AF616831

Capillary Electrophoresis Technology

CHROMATOGRAPHIC SCIENCE SERIES

A Series of Monographs

Editor: JACK CAZES
Sanki Laboratories, Inc.
Mount Laurel, New Jersey

1. Dynamics of Chromatography, *J. Calvin Giddings*
2. Gas Chromatographic Analysis of Drugs and Pesticides, *Benjamin J. Gudzinowicz*
3. Principles of Adsorption Chromatography: The Separation of Nonionic Organic Compounds, *Lloyd R. Snyder*
4. Multicomponent Chromatography: Theory of Interference, *Friedrich Helfferich and Gerhard Klein*
5. Quantitative Analysis by Gas Chromatography, *Josef Novák*
6. High-Speed Liquid Chromatography, *Peter M. Rajcsanyi and Elisabeth Rajcsanyi*
7. Fundamentals of Integrated GC-MS (in three parts), *Benjamin J. Gudzinowicz, Michael J. Gudzinowicz, and Horace F. Martin*
8. Liquid Chromatography of Polymers and Related Materials, *Jack Cazes*
9. GLC and HPLC Determination of Therapeutic Agents (in three parts), *Part 1 edited by Kiyoshi Tsuji and Walter Morozowich, Parts 2 and 3 edited by Kiyoshi Tsuji*
10. Biological/Biomedical Applications of Liquid Chromatography, *edited by Gerald L. Hawk*
11. Chromatography in Petroleum Analysis, *edited by Klaus H. Altgelt and T. H. Gouw*
12. Biological/Biomedical Applications of Liquid Chromatography II, *edited by Gerald L. Hawk*
13. Liquid Chromatography of Polymers and Related Materials II, *edited by Jack Cazes and Xavier Delamare*
14. Introduction to Analytical Gas Chromatography: History, Principles, and Practice, *John A. Perry*
15. Applications of Glass Capillary Gas Chromatography, *edited by Walter G. Jennings*
16. Steroid Analysis by HPLC: Recent Applications, *edited by Marie P. Kautsky*
17. Thin-Layer Chromatography: Techniques and Applications, *Bernard Fried and Joseph Sherma*
18. Biological/Biomedical Applications of Liquid Chromatography III, *edited by Gerald L. Hawk*
19. Liquid Chromatography of Polymers and Related Materials III, *edited by Jack Cazes*
20. Biological/Biomedical Applications of Liquid Chromatography, *edited by Gerald L. Hawk*
21. Chromatographic Separation and Extraction with Foamed Plastics and Rubbers, *G. J. Moody and J. D. R. Thomas*

22. Analytical Pyrolysis: A Comprehensive Guide, *William J. Irwin*
23. Liquid Chromatography Detectors, *edited by Thomas M. Vickrey*
24. High-Performance Liquid Chromatography in Forensic Chemistry, *edited by Ira S. Lurie and John D. Wittwer, Jr.*
25. Steric Exclusion Liquid Chromatography of Polymers, *edited by Josef Janča*
26. HPLC Analysis of Biological Compounds: A Laboratory Guide, *William S. Hancock and James T. Sparrow*
27. Affinity Chromatography: Template Chromatography of Nucleic Acids and Proteins, *Herbert Schott*
28. HPLC in Nucleic Acid Research: Methods and Applications, *edited by Phyllis R. Brown*
29. Pyrolysis and GC in Polymer Analysis, *edited by S. A. Liebman and E. J. Levy*
30. Modern Chromatographic Analysis of the Vitamins, *edited by André P. De Leenheer, Willy E. Lambert, and Marcel G. M. De Ruyter*
31. Ion-Pair Chromatography, *edited by Milton T. W. Hearn*
32. Therapeutic Drug Monitoring and Toxicology by Liquid Chromatography, *edited by Steven H. Y. Wong*
33. Affinity Chromatography: Practical and Theoretical Aspects, *Peter Mohr and Klaus Pommerening*
34. Reaction Detection in Liquid Chromatography, *edited by Ira S. Krull*
35. Thin-Layer Chromatography: Techniques and Applications. Second Edition, Revised and Expanded, *Bernard Fried and Joseph Sherma*
36. Quantitative Thin-Layer Chromatography and Its Industrial Applications, *edited by Laszlo R. Treiber*
37. Ion Chromatography, *edited by James G. Tarter*
38. Chromatographic Theory and Basic Principles, *edited by Jan Åke Jönsson*
39. Field-Flow Fractionation: Analysis of Macromolecules and Particles, *Josef Janča*
40. Chromatographic Chiral Separations, *edited by Morris Zief and Laura J. Crane*
41. Quantitative Analysis by Gas Chromatography, Second Edition, Revised and Expanded, *Josef Novák*
42. Flow Perturbation Gas Chromatography, *N. A. Katsanos*
43. Ion-Exchange Chromatography of Proteins, *Shuichi Yamamoto, Kazuhiro Nakanishi, and Ryuichi Matsuno*
44. Countercurrent Chromatography: Theory and Practice, *edited by N. Bhushan Mandava and Yoichiro Ito*
45. Microbore Column Chromatography: A Unified Approach to Chromatography, *edited by Frank J. Yang*
46. Preparative-Scale Chromatography, *edited by Eli Grushka*
47. Packings and Stationary Phases in Chromatographic Techniques, *edited by Klaus K. Unger*
48. Detection-Oriented Derivatization Techniques in Liquid Chromatography, *edited by Henk Lingeman and Willy J. M. Underberg*
49. Chromatographic Analysis of Pharmaceuticals, *edited by John A. Adamovics*
50. Multidimensional Chromatography: Techniques and Applications, *edited by Hernan Cortes*
51. HPLC of Biological Macromolecules: Methods and Applications, *edited by Karen M. Gooding and Fred E. Regnier*

52. Modern Thin-Layer Chromatography, *edited by Nelu Grinberg*
53. Chromatographic Analysis of Alkaloids, *Milan Popl, Jan Fähnrich, and Vlastimil Tatar*
54. HPLC in Clinical Chemistry, *I. N. Papadoyannis*
55. Handbook of Thin-Layer Chromatography, *edited by Joseph Sherma and Bernard Fried*
56. Gas–Liquid–Solid Chromatography, *V. G. Berezkin*
57. Complexation Chromatography, *edited by D. Cagniant*
58. Liquid Chromatography–Mass Spectrometry, *W. M. A. Niessen and Jan van der Greef*
59. Trace Analysis with Microcolumn Liquid Chromatography, *Miloš Krejčí*
60. Modern Chromatographic Analysis of Vitamins: Second Edition, *edited by André P. De Leenheer, Willy E. Lambert, and Hans J. Nelis*
61. Preparative and Production Scale Chromatography, *edited by G. Ganetsos and P. E. Barker*
62. Diode Array Detection in HPLC, *edited by Ludwig Huber and Stephan A. George*
63. Handbook of Affinity Chromatography, *edited by Toni Kline*
64. Capillary Electrophoresis Technology, *edited by Norberto A. Guzman*

ADDITIONAL VOLUMES IN PREPARATION

Thin-Layer Chromatography: Techniques and Applications, Third Edition, Revised and Expanded, *Bernard Fried and Joseph Sherma*

Capillary Electrophoresis Technology

edited by
Norberto A. Guzman
*The R.W. Johnson Pharmaceutical Research Institute
Raritan, New Jersey*

Marcel Dekker, Inc. New York • Basel • Hong Kong

Library of Congress Cataloging-in-Publication Data

Capillary electrophoresis technology / edited by Norberto A. Guzman.
p. cm. -- (Chromatographic science series ; 64)
Includes bibliographical references and index.
ISBN 0-8247-9042-1 (acid-free paper)
1. Capillary electrophoresis. I. Guzman, Norberto A.
II. Series: Chromatographic science ; v. 64.
QP519.9.C36C36 1993
574.19'285--dc20 93-14091
CIP

The publisher offers discounts on this book when ordered in bulk quantities. For more information, write to Special Sales/Professional Marketing at the address below.

This book is printed on acid-free paper.

MARCEL DEKKER, INC.
270 Madison Avenue, New York, New York 10016

Current printing (last digit):
10 9 8 7 6 5 4 3 2 1

PRINTED IN THE UNITED STATES OF AMERICA

Foreword

The technique of capillary electrophoresis is at the crossroads of the more established analytical separation techniques of conventional electrophoresis, liquid chromatography, and gas chromatography. Capillary electrophoresis owes most of its separation modes (e.g., free zone, isoelectric focusing, gel sieving) to conventional electrophoresis. It owes much of the necessary detection technology to the field of liquid chromatography. Finally, the capillaries themselves, as well as much of their surface modification chemistry, is owed to gas chromatography. This "crossroads of chemical analysis" has spawned a health interchange among scientists of diverse backgrounds. In what other field can we find biochemists, molecular biologists, pharmaceutical chemists, chromatographers, laser spectroscopists, mass spectroscopists, and electrochemists all intensely interested in and aware of one another's work?

All of this cross-fertilization is producing a powerful and versatile analytical tool, one whose capabilities in many ways are beginning to surpass those of its "parent" techniques. The lowest detection limits in the field of separations, for example, are being obtained in capillary electrophoresis with laser-induced fluorescence detection, where single-molecule detection has recently been achieved. Capillary electrophoresis, in the mode of micellar electrokinetic chromatography (MEKC), even permits the separation of uncharged molecules, through the clever use of charged micelles as a "pseudophase" in which to partition analyte molecules. What other technique can claim to be highly effective at separating substances ranging in size from lithium ions through proteins and polynucleotides, all the way to cellular and subcellular fragments? And how many techniques offer the capability to measure the chemical composition of single cells?

It is nonetheless important to recognize that capillary electrophoresis is actually a complementary technique to GC, LC, and conventional electrophoresis. GC is still the method of choice for analysis of volatile compounds. LC is still a highly successful separation technique for nonvolatile molecules, and is much better suited to preparative work and scale-up than capillary electrophoresis. Conventional electrophoresis is still a less expensive, albeit more labor-intensive, way to accomplish analysis of biopolymers.

It seems that almost everyone involved in chemical analysis is either doing or considering doing capillary electrophoresis. Although this is certainly an overstatement, it is surprisingly not that far from the truth. A glance at the table of contents of any journal on chemical analysis is highly likely to turn up more than one article on capillary electrophoresis. All this interest creates a need for an accurate and coherent presentation of the field as it currently stands. The days are over when one could become acquainted with the field by reading a few dozen published articles. The total number of publications on capillary electrophoresis is now approaching 1000. There are many new ideas that the practitioner of capillary electrophoresis must master. These include the unusual phenomenon of electrostatic flow, the relationship between electric field strength and speed of analysis, the limitations on heat dissipation and its relationship to applied voltage and buffer conductivity, and many others. Despite this, capillary electrophoresis is still a comparatively simple technique at heart, one that is capable of supplying the user with good results after a relatively short time of working with it.

The book before you is a large and impressive volume written by experts in the field of capillary electrophoresis. It is a thorough treatment of the field as it exists, and offers many insights into where the field is heading. The book covers all important aspects of capillary electrophoresis, from the basics of the electrophoretic process through the various operational modes of capillary electrophoresis, details of hardware and instrumentation, and, finally, a generous treatment of applications. It is the intention of the various authors to acquaint you with all aspects of capillary electrophoresis, the good as well as the bad, the advantages as well as the flaws. I believe that the authors have succeeded, and have produced a marvelous, up-to-date, and well-organized guide and reference work to the exciting field of capillary electrophoresis.

James W. Jorgenson, Ph.D.
Department of Chemistry
University of North Carolina
Chapel Hill, North Carolina

Preface

Perhaps the most rapidly developing area in separation science today is the burgeoning field of capillary electrophoresis. The advent of this technology was confined primarily to analytical chemistry, but the many unique features of capillary electrophoresis have enabled this technology to permeate all areas of chemistry, biology, and medicine. With the availability of a variety of commercial instruments, capillary columns, additives, and buffer conditions, capillary electrophoresis has become suitable for a wide variety of applications. Its use has resulted in a new understanding of the modern concepts of resolution, sensitivity analysis, and characterization of numerous types of chemical and biological substances.

Among the many unique features of capillary electrophoresis are its high power of resolution, high mass sensitivity, low sample volume requirements, and overall versatility. It is possible now to analyze substances present in picoliter-femtoliter volumes and to reach zeptomole sensitivity. Basically, we have achieved an important goal in reaching the molecular counting level of substances by using capillary electrophoresis.

The need to address the explosive growth in knowledge and interest in this subject and to bring capillary electrophoresis out of the analytical chemistry laboratories and into the biological and clinical arena has led to the development of this book. The task required the full cooperation and dedication of many outstanding contributors. These nationally and internationally recognized authorities on the subject of capillary electrophoresis, with special competence in their respective fields, have contributed superb, up-to-date discussions of the topics assigned to them.

This book is intended for a broad audience. It is aimed both at newcomers to the field of capillary electrophoresis, as a self-study guide for investigators unfamiliar with the subject, and at more advanced researchers in separation science working in basic research, pharmaceutical and food science, and biotechnology who are seeking an in-depth update on separation, detection, and characterization using this technology. In addition, this book should be of interest to physicians and investigators working in the diagnostic, environmental, forensic, and general clinical fields.

The objectives of this project, since its inception, were to digest the mass of literature and to trace the sequence of new observations that have led to our present knowledge, to define the present concepts of capillary electrophoresis technology, and to broaden the scope of applications for the future, ideally in such areas as human genome research and other biological and medical applications. The book has been organized in such a way that, in some cases, more than one chapter deals with a particular topic. While this organization results in some overlap, it serves to present different opinions and avoids possible information gaps in the various subjects. A further advantage of this approach is that, in general, each chapter may be read as a separate unit while remaining an integrated part of the total text. The interested reader will find an abundance of references at the end of every chapter. It is my hope that in this form the book will prove to be of great value in helping chemists and biologists make capillary electrophoresis a useful tool in their work.

The book attempts to give a general picture of the theoretical concepts and mathematical expressions of the technology, of the various advances in instrumentation hardware, and of some of the most recent and important technological applications.

This volume is not intended to be an atlas of operative technique, although the major technical maneuvers of most routine operations and applications are illustrated. It is my hope that it may guide the student and the scholar through our present-day knowledge of the technology of capillary electrophoresis and inspire its further development. Capillary electrophoresis can improve our ability to determine the quality of the food that we eat, to analyze the environment in which we live, and to improve the accuracy of the diagnosis and prognosis of disease. Advances made in this and other analytical techniques can bring immediate results for improved health and a better life.

Norberto A. Guzman

Contents

Contributors

Juan P. Advis, D.V.M., Ph.D. Professor, Department of Animal Sciences, Rutgers University, New Brunswick, New Jersey

Clifford M. Berck, M.D. Clinical Assistant Professor, SUNY–Health Science Center at Brooklyn, Brooklyn, New York

Petr Boček, Ph.D., D.Sc. Head, Institute of Analytical Chemistry, Czech Academy of Sciences, Brno, Czech

Ernst E. Brueggemann Toxinology Division, United States Army Medical Research Institute of Infectious Diseases, Fort Detrick, Maryland

Da Yong Chen, Ph.D. Department of Chemistry, University of Alberta, Edmonton, Alberta, Canada

Marcella Chiari, Ph.D. Division of Biomedical Sciences and Technologies, Department of Organic Chemistry and Biochemistry, University of Milan, Milan, Italy

Ann M. Dougherty, M.S.* Research Biochemist, BioPharm Separations Group, Supelco, Incorporated, Bellefonte, Pennsylvania

Norman Dovichi, Ph.D. Professor, Department of Chemistry, University of Alberta, Edmonton, Alberta, Canada

---

**Current affiliation*: Beckman Instruments, Inc., Somerset, New Jersey

Ziad El Rassi, Ph.D. Associate Professor, Department of Chemistry, Oklahoma State University, Stillwater, Oklahoma

Heinz Engelhardt, Ph.D. Professor, Angewandte Physikalische Chemie, University of Saarlandes, Saarbrücken, Germany

Andrew G. Ewing, Ph.D. Professor, Department of Chemistry, The Pennsylvania State University, University Park, Pennsylvania

Salvatore Fanali, Ph.D. Istituto di Cromatografia, Consiglio Nazionale delle Ricerche, Monterotondo Scalo, Italy

Frantisek Foret, Ph.D. Staff Scientist, Barnett Institute of Chemical Analysis and Materials Science, Northeastern University, Boston, Massachusetts

Chuzo Fujimoto, Ph.D. School of Materials Science, Toyohashi University of Technology, Toyohashi, Japan

Petr Gebauer, Ph.D. Scientist, Institute of Analytical Chemistry, Czech Academy of Sciences, Brno, Czech

S. Douglass Gilman Research Assistant, Department of Chemistry, The Pennsylvania State University, University Park, Pennsylvania

Carmen L. Gonzalez, B.Sc. Scientist, Protein Research Unit, Princeton Biochemicals, Inc., Princeton, New Jersey

András Guttman, Dr.Sc. Principal Research Scientist, P/ACE Research and Development, Beckman Instruments, Inc., Fullerton, California

Norberto A. Guzman, Ph.D.* Senior Scientist, Protein Research Unit, Princeton Biochemicals, Inc., Princeton, New Jersey

Heather R. Harke, M.S. Department of Chemistry, University of Alberta, Edmonton, Alberta, Canada

Mark A. Hayes Research Assistant, Department of Chemistry, The Pennsylvania State University, University Park, Pennsylvania

Luis Hernandez, M.D. Professor, Laboratory of Behavioral Physiology, Department of Physiology, Los Andes University, Mérida, Venezuela

Harry B. Hines Toxinology Division, United States Army Medical Research Institute of Infectious Diseases, Fort Detrick, Maryland

---

**Current affiliation*: Principal Scientist, Biotechnology Research and Development, The R. W. Johnson Pharmaceutical Research Institute, a Johnson & Johnson Company, Raritan, New Jersey

Arthur M. Hoyt, Jr., Ph.D. Professor, Department of Chemistry, University of Central Arkansas, Conway, Arkansas

Haleem J. Issaq, Ph.D. Senior Scientist, Chemical Synthesis and Analysis Laboratory, Program Resources, Inc./DynCorp, Frederick Cancer Research and Development Center, National Cancer Institute, National Institutes of Health, Frederick, Maryland

George M. Janini, Ph.D. Scientist, Chemical Synthesis and Analysis Laboratory, Program Resources, Inc./DynCorp, Frederick Cancer Research and Development Center, National Cancer Institute, National Institutes of Health, Frederick, Maryland

Kiyokatsu Jinno, Ph.D. School of Materials Science, Toyohashi University of Technology, Toyohashi, Japan

Narahari Joshi Department of Physiology, School of Medicine, Los Andes University, Mérida, Venezuela

Barry L. Karger, Ph.D. Director, Barnett Institute of Chemical Analysis and Materials Science, and James L. Waters Professor of Analytical Chemistry, Northeastern University, Boston, Massachusetts

Ernst Kenndler, Ph.D. Professor, Institute of Analytical Chemistry, University of Vienna, Vienna, Austria

Morteza G. Khaledi, Ph.D. Associate Professor, Department of Chemistry, North Carolina State University, Raleigh, North Carolina

Jörg Kohr, Ph.D. Angewandte Physikalische Chemie, University of Saarlandes, Saarbrücken, Germany

Ira S. Krull Associate Professor, Department of Chemistry, Northeastern University, Boston, Massachusetts

Cheng S. Lee, Ph.D. Assistant Professor, Department of Chemical and Biochemical Engineering, University of Maryland, Baltimore, Maryland

J. R. Mazzeo* Graduate Student, Department of Chemistry, Northeastern University, Boston, Massachusetts

Wassim Nashabeh, Ph.D.* Graduate Research Assistant, Department of Chemistry, Oklahoma State University, Stillwater, Oklahoma

---

**Current affiliation*: Research Scientist, Chemical Products Group, Research and Development Department, Waters Chromatography Division, Millipore Corporation, Milford, Massachusetts

David M. Northrop, M.S.F.S., Ph.D. Forensic Scientist, Washington State Patrol Crime Laboratory, Kennewick, Washington

Koji Otsuka, Ph.D. Associate Professor, Department of Industrial Chemistry, Osaka Prefectural College of Technology, Neyagawa, Osaka, Japan

Changyu Quang Graduate Student, Department of Chemistry, North Carolina State University, Raleigh, North Carolina

Fred E. Regnier, Ph.D. Professor, Department of Chemistry, Purdue University, West Lafayette, Indiana

Pier Giorgio Righetti, Ph.D. Professor of Biochemistry, Division of Biomedical Sciences and Technologies, Department of Organic Chemistry and Biochemistry, University of Milan, Milan, Italy

Rachhpal S. Sahota Analytical Drug Development, Proctor & Gamble Pharmaceuticals, Norwich, New York

Gerhard Schomburg, Prof.Dr., Dr.h.c. Department of Chromatography, Max-Planck-Institut für Kohlenforschung, Mülheim an der Ruhr, Germany

Mark R. Schure, Ph.D. Theoretical Separation Science Laboratory, Rohm and Haas Company, Spring House, Pennsylvania

Richard D. Smith, Ph.D. Senior Staff Scientist and Group Leader, Chemical Sciences Department, Pacific Northwest Laboratory, Richland, Washington

Scott C. Smith, Ph.D. Senior Scientist, Magellan Laboratories, Inc., Research Triangle Park, North Carolina

Joost K. Strasters, dr.ir. Research Investigator, Department of Analytical Sciences, Sterling Winthrop Pharmaceuticals Research Division, Malvern, Pennsylvania

Shigeru Terabe, Ph.D. Professor, Faculty of Science, Himeji Institute of Technology, Kamigori, Hyogo, Japan

Wolfgang Thormann, Ph.D. Head, Drug Assay Laboratory, and Associate Professor, Department of Clinical Pharmacology, University of Bern, Bern, Switzerland

---

**Current affiliation*: Postdoctoral Research Associate, Barnett Institute, Northeastern University, Boston, Massachusetts

Kenneth B. Tomer, Ph.D. Research Chemist, Laboratory of Molecular Biophysics, National Institute of Environmental Health Sciences, Research Triangle Park, North Carolina

Maria A. Trebilcock, Ph.D. Senior Scientist, Protein Research Unit, Princeton Biochemicals, Inc., Princeton, New Jersey

Pei Tsai, Ph.D. Postdoctoral Fellow, Department of Chemical and Biochemical Engineering, University of Maryland, Baltimore, Maryland

Takao Tsuda, Ph.D. Associate Professor, Department of Applied Chemistry, Nagoya Institute of Technology, Nagoya, Japan

Harold R. Udseth, Ph.D. Senior Research Scientist, Department of Chemical Sciences, Pacific Northwest Laboratory, Richland, Washington

Phillipe Verdeguer, Ph.D. Analytical Department Manager, Europhor Instruments, Toulouse, France

Dan Wu Department of Chemistry, Purdue University, West Lafayette, Indiana

Edward S. Yeung, Ph.D. Distinguished Professor in Liberal Arts and Sciences, Department of Chemistry and Ames Laboratory—USDOE, Iowa State University, Ames, Iowa

Jianzhong Zhang Graduate Student, Department of Chemistry, University of Alberta, Edmonton, Alberta, Canada

I
OVERVIEW

1

Capillary Electrophoresis: *Introduction and Assessment*

Barry L. Karger and Frantisek Foret

Northeastern University
Boston, Massachusetts

INTRODUCTION

High-performance capillary electrophoresis (HPCE) has achieved remarkably rapid development since its introduction in the early to middle 1980s. From a technique for which only homemade systems were available up until 1989, today approximately ten companies offer automated instrumentation and total sales of 30–35 million dollars (1992) are expected to grow to approximately 150 million dollars in 1996 [1].

Capillary electrophoresis actually has had a long history, if one considers electrophoresis conducted in columns as its ancestor. Column electrophoresis was exploited over 30 years ago by several workers, especially Hjertén [2] and Catsimpoolas [3]. Separately, Martin and Everaerts developed capillary isotachophoresis [4], which was commercialized in the mid-1970s. Although the instrumentation was not a significant commercial success, capillary isotachophoresis is still practiced in research laboratories, particularly in Europe.

In the early days, tube diameters of 1–3 mm, followed by 200–400 μm [5], were employed. In 1971, Everaerts explored the separation potential of capillary zone electrophoresis (CZE) performed in 200-μm Teflon capillaries with on-column UV detection [6], and in 1981, Jorgenson and co-workers convincingly demonstrated the analytical potential of CZE using 75-μm fused silica capillaries [7–9]. The experience of this laboratory in capillary chromatography was highly useful in designing instrumentation for this new electrophoretic system. The high-resolution capability of the method was clearly shown, as illustrated in Fig. 1. The importance of capillary operation to reduce the joule heating effects caused by the desirable high fields was emphasized. This work is rightly credited with the initiation of high-performance capillary electrophoresis.

As is typical of new methodology, an induction period followed the foregoing pioneering work, with only a few academic laboratories joining in the development; noteworthy were the studies of Hjertén on capillary zone electrophoresis [10] and Terabe on micellar electrokinetic chromatography (MEKC) [11]. However, by the late 1980s, it was clear to a number of investigators and instrument companies that capillary electrophoresis was a major field, especially in the biological sciences. The introduction of the first-generation equipment in 1989 opened up the possibility for several workers to explore capillary electrophoresis for their specific applications. Instrumentation continues to advance toward a fully automated tool, and there is general agreement that HPCE is currently in a rapid growth stage. This book details advances that convincingly demonstrate the outstanding potential of the technique.

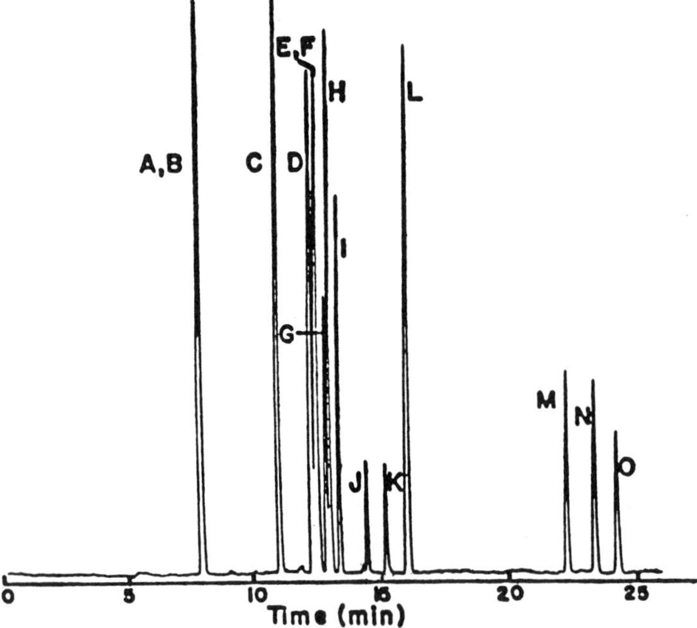

Figure 1 Zone electrophoretic separation of dansyl amino acids: A, unknown impurity; B, ε-labeled lysine; C, dilabeled lysine; D, asparagine; E, isoleucine; F, methionine; G, serine; H, alanine; I, glycine; J and K, unknown impurities; L, dilabeled cysteine; M, glutamic acid; N, aspartic acid; O, cysteic acid. The concentration of each derivative is approximately 5×10^{-4} M, dissolved in operating buffer. (From Ref. 7)

It is interesting that capillaries have already been widely used with slab gel electrophoresis (e.g., two-dimensional gel electrophoresis [12]). Here, a capillary column is filled with a gel, such as agarose, then isoelectric focusing (separation based on pI differences) is conducted in the capillary column. After focusing, the gel is extruded from the column into a trough on a slab for the second dimension of sodium dodecyl sulfate (SDS) gel electrophoresis (separation based on molecular weight [relative molecular mass] differences). The difference in this operation from the current technique is that in HPCE, detection is typically on-column or on-line (detector attached after the column). As such, capillary electrophoresis shares much in common with liquid chromatography. Indeed, data systems developed originally for liquid chromatography (LC) can be easily incorporated into capillary systems.

From its very beginning, workers have emphasized that HPCE is a complementary tool to HPLC. In some respects this has been one of the greatest advantages of HPCE, and yet one of its greatest challenges. Capillary electrophoresis has advanced rapidly, in large part because of the accumulated instrument developments in liquid chromatography and the theoretical understanding of column performance. At the same time, however, HPLC is a well-developed tool, rugged and fully automated. Therefore, capillary electrophoresis is presented with significant challenges relative to replacement of liquid chromatography in those areas for which HPLC is successful. However, liquid chromatography is not so useful in certain applications (e.g., high molecular weight separations) and capillary electrophoresis may offer simpler approaches to optimization and to rapid separation of complex mixtures. It is undoubtedly true that both methods will find their niches, and both will be widely used in a complementary fashion in the years ahead.

Capillary electrophoresis offers some clear advantages over slab gel electrophoresis for automation, speed, and quantitative capability. Slab gel methods, on the other hand, are simple, inexpensive, and allow parallel processing (i.e., simultaneous multisample analysis). Nevertheless, for many applications CE may be expected to replace slab gel electrophoresis. As examples, the molecular weight separation of SDS–protein complexes, isoelectric focusing, and DNA separations are being implemented by HPCE. Another major potential field of application is in oligosaccharide separation, used, for example, for carbohydrate mapping. The very high resolution possible with capillary electrophoresis would appear to make this method appealing.

Beyond comparisons with other separation methods, capillary electrophoresis is becoming recognized in its own right as a general analytical tool. The key characteristic in this regard is its miniature size. Obviously, environmental issues (e.g., solvent removal) are substantially reduced in CE relative to LC. Secondly, reagents (e.g., biospecific agents) that might normally be expensive to utilize become accessible when the total column volume is only 1–2 μL. Thirdly, chemical reactions such as complexation (e.g., binding studies) can be routinely studied on small amounts of material. Furthermore, such studies are performed in solution, with no intervening surfaces. Indeed, the complexing species can be mixed by use of an electric field, rather than by volume mixing.

Capillary electrophoresis is an evolving technique. Not all issues are resolved at the present time; for example, detection sensitivity, ruggedness, and a basic understanding of the role of buffer and buffer additives. Nevertheless, the accumulative experience to date has provided substantial opportunities for the solution of analytical problems. We can anticipate further rapid development in the next few years.

The purpose of this chapter is to introduce HPCE and to provide an assessment of the current status. Our goal is to provide an overview of the field and to set the stage for the following chapters.

FUNDAMENTALS

Electrophoretic Migration

High-performance capillary electrophoresis (HPCE) stands for all modes of electrophoretic separations performed in capillaries [i.e., capillary zone electrophoresis (CZE), capillary isotachophoresis (CITP), capillary isoelectric focusing (CIEF), and micellar electrokinetic capillary chromatography (MEKC)]. Capillary zone electrophoresis can be further divided into open-tube, with or without electroosmotic flow (EEO), and gel- or polymer network-filled capillaries.

The *background electrolyte* (BGE) is an electrolyte solution used as a liquid medium for the electrophoretic separation. Where relevant, the BGE should assure constant conditions, such as ionic strength and pH, during the separation. Constituents of the background electrolyte with the same sign as the separated ions are called *coions*, and those with the opposite charges *counterions*. The choice of buffer may also be dictated by the conductivity of the solution, which should be low to maintain low currents (and thus minimal joule heat).

Upon the application of a constant electric field E ($E = V/L$ where V is the voltage applied across the capillary of the length L), ionic species undergo an electrostatic force, F_e, which is proportional to the electric field strength and the charge (q) of the particular ion,

$$F_e = qE \tag{1}$$

This force causes the acceleration of ions toward the oppositely charged electrode. As the velocity of the ions increases, the counteracting frictional force (F_f) caused by the surrounding solution slows down the species. Assuming a spherical molecule of radius a, the frictional force can be expressed in terms of Stokes' law as

$$F_f = 6\pi\eta a v \tag{2}$$

where η = solution viscosity. After steady state is reached (practically immediately), the ions move with a constant velocity v, proportional to the applied electric field

$$v = \mu E \tag{3}$$

where the proportionality constant μ, the electrophoretic mobility, is a characteristic property of a given ion in a given medium and at a given temperature. Since $F_e = F_f$, through combination of Eqs. 1, 2, and 3, μ can be expressed as

$$\mu = \frac{q}{6\pi\eta a} = \frac{v}{E}\ (\mathrm{cm^2\ V^{-1}\ s^{-1}}) \tag{4}$$

The viscous drag of the solvent and the charge and size of the solute thus control the migration of a species in an applied electric field.

Equation 4 expresses the critical role of solvent viscosity on electrophoretic migration. As has been noted many times, viscosity is quite temperature-dependent, ~2%/°C, which means that temperature control is mandatory for high reproducibility in capillary electrophoresis (see later discussion on joule heat removal).

For a given charge q, the larger the size of the ion, the slower the migration. If spherical ions are assumed, the mobility will scale to molecular weight (MW) by $\mathrm{MW}^{-2/3}$. For complex protein structures, this dependence changes to $\mathrm{MW}^{-1/3}$. In general, for molecules of moderate size, spherical shape is a reasonable approximation in free solution; however, for large molecules (e.g., virus particles) shape and orientation must be taken into account [13].

Table 1 presents a list of mobilities of typical small ions at room temperature and zero or low ionic strength of the solvent medium, μ_0. In the presence of buffer or salt ions, the actual mobility will be lower than the limiting ionic mobility because of the mutual interaction of ions. This dependence of mobility on ionic strength in dilute solutions of univalent electrolytes is typically expressed in terms of the Debeye–Hückel–Onsager equation [14a]; however, as the ionic strength is significantly increased or if polyvalent electrolytes are used, the estimation of μ can become quite complex. Thus, as in chromatography, measured values of mobility in specific solvents, buffers, and temperatures are required for precise results. Nevertheless, the values in Table 1 and more extensive compilations of limiting ionic mobilities [14b] provide useful information in the selection of electrolytes to manipulate solute migration in HPCE.

Effective Electrophoretic Mobility

A substance that is partially ionized in solution moves under the influence of an electric field, with a velocity proportional to its degree of ionization. The corre-

Table 1 Ionic Mobilities, μ, and Acidity Constants, p*K*, of Some Selected Ions.

Substance[a]	$\mu(10^{-5}\ cm^2\ V^{-1}\ s^{-1})$	p*K*
ACES	−31.3	6.84
Acetate	−42.4	4.76
Benzoate	−32.9	4.17
BES	−24.0	7.16
Chloride	−79.1	—
HEPES	−21.8	7.51
Histidine	−28.3	9.33
2-Hydroxyisobutyrate	−33.5	3.97
MES	−26.8	6.13
MOPS	−24.4	7.16
Phosphate	−34.1	2.15
	−58.3	7.22
	−71.5	11.50
TAPS	−25.0	8.30
TES	−22.4	7.43
Cations		
β-Alanine	37.5	3.55
ε-Aminocaproic acid	30.0	4.44
AMPD	32.0	8.78
Ammonium	72.2	9.25
Histidine	29.6	6.04
Imidazole	52.0	7.15
Lithium	36.0	—
Potassium	76.0	—
Sodium	53.0	—
Tris (hydroxymethyl) aminomethane	29.5	8.08

[a]Block letters abbreviations correspond to respective Good's buffers.
Source: Ref. 14b.

sponding electrophoretic mobility is called the *effective electrophoretic mobility*, μ′; a simple relation between the effective and actual ionic mobilities exists [15]:

$$\mu' = \sum_i \mu_i x_i \tag{5}$$

where μ_i and x_i are the corresponding ionic mobilities and mole fractions of ions formed in solution. As in chromatography, buffer pH manipulation thereby provides a powerful tool for selectivity control.

It is straightforward to quantitate the role of pH on electrophoretic migration. The effective electrophoretic mobility of an acid, HA, which can exist in solution as either uncharged HA or its conjugate base A^-, is given by the following relationship.

$$\mu'_{HA} = \{\mu_{A-}\}\{x_{A-}\} + \mu_{HA}x_{HA} \tag{6}$$

Since the mobility of the nonionized acid μ_{HA} is zero ($q = 0$) and the mole fraction of ionized conjugate base molecules can be expressed in terms of the pK of the acid, the effective mobility can be written as [16]

$$\mu'_{HA} = (\mu_{A-})\left[\frac{10^{-pK}}{(10^{-pK} + 10^{-pH})}\right] \tag{7}$$

From Eq. 7, it follows that at low pH (pH < pK − 2) at which the acid is essentially nonionized, the mobility of HA will be zero. At pH = pK, half of the molecules will be ionized, and the corresponding effective electrophoretic mobility will be 50% of actual ionic mobility of the anion. At pH > (pK + 2), all molecules will be ionized, and the effective mobility will be equal to the actual ionic mobility. Figure 2a illustrates this behavior. For proteins and related molecules that possess multiple charged groups with different pKs, μ'−pH curves will be quite complex (see Fig. 2b). Nevertheless, pH control represents one of the most effective means to influence separation in electrophoresis, not only in terms of control of ionization, per se, but also in terms of complex formation or ion-pairing equilibria. Buffer pH change during a run has also been implemented by several workers [17–19].

Electroosmotic Flow

An important phenomenon that can occur in capillary electrophoresis is electroosmotic transport of ions during analysis. Electroosmosis originates from the electrophoretic movement of the hydrated part of the electric double layer at the capillary wall. In uncoated fused silica capillaries, the electric double layer is formed from the ionization of silanol groups present on the surface of the capillary. The negative charge on the capillary surface is balanced by the positively charged layer of hydrated cations. Upon application of the electric field, this layer begins to move toward the cathode and, owing to viscous drag, transports the bulk liquid inside the capillary. The velocity of the electroosmotic flow v_{eo} is given by the Helmholtz–Smoluchowski equation:

$$v_{eo} = \frac{-\varepsilon\zeta}{\eta' E} \tag{8}$$

where ε is the dielectric constant of the solvent, ζ the zeta potential at the capillary wall, and η' the viscosity of the double layer (often approximated as the bulk solution viscosity). As the term $\varepsilon\zeta/\eta'$ has the dimensions of electrophoretic mobility, it is frequently denoted as an electroosmotic mobility μ_{eo}.

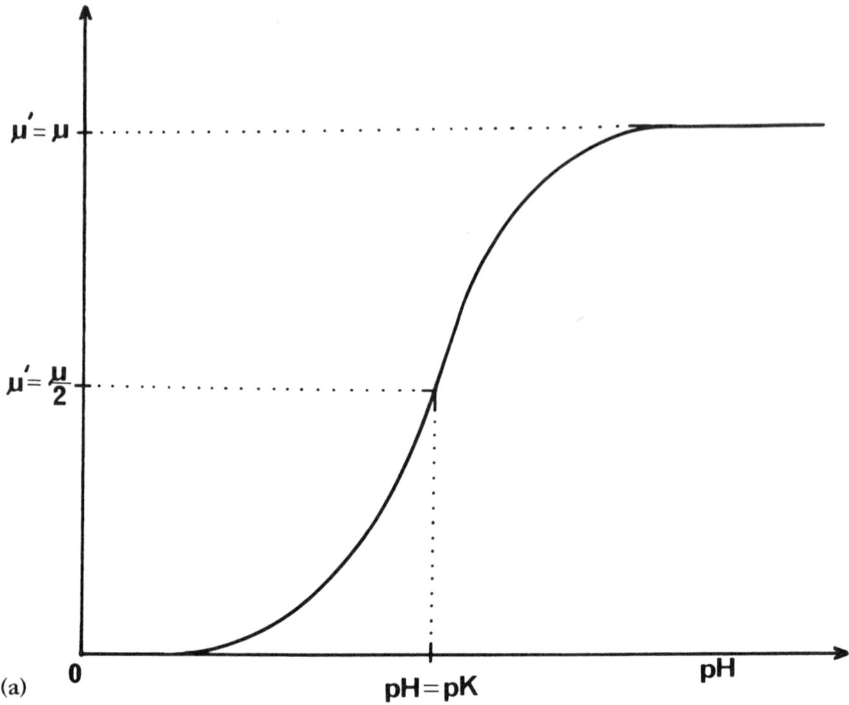

Figure 2 (a) Dependence of the effective mobility μ of a weak monovalent acid on pH. (b) Dependence of mobility on pH for a hypothetical protein.

An important feature of the electroosmotic flow is its velocity flow profile that, unlike the parabolic hydrodynamic flow caused by pressure drop [20], is almost pluglike and does not significantly contribute to the dispersion of zones during the separation (Fig. 3a). When no electroosmotic flow occurs, ion movement will again be pluglike. It is this plug flow that permits the use of an order of magnitude-wider tube diameter for HPCE than for capillary HPLC (e.g., 50–100 μm vs 5 μm). Very narrow-bore capillaries are needed in LC to minimize the radial distance for diffusion to relax the concentration inhomogeneities arising from the parabolic flow profile. The use of wider-diameter capillaries in HPCE substantially reduces the instrumental problems of miniaturization.

Two limiting modes of open-tube capillary electrophoresis can be distinguished, separation with and without electroosmotic flow (EOF). In the case of little or no electroosmotic flow, the surface of the capillary is coated with a neutral layer to reduce the zeta potential. Also, the viscosity of the coated layer is generally increased to reduce v_{eo} [2]. Both modes, with and without electroosmotic flow, are used in current practice,

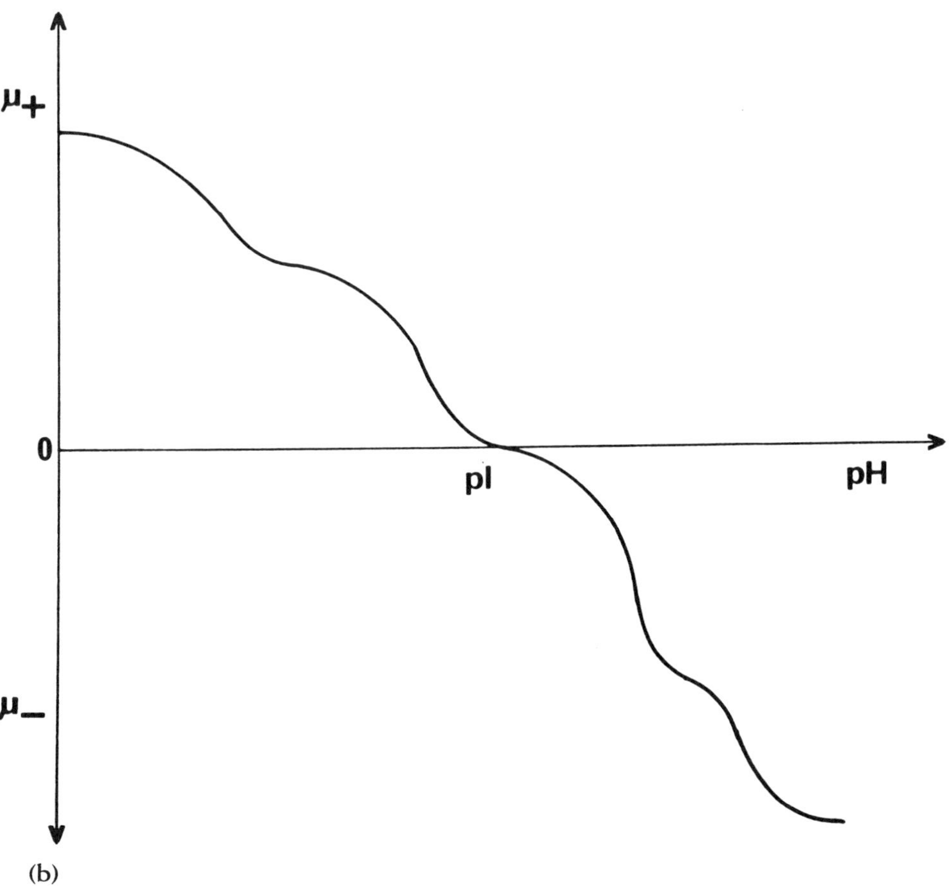

(b)

each with its advantages and limitations. In electroosmotic flow, the separation of both cations and anions is, in principle, possible, provided that the electroosmotic mobility is higher than the electrophoretic mobility of ions moving in the opposite direction and provided that ions of charge opposite that of the surface of the wall do not adsorb. Electroosmotic flow can also be used as a means to speed up specific separations.

With bare silica, which is negatively charged above pH of about 3, cations will move faster than the bulk flow velocity, whereas anions will move slower. The charge, and thus zeta potential of the silica, is quite pH dependent, as shown in Fig. 3b; hence, careful pH control is necessary, particularly near neutral pH. This pH effect can be reduced by coating the surface with pH-independent negative- or

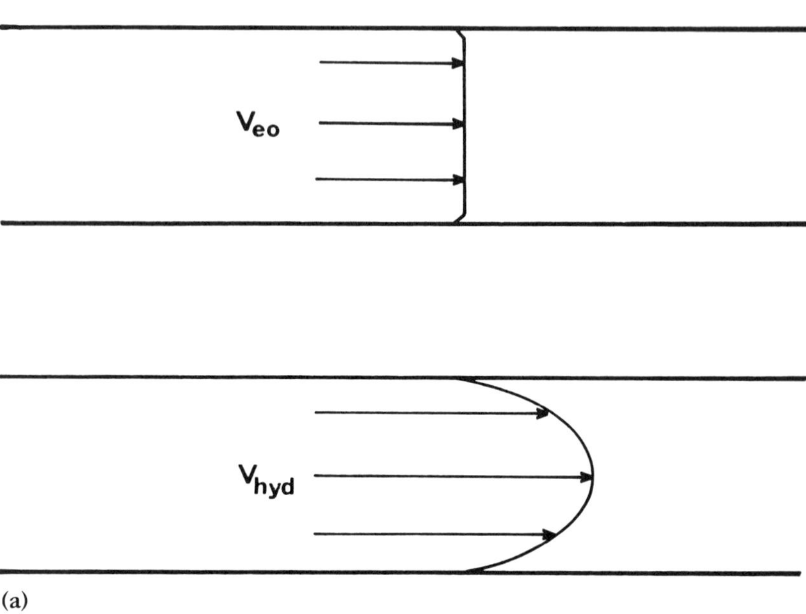

(a)

Figure 3 (a) Velocity profiles of liquid flowing in a capillary under the action of electroosmosis (v_{eo}) and hydrostatic pressure (v_{hyd}). (b) Electroosmotic flow as a function of buffer pH. Flow rate was measured in a 110-cm length of 53-μm–id fused silica filled with 150 mM phosphate buffer at the specified pH. The applied potential was 30 kV; the separation was monitored 75 cm from the anode. Phenol and acetone dissolved in the separation buffers were used as markers. (From Ref. 42)

positive-charged groups (e.g., quaternary ammonium ions). In the latter event, the positively charged surface reverses the bulk flow relative to bare silica. A second approach is to control the zeta potential on the wall by the use of a radial potential gradient imposed by an external field [21,22]. The use of this approach in the separation of biological substances has been demonstrated by Lee and co-workers [23]. Since this method is under development, its relation to other schemes for manipulating electroosmotic flow will need further study (e.g., in pH regions where strong electroosmotic flow exists—pH > 6 for bare silica).

A significant potential problem with charged surfaces is electrostatic adsorption of oppositely charged species. This is especially true for biopolymers (e.g., proteins). (Note that hydrophobic adsorption at the wall can also occur with macromolecules.) The use of similarly charged or neutral hydrophilic coatings is often necessary to reduce adsorption. For example, polyamines can be adsorbed (and cross-linked if desired) to the fused silica surface to create a positively charged surface for the separation of very basic proteins, such as histones (Fig. 4a) [24,25]. A second

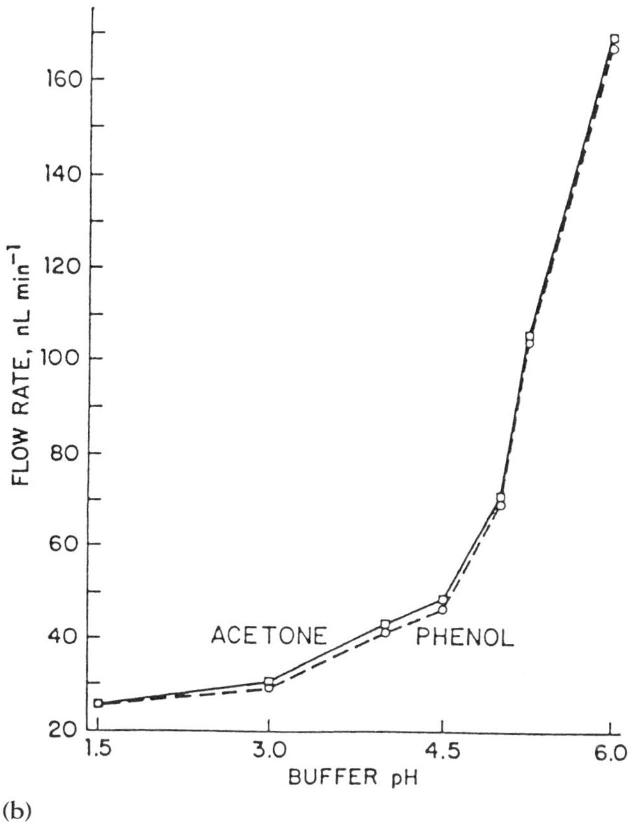

(b)

approach involves the use of thin films of neutral hydrophilic coatings. Since the electric force field extends to approximately 30Å, films of this thickness or lower will reduce, but not eliminate, electroosmotic flow. Regnier and co-workers have adopted this latter strategy of a protective-charged surface in their use of a detergent-adsorbed coating [26]. In the example in Fig. 4b, the separation of acidic and basic proteins is illustrated using an *n*-alkyl C-18–bonded phase covalently attached to the wall, to which a buffer solution containing low concentrations of Brij 35 is added. The detergent equilibrates by adsorption with the C-18 layer to create a dynamically controlled hydrophilic surface with electroosmotic flow. The detergent can also help solvate biopolymers to reduce adsorption. Covalently attached thin layer polyether phases can also be used with electroosmotic flow [27,28].

An alternative approach for CZE is to eliminate electroosmosis by the use of a thicker neutral coating or a viscous additive to the running buffer. Note that electroosmosis is frequently eliminated when employing isotachophoresis or iso-

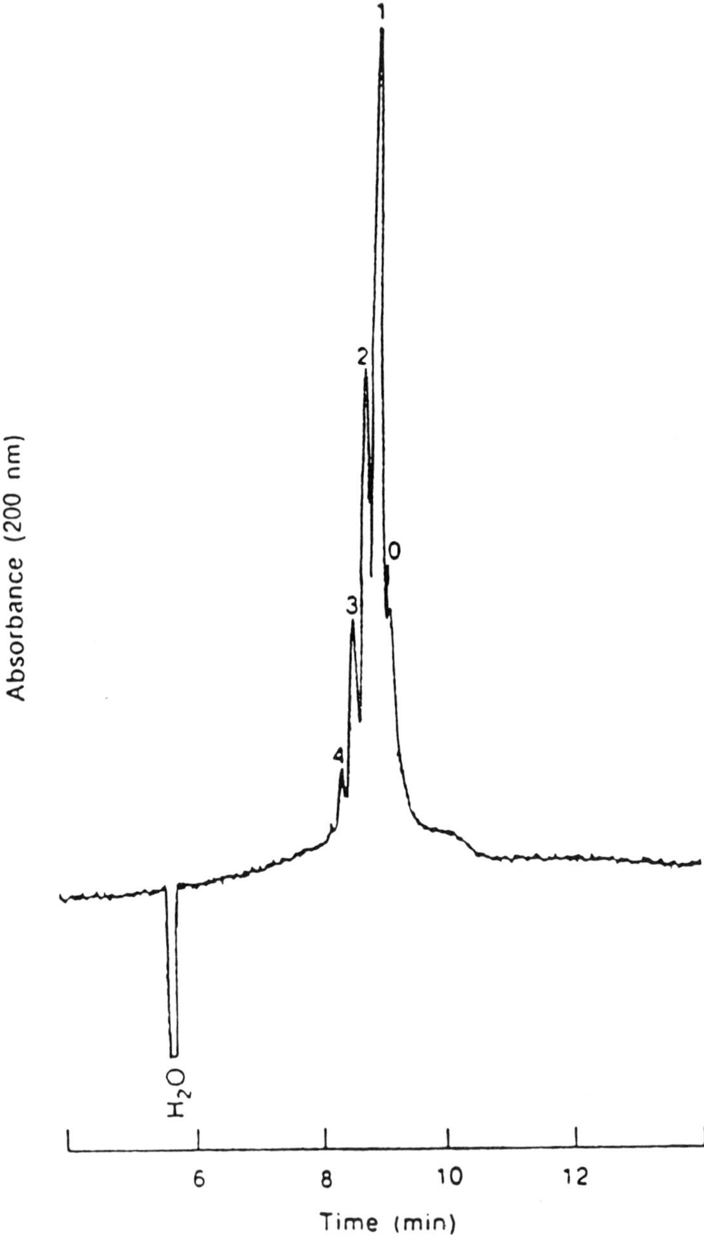

(a)

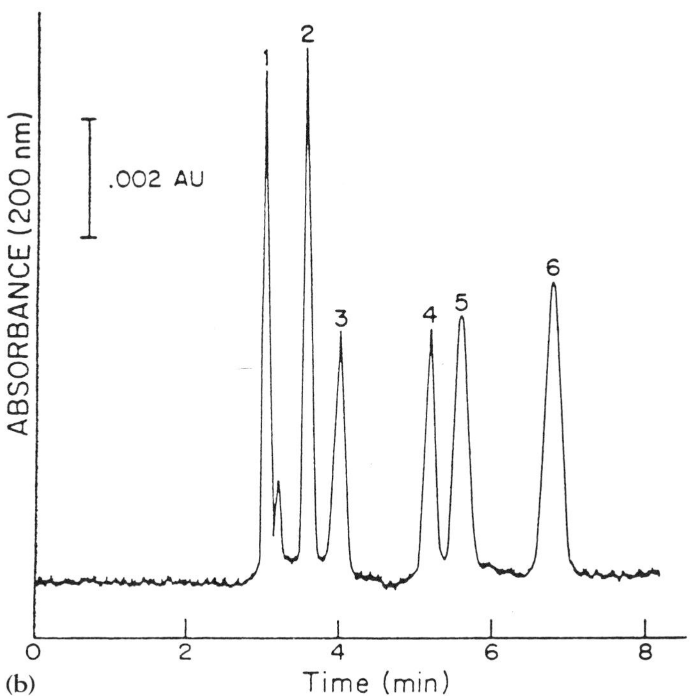

Figure 4 (a) Separation of multiacetylated cuttlefish testes histone H4. Values above peaks represent the number of modified lysines per H4 molecule. Conditions: field strength, 215 V cm^{-1}; current, 2 μA; buffer, 10 mM sodium acetate; capillary length, 70 cm (48 cm to detector); capillary diameter, 50 μm; separation temperature, 30°C. Horizontal scale in minutes. (b) Electropherogram of proteins at pH 7.0: (1) myoglobin, (2) conalbumin, (3) transferrin, (4) β-lactoglobulin B, (5) β-lactoglobulin A, and (6) ovalbumin. Detection was at 200 nm, 300 V cm^{-1}, and 26 μA on a 75-μm × 30-cm modified capillary, using 0.01 M phosphate/0.001% (w/w) Brij 35 buffer. (From Refs. a, 25; b, 26)

electric focusing; however, it can also at times be advantageous to eliminate electroosmotic flow in CZE. Historically, the addition of a neutral hydrophilic polymer, such as methylcellulose, was employed to minimize electroosmotic flow. This approach is used today in CITP [29] and with polymer networks (see later). Presumably, some methylcellulose adsorbs to the capillary wall to reduce protein adsorption and increase viscosity in the electrical double-layer region.

A popular neutral covalent coating to minimize electroosmosis is that designed by Hjerten [30]. As originally developed, a bifunctional agent such as methacryloyloxypropyltrimethoxysilane is attached to the silica surface, followed by free radical polymerization of linear (i.e., non–cross-linked) polyacrylamide. Long chains of polyacrylamide are thereby produced to yield a fuzzy viscous hydrophilic surface. This coating was modified by Novotny and co-workers [31] by using the Grignard reaction to attach the bifunctional silane reagent to the wall. This created an Si—C bond, rather than an Si—O—Si bond, the latter being less hydrolytically stable. Although successful, this approach is synthetically cumbersome and requires 48 hours for completion. More recently, Lee and co-workers have used a cross-linking step with the bifunctional agent for added stability [32].

The linear polyacrylamide coating has been successfully employed for protein separations by CZE and by CIEF (Fig. 5). Although effective, the coating is not fully stable, especially at a pH higher than 9, both from the point of view of hydrolysis of the Si—O—Si bond as well as hydrolysis of acrylamide to acrylic acid. Nevertheless, modifications of the coating or additives did permit in excess of 45 runs at pH 9 [33].

Other coatings and additives have also been employed [34–37]. However, it is fair to say that work is not yet complete in this field. The history of reversed-

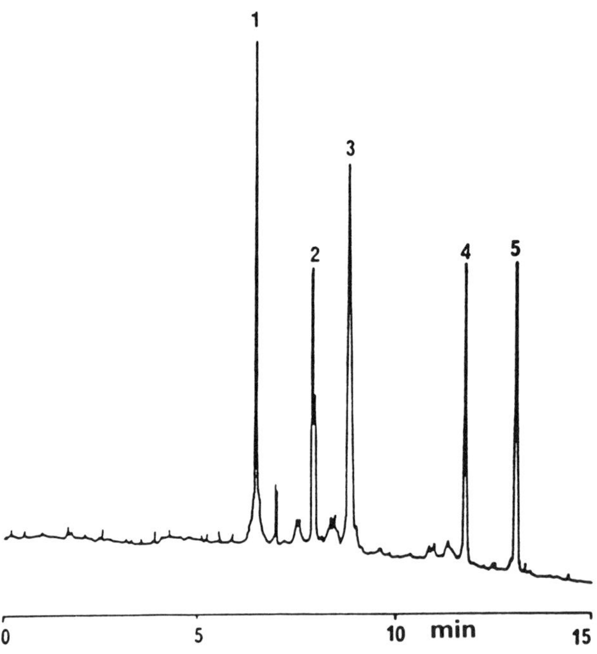

Figure 5 CZE separation of acidic proteins in a coated capillary at pH 8.8. (1) Glucose 6-phosphate dehydrogenase, (2) trypsin inhibitor, (3) β-lactoglobulin B, (4) 1-asparaginase, (5) α-lactalbumin. Detection was at 214 nm, 300 V cm^{-1}, and 7 μA on a 75-μm, effective length 30-cm coated capillary. BGE: 0.02 M TAPS-AMPD, pH 8.8. (From Ref. 194)

phase liquid chromatography is useful to cite. In the early days of HPLC (mid-1970s), *n*-alkyl-bonded phase columns were not fully stable, and many substances adsorbed because of silanol effects (e.g., basic drugs). Over the years, with accumulated experience and knowledge, reversed-phase columns today are rugged and routinely used. After having said this, it needs to be emphasized that successful HPCE operation is possible today for a wide variety of problems. Successful approaches will be discussed in the open-tube column section.

Achievement of Separation

As in chromatography [20], we are interested in the band disengagement or resolution of two or more components and the speed with which separation can occur. Following Giddings [38], resolution, R_s can be expressed as

$$R_s = \frac{1}{4}\left[\frac{\Delta\mu}{\bar{\mu}}\right]N^{1/2} \tag{9}$$

where $\Delta\mu$ is the difference in mobilities of the two species, $\bar{\mu}$ is the average mobility, and N is the number of theoretical plates.

If electroosmotic flow is included in Eq. 9, the resolution expression becomes

$$R_s = \frac{1}{4}\left[\frac{\Delta\mu}{\bar{\mu} \pm \mu_{eo}}\right]N^{1/2} \tag{10}$$

Clearly, the relation of $\bar{\mu}$ to μ_{eo} will affect the resolution. The electroosmotic flow in the direction of the migration will decrease the resolution and vice versa.

The separation factor α can be defined as

$$\alpha = \frac{\mu_1}{\mu_2} \tag{11}$$

Equation 11 can be included in Eq. 10, and with the simplication of $\bar{\mu} \simeq \mu_1$, we can write

$$\frac{\Delta\mu}{\mu_1 \pm \mu_{eo}} \simeq \left[\frac{\alpha - 1}{\alpha \pm \dfrac{\mu_{eo}}{\mu_2}}\right] \tag{12}$$

when $\mu_{eo} = 0$, Eq. 12 becomes $(\alpha - 1/\alpha)$ which is identical with the relation of the separation factor to resolution in chromatography [20]. Because of the $\alpha - 1$ dependency, as α becomes small, the number of theoretical plates required to achieve a particular separation increases substantially. For example, with no electroosmotic flow, for $R_s = 1$ and $\alpha = 1.1$, $N_{req} = 1600$; $\alpha = 1.01$, $N_{req} = 160{,}000$; and for $\alpha = 1.004$, $N_{req} = 1$ million. One of the advantages of HPCE is that plate counts of 10^5-10^6 or higher are possible; in HPLC, plate counts of 10^3-10^4 are more typical.

The very high efficiency means that column conditions to achieve separation are frequently easy to find. Indeed, often after only one or two scouting runs, separation is attained. That the separation factor α may be relatively of lower importance for the separation of closely related species in HPCE than in HPLC can be significant in the separation of chiral substances [39]. Second, the very high efficiency means that short column lengths can be used for fast separation. Indeed, this aspect, along with the high electric fields (roughly one order of magnitude higher than in slab gel operation), indicates that HPCE can be a fast method (separations in general take 10–15 minutes, but can be made in under 1 minute). Interestingly, in the pioneering days of HPCE, HPLC was thought to be a slower method, but advances in this latter field [40,41] have made separation times comparable for HPCE and HPLC.

As mentioned, Eq. 10 shows that resolution can be affected by the relative value of $\bar{\mu}$ to μ_{eo}. When $\bar{\mu}$ is small in comparison with μ_{eo} (independent of the direction of flow), resolution is small, since the denominator will be controlled by μ_{eo}. This means that solutes that migrate close to an uncharged species will have a substantial loss in resolution. This result is analogous to that in chromatography [20]; when retention is close to the unretained peak time (i.e., when k', the capacity factor, is close to zero), resolution suffers. In both events, transport is mainly by bulk flow, not by the separation method–electrophoresis or chromatography.

Separation thus requires that $\bar{\mu}$ be significant relative to μ_{eo}. When the solutes electrophoretically migrate in the same direction as the bulk flow, migration times will be short without substantial change in resolution. This approach can be useful for rapid analysis; however, there are limits. The solutes and the capillary wall will have opposite charges and adsorption may occur. Furthermore, charge differences may not be so critical (e.g., peptides run under acidic conditions with a bare silica column). Nevertheless, low-pH buffer operation can be a useful approach, particularly when it is realized that the extent of negative charge on the fused silica surface is limited, thereby reducing adsorption [42].

Finally, μ and μ_{eo} can have opposite signs, in which case the charge on the fused silica wall would be the same as that of the solutes. Interestingly, whereas solutes with the highest mobilities migrate fastest when μ and μ_{eo} have the same charge signs, the opposite is true when μ and μ_{eo} have opposite signs.

Returning to Eq. 10, the second factor influencing R_s is the number of theoretical plates or efficiency of the system. There are several factors that can influence band broadening in HPCE, and clearly an understanding of how these factors can be minimized is critical to successful operation. Given the miniaturization of the system (the column volume is typically $< 2\ \mu L$), extracolumn effects, arising from injection and detection, are important to consider. Most injection procedures use 5- to 25-nL sample volumes, with focusing before separation to minimize extracolumn effects; however, sample volumes as high as 1 μL are possible (see later discussion on sample preconcentration). On-column detection results in 1 to 5 nL cell volumes, which are generally satisfactory for performance. Thus, it can be concluded that, for the most part, systems and procedures have been properly designed to minimize the influence of extracolumn effects on band broadening.

A second factor that can influence efficiency is the amount of sample on the column. Sample overloading arises from distortion of the conductivity in the sample zone relative to other regions in the column [43]. As a consequence of the conductivity differences, the local field within the zone will differ from other buffer regions. Depending on the relative conductivities, either band fronting or tailing can result. However, proper buffer selection can minimize band asymmetry so that trace analysis can be conducted without sample splitting (see later discussion on trace analysis). Consequently, lower sample amounts generally lead to higher efficiencies.

A third factor that can cause band broadening is sample adsorption to the column wall. Generally, adsorption is diffusion-controlled and rapid; thus, the rate of desorption will depend on the strength of adsorption. With the very high efficiencies possible in HPCE, even a minimal amount of adsorption can lead to a reduction in performance. Furthermore, adsorption can lead to losses of sample, particularly given the high surface area/volume ratios in the capillary system. Adsorption of any component in a mixture can alter the performance of all species by changing the local zeta potential at the wall [44,45]. Such changes can lead to bulk solution-mixing effects within the altered zeta potential region, thereby broadening all components. Indeed, the open-tube separation of proteins requires that adsorption be minimized for all species present, to obtain the highest efficiencies.

Assuming the foregoing effects are minimized, band broadening in open-tube operation will be mainly due to axial diffusion for which N can be expressed as [7]

$$N = \frac{\mu E \ell}{2D} = \frac{\mu V}{2D}\left[\frac{\ell}{L}\right] \tag{13}$$

where D = the diffusion coefficient of the species in the running buffer at the column temperature T, ℓ = column length to the detector, and L = total column length between the reservoirs. The higher the electric field, the less time the substance will spend in the column and the less time that will be available for diffusion. Thus, faster separations lead to higher plate counts, as seen from Eq. 13.

The field cannot be taken to extremely high levels because of the effect of joule heating. A temperature difference ΔT between the center of the capillary and the wall can result, leading to different migration rates across the capillary and, accordingly, to band broadening [46]. The effect of joule heating can be seen in the following proportionality:

$$\Delta T \sim E^2 \lambda r^2 \tag{14}$$

where λ = molar conductivity of the solution, and r = tube radius. This expression shows that high electric fields, as well as high conductivity buffers, cause increased joule heating. Critical to the effect of joule heating is the tube diameter. In one sense, the use of capillaries can be viewed as the necessary ingredient to employ high electric fields for fast separation. This principle has been taken by Jorgenson in the use of 5-μm diameter capillaries with electric fields of 1000–2000 V cm^{-1}. By this approach, separations of 1 second or less have been demonstrated [47].

Another significant consideration for joule heating is the ability to rapidly remove the heat generated. It is widely accepted today that active cooling of the capillary is important to permit a wide range of conditions for operation (e.g., high-electric fields, high-conductivity buffers, wide-diameter columns). In the future, it can be assumed that improved cooling systems will be developed and that this will permit even faster separations or the use of wider-diameter capillaries.

INSTRUMENTATION AND OPERATION

Apparatus

The basic instrumentation for HPCE for all modes of capillary electrophoresis is shown in Fig. 6. The separation capillary is placed between two electrode reservoirs filled with the background electrolyte. Platinum electrodes E serve to connect the high voltage power supply HVPS, which delivers approximately up to 100 μA and

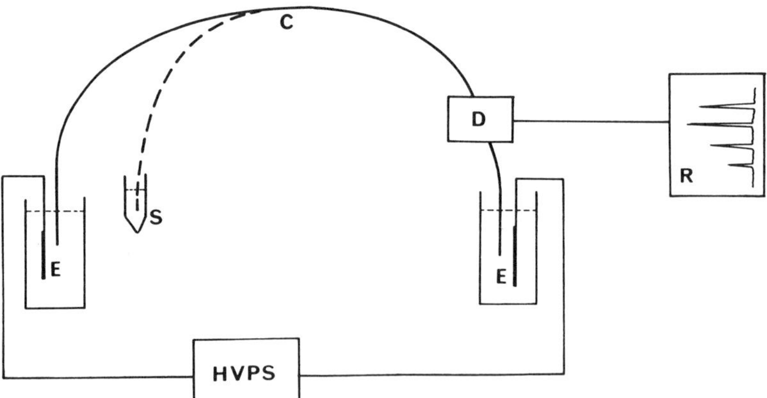

Figure 6 Basic scheme of the CE instrument. The separation capillary (C) is connected to the high-voltage power supply (HVPS) by electrolyte reservoirs equipped with electrodes (E). The sample (S) is injected either by flow or by electromigration, and separated zones are detected on the opposite end of the capillary by a suitable detector (D). The detection signal is either recorded by a line recorder (R), or can be further evaluated by a computer.

30 kV. The sample S is injected by pressure or electrophoretically into the capillary C, and zones of separated substances are analyzed by a suitable detector D (mainly UV) at the opposite end of the capillary. A typical instrument for CZE is now equipped with a fused silica separation capillary 25- to 100-μm id × 20–100 cm. The polyimide that surrounds the fused silica is removed by heat or by cutting (~ 2 mm) at the detection point. It is generally accepted that the temperature should be controlled in the capillary for good migration time reproducibility. Furthermore, as noted, active cooling (i.e., efficient heat removal from the outside walls) permits a wide range of conditions to be used [48].

Detection

Ultraviolet

The most widely used detector, currently, is UV absorbance, with the fused silica capillary acting as the on-column detector cell. Essentially, the commercial detectors represent modified UV detectors for HPLC. As such, many of the advances learned from the early days of HPLC have been incorporated, so that noise levels are often as low as 10^{-5} absorbance units. Commercially available detectors are single-wavelength, multiple-wavelength, or rapid-scanning spectrometers, and there has been initial commercialization of photodiode array detection. As with HPLC, the selection of the detector will be based on the specific problem of the analyst.

Equation 3 demonstrates that the velocity of ions in an electric field will depend on their mobility. Thus, the velocity of the various peaks migrating through the fixed detector cell will not be constant, but will decrease with longer migration times. Since the UV detector is concentration-sensitive, this velocity difference will have no effect on detector signal (except for a minor change in band broadening); however, it will have a significant effect on the peak width. Peak area measurements, therefore, require correction for migration velocity to be able to accurately measure relative peak areas. This point has been made many times in the literature [49,50]. Note that, since theoretical plates are a measure of the ratio of migration time to peak width in time units, no correction is necessary for this calculation.

Optimum detection for nucleic acid components is at 260 nm, whereas for peptides or proteins, low wavelength (210 nm or lower) is most sensitive. For carbohydrates, some derivatization is usually thought to be necessary to be able to sense in the UV region (e.g., with 2-aminopyridine) [51]. Other samples require appropriate wavelengths, as is well-known in HPLC. With filter UV detectors, operation at 191 nm is possible, at which, for peptides, a factor of two to three improvement in detector signal over 210 nm is obtained. Recently, high-sensitivity CE analysis of underivatized carbohydrates was also achieved at 191 nm [52]. In one commercial system, operation at 185 nm is possible [53]. These higher-sensitivity wavelengths have the disadvantage of detection of more interferences in the system, and buffers must be carefully selected. Nevertheless, the use of absorbance at wavelengths below 200 nm is important and should see wider application in the future.

One of the main issues with UV detection is that of insufficient detection limits. It is interesting that HPLC is often viewed as a more sensitive method than HPCE. Concentration detection limits, where concentration is defined in terms of the sample, are of the order of $10^{-5}-10^{-6}$ M for HPCE, whereas in HPLC, two orders of magnitude lower are possible. Interestingly, some workers emphasize mass detection limits for which, because of miniaturization, HPCE appears quite favorable (e.g., picomole to femtomole detection with UV). However, the amount of sample used is quite small and, as a result, individual species may not be detected. Sample concentration detection levels represent a more meaningful measure of limits of analysis.

The main reason for the difference in concentration detection limits is the relative volume of sample used in the two methods. In HPCE, typical injection volumes are 1–10 nL whereas in HPLC, 1–10 μL or even as much as 100–1000 μL can be readily employed. The critical point is the use of a trace enrichment or a preconcentration step at the head of the column. In HPLC, it is typical to operate initially under weak solvent conditions, such that sample components will adsorb and preconcentrate at the beginning of the column. Subsequently, gradient elution or step gradients are employed for migration of the components through the LC column. The need for the preconcentration step prior to the column and/or at the head of the column in HPCE is self-evident.

Several approaches have been developed for preconcentration in HPCE. In one procedure, in direct analogy to HPLC, a small amount of packing is placed between two frits at the head of the HPCE column. This packing can either be reversed-phase or a biospecific adsorbent (e.g. antibody) [54]. After adsorption of the sample on the surface of the packing, a small amount of displacing solvent (e.g., methanol in reversed-phase) is used for desorption. Although successful, this approach carries with it the danger of nonadsorption of specific components as well as the inability to desorb components completely from the packing. Furthermore, since HPCE is fundamentally a solution technique, the incorporation of a surface adsorption step is often undesirable.

A more typical approach is to employ a sample matrix in which the solvent has a lower conductivity than the running buffer in the column. In many cases, water or dilute buffer is used as the sample solvent. Since the voltage will then drop mainly over the low conductivity region, a large field will be imparted to the sample, driving ionic components onto the capillary where they will be essentially stopped owing to the low field across the column. The approach can be used with electrophoretic or pressure injection, and has been amply discussed by Chien and Burgi [55]. The method is widely used and can yield injection volumes in the 50 nL level without much difficulty.

An alternative approach is to employ isotachophoresis as a means of focusing the sample. Even though capillary isotachophoresis as a separation tool has not been a commercial success, nevertheless, the method is a powerful preconcentrating procedure. As first developed for HPCE, a coupled column arrangement was used. Wide-diameter Teflon tubing (e.g., 300 μm) permitted the use of as much as 10–50 μL for isotachophoretic focusing [56]. Detection of ITP-focused zones in the first column could be conveniently achieved with conductivity detection, and at the appropriate times, concentrated zones could be transported to the analytical capillary (50–100 μm) for separation purposes. This coupled column approach has been used not only with UV detection, but also as a means of preconcentrating before CE–mass spectrometry [57]. The coupled column arrangement has the advantage of being able to use large volumes as well as to provide electrophoretic cleanup of the sample before injection onto the analytical column. However, it adds instrumental complexity to the capillary zone electrophoretic system. In most cases where approximately a 1-μL volume will be satisfactory, the appropriate volume can be simply pressure-injected into the column itself [58]. Isotachophoretic focusing is first employed followed by CZE separation (i.e., transient on-column ITP preconcentration); all this can be readily accomplished on present-day commercial equipment.

The principle of preconcentration and self-sharpening is well-known in isotachophoresis. The concentrations and thus peak widths of the migrating zones are determined by the composition of the leading electrolyte. In the most simple case of monovalent ions, the concentration in the sample zone C_S is related to the concentration of the leading electrolyte C_L and mobilities of the sample components μ_S, leading ion μ_L, and counter ion μ_R by Eq. 15

$$C_S = \frac{C_L \mu_S (\mu_L + \mu_R)}{\mu_L (\mu_S + \mu_R)} \tag{15}$$

In practice, the concentration in the ITP zones are in the range of 0.1–0.8 C_L. From Eq. 15 it also follows that the concentration in the separated zones is independent of the original sample concentration. This property is essential for the analysis of minor sample components, which can easily be preconcentrated by at least several orders of magnitude. Thus, to achieve preconcentration, a leading electrolyte, with mobility higher than the sample components, and a terminating electrolyte, with mobility lower than the sample components, are necessary.

Transient on-column ITP preconcentration followed by CZE separation in the same column can be accomplished in several ways [58]. In one example, the column can first be filled with a running buffer in which a coion buffer component has high mobility (e.g., ammonium ion for cationic solutes and chloride ion for anionic solutes). The sample is first injected into the column, preferably by pressure, and then the injection side of the capillary is dipped into a terminating electrolyte buffer reservoir. With the application of the electric field, isotachophoretic focusing is established. The time for focusing is a function of the concentration of the salts in the sample, the higher its concentration, the longer the necessary time. In normal cases, 2–10 minutes will be satisfactory. After this period, the injection side of the capillary is placed in a buffer reservoir containing the leading electrolyte. The leading electrolyte ions then move through the sample and relax the focused zones into CZE separation. This change in buffer reservoirs can be easily accommodated by commercial instruments.

A second approach is to add leading ions to the sample itself. Here, the terminating buffer electrolyte is initially in the column. The sample–salt matrix is then injected into the column, and the injection side of the capillary is placed in the buffer reservoir containing the terminating electrolyte. Upon the application of the electric field, the leading electrolyte, here the salt, stacks and provides an ITP zone behind which the sample components can focus. Since the amount of salt is limited, the stacked zone will collapse in a short time, relaxing the sample com-

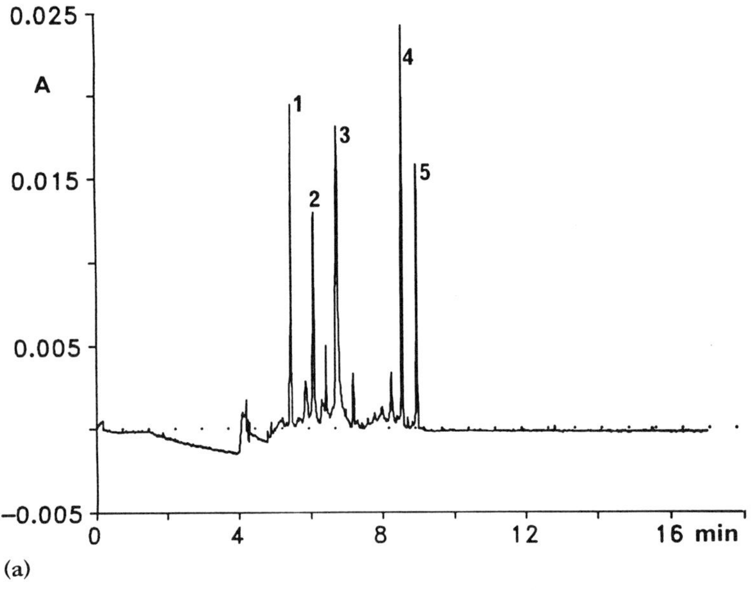

Figure 7 (a) CZE separation of acidic proteins with ITP preconcentration. BGE: 0.02-M TAPS-Tris, pH 8.3. Sample: (1) glucose-6-phosphate dehydrogenase, (2) trypsin inhibitor, (3) β-lactoglobulin B, (4) asparaginase, (5) α-lactalbumin, about 10^{-7} M each dissolved in 5 mM Tris-HCl buffer. Capillary: coated, effective length, 1, was 47 cm × 75-μm id, A-UV absorbance at 214 nm (b) CZE separation of basic proteins with ITP preconcentration. BGE: 0.02-M β-alanine-acetic acid, pH 4.3. Sample: (1) lysozyme, (2) cyctochrome *c*, (3) ribonuclease A, (4) myoglobin, (5) α-chymotrypsinogen, about 3×10^{-7} M each, dissolved in 5-mM sodium acetate. Other conditions as in (a).

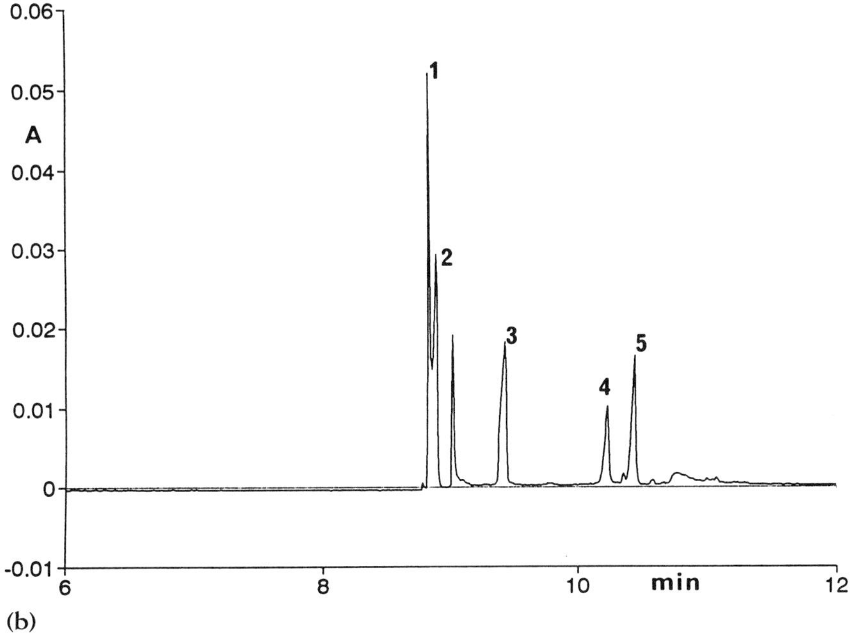

(b)

ponents for CZE separation. A theoretical discussion of this approach has been presented by Bocek and co-workers [59].

To illustrate this approach, Fig. 7a shows the separation of acidic proteins at pH 8.8, using a 500-nL injection volume in which focusing with added salt (Tris-HCl) is employed. The concentration of the samples are in the range of 10^{-7}–10^{-8} M, roughly two orders of magnitude below that typically seen with UV detection. Figure 7b shows a similar separation of basic proteins at pH 4.4, here, using sodium acetate as the salt. In both cases, a hydrophilic coating on the capillary wall was employed to minimize adsorption, and current was maintained constant. Reproducibilities of peak areas (2–3% RSD) and migration times (0.5% RSD), typical for CZE were found using standard mixtures [60]. Where necessary, a desalting may be used prior to ITP prefocusing. An added advantage of this ITP-focusing procedure over low-conductivity sample matrices is that the temperature in the sample zone will not be significantly elevated. A previous paper showed that the elevated temperature in the low conductivity region can sometimes reach 90°C or higher, leading to sample decomposition [61]. In addition, proteins are generally more stable with salt present than in pure water. In summary, with proper care, injections as high as 1 μL can be frequently achieved with the present day HPCE instrumentation.

Fluorescence

In contrast with UV detection, fluorescence is more selective and potentially more sensitive. Commercially available detection systems using conventional light sources, such as high-pressure mercury arc lamps, are available. For strongly fluorescent substances, two orders of magnitude lower detection limits are typically possible relative to UV.

In the past few years, there has been much interest in using laser-induced fluorescence (LIF) for ultratrace detection of substances in capillary electrophoresis [62]. This arises from the high intensity of the laser source, along with the ability to focus the coherent light on a small spot within the capillary. A helium–cadmium laser is used for excitation at 325 nm, an argon ion laser at 488 nm and 514 nm, and a helium–neon green laser at 543 nm, 594 nm, and above 600 nm. In the future, very inexpensive diode lasers will find wide application. Commercial argon ion laser-induced fluorescence detectors attached to CE instruments are now available [63].

Remarkably low detection limits are, in principle, possible in LIF. Indeed, given the small volume of the band in the detector cell, detection limits in the attomole and zeptomole ranges have been achieved [64–66] (zeptomole = less than 1000 molecules). This translates into concentration detection limits of 10^{-12} M or lower. However, it needs to be emphasized that such concentration detection limits represent the potential of LIF and are not presently attainable with real samples. The difficulties in analyzing real samples at the ultratrace level are well-known. Losses from adsorption on surfaces, matrix effects, and difficulties in constructing calibration curves, other than by dilution from higher concentrations, lead to detection limits more typically in the 10^{-8}–10^{-10} M level, still lower than UV detection. Furthermore, in many examples, derivatization to fluorescently active species is a necessary step [67]. Such procedures have potential problems in formation of extra products, nonquantitative reaction at the ultratrace level, and matrix interferences of reaction conditions. Thus, it must be emphasized that the LIF detection system is far ahead of the chemistry and sample-handling procedures for analysis of biological samples at the ultratrace level. With further study and with the development of derivatizing agents, there is the expectation that detection limits will continue to decrease over a period of time.

One interesting recent development involves the use of a high-powered argon ion laser, which is frequency doubled to provide excitation at roughly 275 nm [68]. Intrinsic fluorescence detection of proteins by tryptophan and tyrosine residues is then possible. This approach eliminates the need for derivatization, and thus in specific cases, laser-induced intrinsic fluorescence can be quite useful. However, it needs to be recognized that the response of individual substances can vary greatly and, therefore, calibration plots will be necessary for all quantitative determinations. Sample handling at the trace level still remains an issue. Nevertheless, a potentially exciting application of this technology is in the analysis of single red blood cells [69].

Indirect Detection

One of the problems with fluorescence detection is the need for fluorescently active species. The same argument can be used, but to a smaller extent, for UV detection. As originally introduced for HPLC [70], indirect detection can be employed for UV or fluorescently inactive species. The scheme of indirect detection is shown in Fig. 8a. Here, the background electrolyte contains either a UV-absorbing or fluorescently active constituent that provides a stable baseline signal. As zones migrate through the detector, they effectively dilute or displace the absorbing constituent to reduce the background signal, leading typically to negative peaks. The linear dynamic range is lower than in UV or fluorescence detection, and since the approach leads to universal detection, all substances, including potential interferences will be sensed. Indirect detection has been used most effectively in the separation and analysis of metal cations and small organic and inorganic anions [71,72]. As shown in Fig. 8b, the approach is rapid and can be competitive with ion chromatography. In the example shown, the background electrolyte is chromate; however, other UV-active substances can be used, depending on the sample. Indirect laser-induced fluorescence detection can also be employed as a trace universal analysis procedure [73]. However, the stability of the laser light is difficult to control.

Capillary Electrophoresis–Mass Spectrometry

Given the zero or very low bulk flow rate (< 1 μL/min), resulting from electroosmosis in the capillary, CE should be a compatible technique for coupling to mass spectrometry (MS). Indeed, routine coupling will ultimately be an important development for the full success of capillary electrophoresis. In many analyses, it is necessary to determine the identity of individual peaks for method validation. Furthermore, MS can serve not only as a universal CE detector, but also as a second dimension of separation. In one approach, samples can be collected after HPCE separation and then analyzed by mass spectrometry [74]. However, this approach is not as desirable as on-line coupling.

The most important issue in on-line CE–MS is the interface for transfer of substances between the column and the vacuum system of the mass spectrometer. Developments in mass spectrometry have greatly aided in the coupling, particularly in terms of atmospheric ionization. Transport directly into the vacuum would lead to hydrodynamic flow in the capillary and loss in performance. Smith and co-workers were the first to develop on-line CE–MS under atmospheric conditions using electrospray ionization [75]. In electrospray, a high voltage (3–6 kV) is applied between the electrospray tip and the entrance of a quadrupole mass spectrometer. This potential drop provides a cloud of charged droplets, with evaporation assisted by a counterflow of warm drying nitrogen gas [76]. Charged ions are released from very small droplets in the vapor state (ion evaporation) and are then brought directly into the vacuum region of the mass spectrometer through a sampling orifice. To optimize flows at the electrospray interface, when using CE, a flowing sheath consisting of typically 0.1–1% acetic acid in 1:1 methanol/water is often used. A related method, pneumatically assisted electrospray (ion spray), originally introduced for LC–MS by Henion and co-workers [77], has been used for CE–MS, as well [78].

Figure 9 shows the CE–MS separation and mass spectra of proteins using the electrospray approach. An important feature of electrospray ionization illustrated in this example is that multiply charged ions are produced, for example, protonation of many basic groups on a protein. This multiple ionization permits high molecular weight compounds to be analyzed with an ordinary quadrupole mass spectrometer,

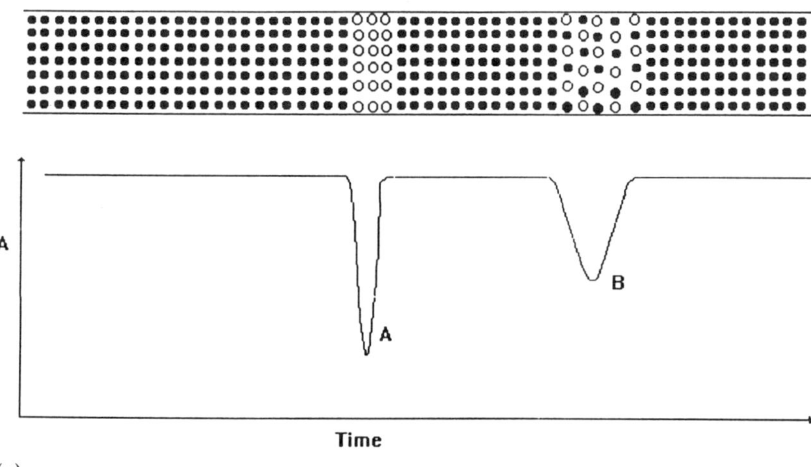

(a)

Figure 8 (a) The principle of the indirect detection in CZE. A constituent of the BGE with the required optical property (black dots) is used to provide a stable background signal. When optically transparent sample ions (white dots) replace the BGE constituent, a decrease of the absorbance A during their transition through the detection cell is recorded as negative peaks A, B. (b) Electropherogram of 36 anions. Peaks, concentrations [ppm]: 1, thiosulfate [1.3]; 2, bromide [1.3]; 3, chloride [0.7]; 4, sulfate [1.3]; 5, nitrite [1.3]; 6, nitrate [1.3]; 7, molybdate [3.3]; 8, azide [1.3]; 9, tungstate [3.3]; 10, monofluorophosphate [1.3]; 11, chlorate [1.3]; 12, citrate [0.7]; 13, fluoride [0.3]; 14, formate [0.7]; 15, phosphate [1.3]; 16, phosphite [1.3]; 17, chlorite [1.3]; 18, glutarate [1.7]; 19, *o*-phthalate [0.7]; 20, galactarate [1.3]; 21, carbonate [1.3]; 22, acetate [1.3]; 23, chloroacetate [0.7]; 24, ethanesulfonate [1.3]; 25, propionate [1.3]; 26, propanesulfonate [1.3]; 27, *dl*-aspartate [1.3]; 28, crotonate [1.3]; 29, butyrate [1.3]; 30, butanesulfonate [1.3]; 31, valerate [1.3]; 32, benzoate [1.3]; 33, *d*-glutamate [1.7]; 34, pentanesulfonate [1.7]; 35, *d*-gluconate [1.7]; 36, *d*-galacturonate [1.7]. The electrolyte is 5 mM chromate and 0.4 mM OFM Anion-BT adjusted to pH 8.0. Applied potential is 30 kV (negative polarity). Capillary dimensions are 60 cm (52 cm to detector) × 50-μm–id fused silica. Indirect UV detection. Injection by electromigration at 1 kV for 15 seconds. (From Ref. 72)

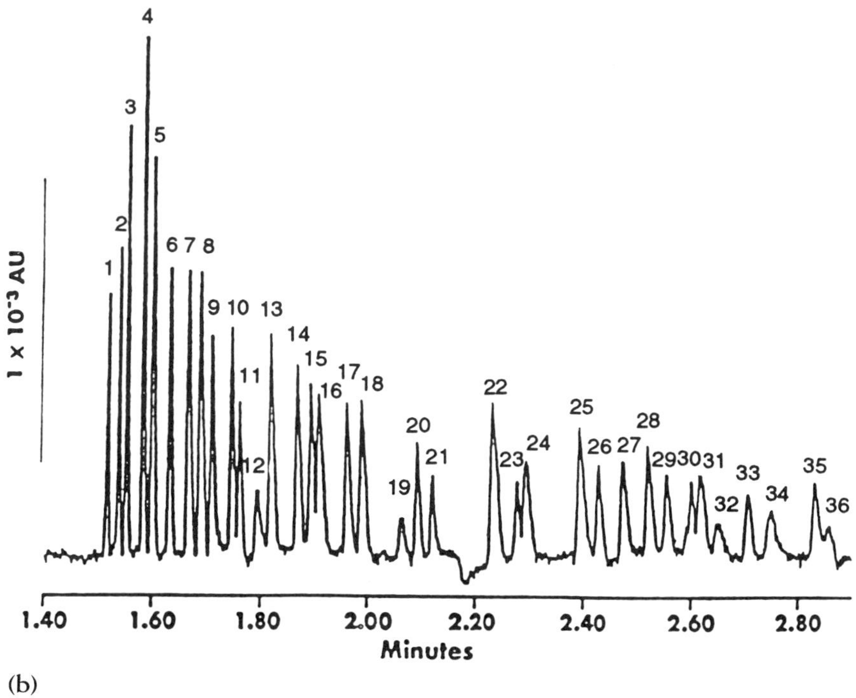

(b)

within the range up to 3000 m/z. Molecular weights as high as 100,000 or more have been determined by this approach [79].

Several issues remain to be resolved for the use of CE–MS. First, a better understanding of the influence of buffer constituents and pH relative to electrospray signal is necessary. In general, acidic buffer systems in the capillary have most often been used. For example, peptides have been analyzed under acidic conditions using capillaries with positively charged surfaces [80]. Some work using basic pH conditions in the capillary has also been undertaken [81].

A second issue is detection levels. Typically, concentration detection limits of roughly 10^{-5} M have been found, and this is certainly not low enough for trace analysis. Coupled column ITP [57] and transient on-column ITP focusing has been used as a means to analyze proteins at 10^{-7} M or lower level [82]. In addition, narrow-diameter capillaries of 20 μm or smaller have been used to reduce the current in the capillary to the nanoampere range, which is similar to that in the electrospray cloud. This permits a more efficient utilization of ions from the capillary. With this approach, subfemtomole analyses have been shown [83]. Another limiting factor that is related to electrospray system is that only a small fraction of the ions formed

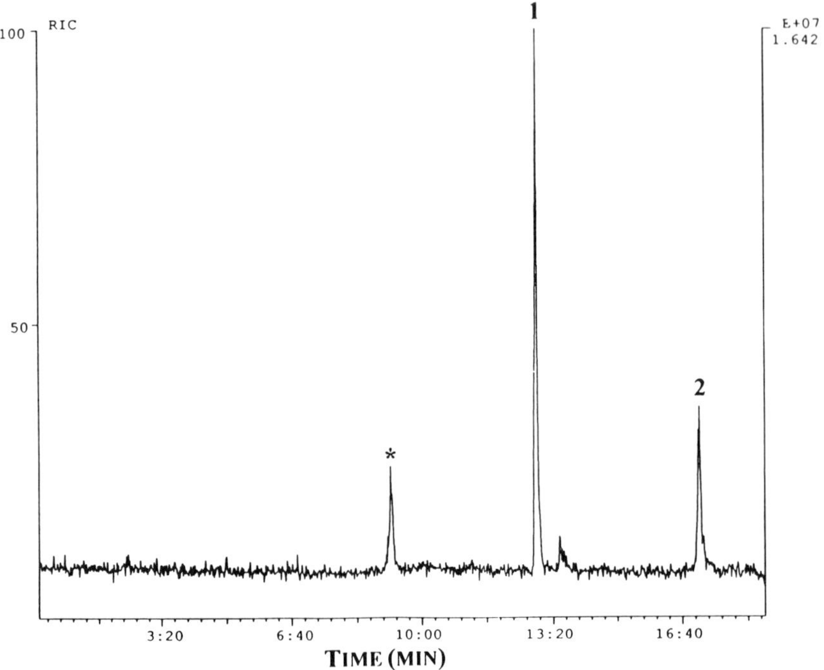

Figure 9 Full scan (m/z 600–m/z 2000) reconstructed ion electropherogram of 450 nM cytochrome *c* (1) and 270 nM myoglobin (2) dissolved in 0.005 M ammonium acetate buffer. The zones were focused by transient isotachophoretic migration. The peak marked (*) is from the rear boundary of the ammonium zone. Injection volume, 750 nL (injected quantity of 326 fmol and 195 fmol, respectively). BGE: 0.02 M 6-aminohexanoic acid + acetic acid, pH 4.4. CE conditions: voltage 18 kV, current 6 μA. Capillary: 75 μm × 50 cm. (From Ref. 82)

in the electrospray are actually collected by the sample orifice and transported into the mass analyzer. Improvements in ion collection efficiency will also greatly aid lowering detection limits.

Rapid developments in CE–MS over the next several years can be anticipated. One present difficulty is that the peak widths of rapid separations of high efficiency in CE are typically only a few seconds. For effective utilization of such separations, rapid-scanning mass spectrometers must be employed. The development of time-of-flight mass spectrometry with electrospray ionization can be expected in the coming years [84,85]. In addition, ion-trap mass spectrometry with electrospray ionization [86] may also be used. Within the next few years, it is expected that CE–MS will become a routine and rugged tool for analysis of samples at the femtomole to attomole levels. As in other areas, it will fully complement LC–MS, which has now become a well-developed tool.

OPEN-TUBE CAPILLARY ELECTROPHORESIS

This chapter has already covered several aspects of capillary electrophoretic operation in the open-tube format (to be contrasted with polymer network-filled capillaries). In this section, the applicability of the methodology will be discussed. Only a small flavor of all the possibilities will be presented. Much more will be covered in the other chapters of this book.

Micellar Electrokinetic Capillary Chromatography

Equation 4 has shown that mobility depends on the solute properties in terms of q/a (i.e., effective charge per unit mass). Thus, both the overall charge and size of the solute can influence migration and separation. If neutral species are considered, $q = 0$ and $\mu = 0$, leading to no separation. Without electroosmotic flow, such neutral solutes remain essentially stationary, whereas with flow, they move at the same velocity as the buffer ions, v_{eo}. Another aspect of Eq. 4 is that hydrophobic or nonpolar characteristics do not directly influence μ and thus separation. Such forces, however, can at times be very critical from a separation point of view (e.g., reversed-phase chromatography).

Considering these issues, Terabe in 1984 [11] introduced the method of micellar electrokinetic chromatography (MEKC). A detergent, such as SDS, is added to the running buffer in sufficient concentration to be above its critical micelle concentration (CMC). Since the detergent SDS is negatively charged, it will migrate slowly in a bare fused silica column as it is dragged toward the detector by the bulk electroosmotic flow of opposite polarity.

It is well-known that solutes can partition within the micelle by hydrophobic interactions. When a solute is in the micelle, it will migrate with the velocity of the micelle v_m. On the other hand, when it is free of the micelle, it will migrate with its own zone velocity v. If the solute is neutral, this latter velocity would equal the bulk electroosmotic flow velocity v_{eo}. In MEKC, the migration time of the solute t_R is transformed from being solely due to electrophoresis to being controlled by the capacity factor k' for solute partitioning in the micelle [11]:

$$t_R = \frac{t_o(1 + k')}{1 + \dfrac{t_o}{t_m} k'} \tag{16}$$

where t_m is the migration time of the micelle. This equation is identical with the retention time equation in chromatography [20], except for $(t_o/t_m)k'$ in the denominator. On this basis, the micelle can be considered as the "stationary phase" of chromatography, moving at a fixed constant rate through the column.

Figure 10 illustrates MEKC with standard drug substances. The sharp bands are a consequence of the rapid formation and breakage of the micelles (millisecond time scale), leading to fast mass transfer. Also, the separation time window is a function of the difference between t_o and t_m, and this ratio affects solute migration time t_R (see Eq. 16). If k' is too large, the solute will essentially migrate with the micelle time t_m, which could be excessive. A large k' may be a consequence of significant hydrophobicity for the solute. The change to a more hydrophilic detergent (e.g., a bile salt instead of SDS) would speed up separation here [87]. One can also control t_m by manipulating the charge on the micelle [88]. On the other hand, if the k' value is too small (0.2 or less), the solute will act as though it is unaffected by the micelle.

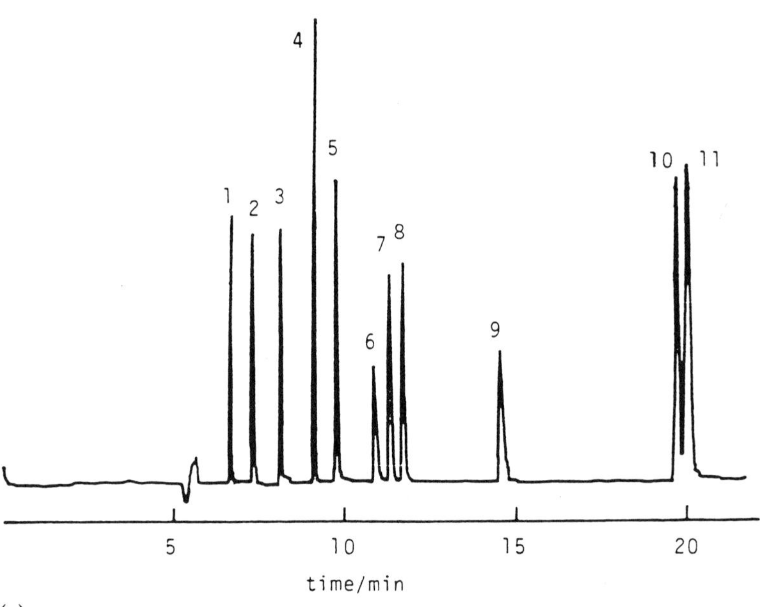

(a)

Figure 10 (a) Micellar EKC analysis of a cold medicine: 1, acetaminophen; 2, caffeine; 3, sulpyrin; 4, naproxen; 5, guaifenesin; 6, impurity; 7, phenacetin; 8, ethenzamide; 9, 4-isopropylantipyrine; 10, noscapine; 11, chlorpheniramine and tipepidine; applied voltage, 20 kV; detection wavelength, 220 nm. Capillary: 50 μm × 650 mm (500 mm to detector). (b) Chromatogram of four enantiomeric solutes and two achiral solutes. Buffer: 0.05 M SDS solution of pH 9.0 containing 40 mM β-CD and 20 mM sodium *d*-camphor-10-sulfonate. (1) Thiopental (sodium); (2) pentobarbital (calcium); (3) 2,2,2-trifluoro-1-(9-anthryl)ethanol, (4) 2,2-dihydroxy-1,1′-dinaphthyl; (5) phenobarbital; (6) barbital. (From Refs. a, 192; b, 193)

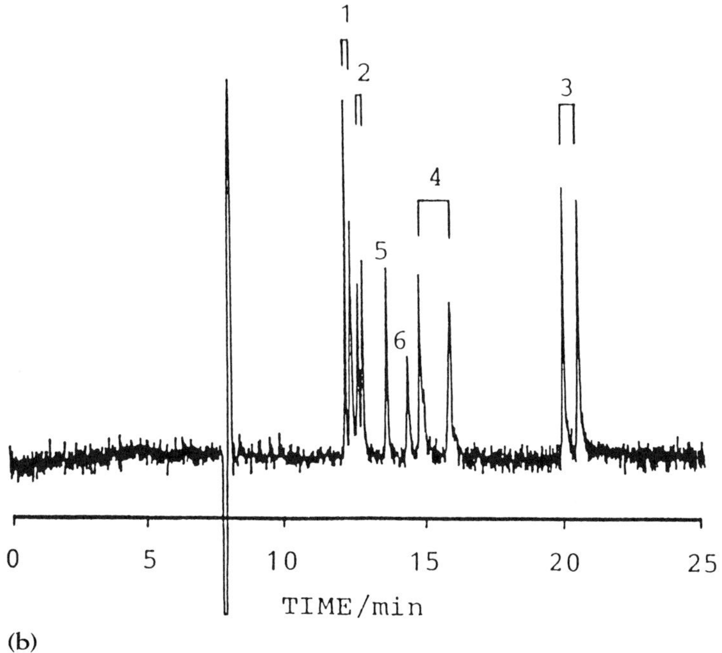

(b)

Micellar electrokinetic chromatography was one of the first rugged methods developed for HPCE, and it is now a well-established approach. Since there is a transformation from electrophoresis to partition chromatography in HPCE, neutral solutes can be separated. Also, hydrophobic interactions can be involved in the separation process by partition within the interior of the micelle. Of course, ion pairing with the surface of the micelle can also play a role in separation, as well as ligand exchange when the micellar surface contains metal ligands [89].

In general, MEKC is limited to small molecules that can penetrate into the micelle. Thus, amino acids, for example, PTH–amino acid derivatives [90], drugs (see Fig. 10), and organic chemicals work well. Proteins, which are larger species, are more difficult to solubilize within the micelle. Some efforts to use larger inclusion bodies are underway [91]; however, this area needs more study.

MEKC can be considered an interacting medium to manipulate and control electrophoretic migration. There are many other possible interaction approaches, and it is easy to predict that this area will expand greatly in the coming years. We can only cite a few examples here.

Cyclodextrins, appropriately modified for electrophoretic migration, can be used as selective complexing systems, based on the ability of species to penetrate the toroidal cavity of the complexing agent. When an inclusion complex forms, the solute electrophoretically migrates at a different rate than when uncomplexed. Modified cyclodextrins have been successfully employed for peptides [92] and chiral separations [93]. It is noteworthy that for chiral species, an alternative separation mechanism to selective inclusion can be provided by differential complexation with specific proteins added to the running buffer (e.g., bovine serum albumin; BSA) [94]. As in HPLC, underivatized enantiomeric pairs can be separated by this latter approach.

Complexation principles lead directly to affinity electrophoresis [95]. In this method, biospecific interaction plays a significant role in selectively modifying electrophoretic mobility. Although the field of affinity capillary electrophoresis (ACE) is in its infancy [96], two significant applications are already beginning to emerge. First, the rapid determination of binding constants has recently been demonstrated for an enzyme with its inhibitor [96a] and for heparin with a small peptide [97]. In the former, both the complexed and uncomplexed enzyme species were rapidly quantitated for the determination of the binding constant and binding characteristics (Scatchard plot). The approach was identical with standard mobility shift assay procedures widely used in slab gel electrophoresis. On the other hand,

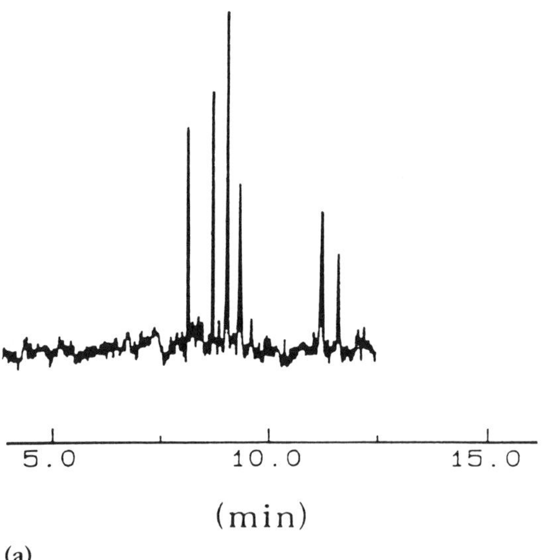

Figure 11 Optimum in separation at pH 8.22. Buffer: tricine/KCl 10/20 mM, pH 8.22; voltage: 2 kV and 6 seconds for injection, 20 kV and 14 μA for electrophoresis. Elution order: myoglobin (whale skeletal muscle), myoglobin (horse heart), carbonic anhydrase B, carbonic anhydrase A, β-lactoglobulin B, β-lactoglobulin A. (b) Electropherogram of hGH digest in pH 8.1 tricine buffer. (From Refs. a, 101; b, 102)

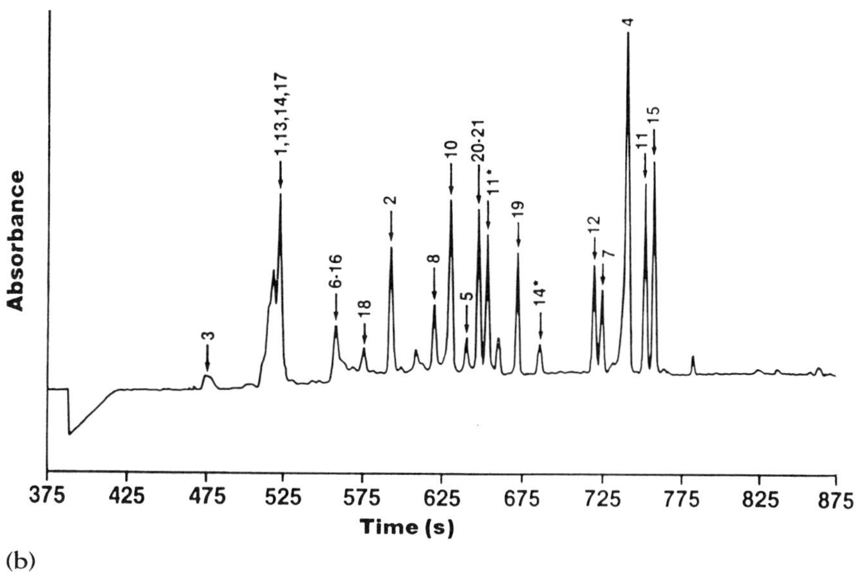

(b)

since the free peptide was positively charged and heparin and peptide–heparin complex were both negatively charged [97], only the free peptide was injected into the capillary for quantitation.

These approaches will become of real value in the drug discovery area. Today, selection of therapeutic agents often involves random searches of numerous related species (e.g., peptides). An important clue in the selection of a potentially active species is the extent to which such a molecule will bind to an antibody, receptor, or other. The use of HPCE offers the advantages of (1) speed and automation, (2) the need for only a small quantity of complexing agent and binding agents, (3) binding studies of multiple species in one run containing arrays of biospecific agents, and (4) minimal purification. Although binding studies on arrays of peptides on chips are being actively developed [98], nevertheless, HPCE would appear to have an important future in this field for antigen–antibody, protein–DNA, drug–receptor, and other important biospecific interactions.

Another area for which ACE can be important in the future is trace analysis. It is well-known that labeled biospecific agents can lead to high-detection specificity at the trace level [99]. Immunoassays are based on this principle. The use of fluorescently labeled fragments with laser detection would appear to offer many potential advantages in CE. The problems of sample derivatization at the trace level have already been discussed. For labeled biospecific agents, many of these problems are overcome, since the derivatization essentially arises from the noncovalent attachment of the labeled substance. Moreover, since only a small quantity of material is used in each assay, complex derivatization and purification procedures to obtain the labeled fragments can be employed, if necessary.

Recently, our laboratory has fluorescently labeled (rhodamine) Fab′ fragments of antihuman growth hormone and achieved quantitative analysis of methionyl-human growth hormone at detection levels below 5×10^{-12} M [100]. Future work with cyanine dyes and inexpensive diode lasers in the 633-nm excitation region should decrease background noise, even beyond that for rhodamine excitation, for potentially improved detection limits. Fluorescently labeled fragments should in the future become important reagents for ultratrace analysis by HPCE.

Capillary Zone Electrophoresis

The separation of peptides and proteins in the CZE format has already been discussed, and this section will simply integrate that material with a few added topics to provide a perspective on the field.

As mentioned, a main concern is to reduce adsorption to the walls of bare silica. In 1986, Lauer and McManigill published an important paper in which they showed separation of basic proteins with plate counts of 8×10^5 using a running buffer of pH 8.8 and added amine (Fig. 11a) [101]. The basic pH reduced the positive charge on the proteins, and as in HPLC, the amine provided an adsorption competitor for the silanol groups on the silica wall. This work convincingly demonstrated the potential of CZE for high resolution of proteins. Basic pH conditions thus represent one strategy to attempt separation of oligopeptides and proteins. For example, Fig. 11b illustrates the separation of peptides from a tryptic digest of human growth hormone at pH 8.1 with a tricine buffer and 0.02 M morpholine [102].

The numbers on top of the electrophoretic peaks in Fig. 11b represent the elution order of the peptide fragments in reversed-phase LC. It is readily apparent that an entirely different pattern is observed in CZE relative to RPLC and, therefore, the two approaches are complementary to each other. The same principles of complementarity for peptide mapping are found when acidic conditions (e.g., pH 2.5) are used for separation. Researchers have collected peaks off an HPLC column and then run them in CZE to ascertain the number of components that are under an LC band [103]. Such an approach obviously argues for on-line coupling (i.e., LC–CE). This coupling has now been successfully accomplished by Jorgenson and co-workers, and impressive separations of complex mixtures have been achieved [104]. The strategy of this procedure is to collect fractions in a sample loop and to analyze an individual fraction by CZE before the full collection of the next LC fraction in the sample loop. Separations in CZE of only a few seconds are required to analyze all the fractions from HPLC in a continuous fashion. An alternative approach to two-dimensional separations is to couple CE (or LC) directly to mass spectrometry (a separation procedure for ions in the vapor state). As previously noted, the full development of CE–MS will be of significance to CE both from the points of view of resolution and peak identification.

Returning to the application of bare silica for peptide and protein separations, additives have been used to reduce adsorption (and control electroosmotic flow). Such strategies include (a) zwitterion or detergent added to the running buffer [105]; (b) 100- to 200-mM salt concentrations (to suppress electrostatic interactions) [106]; (c) polymer additives (e.g., methylcellulose [107]), and (d) polyamines [25]. Each strategy has advantages in specific cases, and all will not be successful for every application—the sample matrix or specific protein could significantly affect the approach taken. The list does, however, show that many problems can be addressed, and there is still the opportunity to use coated capillaries.

Coated capillaries offer the advantage of good reproducibility and the minimization of adsorption; however, above pH 9–10, such columns may not be stable for long periods owing to either polyacrylamide hydrolysis or hydrolysis of the Si—O—Si bond, whereas bare silica may be easier to use at basic pH. As will be seen, pH stability is especially important in CIEF.

In general, the highest peak efficiencies will be achieved when electroosmotic flow is negligible (i.e., by the use of neutral coating layer thicknesses of 100 Å or higher). However, electroosmotic flow allows both acidic and basic proteins to be separated. Thus, it seems reasonable to conclude that both approaches will be of value in the future.

Finally, it is worth emphasizing that the very high efficiencies of CZE for proteins make this method potentially powerful for analyzing biopolymers. The free-solution aspects of the method are also worth noting in that recoveries, in general, should be high, assuming wall adsorption is minimized. One interesting application of CZE is in the process monitoring of various stages of purification in biotechnology [108]. The cited paper emphasized the direct analysis of samples from complex matrices such as fermentation broths. Given the high resolution of CZE, it is also ideally suited for analysis of final purification or polish stages. Applications such as these are appearing in the literature, and this leads to the conclusion that in the future, CZE will become an important separation tool for peptides and proteins.

Capillary Isoelectric Focusing

Isoelectric focusing (IEF) is one of the most powerful techniques for the high resolution of protein mixtures, with separation based on differences in isoelectric points (pI) of the species [109,110]. The technique depends on establishment of a stable pH gradient along the separation path generated either by a suitable liquid mixture of zwitterionic components—ampholytes—or by similar groups covalently attached to the structure of a stabilizing gel—Immobilines [111]. During focusing, proteins move toward the position at which the pHs equal their respective pIs. Since the proteins possess no net charge at their pI, they remain focused at pH = pI. Conventional IEF is performed using a gel matrix, such as polyacrylamide or agarose, either in slabs or in tubes. It is a widely used method of separation and analysis, well understood by the biochemist.

The resolving power in IEF can be expressed as Δ(pI) for two species as [112]:

$$\Delta(\text{pI}) = 3\left(\frac{D[d(\text{pH})/dx]}{E[-d\mu/d(\text{pH})]}\right)^{1/2} \tag{17}$$

Resolution is increased as Δ(pI) is increased. Thus, the shallower the pH gradient, $d(\text{pH})/dx$, the better the separation after focusing. Since the bands are in a sharp zone at their pI, there is a significant concentration gradient that drives the molecules away from the pI region by diffusion D. However, the higher the field E and the more sensitive the electrophoretic mobility of the molecule to pH, $d\mu/d(\text{pH})$, the sharper the bands will be. Hence, high fields improve resolution, and capillaries permit high E values, without the joule heating effect. Thus, CIEF is a logical means of conducting IEF separations. Furthermore, a stabilizing gel, necessary in the slab format, is not required in the capillary, for which the walls serve the purpose of an anticonvective medium. In addition, as with all modes of CE, CIEF offers the possibility of automation, speed, and the ability to quantitate bands.

Hjertén is generally recognized as the pioneer in CIEF, initially in capillaries of 200 μm [113] and then in more typical 50–100 μm [114]. In all his applications, a coated capillary has been employed to minimize electroosmotic flow to a negligible level. Typically, the sample and ampholytes of the appropriate pH range (3–10, 5–8) are mixed together and are injected into the column. Focusing is established, as measured by the drop of the current to a minimal value. Note that in this approach, the sample is injected into the whole column and then focused. Thus, in analogy to the previous discussion of ITP preconcentration, CIEF is suited for trace analysis of proteins.

Once focused, the CIEF procedure requires some procedure for determination of the zones. One approach, at an early developmental stage, is to perform whole-column detection, in which the refractive index gradient of the concentrated zones is measured by the scattering of laser light [115]. This method would be rapid and would indeed not require complete focusing for quantitative analysis; however, whole-column detection is far from being commercialized and, consequently, it is now difficult to predict its future.

The current practice of CIEF is to mobilize the focused bands past the detector in one of two ways. First, the composition of either the anolyte or catholyte can be changed to create a pH gradient in the column [114]. This can be most practically accomplished by adding salt to either buffer reservoir [116]. Addition of salt (e.g., NaCl or zwitterion) to the cathode (basic pH) reservoir would, e.g., create a competition between Cl^- and OH^- from entering the column, thereby reducing the

OH^- concentration and the pH. A pH gradient is then established to move the ampholytes and the focused zones by the detector with the electric field applied. Figure 12a illustrates this approach in the separation of hemoglobin variants [117]. Note that hemoglobin A and F differ by only 0.05 in pI units, and there is still baseline separation. A good discussion of this method can be found in a paper by Wehr et al. [118].

One issue with the salt mobilization approach, however, is that the pH or pI versus time plot is not linear. Thus, resolution as a function of pI differences will not be constant over the whole pH range, and accurate determination of pI values in nonlinear regions would be difficult. An alternative mobilization procedure involves flow of the contents of the capillary by the detector with the electric field applied [114,119]. In the method that has been published, the sample and ampholytes are focused first, followed by the flow of the focused zones to the detector, with application of the high voltage. The resolution does not appear quite as high as the salt mobilization procedure; however, the method yields linear pH (pI) versus time plots over the whole range of pH 3–10. Figure 12b shows the separation of four peptides, with pI ranging from 2.75 to 3.3, a pH region that can be difficult to achieve without linear pI versus time.

One general issue in CIEF is that when operating over the pH range of 3–10, the capillary is exposed to high pH from the cathode. This creates severe demands on the coating [120]; for example, the previously described linear polyacrylamide coating would not be successful. A proprietary coating is used in the salt mobilization procedure [117,118]. A GC immobilized coating (DB-1 wax), which should be pH stable, has been used [119]. Also, 0.4% methylcellulose is added to the running buffer/ampholytes to reduce adsorption on the coating and to suppress any electroosmotic flow.

An alternative procedure to the foregoing approaches is to conduct CIEF in bare silica in which mobilization starts at the beginning of the analysis owing to electroosmotic transport [121,122]. To control electroosmotic flow and potential adsorption, a polymer, such as methylcellulose, is added to the buffer. A thin covalent coating that allows electroosmotic flow can also be employed. This method has the potential to work over the pH range of 3–10 and is simple to operate. Two potential issues, however, are that the pH (pI) versus time plots are nonlinear as a consequence of the change in zeta potential at the wall with pH, and the possibility of adsorption of matrix components with "real" samples affecting electroosmotic flow. The use of a coated capillary that still allows electroosmotic flow may help in this regard.

A detailed discussion of the CIEF approaches has been presented to emphasize that successful results can be achieved in terms of reproducibility of migration and quantitation. CIEF is a significant procedure because there are important applications currently being performed with slab gels, for example, antibody characterization, hemoglobin variant analysis, and protein degradation (e.g., deamidation) analysis.

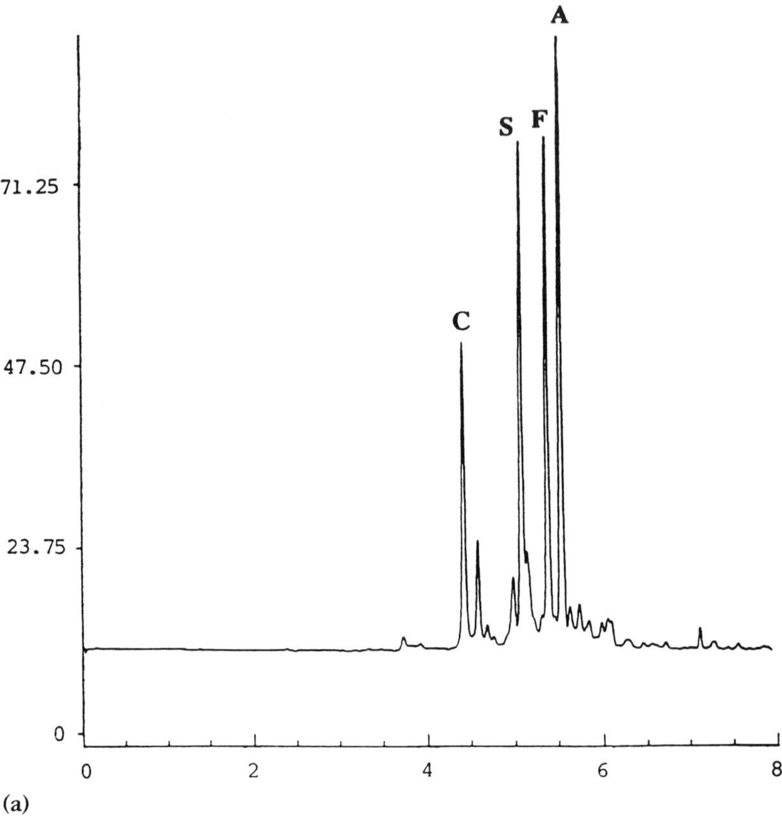

Figure 12 (a) Separation of hemoglobin variants by capillary IEF in a 12-cm × 25-μm coated capillary using three to ten ampholytes. Focusing and mobilization were carried out at 8-kV constant voltage. Protein concentration was approximately 250 μg/mL for each protein. Isoelectric points are (1) hemoglobin A, pI 7.10; (2) hemoglobin F, pI 7.15; (3) hemoglobin S, pI 7.25; (4) hemoglobin C, pI 7.50. (b) Capillary isoelectric focusing of RNase T and RNase ba. Sample 200 μg/mL, 4 ng injected. Capillary: 50 μm × 72 cm (50 cm to detector). (From Refs. a, 117; b, 119)

Use of CIEF should have clear advantages over slab gel operation in terms of quantitation, speed, and automation, so that with rugged procedures and columns, it is anticipated that CIEF will be widely used.

It is anticipated that CIEF will rapidly develop over the next few years in terms of new columns, optimized mobilization procedures and, most especially, broad applications to validate the technique. It is expected that CIEF will be used in conjunction with CZE and size separation of SDS–protein complexes as the main tools for protein separation and analysis by HPCE.

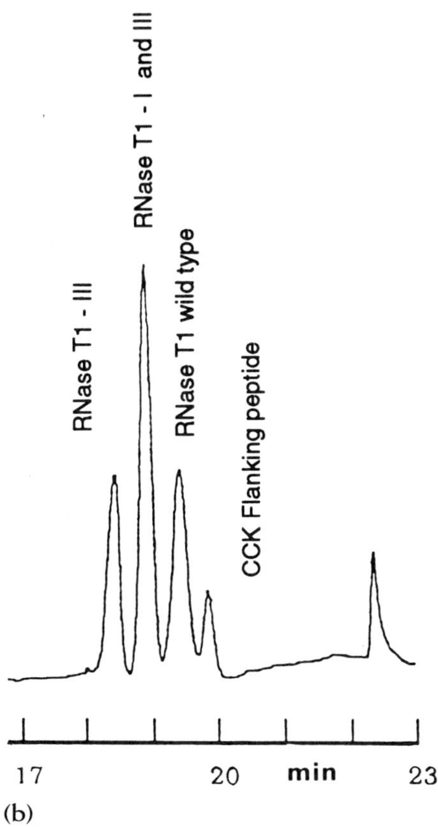

(b)

POLYMER NETWORKS

At present, separation in slab gel electrophoresis is based primarily on solute size differences, with migration electrically driven. For example, in free solution, single- and double-stranded DNA molecules, of roughly more than ten bases in length, migrate with a constant mobility, independently of base length [123]. All DNA separations in electrophoresis are based on molecular size differences by the restricted migration of DNA molecules through porous gel media under the influence of an applied electric field [124]. Indeed, most separations of DNA are conducted to determine molecular mass or base number.

A second important application of slab gel electrophoresis is the sodium dodecyl sulfate–polyacrylamide gel electrophoresis (SDS–PAGE) separation of proteins on the basis of relative molecular mass (M_r) [125]. As with DNA, SDS–protein complexes in free solution migrate with the same mobility, independent of M_r. However, when electrophoresis is conducted through a porous gel network, separation is based on sieving, with mobility generally proportional to log M_r. A third significant biopolymer class is that of carbohydrates. Until recently, carbohydrates have generally been separated by electrophoresis as frequently as by chromatography. When separation has been by electrophoresis, borate complexes or affinity electrophoresis has been used [95]. Molecular mass separation can also be important for carbohydrates. Thus, sieving behavior can be used in the separation of carbohydrates, and this is being developed for HPCE [126].

Because HPCE uses an electric field of at least one order of magnitude higher than slab gel electrophoresis, the transfer of matrices directly from slabs to capillaries has required some study. Nevertheless, much progress has been made, and successful separations can be achieved in many applications. It is thus possible to conduct high-speed–high-resolution size separations for DNA, SDS–protein complexes, and carbohydrates. The next section deals with sieving matrices, and this will be followed by an overview of applications and future directions.

Sieving Matrices

The most frequently used matrices in slab gel electrophoresis are cross-linked polyacrylamide and agarose. Depending on the monomer and cross-linker concentrations, pores of 50–250 Å are typically produced for polyacrylamide gels [127]. Gelation of agarose leads to pore sizes of about 500–2000 Å or higher [128]. Thus, cross-linked polyacrylamide is employed for the separation of molecules with M_r of up to a few hundred thousand, whereas agarose is selected for separations beyond this range. In HPCE, little work has yet been done with agarose gels; therefore, this topic will not be discussed here.

The first reports of DNA capillary gel electrophoresis were from Cohen et al. [129], who showed the remarkable resolving power of cross-linked polyacrylamide. Figure 13a shows the size separation at 300 V cm^{-1} of polydeoxyadenylic acid [$p(dA)_{40-60}$] using a 3% T, 5% C (T = monomer concentration, C = cross-linker, *bis*, concentration [130]) column, from which 20 peaks differing in only one nucleotide each are baseline-resolved between 7 and 8 minutes running time. In a later paper dealing with longer oligonucleotides, the high-resolving power was emphasized when plate counts as high as 3×10^7/m were obtained (Fig. 13b) [131]. The speed and resolution of such columns led immediately to possible application of capillary electrophoresis to many areas of genetics, such as (a) DNA sequencing, (b) assessment of purity of oligonucleotide synthesis, (c) PCR product analysis, and (d) DNA fingerprinting, to mention only a few topics. Currently, DNA analysis by capillary electrophoresis is rapidly developing, and some applications will be discussed in the next section.

The preparation of cross-linked polyacrylamide gels has evoked much discussion in the literature. First, if one prepares a gel in a capillary of bare silica, upon application of the electric field, the gel is observed to extrude from the column because of electroosmotic pressure. A neutral coating on the walls that reduces electroosmotic flow [e.g., 30], can eliminate this problem. A second approach, used by this laboratory, is to covalently attach the gel to the wall by a bifunctional agent [30]. Thus, rather than a linear polyacrylamide coating, a cross-linked gel is attached throughout the column. This approach eliminates several steps in the preparation of the matrix and, in addition, prevents movement of the gel in the detector region, leading to background noise. We have found that bubble-free columns can routinely be obtained by this approach.

A second issue in the preparation of cross-linked polyacrylamide is the condition of polymerization. The free-radical reaction has been well characterized in the literature [127]. Polymerization is affected by temperature (exothermic reaction), oxygen (quencher), and the amount of initiator and catalyst. In addition, alcohols, which are sometimes used as stabilizers, can also quench the polymerization. It is clear that reproducible results require careful control of conditions. Also, it is important to emphasize that two gels that are made under different conditions can yield significantly different sieving behavior, in spite of having the same nominal composition (%T, %C). Therefore, care must be exercised in comparing columns from different laboratories.

A third issue is the substantial volume reduction that occurs on polymerization. The use of low-monomer concentrations (up to 5–6% T), independent of cross-linker concentration, allows successful gel preparation within the column; however, higher %T compositions become more difficult. One approach to counteract the volume reduction is to polymerize under pressure [132]. A better approach was introduced by Novotony and co-workers, in which the initiator, ammonium persulfate, and the catalyst, TEMED, were electrophoretically introduced at opposite ends of the capillary, with TEMED also in the buffer. In this fashion, polymerization does not take place instantaneously throughout the column, but only in those areas where TEMED and persulfate mix. The gel can then adjust to the volume changes, and the exothermicity of the reaction is controlled. With this approach, gels of 20% T or higher could be made, and such columns provide high-resolution separation of carbohydrates (see later). Others have also contributed to the advancement of gel preparation. Special note should be given to the laboratories of Schomburg [133], Poppe [134], Baba [135], and the early work of Hjerten [136].

There has been some confusion in the literature concerning the differentiation between the preparation and use of cross-linked columns. Even if the columns are well prepared, their use in practice can lead to gel collapse and bubble formation, with a concomitant complete loss in current. Following Tanaka [137], the simple picture to understand the collapse of gels under stress is to consider that the pore structure results from a steady state in osmotic pressure between the pores. If the osmotic pressure increases in one region, the polymer fibers will try to respond by pore expansion (or shrinkage); however, because of covalent cross-linkage, reversible change in structure can occur only over a limited stress range, beyond which collapse

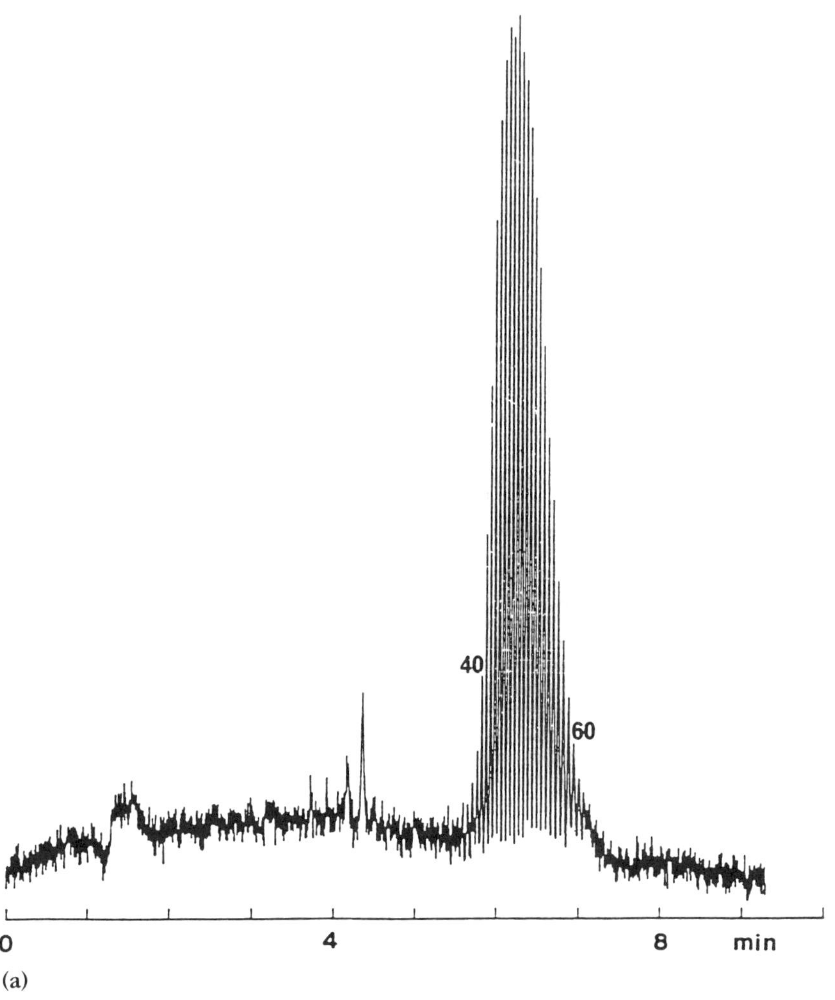

Figure 13 An HPCE separation of the polydeoxyadenylic acid test mixture, $p(dA)_{40-60}$. Capillary dimensions were 270 × 0.075-mm id (effective length, 600 mm); running buffer: 0.1 M Tris/0.25 M borate/7 M urea, pH = 8.3. (b) HPCE separation of polydeoxythymidylic acid mixture, $p(dT)_{20-160}$ in 3% T, 5% C polyacrylamide gel. Capillary dimensions, 700 × 0.075-mm id (effective length, 500 mm); running buffer as in (a); applied field, 300 V cm^{-1}, 13 μA. (From Refs. a, 129; b, 131)

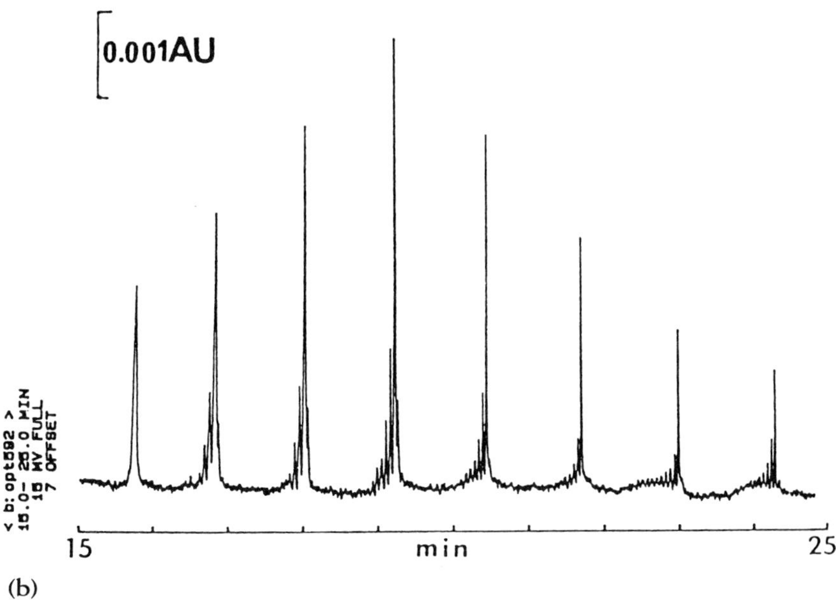

(b)

occurs. Additives, such as ethylene glycol or lower alcohols [128], can help make the gel more flexible (i.e., be more resistant to stress), but ultimately, collapse may result.

There are several possible stresses that can take place in gels. First, elevated temperature will result in irreversible gel alteration. Thus, 3% T, 5% C columns with 7 M urea produce bubbles above 25–30°C [138]; hence, careful temperature control is required. Second, it is well-known that ion mobility within the gel can differ from free solution, leading to a low-conductivity zone migrating through the gel [139]. This low-conductivity zone will become hot owing to a voltage drop across the zone, and this temperature increase can lead to bubble formation. Placing gel or noncross-linked polymer in the buffer reservoirs can help reduce this effect [140].

Third, injection of salts (or template in DNA sequencing) can cause stress (osmotic pressure changes) at the head of the column, also leading to collapse in this region. Desalting the sample using miniaturized membranes and removing template can be useful here. In SDS–PAGE, injection of sample matrices with high SDS concentration will also create ionic stress. Therefore, injection must be carefully performed when using cross-linked polyacrylamide for the separation of SDS–protein complexes. Fourth, the high fields used in HPCE will often stretch the gels, leading ultimately to collapse. Relaxation (low-field operation) between runs can extend the column's lifetime; however, this approach will reduce analytical throughput. In summary, some care must be exercised when using cross-linked gels; however, remarkable separations are possible with such columns.

An alternative, and, if feasible, preferable approach, is to select noncross-linked polymer matrices for electrophoretic separation by sieving. The use of such matrices was demonstrated on slab gels by Bode, using, for example, linear polyacrylamide and polyethylene glycol [141], and by Chrambach and co-workers using linear polyacrylamide for SDS–PAGE [142]. Interestingly, Chrambach et al. found that polymer concentrations lower than 10% T could not be maintained on the slab. On the other hand, the anticonvective nature of the capillary walls can permit use of compositions lower than 10% T.

Grossman and Soane have explored low-viscosity polymer-sieving matrices in terms of the principles of entangled polymer solutions [143]. Others have used similar matrices and have described the method in terms of nongel-sieving [144] or buffer additives. In many respects, these noncross-linked polymer matrices follow behavior similar to that in cross-linked gels. To emphasize that similarity, the name *polymer network*, as originally suggested by Bode [141], is recommended as a common term for the cross-linked and noncross-linked polymer-sieving matrices.

The noncross-linked matrices have specific advantages over cross-linked gels in terms of practical operation. First, greater stress can be applied to the polymer network before collapse, because the polymer chains have greater flexibility to relieve the stress. Thus, in a recent paper, linear polyacrylamide was shown to have a much longer column lifetime than cross-linked polyacrylamide for capillary SDS–PAGE [145]. Second, at lower viscosities (e.g., even up to ~3000 cp) the matrix can be replaced after each run [146]. Replaceable linear polyacrylamide has been useful for high-resolution DNA separations [147] and, more recently, DNA sequencing [148]. Lower viscosities (a few hundred cp) can be easily replaced on commercial instruments. This replaceable characteristic permits long column lifetimes, often with less sample cleanup than required for fixed column matrices.

Third, the low-viscosity matrices can use pressure injection. For cross-linked gels, only electrokinetic injection is possible. Pressure injection may not require desalting of the sample, which is often necessary for electrokinetic injection, and there is no solute bias, as can occur with electrokinetic injection [149]. Currently, a variety of polymer networks are used, including methylcellulose, polyethylene glycol, and dextran.

DNA Analysis

Figure 13 demonstrates that HPCE has sufficient resolution and speed to be used to determine the purity of oligonucleotide synthesis. Failure sequences can be easily observed [129]. The speed and resolution of capillary electrophoresis is such that it should be a method of choice. A potentially important application for single-stranded oligonucleotide will be in antisense DNA analysis [150]. Here, modified oligonucleotides are synthesized in the range of 25–30 bases in length. Since these species are to be used for therapeutic purposes, the need to assess purity in a rapid manner is self-evident.

Capillary electrophoresis has created much interest in the field of DNA sequencing. One of the stated needs of the Human Genome Project is to increase substantially the throughput of DNA sequence analysis while reducing the cost per base of determined sequence. The strategy is to increase the rate of separation and, through parallel processing, determine many sequences at the same time. The high-electric field capabilities of capillary electrophoresis offers a means of speeding up separation. An overview of DNA-sequencing procedures has been prepared [151].

As seen in Fig. 13 and corroborated by other studies [152–154], HPCE has the requisite efficiency and resolving power to be used for DNA sequencing. In this application, single-stranded DNA molecules must be resolved with single-base differences. It is interesting to explore how powerful the approach is. Since migration versus base number is linear in the region at which sequences can be read, the separation factor α (Eq. 11) will, as a consequence, decrease as the base length increases. Indeed, quite small α values may occur, and yet at least partial resolution may still result, because of the high efficiency available. Assuming, for example, $N = 4 \times 10^6$ and $R_s = 0.6$, the combination of Eqs. 10 and 12, with $\mu_{eo} = 0$, leads to an α value of only 1.0012. This value can be contrasted with chromatography for which α values of 1.02 would be quite low ($N \sim 10^4$) [20]. In principle, the greater the column efficiency in HPCE, the longer the base number that can be sequenced. The pore size of the polymer network and the applied electric field will also influence resolution through the band-spacing; this topic will be discussed shortly.

In the typical HPCE DNA-sequencing experiment, primer labeled with four different dyes is chain-extended on a template with DNA polymerase, using the Sanger strategy of dideoxy-chain termination [151]. The four dyes are commercially available fluorescein and rhodamine derivatives (FAM, JOE, ROX, TAMRA), as used in the standard slab gel laser-induced fluorescence approach [155]. Each dye represents chain termination of a specific base. From the migration times of the fragments, the base number can be obtained, and from the appropriate excitation and emission wavelengths, the specific base associated with a dideoxy-terminator can be ascertained. The four separate dyes allow all four termination reactions to be separated in one column (or one lane). In addition, the fluorescent dyes result in low-level detection using laser-induced fluorescence.

A capillary gel system for DNA sequencing was initially presented employing a single low-power argon ion laser with four photomultiplier tubes (PMTs) and appropriate beam splitters and filters [156]. By using a 3% T, 3% C gel, sequencing could be read to approximately 300 bases in length. Relative to individual lanes on a slab gel, it was noted that a 25-fold increase in separating speed was possible. Other instrumental approaches with four dyes include two lasers and a filter wheel with four PMTs [157] and a charge-coupled device (CCD) for increased spectral resolution [158]. Our laboratory has used a two-laser–two-window approach with an intensified diode array for high sensitivity and high spectral resolution [159].

Although the four-dye approach is general for sequencing long ranges of base lengths, other strategies using one or two dyes are possible for stretches of approximately 300–400 bases in length. Of course, this simplifies the instrumentation necessary for sequencing. Tabor and Richardson recently optimized the Sanger chain termination-sequencing reaction such that approximately equal amounts of each reaction product can be obtained [160]. With this new chemistry, it is possible to determine sequencing reaction products not only in terms of specific dyes, but also in terms of peak height (or area) ratios [161–163].

As noted earlier, gel stability has been an issue with DNA sequencing [140]. Although improvements in the sample cleanup step can be used, our laboratory has adopted the replaceable linear polyacrylamide approach. The approach involves the separation of sequencing products using a 6% T column and a labeling strategy of two dyes (FAM and JOE) and two concentration ratios (3:1), as used by Dovichi and co-workers [163]. Successful sequencing up to 300 bases in less than 30 minutes is possible [148]. Figure 14 shows the separation of the FAM channel of a sequencing run in which T/C stop ratios are 3:1. Good separation is obtained for up to 300 bases, and the identification of dideoxy-T and C reaction products is straightforward. After each run, the matrix is removed, and the column filled with new polymer network polyacrylamide. By this simple approach, hundreds of sequencing runs are possible from a single column.

Parallel processing is a necessary feature of DNA sequencing to increase throughput. One approach, advanced by Mathies and co-workers [162], is to move an array of columns past a fixed laser beam and, by confocal optics, transport the emitted light to the appropriate PMT. The second approach, advanced by Smith and co-workers, uses ultrathin slab gels ($< 100\ \mu m$), so that high electric fields can again be employed [164]. The parallel processing is achieved by multiple lanes in the ultrathin slabs. It is not yet clear which method will be more acceptable, as this will obviously depend on the ease of commercialization and automation. Finally, the use of capillary gel electrophoresis for DNA sequencing may ultimately be replaced by newer methods, such as hybridization on oligonucleotide arrays attached to chips [98] or mass spectrometry [165,166].

Besides single-stranded DNA analysis, the other important area for HPCE with sieving media involves separation of double-stranded molecules. This approach works quite well in the resolution of fragments up to about 1 kbp. Figure 15a shows the separation of φX174 *Hae* III restriction fragments on a 3% T, 0.5% C polyacrylamide column, and Fig. 15b illustrates a separation of a 1-kbp ladder on the same column [167]. Similar resolution can be obtained on replaceable linear polyacrylamide and on other matrices [143,144,168,169]. Occasionally, resolution is increased by adding an intercalating agent to the sample (e.g., ethidium bromide [170] or thiazole orange [171]). In the later, trace analysis using the argon ion laser is possible.

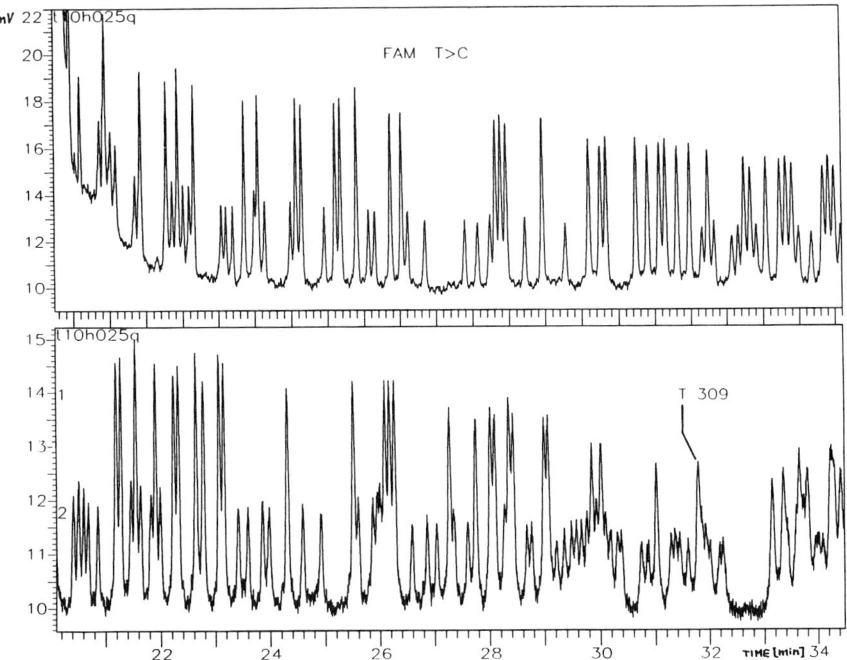

Figure 14 M13mp18 sequencing. FAM dye-labeled primer (M13-21, ABI); base coding: T/C = 2.6:1 peak height ratio. Conditions: Capillary: 75-μm id, 360-μm od; effective length, l, was 18, total length, L, was 33 cm; 6% replaceable linear polyacrylamide; 1 × TBE/4 M urea/30% formamide; E = 250 V cm^{-1}.

The high-speed, high-resolution capabilities of double-stranded DNA molecules of HPCE using replaceable polymer networks opens the possibility for several important applications, including diagnostics, PCR product analysis, DNA fingerprinting, and plasmid analysis. Interest in this area continually grows, especially as the knowledge base of genetics increases.

Figure 15b shows the interesting result that band-spacing, as a function of molecular mass, decreases significantly above 1 kbp. This reduction in band-spacing arises from the stretching of the DNA molecules in the high applied electric field, which results in the species having similar mobilities [172]. As evidence of this behavior, it was observed that improved resolution above 1 kbp occurred as the field decreased [167]. This field effect was recently used for enhanced resolution in a field-programming experiment [173].

It is not possible within the space constraints of this chapter to discuss the very important and interesting modeling of DNA electric field migration through polymer networks. A generally accepted model for DNA is based on biased reptation

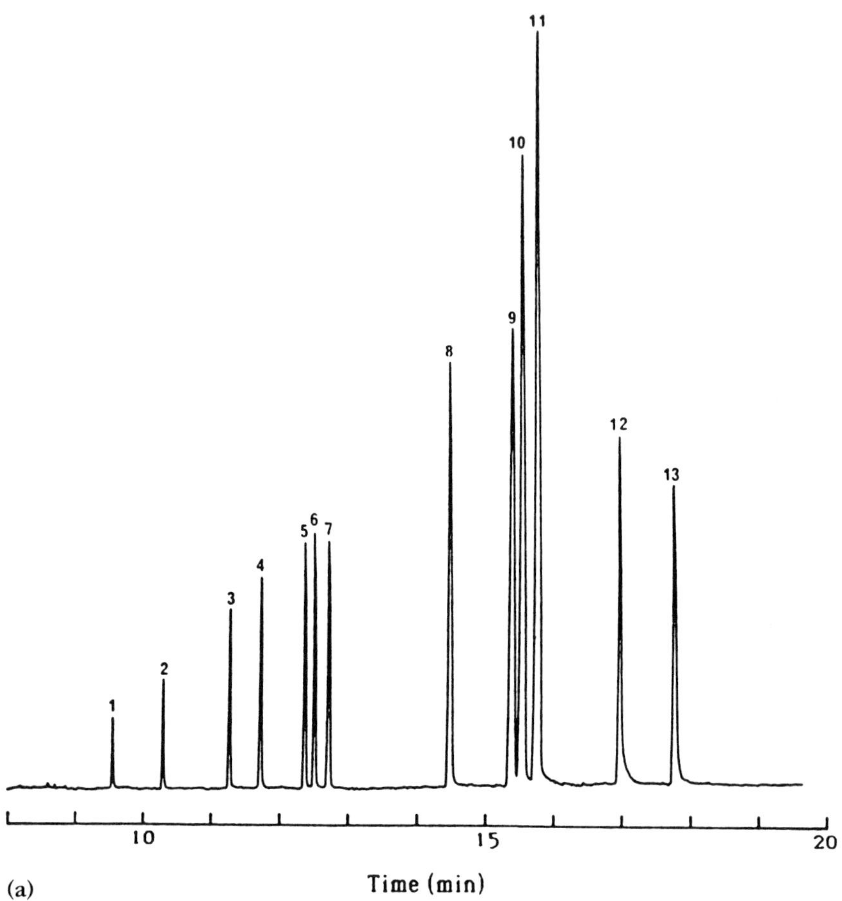

Figure 15 (a) Separation of *Hae* III digest of ϕX174 DNA, *Eco*R I digest of pBR322 and *Eco*R I digest of M13mp18 on a 3% T, 0.5% C polyacrylamide capillary column. The 11 fragments of the ϕX174 digest have been identified as the first 11 peaks in the electropherogram, followed by pBR322 and M13mp18. Identification was made by spiking each peak with slab gel-isolated species. Peaks: 1 = 72; 2 = 118; 3 = 194; 4 = 234; 5 = 271; 6 = 281; 7 = 310; 8 = 603; 9 = 872; 10 = 1078; 11 = 1353; 12 = 4363 and 13 = 7253 base pairs. Conditions: effective length, 1, was 30 cm; total length, L, was 40 cm; applied field, E, was 250 V cm^{-1}; current generated, i, was 12.5 μA; and sample was injected electrophoretically at 10 kV for 0.5 seconds. Buffer: 100 mM Tris-borate (pH 8.3), 2 mM EDTA. (b) Separation of 1000-base pair DNA ladder on a 3% T, 0.5% C polyacrylamide capillary. Fragment lengths range from 75 to 12216 bp. All peaks were identified by spiking with slab gel-isolated species. Peak legend: 1 = 75; 2 = 142; 3 = 154; 4 = 200; 5 = 220; 6 = 298; 7 = 344; 8 = 394; 9 = 506; 10 = 516; 11 = 1018; 12 = 1635; 13 = 2036; 14 = 3054; 15 = 4072; 16 = 5090; 17 = 6108; 18 = 7126; 19 = 8144; 20 = 9162; 21 = 10180; 22 = 11198; 23 = 12216 base pairs. (From Ref. 167)

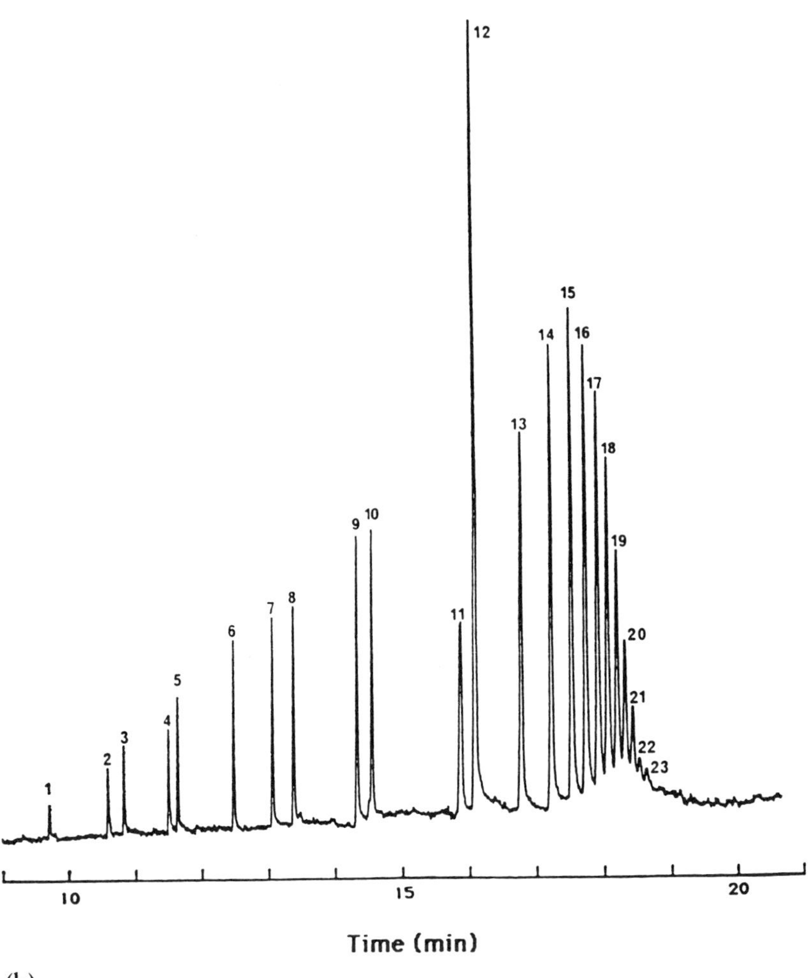

(b)

with regions of mobility behavior as a function of base number—(a) normal sieving—Ogston, for small base sizes; (b) intermediate-sized species of biased reptation in which the molecules are constrained in the paths they take for migration by virtue of the matrix and pore size; and (c) larger molecules of biased reptation with stretching [174]. The base size at which molecules change from one region to another is a function of the applied field, pore size, and molecular size. Higher fields lead to stretching at lower base numbers. Also, it is possible to find inversion in migration order in the regions between biased reptation and plateau stretching [174].

Ultimately, stretching limits the range of single-stranded DNA-sequencing product fragments that can be resolved [175,176]. Indeed, predictions of limits of roughly 800 bases, beyond which sequences can no longer be read, have been advanced. The ultimate limit is now uncertain, but will depend on labeling strategy, gel structure, and applied field. Obviously, high fields are desirable for rapid separation, but this can lead to shorter-reading sequences [175]. Longer-range sequencing will require longer analysis time, but will also reduce the number of sequencing reactions. The strategy taken will obviously depend on the goal and requirements of the research.

Pulsed field electrophoresis has become an important tool in slab gel operation, particularly for the separation of very large and megabase pair DNA [177]. Pulsing in HPCE can influence separation, as well, even for sizes of 1 kbp [167,178]. The issue of waveform fidelity in pulsed field capillary electrophoresis has also been fully addressed [179]. Pulsed field electrophoresis is at an early stage in development in HPCE, and it is anticipated that it may be a useful tool in manipulating separation, not only for DNA, but for other biopolymers, as well.

In summary, DNA analysis is an important application of HPCE using polymer network-filled capillaries. When feasible, replaceable matrices are the method of choice. Beyond the examples discussed in this chapter, there are several other areas for which DNA analysis will be useful. These include mutational analysis, DNA adduct analysis, hybridization assays, and diagnostic testing. It is thus anticipated that DNA analysis by HPCE will grow substantially in the next few years and become an important tool to the life scientist.

Protein Analysis

Protein separations under native conditions are possible by HPCE using gel-filled capillaries [180]. In general, however, open-tube CZE is also a powerful tool and will often be used. Thus, beyond the added size selectivity, it would appear that the use of gel-filled capillaries may find greatest value in the analysis of polyamino acids [181]. On the other hand, the molecular mass separation of SDS–protein complexes by HPCE would appear to be a method that may be widely used by protein chemists. As in CIEF, this would involve transferring an already-existing slab gel method (SDS–PAGE) to the capillary format, with the added advantages of speed, quantitation, and resolution.

In the late 1980s, this laboratory demonstrated SDS–PAGE separation of proteins using cross-linked polyacrylamide gel-filled capillaries [182]. The expected linear behavior of mobility versus log M_r was observed, and Ferguson plots [127] revealed true sieving to be occurring. Because of the instability of cross-linked gels to osmotic shock (and perhaps adsorption of SDS on the polymer chains), care had to be exercised to obtain usable column lifetimes. In the work, 7 M urea was added to the running buffer as a stabilizer and to reduce pore size. In addition, the column was cooled with a water jacket to actively remove heat. A later paper also demonstrated SDS–PAGE using cross-linked gels with ethylene glycol as a stabilizer [183]. A total of 40 runs was claimed for the lifetime of the column.

To improve column lifetime, noncross-linked linear polyacrylamide can be employed as a sieving medium, in which, as discussed earlier, the polymer fibers are able to respond to stress better than the cross-linked gels [145,184–186]. However, another problem remains with polyacrylamide as matrix. Ultraviolet detection must be at 230 nm or above (e.g., 280 nm) because of the absorptivity of the polymer. Such wavelengths are significantly less sensitive for proteins than 210 nm, or lower. Therefore, selection of UV-transparent polymers that allow sieving is a preferable approach. Moreover, as with DNA analysis, the use of replaceable polymer networks is desirable.

This laboratory has utilized dextran, M_r 2×10^6 Da, as a sieving medium for the separation of SDS–protein complexes [145]. Figure 16a shows the separation of a mixture of standard proteins from 14×10^3 to 220×10^3 Da with such a column. The method is directly transferable from slab gel, with detection limits of approximately 0.5 μg/mL. Rapid and reproducible quantitative analysis and estimation of M_r were possible. In addition, after each run, the polymer network was replaced, leading to hundreds of runs on a single column. Indeed, significantly higher precision in migration time was obtained by replacing the matrix after each run than by injecting the next sample in the same column [145]. A washing step was also added for particularly dirty samples; however, it should be noted that this is often unnecessary, since SDS is a strong solubilizing agent. Figure 16b illustrates the separation of an SDS–*Escherichia coli* crude extract, using polyethylene glycol, M_r 10^5 Da, as the UV-transparent sieving medium [145]. A proprietary matrix has been used by other workers, as well [187]. It can be noted that those proteins the M_r of which cannot be accurately determined by slab gels (e.g., glycoproteins, histones), will also be inaccurate in this method.

This SDS–protein complex approach with a UV-transparent matrix is currently being employed by biochemists to assess purity and to determine protein M_r. The use of HPCE of SDS–protein complexes is not as precise or accurate as mass spectrometry, but it is less expensive. In addition, HPCE finds real use in the analysis of high-M_r species (e.g., $> 10^5$ Da), since usually, the proteins of such large species will be heterogeneous, leading to potential complexity in the mass spectrum. An important application of the HPCE of SDS–protein complexes is thus the analysis of antibodies and antibody fragments (CIEF is important in this application, as well).

Carbohydrate Analysis

The analysis of carbohydrates by HPCE is a relatively new area that has much potential [188]. The high-resolving power of the CE method and the possibility for low-detection limits makes the approach particularly appealing. As noted earlier, most work to date has involved derivatization of the carbohydrate to form fluorescently or UV-active complexes. In laser-induced fluorescence, quite low levels of detection are, in principle, possible (subattomole).

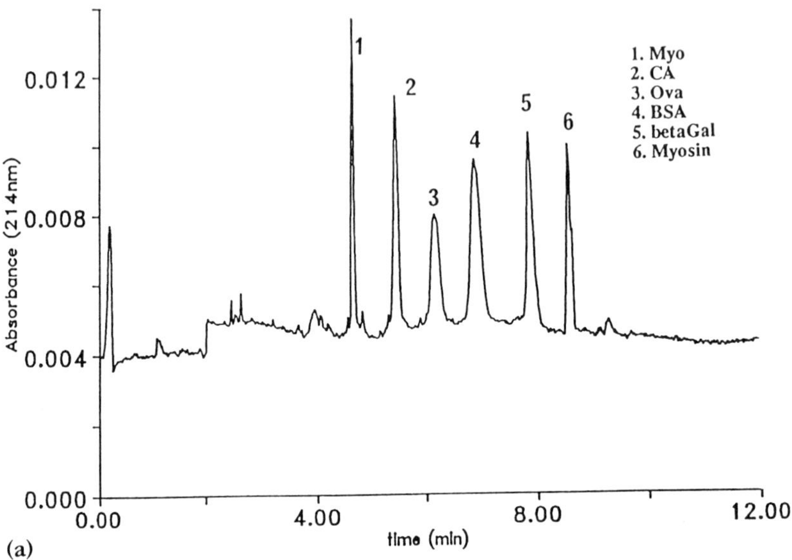

Figure 16 (a) Rapid separation of SDS–proteins. Buffer: 0.06 M AMPD-CaCO pH 8.8, 0.1% SDS, 10% dextran (M_r 2 × 10^6). Column: 23 cm in length with an effective length of 18 cm. Applied electric field: 400 V cm^{-1}, resulting in 30 μA. Injection at 10 kV for 2 seconds. (b) Electropherogram of an SDS–*E. coli* crude extract using capillary PEG polymer network. Conditions: buffer: 0.1 M Tris-CHES, 0.1% SDS, pH 8.8; polymer: 3% w/V PEG 100,000; capillary: 40 cm, 100-μm id; applied electric field: 300 V cm^{-1}; detection: 214 nm; OG: internal standard—2 Orange G. (From Ref. 145)

Although open-tube separations have been achieved, gels will also play an important role in this area, particularly for larger carbohydrates. Figure 17, from the laboratory of Novotny, illustrates the significant resolving power of this approach [189]. In this example, a 10% T, 3% C polyacrylamide gel column is used for the separation of a sample of partially hydrolyzed dextrin 15. The inset shows that migration time (1/μ) is proportional to molecular mass (degree of polymerization) of the polymer. Polymer networks offer high-resolution capabilities and a unique mode of separation based on size. It is likely that this approach, when fully developed, will be of real value to carbohydrate chemists.

In summary, it is clear that capillary electrophoretic separations based on differential sieving behavior is a major mode of resolution. It is most applicable to biopolymers and other polymeric species [190]. We can anticipate further growth in this area in the forseeable future.

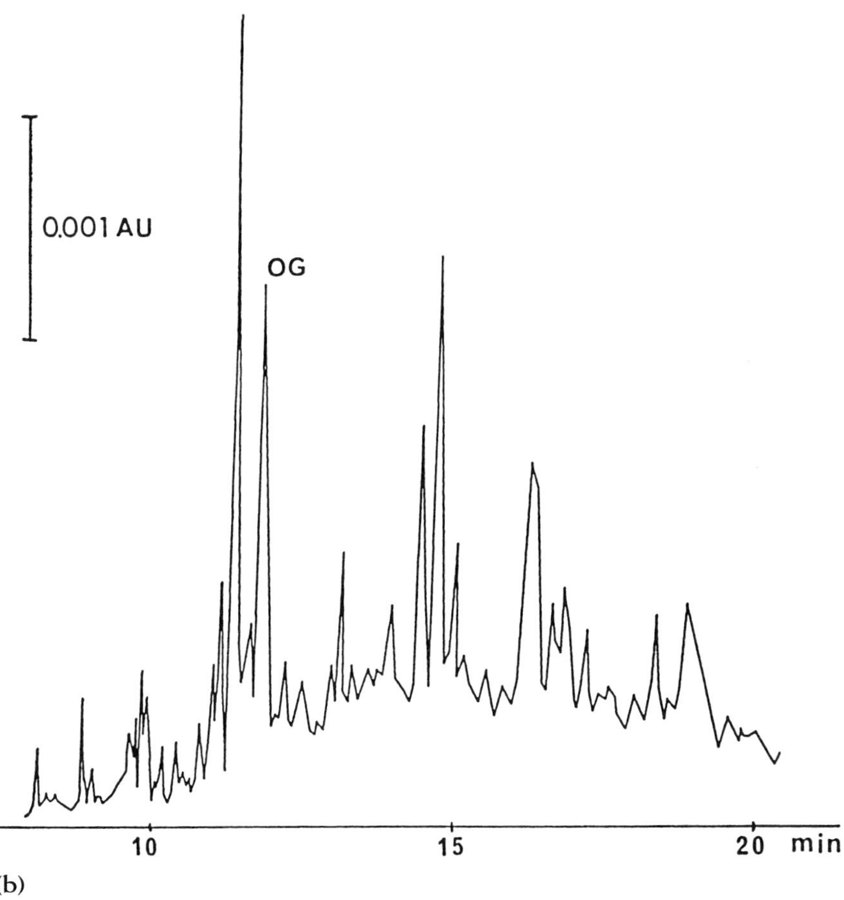

(b)

CONCLUSION

This overview of high-performance capillary electrophoresis has set the stage for the following chapters, which provide significantly more in-depth discussion of individual topics. It is, however, appropriate to conclude this chapter with a final assessment of the current and future potential of HPCE.

Within the past several years, capillary electrophoresis has grown from a laboratory curiosity to an established separation and analytical tool. As one indication of this change, the Fifth International Symposium on HPCE, held in Orlando, Florida in January 1993, included well over 250 posters and oral presentations, with many papers dealing with the application of the methodology to meaningful analytical problems. With new chemistries and separation procedures continually being produced, even broader applicability can be anticipated in the next several years.

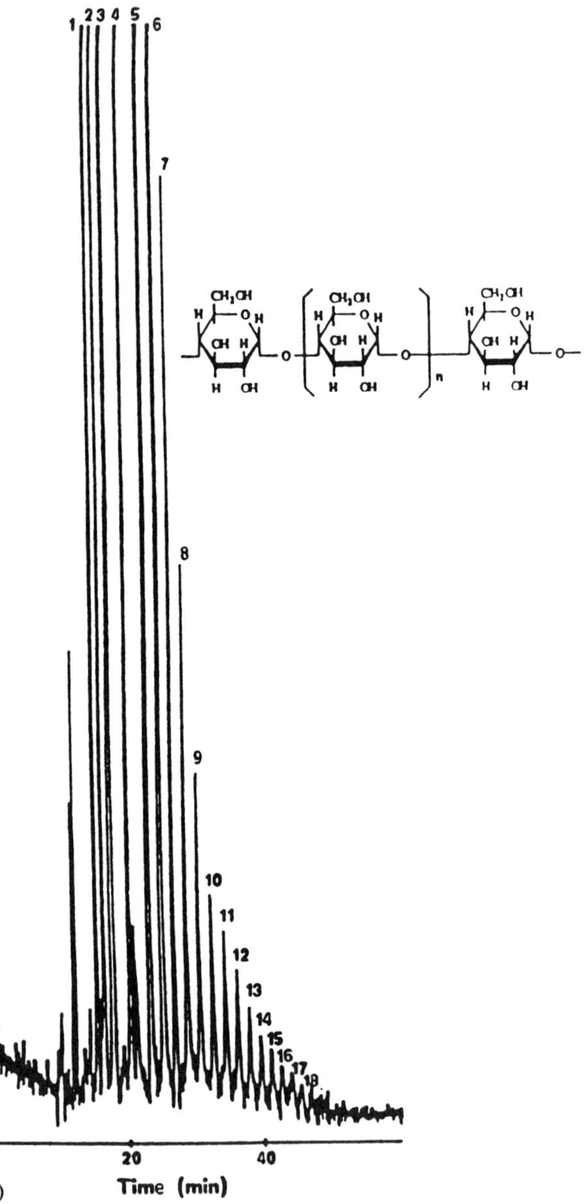

Figure 17 Electrophoretic separation of a partially hydrolyzed dextrin 15 in a polyacrylamide gel-filled column. Correlation between molecular mass–oligomers from maltodextrin and their corresponding migration times. Gel concentration, 10% T, 3% C; capillary, 26 cm (19 cm effective length) × 50 μm; buffer, 0.1 M Tris-0.25 M borate-7 M urea (pH 8.33); applied field, 269 V cm^{-1} (20 μA). (From Ref. 126)

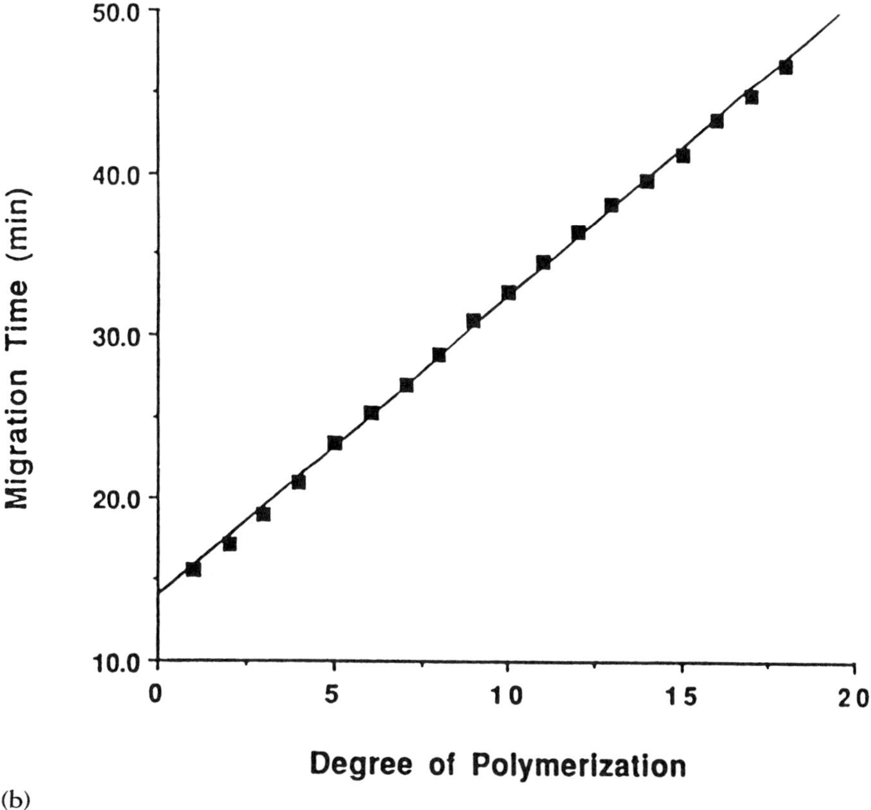

(b)

High-performance capillary electrophoresis fits in with a trend in science toward miniaturization. Indeed, researchers are even developing capillary electrophoresis on planar chips [191]. Furthermore, the method is quite compatible with biological samples. The speed and efficiency of CE mean that a number of small-molecule applications will also emerge in the years ahead, beyond those already demonstrated (e.g., ion and drug analysis).

Several developments are needed to make HPCE an even more important separation tool in the 1990s. First, an increasing number of applications broadening the methods are required. Second, the instrumentation and chemistries need to become more rugged for broad and routine use. Third, CE–MS must emerge as a practical method, as LC–MS has become. Fourth, automated sample-handling methods, compatible with CE volume sizes, must be developed. Fifth, the full use of CE as a miniaturized analytical tool must be demonstrated.

These and other advances are only a matter of time. In many respects, HPCE is following closely the development of HPLC in the 1970s and 1980s. Over the years, HPLC emerged as a rugged tool that is now routinely used by scientists around the world. Indeed, today it is the most valuable separation method for organic and biological substances. One difference, however, between the development of HPCE relative to HPLC is that the rate of growth will be much faster for HPCE, in large part because of the accumulated experience from HPLC.

The prediction for the future advancement of HPCE is thus highly positive. The next few years should see rapid growth with diverse applications. A large variety of problems will best be solved using this technique. The following chapters should convincingly demonstrate that HPCE will take its rightful place as a major separation and analytical method.

ACKNOWLEDGMENTS

The authors thank NIH under grant GM15847 for support of the work in preparing this overview. Contribution No. 553 from the Barnett Institute.

REFERENCES

1. Strategic Directions Intl., *R & D Magazine*, Aug: 2 (1992).
2. S. Hjertén, *Chromatogr. Rev.*, 9:122 (1967).
3. N. Catsimpoolas, *Sep. Sci.*, 6:435 (1971).
4. A. J. P. Martin, and F. M. Everaerts, *Anal. Chim. Acta*, *38*:223 (1967).
5. R. Virtanen, *Acta Polytech. Scand.*, *123*:1 (1974).
6. F. E. P. Mikkers, F. M. Everaerts, and T. P. E. M. Verheggen, *J. Chromatogr.*, *169*:11 (1979).
7. J. W. Jorgenson, and K. D. Lukacs, *Anal. Chem.*, *53*:1298 (1981).
8. J. W. Jorgenson, and K. D. Lukacs, *J. Chromatogr.*, *218*:209 (1981).
9. J. W. Jorgenson, and K. D. Lukacs, *Clin. Chem.*, *27*:1551 (1981).
10. S. Hjertén, *J. Chromatogr.*, *270*:1 (1983).
11. S. Terabe, K. Otsuka, K. Ichikawa, A. Tsuchiya, and T. Ando, *Anal. Chem.*, *56*:113 (1984).
12. A. T. Endler, and S. Hanash, eds., *Two-Dimensional Electrophoresis*, VCH, Weinheim (1989).
13. P. D. Grossman, and D. S. Soane. *Anal. Chem.*, *62*:1592 (1990).

14a. J. Vacík, *Electrophoresis, Part A: Techniques* (Z. Deyl, ed.) *J. Chromatogr. Libr.*, *18*:11 (1979).

14b. J. Pospíchal, P. Gebauer, P. Boček, *Chem. Rev.*, *89*:419 (1989).

15. A. Tiselius, *Nova Acta Reg. Soc. Sci. Upps.*, 7:1 (1930).
16. P. Boček, M. Deml, P. Gebauer, and V. Dolník, *Analytical Isotachophoresis*, VCH Verlagsgesellschaft, Weinheim (1988).
17. V. Šustáček, F. Foret, and P. Boček, *J. Chromatgr.*, *480*:271 (1989).
18. V. Purghart, and D. E. Games, *J. Chromatogr.*, *605*:139 (1992).

19. C. Whang, and E. S. Yeung, *Anal. Chem.*, *64*:502 (1992).
20. B. L. Karger, L. R. Snyder, C. Horvath, eds., *An Introduction to Separation Science*, Wiley, New York, 1973.
21. C. S. Lee, W. C. Blanchard, and C. T. Wu, *Anal. Chem.*, *62*:1550 (1990).
22. M. A. Hayes, and A. G. Ewing, *Anal. Chem.*, *64*:512 (1992).
23. C. T. Wu, T. Lopes, B. Patel, and C. S. Lee, *Anal. Chem.*, *64*:886 (1992).
24. J. K. Towns, and F. E. Regnier, *J. Chromatogr.*, *516*:96 (1990).
25. J. E. Wiktorowicz, and J. C. Colburn, *Electrophoresis*, *11*:769 (1990).
26. J. K. Towns, and F. E. Regnier, *Anal. Chem.*, *63*:1126 (1991).
27. J. K. Towns, J. Bao, and F. E. Regnier, *J. Chromatogr.*, *559*:227 (1992).
28. W. Nashabeh, and Z. Rassi, *J. Chromatogr.*, *559*:367 (1991).
29. P. Gebauer, and W. Thormann, *J. Chromatogr.*, *558*:423 (1991).
30. S. Hjertén, *J. Chromatogr.*, *347*:191 (1985).
31. C. A. Cobb, V. Dolník, and M. V. Novotny, *Anal. Chem.*, *62*:2478 (1990).
32. M. Huang, W. P. Vorkink, and M. L. Lee, *J. Microcolumn Sep.*, *4*:233 (1992).
33. M. Zhu, R. Rodriguez, D. Hansen, and T. Wehr, *J. Chromatogr.*, *516*:123 (1990).
34. J. Kohr, and H. Engelhardt, *J. Microcolumn Sep.*, *3*:491 (1991).
35. M. Huang, W. P. Vorkink, and M. L. Lee, *J. Microcolumn Sep.*, *4*:135 (1992).
36. G. J. M. Bruin, J. P. Chang, R. P. Kuhlman, K. Zegers, J. L. Kraak, and H. Poppe, *J. Chromatogr.* *471*:429 (1989).
37. W. G. H. M. Muijselaar, and C. H. M. M. Bruijn, and F. M. Everaerts, *J. Chromatogr.*, *605*:115 (1992).
38. J. C. Giddings, *Sep. Sci.*, *4*:181 (1969).
39. P. Gozel, E. Gassmann, H. Nickelsen, and R. N. Zare, *Anal. Chem.*, *59*:44 (1987).
40. K. Kalghati, and C. Horvath, *J. Chromatogr.*, *443*:343 (1988).
41. N. B. Afeyan, S. P. Fulton, and F. E. Regnier, *J. Chromatogr.*, *544*:267 (1991).
42. R. M. McCormick, *Anal. Chem.*, *60*:2322 (1988).
43. F. E. P. Mikkers, F. M. Everaerts, and T. P. E. M. Verheggen, *J. Chromatogr.*, *169*:1 (1979).
44. F. Foret, and P. Boček, *Advances in Electrophoresis*, Vol. 3 (A. Chrambach, M. J. Dunn, and B. J. Radola, eds.), VCH, Weinheim, p. 297 (1989).
45. J. K. Towns, and F. E. Regnier, *Anal. Chem.*, *64*:2473 (1992).
46a. J. H. Knox, *Chromatographia*, *26*:329 (1988).
46b. S. Hjertén, *Electrophoresis*, *11*:665 (1990).
47. C. A. Monning, and J. W. Jorgenson, *Anal. Chem.*, *63*:802 (1991).
48. R. J. Nelson, A. Paulus, A. S. Cohen, A. Guttman, and B. L. Karger, *J. Chromatogr.* *480*:111 (1989).
49. X. Huang, W. F. Coleman, and R. D. Zare, *J. Chromatogr.*, *480*:95 (1989).
50. P. D. Grossman, et al., *Am. Biochem. Lab.*, *2*:35 (1990).
51. W. Nashabeh, and Z. E. Rassi, *J. Chromatogr.*, *600*:279 (1992).
52. P. Hermentin, R. Witzel, R. Doenges, R. Bauer, H. Haupt, T. Patel, R. B. Parekh, and D. Brazel, *Anal. Biochem.*, *206*:419 (1992).
53. Millipore Corp., Waters Chromatography Div., Milford, Massachusetts.
54. N. A. Guzman, M. A. Trebilcock, and J. P. Advis, *J. Liquid Chromatogr.*, *14*:997 (1991).

55. R. Chien, and D. S. Burgi, *Anal. Chem.*, *64*:489A (1992).
56. D. Kaniansky, and J. Marák, *J. Chromatogr.*, *498*:191 (1990).
57. A. P. Tinke, N. J. Reinhoud, W. M. A. Niessen, U. R. Tjaden, and J. van der Greef, *Rapid Commun. Mass. Spectrom.*, 6:560 (1992).
58. F. Foret, E. Szoko, and B. L. Karger, *J. Chromatogr.*, *608*:3 (1992).
59. P. Gebauer, W. Thormann, and P. Boček, *J. Chromatogr.*, *608*:47 (1992).
60. F. Foret, E. Szoko, and B. L. Karger, *Electrophoresis*, (1993) (in press).
61. A. Vinther, A. Soeberg, L. Nielsen, J. Pedersen, and K. Biedermann, *Anal. Chem.*, *64*:187 (1992).
62. E. Gassman, J. E. Kuo, and R. Zare, *Science*, *230*:813 (1985).
63. Beckman Instruments, Fullerton, California.
64. F. Y. Cheng, *Science*, *242*:562 (1988).
65. L. Hernandez, J. Escalona, N. Joshi, and N. Guzman, *J. Chromatogr.*, *559*:183 (1991).
66. J. V. Sweedler, J. B. Shear, H. A. Fishman, and R. N. Zare, *Anal. Chem.*, *63*:496 (1991).
67. J. Liu, O. Shirota, and M. Novotny, *Anal. Chem.*, *63*:413 (1991).
68. T. T. Lee, and E. S. Yeung, *J. Chromatogr.*, *595*:319 (1992).
69. B. L. Hogan, and E. S. Yeung, *Anal. Chem.*, *64*:2841 (1992).
70. H. Small, T. Stevens, and W. Bauman, *Anal. Chem.*, *47*:1801 (1975).
70a. G. J. Bruin, A. C. Asten, X. Xu, and H. Poppe, *J. Chromatogr.*, *608*:97 (1992).
71. F. Foret, S. Fanali, A. Nardi, and P. Boček, *Electrophoresis*, *11*:780 (1990).
72. W. R. Jones, and P. Jandik, *J. Chromatogr.*, *608*:385 (1992).
73. E. S. Yeung, and W. G. Kuhr, *Anal. Chem.*, *63*:275A (1991).
74. K.-O. Eriksson, A. Palm, and S. Hjertén, *Anal. Biochem.*, *201*:211 (1992).
75. J. A. Olivares, N. T. Nguyen, C. R. Yonker, and R. D. Smith, *Anal. Chem.*, *59*:1232 (1987).
76. J. B. Fenn, M. Mann, C. K. Meng, S. F. Wong, and C. M. Whitehouse, *Science*, *246*:64 (1989).
77. A. P. Bruins, T. R. Covey, and J. D. Henion, *Anal. Chem.*, *59*:2642 (1987).
78. E. D. Lee, W. Mück, J. D. Henion, and J. R. Covey, *J. Chromatogr.*, *458*:313 (1988).
79. J. A. Loo, H. R. Udseth, and R. O. Smith, *Anal. Biochem.*, *197*:404 (1989).
80. P. Thibault, C. Paris, and S. Pleasance, *Rapid Commun. Mass Spectrom.*, *5*:484 (1991).
81. R. D. Smith, J. A. Loo, C. J. Barinaga, C. G. Edmonds, and H. R. Udseth, *J. Chromatogr.*, *480*:211 (1989).
82. T. J. Thompson, F. Foret, P. Vouros, and B. L. Karger, *Anal. Chem.*, (1992) (in press).
83. J. H. Wahl, D. R. Goodlett, H. R. Udseth, and R. D. Smith, Proceedings 40th ASMS Conference on Mass Spectrometry and Allied Topics, May 31–June 5, Washington, D.C., 404 (1992).
84. E. D. Lee, Z. Liu, and M. L. Lee, *Anal. Chem.*, *63*:2897 (1991).
85. J. G. Boyle, and C. M. Whitehouse, *Anal. Chem.*, *64*:2084 (1992).
86. J. C. Schwartz, and I. Jardine, Proceedings 40th ASMS conference on mass spectrometry and allied topics, May 31–June 5, Washington, D.C., p. 709 (1992).

87. H. Nishi, T. Fukuyama, M. Matsuo, and S. Terabe, *Anal. Chim. Acta*, *236*:281 (1990).
88. J. Cai, and Z. El Rassi, *J. Chromatogr.*, *608*:31 (1992).
89. A. S. Cohen, S. Terabe, J. A. Smith, and B. L. Karger, *Anal. Chem.*, *59*:1021 (1987).
90. K. C. Waldron, and N. J. Dovichi, *Anal. Chem.*, *64*:1396 (1992).
91. S. Terabe, N. Matsubara, Y. Ishihama, and Y. Okada, *J. Chromatogr.*, *608*:23 (1992).
92. J. Liu, K. A. Cobb, and M. Novotny, *J. Chromatogr.*, *519*:189 (1990).
93. T. Ueda, F. Kitamura, R. Mitchell, T. Metcalf, T. Kuwana, and A. Nakamoto, *Anal. Chem.*, *63*:2979 (1991).
94. J. Yang, and D. S. Hage, Fifth International Symposium on HPCE, Orlando, January 25–28, 1993.
95. K. Shimura, *J. Chromatogr.*, *510*:251 (1990).
96. J. C. Kraak, S. Busch, and H. Poppe, *J. Chromatogr.*, *608*:257 (1992).
96a. Y. H. Chu, L. Z. Avila, H. A. Bieluyck, and G. M. Whitesides, *J. Med. Chem.*, *35*:2915 (1992).
97. N. H. H. Heegaard, and F. A. Robey, *Anal. Chem.*, *64*:2479 (1992).
98. S. P. A. Fodor, et al., Science, *251*:767 (1991).
99. I. A. Hemmila, ed., *Applications of Fluorescence in Immunoassays*, John Wiley & Sons, New York (1991).
100. K. Shimura, and B. L. Karger, Fifth International Symposium on HPCE, Orlando, January 25–28, 1993.
101. H. H. Lauer, and D. McManigill, *Anal. Chem.*, *58*:166 (1986).
102. R. G. Nielsen, and E. C. Rickard, *J. Chromatogr.*, *516*:99 (1990).
103. P. D. Grossman, et al., *Anal. Chem.*, *61*:1186 (1989).
104. M. M. Bushey, and J. W. Jorgenson, *Anal. Chem.*, *62*:978 (1990).
105. M. M. Bushey and J. W. Jorgenson, *J. Chromatogr.*, *480*:301 (1989).
106. F. A. Chen, *J. Chromatogr.*, *559*:445 (1991).
107. S. Hjertén, *Arkiv. Kemi*, *13*:151 (1958).
108. W. M. Hurni, and W. J. Miller, *J. Chromatogr.*, *559*:337 (1991).
109. P. G. Righetti, *Isoelectric Focusing: Theory, Methodology and Applications*, Elsevier, Amsterdam, 1983.
110. R. A. Mosher, D. A. Saville, and W. Thormann, *The Dynamics of Electrophoresis* (B. J. Radola, ed.), VCH, New York (1992).
111. M. Chiari, and P. G. Righetti, *Electrophoresis*, *13*:187 (1992).
112. O. Vesterberg, and H. Svensson, *Acta Chem. Scand.*, *20*:820 (1966).
113. S. Hjertén, and M. D. Zhu, *J. Chromatogr.*, *346*:265 (1985).
114. F. Kilar, and S. Hjertén, *J. Chromatogr.*, *480*:351 (1989).
115. J. Wu, and J. Pawliszyn, *Anal. Chem.*, *64*:(1992) (in press).
116. S. Hjertén, L. Liao, and K. Yao, *J. Chromatogr.*, *387*:127 (1987).
117. M. Zhu, R. Rodriguez, and T. Wehr, *J. Chromatogr.*, *559*:479 (1991).
118. T. Wehr, M. Zhu, R. Rodriguez, D. Burke, and K. Duncan, *Am. Biochem. Lab.*, *8*:23 (1990).
119. M. Chen, and J. E. Wiktorowicz, *Anal. Biochem.*, *206*:84 (1992).
120. T. J. Nelson, *J. Chromatogr.*, *623*:357 (1992).
121. J. R. Mazzeo, and I. S. Krull, *Anal. Chem.*, *63*:2852 (1991).

122. W. Thormann, J. Caslavska, S. Molteni, and J. Chmelík, *J. Chromatogr.*, *589*:321 (1992).
123. B. M. Olivera, P. Baine, and N. Davidson, *Biopolymers*, *2*:245 (1964).
124. N. C. Stellwagen, *Biochemistry*, *22*:6180 (1983).
125. A. T. Andrews, *Electrophoresis*, 2nd ed., Clarendon Press, Oxford, p. 117 (1986).
126. J. Liu, O. Shirota, and M. Novotny, *J. Chromatogr.*, *559*:223 (1991).
127. A. Chrambach, *The Practice of Quantitative Gel Electrophoresis*, VCH, Deerfield Beach, Fl., p. 64 (1985).
128. I. T. Norton, D. M. Goodall, K. R. J. Ansten, E. R. Morris, and D. A. Rees, *Biopolymers*, *25*:1009 (1986).
129. A. S. Cohen, D. R. Najarian, A. Paulus, A. Guttman, J. A. Smith, and B. L. Karger, *Proc. Natl. Acad. Sci. USA*, *85*:9660 (1988).
130. S. Hjertén, *Arch. Biochem. Biophys.*, Suppl. 1: 147 (1962).
131. A. Guttman, A. S. Cohen, D. N. Heiger, and B. L. Karger, *Anal. Chem.*, *62*:137 (1990).
132. P. F. Bente, and J. Myerson, Hewlett-Packard, European Patent EP0272925 A2, June 29, 1988.
133. H. G. Yin, J. A. Lux, and G. Schomburg, *J. High Resolut. Chromatogr.*, *13*:436 (1990).
134. T. Wang, G. J. M. Bruin, J. C. Kraak, and H. Poppe, *Anal. Chem.*, *63*:2207 (1991).
135. Y. Baba, and M. Tsuhako, TRAC, *11*:280 (1992).
136. S. Hjertén, and M. D. Zhu, *J. Chromatogr.*, *327*:157 (1985).
137. T. Tanaka, *Sci. Am.*, *244*:124 (1981).
138. A. Guttman, A. S. Cohen, and B. L. Karger, unpublished results.
139. M. Spencer, *Electrophoresis*, *4*:36 (1983).
140. H. Swerdlow, K. E. Dew-Jager, K. Brady, N. J. Dovichi, and R. Gesteland, *Electrophoresis*, *13*:475 (1992).
141. H. J. Bode, *Electrophoresis '79*, Walter de Gruyter & Co., New York, p. 39 (1980).
142. D. Tietz, M. H. Gottlieb, J. S. Fawcett, and A. Chrambach, *Electrophoresis*, 7:217 (1986).
143. P. D. Grossman, and D. S. Soane, *J. Chromatogr.*, *559*:257 (1991).
144. M. Zhu, D. L. Hansen, S. Burd, and F. Gannon *J. Chromatogr.*, *480*:311 (1989).
145. K. Ganzler, K. S. Greve, A. S. Cohen, and B. L. Karger, *Anal. Chem.*, *64*:2655 (1992).
146. J. Sudor, F. Foret, and P. Boček, *Electrophoresis*, *12*:1056 (1991).
147. B. L. Karger, *HPCE '91*, San Diego, Calif.
148. M. Ruiz-Martinez, J. Berka, A. Belenkii, F. Foret, and B. L. Karger, *Anal. Chem.*, (submitted).
149. X. Huang, J. A. Luckey, M. J. Gordon, and R. N. Zare, *Anal. Chem.*, *611*:766 (1989).
150. A. S. Cohen, Lecture 108, Eighth International Symposium on Capillary Electrophoresis and Isotachophoresis, Rome, October (1992).
151. G. L. Trainor, *Anal. Chem.*, *62*:418 (1990).
152. H. Drossman, J. A. Luckey, A. J. Kostichka, J. D'Cunha, and L. M. Smith, *Anal. Chem.*, *62*:900 (1990).

153. H. Swerdlow, and R. Gesteland, *Nucleic Acids Res.*, *18*:1415 (1990).
154. A. S. Cohen, D. R. Najarian, and B. L. Karger, *J. Chromatogr.*, *516*:49 (1990).
155. Applied Biosystems, Inc., Foster City, California.
156. J. A. Luckey, H. Drossman, A. J. Kostichka, D. A. Mead, J. D'Cunha, T. B. Norris, and L. M. Smith, *Nucleic Acids Res.*, *18*:4417 (1990).
157. H. Swerdlow, J. Z. Zhang, D. Y. Chen, H. R. Harke, R. Greg, S. Wu, and N. J. Dovichi, *Anal. Chem.*, *63*:2835 (1991).
158. A. E. Karger, J. M. Harris, and R. F. Gesteland, *Nucleic Acids Res.*, *19*:4955 (1991).
159. A. S. Cohen, S. Carson, A. Belenkii, and B. L. Karger, *Anal. Chem.*, (1992) (in press).
160. S. Tabor, and C. C. Richardson, *J. Biol. Chem.*, *256*:8322 (1990).
161. S. L. Pentoney, Jr., K. D. Konrad, and W. Kaye, *Electrophoresis*, *13*:467 (1992).
162. X. C. Huang, M. A. Quesada, and R. A. Mathies, *Anal. Chem.*, *64*:2149 (1992).
163. D. Y. Chen, H. R. Harke, and N. J. Dovichi, *Nucleic Acids Res.*, *20*:4873 (1992).
164. A. J. Kostichka, M. L. Marchbanks, R. L. Brumley, H. Drossman, and L. M. Smith, *Bio/Technology*, *10*:78 (1992).
165. T. Hunkapiller, R. J. Kaiser, B. F. Koop, and L. Hood, *Science*, *254*:59 (1991).
166. K. B. Jacobson, and H. F. Arlinghaus, *Anal. Chem.*, *64*:315A (1992).
167. D. N. Heiger, A. S. Cohen, and B. L. Karger, *J. Chromatogr.*, *516*:33 (1990).
168. H. E. Schwartz, K. Ulfelder, F. J. Sanzeri, M. P. Busch, and R. G. Brownlee, *J. Chromatogr.*, *559*:267 (1991).
169. P. Boček, and A. Chrambach, *Electrophoresis*, *13*:31 (1992).
170. A. Guttman, and N. Cooke, *Anal. Chem.*, *63*:2038 (1991).
171. H. E. Schwartz, and K. Ulfelder, *Anal. Chem.*, *64*:1737 (1992).
172. M. Lalande, J. Noolandi, C. Turmel, R. Brousseau, J. Rousseau, and G. W. Slater, *Nucleic Acids Res.*, *16*:5427 (1988).
173. A. Guttman, B. Vanders, and N. Cooke, *Anal. Chem.*, *64*:2248 (1992).
174. G. W. Slater, and J. Noolandi, *Biopolymers*, *28*:1781 (1989).
175. G. W. Slater, and G. Drouin, *Electrophoresis*, *13*:574 (1992).
176. P. D. Grossman, S. Menchu, and D. Hershey, Genet. Anal. Tech. Appl., 9:9 (1992).
177. L. Ulanovsky, and M. Burmeister, eds., *Methods in Molecular Biology, Vol. 6, Pulsed Field Gel Electrophoresis Techniques*, Humuna Press, Clifton, N.J. (1991).
178. T. Demana, M. Lanan, and M. D. Morris, *Anal. Chem.*, *63*:2795 (1991).
179. D. N. Heiger, S. M. Carson, A. S. Cohen, and B. L. Karger, *Anal. Chem.*, *64*:192 (1992).
180. A. S. Cohen, A. Paulus, and B. L. Karger, *Chromatographia*, *24*:15 (1987).
181. V. Dolník, and M. V. Novotny, *Anal. Chem.*, *64*:(1992) (in press).
182. A. S. Cohen, and B. L. Karger, *J. Chromatogr.*, *397*:409 (1987).
183. K. Tsuji, *J. Chromatogr.*, *550*:823 (1991).
184. A. Widhalm, C. Schwer, D. Blass, and E. Kenndler, *J. Chromatogr.*, *546*:446 (1991).
185. T. Manabe, and S. Terabe, Poster PT-23, *HPCE '91*, San Diego.
186. D. Wu, and F. E. Regnier, *J. Chromatogr.*, *608*:349 (1992).
187. D. M. Demorest, S. Chen, W. E. Werner, R. Dubrow, J. L. Colburn, and J. E. Wiktorowicz, Poster M28, 4th Intl. Symp. HPCE, Amsterdam, February (1992).
188. M. V. Novotny, and J. Sudor, *Electrophoresis*, (in press).

189. J. Liu, O. Shirota, and M. Novotny, *J. Chromatogr.*, 559:223 (1991).
190. J. B. Poli, and M. R. Schure, *Anal. Chem.*, 64:896 (1992).
191. J. D. Harrison, A. Manz, Z. Fan, H. Ludi, and H. M. Widmer, *Anal. Chem.*, 64:1926 (1992).
192. S. Terabe, *TRAC*, 8:129 (1989).
193. H. Nishi, T. Fukugama, S. Terabe, *J. Chromatogr.*, 553:503 (1991).
194. D. Schmalzing, F. Foret, C. Piggee, and B. L. Karger, in preparation.

2
Micellar Electrokinetic Chromatography

Shigeru Terabe

Himeji Institute of Technology
Kamigori, Hyogo, Japan

Capillary zone electrophoresis (CZE) is a highly efficient separation method that employs a fused silica capillary of less than 100 μm internal diameter (i.d.) and 30–100 cm length. The separation principle of CZE is based on the differential electrophoretic mobility and, therefore, only charged compounds or ions can be separated by this method. Under conventional conditions, the electroosmotic flow transports the bulk solution in the capillary with a flat velocity profile from the positive to negative electrode, and it is stronger than the electrophoretic migration of charged substances at a pH above 6 [1,2]. Therefore, most solutes, even if they are anionic, migrate toward the negative electrode.

Although CZE has a high separation power, it is not applicable to the separation of uncharged compounds, because neutral compounds have no electrophoretic mobility. To overcome this difficulty, we have developed electrokinetic chromatography (EKC) [3], which is named after electrokinetic phenomena and chromatography. The separation principle of EKC is the same as that of chromatography, but EKC utilizes electrokinetic phenomena to perform the chromatography instead of a liquid delivery pump. For chromatographic separation, we need two phases, between which analyte molecules are distributed, and one of the two phases must migrate at velocities different from the other, which may not migrate. Since charged substances have electrophoretic mobilities, they can migrate at a velocity different from the surrounding bulk solution. Therefore, a charged substance that can incorporate analyte molecules is appropriately usable as a "phase" in EKC. We named this electrophoretically migrating phase a *separation carrier*.

Micellar EKC (MEKC) [4,5] uses ionic micelles as the separation carrier. MEKC is the most popular and useful among other EKC techniques. Cyclodextrin (CD) derivatives having ionic groups are also useful separation carriers used in CDEKC [6], which is especially helpful for the separation of aromatic isomers and enantiomers. Two other EKC techniques have now been published: ion-exchange EKC [7], which employs a polymer ion as the separation carrier, is effective for the separation of isomeric ions having close or identical electrophoretic mobilities; microemulsion EKC [8] is similar to MEKC, but it is closer to pure liquid–liquid partition chromatography than to MEKC, although only one paper has been published. Although the term *micellar electrokinetic capillary chromatography* (MECC) [9] is widely used for MEKC, the reason we use MEKC instead of MECC is that MEKC is only a mode of EKC, and EKC is the more versatile technique, as mentioned above.

In this chapter, fundamental characteristics of MEKC are explained from the viewpoint of separation analysis. This is not a comprehensive review of MEKC, and most examples cited are from the author's work.

SEPARATION PRINCIPLE

Qualitative Explanation

A schematic diagram of the separation principle is shown in Fig. 1. A capillary is filled with an ionic surfactant solution of a concentration higher than its critical

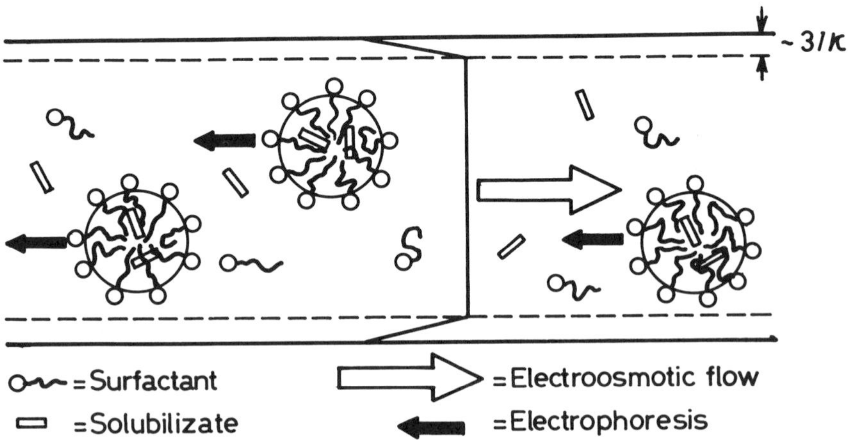

Figure 1 Schematic of the separation principle of MEKC.

micelle concentration (CMC), above which the micelle is formed, by the aggregation of surfactant molecules, as an electrophoretic solution instead of the simple buffer solution used in CZE. The ionic micelle works as the separation carrier, as shown in Fig. 1, although the micelle exists only in a dynamic equilibrium state in the presence of the monomeric surfactant. The micelle is usually depicted as a sphere (see Fig. 1), but its size and shape depend on the individual surfactant. The aggregation number has a distribution, and the average lifetime of the micelle is about 1 second.

When sodium dodecyl sulfate (SDS), one of the most popular surfactants, is employed, the micelle consists of about 60 molecules and is negatively charged. The SDS micelle, therefore, migrates toward the positive electrode by electrophoresis. The electroosmotic flow is in the direction of the negative electrode, and it is stronger than the electrophoretic migration of the SDS micelle under conditions of pH above 5 [10]. Thus, the SDS micelle also migrates toward the negative electrode at a velocity equal to the difference between the electroosmotic and electrophoretic velocities.

A neutral analyte injected into the micellar solution will migrate at the electroosmotic velocity when it is free from the micelle, and at the velocity of the micelle when it is incorporated into the micelle. Provided the equilibrium distributing the analyte between the micelle and the surrounding aqueous phase is quickly established, the analyte will migrate at the velocity between the two extremes, the electroosmotic velocity, v_{eo}, and the velocity of the micelle, v_{mc}, as shown in Fig. 2A, where the analyte is assumed to be equally distributed between the two.

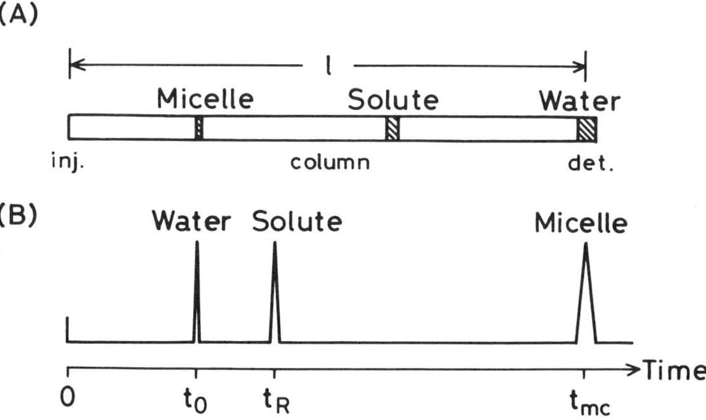

Figure 2 (A) Schematic of the zone separation in MEKC and (B) chromatogram. (From Ref. 5)

Description of the Migration With the Chromatographic Capacity Factor

We can define the capacity factor, k', as the ratio of the number of the analyte molecules incorporated into the micelle, n_{mc}, over that in the aqueous phase, n_{aq}, as in Eq. 1.

$$k' = \frac{n_{mc}}{n_{aq}} \tag{1}$$

Then, $1/(1 + k')$ becomes equal to the fraction of the analyte in the aqueous phase, because

$$\frac{1}{1 + k'} = \frac{n_{aq}}{n_{aq} + n_{mc}} \tag{2}$$

Similarly, $k'/(1 + k')$ is equal to the fraction of the analyte in the micelle, as

$$\frac{k'}{1 + k'} = \frac{n_{mc}}{n_{aq} + n_{mc}} \tag{3}$$

The migration velocity of the analyte, v_s, can be given as

$$v_s = \frac{1}{1 + k'} v_{eo} + \frac{k'}{1 + k'} v_{mc} \tag{4}$$

Substituting each velocity in Eq. 4 by the following equations:

$$v_s = \frac{l}{t_R} \tag{5}$$

$$v_{eo} = \frac{l}{t_0} \tag{6}$$

$$v_{mc} = \frac{l}{t_{mc}} \tag{7}$$

where t_R, t_0, and t_{mc} are the migration times of the analyte, the aqueous phase, and the micelle, respectively (see Fig. 2B) and l is the effective length of the capillary from the injection end to the detector cell, we can obtain the relation between the migration time and the capacity factor,

$$t_R = \frac{1 + k'}{1 + (t_0/t_{mc})k'} t_0 \tag{8}$$

When t_{mc} is infinity or the micelle stays still in the capillary, which can be attained only when the electroosmotic velocity is identical with the electrophoretic velocity of the micelle in opposite directions, Eq. 8 becomes

$$t_R = (1 + k')t_0 \tag{9}$$

which is the same as the corresponding equation of conventional chromatography. Under these conditions, the micelle will work as an ideal stationary phase of liquid chromatography, because it does not require the supporting material, such as silica gel, in addition to the homogeneous size and distribution in the column.

On the contrary, when t_0 is infinity or electroosmosis is completely suppressed, Eq. 8 gives

$$t_R = (1 + 1/k')t_{mc} \qquad (10)$$

Here, the aqueous phase stays still in the capillary and the micelle will migrate through the aqueous phase. Thus, the electroosmotic flow is not essential in MEKC, although all the analytes migrate toward the positive electrode in the absence of electroosmosis, if SDS is used. In this instance, the migration velocity of the analyte that has a larger capacity factor is faster than that of one having a smaller capacity factor.

In general, both the micelle and the surrounding aqueous phase migrate in the capillary; therefore, every analyte, if it is electrically neutral, must reach the detector during the migration time window between t_0 and t_{mc}. As the ratio t_0/t_{mc} decreases, the migration time window expands.

To obtain the capacity factor according to Eq. 8, we need to know the three migration times: t_0, t_R, t_{mc}. The marker of the electroosmotic flow must be neutral, free from the micelle, and have UV absorbance. No such compounds are available, and methanol is used because the distribution coefficient is negligibly small and can be detected as a baseline deflection with a UV detector owing to a refractive index change. The tracer of the micelle must be completely incorporated into the micelle. Sudan III or IV is often employed. Timepidium bromide [11] is also a good tracer for an anionic micelle. Also, it is not always easy to obtain the values of t_0 and t_{mc}.

Figure 3 shows a typical example of MEKC separation, in which eight electrically neutral compounds are successfully resolved in 17 minutes. The capacity factor scale is inserted in the figure to indicate the relation between migration time and capacity factor. The capacity factor of infinity means that the analyte has the same migration time as the micelle or that it is totally incorporated into the micelle. Theoretical plate numbers calculated from the peak widths are 200,000–250,000.

Description of the Migration With Electrophoretic Mobility

The foregoing explanation is from a chromatographic standpoint. The migration of the analyte by MEKC can also be described from the viewpoint of electrophoresis. In CZE, the migration velocity of the analyte is expressed as

$$v_s = [\mu_{eo} + \mu_{ep}(s)]E \qquad (11)$$

where μ_{eo} and $\mu_{ep}(s)$ are the electroosmotic mobility and electrophoretic mobility

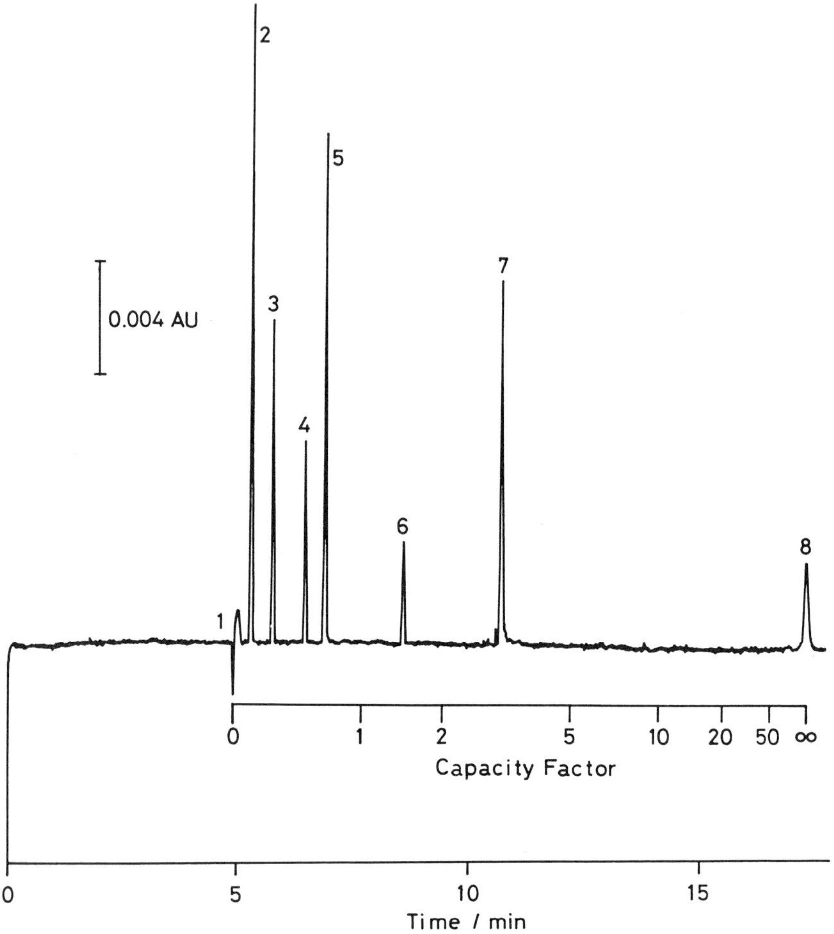

Figure 3 Micellar electrokinetic chromatogram of a mixture: 1, methanol; 2, resorcinol; 3, phenol; 4, *p*-nitroaniline; 5, nitrobenzene; 6, toluene; 7, 2-naphthol; 8, Sudan III. Conditions: capillary, 50-μm i.d. × 650 mm (effective length 500 mm); separation solution, 30 mM SDS in 50 mM phosphate–100 mM borate buffer (pH 7.0); applied voltage, 15 kV; current 33 μA; detection, absorbance at 210 nm. (From Ref. 3)

of the analyte, respectively. We can apply this equation to MEKC by assuming the effective electrophoretic mobility, $\mu^*_{ep}(s)$, for the neutral analyte [12] as

$$\mu^*_{ep}(s) = \frac{k'}{1 + k'} \mu_{ep}(mc) \tag{12}$$

where $\mu_{ep}(mc)$ is the electrophoretic mobility of the micelle. Thus, the velocity of the neutral analyte in MEKC is given as

$$v_s = [\mu_{eo} + \mu^*_{ep}(s)]E \tag{13}$$

The difference between Eq. 4 and Eq. 13 should be noted, because $\mu_{ep}(mc)E$ is not equal to v_{mc} (see Eq. 21). The effective mobility means that even the neutral analyte has an apparent mobility equal to the product of the fraction of the analyte incorporated into the micelle and the mobility of the micelle. The effective mobility can be calculated according to Eq. 13, provided v_{eo} or μ_{eo} is known. This may be an advantage for use of the effective mobility over the capacity factor, which requires both t_0 and t_{mc} to calculate. However, the capacity factor provides quantitative information about the distribution equilibrium, but the effective mobility gives only qualitative information about it. In other words, the capacity factor cannot be obtained according to Eq. 12, unless $\mu_{ep}(mc)$ is known.

RESOLUTION

Resolution Equation

The resolution equation of MEKC [5] is given as

$$R_s = \frac{\sqrt{N}}{4}\left(\frac{\alpha - 1}{\alpha}\right)\left(\frac{k'_2}{1 + k'_2}\right)\left(\frac{1 - t_0/t_{mc}}{1 + (t_0/t_{mc})k'_1}\right) \tag{14}$$

Where N is the theoretical plate number, α the separation factor equal to k'_2/k'_1, and k'_1 and k'_2 are capacity factors of the analyte 1 and 2, respectively. The equation is similar to that of conventional chromatography, except for the last term of the right-hand side. The dependence of R_s on the plate number and the separation factor is the same as in conventional chromatography, as shown in Eq. 14, but the effect of the capacity factor on R_s is significantly different. The optimum value of the capacity factor is dependent on the ratio t_0/t_{mc} as discussed later. Each term in the right-hand side of Eq. 14 is discussed in the following.

Effect of the Plate Number

A distinct characteristic advantage of MEKC over high-performance liquid chromatography (HPLC) is in its high efficiency. The plate number obtained by MEKC easily reaches approximately 200,000, more than ten times higher than that of HPLC. The main causes of band broadening in MEKC are molecular diffusion, the kinetics of partition equilibrium, and microheterogeneity of the micelle. Molecular diffusion is the most important for most analytes. The other effects become considerable only for those analytes having large capacity factors. Therefore, the higher the applied voltage, in general, the higher plate number we can expect. However, since the micellar solution has a high conductivity, less than 25 kV is employed for a 50-μm i.d. × 50-cm capillary. The separation factor will be discussed in the following section.

Effect of the Capacity Factor

To discuss the capacity factor term, the product of the last two terms of the right-hand side of Eq. 14 is written as

$$f(k') = \left(\frac{k'}{1 + k'}\right)\left(\frac{1 - t_0/t_{mc}}{1 + (t_0/t_{mc})k'}\right) \qquad (15)$$

where k'_1 is assumed to be equal to k'_2, because we are discussing the optimum capacity factor for the separation of two close peaks. The dependence of $f(k')$ on k' is given in Fig. 4 for different values of t_0/t_{mc}. To find the optimum k' that gives

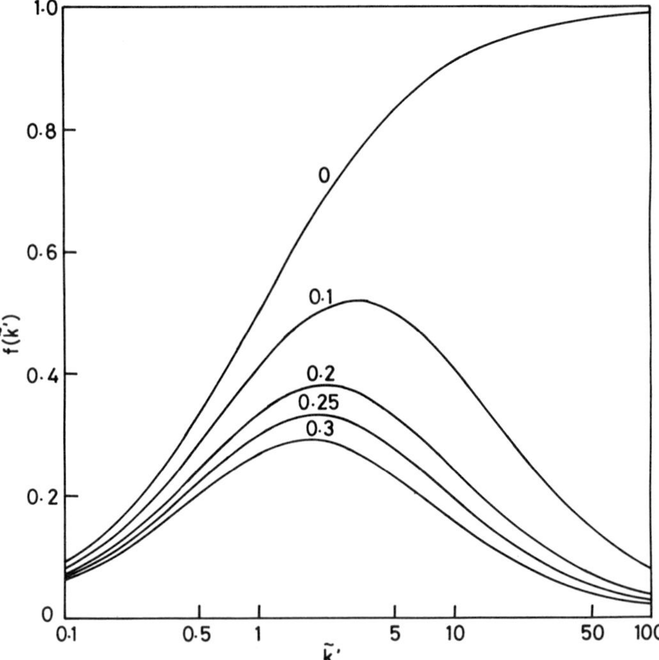

Figure 4 Dependence of $f(k')$ on the capacity factor, k', for different t_0/t_{mc} values, which are given on each curve. (From Ref. 5)

the highest resolution, we can differentiate Eq. 15 and find k' that makes the first derivative of Eq. 15 equal zero [13].

$$k'_{opt} = \sqrt{t_{mc}/t_0} \tag{16}$$

Under the conditions of pH 7, t_{mc}/t_0 is about 4; hence, k'_{opt} is close to 2. Large capacity factors mean that most fraction of analytes are incorporated into the micelle, and small values mean the reversed situation or that most analytes exist in the aqueous phase. The result given by Eq. 16 suggests that the capacity factors should not be close to either extreme for separation by MEKC. In general, capacity factors between 1 and 5 are recommended for the optimal resolution.

The capacity factor is related to the distribution coefficient, K, by

$$k' = K\frac{V_{mc}}{V_{aq}} \tag{17}$$

where V_{mc}/V_{aq} is the phase ratio and V_{mc} and V_{aq} are volumes of the micelle and the remaining aqueous phase. The micellar volume is related to the surfactant concentration, C_{srf}, by

$$V_{mc} = \bar{v}(C_{srf} - \text{CMC}) \tag{18}$$

where $\bar{v}$ is the specific volume of the micelle. The volume of the aqueous phase is the rest of the solution. Therefore,

$$\frac{V_{mc}}{V_{aq}} = \frac{\bar{v}(C_{srf} - \text{CMC})}{1 - \bar{v}(C_{srf} - \text{CMC})} \tag{19}$$

Since the volume of the micelle is usually negligible in comparison with that of the aqueous phase, Eq. 20 is obtained.

$$k' = K\bar{v}(C_{srf} - \text{CMC}) \tag{20}$$

This equation suggests that the capacity factor increases linearly with the surfactant concentration; hence, it is straightforward to obtain the desired value of the capacity factor by adjusting the concentration of the surfactant. This is a characteristic feature of MEKC that is different from reversed-phase (RP)-HPLC, for which the alteration of the phase ratio is seriously limited.

Effect of the Electroosmotic Velocity

The effect of the electroosmotic velocity on resolution can also be discussed on the basis of Eq. 14. The velocity of the micelle is written as

$$v_{mc} = [\mu_{eo} + \mu_{ep}(\text{mc})]E \tag{21}$$

Then,

$$t_0/t_{mc} = 1 + \mu_{ep}(\text{mc})/\mu_{eo} \tag{22}$$

The mobility μ_{eo} and $\mu_{ep}(mc)$ usually have different signs, and the ratio $\mu_{ep}(mc)/\mu_{eo}$ is smaller than zero and larger than -1. Therefore, t_0/t_{mc} is smaller than 1. When $\mu_{eo} = -\mu_{ep}(mc)$, the migration time of the micelle, t_{mc}, becomes infinity. The smaller t_0/t_{mc} gives the higher resolution, as shown in Fig. 4. If we assume a negative sign for t_0/t_{mc}, when $\mu_{ep}(mc)/\mu_{eo}$ is smaller than -1, Eq. 14 suggests possible extremely high resolution for the analyte having the capacity factor close to $-(t_{mc}/t_0)$ at the expense of a long time [10]. Some examples of the effect of the electroosmotic velocity on resolution have been published [14,15].

SELECTIVITY MANIPULATION

The separation factor can be manipulated through chemical consideration. The micelle corresponds to the stationary phase in RP-HPLC relative to selectivity manipulation, as does the surrounding aqueous phase to the mobile phase. Both the micelle and the aqueous phase are easily modified with additives for improved separation. In this section, general consideration about the factors affecting selectivity are described.

Effect of the Micellar Structure

Surfactants

Surfactants have a hydrophobic and a hydrophilic group within each molecule, and both groups affect selectivity in MEKC. The hydrophilic, or ionic group, is generally more important in determining selectivity than is the hydrophobic group. Figure 5 and Table 1 show examples of the effect of the surfactant structure. The structural difference between SDS and cetyltrimethylammonium bromide (CTAB) is mainly in the ionic groups: sulfate in SDS and quarternary ammonium in CTAB. The hydrophobic groups are similar and different only in the length of the alkyl chains: dodecyl in SDS and hexadecyl in CTAB. The migration order of nitrobenzene shown in Figure 5 was drastically altered from SDS to CTAB, probably because of the difference in ionic groups between the two surfactants. Similar examples have also been published [16]. Table 1 demonstrates differences of distribution coefficients of the test samples, shown in Fig. 3, among three related surfactants [5]. For resorcinol, phenol, *p*-nitroaniline, and nitrobenzene, the distribution coefficients are virtually equal between SDS and sodium tetradecyl sulfate (STS), which have identical ionic groups, but different lengths of alkyl chains, but the distribution coefficients are significantly different between SDS and sodium dodecanesulfonate (SDDS), which have identical alkyl groups, but different ionic groups. These results shown in Fig. 5 and Table 1 strongly suggest that many compounds are adsorbed on the surface of the micelle or at least interact considerably with the surface of the micelle.

Although MEKC is capable of separating neutral compounds, it also allows an improved separation of ionic analytes when they are not successfully separated by

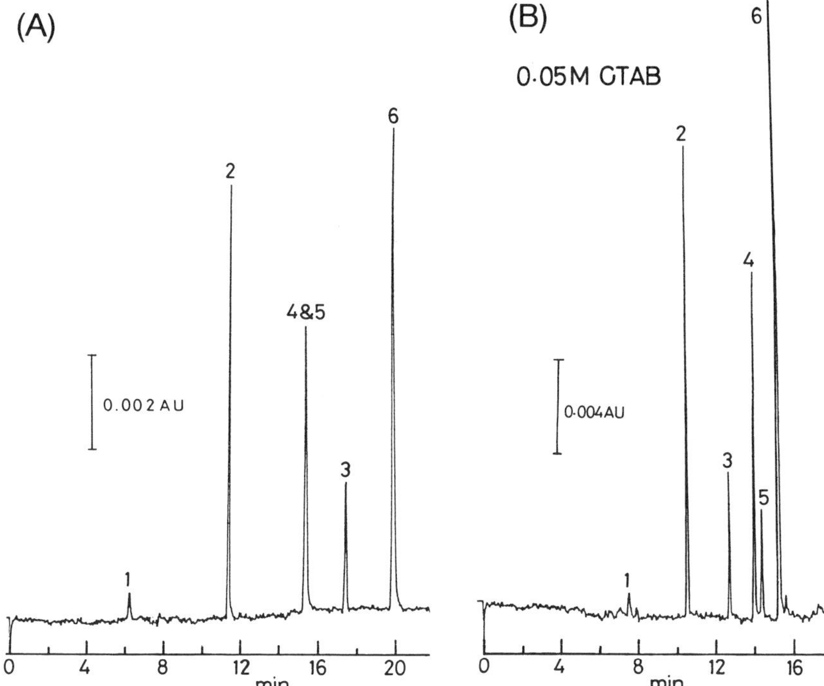

Figure 5 Effect of the surfactant structures on selectivity: 1, water; 2, aniline; 3, nitrobenzene; 4, *m*-nitroaniline; 5, *p*-nitroaniline; 6, *o*-nitroaniline. Conditions: capillary, 50 μm i.d. × 650 mm (effective length 500 mm); separation solution, (A) 100 mM SDS in 50 mM phosphate–100 mM borate buffer (pH 7.0), (B) 50 mM CTAB in 100 mM Tris–HCl (pH 7.0); applied voltage, 15 kV; detection, absorbance at 230 nm.

CZE alone. Since the micelle used in MEKC has a charge on the surface, the analyte having the opposite charge will strongly interact with the micelle through electrostatic force, and the analyte having the same charge will interact weakly owing to the electrostatic repulsion. Many excellent examples that demonstrate improved resolution by MEKC compared with CZE have been published [17]. Micellar ECK is also a powerful technique for the separation of mixtures containing both ionic and neutral analytes.

Bile salts, such as sodium cholate, sodium deoxycholate, sodium taurodeoxycholate, which are supposed to form helical micelles [18], show selectivity significantly different from the long-alkyl–chain surfactants [19]. Bile salts are natural surfactants and have chirality. Therefore, a bile salt micelle can separate enantiomers [18,20–22].

Table 1 Distribution Coefficients at 35°C

Solute	Distribution coefficient		
	SDS[a]	STS[b]	SDDS[c]
Resorcinol	21.6	20.8	27.7
Phenol	52.1	52.3	56.1
p-Nitroaniline	103	100	84.3
Nitrobenzene	135	138	111
Toluene	318	345	288
2-Naphthol	656	789	698

[a]Sodium dodecyl sulfate.
[b]Sodium tetradecyl sulfate.
[c]Sodium dodecanesulfonate.
Source: Ref. 5.

Mixed Micelle

The micelle used in MEKC can be modified by forming a mixed micelle with ionic or nonionic surfactants, or by replacing the counterion with a different ion. The mixed micelle consisting of an ionic and a nonionic surfactant is usually larger and has a lower charge density on the surface and, hence, a lower electrophoretic mobility in comparison with the original ionic micelle. These changes in the micelle structures cause the narrower migration time window [23]. Since the surface of the mixed micelle is different from that of the original ionic micelle, the mixed micelle naturally shows different selectivity [23], as expected from the results given in Fig. 5 and Table 1.

Effect of pH

The constituent of the buffer dissolving the surfactant does not significantly affect selectivity, but the pH of the buffer is an important factor to manipulate selectivity of ionizable analytes. As mentioned earlier, an ionized form of the analyte having the same charge as the micelle will be incorporated into the micelle less than its un-ionized form. In case of acids, the increase in pH will promote ionization and, hence, the distribution to the anionic micelle will be retarded. Thus, the distribution coefficient, migration time, and selectivity can be substantially altered by changing the pH for ionizable analytes [24,25].

Effect of Temperature

The distribution coefficient is dependent on temperature, and an increase in temperature causes reduction of the migration time because of the decrease in the distribution coefficient. The temperature rise also results in increases of v_{eo} and

v_{ep}(mc) by the same extent owing to the lowered viscosity. The dependence of the distribution coefficient on temperature is different among analytes; therefore, temperature will also affect selectivity as shown in Fig. 6 [26]. The temperature effect on selectivity is unremarkable, but temperature seriously affects the migration time. Therefore, it is essential to the reproducibility of results to keep temperature precisely constant.

Effect of Additives to the Aqueous Phase

The most versatile techniques for manipulating selectivity in MEKC are the use of additives to the micellar solution to modify the aqueous phase as well as the choice of surfactants. Modification of the mobile phase in HPLC has been highly developed, and many of the additives can also be applicable to MEKC. However, we must consider the difference between the micelle and the stationary phase in HPLC to

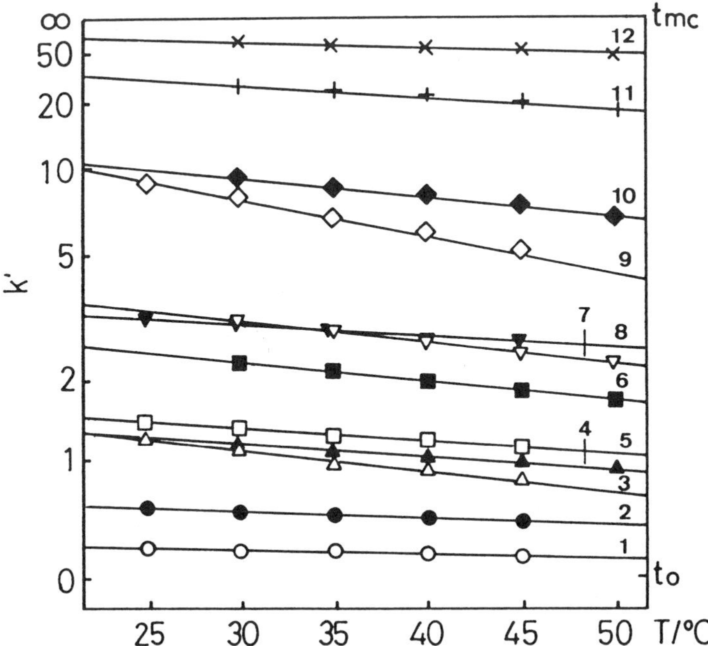

Figure 6 Dependence of the capacity factor, k', on temperature, T. 1, resorcinol; 2, phenol; 3, *p*-nitroaniline; 4, *o*-cresol; 5, nitrobenzene; 6, 2,6-xylenol; 7, 2,4-xylenol; 8, toluene; 9, 2-naphthol; 10, *p*-propylphenol; 11, *p*-butylphenol; 12, *p*-amylphenol. Conditions: capillary, 50 μm i.d. × 570 mm (effective length 500 mm); separation solution, 50 mM SDS in 50 mM phosphate–100 mM borate buffer (pH 7.0); applied voltage, 15 kV; detection, absorbance at 214 nm. (From Ref. 26)

take advantage of such additives. So far, four groups of additives have been useful in MEKC: cyclodextrins, ion-pair reagents, urea, and organic solvents.

Cyclodextrins

Cyclodextrin (CD) is now popular in the field of chromatography. Most of the techniques using CD are based on its capability to recognize specific molecules that fit its hydrophobic cavity. Cyclodextrin is especially effective for the separation of aromatic isomers and aromatic enantiomers having the chiral center close to the aromatic ring. From the viewpoint of MEKC, CD is electrically neutral and has no electrophoretic mobility. Although the cavity of CD is hydrophobic, the surface is hydrophilic and, hence, CD is assumed not to be incorporated into the micelle. A surfactant molecule may be included into the CD cavity.

The separation principle of cyclodextrin-modified MEKC (CD/MEKC) is schematically shown in Fig. 7. The analyte molecule included by CD migrates at the same velocity as the electroosmotic flow, because CD behaves electrophoretically as the aqueous phase. Therefore, the addition of CD reduces the apparent distribution coefficient and enables the separation of highly hydrophobic analytes, which will be almost totally incorporated into the micelle in the absence of CD. The CD addition also greatly alters selectivity.

Figure 8 shows an example of the effect of CD addition on the separation of highly hydrophobic and isomeric compounds [27]. Eleven isomeric trichlorobiphenyls migrated together at a velocity equal to v_{mc} in the absence of CD, because they are totally incorporated into the micelle. As shown in this example, CD/MEKC is powerful enough to resolve the isomeric mixture. Urea was added to increase the

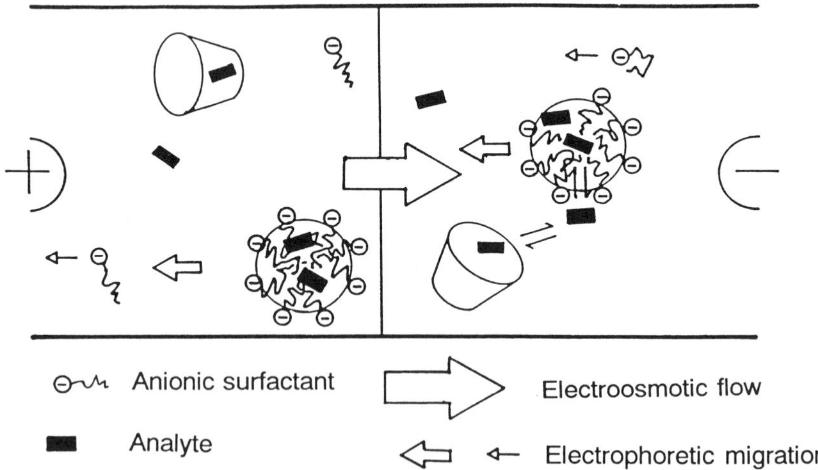

Figure 7 Schematic of the separation principle of CD/MEKC.

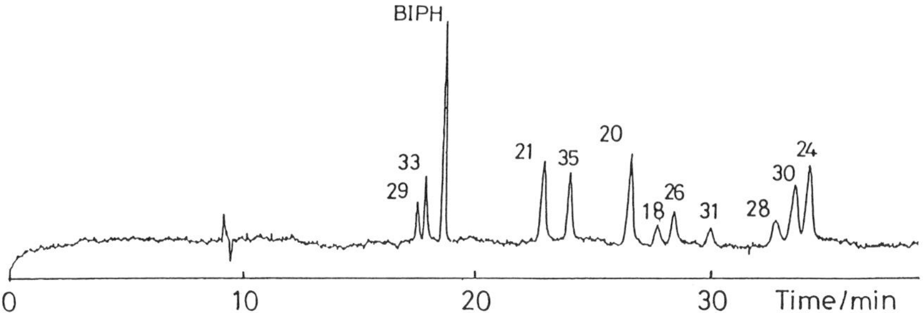

Figure 8 Separation of 11 trichlorobiphenyl isomers. Peaks are identified with the IUPAC number: 18 = 2,2′,5-; 20 = 2,3,3′-; 21 = 2,3,4-; 24 = 2,3,6-; 26 = 2,3′,5-; 28 = 2,4,4′-; 29 = 2,4,5-; 30 = 2,4,6-; 31 = 2,4′,5-; 33 = 2′,3,4-; and 35 = 2′,3,4-trichlorobiphenyl; BIPH = biphenyl. Conditions: capillary, 50 μm i.d. × 650 mm (effective length 500 mm), separation solution; 60 mM γ-CD, 100 mM SDS, and 2 M urea solution in 100 mM borate–50 mM phosphate buffer (pH 8.0); applied voltage 15.4 kV; detection, absorbance at 220 nm. (From Ref. 27)

solubility of CD, although it was not always necessary in this example. The use of β-CD instead of γ-CD was not particularly successful, probably because the cavity of β-CD was a little too small to include the analyte together with an SDS molecule.

Cyclodextrin/MEKC is also very helpful for chiral separation because of the chirality of CD [28,29; see also Chap. 21]. One advantage of CD/MEKC over RP-HPLC using CD as a modifier of the mobile phase is a small amount of CD required in CD/MEKC. This means that even an expensive CD derivative is available in CD/MEKC, because the amount necessary to prepare the separation solution is from 0.1 to 1 mmol or less.

Ion-Pair Reagent

Ion-pair reagents are frequently employed in RP-HPLC for the separation of ionic analytes. There is an essential difference in the separation of ionic compounds between MEKC and RP-HPLC, although both have close similarities for the separation of nonionic compounds. The difference is due to the electrical charge on the micelle. Figure 9 schematically illustrates the effect of the addition of the tetraalkylammonium salt to the SDS solution for the separation of anionic and cationic analytes [11].

For an anionic analyte, such as a carboxylate shown in Fig. 9A, the carboxylate anion will not combine with the anionic SDS micelle owing to the electrostatic repulsion in the absence of the ammonium salt additive. The ammonium cation will couple with the carboxylate anion to form an ion pair. The paired ions thus formed are electrically neutral and will be incorporated into the micelle more easily than the carboxylate form. An increase in the concentration of the ammonium salt brings about an increase in the migration time. The sodium ion on the surface of

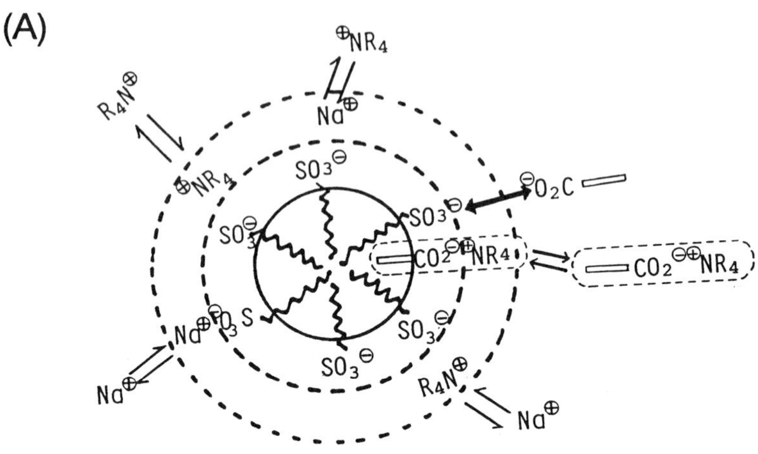

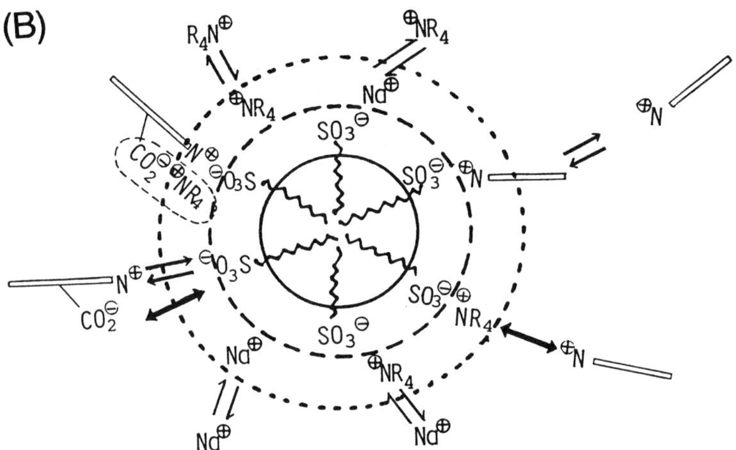

Figure 9 Schematic illustration of the effect of ion-pair reagents (tetraalkylammonium salt) in MEKC with SDS micelle. (A) Anionic analyte; (B) cationic and zwitterionic analyte. Solid circle shows hydrophobic core and broken circle the Stern layer of the micelle. (From Ref. 11)

the SDS micelle will be partly substituted by the ammonium ion, as shown in Fig. 9A, and it will change the surface characteristics of the SDS micelle.

For cationic analytes, the addition of the ammonium salt causes a decrease in the migration time, because the ammonium salt competes with the cationic analyte in combining with the anionic SDS micelle, as shown in Fig. 9B, and reduces the incorporation of the analyte into the micelle. The zwitterionic analytes have both anionic and cationic groups in a molecule, and the effect of an ammonium salt is the sum of both aforementioned mechanisms. An example of the addition of tetramethylammonium bromide to an SDS solution for the separation of cephalosporin antibiotics is given in Fig. 10 in comparison with CZE and MEKC [11]. Although the separation of these antibiotics was successful using a 150-mM SDS solution without the ammonium salt addition [30], Fig. 10 is given to demonstrate the effect of the ion-pair reagent.

Urea

A high concentration of urea is known to increase the solubility of hydrophobic compounds into water as well as to hinder hydrogen bond formation in the aqueous phase. The addition of a high-concentration urea to the SDS solution enabled the MEKC separation of highly hydrophobic compounds, as shown in Fig. 11 [31], in which a chromatogram obtained using the SDS solution without urea is also given for comparison. As shown in Fig. 11, these steroidal compounds were not successfully resolved with the SDS solution alone because their capacity factors were much larger

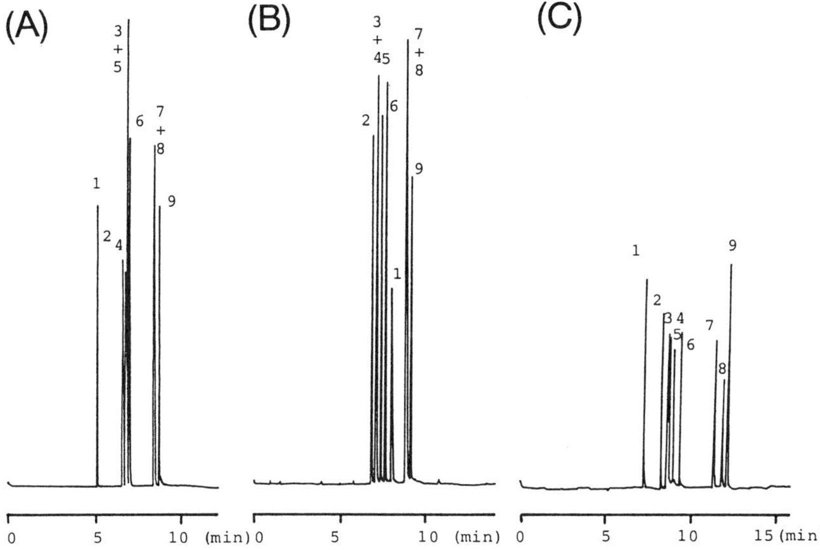

Figure 10 (A) Separation of cephalosporin antibiotics by CZE, (B) MEKC, and (C) MEKC using the ion-pair reagent. 1, C-TA; 2, ceftazidime; 3, cefotaxime; 4, cefmenoxime; 5, cefoperazone; 6, cefpiramide; 7, cefpimizole; 8, cefminox; 9, ceftriaxone. Conditions: capillary, 50 μm × 650 mm (effective length 500 mm); separation solution, (A) 20 mM borate–20 mM phosphate buffer (pH 9.0); (B) 50 mM SDS in the same buffer as in (A); (C) the same SDS solution as in (B), but containing 40 mM tetramethylammonium bromide; applied voltage 20 kV; detection, absorbance at 210 nm. (From Ref. 11)

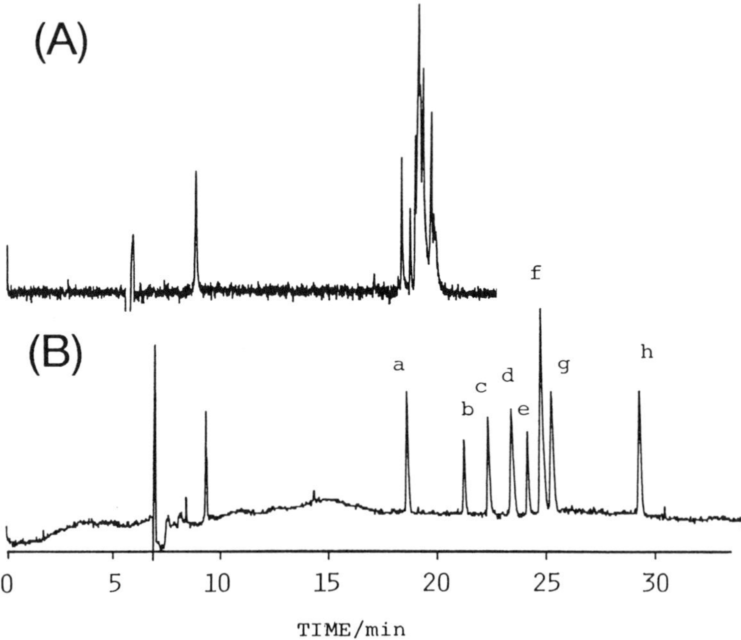

Figure 11 The effect of urea addition to the SDS solution. a, hydrocortisone; b, hydrocortisone acetate; c, betamethasone; d, cortisone acetate; e, triamcinolone acetonide; f, fluocinolone acetonide; g, dexamethasone acetate; h, fluocinonide. Conditions: separation solution, (A) 50 mM SDS in 20 mM borate–20 mM phosphate buffer (pH 9.0); (B) the same SDS solution as in (A), but containing 6M urea. Other conditions are the same as in Fig. 10. (From Ref. 31)

than the optimal value. The high-concentration urea slightly reduced the electroosmotic velocity and considerably reduced the migration velocity of the micelle. Although the urea addition caused a slight increase in viscosity, the decrease of v_{mc} was more significant than expected from the viscosity change [31]. However, the reason of the reduced velocity of the micelle is not obvious. Interestingly, the steroidal compounds given in Fig. 11 were also successfully resolved by CD/MEKC [32], which gave different migration orders.

Organic Solvents

Organic solvents miscible with water are widely accepted as mobile-phase modifiers in RP-HPLC. The main objective of using such organic solvents is to adjust the capacity factor close to the optimum value in RP-HPLC, because the capacity factor decreases as the concentration of the organic solvent in the mobile phase increases. We may also expect the same effect for the organic solvent in MEKC, but a high concentration of the organic solvent may break down the micelle. To reduce the capacity factor in MEKC, we are able to use CD or urea for highly hydrophobic analytes and the ion-pair reagent for ionic analytes, rather than a low concentration of the micelle. However, the use of organic solvents may contribute to improvement of the resolution or to alteration of the selectivity. Some papers have been published on the use of an organic solvent [33,34], such as methanol, 2-propanol, and acetonitrile. In general, the addition of methanol or 2-propanol reduces the electroosmotic velocity and thereby expands the migration time window.

APPLICATIONS

The scope of MEKC separation is similar to that of RP-HPLC. Reversed-phase HPLC is a widely accepted, well-established, and reliable analytical method, and numerous applications in various fields have been developed for routine analysis. An obvious advantage of RP-HPLC over MEKC is the capability of fractionating a relevant peak for preparative purposes.

Micellar EKC is a relatively new technique compared with RP-HPLC, and its instrumentation is far behind that of HPLC. The main advantage of MEKC over RP-HPLC is in its high separation efficiency. Various examples that show better resolution than with RP-HPLC have been published (see e.g., 17). Micellar EKC has also some other advantages over RP-HPLC (e.g., minute amounts of sample and separation solutions, a short analysis time, an easy cleaning of the capillary). It is also noted that MEKC is a branch of capillary electrophoresis and it is complementary to CZE.

Applications of MEKC are as yet mainly confined to the separation of small molecules, because the size of the micelle is not large enough to incorporate big molecules such as proteins and DNA. The most promising application of MEKC will be in pharmaceutical and biomedical fields, especially when the sample volume is limited. Since this chapter does not aim for a comprehensive review of MEKC, a few examples of separation by MEKC are given in the following.

Jorgenson and Lukacs [1] have shown a CZE separation of some dansylated amino acids in their first paper on capillary electrophoresis. Although an extremely high efficiency of the CZE separation was clearly demonstrated, the resolution of many dansylated amino acids was not particularly successful because many dansylated amino acids have close electrophoretic mobilities. Figure 12 demonstrates the separation of 21 dansylated amino acids by MEKC, using a 100-mM SDS solution at pH 8.3. It is apparent from this example that MEKC is more powerful in separating complex mixtures such as this example than is CZE, because of the various factors that are available for the manipulation of selectivity. Some other examples of separations of complex mixtures by MEKC have been reported other than those shown in Figs. 8, 9, and 11: alkylphenols [4], other aminoacid derivatives [24,31,35], chlorinated phenols [25], nucleic acid constituents [36,37], oligonucleotides [37],

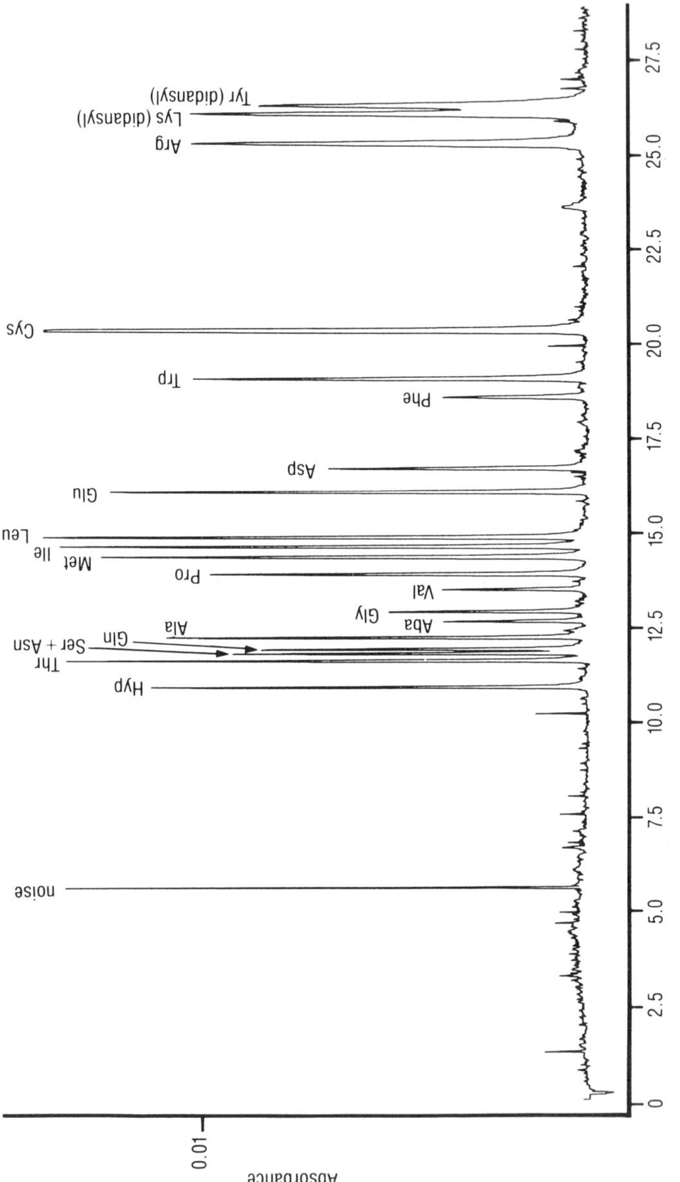

Figure 12 Separation of dansylated amino acids. Conditions: capillary, 75 μm i.d. × 570 mm (effective length 500 mm); separation solution, 100 mM SDS in 100 mM borate buffer (pH 8.3); temperature, 20°C; detection, absorbance at 200 nm. (From Ref. 28)

and pharmaceuticals [many examples in Refs. 17,38]. There are increasing numbers of examples of the separation of complex mixtures by MEKC.

A potential application field for MEKC is that of drug analysis in serum or plasma [39,40]. Figure 13 shows an example of the direct assay of an antibiotic in plasma [40]. The separation of antipyrine (AP) and cefpiramide (CPM) was easily attained, even without the SDS addition, if the samples were dissolved in the buffer as given in Fig. 13A. However, the drug peaks were barely observed under the same conditions given in Fig. 13A when samples were spiked into plasma because the drug peaks were obscured by those of plasma proteins. The use of the SDS solution permitted the separation of the drug peaks from the protein peaks, probably owing to the binding of SDS molecules to the proteins. For the purpose of drug assay, detection of protein is unnecessary, and the proteins in the capillary can easily be washed out by a pressurized flow after the detection of the drug peaks. Nakagawa et al. [40] also observed that protein-bound drug was rapidly released by this technique. To make use of this technique, we must optimize conditions, as the drug peaks migrate faster than the protein peaks [41].

Reproducibility of the migration time and precision of quantitation by MEKC have been reported to be comparable with, or sightly less than, those of RP-HPLC, provided the temperature is kept constant [16,42]. Viscosity is sensitive to temperature and, therefore, the migration velocity is susceptible to temperature. Most commercial instruments have thermostated ovens for the capillary, but it should be remembered that the real temperature in the capillary is usually higher than the oven temperature, depending on current, because heat dissipation is not complete when joule heating is high [43]. The peak area depends on the migration velocity of each compound, since the residence time in the detector cell depends on the

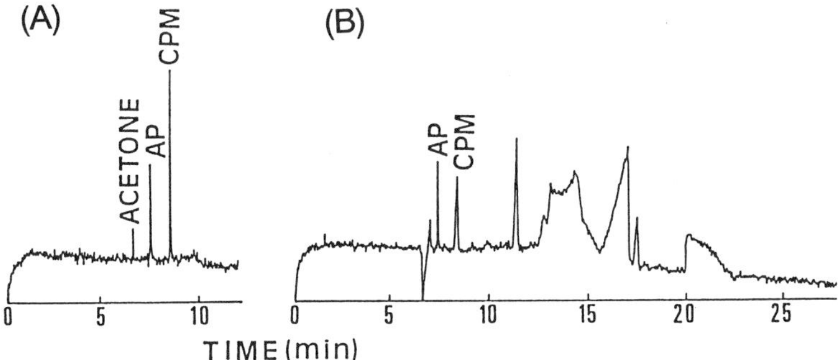

Figure 13 Analysis of drugs in buffer (A) and in plasma (B) by MEKC. AP, antipyrine; CPM, cefpiramide. Conditions: capillary, 50 μm i.d. × 750 mm (effective length 550 mm); separation solution, 10 mM SDS in 50 mM phosphate buffer (pH 8.0); applied voltage, 15 kV; detection, absorbance at 280 nm; temperature, 37°C. (From Ref. 40)

migration velocity in capillary electrophoresis; therefore, the precision of the peak area is affected by the reproducibility of the migration time.

ACKNOWLEDGMENT

The author is most grateful to his co-workers, whose names are cited in the references, especially to Drs. Koji Otsuka and Hiroyuki Nishi, for their cooperation in developing EKC techniques. He also greatly appreciates Dr. Toshio Nakagawa's original idea of using the micelle in electrophoresis, as described in the author's first paper on MEKC [4]. Part of the research has been supported by the Grant-in-Aid for Scientific Research from the Ministry of Education, Science and Culture. The author also thanks Nissan Science Foundation, Yokogawa Electric Corp., Shimadzu Corp., Hitachi, Ltd., and Sumitomo Chemical Co. for their financial support to the research.

REFERENCES

1. J. W. Jorgenson, and K. D. Lukacs, *Anal. Chem.*, *53*:1298 (1981).
2. K. D. Lukacs, and J. W. Jorgenson, *J. High Resolut. Chromatogr. Chromatogr. Commun.*, *8*:407 (1985).
3. S. Terabe, *Trends Anal. Chem.*, *8*:129 (1989).
4. S. Terabe, K. Otsuka, K. Ichikawa, A. Tsuchiya, and T. Ando, *Anal. Chem.*, *56*:111 (1984).
5. S. Terabe, K. Otsuka, and T. Ando, *Anal. Chem.*, *57*:834 (1985).
6. S. Terabe, H. Ozaki, K. Otsuka, and T. Ando, *J. Chromatogr.*, *332*:211 (1985).
7. S. Terabe and T. Isemura, *Anal. Chem.*, *62*:650 (1990).
8. H. Watarai, *Chem. Lett.*, 391 (1991).
9. D. E. Burton, M. J. Sepaniak, and M. P. Maskarinec, *J. Chromatogr. Sci.*, *24*:347 (1986).
10. K. Otsuka and S. Terabe, *J. Microcol. Sep.*, *1*:150 (1989).
11. H. Nishi, N. Tsumagari, and S. Terabe, *Anal. Chem.*, *61*:2434 (1989).
12. K. Ghowsi, J. P. Foley, and R. J. Gale, *Anal. Chem.*, *62*:2714 (1990).
13. J. P. Foley, *Anal. Chem.*, *62*:1302 (1990).
14. S. Terabe, H. Utsumi, K. Otsuka, T. Ando, T. Inomata, S. Kuze, and Y. Hanaoka, *J. High Resolut. Chromatogr. Chromatogr. Commun.*, *9*:666 (1986).
15. J. A. Lux, H. Yin, and G. Schomburg, *J. High Resolut. Chromatogr.*, *13*:145 (1990).
16. H. Nishi, T. Fukuyama, M. Matsuo, and S. Terabe, *J. Pharm. Sci.*, *79*:519 (1990).
17. H. Nishi and S. Terabe, *Electrophoresis*, *11*:691 (1990).
18. R. O. Cole, M. J. Sepaniak, W. C. Hinze, J. Gorse, and K. Oldiges, *J. Chromatogr.*, *557*:113 (1991).
19. H. Nishi, T. Fukuyama, M. Matsuo, and S. Terabe, *J. Chromatogr.*, *513*:279 (1990).
20. H. Nishi, T. Fukuyama, M. Matsuo, and S. Terabe, *J. Chromatogr.*, *515*:233 (1990).
21. S. Terabe, M. Shibata, and Y. Miyashita, *J. Chromatogr.*, *480*:403 (1989).
22. H. Nishi, T. Fukuyama, M. Matsuo, and S. Terabe, *Anal. Chim. Acta*, *236*:281 (1990).
23. H. T. Rasmussen, L. K. Goebel, and H. M. McNair, *J. High Resolut. Chromatogr.*, *14*:25 (1991).

24. K. Otsuka, S. Terabe, and T. Ando, *J. Chromatogr.*, *332*:219 (1985).
25. K. Otsuka, S. Terabe, and T. Ando, *J. Chromatogr.*, *348*:39 (1985).
26. S. Terabe, T. Katsura, Y. Okada, Y. Ishihama, and K. Otsuka, *J. Microcol. Sep.*, *5*:23 (1993).
27. S. Terabe, Y. Miyashita, O. Shibata, E. R. Barnhart, L. R. Alexander, D. G. Patterson, B. L. Karger, K. Hosoya, and N. Tanaka, *J. Chromatogr.*, *516*:23 (1990).
28. Y. Miyashita and S. Terabe, *Beckman P/ACE System 2000 Applications Data*, DS-767 (1990).
29. H. Nishi, T. Fukuyama, and S. Terabe, *J. Chromatogr.*, *553*:503 (1991).
30. H. Nishi, N. Tsumagari, T. Kakimoto, and S. Terabe, *J. Chromatogr.*, *477*:259 (1989).
31. S. Terabe, Y. Ishihama, H. Nishi, T. Fukuyama, and K. Otsuka, *J. Chromatogr.*, *545*:359 (1991).
32. H. Nishi and M. Matsuo, *J. Liquid Chromatogr.*, *14*:973 (1991).
33. K. Otsuka, S. Terabe, and T. Ando, *Nippon Kagaku Kaishi*, 950 (1986).
34. A. T. Balechunas and M. J. Sepaniak, *Anal. Chem.*, *59*:1466 (1987).
35. H. Nishi, T. Fukuyama, and M. Matsuo, *J. Microcol. Sep.*, *2*:234 (1990).
36. A. S. Cohen, S. Terabe, J. A. Smith, and B. L. Karger, *Anal. Chem.*, *59*:1021 (1987).
37. K. H. Row, M. P. Griest, and M. P. Mascarinec, *J. Chromatogr.*, *409*:193 (1987).
38. S. Fujiwara, S. Iwase, and S. Honda, *J. Chromatogr.*, *447*:133 (1988).
39. T. Nakagawa, Y. Oda, A. Shibukawa, and H. Tanaka, *Chem. Pharm. Bull.*, *36*:1622 (1988).
40. T. Nakagawa, Y. Oda, A. Shibukawa, H. Fukuda, and H. Tanaka, *Chem. Pharm. Bull.*, *37*:707 (1989).
41. H. Nishi, T. Fukuyama, and M. Matsuo, *J. Chromatogr.*, *515*:245 (1990).
42. K. Otsuka, S. Terabe, and T. Ando, *J. Chromatogr.*, *396*:350 (1987).
43. R. J. Nelson, A. Paulus, A. S. Cohen, A. Guttman, and B. L. Karger, *J. Chromatogr.*, *480*:111 (1989).

3

Conventional Isoelectric Focusing and Immobilized pH Gradients: *An Overview*

Pier Giorgio Righetti and Marcella Chiari

University of Milan
Milan, Italy

In this chapter we will review the mathematical models pertaining to separations by isoelectric focusing (IEF) in carrier ampholyte (CA) buffers and in immobilized pH gradients (IPG). In CA-IEF, the steady-state distribution profile of amphoteric compounds predicts gaussian distributions only if and when the background conductivity remains constant within the focused zone. In practice, this is achieved only when focusing across neutrality. In all other pH regions, the tendency is to produce asymmetric peaks, with negatively (in acidic regions) or positively (in basic gradients) skewed profiles. The transient state, in uniformly loaded samples, is seen as two discernible peaks migrating from the two ends of the column and merging into one at the pI position. The decay state seems to be generated by an isotachophoretic phenomenon of anodic, cathodic, or symmetric loss of carrier ampholytes. The IPGs are characterized by indefinitely stable pH gradients and by an extremely high resolution. For producing wide pH gradients, computer algorithms are described that optimize any mixture of Immobiline chemicals for producing linear or nonlinear pH gradients, such as convex, concave exponential, and sigmoidal. Focusing systems adapted to capillary zone electrophoresis are also described.

THE EARLY DAYS

There are rumors that when H. Svensson was a young assistant of A. Tiselius at the Institute of Physical Chemistry in Uppsala, during the years 1939–1945, he was already engaged in studying the possibility of forming pH gradients by convection-free electrolysis of salts. In fact, when a salt is electrolyzed under convection-free conditions, acid and base are completely separated and acquire concentration courses characterized by an exponential decline from the respective electrode toward the middle of the apparatus (Fig. 1). Because of the very high resistance of pure solvent, we can visualize a steady state in which acid and base attain vanishing concentrations at the same point, the position of which will depend on the transference numbers and diffusion coefficients of acid and bases. For sodium sulfate (Na_2SO_4) in Fig. 1, the point of neutrality will lie just about at the center of the electrolyzer, between NaOH and sulfuric acid. As a result of this decaying concentration profile, a huge pH drop will occur at the center of the electrolyzer, with very shallow pH gradients occupying all the other space (a pH 1.7–2.6 gradient to the left and a pH 11.5–12.3 to the right). According to Rilbe [1], such gradients are quite useful (for what, however, was not specified). Not a minor inconvenience of such a system was that, at the center of the apparatus, water would boil over owing to huge ohmic resistance, forcing young Svensson to stand night guard at the apparatus judiciously dropwise adding some Na_2SO_4 in the central chamber. The same rumors have him trying to escape to Marseilles or to Calvi (Sardinia) endeavoring to enroll in the Foreign Legion, but during World War II this was a somewhat difficult proposition. He managed, nevertheless, to discuss his thesis (by due permission of the Philosophical Faculty of the University of Uppsala) on March 9, 1946, at 10 AM, on *The Moving*

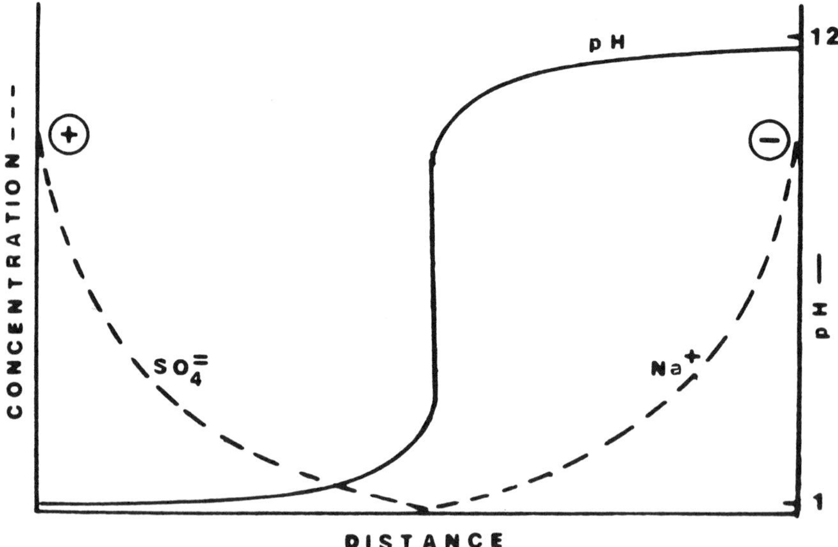

Figure 1 Simulated pH profile and ion distribution during steady-state electrolysis of Na_2SO_4.

Boundary Method. How such an early, disastrous attempt left him with such a vivid love for pH gradients, to the point of leading him to the discovery of isoelectric focusing, remains a mystery.

In the mid 1950s, Kolin explored other ways for producing pH gradients. Under the spell of Tiselius, too, he built a U-tube apparatus and filled the "separation arm" with a discontinuous buffer made by titrating the same buffering ion (say phosphate) to two different pH values (say pH 6 and 7; Fig. 2). He then injected the sample at the interface between these two limit solutions and allowed a pH gradient to develop in the sample zone by diffusion. As the electric circuit was closed on the separation cell, a very quick focusing ensued, dubbed by him *isoelectric line spectra*. This was also a dead end, since there was no way to control the slope of the pH gradient over the separation axis, as the nonamphoteric buffers and titrants quickly migrated to the respective electrodes. Some scientists at the National In-

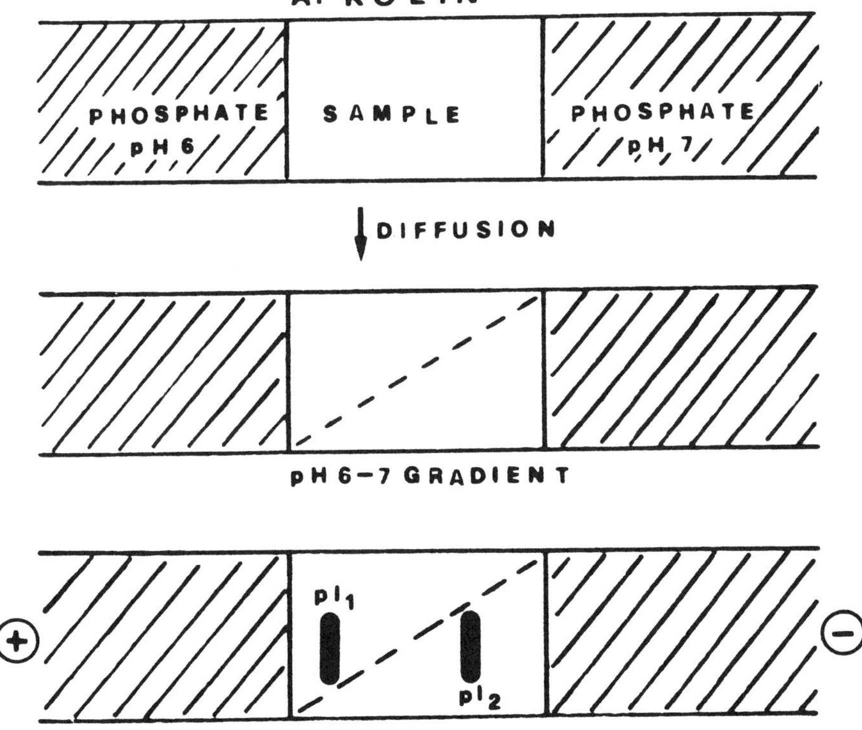

Figure 2 Isoelectric focusing according to Kolin. Top: two juxtaposed layers of buffer with a sample zone inbetween; middle: diffusion to steady state with formation of a pH gradient; bottom: focusing of protein ions under the influence of an electric field.

stitutes of Health (NIH) accused him of "black magic" and sent magicians to his laboratory trying to exorcise evil spirits. Yet, it was a magnificent invention that indicated the right path to Svensson, eager to return to the love of his youth.

CONVENTIONAL ISOELECTRIC FOCUSING

Isoelectric focusing (IEF) in carrier ampholyte (CA) buffers was born through the stubborn efforts of Svensson [2,3], who first understood that, to create and maintain a pH gradient (i.e., a proton concentration gradient) in an electric field, one had to resort to a set of peculiar buffers that would have to be, at the very least (a) amphoteric and (b) carriers (of buffering power and conductivity) at the isoelectric point (pI). Such a gradient, generated by electrolysis of an initially homogeneous mixture of the CA-buffers, is called a *natural pH gradient*, as opposed to *artificial pH gradients* obtained by diffusion of two limiting solutions in the absence of an electric field [4]. Svensson first perceived, what now appears obvious, that an ampholyte in the isoelectric state may have a conductivity, despite its zero mobility and that the overall charge is close to zero (let us recall here that the values of the isoelectric and isoionic points differ slightly). Svensson also first used the concept of an ampholyte with two dissociation stages close to the isoelectric point (an absolute requirement for buffering and conducting at the pI).

The Steady State

Distribution Profile of an Ampholyte About Its Isoelectric Point

Under steady-state conditions (as obtained by balancing the simultaneous electrophoretic and diffusional mass transports) Svensson [2] has derived the following differential equation describing the concentration profile of a focused zone:

$$C\mu i/qk = D(dC/dx) \tag{1}$$

where:

C = concentration of a component in arbitrary mass units per arbitrary volume unit.

μ = electric mobility in $cm^2\ V^{-1}\ s^{-1}$ of ion constituent except H^+ and HO^-, with positive sign for cationic and negative sign for anionic migration.

i = electric current in A.

q = cross-sectional area in cm^2 of electrolytic medium, measured perpendicularly to the direction of current.

k = conductance of medium, in $ohm^{-1}\ cm^{-1}$.

D = diffusion coefficient in $cm^2\ s^{-1}$ of component corresponding to the ion constituent with mobility μ.

x = coordinate along the direction of current, x increasing from zero at the anode toward the cathode.

Each term in Eq. 1 expresses the mass flow per second and area in square centimeters of the cross section; that to the left being the electric, that to the right the diffusional mass flows. If Eq. 1 is written in the form

$$(i\mu/q)(dx/k) = D(dC/C) \tag{2}$$

it is seen that it is possible to integrate it if μ is known as a function of pH and D as a function of C. Specifically, if the conductance, the diffusion coefficient, and the derivative

$$p = -d\mu/dx = -\left[\frac{d\mu}{d(\mathrm{pH})}\right]\left[\frac{d(\mathrm{pH})}{dx}\right] \tag{3}$$

(where p is the ratio between the protein titration curve and the slope of the pH gradient over the separation axis) can be regarded as constant within the focused zone, then $\mu = -px$, and one obtains the following analytical solution:

$$C = C_0 \, exp\left[\frac{-(pix^2)}{2qkD}\right] \tag{4}$$

where x is now defined as being $= 0$ at the concentration maximum C_0. This is a gaussian concentration distribution with the inflection points at:

$$x_i = \pm\sqrt{\frac{qkD}{pi}} \tag{5}$$

where x_i denotes the width of the gaussian distribution of the focused zone measured from the top of the distribution of the focused ampholyte to the inflection point (1 standard deviation). The course of the pH gradient is $d(\mathrm{pH})/dx$, and $d\mu/d(\mathrm{pH})$ represents the titration curve of the ampholyte. It should be remembered that this gaussian profile holds only if and as long as the conductivity of the bulk solution comprised within the zone is constant. Constant conductivity along a pH gradient is quite difficult to maintain, especially as one approaches pH extremes (below pH 4 and above pH 10), if for no other reason than because the nonnegligible concentration of H^+ and OH^- present in the bulk liquid begins to strongly contribute. We have simulated such focusing conditions in two extreme pH intervals (pH 3–4 and pH 10–11) and across neutrality (pH 6.5–7.5) [5]. For simplicity, the gradients have been assumed to be formed by immobilized buffering species, which would contribute minimal conductivity to the liquid phase, and of which the diffusion coefficients, transport numbers, and mobilities would be zero. The gradients within the stated pH intervals are assumed to be linear. In such simulations we have computed the steady-state distribution of three different ampholytes, all having pI values in the middle of each gradient (thus pI 3.5, pI 7.0, and pI 10.5, respectively), but with different ΔpKs of 1, 2, or 3 (thus, in the example of the pH 3–4 interval, $\Delta pK = 1$ is an ampholyte with $pK_1 = 3.0$ and $pK_2 = 4.0$; $\Delta pK = 2$ has a $pK_1 = 2.5$ and $pK_2 = 4.5$; and $\Delta pK = 3$ has $pK_1 = 2.0$ and $pK_2 = 5.0$; the same applies to the pI 7.0 and 10.5 ampholytes in the other two pH ranges). All ampholytes selected for simulation are simple biprotic species. The steady-state profiles have been computed for three different current densities: 1, 5, and 15 A m^{-2}, except for simulations across neutrality (pH 6.5–7.5) for which much lower values had to be used (respectively, 0.02, 0.1, and 0.3 A m^{-2}). The results are summarized in Fig. 3 (A–C). The steady-state profiles for the three different ΔpK species are shown (a) in the pH 3–4 gradient (left panel), (b) in the pH 6.5–7.5 interval (central section), and (c) in the pH 10–11 span (right panel). As shown in the left and right panels, at low current densities (1 A m^{-2}), the three ampholytes exhibit negatively (pH 3–4) and positively (pH 10–11) skewed profiles, the degree of skewness increasing substantially from the $\Delta pK = 1$ to the $\Delta pK = 3$ species. As the current densities are increased to 5 A m^{-2} (B panels) or to 15 A m^{-2} (C panels), two phenomena are apparent: the kurtosis is reduced for the species with larger ΔpKs (i.e., the zones sharpen up into taller and narrower bands) and also the skewness is quite decreased. However, neither is completely eliminated, and it is seen that even at the highest current densities the $\Delta pK = 1$ (considered to be a good carrier ampholyte) is still distinctly skewed. In the pH 10–11 interval, both phenomena, skewness and kurtosis, are considerably reduced compared with the pH 3–4 interval. This is because the conductivity profile is not a steep exponential as in the acidic gradient (in general, the ratio of background conductivities is about 2:1). In turn, this is linked to the mobility of H^+ and OH^- ions, the former being 3.625×10^{-4}, the latter 1.985×10^{-4} ($cm^2 V^{-1} s^{-1}$ at 25°C in free solution). As a result of this, at any current density, the peaks focused in the pH 10–11 gradient are sharper than the corresponding ones in the pH 3–4 span. Conversely, only focusing across neutrality generates symmetric steady-state profiles (see middle panels of Fig. 3) and this occurs at any current density and for the three pI 7.0 species, irrespective of their ΔpKs.

This does not pose a major problem for proteins, which usually have a rather small diffusion coefficient and remarkably good titration curves (i.e., with steep slopes about the pI value), but it could severely decrease resolution for peptides or small amphoteric molecules, such as amino acids.

The Law of pH Monotony

As formulated in 1967 by Svensson, this law states that a natural pH gradient, developed by the current itself, is positive throughout the gradient, as the pH increases steadily and monotonically from anode to cathode. Reversal of the pH gradient at any position between the electrodes is, thus, incompatible with the steady state. Accordingly, in stationary electrolysis, two ampholytes can never be completely separated from one another, unless a third ampholyte, with intermediate pI, is present in the system. Let us consider Fig. 4: two adjacent CAs, with pI 3.1

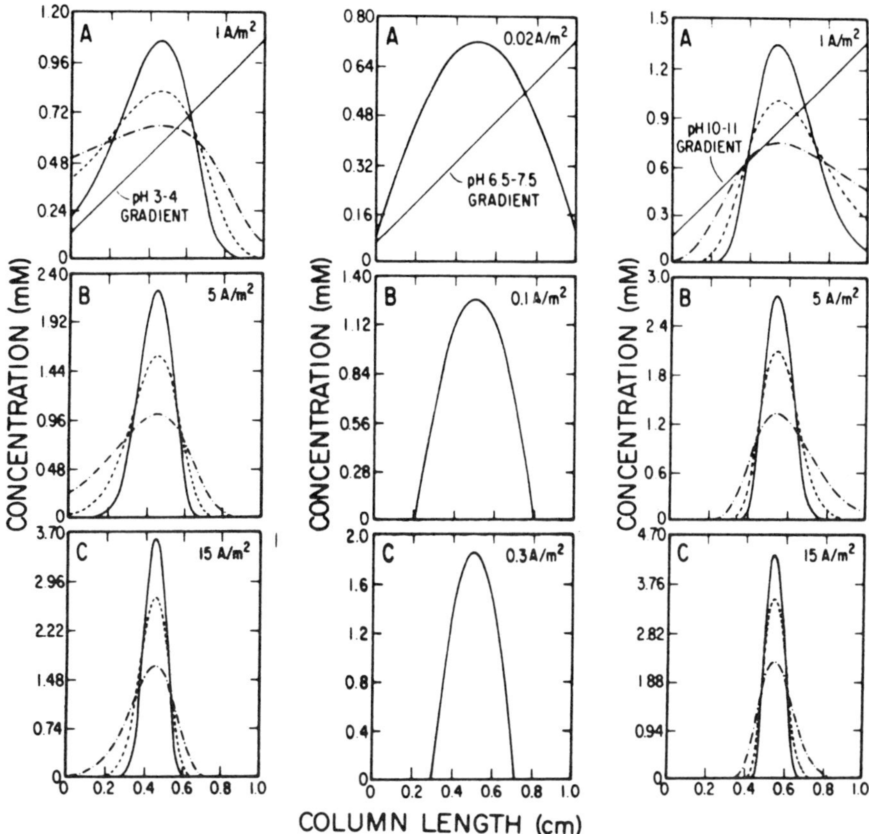

Figure 3 Computer simulation of focusing ampholytes at neutral and pH extremes. Left three panels: focusing in the pH 3–4 interval. Three pI 3.5 ampholytes, with Δp*K*s 1 (solid curve), 2 (broken), and 3 (broken and dotted curve) are focused in a pH 3–4 range formed by linear concentration gradients of an imaginary acidic Immobiline (p*K* 3.0) and a basic one (p*K* 4.0), ranging from 10 to 5 mM. The steady state profiles are reported for current densities of 1 (A), 5 (B) and 15 (C) A m^{-2}, with a fourth curve reporting the pH gradient. Central three panels: focusing in the pH 6.5–7.5 interval. All conditions as in the pH 3–4 range, except that the three ampholytes with Δp*K*s 1, 2, and 3 have a pI of 7.0 and that the two background Immobilines have a p*K* of 6.5 (acidic species) and of 7.5 (basic component). Here the three current densities are 0.02 (A), 0.1 (B) and 0.3 (C) A m^{-2}. Right 3 panels: focusing in the pH 10–11 range. All conditions as in the pH 3–4 interval, except that the three ampholytes with Δp*K*s 1, 2, and 3 have a pI of 10.5 and that the two background Immobilines have a p*K* of 10.0 (acidic species) and of 11.0 (basic component). (From Ref. 5)

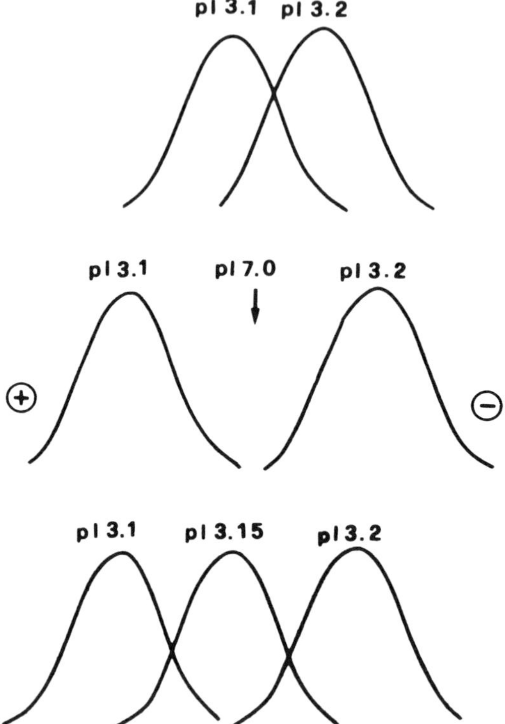

Figure 4 The law of pH monotony. Top: overlapping gaussians of two adjacent ampholytes; center: separation of these gaussians for formation of a zone of pure water in between; bottom: a third, intermediate ampholyte focusing between two adjacent species.

and 3.2, have to overlap to some extent. If complete separation were ever to be achieved, a zone of pure water (pI 7.0) would develop between them. This situation is excluded by the foregoing law; thus, if separation of the two has occurred between the pI 3.1 and pI 3.2 species, a third ampholyte, with pI 3.05, should focus between them. Ideally, the CAs should be evenly distributed over the entire pH range and should also be present at about neutrality to avoid formation of a zone of pure water, with highly deleterious effects (see later).

Description of Infinite-Component Mixtures

In this section we construct a mathematical model for a mixture of chemical subsystems when the number of mixture components is sufficiently high [6]. We can consider a mixture composed of a solvent (water), its ions, and an infinite number of closed chemical subsystems. We will characterize the individual closed chemical subsystems by the sorting parameter s. To describe an infinite component mixture, we will introduce the concentration distribution function for the components of the mixture, $X(s, x, t)$, depending on the multidimensional sorting parameter s. The composition of the mixture occupying the volume at the initial instant of time is specified by the global distribution function

$$M(s) = \int R\, X(s, x, t)dv \tag{6}$$

where M is the mass concentration in a zone (mol/L) and dv is an infinitesimal incremental volume. The physicochemical properties of such a mixture will be described using the distribution functions of the diffusion coefficient $D[s, \psi(x, t)]$ the mobilities $\mu[s, \psi(x, t)]$, the conductivities $\sigma(s, \psi(x, t)]$, the diffusional conductivity $\sigma^D(s, \psi(x, t)]$, and the molar charge $\varepsilon[s, \psi(x, t)]$. We will recall that, for closed chemical subsystems, D, μ, σ, σ^D, and ε depend only on the acidity of the mixture $\psi(x, t)$. We shall restrict ourselves to the case in which the mixture as a whole is at rest, its density is constant, there are no barodiffusion and thermal diffusion effects, the dielectric constant of the mixture equals that of the solvent, and the D and μ values of individual components do not depend on temperature. We shall additionally assume that the mixture of chemical subsystems is electrically neutral. For a specified monotonic acidity profile $\psi(x)$, for a known set of amphoteric substances, characterized by the quantities D, μ, σ, σ^D, and ε above, it is possible to calculate the steady-state distribution profile $M(s)$ if the potential V or the electric current density j are specified.

Let us now proceed with a practical example. In 1962, Svensson, after a 6-month search in chemical catalogues, listed a series of 37 amphoteric substances potentially useful in IEF. The quality mark of such substances is the $|\mathrm{pI} - \mathrm{pK}|$ value, which should be as small as possible (ideally, about 0.5). By simulating the steady-state distribution profile in the pH 2.7–10.7 range ($j = 102$ A cm^{-2}; $V = 3400$) one obtains the results of Fig. 5: it is seen that it is possible to obtain a fairly linear pH gradient, followed by a more hectic pK variation. The $M(s)$ profile (in mol L^{-1}) is clearly uneven, and it follows the relative distribution of pIs of the 37 compounds (note that there are 12 amphoteres in the pH 3–4, 5 in the pH 4–5, and 5 in the pH 5–6 ranges) [3]. Since a quite linear pH gradient has been obtained, one might think that such a mixture of 37 amphoteric buffers is sufficient for good focusing. In fact, it is not, since the field intensity (E, in V cm^{-1}) is highly irregular, varying from as low as a few volts in regions of high σ up to $>$ 1200 V cm^{-1} across neutrality. Such an irregular V profile will never allow proper sample focusing: in addition, owing to σ minima at about pH 7, with production of a tremendous joule heat, the liquid could literally boil over in such a region [1]. There are, in fact, huge pI gaps among tyrosyltyrosine (pI 5.85), isoglutamine (pI 6.10), lysylglutamic acid (pI 6.81), and histidylglycine (pI 7.30) (see Table I in ref. 3). Nature has been quite parsimonious in creating amphoteric species that

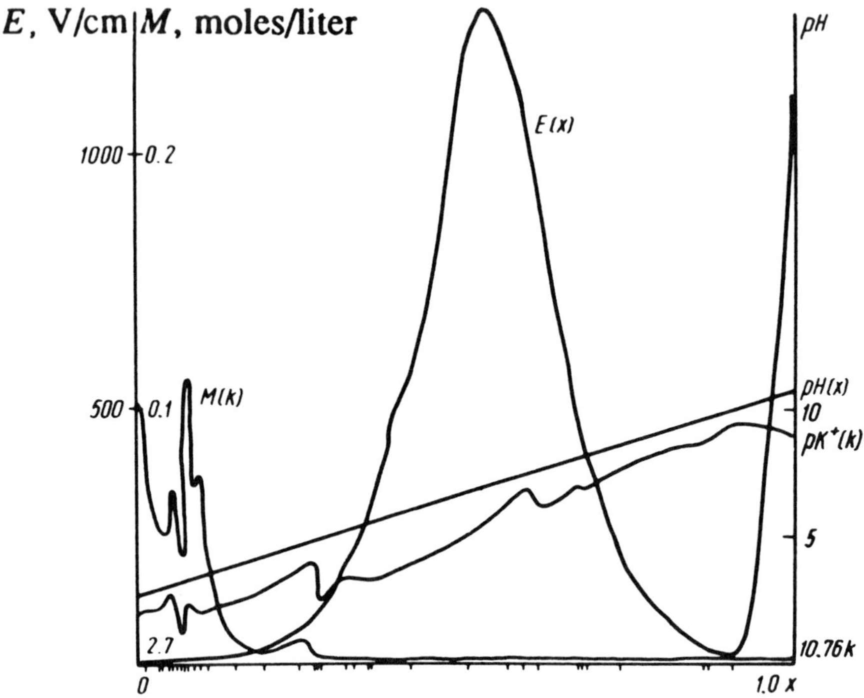

Figure 5 Results of the calculation of the total concentration $M(k)$ and field intensity $E(x)$ for the set of 37 amphoteric substances given by Svensson, with which a linear pH profile is created in the range $2.77 \leq \text{pH} \leq 10.76$ (From Ref. 6)

buffer near neutrality. Even protein lysates, used as background electrolytes for IEF, did not help much: very few peptides focus and buffer close to neutrality. The problem was solved by Vesterberg [7], who advanced the brilliant idea of producing a highly heterogeneous mixture of amphoteres by reacting oligoamines (up to six nitrogen long) with acrylic acid: a "chaotic" organic chemistry, able to produce several hundred amphoteres with the required property of closing the gaps near neutrality and producing a fairly even buffering power all along the pH scale. It might be asked, then, how many different species of carrier ampholytes should be present per pH unit for generation of a stepless pH course. This question can be answered by formulating a condition for completely unresolved double zones of adjacent carrier ampholyte peaks [8].

Could one actually calculate how many different species of carrier ampholytes should be present per pH unit for generation of a stepless pH course? In a real system, this might be an impossible task. However, for an ideal system, comprising monovalent ampholytes with equal diffusion coefficients and electric mobilities, equal relative concentrations, and evenly spaced pI values on the pH scale, and exhibiting the same difference between their two pK values, Almgren [9] has been able to simulate the steady-state profiles. To solve this problem, we need to know the following parameters: ΔpI and ΔpK and the width of the gaussian distribution curve at the steady state (expressed in the abscissa of Fig. 6A,B as yM: because it is a column coordinate, but it has the dimensions of moles per liter). As shown in Fig. 6A, a system comprising eight carrier ampholytes, distributed over 0.25 pH units (from pH 7.0 to 7.25), in which each individual species has a $\Delta pK = 2$, and spaced at $\Delta pI = 0.05$, is able to generate smooth pH and conductivity (H) courses. According to these data, it would appear that at least 30 carrier ampholytes per pH unit are needed for production of a stepless pH gradient. Conversely, if the same CA species ($\Delta pK = 2$) are now spaced at $\Delta pI = 0.1$ (thereby halving the number of amphoteres to 15/pH unit) the pH gradient will increase by steps (Fig. 6B). These plateaus are centered in the middle of each peak, whereas the pH grows linearly in the boundary between each pair of ampholytes. Thus, the notion that good focusing can be obtained by using an artificial mixture of 47 buffers, as proposed in 1982 by Cuono and Chapo [10], goes against the elementary rules of IEF. Needless to say, this idea of the 47 buffers was abandoned as soon as it was proposed.

The Transient State

It sounds odd that one should mention the transient after the steady state, but this is historically the evolution of the field. Nobody elaborated on this topic, having the steady state so elegantly discussed by Svensson (as well as, as it turned out later, by Kaumann [11] and Schumaker [12]). Only much later Catsimpoolas [13] started measuring the approach to the pI position. He devised a protocol in which an ampholyte (histidyltyrosine in Fig. 7) was not pulse-loaded, but uniformly distributed between the two electrodic solution. It is seen from Fig. 7 that two discernible peaks migrate from the two ends of the column (positive and negative) until they merge into one at the pI position. These data illustrate one of the fundamental aspects of the transient state in IEF: the changes that occur begin at the boundaries (electrodes) and propagate inward. If one plots the peak position difference (Δx) versus time (Fig. 8) one can appreciate another phenomenon: at any given time the cations (the positively charged species moving away from the anode) move faster toward the pI position than anions (i.e., the species repelled from the cathode). For simple amphoteric species, this might be because anions are, in general, more hydrated than cations [14], so that their radius is somewhat larger; thus, the velocity (which depends on the charge/radius ratio) is correspondingly smaller. For proteins, there is usually another added cause: the two branches of the titration curve about the pI value might have quite different slopes, so that the net positive or negative

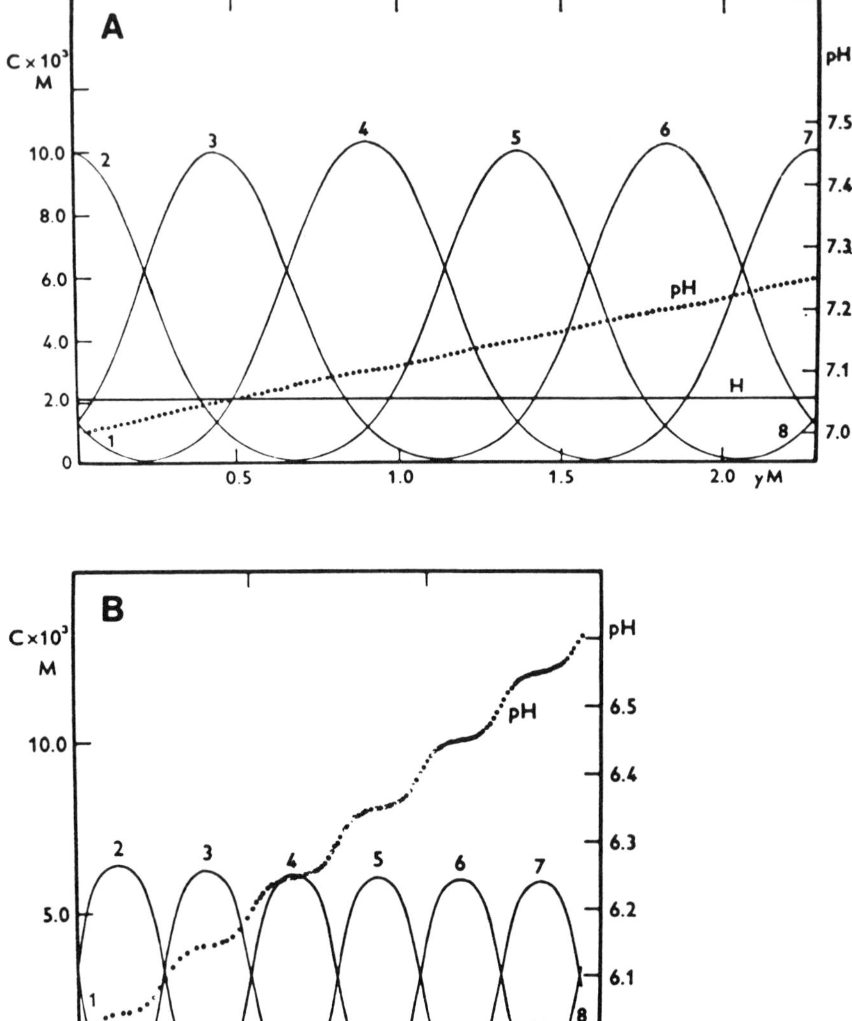

Figure 6 Computer simulation of focusing of ideal, monomonovalent ampholytes with equal diffusion coefficients and electric mobilities. The abscissa (*yM*) has the dimensions of moles per liter; the left ordinate is in moles, H is the conductivity and the dotted line is the pH gradient. (A) focusing of eight carrier ampholytes over 0.25 pH units, having $\Delta pK = 2$ and spaced at $\Delta pI = 0.05$. (B) focusing of eight carrier ampholytes over 0.5 pH units, having $\Delta pK = 2$ and spaced at $\Delta pI = 0.1$. (From Ref. 9)

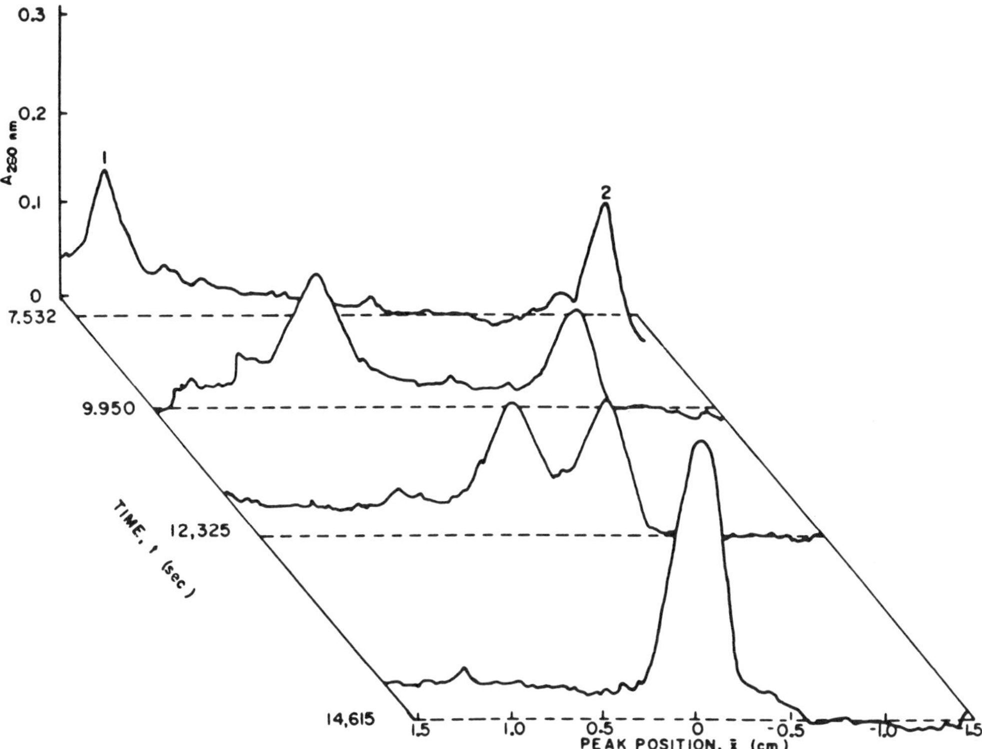

Figure 7 Scanning electrophoretic patterns of histidyltyrosine peaks migrating toward the pI position. Peak 1 migrates away from the negative electrode and peak 2 from the positive. The pI position is arbitrarily set at $x = 0$. The original sample distribution was uniform and the ampholine concentration 2%. (From Ref. 13)

charges away from the pI value are unequal: this calls for markedly different rates of approach to the pI position.

The Decay State

Equilibria are the most difficult thing to maintain here on Earth (and perhaps in the universe). Everything decays, even for our funambulist Italian politicians, who have found incredible metastatic equilibria in their fantastic jargon, which includes "parallel convergences" and soon will be extended to "diverging convergences." Therefore, it is no surprise that IEF, guaranteed by mathematical formulas to be indefinitely stable, would be subject to decaying in real life. But the causes of this degradation have been quite obscure for many years. Murel et al. [15], by computer simulation, have suggested that pH gradient stability may depend on a balance

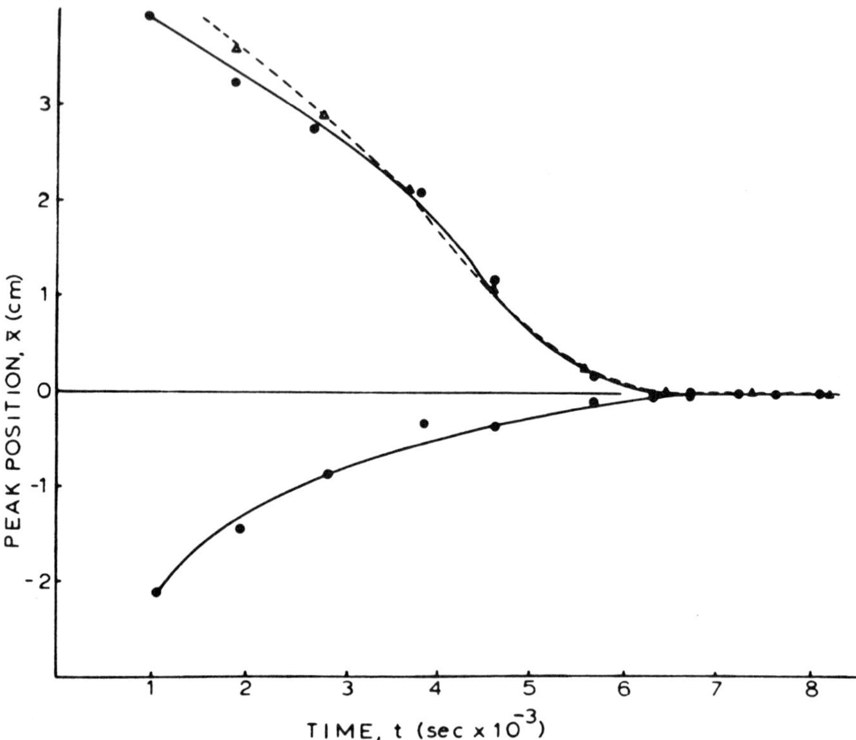

Figure 8 Plot of peak position difference (Δx) versus time. Histidyltyrosine, 2% ampholine concentration. (From Ref. 13)

between mobilities and concentrations of protons and hydroxyl ions migrating in opposite directions. In other words, according to these authors, the "proton cloud" generated at the anode travels toward the cathode faster than the "$-OH^-$ cloud" moving in the opposite direction; hence, a net drift toward the cathode. They suggest a mechanism of "pressure" of the proton cloud, just as in a thermoionic valve: such cloud has the effect of protecting the anodic end and pushing the cathodic end of the pH gradient out of the electrophoretic space. If this hypothesis were correct, equalization of the product (H^+ molarity $\times$ H^+ mobility) with the product (OH^- molarity $\times$ OH^- mobility) should stop the cathodic drift, but this does not seem to be true [16]. Probably, the *cathodic drift* (so-called because the pH gradient decay is most often at the cathodic end) has a multifactorial origin. One of the possible causes could be ingrained in the anticonvective support itself. The most popular matrix used today (a polyacrylamide gel), although believed to be devoid of ionizable groups, is indeed not quite neutral. We have found that, on the contrary, polyacrylamides are negatively charged owing to (a) trace impurities of acrylic acid in the gel; (b) covalent incorporation of catalysts (persulfate) as terminal groups in the matrix chains; and (c) hydrolysis of amide groups to acrylic acid in the gel layer close to the cathodic gel end [17]. Thus, the only way that we found effective in blocking pH gradient decay in IEF was the production of *balanced matrices* [i.e., the covalent incorporation of tertiary (or better quaternary) amines in stoichiometric amounts compared with the gel negative charges]. To this end we have used 3-dimethylaminopropylmethacrylamide (DMAPMA) or methacrylamidopropyltrimethyl ammonium chloride (MAPTAC), at such levels (in general, micromolar) that the autochthonous negative charges are balanced. As shown in Fig. 9, incorporation of DMAPMA in increasing amounts in polyacrylamide gels progressively quenches the cathodic decay down to *zero drift* conditions at the equivalence point. When an excess of DMAPMA is incorporated into the gel, the drift reverses its polarity and becomes anodic. Thus, the fact that, according to the net charge incorporated into the gel, a cathodic drift, zero drift (balanced gels), or anodic drift can be generated, speaks in favor of electroosmosis being the major cause of gradient decay in polyacrylamide gels. Yet, even this might not be the ultimate cause for pH gradient decay. Svendsen and Schafer-Nielsen [18] have suggested that pH gradients in IEF might decay by an isotachophoretic (ITP) mechanism. They additionally postulated that the nature of the drift could be controlled by appropriate choice of electrolyte type and concentration as well as the volume of the electrode reservoir. This model was further expanded, recently, by Mosher and Thormann [19]. These authors have used for their simulation work the generalized model for transient electrophoretic processes of Bier et al. [20]. This model is isothermal and one-dimensional and assumes the absence of liquid flows. It predicts the evolution of concentration, pH, and conductivity profiles as a function of time. Inputs required include the pK and mobility values of each component; the length of the separation space and its segmentation; and the current density, amount of electrophoresis time, and the initial distribution of each component. In the particular case of Figs. 10 and 11, a mixture of 15 biprotic ampholytes with equally spaced pIs (0.5 pH units) between pH 3 and 10 was used for simulation. The difference between pK_1 and pK_2 for each was 2 units, and each ionic mobility was 3×10^{-8} m^2 V^{-1} s^{-1}. The ionic mobility of Na^+ was 5.19×10^{-8} m^2 V^{-1} s^{-1}, whereas the anolyte (phosphoric acid) was assumed to have a pK = 2 and a mobility of 3.67×10^{-8} m^2 V^{-1} s^{-1}. Current density was a constant 20 A m^{-2}, and the boundary conditions employed permitted free mass transport into and out of the separation space. By these simulations it was found that both anodic and cathodic drifts (i.e., progressive loss of the acidic and basic ends of the pH gradient, respectively) were possible, according to the initial concentration in the electrolyte reservoirs. The six panels of Fig. 10 illustrate the progressive decay of the anodic end of the gradient. The initial distribution of all components is shown in the first panel (T = 0). The ampholytes

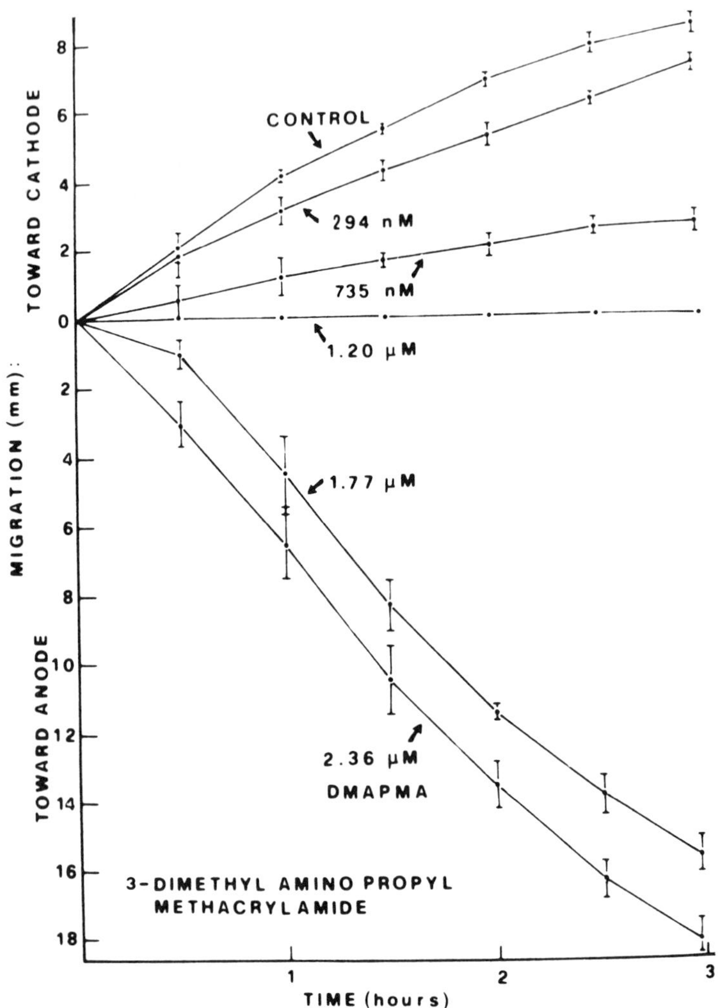

Figure 9 Effect of a tertiary base on the cathodic drift. The gel was polymerized either as such (control, no DMAPMA added) or in presence of increasing amounts of 3-dimethylaminopropylmethacrylamide (DMAPMA), from a minimum of 294 nM up to 2.36 μM. Notice that, above the equivalence point (1.20 μM DMAPMA added) the focused hemoglobin zones start drifting toward the anode (drift reversal). The gel was a 6%T, 4%C bis matrix, 360 μm thick, and contained 2% Ampholine, pH 3.5–10. Focusing was at 10°C, 20 minute prerun, followed by 1-hour sample run (normal human adult hemoglobin). The time period marked on the abscissa begins after this total time of 80 minutes. All cathodic (or anodic) drift measurements were made at a constant voltage of 1000 V (100 V/cm). (From Ref. 17)

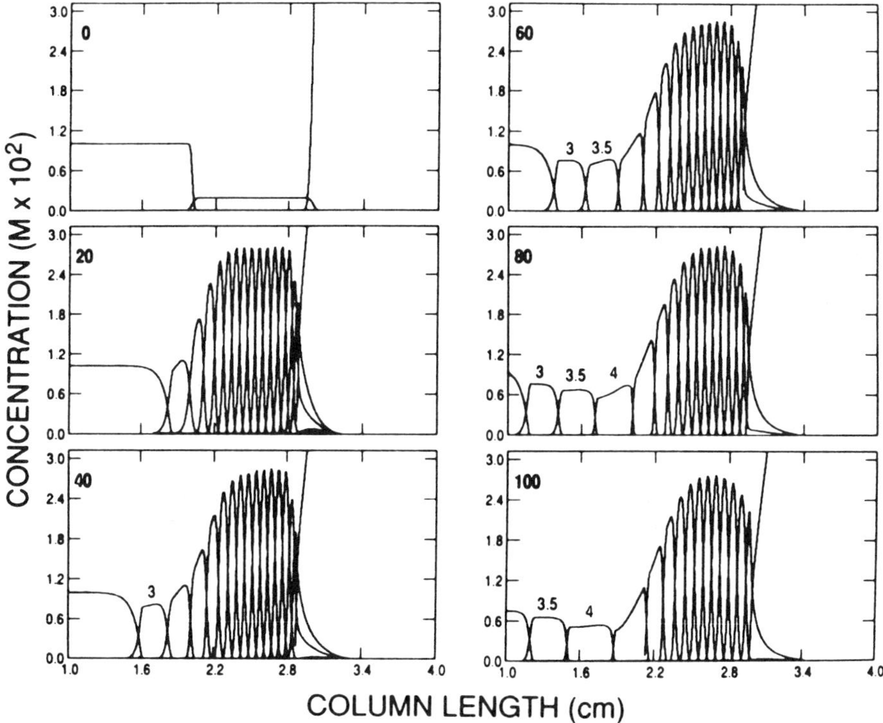

Figure 10 Computer simulation of the focusing of 15 ampholytes with 10 mM phosphoric acid as anolyte and 90 mM NaOH as catholyte. The anode is to the left. The ampholytes are initially distributed uniformly, 2 mM each, at time zero (upper left-hand panel) over the central 1 cm. Time points presented are after 20, 40, 60, 80, and 100 minutes of current flow. At this current density the ampholytes are focused by 20 minutes. The migration of phosphoric acid toward the anode causes the acidic ampholytes to form an isotachophoretic stack, which results in those ampholytes being lost from the focusing space. The ampholytes zones are identified by their pIs. (From Ref. 19)

are uniformly distributed at a concentration of 2 mM across the central 1 cm of the separation axis (the focusing space). The anolyte (phosphoric acid, 10 mM) occupies the left side (anodic portion) of the separation space. The catholyte (NaOH, 90 mM) is located in the right side. The additional five panels present the concentration distributions of all components at 20-minute intervals up to 100 minutes. In the second panel ($T = 20$), it is seen that the ampholytes have focused and the H_3PO_4 boundary has been displaced toward the anode, whereas the Na^+ boundary remains stationary. The 40-minute profile shows ampholyte pI = 3 (the most acidic species) clearly forming an isotachophoretic zone, with a characteristic plateau, behind the

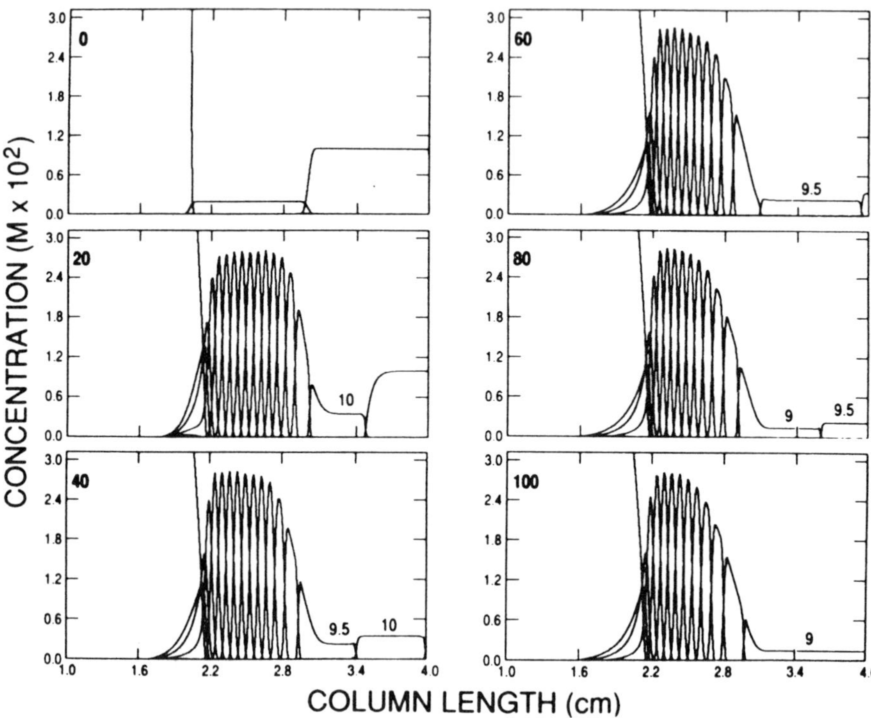

Figure 11 Computer simulation of the focusing of 15 ampholytes with 100 mM H_3PO_4 and 10 mM NaOH as anolyte and catholyte, respectively. The anode is to the left. All other conditions as in Fig. 7. The NaOH acts as a leader in the isotachophoretic sense for the basic ampholytes, causing their progressive loss from the focusing space (the central 1 cm of the column). The ampholyte zones are identified by their pIs. None of the acidic ampholytes have been lost from the focusing space by 100 minutes. (From Ref. 19)

phosphate boundary. In the 60- and 80-minute profiles, it is seen that the adjacent species (pI = 3.5 and pI = 4) are also mobilized by forming an isotachophoretic train; in addition, the phosphoric acid zone has nearly left the electrophoretic space. In the last panel (T = 100 minutes) a stack of ampholytes has formed and the most acidic species (pI = 3) has now also vacated the column. Note that, according to the Kohlrausch autoregulating function, the adjusted concentrations decrease from the pI = 2 to the pI = 4 zones, as typical of an isotachophoretic stack. Conversely, at this current density and time interval, the NaOH boundary has hardly moved and, as a consequence, no ampholytes are lost from the cathodic end of the gradient. If now one reverses the molarity of the two end electrolytes, one should then observe a cathodic drift. This situation is simulated in Fig. 11: here, the anolyte is 100 mM H_3PO_4, and the catholyte is reduced to 10 mM NaOH. The same mixture of ampholytes of Fig. 10 is used, with the same initial distribution (T = 0, upper left panel). At T = 20 minutes, the ampholytes are focused and the NA^+ boundary has migrated a substantial distance toward the cathode. This boundary is now acting as a leading ion for ampholyte pI = 10, which is forming an isotachophoretic zone immediately behind. At T = 40 and 60, additional ampholytes are seen stacking behind the leading ion, and again adjusting their plateau concentration according to their mobility. At T = 60, the entire Na^+ zone has vacated the electrophoretic space, and also the pI = 10 ampholyte has almost completely disappeared. The decay progresses as time goes by and new stacked zones begin to form and move out. According to these data, it would appear that, if one were to equalize the product of the concentration × mobility of the two end species, one might hope to block such drift. In fact, equalization of the velocities of the two boundaries (which occurs when the NaOH concentration is 2.25-fold the H_3PO_4 molarity) has merely the effect of producing a symmetric drift (i.e., a pH gradient that decays at both extremes).

IMMOBILIZED pH GRADIENTS

We have seen, in the first part of this discussion, the basic aspects of conventional IEF; now, in this second part, let us examine its evolution. Conventional IEF, in addition to the decay problems outlined earlier, suffers from several drawbacks, such as (a) uneven buffering capacity, (b) uneven conductivity, (c) unknown chemical environment, (d) very low and unknown ionic strength, (e) cathodic drift (or pH gradient instability), and (f) unwieldy to pH gradient engineering. Most of these vexatious problems have never been solved in 25 years of focusing with the carrier ampholyte buffers. For all these reasons, as a result of an intensive collaboration with Dr. Bjellqvist group in Stockholm and Dr. Görg's group in Munich, in 1982 we launched the technique of immobilized pH gradients (IPG) [21]. The IPGs are based on the principle that the pH gradient, which exists before the IEF run itself, is copolymerized, and thus insolubilized, within the fibres of a polyacrylamide matrix. This is achieved by using, as buffers, a set of nonamphoteric, weak acids and bases, having the following general chemical composition: $CH_2{=}CH{-}CO{-}NH{-}R$, where R denotes either weak carboxyl or weak tertiary amino groups. Six of these chemicals are available, from Pharmacia-LKB, under the trade name Immobiline; however, we have described a total of 16 of these compounds, the formulas and general properties of which can be found in a recent survey [22]. During gel polymerization, these buffering species are efficiently incorporated into the gel (84–86% conversion efficiency at 50°C for 1 hour). Immobiline-based pH gradients can be cast in the same way as conventional polyacrylamide gradient gels, using a density gradient to stabilize the Immobiline concentration gradient, with the aid of a standard, two-vessel gradient mixer. As shown by their chemical formulas, these buffers are no longer amphoteric, as in conventional IEF, but are bifunctional. At one end of the molecule is located the buffering (or titrant) group, and at the other end is an acrylic double bond, which disappears during immobilization of the buffer in the gel matrix. The IPGs have proved most successful in protein fractionation, and today are believed to represent the electrokinetic technique with the highest resolving power. The technique is straightforward in its use; however, the complexity of the system came when formulating extended pH gradients. It is thus our pleasure to recollect here the history of development of our computational approach to IPGs and to give new examples of this method. In the cornerstone publication [21], only modified Handerson–Hasselbalch equations were given for computing narrow to ultranarrow pH gradients, extending to a maximum of 1 pH unit. Such equations were based on the *tandem principle* (i.e., on the use of only two Immobiline species, one buffering and the other titrating within the desired pH interval). However, when several buffering species had to be mixed and titrated to form an extended pH gradient, the problem became quite complex. We adopted the basic concepts of gradient elution as first proposed in 1959 by Peterson and Sober [23]. In our first computer simulation of extended pH gradients, we introduced:

1. The dissociations constants of acidic and basic Immobilines
2. The law of electroneutrality
3. A polynomial calculating the value of $[H^+]$ (and thus pH) along the titration as a function of all dissociated species present
4. The generalized Peterson-Sober equation [23], for solving the foregoing polynomial by calculating the actual concentration of each Immobiline in the output flow
5. An equation for calculating the ionic strength along the pH gradient
6. An equation for computing the buffering power (β) during titration

In practice, this procedure proved too laborious: we had to prepare five different solutions (as we used a five-chambered mixer) pretitrated to given pH values, increasing from anode to cathode. Consequently, this approach was abandoned in favor of a two-vessel gradient mixer [24]: the same set of equations was used; in addition we introduced an optimization algorithm (based on Cauchy's minimum slope criterion) that had as a goal the linearization of the pH gradient by keeping the β power constant along the titration. This version proved to be quite popular and allowed us to publish a series of wide pH gradients (spanning 2–6 pH units) optimized in terms of linearity of pH gradient and constant β profile [25]. The set goal: within each formulation, the deviation from linearity had to be maintained within 1% of the stated pH interval. In this paper, we described two approaches to the generation of extended pH gradients: (a) in one case, the same concentration of buffering Immobilines would be placed in the two vessels of the mixer; (b) in the other, different concentrations of buffering groups could be present in each chamber. The latter case (a sort of unorthodox titration) allows the shift of the apparent p*K* of a given buffer, thereby filling-in holes of buffering power produced by too wide Δp*K* values.

Even this last approach would not have been optimal in quite a few instances: for example, in developing two-dimensional (2-D) maps, best resolution in the focusing dimension would clearly be obtained by nonlinear pH gradients, following the relative abundance of isoelectric proteins along the pH scale. Thus, we soon recalculated wide, nonlinear IPG recipes for use in 2-D maps and in all cases requiring analysis of highly heterogeneous samples [26]. This approach was particularly interesting, as it proved for the first time, the possibility of having in the gel cassette a nonlinear pH gradient supported by a linear density gradient (the latter is always present when pouring IPGs, as the liquid elements, during delivery, have to be stabilized against convection and remixing). Since, at that time, we had given extended recipes (especially the 5- and 6-pH unit intervals) utilizing, in addition to the set of six weak acidic and basic Immobilines, also two titrants, one strong acid and a strong base (not commercially available) we had to publish a new set of formulations, optimized in the absence of such titrants [27]. During that period, other types of focusing in nonamphoteric buffers were introduced: Chrambach's "arrested stack" [28], Rilbe's [29] "steady-state rheoelectrolysis," and Bier's "physically" immobilized pH gradients [30]. We were able to use our IPG computer program to resimulate their data at the light of our acquired knowledge and find the merits and limits of each alternative route to *nonamphoteric buffer focusing* [31].

At the beginning of 1986, we started thinking of expanding the fractionation capability of IPGs: up to this moment, the most extended pH interval described was a pH 4–10. For this reason, we had not included the dissociation products of water (H^+ and OH^-) in our simulations, since within the pH 4–10 range their concentration is negligible. At that time, we started focusing dansylated amino acids (which exhibited pIs in the pH 3–4 interval) [32], and we realized that there was a strong divergence between calculated and experimental pH gradients: thus, our computer program was expanded to include the effects of H^+ (pK = −1.74) and OH^-(pK = 15.75) on β, ionic strength, and pH profile [33]. In fact, simulations were not only limited to acidic, but included also quite basic (pH 10–11) intervals: in this latter instance, we were able to model IPG behavior by modifying the more general program for steady-state IEF, as developed by Bier et al. [20]. All the work produced up to the end of 1986 was summarized in an article [34] that gave the overall view of the computational method and described the architecture of the program.

In 1988, we started a long-range program on the characterization of existing Immobilines and on the synthesis of new species. In fact, we wanted to expand to the utmost the capabilities of IPGs, so new chemicals were sorely needed. In a series of papers, we could in fact

1. Describe the synthesis and purification of existing Immobilines [35,36]

2. Produce a new, more acidic acrylamido buffer with a p*K* of 3.1 [37,38] for fractionations extending down to pH 2.5
3. Prepare two analogues of the two morpholino species (p*K* 6.2 and 7.0) having slightly increased p*K* values (6.6 and 7.4, respectively) by substituting the morpholino with a thiomorpholino ring [39]
4. Synthesize a new hydrophilic Immobiline with a p*K* = 8.05 [40].

At this point, the family of Immobilines had considerably expanded and our former program (which was limited to mixtures of no more than ten species, including buffers and titrants) could no longer handle the new generation. In addition to this, some authors [41] had described oligoprotic acrylamido buffers, and these species too were not contemplated in our computational approach. Thus, we completely changed our pH simulation program and we described a new, powerful system, having the following characteristics:

1. It is under MS-DOS, thus it is fully compatible with IBM hardware.
2. The modules are written in FORTRAN 77 subset implemented by the Microsoft FORTRAN 3.2 compiler, with the graphic functions in GW-BASIC.
3. It allows computation of pH gradient, β and ionic strength, not only of monoprotic, but also of polyprotic (polyacids, polybases, zwitterions) buffers, each with up to ten protolytic groups.
4. It can handle a much more complex mixture of up to 48 polyprotic buffers plus two titrants (as opposed to ten monoprotic species in the former program).
5. It includes automatically in the computation the two Immobilines generated by water dissociation [inserted in the formulations as a base, p*K* = 0, and an acid, p*K* = 14, taken at unit activity (i.e., 1 mol L^{-1}).

As the accompanying equations describing the system are quite complex, we are here only quoting this recent development and refer the interested readers to the original articles [34,38,42,43]. As a result of this new program, we have recently given new formulations for IPGs, including pH extremes: the widest possible pH range that could be formulated is a pH 2.5–11 interval, spanning 8.5 pH units [44].

More recently, we have modified our monoelectrolyte gradient simulator (MGS) program, by introducing new optimization algorithms, with better target functions. We shall give here an example on the performance of this new program. As just stated, we were able to calculate, with the Celentano et al. [43] program, new extended recipes, spanning as much as 8.5 pH units [44]. Figure 12 gives the profile and the statistics of such a wide pH interval: first of all, it is seen that, although reasonably linear in the pH 3.5–10 range, it deviates substantially from linearity outside these boundaries, with a maximum negative deviation of −0.25 units (at near pH 3) and a maximum positive of +0.2 units (at near pH 10.2). The reasons: smoothing of β works at its best only when, for each buffering Immobiline, the same molarity is used in each of the two vessels; in addition, such a

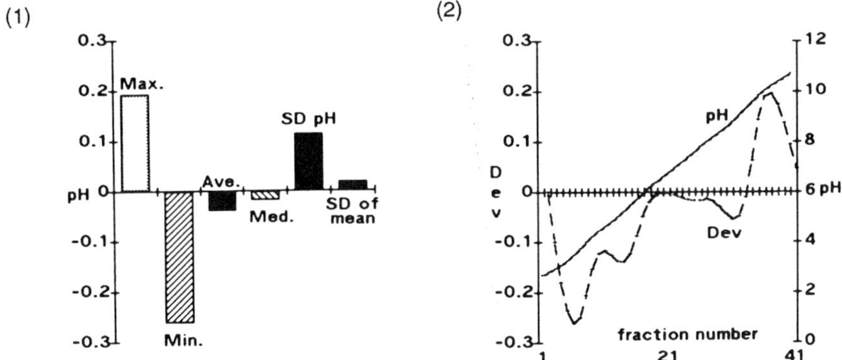

Figure 12 (1) pH profile and (2) deviation from linearity of the pH 2.5–11 recipe according to Gianazza et al. [44]. Max. and Min., maximum and minimum deviation; Ave., average; SD, standard deviation.

target cannot function outside a pH 4–10 interval, owing to the buffering power of water, represented by two branches of an hyperbole. We have resimulated the pH 2.5–11 interval with the new program MGS-2: this program has a new target function to be minimized, namely the standard deviation of the residues from linearity, taking as ideal shape a straight pH line joining the two pH extremes in the limiting solutions of the two starting recipes (in the mixing chamber and reservoir). Figure 13 offers the "pedigree" of such a new optimized interval. It is

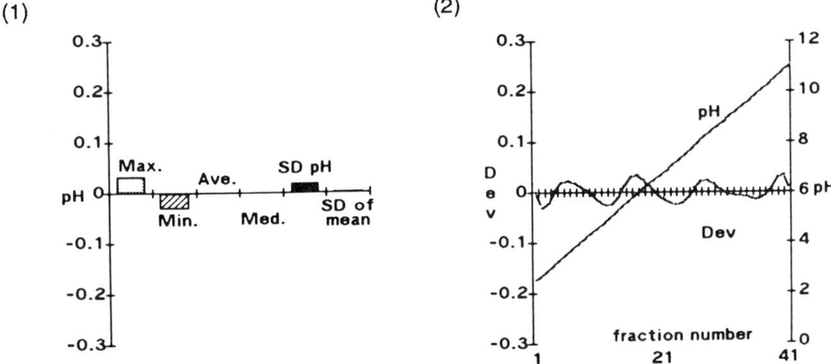

Figure 13 (1) pH profile and (2) deviation from linearity of the pH 2.5–11 recipe calculated according to Tonani and Righetti [54]. Max. and Min., maximum and minimum deviation; Ave., average; SD, standard deviation.

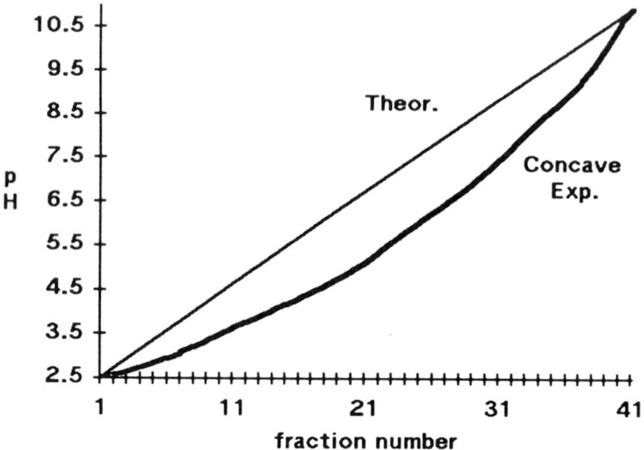

Figure 14 Concave exponential (Exp.) gradient (pH 2.5–11) obtained with the recipe of Righetti and Tonani [55]. Theor., theoretical linear curve.

immediately apparent how, by using the same set of nine chemicals (plus the p*K* 3.6), the new optimized gradient is remarkably linear. In the points of maximum deflection, the positive and negative deviations from linearity are never >0.03 pH units (compare with 0.2 and 0.25, respectively, in the previous recipe). Thus, it is seen that this new optimization algorithm can perform almost ten times better than the previous one. The philosophy of the new algorithm can be appreciated on the profile of the deviation curve in Fig. 13A: it is seen that such deviations oscillate constantly around zero, in an almost sinusoidal way. In other words, each positive deviation is counteracted immediately by a negative one, the sum of the areas of positive and negative deviations tending to zero (see the bar SD of mean in Fig. 13B). However, our new program can do even better than that: in fact, when working with 2-D maps or with very complex samples, it would not pay to use a linear wide gradient, since 70% of the proteins have pIs below pH 7, and only 30% are alkaline. Thus, ideally, the pH 2.5–11 IPG described previously should be nonlinear, in fact, it should be a concave exponential, with a pH shape following the relative abundance of proteins along the pH scale. Figure 14 gives the profile for such nonlinear pH 2.5–11 interval, as calculated by our new program. This new recipe has exactly the required properties, since it gives two-thirds of the separation space to acidic proteins and only one-third to basic species. It should be the ultimate in 2-D maps of highly heterogeneous samples. For a more general survey on both topics, conventional isoelectric focusing and immobilized pH gradients, refer to two manuals, as reported in refs. [56] and [57], respectively.

ISOELECTRIC FOCUSING IN CAPILLARY ZONE ELECTROPHORESIS

Conventional IEF in CA buffers can also be efficiently performed in capillary zone electrophoresis (CZE), provided some stringent requirements are met. First, the inner surface of the fused silica capillary should be coated to eliminate or at least quench the negative surface charge caused by ionization of silanol groups (having a range of p*K* values between pH 3 and 5). Such ionization not only induces a strong electroosmotic flow (which would wash away at the cathode-focused proteins, if any), but would prevent pH gradient formation, owing to the buffering power of the ionizing silanols on the wall and, finally, would favor strong binding of proteins to the capillary wall. An efficient way of suppressing these wall charges would be to bury them under a fluffy coating. Hjertén [45] first proposed coating the wall with strings of linear polyacrylamide, covalently bound by a divalent "bind silane." Unfortunately, the siloxane bridge thus formed is susceptible to alkaline hydrolysis: based on such evidence, Cobb et al. [46] have developed a hydrolytically stable coating, based on producing a direct silicium–carbon linkage with the aid of Grignard reagents. Alternatively, Swedberg [47] has proposed an aryl pentafluoro-modified capillary surface (a review on inner surface coating has recently appeared [48]).

The first report on CZE-IEF was from Hjertén and Zhu [49]. Subsequently, Hjertén et al. [50,51] offered a thorough theoretical study on the focusing and mobilization conditions in CZE. Usually, the capillary is filled with 2.5% carrier ampholytes (in the desired pH interval) and focusing is performed between 20 mM phosphoric acid (anolyte) and 20 mM NaOH (catholyte). In narrow tubes (e.g., 50 μm inner diameter, but in some cases as low as 25 μm) at least 300 V cm^{-1} can be applied with barely 4–6 minutes for reaching steady state (an indicator is that the current drops from about 15 to as low as 1 μA, in 50-μm columns). Mobilization is, in general, achieved by adding to the anolyte 80 mM NaCl: as the Na^+ boundary sweeps through the column it causes a drastic pH increase, which drives the proteins past the cathodic detector (set at 280 nm). During mobilization the current rises from 1 to as high as 200 μA and the process lasts for about 10 minutes at 400 V cm^{-1}. Mobilization can also be achieved, after attainment of steady state, by replacing the catholyte with 10 mM glycine (pH 6) and emptying the tube (still under applied voltage) by pumping anolyte (20 mM phosphoric acid) into the capillary from the anodic side (at a rate of about 0.12 μL min^{-1}). After recording past the cathodic detector, the protein zones can be recovered with the aid of a fraction collector. Kilár and Hjertén [52] have shown that, in narrow pH ranges, transferrin can be resolved into the full spectrum of iron-free, partially saturated, and fully saturated iron forms. In addition, Kilár [53] has proposed that the current value associated with the transit of as protein peak during mobilization could be linked to the pI value of the protein under analysis.

Isoelectric focusing in CZE might have to be performed at your own risk: of isoelectric protein precipitation. At such ionic strength extremes (an ion concentration of barely 1 mEq L^{-1}) and at zero net surface charge, proteins tend to form aggregates that precipitate out. Precipitation in CZE–IEF is evidenced by losses or fluctuations of current. Passage of precipitated particles through the detector is evidenced by spikes in the UV tracing. Several strategies can be employed to minimize precipitation, such as (a) reducing the sample concentration (often at less than 1 mg mL^{-1} no precipitation occurs); (b) reducing the focusing or mobilization times; (c) incorporation of additives. Use of IEF is compatible with any neutral or zwitterionic additives. Two classes are generally employed: (a) organic modifiers (e.g., glycerol, ethylene glycol) and (b) nonionic and zwitterionic surfactants (e.g., Triton X-100, Nonidet P-40, Brij-35, CHAPS, sulfobetaines). Of course, urea is an excellent protein solubilizer, but it unfolds them, and it might carbamylate them at a pH-dependent rate (so, as you are working in a pH gradient, we let you guess the sequel of this chronicle).

ACKNOWLEDGMENTS

Supported in part by Agenzia Spaziale Italiana (Roma), by Progetti Finalizzati FATMA and Biotecnologia e Biostrumentazione (CNR, Roma).

REFERENCES

1. H. Rilbe, *Isoelectric Focusing* (N. Catsimpoolas, ed.), Academic Press, New York, pp. 14–52 (1976).
2. H. Svennson, *Acta Chem. Scand.*, *15*:325–341 (1961).
3. H. Svensson, *Acta Chem. Scand.*, *16*:456–466 (1962).
4. A. Kolin, *Isoelectric Focusing*, (N. Catsimpoolas, ed.), Academic Press, New York, pp. 1–11 (1976).
5. R. A. Mosher, M. Bier, and P. G. Righetti, *Electrophoresis*, *7*:59–66 (1986).
6. V. G. Babskii, M. Y. Zhukov, and V. I. Yudovich, *Mathematical Theory of Electrophoresis*, Plenum Press, New York, (1989).
7. O. Vesterberg, *Acta Chem. Scand.*, *23*:2653–2666 (1969).
8. H. Svensson, *J. Chromatogr.*, *25*:266–273 (1966).
9. M. Almgren, *Chem. Scripta*, *1*:69–75 (1971).
10. C. B. Cuono and G. A. Chapo, *Electrophoresis*, *3*:65–70 (1982).
11. W. G. Kauman, *Classe Sci. Acad. R. Belg.*, *43*:854–868 (1957).
12. E. Schumacher, *Helv. Chim. Acta*, *40*:2322–2340 (1957).
13. N. Catsimpoolas, *Isoelectric Focusing* (N. Catsimpoolas, ed.), Academic Press, New York, pp. 1–11 (1976).
14. J. T. Edward and D. Waldron-Edward, *J. Chromatogr.*, *20*:563–573 (1965).
15. A. Murel, I. Kirjanen, and O. Kirret, *J. Chromatogr.*, *174*:1–11 (1979).
16. H. Delincée, *Electrophoresis '89* (B. J. Radola, ed.), de Gruyter, Berlin, pp. 165–171 (1980).

17. P. G. Righetti and C. Macelloni, *J. Biochem. Biophys. Methods*, *6*:1–15 (1982).
18. P. J. Svendsen and C. Schafer-Nielsen, *Electrophoresis '82* (D. Stathakos, ed.), de Gruyter, Berlin, pp. 83–89 (1983).
19. R. Mosher and W. Thormann, *Electrophoresis*, *11*:717–723 (1990).
20. M. Bier, O. Palusinski, R. A. Mosher, and D. A. Saville, *Science 219*:1281–1287 (1983).
21. B. Bjellqvist, K. Ek, P. G. Righetti, E. Gianazza, A. Görg, R. Westermeier, and W. Postel, *J. Biochem. Biophys. Methods*, *6*:317–339 (1982).
22. M. Chiari, C. Ettori, and P. G. Righetti, *J. Chromatogr.*, *559*:119–131 (1991).
23. G. Dossi, F. Celentano, E. Gianazza, and P. G. Righetti, *J. Biochem. Biophys. Methods*, *7*:123–142 (1983).
24. E. Gianazza, G. Dossi, F. Celentano, and P. G. Righetti, *J. Biochem. Biophys. Methods*, *8*:109–133 (1983).
25. E. Gianazza, F. Celentano, G. Dossi, B. Bjellqvist, and P. G. Righetti, *Electrophoresis*, *5*:88–97 (1984).
26. E. Gianazza, P. Giacon, B. Sahlin, and P. G. Righetti, *Electrophoresis*, *6*:53–56 (1985).
27. E. Gianazza, S. Astrua-Testori, and P. G. Righetti, *Electrophoresis*, *6*:113–117 (1985).
28. A. Chrambach and L. M. Hjelmeland, *Electrophoresis '83* (H. Hirai, ed.), de Gruyter, Berlin, pp. 81–97 (1984).
29. H. Rilbe, *J. Chromatogr.*, *159*:193–205 (1978).
30. M. Bier, R. A. Mosher, W. Thormann, and A. Graham, *Electrophoresis '83* (H. Hirai, ed.), de Gruyter, Berlin, pp. 99–107 (1984).
31. P. G. Righetti and E. Gianazza, *J. Chromatogr.*, *334*:71–82 (1985).
32. A. Bianchi-Bosisio, P. G. Righetti, N. B. Egen, and M. Bier, *Electrophoresis*, *7*:128–133 (1986).
33. P. G. Righetti, E. Gianazza, and F. Celentano, *J. Chromatogr.*, *356*:9–14 (1986).
34. F. Celentano, E. Gianazza, G. Dossi, and P. G. Righetti, *Chemometr. Intel. Lab. Syst.*, *1*:349–358 (1987).
35. M. Chiari, E. Casale, E. Santaniello, and P. G. Righetti, *Theor. Appl. Electr.*, *1*:99–102 (1989).
36. M. Chiari, E. Casale, E. Santaniello, and P. G. Righetti, *Theor. Appl. Elect.*, *1*:103–107 (1989).
37. P. G. Righetti, M. Chiari, P. K. Sinha, and E. Santaniello, *J. Biochem. Biophys. Methods*, *16*:185–192 (1988).
38. P. G. Righetti, M. Fazio, C. Tonani, E. Gianazza, and F. C. Celentano, *J. Biochem. Biophys. Methods*, *16*:129–140 (1988).
39. M. Chiari, P. G. Righetti, P. Ferraboschi, T. Jain, and R. Shorr, *Electrophoresis*, *11*:617–620 (1990).
40. M. Chiari, L. Pagani, P. G. Righetti, T. Jain, R. Shorr, and T. Rabilloud, *J. Biochem. Biophys. Methods*, *21*:165–172 (1990).
41. R. Charlionet, R. Sesboüé, and C. Davrinche, *Electrophoresis*, *5*:176–178 (1984).
42. C. F. Celentano, E. Gianazza, C. Tonani, and P. G. Righetti, *Electrophoresis '88*, (C. Schafer-Nielsen, ed.) VCH, Weinheim, pp. 15–27 (1988).
43. F. C. Celentano, C. Tonani, M. Fazio, E. Gianazza, and P. G. Righetti, *J. Biochem. Biophys. Methods*, *16*:109–128 (1988).

44. E. Gianazza, F. Celentano, S. Magenes, C. Ettori, and P. G. Righetti, *Electrophoresis*, *10*:806–808 (1989).
45. S. Hjertén, *J. Chromatogr.*, *347*:191–198 (1985).
46. K. A. Cobb, V. Dolnik, and M. Novotny, *Anal. Chem.*, *62*:2478–2483 (1990).
47. S. A. Swedberg, *Anal. Biochem.*, *185*:51–56 (1990).
48. J. R. Mazzeo and I. S. Krull, *Bio Techniques*, *10*:638–645 (1991).
49. S. Hjertén and M. Zhu, *J. Chromatogr.*, *346*:265–272 (1985).
50. S. Hjertén, K. Elenbring, F. Kilar, J. L. Liao, A. J. C. Chen, J. C. Siebert, and M. D. Zhu, *J. Chromatogr.*, *403*:47–61 (1987).
51. S. Hjertén, J. L. Liao, and K. Yao, *J. Chromatogr.*, *387*:127–138 (1987).
52. F. Kilar and S. Hjertén, *Electrophoresis*, *10*:23–29 (1989).
53. F. Kilár, *J. Chromatogr.*, *545*:403–406 (1991).
54. C. Tonani and P. G. Righetti, *Electrophoresis*, *12*:1011–1021 (1991).
55. P. G. Righetti and C. Tonani, *Electrophoresis*, *12*:1021–1027 (1991).
56. P. G. Righetti, *Isoelectric Focusing: Theory, Methodology and Applications*, Elsevier, Amsterdam (1983).
57. P. G. Righetti, *Immobilized pH Gradients: Theory and Methodology*, Elsevier, Amsterdam (1990).

II
BUFFER SYSTEM

4

The Buffer in Capillary Zone Electrophoresis

George M. Janini and Haleem J. Issaq

Program Resources, Inc./DynCorp
Frederick Cancer Research and Development Center
National Cancer Institute
National Institutes of Health
Frederick, Maryland

The buffer plays a central role in capillary zone electrophoresis (CZE). In analogy to chromatography, the buffer in CZE assumes the role of the mobile phase and the stationary phase. It is the semiconducting property of the buffer that makes the migration of analyte ions possible. Electroosmosis, which is an integral part of CZE, is mainly driven by the residual charge on the inner wall of the capillary, but electroosmotic flow is also controlled by judicious selection of buffer composition. In preparing this chapter, the objective is to highlight the role of the buffer in CZE to help the reader obtain an understanding of the underlying principles and provide a useful groundwork for further study. Mathematical rigor is sacrificed for simplicity and clarity of presentation.

BUFFER SYSTEMS

Buffer pH, type, concentration, and ionic strength influence many solution-state properties such as solubility, reaction rate, and mechanism. In CZE, solute migration velocity, separation, column efficiency, and peak shape are sensitive to changes in buffer characteristics. In particular, the pH is of crucial importance, creating the need for stringent buffer control. Buffer capacity, which is a quantitative measure of buffering ability, is another property that needs to be addressed. The buffer capacity must be high enough such that the local pH and conductivity will not change as a result of sample injection.

In theory, the amounts of material needed to prepare a buffer of defined properties can be calculated. However, approximations in the equation used for buffer pH calculations and in the values of the weak acid or base dissociation constants (p*K*s) give calculated pH values that are slightly different from measured values. Because of the laborious calculations involved, several authors have published tables and practical procedures for the preparation of buffers [1–6]. Others have presented programs for calculating the amounts of material needed to prepare buffers of any specifications [7,8].

A *buffer*, by definition, is a solution that maintains a constant pH by resisting changes in pH as a result of dilution or addition of small amounts of acids and bases. Many buffer solutions are multicomponent systems, and some are mixtures of two or more buffer systems. A simple buffer system is a solution that contains a weak acid and one of its salts (or a weak base and one of its salts). A rule of thumb is that the concentration of the two components should be within a factor of 10 of each other. To illustrate the elementary principles of buffer action, consider a solution containing a weak acid (HA) and its salt (MA). The equilibria involved are as follows:

$$HA + H_2O \rightleftarrows H_3O^+ + A^-; \qquad K_a \tag{1.a}$$

$$A^- + H_2O \rightleftarrows OH^- + HA; \qquad K_h \tag{1.b}$$

where K_a is the ionization constant of HA and K_h is the hydrolysis constant of A^-. The concentration of the weak acid [HA] and its conjugate base $[A^-]$ at equilibrium are calculated as follows:

$$[HA] = C_{HA} - [H_3O^+] + [OH^-] \tag{2.a}$$

$$[A^-] = C_{A-} + [H_3O^+] - [OH^-] \tag{2.b}$$

where C_{HA} and C_{A-} are the analytical concentrations of the acid and the salt respectively and [ ] denotes concentration.

In acidic solution $[OH^-]$ is negligibly small and, except for very dilute solutions, $[H_3O^+]$ to a very good degree of approximation, is much smaller than both [HA] and $[A^-]$. Therefore, it could be assumed that

$$[HA] \simeq C_{HA} \tag{3.a}$$

$$[A^-] \simeq C_{A-} \tag{3.b}$$

By substituting these values in the equilibrium expression presented by Eq. 1.a and rearranging we obtain

$$[H_3O^+] = K_a \frac{C_{HA}}{C_{A-}} \tag{4}$$

By using the definitions, $pH = -\log [H_3O^+]$ and $pK_a = -\log K_a$, we get the expression that is well known as the Henderson–Hasselbalch equation:

$$pH = pK_a + \log \frac{C_{A-}}{C_{HA}} \tag{5}$$

Equation 5 is used for calculation of the concentrations of HA and A^- needed to prepare buffers of any desired pH value, provided that (a) the pK_a of the acid is accurately known, and (b) assumptions Eqs. 3.a and 3.b are valid. Assumptions 3.a and 3.b fail if K_a is large ($K_a > 1 \times 10^{-3}$) or the concentration of either HA or A^-, or both, are small and comparable with $[H_3O^+]$. Table 1 lists a compilation of weak acids, bases, and zwitterions that are commonly used as buffer material [2,9].

Buffer solutions are resistant to changes in pH as a result of dilution and the addition of small amounts of acids and bases, as clearly demonstrated in Figs. 1 and 2, respectively. Figure 1 shows the effect of dilution on the pH of buffered and unbuffered systems. Although the pH of unbuffered systems (HCl and CH_3COOH) increases continuously with dilution, the pH of the buffer ($CH_3COOH + CH_3COONa$) remains constant over a wide range of concentration. Figure 2 shows the effect of addition of small amounts of acids or bases to buffered and unbuffered solutions. Whereas the addition of small amounts of an acid or a base to an unbuffered solution drastically changes its pH, such additions result in negligible changes in the pH of buffered solutions.

Buffer capacity, sometimes referred to as *buffer value* or *buffer unit* is a quantitative measure of the potential of the buffer solution. It is defined as the number of equivalents of strong acid or base needed to cause a 1 unit change in the pH of 1 L of a buffer solution.

The amount of acid or base that could be tolerated by a buffer, without significant change in pH, is directly proportional to the concentration of the species that constitute the buffer as well as their concentration ratio. Buffers are most effective within a 2-unit pH range bracketing the pK_a of the acid (buffer range $pK_a - 1$ to $pK_a + 1$; i.e., $10 \geq C_{A-}/C_{HA} \geq 0.1$). For example, an acetic acid–acetate buffer is most effective in stabilizing the pH of the solution in the pH range 3.74–5.74. It is, therefore, apparent that to prepare a buffer at a desired pH, the acid chosen should have a pK_a that is reasonably close to the desired pH. Buffering is not as effective if C_{A-}/C_{HA} is very large (>10) or very small (<0.1). In such cases, small changes in molarity can have a large effect on the C_{A-}/C_{HA} ratio and thus on the pH.

Temperature changes affect the equilibrium constants of buffer systems and, consequently, change the pH of the solutions. For acidic buffers, pH generally decreases with increasing temperature. Most buffers have $\Delta pK_a/\Delta T$ (change in pK_a per degree Kelvin) about -0.01, but some, as for example, Tris, have higher values [9]. A $\Delta pK_a/\Delta T$ value of -0.01 translates to a decrease of about 0.1 pH units for every 10° rise in temperature.

Table 1 Selected Weak Acids, Bases, and Zwitterionic Compounds and Their p*K* Values at 25°C

Compound	p*K*
Oxalic acid	1.23 (pK_1)
	4.19 (pK_2)
Phosphoric acid	2.12 (pK_1)
	7.21 (pK_2)
	12.32 (pK_3)
Malonic acid	2.9 (pK_1)
	5.7 (pK_2)
Citric acid	3.06 (pK_1)
	4.74 (pK_2)
	5.40 (pK_3)
Formic acid	3.75
Succinic acid	4.19 (pK_1)
	5.57 (pK_2)
Acetic acid	4.74
MES [2-(*N*-morpholino)ethanesulfonic acid]	6.15
ADA (*N*-2-acetamidoiminodiacetic acid)	6.60
PIPES [piperzine-*N*,*N*′-bis(2-ethanesulfonic acid)]	6.80
ACES (*N*-2-acetamido-2-aminoethanesulfonic acid)	6.90
Imidazole	7.00
MOPS [3-(*N*-morpholino)propanesulfonic acid]	7.20
TES [2-{[tris-(hydroxylmethyl)methyl]amino}ethanesulfonic acid]	7.50
HEPES (*N*-2-hydroxyethylpiperazine-*N*′-2-ethanesulfonic acid)	7.55
Hydrazine	7.99
HEPPS (*N*-2-hydroxyethylpiperazine-*N*′-3-propanesulfonic acid)	8.00
TRICINE {*N*-[tris-(hydroxymethyl)methyl]glycine}	8.15
TRIS [tris-(hydroxymethyl)aminomethane]	8.30
BICINE [*N*,*N*-bis(2-hydroxyethyl)glycine]	8.35
Glycylglycine	8.40
Boric acid	9.24
Ammonia	9.26
CHES [2-(cyclohexylaminoethanesulfonic acid)]	9.50
Trimethylamine	9.87
CAPS [3-(cyclohexylamino)propanesulfonic acid]	10.40

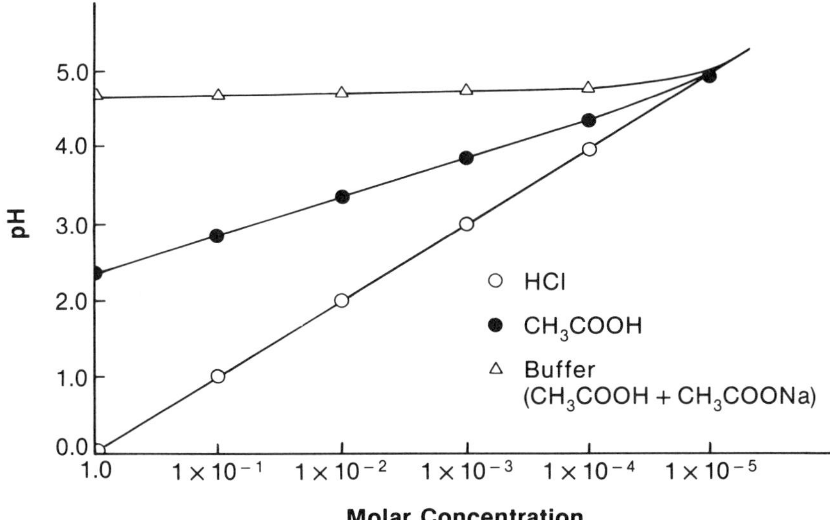

Figure 1 Effect of dilution on the pH of buffered and unbuffered acidic solutions. Initial concentrations: 1 M each; temperature: 25°C, $K_a(CH_3COOH) = 1.85 \times 10^{-5}$.

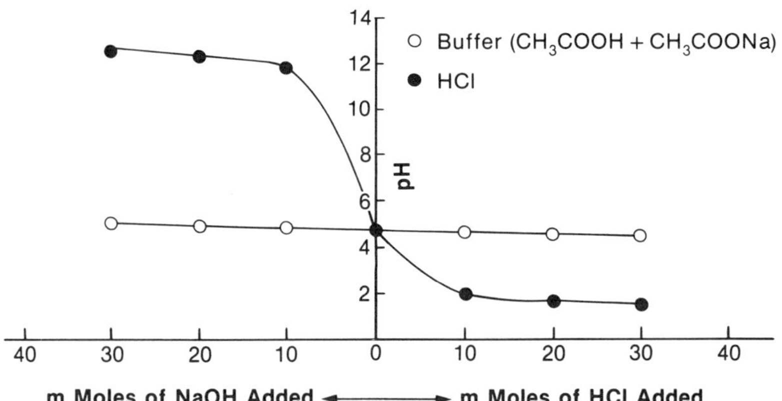

Figure 2 Effect of the addition of small amounts of acids and bases to buffered and unbuffered solutions. Initial concentrations: $[CH_3OOH] = [CH_3COONa] = 0.5$ M, pH 4.74; $[HCl] = 1.85 \times 10^{-5}$, pH 4.74; $K_a(CH_3COOH) = 1.85 \times 10^{-5}$.

Other practical factors that influence the selection of buffers were discussed by Gueffroy [9]. These include stability to oxidation and light, complexation and metal ion-binding properties, and UV and visible light-absorbing ability.

UNDERLYING THEORETICAL CONSIDERATIONS

The theory of electrophoresis deals with the migration of charged particles in a semiconducting fluid under the influence of an electric field. In CZE the semiconducting fluid is contained in a narrow capillary tube, and ionic and ionizable solutes of interest are separated into zones, based on differences in charge, size, and shape. The quantitative treatment of ionic motion is complex, but as long as we do not inquire too deeply into the details, CZE theory can be discussed very simply. To facilitate this discussion, we present abbreviated reviews of the relevant underlying principles of ionic conductivity and mobility, joule heating, the electric double layer, electroosmosis in capillary tubes, and electrophoretic and electroosmotic mobilities. Readers who are interested in more details may consult the referenced literature.

Electrolytic Conductivity and Ionic Mobility

The electric conductance ($1/R$; where R = resistance) of a uniform conductor is directly proportional to its cross-sectional area (A) and inversely proportional to its length (L).

$$\frac{1}{R} = \frac{KA}{L} \text{ or } R = \frac{L}{K_A} \tag{6}$$

where the proportionality constant K = conductivity.

The *electrical conductivity* of an electrolyte solution is the sum of contributions from each of the types of ions present. When two electrodes a distance L apart are at a potential difference (V), a uniform electric field of magnitude $E = V/L$ is created. When an ion of charge q is present in the field, it experiences a force of magnitude $F = qE$. Cations respond by accelerating toward the negative electrode, whereas anions accelerate in the direction of the positive electrode. The translational movement of the ions is opposed by a viscous drag force of magnitude $F' = 6\pi\eta rv$, where v is the ion's velocity, r is the ion's radius, and η is the medium viscosity. When the two forces are counterbalanced the particles move with a steady-state velocity:

$$v = \frac{qE}{6\pi\eta r} = \mu E \tag{7}$$

where μ is the ionic mobility.

Since the velocity of the ions controls the rate at which charge is transported, a direct relation is expected between ionic mobility and conductivity. As a consequence of this theoretical relation, the current generated in a CZE setup is expected to increase with (a) decreasing solution viscosity, (b) increasing charge on the ions constituting the buffer, and (c) decreasing ionic size. Although this is largely true for bulky ions, some discrepancy is noted for small ions. The subject will be discussed in more detail and specific examples will be given later in this chapter. Values for ionic mobilities and limiting ionic conductivities for some ions commonly used in CZE are given in Table 2.

Joule Heating

When a potential difference is applied across a buffer-filled capillary of finite resistance, energy is supplied to the system at a rate governed by the following expression:

$$W = VI \tag{8}$$

where W is the rate of heat generation in watts or joules per second, and I is the current. Most of this energy is lost as heat which, if not efficiently dissipated, tends to (a) raise the buffer temperature, (b) cause convection currents, and (c) cause parabolic temperature variation across the column. The heat so generated is dissipated into the column surroundings through the column wall. Minimizing heat generation and maximizing heat dissipation is critical in CZE because the presence of thermal gradients and other heat-related effects could adversely affect the very high efficiency of this technique. The quantitative treatment of heat dissipation in relation to column

Table 2 Ionic Mobilities and Limiting Ionic Conductivities of Selected Ions in Water at 25°C

Ion	Ionic mobility (10^4 cm^2 S^{-1} V^{-1})	Limiting ionic conductivity (S cm^2 mol^{-1})
H^+	36.23	349.6
Li^+	4.01	38.7
Na^+	5.19	50.10
K^+	7.62	73.50
Rb^+	7.92	77.8
Cs^+		77.2
Zn^{2+}	5.47	105.6
OH^-	20.64	199.1
Cl^-	7.91	76.35
Br^-	8.09	78.1
I^-	7.96	76.8
CH_3COO^-	4.24	40.9
$(COO)_2^{2-}$		148.2

Source: Data from Ref. 10.

geometry and instrumental design is very complex. On the other hand, estimation of the heat generated inside the column could be made very easily.

The amount of heat generated per unit time could be related to relevant CZE experimental parameters by substituting Ohm's Law ($V = IR$) into Eq. 8:

$$W = IV = \frac{V^2}{R} \tag{9}$$

Substituting for R from Eq. 6 and realizing that (a) $K = \Lambda C$, where Λ is the molar conductivity of the solution, and (b) $A = \pi d^2/4$, where d is the column diameter we get:

$$W = \frac{\pi d^2 \Lambda C V^2}{4L} \tag{10}$$

Thus, for a given system the rate of heat generation is directly proportional to (a) the molar conductivity of the solution (i.e., buffer type), (b) the buffer concentration, (c) the applied voltage squared, and (d) the column diameter squared, and is inversely proportional to column length. Others choose to represent the heat generation in terms of power per unit length [11] or power per unit volume [12]. No matter how the power is reported, the numbers are meaningless unless column diameter, length, and material are also reported. Temperature rise inside the column is a function of the quantity of fluid it holds, and efficient heat dissipation is a function of column geometry and cooling system design. For a typical CZE experiment with $V = 25$ kV, $C = 10$ mM, $\Lambda = 126.45$ S cm^{-2} mol^{-1} (NaCl, from Table 2), $L = 50$ cm, and $d = 75$ μm, the quantity of heat generated per second is estimated to be 0.7 W (i.e., 1.4 W m). If this amount of heat is absorbed by the small amount of buffer inside the column (2.2 μL), it will result in a considerable rise in temperature. Clearly, efficient heat dissipation is critical for the success of CZE experiments. As early as 1969, it was suggested by Giddings [13] that heat dissipation is greatly enhanced by using capillary columns of small diameter (100 μm or less). Under such conditions, thermal gradients (which are proportional to column diameter squared) are reduced, resulting in theoretical efficiencies on the order of several million plates [13].

It is estimated that the temperature coefficients of mobility are nearly the same for all ions in a given solvent and are of the same order of magnitude as the temperature coefficient of viscosity [10]. Therefore, it is, expected that the mobility increases by about 2% per kelvin rise in temperature in the neighborhood of 25°C. Thermal gradients across the column cause the ions in the center of the column to move faster than the ions closer to the wall, resulting in zone spreading and distortion. Wieme (14) gave the following expression for the temperature profile inside the column:

$$t_c - t_r = \left(\frac{0.239\ w}{4K}\right) r^2 \tag{11}$$

where t_c = temperature in the center of the column; r = distance from center, and K = thermal conductivity (1.45×10^{-3} cal s^{-1} cm^{-1} K^{-1} for water); and w = power per cubic centimeter. From this equation, a value of 0.18°C is calculated for $(t_c - t_r)$ for the CZE experimental setup described earlier in this section. This translates to a 0.4% difference in mobility between ions in the center and ions close to the column wall.

Heat effects in CZE were studied by several groups [15–19]. Terabe et al. [15] attempted to measure the surface temperature with a thermocouple for a micellar electrokinetic setup. The temperature rise was proportional to the power generated and amounted to about 2.8°C mW^{-1} cm^{-1} for 0.05 M sodium dodecyl sulfate (SDS) solution. Column temperature as measured by Terabe et al. exceeded 70°C. Grushka et al. [17] derived a plate height equation for CZE that described the effect of column temperature and thermal gradients on column efficiency. They observed that, even with thermostating, the temperature of a quartz capillary wall can still reach 38°C under reasonable operating conditions. Nevertheless, their results suggest that thermostating at elevated temperatures may lead to improved column efficiency, and a temperature differential of as much as 1.5°C inside the column will not cause a significant increase in the plate height of the system. Knox and Grant [16] studied the temperature gradient effects on plate height in CZE using a simple expression and essentially arrived at similar conclusions.

The influence of temperature inside the capillary on CZE parameters was evaluated using different cooling systems by Kurosu et al. [18], and Burgi et al. [19] who reported an indirect method for the measurement of column temperature in CZE. They observed that column temperature increases about 1.1°C/0.1 W of power for natural convection cooling and 0.6°C/0.1 W of power for forced-air cooling.

In conclusion, the importance of minimizing heat generation by careful selection of the buffer and CZE experimental parameters, and the importance of efficient heat dissipation by careful instrumental design cannot be overemphasized. It is not the rise in buffer temperature, as a result of heat generation, that is detrimental to CZE column efficiency, but rather, convection currents and thermal gradients inside the column.

The Electric Double Layer

Although ionic solutions are electrically neutral overall, yet surrounding any given ion, there exists an excess of counterions (ions of opposite charge). This time-averaged spherical cloud of opposite charge surrounding a given ion is called the *ionic atmosphere*.

From basic electrostatics, Coulomb law states that the potential (ϕ) at a distance r from an ion of charge q is equal to:

$$\phi = \frac{q}{4\pi\varepsilon_0 D}\left(\frac{1}{r}\right) \tag{12}$$

where ε_0 is the electric permittivity in vacuum and the dielectric constant $D = \varepsilon/\varepsilon_0$ where ε = the electric permittivity of the medium. In vacuum $D = 1$, but in liquid $D > 1$ and, therefore, the potential at any r is lower than its value in vacuum. For example, $D = 78.54$ for water at 25°C and, therefore, at any distance r, the potential is reduced from the vacuum value by about two orders of magnitude. The D for organic solvents is much lower than that of water and, therefore, they are less effective in reducing the coulombic potential. This factor should be taken into consideration when discussing the effect of organic solvent modifiers on CZE parameters. In addition to the role of the solvent we also have to consider the role of the ionic atmosphere. The pursuit of rigorous analytical solutions to this problem is one of the most complex undertakings in solution chemistry; however, approximate solutions could be derived quite easily if the rigor of analysis is not pushed too far.

The diffuse ion cloud surrounding any given ion shields its potential by a space charge of opposite polarity and effectively reduces the potential at any distance relative to its value according to Eq. 12. Debye and Huckel [20,21] derived the following empirical expression for the coulombic potential, taking into consideration the solution dielectric and the shielding effect of the ionic atmosphere:

$$\phi = \frac{q}{4\pi\varepsilon_0 D} \frac{1}{r} \, exp\left(-\frac{r}{r_D}\right) \tag{13}$$

The parameter r_D, which is called the Debye length or Debye radius, determines how strongly the potential is damped as a result of the ionic atmosphere. When r_D is very large, $r/r_D = 0$ and the ionic atmosphere is insignificant. Under such conditions Eq. 13 reduces to Eq. 12. When r_D is small, the shielded potential is much smaller than the unshielded potential and the shielded potential decreases with decreasing r_D. The central ion and the surrounding ionic atmosphere are thought of as an electric double layer. The central ion is rigid, and its charge and radius (radius of shear) are the main factors determining its mobility. The diffuse cloud (mobile) behaves like a shell, with a net charge equal, but opposite to, that of the central ion and with a radius equal to r_D. Qualitatively, r_D is expected to decrease with increasing buffer concentration or ionic strength, since as the concentration is increased a smaller shell will be required to neutralize the charge of the central ion. Theoretically, the parameter r_D is related to experimental parameters by the Debye–Huckel expression [20,21]:

$$r_D = \frac{DkT}{8\pi e^2 \mu} \tag{14}$$

where e = the proton charge, k = Boltzmann constant, T = absolute temperature, and μ = ionic strength of the solution. As we shall see later, the inverse proportionality between r_D and μ and, consequently, between r_D and the square root of buffer concentration is relevant to the discussion of the effect of buffer type and concentration on mobility and solute migration in CZE.

The electrical potential at the radius of shear relative to its value in the distant bulk medium is called the zeta potential or electrokinetic potential. The Debye radius is directly related to the zeta potential through Eq. 13, where the zeta potential is equal to ϕ when r equals the radius of shear. As r_D decreases, so does the zeta potential. An approximate, but qualitatively sound, relation between r_D and the zeta potential is given by Tsuda et al. [22]:

$$\zeta = \frac{4\pi r_D \varepsilon'}{D} \tag{15}$$

where ε' is the total excess charge in solution per unit area. Hence, it could be concluded that both the Debye radius and the zeta potential decrease with increasing ionic strength (concentration) and with increasing charge on the ions constituting the buffer. These conclusions, based on the Debye–Huckel theory of ionic solutions will be the basis for the correlation of solute migration in CZE with buffer type and concentration.

Electroosmosis in Capillary Tubes

Electroosmosis in a capillary tube refers to the propulsion of the bulk solvent in the tube under the influence of an applied electric potential. The underlying cause of electroosmosis is the existence of an electric charge on the inside wall that is in direct contact with the buffer. For example, the surface of fused silica and silica-based glasses contains Si—OH groups that are ionized to SiO^- in alkaline and slightly acidic media (pH > 2). The negatively charged surface is counterbalanced by positive ions from the buffer. A diffuse ionic atmosphere and a double layer are formed, as described in the previous section. The direction of electroosmotic flow is determined by the sign of the charge on the surface of the inner wall of the capillary column. The sign, as well as the density of the surface charge, depends on the column material, surface treatment, and solvent properties [23–25]. Several models have been advanced to represent the double layer including the Helmholtz and Gouy–Chapman models [10] and their combination in the Stern model [26]. All models are qualitatively similar, but differ in the details of the structure of the double layer. Two regions of charge are distinguished: (a) a stationary sheet of negative charge located at the surface of fused silica tubes; and (b) a sheet of negative charge facing the surface in the solution that is mobile. Under the influence of an applied potential, the positive ions in the diffuse region migrate toward the cathode and, in so doing, they entrain the water of hydration, resulting in electroosmotic flow of the bulk fluid inside the tube. Electroosmosis is a bulk flow identical for all ionic and neutral particles, irrespective of their electric charge.

Silica–water systems are reported to have a zeta potential of approximately 150 mV, and the estimated double-layer thickness (r_D) in pure water is about 1 μm [26]. A flat-fronted flow velocity profile is expected with electroosmosis if the capillary tube radius is greater than about seven times r_D [27]. The flat-fronted flow profile is the most celebrated advantage to using electroosmotic flow in capillary liquid chromatography and in CZE. The resulting reduction in solute diffusion, in comparison with pressure-driven flow, greatly improves column efficiency. As explained in the previous section, r_D and, consequently, the electroosmotic zeta potential, decrease with increasing ionic concentration (ionic strength). For example, r_D for a 1×10^{-6} M solution of a monovalent cation–anion salt in water was estimated to be about 0.3 μm compared with 0.003 μm for a 1×10^{-2} M solution of the same salt [26].

Electrophoretic and Electroosmotic Mobilities

The basic theoretical concept of electrophoresis is relatively simple. It involves the migration of charged particles in a semiconducting fluid under the influence of an electric field. Ionic and ionizable solutes are separated based on differences in charge, size, and shape. As described in detail in the earlier section on electrolytic conductivity and ionic mobility when a charged particle is placed in an electric field (E) it experiences a force that is proportional to its effective charge (q) and the electric field strength. The translational movement of the particle is opposed by a viscous drag force that is proportional to the particle velocity (v), hydrodynamic radius (r) and medium viscosity (η). When the two forces are counterbalanced, the particle moves with a steady-state velocity [14]:

$$v_{ef} = \mu_{ef} E \tag{16}$$

where μ_{ef}, the electrophoretic mobility is given by:

$$\mu_{ef} = \frac{q}{6\pi\eta r} \tag{17}$$

Equation 16 is identical with Eq. 7 because it basically describes the same phenomenon. According to Huckel [28], the electrophoretic mobility is related to the zeta potential by the following expression:

$$\mu_{ef} = \frac{D\zeta_{ef}}{6\pi\eta} \tag{18}$$

where ζ is the electrophoretic zeta potential and D is the dielectric constant of the medium.

Equation 18 differs from the Helmholtz von Smoluchowski equation [26] only by a constant factor of 2/3:

$$\mu_{ef} = \frac{D\zeta_{ef}}{4\pi\eta} \tag{19}$$

which, according to Henry [29], is a better representation of the electrophoretic mobility.

The direct relation between the zeta potential and the Debye radius (Eq. 15) was presented earlier in the section on electric double layer. Both are inversely proportional to the square root of concentration (ionic strength) of the buffer, thus providing a theoretical relation between the electrophoretic mobility and buffer type and concentration.

The equations of flow for electroosmosis are identical with those developed for electrophoresis, since both phenomena are complementary. The electroosmotic velocity (v_{eo}) is given by:

$$V_{eo} = \mu_{eo} E \tag{20}$$

where μ_{eo} (the electroosmotic mobility) is represented by an expression similar to Eq. 19:

$$\mu_{eo} = \frac{D\zeta_{eo}}{4\pi\eta} \tag{21}$$

where ζ_{eo} is the electroosmotic zeta potential, and all other terms are as defined earlier. When both electrophoretic and electroosmotic forces simultaneously act on an ion, its net mobility is the algebraic sum of Eqs. 16 and 20:

$$V_{net} = \mu_{net} E = (\mu_{ef} + \mu_{eo}) E \tag{22}$$

The ion's migration time (t_m) is given by:

$$t_m = \frac{L}{(\mu_{ef} + \mu_{eo}) E} = \frac{L^2}{(\mu_{ef} + \mu_{eo}) V} \tag{23}$$

where L is the column length.

FACTORS THAT INFLUENCE MOBILITY, SELECTIVITY, AND RESOLUTION

Buffer Type

Inorganic buffers are made of two parts: a cation (metal ion) and an anion. In this section we will probe the influence of the cation and the anion on migration time, resolution, and selectivity.

Role of the Buffer's Cation

The major factor that influences the electrophoretic mobility of an ion according to Eq. 17 is its charge/radius ratio. However, μ_{ef} is also dependent on the type and concentration of the buffer in as much as the presence of the double layer shields the ion and reduces its effective charge. According to Eq. 18, μ_{ef} is directly proportional to the ion's electrophoretic zeta potential. To gauge the effect of the buffer's cation on μ_{ef} Issaq and co-workers [30,31] conducted an experiment whereby

v_{ef} for dansylalanine probe solute was measured with 0.1 M each of lithium, sodium, potassium, rubidium, and cesium acetates. All five experiments were conducted at a constant electric field strength of 400 V cm^{-1} with the expectation that all other factors that affect μ_{ef} remain constant or approximately so. Figure 3(a) gives a plot of μ_{ef} versus the reciprocal of the cation's crystal radii. The data was obtained from Table 1, Ref. 30. The plot is linear and the highest mobility is attained with the lithium salt, which has the shortest ionic crystal radius. It is to be noted that r in the Helmholtz–von Smoluchowski equation is the hydrodynamic radius (the effective radius of the ion in solution, taking into account all the solvent molecules the ion carries in its inner shell) and not the ion's crystal radius. Small ions are more extensively solvated than larger ions of the same charge because the electric field at the surface of an ion is proportional to q/r^2 [see Table 1, Ref. 30].

Issaq and co-workers also measured μ_{eo} for the same systems under identical conditions (using mesityl oxide as a marker for electroosmotic mobility) and plotted the results as shown in Fig. 3(b). The same trend is observed; however, μ_{eo} data for rubidium and cesium are significantly lower than expected from the theoretical linear relation. This is most likely due to the adsorption of these large-sized cations on the capillary wall, thereby altering its charge and effectively reducing the electroosmotic zeta potential. Green and Jorgenson [32] studied the role that alkali metal salts play in minimizing adsorption of proteins on fused silica and obtained

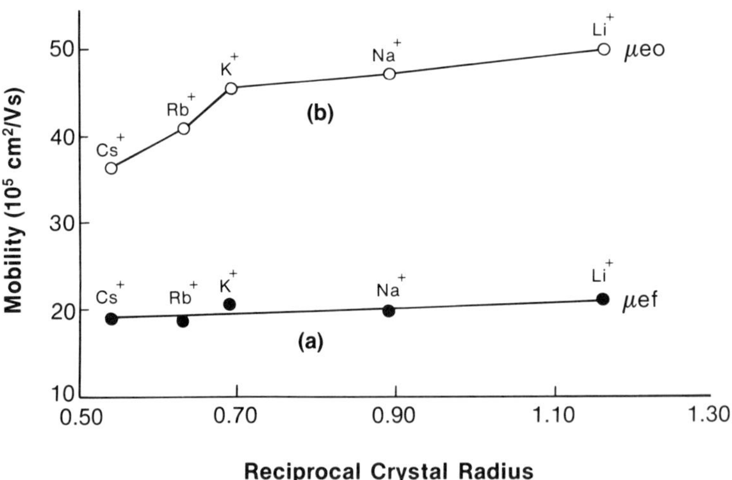

Figure 3 Effect of buffer cation on mobility. (a) Electrophoretic mobility versus reciprocal crystal radius; (b) Electroosmotic mobility versus reciprocal crystal radius. Solutes, dansyl alanine and mesityl oxide; column, 50 cm × 75 μm fused silica; instrument, Beckman Model P/ACE System 2000. (From Ref. 31)

results in agreement with these findings. They observed that the larger the cation's size, the more effective it is in covering the surface of the capillary tube. Figure 4 shows the effect of cation type on the separation and mobility of a test mixture of four dansylated amino acids. It is clear that baseline resolution of the four test solutes was achieved only in the CsAC buffer. The figure also shows that the mobility of the solutes decreased with an increase in the cation's crystal radius. Table 3 shows that at 20-kV applied voltage the resulting current increases with increasing cation size. For example, LiAc produced a current half that produced by CsAc at the same molar concentration and applied voltage [30]. Thus, in comparing the CZE properties of alkali salts (same concentration and same anion), Cs provided the best resolution at the expense of longer analysis time and higher current.

Role of the Buffer's Anion

Green and Jorgenson [32] studied a series of four potassium salts (KCl, KNO_3, KBr, and K_2SO_4) to see what effect the anion would have on the capacity factor (k') of proteins. When these salts were used as buffer additives at high concentrations,

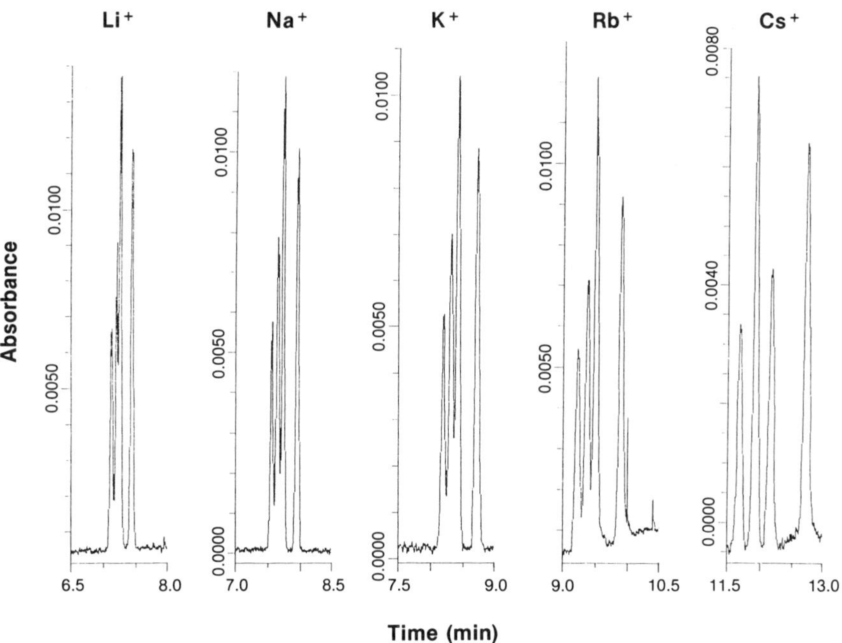

Figure 4 Effect of the cation type on the resolution and migration times of dansylated leucine, proline, methionine, and alanine, at constant voltage of 20 kV and constant pH 6.45. The solutions' concentrations are 0.05 M; temperature constant at 20°C, detection at 254 nm. (From Ref. 30)

Table 3 pH and Current Measurements at 20 kV for the Acetate Salts of the Alkali Metals

Salt	Molarity	pH	Observed current (μA)
LiAc	0.1	8.09	111
NaAc	0.1	8.11	147
KAc	0.1	7.59	195
RbAc	0.1	9.78	204
CsAc	0.1	6.45	215

Table 4 pH and Current Measurements at 20 kV of Seven Sodium Salt Solutions (0.05 M)

Salt Compound	Measured pH	Measured current (μA)	Measured current for the solutions adjusted to pH 8
Sodium acetate	8.0	74.5	74.0
Sodium tetraborate	9.3	137.4	118.0
Sodium citrate	8.9	246.5	243.3
Sodium phosphate dibasic	8.8	162.0	160.7
Sodium hydrogen carbonate	8.7	69.0	77.3
Sodium nitrate	6.4	121.0	
Sodium nitrite	6.9	107.0	

they yielded equivalent k' values, suggesting that, under these experimental conditions, the anion does not have a measurable effect on k'. Van Orman et al. [33] determined "that buffer composition is of a little importance, while ionic strength is a primary factor in the electroosmotic flow."

It is clear from the previous section that the buffer's cation influences the mobility, migration time, and resolution. Issaq and his co-workers also studied the effect of the buffer's anion on these CZE parameters, by using seven sodium salts [34]. Table 4 lists the seven salts selected for this study, along with their pH (0.05 M solution) and resulting current, at an applied voltage of 20 kV. The unadjusted pH (column 2, Table 4) of the salt solutions is mainly in the basic range and can be adjusted to pH 8 with the corresponding acid. Neither sodium nitrate nor sodium nitrite were adjusted to pH 8 because this would have required the use of a base, which would have altered the concentration of the sodium ion in solution and would have led to erroneous results. Table 4 shows that at 20-kV applied voltage, the resulting current varies widely depending on the anion used. Sodium citrate produced a current 3.6 times larger than that produced by sodium hydrogen carbonate having the same concentration. This means that (a) the power produced by using sodium citrate is approximately 3.6 times larger than that produced by sodium hydrogen carbonate or sodium acetate; and (b) the heat generated by the citrate buffer is much larger than that generated by the other two buffers. Sodium phosphate, one of the buffers most widely used for the separation of biological compounds, produces a current that is twice as high as that of sodium acetate. When the pH of these sodium salt solutions was adjusted to 8, there was little or no change in the current produced (compare columns 3 and 4, Table 4) except with sodium tetraborate because the pH was adjusted from pH 9.25 to 8. This shows that slight adjustments in pH (less than 1 pH unit) may not significantly affect the current produced. The effect of the buffer anion on μ_{eo} is given in Table 5. Changes in μ_{eo} are less than 10% for mesityl oxide in phosphate, acetate, citrate, and bicarbonate, but μ_{eo} is faster (~25%) in tetraborate buffer than in the other four buffers under the same experimental conditions. Migration times of eight dansylated amino acids in the five test buffers (see Table 5) follow the μ_{eo} trend. Examination of this table shows that differences in t_m are observed from buffer to buffer. For example, dansylated cysteic acid, which eluted last in each buffer, has a difference of more than 50% in t_m; the fastest t_m was obtained in sodium bicarbonate (largest μ_{eo}), whereas the slowest was achieved in sodium tetraborate (smallest μ_{eo}). Also, dansylated arginine and lysine coeluted in sodium acetate, but were well resolved in sodium citrate. Table 6 lists the separation factors (α) for the eight amino acids in the five sodium buffers. The table shows that the optimum separation for each pair was not realized in one specific buffer, but rather, in combinations of more than one buffer. Overall, sodium citrate gave the best α values, whereas sodium bicarbonate gave the worst α values. Also, some pairs were resolved better in one buffer (arginine–lysine, alanine–cystine),

Table 5 Migration Times (min) of Eight Dansylated Amino Acids and μ_{eo} of Mesityloxide ($10^5 cm^2/Vs$) in Five Different Sodium Phosphate Buffers Having the Same Concentration (0.05 M) and pH (8.0), at an Applied Voltage of 20 kV

Amino acid	Phosphate	Acetate	Citrate	Bicarbonate	Tetraborate
Dans-arginine	4.25	4.13	4.48	4.04	5.00
Dans-lysine	4.32	Coeluted	5.36	4.11	5.12
Dans-tyrosine	5.60	5.39	5.93	5.23	6.83
Dans-methionine	6.19	5.88	6.68	5.70	7.62
Dans-alanine	6.42	6.05	6.94	5.87	7.92
Dans-cystine	7.66	7.10	8.69	6.89	9.52
Dans-glutamic acid	10.11	8.82	11.81	8.61	12.86
Dans-cysteic acid	10.80	9.40	13.31	9.20	14.02
Mesityl oxide	49.7	49.0	47.7	51.8	41.2

Table 6 Separation Factor (α) Values for Eight Dansylated Amino Acids in Five Sodium Buffers (0.05 M, pH 8) at an Applied Voltage of 20 kV

Amino acid pairs	Phosphate	Acetate	Citrate	Bicarbonate	Tetraborate
Lys–Arg	1.02		1.20	1.02	1.02
Tyr–Lys	1.30	1.31	1.11	1.27	1.33
Met–Tyr	1.11	1.09	1.13	1.09	1.12
Ala–Met	1.04	1.03	1.04	1.03	1.04
Cys–Ala	1.19	1.17	1.25	1.74	1.20
Glu–Cys	1.32	1.24	1.36	1.25	1.35
Cyst–Glu	1.07	1.08	1.13	1.07	1.09

whereas others were resolved equally well in all of the buffers (methionine–alanine). This indicates that selectivity differences exist, which means that the buffer's anion affects selectivity under the present experimental conditions.

The results clearly show that the buffer's anion influences not only electroosmotic flow and mobility times, but also resolution, selectivity, and the power produced at an applied voltage. Therefore, careful attention should be paid to the selection of the buffer, to produce optimum results and generate minimum heat.

Effect of Buffer Concentration on Mobility

Several workers [35–38] have reported data on the effect of buffer concentration on mobility. Altria and Simpson [36,37] observed that the mobility is inversely proportional to concentration and that a plot of the logarithm of concentration versus mobility is linear. Bruin et al. [35] studied overall mobility versus buffer concentration. Figure 4, Ref. 35 shows that the relation is linear, whereas Fig. 5 of the same reference gives data that fit a curve of decreasing mobility with increasing concentration. Nashabeh and El-Rassi [38] also presented data on the effect of buffer concentration on mobility. In all, none of the aforementioned literature provided theoretical justification for their plotting procedures. As early as 1975 Wieme [14] suggested that electrophoretic mobility and its complementary electroosmotic mobility should be directly proportional to the reciprocal of the square root of ionic strength. His conclusion was based on the same theoretical arguments presented in the previous section.

To test the validity of this approach Issaq et al. [31] measured μ_{ef} for dansyl alanine and μ_{eo}, with mesityl oxide as a marker, at different concentrations of an acetate and a phosphate buffer system. All experiments were conducted at an electric field strength of 90 V cm^{-1} to ensure the absence of joule-heating complications. Figure 5 shows the data plotted as mobility versus reciprocal of the square root of buffer concentration. All plots were linear with correlation coefficients in excess of 0.995. As the buffer concentration is increased, the electrophoretic and electro-

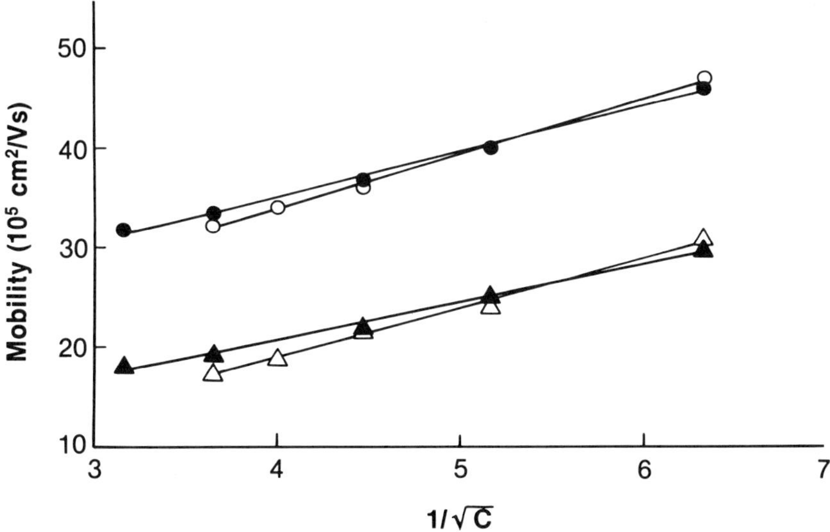

Figure 5 Mobility as a function of buffer concentration. ○, μ_{eo}, acetate; ●, μ_{eo}, phosphate; △, μ_{ef} (dansyl alanine, acetate); ▲, μ_{ef} (dansyl alanine, phosphate). Column, 50 cm × 75 μm fused silica; instrument, Beckman Model P/ACE System 2000. (From Ref. 31)

osmotic mobilities are both reduced. The net mobility ($\mu_{ef} + \mu_{eo}$) is, therefore, expected to follow the same trend.

In accordance with the results presented in Fig. 5 and the relation expressed in Eq. 23, a linear relation is expected if one plots migration time (at constant field strength) versus square root of concentration. Figure 6 shows representative plots with an acetate buffer at pH = 5.0 and a field strength of 90 V cm^{-1}. The plots in the figure converge to a single point, representing the migration times of the solutes in pure water. The figure shows that the migration time of any solute increases and the separation factor continuously improves with increasing concentration. The linear relation between solute migration time and the square root of concentration suggests that the migration time will double for every fourfold increase in buffer concentration. The larger the difference in the net mobility of two neighboring solutes, the larger their separation factor will be at higher concentrations. The same data were plotted as t_m versus concentration and t_m versus the logarithm of concentration (figures not shown), and the plots appeared linear. However, the plots did not converge at infinite dilution, and the correlation coefficients were not as good as reported in Fig. 6. This, in conclusion, could be taken as experimental verification that the mobility is inversely proportional to the square root of the buffer concentration.

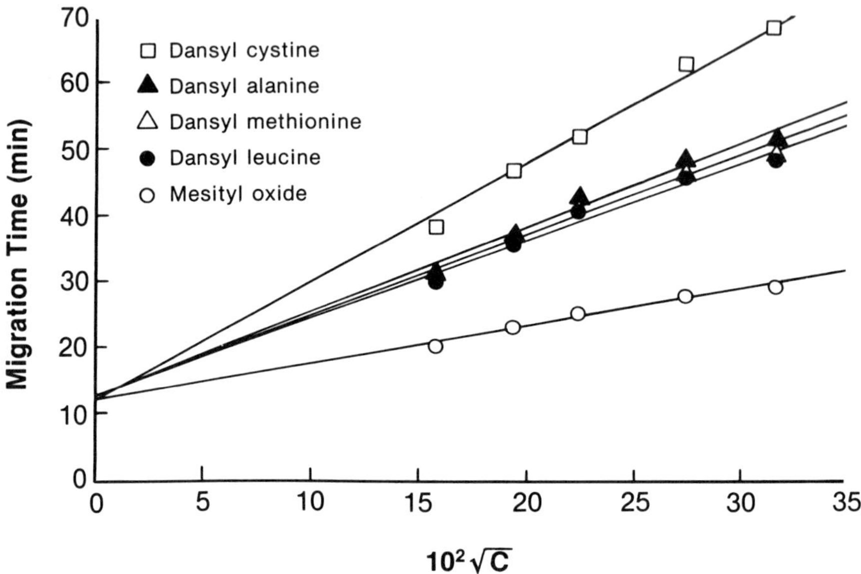

Figure 6 Migration time as a function of buffer concentration. Buffer, acetate at pH 5.0; field strength, 90 V cm^{-1}; column, as in Fig. 5. (From Ref. 31)

Effect of Applied Voltage on Mobility

The effect of applied voltage on solute mobility and separation parameters in CZE was thoroughly investigated by several groups [22,35,36,39–45]. According to Eq. 16, the solute velocity (or the reciprocal of migration time) is linearly proportional to the applied voltage. In practice, however, the solute velocity is not a linear function of voltage [39]. It has been argued that, as voltage is increased, the heat generated results in an increase in temperature and the formation of a radial temperature gradient, with the temperature near the wall (where heat is dissipated to the surroundings) being lower than that at the center. Higher temperatures result in increased buffer conductivity, solute diffusion coefficient and double-layer thickness, and decreased buffer density and viscosity [14]. The net effect of these changes is an increase in solute mobility and a decrease in column efficiency.

Issaq et al. [31] measured μ_{ef} for dansyl alanine and dansyl methionine and μ_{eo} (with mesityl oxide as marker), at different voltages with different concentrations of an acetate buffer (pH = 5.0). Figure 7 shows the effect of voltage on the electrophoretic mobility of dansyl alanine and Fig. 8 shows the effect of applied voltage on electroosmotic mobility. The figures demonstrate that the effect of applied voltage on electrophoretic and electroosmotic mobility is qualitatively similar. In both cases, the mobility increases with increasing applied voltage, in agreement with

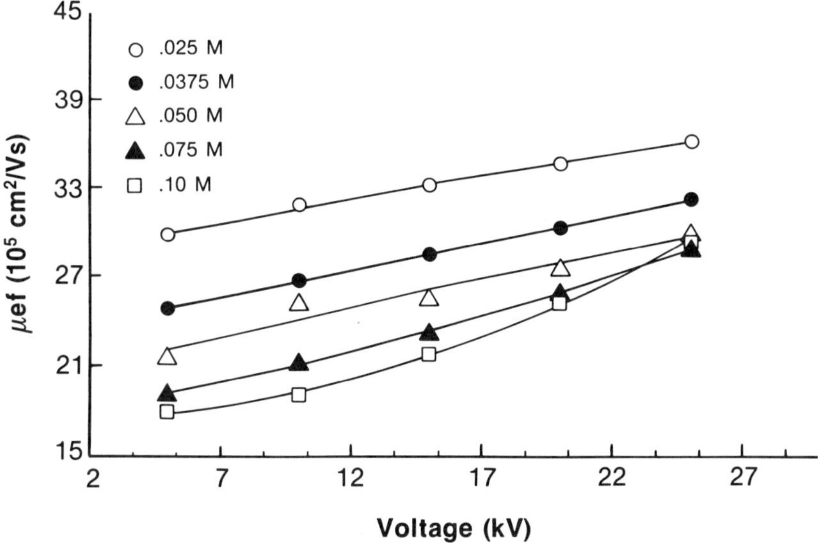

Figure 7 Electrophoretic mobility as a function of applied voltage. Buffer, acetate at pH 5.0; solute, dansyl alanine; column, as in Fig. 5. (From Ref. 31)

Altria and Simpson's findings [36]. At low buffer concentration (25 mM), the mobility increases linearly with increasing applied voltage. Since the mobility is not a constant, even under conditions for which joule heating may not be a disruptive factor, plots of migration velocity versus voltage (Eq. 22) or migration time versus reciprocal voltage (Eq. 23) are not expected to be linear, as was originally believed [44].

As an alternative plotting procedure that takes into consideration the linear dependence of mobility on voltage, Atamna et al. [45] showed that

$$\frac{1}{t_m V} = m + n\,V \tag{24}$$

where m and n are constants. Equation 24 is derived from Eq. 23, assuming that μ_{ef} and μ_{eo} are linear functions of voltage (see Figs. 7 and 8). Figure 9 shows that Eq. 24 represents an excellent fit for t_m versus V data where the plots were linear with correlation coefficients exceeding 0.995 [45].

The observable effect of increasing mobility with increasing voltage is mainly attributed to temperature increases inside the column, as a result of joule heating. Specifically, this is a direct consequence of the larger temperature coefficient of viscosity (2%/°C). There are other factors besides temperature that, if operative, may increase solute mobility with increasing applied voltage. For example, μ increases with increasing voltage owing to the Wien effect [43]; however, this effect

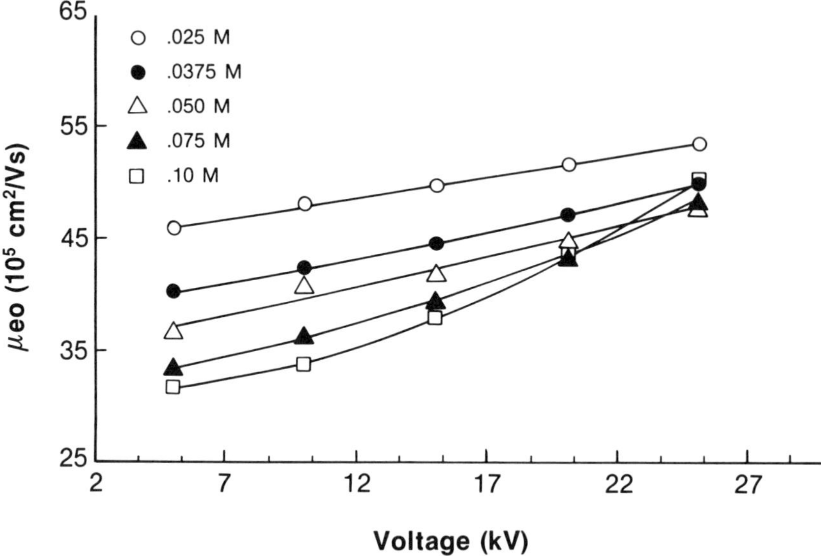

Figure 8 Electroosmotic mobility as a function of applied voltage. Buffer, acetate at pH 5.0; solute, mesityl oxide; column, as in Fig. 5. (From Ref. 31)

is significant only at a field strength in excess of 1×10^5 V cm^{-1}, which is far beyond what is normally used in CZE. Furthermore, rodlike molecules tend to increase their alignment in the direction of flow with increasing field strength; however, this effect is operative only if the shape of the solute molecules deviates significantly from spherical symmetry [43].

Effect of Buffer Concentration and Applied Voltage on Column Efficiency, Selectivity, and Resolution

Column efficiency in CZE is gauged by the number of theoretical plates (N) generated by the column. In practice, many experimental parameters, such as solute diffusion, finite sample injection volume, distortion of the flat flow profile by adsorption to capillary wall, and joule heating, can adversely affect column efficiency. A detailed examination of the factors that affect column efficiency is beyond the scope of this chapter. The subject has been extensively covered elsewhere [13,46–50].

In this section, we discuss the effect of applied voltage and buffer concentration on column efficiency, selectivity, and resolution under ideal experimental conditions for which the only operative zone-broadening mechanism is solute diffusion.

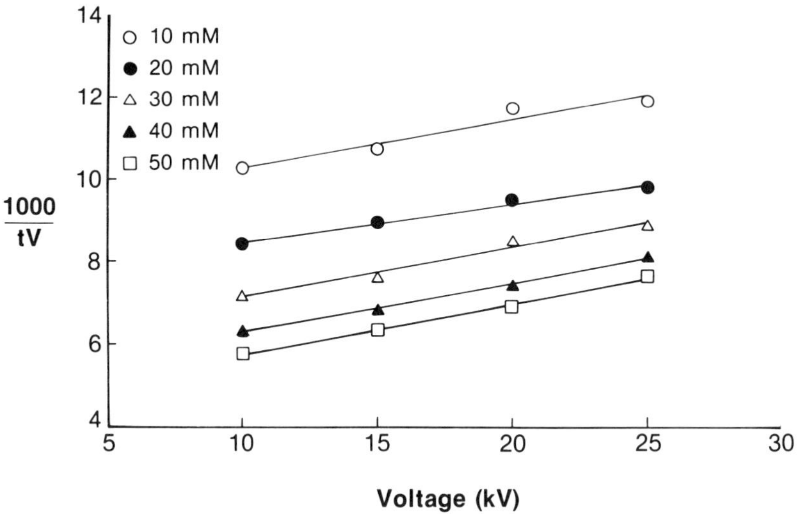

Figure 9 Migration time for dansyl leucine as a function of applied voltage at different buffer concentrations. Buffer, acetate at pH 5.0; field strength, 400 V cm^{-1}; column, 50 cm × 75 μm i.d. fused silica; instrument, Beckman Model P/ACE System 2000. (From Ref. 45)

If solute diffusion is solely responsible for zone broadening, the number of theoretical plates (N) is given by [13]:

$$N = \frac{L^2}{\sigma^2} \tag{25}$$

where L is the column length and σ^2 is the zone variance, which is given by the Einstein equation [13]:

$$\sigma^2 = 2D't_{\mathrm{m}} \tag{26}$$

where D' is solute diffusion coefficient and t_{m} is solute migration time. Substituting Eq. 26 in Eq. 25 yields an expression for N in terms of experimental parameters:

$$N = \frac{L^2}{2D't_{\mathrm{m}}} = \frac{\mu V}{2D'} \tag{27}$$

where $t_{\mathrm{m}} = L/v = L^2/\mu V$, and $\mu = (\mu_{\mathrm{ef}} + \mu_{\mathrm{eo}})$.

The resolution (R_{s}) of two close-lying zones in an electropherogram is given by [13]:

$$R_{\mathrm{s}} = \frac{1}{4}N^{1/2}\frac{\Delta v}{\bar{v}} \tag{28}$$

where $\Delta v/\bar{v}$ is the relative migration velocity difference of the two zones, which in terms of solute mobilities is equal to $(\mu_{ef}(1) - \mu_{ef}(2))/(\bar{\mu}_{ef} + \mu_{eo})$.

By substituting the expression for the relative velocity difference and the expression for the number of theoretical plates into Eq. 28, Jorgenson and Lukacs arrived at the following expression for resolution [44]:

$$R_s = 0.177(\mu_{ef}(1) - \mu_{ef}(2))\left[\frac{V}{D'(\bar{\mu}_{ef} + \mu_{eo})}\right]^{1/2} \quad (29)$$

According to Jorgenson and Lukacs, Eq. 27 suggests a misleading approach to improved separation efficiency because it promotes large values of μ_{eo} in the same direction as the electrophoretic mobility, whereas it is apparent from examination of Eq. 29 that this will decrease the actual resolution of the two zones. In conclusion, they argue that the best resolution is obtained when the electroosmotic flow just balances the electrophoretic migration ($\mu_{eo} = -\mu_{ef}$). At this point solutes with extremely small differences in mobility may be resolved, albeit, at a large expense in time. In support of this argument, Jorgenson and Lukacs presented Figs. 5 and 6 [44], showing that N improves with increasing voltage (i.e., shorter analysis time) and at the same time presented Fig. 7 of the same reference to show that the resolution improves with decreasing electroosmotic flow (longer analysis time) as a result of selectivity enhancement (larger relative velocity differences).

As is evident from Eq. 29, resolution in CZE may be optimized by controlling the electroosmotic flow, provided that the procedures employed do not adversely affect the electrophoretic mobility or column efficiency. Coating or chemically treating the capillary is one way of achieving this optimization [35,51], and applying an external electric field is another [52]. However, other means that are commonly used to affect electroosmotic flow, such as varying the buffer pH [36,40], applying buffer additives [36,53], changing the concentration of the buffer [31,36,41,42,45], and changing the applied voltage, will almost certainly affect electrophoretic mobility and column efficiency.

The effect of applied voltage on resolution was investigated by several groups [35,36,39,40,42,43]. By all accounts, experimentally determined resolution improves with increasing voltage, up to a certain point, beyond which the voltage is so high it contributes to zone broadening by the joule-heating effect. The effect of buffer concentration is not as dramatic as that of applied voltage [31,36,41,42,45]; however, it has been reported that resolution slightly increases with increasing buffer concentration [31,42,45].

To analyze the effect that changes in applied voltage and buffer concentration have on resolution, we derived the following expression, assuming that solute diffusion is the only zone-broadening mechanism [45]:

$$R_s = \left(\frac{L^2}{32D'}\right)^{1/2} \frac{\Delta t_m}{(t_m(2))^{3/2}} \quad (30)$$

where Δt_m is the migration time difference of the two zones, $t_m(2)$ is the migration time of the slower migrating solute, and the other terms are as defined earlier. Equation 30 expresses the resolution as a function of solute migration time. It is identical with Eq. 29 in substance, but differs in the choice of parameters. The two forms reduce to each other by simple algebraic manipulations considering the relation between the parameters as given in the previous sections.

Equation 30 was used to analyze the data for dansyl methionine–dansyl alanine (Table 1, Ref. 45). R_s values for this solute pair were calculated at different applied voltages and buffer concentrations as detailed [45]. The resolution was then plotted as a function of concentration and voltage, resulting in the resolution surface shown in Fig. 10. Isochrons (i.e., lines of constant analysis time) were also calculated and plotted in Fig. 10 (dark lines). These lines are used to predict the behavior of the resolution under constant analysis time. The $R_s - V$ plane in this figure is featureless, since R_s continuously increases with an increase in V at all concentrations. Buffer concentration effects are noted in the following observations. At the high-voltage

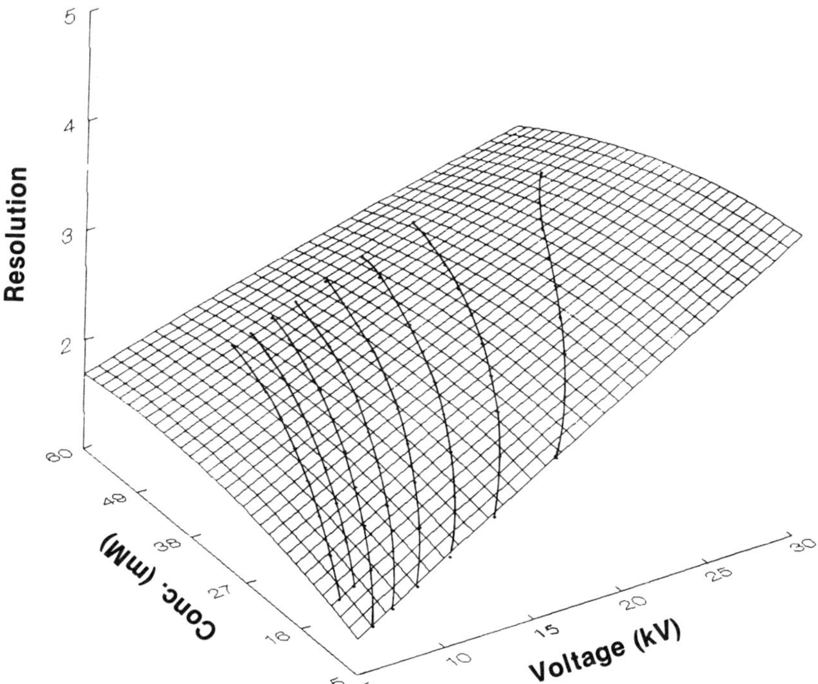

Figure 10 Calculated resolution surface for dansyl methionine–dansyl alanine in applied voltage and buffer concentration space. The dark nongrid lines on the surface are isochrons, from 300s (lower border) to 1000s (higher border) in increments of 100s. (From Ref. 45)

border, R_s decreases continuously with increasing solution concentration because the small, favorable increase in differential migration is overwhelmed by the large, unfavorable decrease in N. However, at the low-voltage border, R_s passes through a maximum as the concentration is increased. For example, at 10 kV, R_s is maximum at $C = 35$ mM, and at 15 kV the maximum R_s value is at $C = 20$ mM. This behavior is observed mainly because ΔT_m is small at high voltages and only marginally increases with increasing concentration, whereas at low voltages, ΔT_m is large and shows a relatively larger increase with increasing concentration.

The isochronal lines show that at constant analysis time, resolution is optimized by simultaneously increasing voltage and buffer concentration. However, a slight increase in voltage necessitates a large increase in buffer concentration to achieve higher resolution under isochronal conditions.

Effect of Buffer's pH on Mobility, Resolution, and Selectivity

The buffer's pH is a very important parameter that affects the quality of the separation. McCormick [54] found that increasing the pH from 1.5 to 6.0 resulted in an increase in μ_{eo} (Fig. 11). Others [55] found that t_m of water-soluble vitamins increased with an increase in pH from 7 to 10 (Fig. 12). Note also that resolution

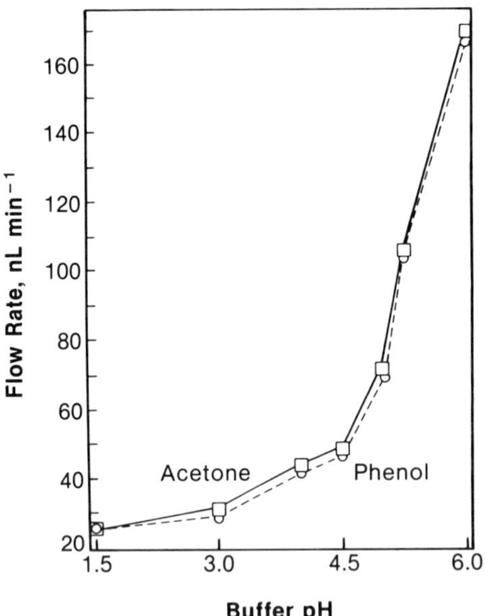

Figure 11 Electroosmotic flow as a function of buffer pH. (From Ref. 54)

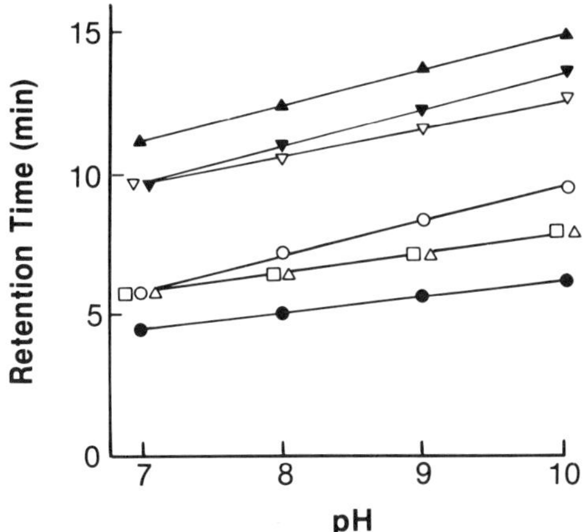

Figure 12 Effect of pH on migration time in CZE. (From Ref. 55)

improved at higher pH. Also, changes in selectivity were observed along with changes in pH. Figure 13 [56] shows the separation of a series of peptides. Note that peaks 8 and 10 shifted positions with changes in pH, whereas other experimental conditions remained constant. This phenomenon of changing selectivity by changing pH can be used to optimize the separation of a mixture. Figure 14 illustrates this point, for which baseline resolution of the mixture was achieved at pH 9.5 [57]. Therefore, it is important, when attempting to resolve a mixture, to verify the effect of pH, because as clearly shown in Figs. 12–14, small changes in pH can result in the separation of closely migrating solutes.

Effect of Temperature on Mobility and Resolution

As mentioned in the section on joule heating, one of the limiting factors in CZE is the heat inside the capillary. However, an experiment may be performed at higher or lower temperatures depending on the stability and the solubility of the solute in the buffer at a certain temperature. In general, at higher temperatures, solutes are more soluble in the buffer. Also, buffer viscosity decreases with an increase in temperature, which means that t_m will be faster at higher temperatures.

Figure 15 shows that t_m decreased by about 25% when the column temperature was increased by 20°C, from 19°C to 39°C [58]. This was also true for the t_m of dipeptides (Fig. 16). In addition, the resolution improves at lower temperature (Figs. 15 and 16). However, note that in Fig. 16, FD is resolved from FF and FY at 25°C. At the higher temperatures FF and FY comigrated, but their resolution from FD improved [59]. This means that in certain cases, resolution may improve with an increase in column temperature.

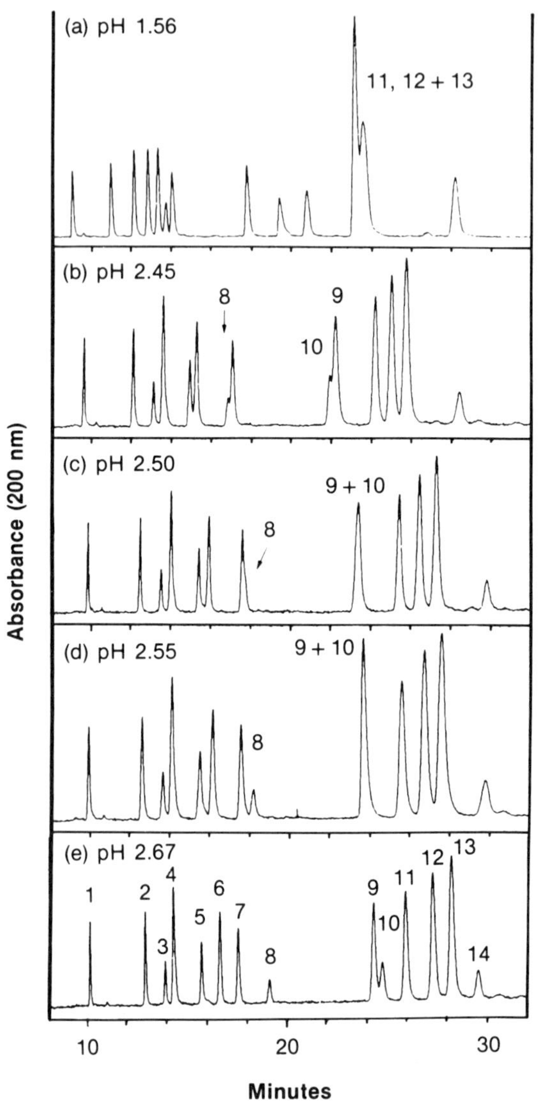

Figure 13 Effect of pH on the resolution of peptide mixture in CZE. (From Ref. 56)

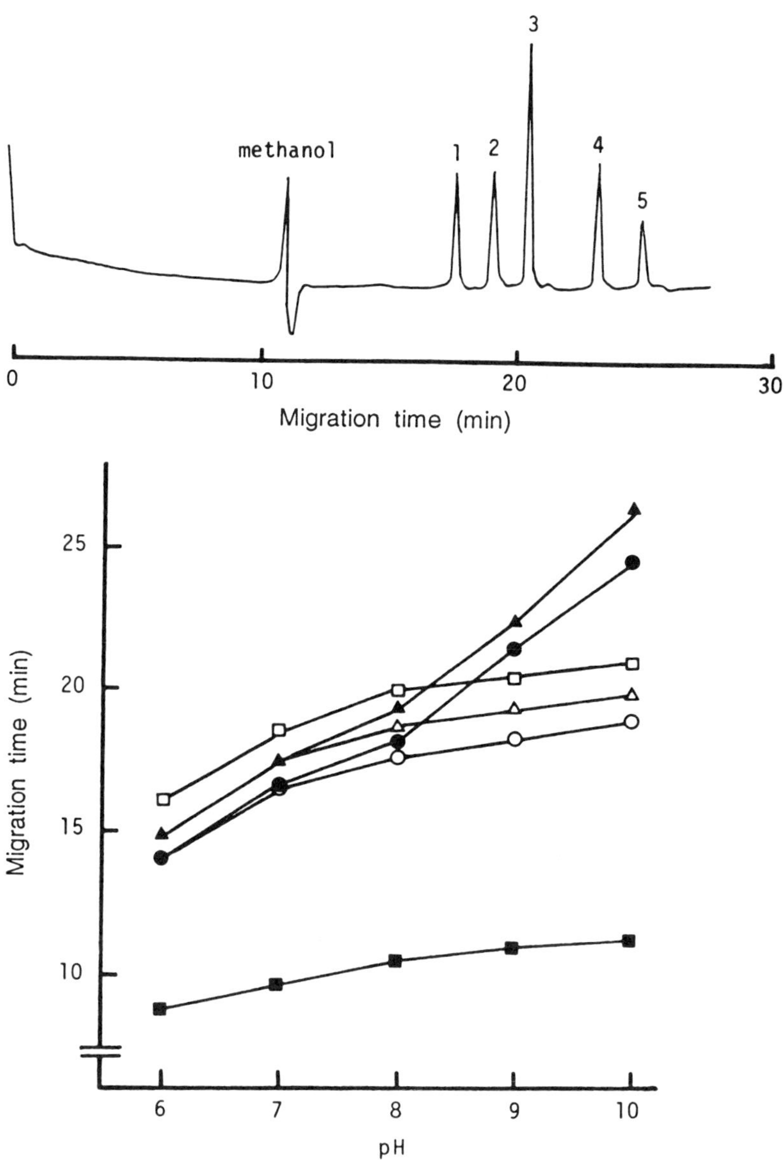

Figure 14 Effect of pH on the separation of cinnamic acid and its analogues. (bottom) effect of pH on retention, (top) electropherogram at pH 9.5. (From Ref. 57)

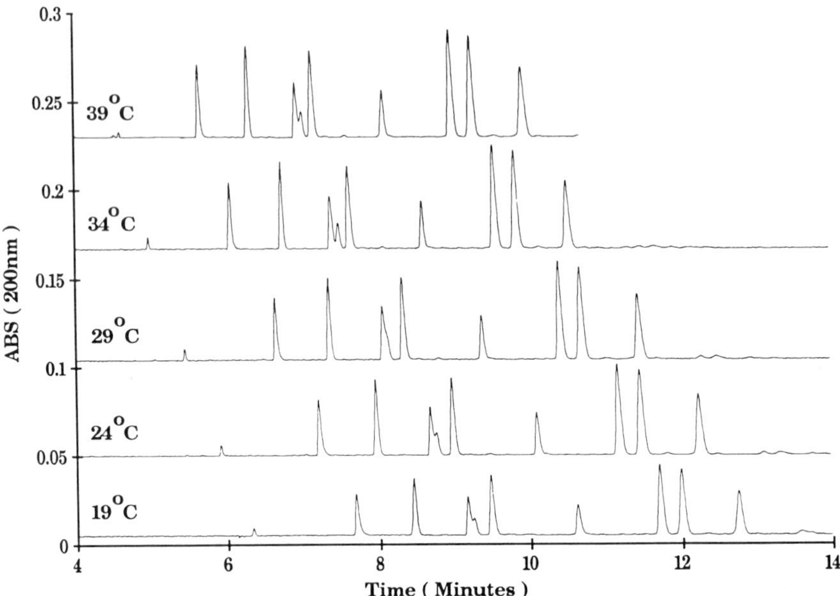

Figure 15 Peptide migration times as a function of capillary temperature. (From Ref. 58)

BUFFER ADDITIVES

In the previous section, the discussion focused on the parameters of unmodified buffers. In this section, the effect of using inorganic and organic additives to the buffer on migration time, resolution, and selectivity will be discussed.

Olechno [60] used a mixed buffer solution composed of two salts, sodium borate and sodium phosphate, in the same solution to resolve mixtures of antibacterials, sulfonamides, B_6 vitamins, and cephalosporins (Fig. 17). Two salts in the same buffer are normally used to extend the buffering capacity of the buffer solution. We have no data to compare for quality of the separation if these compounds were to be run in a single salt buffer. Also, a series of 11 anions were resolved by CZE in less than 6 minutes in a buffer made of $K_2Cr_2O_7$, $Na_2B_4O_7$, H_3BO_3, and diethylenetriamine at pH 7.8 [60].

The first chiral separation in CZE was reported by Gassman et al. [61] by whom a mixture of amino acid enantiomers was resolved by adding Cu(II)–L-histidine to the buffer solution. Also, the addition of copper sulfate and aspartame to an ammonium acetate buffer allowed the separation of four pairs of dansylated DL-amino acids (Fig. 18) [62].

The addition of organic solvents to the buffer solution to improve resolution and control mobility was studied [53,58,63]. It was observed that the addition of

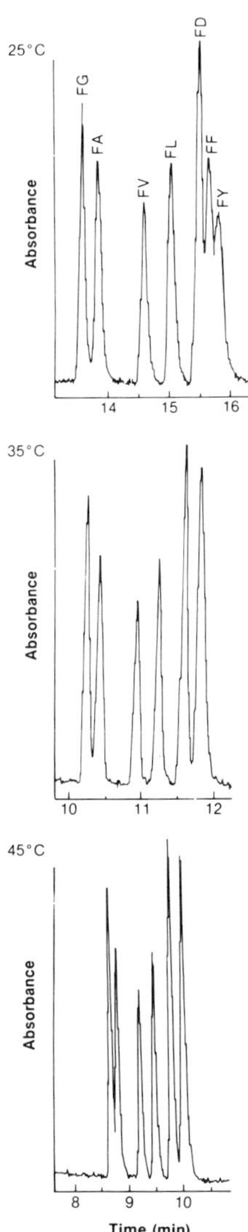

Figure 16 Separation of dipeptides as a function of temperature. (From Ref. 58)

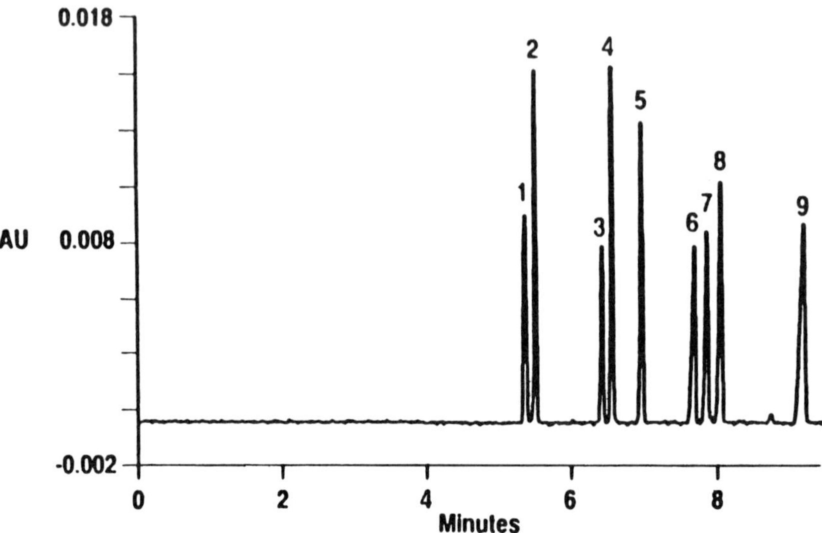

Figure 17 Electropherogram of the separation of cephalosporins and their precursors using a phosphate–borate buffer. (From Ref. 60)

methanol increased t_m (Fig. 19), whereas the addition of acetonitrile [Fig. 20] decreased it [58]. The change in mobility as a result of the addition of organic modifiers was mainly attributed to viscosity, assuming that viscosity has more influence on mobility than the dielectric constant or the zeta potential [63]. In another study, Fujiwara and Honda [53] found that μ_{eo} increased with the addition of acetonitrile, whereas it decreased with the addition of methanol. They also reported that both methanol and acetonitrile raised electrophoretic mobility and, as a result, better resolution was obtained. Figure 21 shows the effect of the addition of these two organic modifiers on retention times [53]. In all cases t_m decreased as the amount of acetonitrile in the buffer solution increased, as in Fig. 20, whereas it remained mainly constant with the addition of methanol. The results of the methanolic solution do not agree with those of Fig. 19. This might be because, although the buffer concentration was constant throughout the experiment (see Fig. 19), they were variable in Fig. 21. Also, although McLaughlin et al. [58] claimed that viscosity plays a role Fujiwara and Honda [53] claim that intramolecular hydrogen bonding plays a role. Figure 22 shows the effect of methanol and acetonitrile addition to the buffer on the resulting voltage at constant applied current [53]. The results show that the voltage increased by an increase in the organic modifier. These results, although accurate, may give an erroneous impression, since the authors, as just mentioned, did not keep the buffer concentration in all the solutions the same. This dilution in buffer concentration would lead to increase in the voltage.

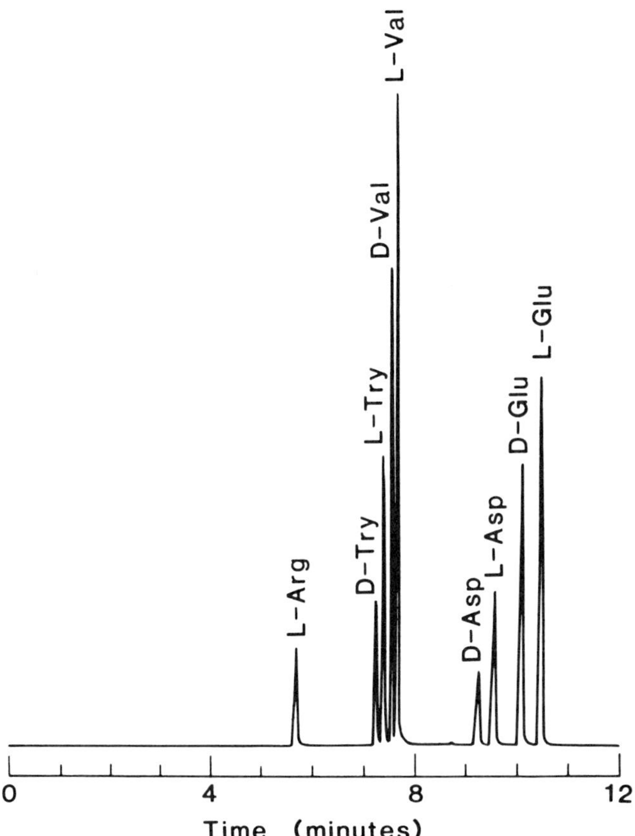

Figure 18 Electropherogram of a mixture of four DNS-DL-AAs. DNS-L-Arg is used as an internal standard. Electrolyte composition is as follows: 2.5 mM $CuSO_4 \cdot 5H_2O$, 5.0 mM aspartame, and 10 mM NH_4–OAc, pH 7.4; capillary, 75 μm i.d., 100 cm (75 cm to the detection zone); applied voltage, −30 kV; current, ~33 μA. (From Ref. 62)

The addition of a surfactant, such as SDS to the CZE buffer will allow the separation of neutral molecules. Terabe et al. [64] first introduced micellar electrokinetic chromatography (MECC, MEKC) in 1984 for the separation of neutral molecules. The theory of MECC is presented in Chapter 2 and will not be discussed here. Anionic, cationic, and nonionic surfactants have been used as pseudophases in MECC [65]. Selectivity in MECC can be manipulated by adding inorganic salts or organic solvents [65]. A few examples are given here.

Sepaniak and his co-workers [66], Fujiwara and Honda [67], and Terabe et al. [23] studied the effect of the addition of an organic solvent to the micellar phase. It was found that the addition of organic solvents to the pseudophase serves a few

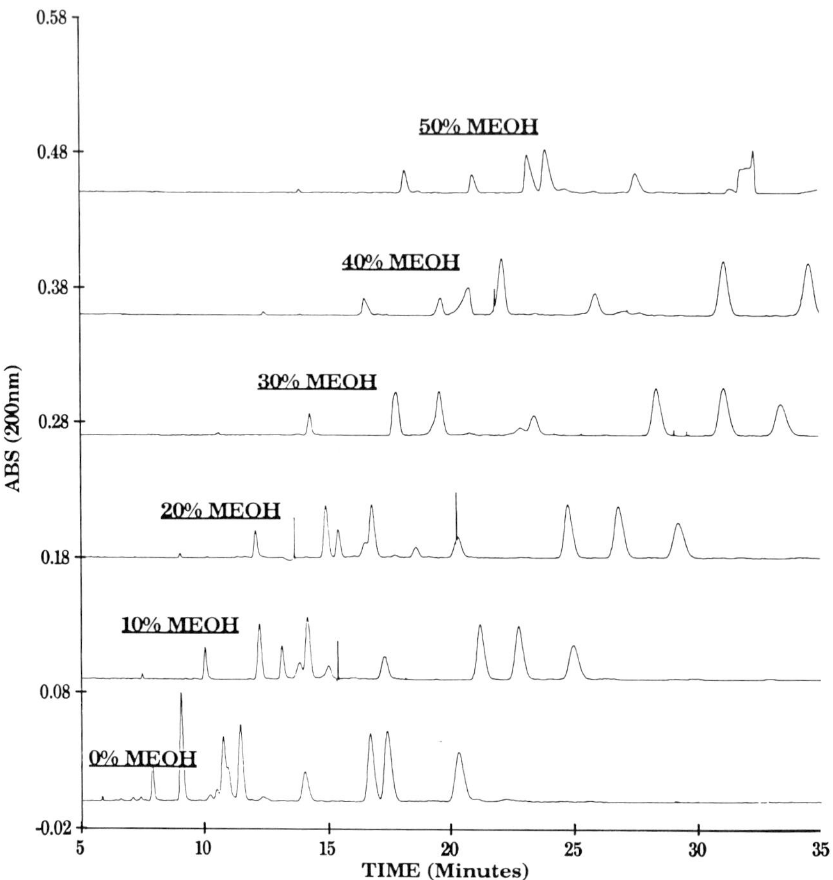

Figure 19 Effect of addition of methanol on migration time of bioactive peptides. (From Ref. 58)

purposes. Organic solvents interact with the capillary wall and, as a result, (a) slow down electroosmotic flow, (b) decrease the polarity of the mobile phase, (c) alter the partition coefficient, (d) may improve the selectivity, (e) extend the usefulness of the technique to more hydrophobic compounds, and (f) allow the development of gradient elution in MECC. Selectivity is enhanced by the addition of organic modifiers because the shift in equilibrium is greater for hydrophobic solutes compared with hydrophilic solutes [68]. However, the addition of an organic solvent at high percentages, above 15% (v/v) drops column efficiency, and migration times become impractical. It was found that analysis time in MECC is dependent on the amount

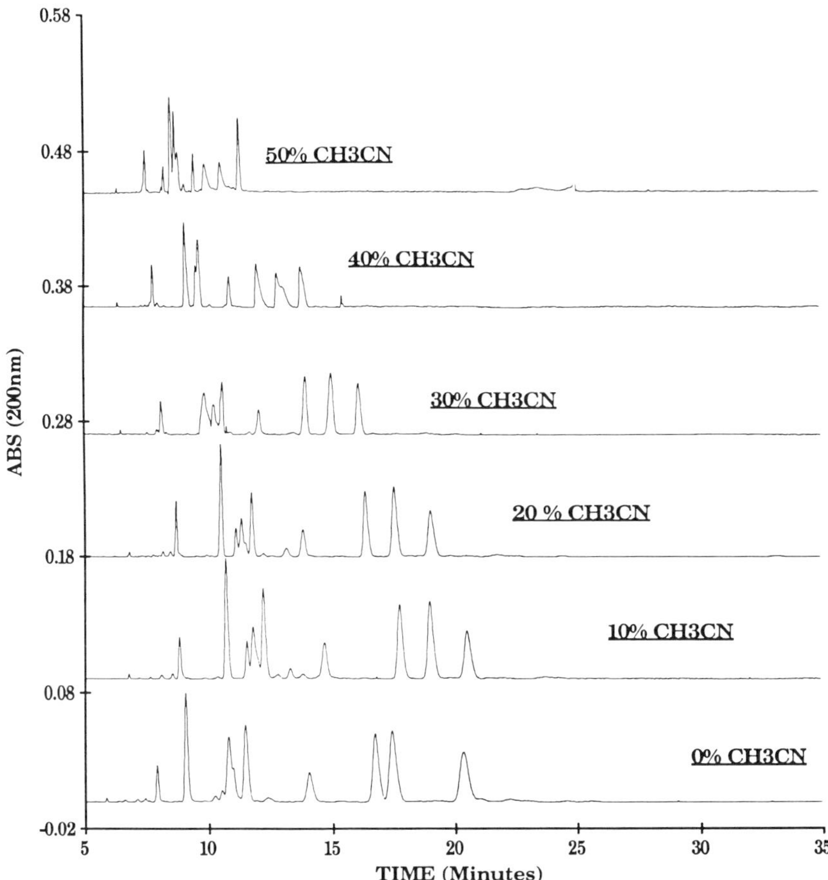

Figure 20 Effect of addition of acetonitrile on migration time of bioactive peptides. (From Ref. 58)

and type of organic modifier; methanol and isopropanol extended the elution range, whereas acetonitrile and dioxane did not appreciably affect the flow [66]. The addition of 20% (v/v) methanol to a phosphate–borate buffer, pH 8, containing 25 mM SDS, allowed the resolution of dansylated methylamine from dansylated methyl-d_3-amine in 83 min [69].

Balchunas and Sepaniak [66] were the first to report the use of gradient elution in MECC to resolve a mixture of ten primary and secondary amines by stepwise gradient elution. Later, Sepaniak et al. [70] reported the building of an apparatus

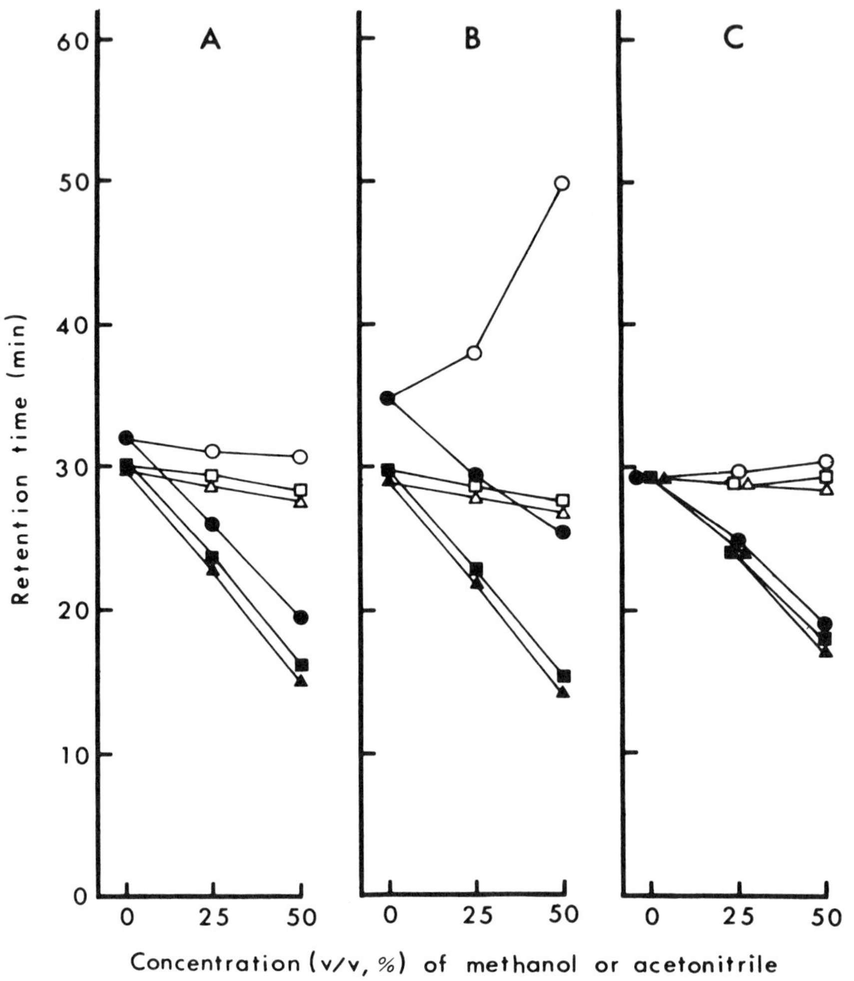

Figure 21 Retention times of substituted benzoic acids as a function of the concentration of methanol (open symbol) or acetonitrile (closed symbol): (○, ●) *ortho*, (□, ■) *meta*, (△, ▲) *para*; (A) aminobenzoic acid, (B) hydroxybenzoic acid, (C) methylbenzoic acid. (From Ref. 53)

for continuous gradient elution, which was used for the resolution of a mixture of alkyl amines.

Brij 35 [polyoxyethylene (23) dodecanol] is a nonionic surfactant that has little use in MECC because it cannot migrate electrophoretically. However, when added to an ionic surfactant, it can affect the separation. Also, Brij 35 can be added to

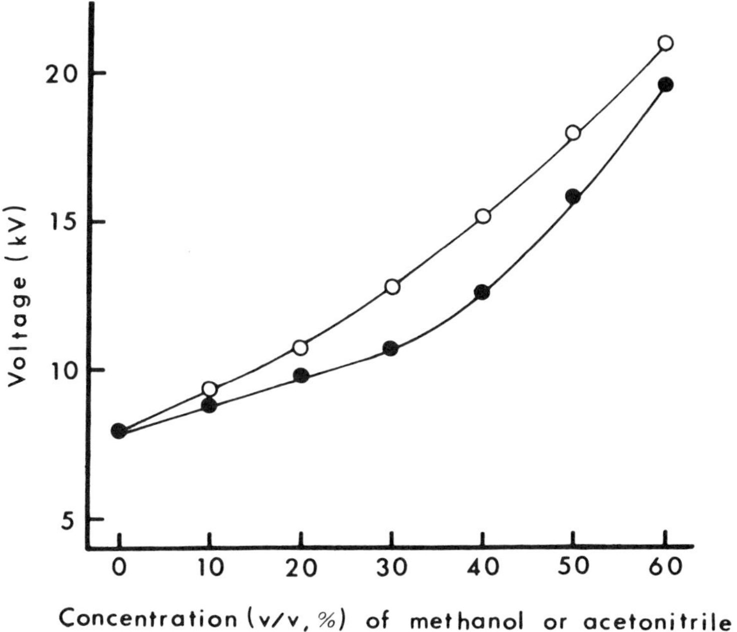

Figure 22 Relationship between voltage and concentration of organic solvent: open symbol, methanol; closed symbol, acetonitrile. (From Ref. 53)

the micellar phase without increasing the joule heating. Rasmussen et al. [71] were able to resolve benzene from benzaldehyde by adding 0.025 M Brij 35 to 0.025 M SDS in 0.01 M Na_2HPO_4.

Towns and Regnier [72] examined the effect of polyoxyethylene surfactant size and structure on protein exclusion and small molecular separation. Brij 35 and other polyoxyethylene surfactants (Tween 20, Tween 40, Tween 80, and Brij 78) were used to coat the polysiloxane inner surface and to create a hydrophilic network, which allows the separation of acidic and basic proteins [68]. When the headgroup size was held constant (Tween series), the hydrocarbon chain length had no influence on the capillary performance. However, there was a large change in capillary performance when headgroup size and structure were varied.

Nishi et al. [73] studied the effect of the addition of tetraalkylammonium salts (TAAS) to an SDS micellar phase on the separation of ionic substances. The results show that the addition of TAAS to the SDS solution resulted in better separation of the solutes in the mixture. For example, the separation of nine closely related antibiotics, having both anionic and cationic groups in a molecule, were best resolved when TAAS (40 mM) was added to the SDS (50 mM) micellar solution. The addition of these salts to the micellar system decreases the migration times of cationic solutes and increases the migration times of anionic solutes. It was suggested that an ion-pair type mechanism is involved whereby the TAAS ions are likely to combine with the anionic SDS micelle as a counterion, instead of the sodium ion, and alter the character of the micelles.

Yu and Dovichi [74] were able to resolve a mixture of dabsyl amino acids using a 1:1 acetonitrile/20-mM phosphate buffer (pH 7.0) that contained 5 mM SDS. Of the 18 amino acids injected two pairs coeluted, isolucine and valine, and cysteine and tyrosine. Cohen et al. [75] resolved a chiral mixture of dansylated amino acids in MECC by complexation with Cu(II) and *N*,*N*-didecyl-L-alanine. Dobashi et al. [76] reported the MECC separation of racemic mixtures of *N*-acetylated amino acids by employing a chiral micelle made of surfactants derivatized with L-amino acids. The best results were obtained when *N*-dodecanoyl-L-valinate was used. Modification of the micellar solution with methanol affected the selectivity of the solutes. Otsuka et al. [77] investigated the use of SDVal for chiral separations in MECC, and reported that the addition of SDS, methanol, and urea improved the peak shapes and resolution, and affected the selectivity.

Gozel et al. [78] were able to use a micellar solution composed of 20 mM sodium tetradecylsulfate (STS; pH 7.81), 2.5 mM copper sulfate, 5.0 mM aspartame, and 10 mM ammonium acetate to resolve a chiral mixture of dansylated D-amino acids from L-amino acids. The addition of STS resulted in the resolution of the enantiomer pairs, which were not completely separated in its absence.

Terabe and Isemura [79] resolved a mixture of naphthalene disulfonates by modifying the buffer with a soluble cationic polymer. The polymer acts as a pseudostationary phase, mimicking an ion exchanger.

Nishi et al. [73] showed that the addition of tetraalkylammonium salts to the micellar solution improved the separation of some ionic substances. Terabe et al. [80] found that the addition of 10 mM SDS to the micellar bile salt solution cut the retention time in half, without considerably affecting the resolution. Karger and his co-workers [81] resolved 14 out of a mixture of 18 oligonucleotides of eight bases, each with a different sequence, in less than 30 minutes by adding zinc and SDS to the buffer system. Their explanation is that the addition of low concentrations of divalent metals and SDS to the buffer system leads to a significant enhancement of the time window and good separation of the oligonucleotides. Metal ions are electrostatically attracted to the surface of the micelle, where they can be selectively complexed with appropriate solutes, which can manipulate selectivity. Zinc(II) was used because it is intermediate in oligonucleotide-binding strength between Mg(II) and Cu(II). Zinc(II) is also transparent in the range of oligonucleotide detection [82]. Copper(II)-L-histidine and Cu(II)-aspartame [78] were used for chiral separations in MECC.

Cyclodextrins (CD) have been used as multimodal phases for the separation of small molecules [83] and chiral compounds in HPLC [84,85], GC [86], isotachophoresis [87], and gel-filled HPCE [88]. Terabe et al. [89] were the first to use an ionic CD derivative as a substitute for the micelle in MECC, which can form inclusion complexes. They modified a 0.1 M phosphate buffer (pH 7.0) with 0.025 M 2-*O*-carboxymethyl-β-cyclodextrin and were able to resolve a series of substituted benzene isomers. Fanali investigated the effect of β-CD and derivatized β-CD in the buffer solution on the resolution of optical isomers [82]. The results show poor resolution of the enantiomers of ephedrine and isoproterenol, although high concentrations of β-CD were added to the buffer solution. However, the addition of 18 mM Heptakis (2,6-di-*O*-methyl-β-CD) to the buffer (10 mM Tris-H_3PO_4, pH 2.4) resulted in the separation of the optical isomers studied. Moreover, Terabe et al. [90] explored the applicability of CD-MECC to the separation of highly hydrophobic and closely related compounds, such as polychlorinated biphenyls (PCBs) tetrachlorodibenzodioxin (TCDD) isomers and polycyclic aromatic hydrocarbons (PAHs). The results show that the addition of 40 mM γ-CD to the separation solution enabled the separation of all chlorinated benzene congeners, which were not resolved in the absence of γ-CD. Eleven trichlorobiphenyl isomers that comigrated with the micelle were completely resolved when 60 mM γ-CD was added. The TCDD isomers were also resolved by γ-CD MECC. Nishi and Matsuo [91] reported that the addition of CD to the SDS solution improved the resolution of lipophilic compounds, corticosteroids, and aromatic hydrocarbons.

It is clear from the foregoing discussion that not only buffer pH, type, ionic strength, and temperature affect resolution and selectivity, but organic additives also play an important role.

ACKNOWLEDGMENT

Research sponsored by the National Cancer Institute, Department of Health and Human Services, under contract N01-CO-74102 with PRI/DynCorp. The content of this publication does not necessarily reflect the views or policies of the Department of Health and Human Services, nor does mention of trade names, commercial products, or organizations imply endorsement by the US Government.

REFERENCES

1. S. P. Colowick and N. O. Kaplan, eds., *Methods in Enzymology*, Vol. 1, Academic Press, New York (1955).
2. C. Long, ed., *Biochemist's Handbook*, Van Nostrand, Princeton, New Jersey (1961).
3. R. G. Bates, *Ann. N.Y. Acad. Sci.*, *92*:341 (1961).
4. P. J. Elving, J. M. Markowitz, and I. Rosenthal, *Anal. Chem.*, *28*:1179 (1956).
5. N. E. Good, G. D. Winget, W. Winter, T. N. Connolly, S. Izawa, and R. M. M. Singh, *Biochemistry*, *5*:467 (1966).
6. D. D. Perrin, *Aust. J. Chem.*, *16*:572 (1963).

7. D. D. Perrin and B. Dempsey, *Buffers for pH and Metal Ion Control*, Holstand, New York, Chap. 5 (1974).
8. W. J. Lambert, *J. Chem. Educ.*, *67*:105 (1990).
9. D. E. Gueffroy, "Buffers, A Guide for the Preparation and Use of Buffers in Biological Systems," Monograph, Behring Diagnostics, Doc. No. 4140-285, La Jolla, California (1985).
10. P. W. Atkins, *Physical Chemistry*, 3rd ed., Oxford University Press, Oxford (1986).
11. R. A. Wallingford and A. G. Ewing, *Adv. Chromatogr.*, *29*:1 (1990).
12. J. H. Knox, *Chromatographia*, *26*:329 (1988).
13. J. C. Giddings, *Sep. Sci*, *4*:181 (1969).
14. R. J. Wieme, *Chromatography, A Laboratory Handbook of Chromatrographic and Electrophoretic Methods*, 3rd ed. (E. Heftman, ed.), Van Nostrand Reinhold, New York, Chap. 10 (1975).
15. S. Terabe, K. Otsuka, and T. Ando, *Anal. Chem.*, *57*:834 (1985).
16. J. H. Knox and I. H. Grant, *Chromatographia*, *24*:135 (1987).
17. E. Grushka, R. M. McCormick, and J. J. Kirkland, *Anal. Chem.*, *61*:241 (1989).
18. Y. Kurosu, K. Hibi, T. Sasaki, and M. Saito, *J. High Resolut. Chromatogr.*, *14*:200 (1991).
19. D. S. Burgi, K. Salomon, and R. L. Chien, *J. Liq. Chromatogr.*, *14*:847 (1991).
20. P. Debye and E. Huckel, *Physik. Z.*, *24*:305 (1923).
21. P. Debye and E. Huckel, *Physik. Z.*, *24*:185 (1923).
22. T. Tsuda, K. Nomura, and G. Nakagawa, *J. Chromatogr.*, *248*:241 (1982).
23. S. Terabe, H. Utsumi, K. Otsuka, T. Ando, T. Inomata, S. Kuze, and Y. Yanaoka, *J. High Resolut. Chromatogr.*, *9*:666 (1986).
24. T. Tsuda, *J. High Resolut. Chromatogr.*, *10*:622 (1987).
25. W. D. Pfeffer and E. S. Yeung, *Anal. Chem.*, *62*:2178 (1990).
26. T. S. Stevens and H. J. Cortes, *Anal. Chem.*, *55*:1365 (1983).
27. V. Pretorius, B. J. Hopkins, and J. D. Schieke, *J. Chromatogr.*, *99*:23 (1974).
28. E. Hückel, *Physik K.*, *25*:204 (1924).
29. D. C. Henry, *Proc. R. Soc. (Lond.)*, *133*:106 (1931).
30. I. Z. Atamna, C. J. Metral, G. M. Muschik, and H. J. Issaq, *J. Liq. Chromatogr.*, *13*:2517 (1990).
31. H. J. Issaq, I. Z. Atamna, G. M. Muschik, and G. M. Janini, *Chromatographia*, *32*:155 (1991).
32. J. S. Green and J. W. Jorgenson, *J. Chromatogr.*, *478*:63 (1989).
33. B. B. Van Orman, T. M. Olefirowicz, G. G. Liversidge, A. G. Weing, and G. L. McIntire, Poster P-116 at the Second International Symposium on High Performance Capillary Electrophoresis, San Francisco, January (1990).
34. I. Z. Atamna, C. J. Metral, G. M. Muschik, and H. J. Issaq, *J. Liq. Chromatogr.*, *13*:3201 (1990).
35. G. J. M. Bruin, J. P. Chang, R. H. Kuhlman, K. Zegers, J. C. Kraak, and H. Poppe, *J. Chromatogr.*, *471*:429 (1989).
36. K. D. Altria and C. F. Simpson, *Chromatographia*, *24*:527 (1987).
37. K. D. Altria and C. F. Simpson, *Anal. Proc.*, *25*:85 (1988).
38. W. Nashabeh and Z. El-Rassi, *J. Chromatogr.*, *514*:57 (1990).

39. T. Tsuda, K. Nomura, and G. Nakagawa, *J. Chromatogr.*, *264*:385 (1983).
40. K. D. Lukacs and J. W. Jorgenson, *J. High Resolut. Chromatogr.*, *8*:407 (1985).
41. K. D. Altria and C. F. Simpson, *Anal. Proc.*, *23*:453 (1986).
42. H. T. Rasmussen and H. M. McNair, *J. Chromatogr.*, *516*:223 (1990).
43. P. D. Grossman and D. S. Soane, *Anal. Chem.*, *62*:1593 (1990).
44. J. W. Jorgenson and K. D. Lukacs, *Anal. Chem.*, *53*:1298 (1981).
45. I. Z. Atamna, H. J. Issaq, G. M. Muschik, and G. M. Janini, *J. Chromatogr.*, *588*:315 (1991).
46. F. Foret, M. Deml, and P. Bocek. *J. Chromatogr.*, *452*:601 (1988).
47. E. Grushka and R. M. McCormick, *J. Chromatogr.*, *471*:421 (1989).
48. X. Huang, W. F. Coleman, and R. N. Zare, *J. Chromatogr.*, *480*:95 (1989).
49. G. O. Roberts, P. H. Rhodes, and R. S. Snyder, *J. Chromatogr.*, *480*:35 (1989).
50. K. Otsuka and S. Terabe, *J. Chromatogr.*, *480*:91 (1989).
51. S. Hjerten, *J. Chromatogr.*, *347*:191 (1985).
52. C. S. Lee, W. C. Blanchard, and C.-T. Wu, *Anal. Chem.*, *62*:1550 (1990).
53. S. Fujiwara and S. Honda, *Anal. Chem.*, *59*:487 (1987).
54. R. M. McCormick, *Anal. Chem.*, *60*:2322 (1988).
55. S. Fujiwara, S. Iwase, and S. Honda, *J. Chromatogr.*, *447*:133 (1988).
56. M. Strickland and N. Strickland, *Am. Lab.*, Nov.: 64 (1990).
57. S. Fujiwara and S. Honda, *Anal. Chem.*, *58*:1811 (1986).
58. G. M. McLaughlin, J. A. Nolan, J. L. Lindahl, R. H. Palmieri, K. W. Anderson, S. C. Morris, J. A. Morrison, and T. J. Bronzert, *J. Liq. Chromatogr.*, *15*: (in press, 1992).
59. H. J. Issaq, G. M. Janini, I. Z. Atamna, and G. M. Muschik, *J. Liq. Chromatogr.*, *15*:1129 (1992).
60. J. Olechno, Dionex Corporation, personal communication (1991).
61. E. Gassman, J. C. Kuo, and R. N. Zare, *Science*, *230*:813 (1985).
62. P. Gozel, E. Gassmann, H. Michelson, and R. Zare, *Anal. Chem.*, *59*:44 (1987).
63. A. S. Cohen, A. Paulus, and B. L. Karger, *Chromatographia*, *24*:15 (1987).
64. S. Terabe, K. Otsuka, K. Ichikawa, A. Tsuchiya, and T. Ando, *Anal. Chem.*, *56*:111 (1984).
65. G. M. Janini and H. J. Issaq, *J. Liq. Chromatogr.*, *15*:927 (1992).
66. A. T. Balchunas and M. J. Sepaniak, *Anal. Chem.*, *60*:617 (1988).
67. S. Fujiwara and S. Honda, *Anal. Chem.*, *59*:2773 (1987).
68. F. P. Tomasella, J. Fett, and L. J. Clinelove, *Anal. Chem.*, *63*:474 (1991).
69. M. M. Bushey and J. W. Jorgenson, *J. Micorcol. Sep.*, *1*:125 (1989).
70. M. J. Sepaniak, D. F. Swaile, and A. C. Powell, *J. Chromatogr.*, *480*:185 (1989).
71. H. T. Rasmussen, L. K. Goebel, and H. M. McNair, *J. Chromatogr.*, *517*:549 (1990).
72. J. K. Towns and F. E. Regnier, *Anal. Chem.*, *63*:1126 (1991).
73. H. Nishi, N. Tsumagari, and S. Terabe, *Anal. Chem.*, *61*:2434 (1989).
74. M. Yu and N. J. Dovichi, *Anal. Chem.*, *61*:37 (1989).
75. A. Cohen, A. Paulus, and B. Karger, *Chromatographia*, *24*:15 (1987).
76. A. Dobashi, T. Ono, S. Hara, and T. Yamaguchi, *J. Chromatogr.*, *480*:413 (1989).
77. K. Otsuka, J. Kawahara, K. Tetekawa, and S. Terabe, *J. Chromatogr.*, *559*:209 (1991).
78. P. Gozel, E. Gassman, H. Michelson, and R. N. Zare, *Anal. Chem.*, *59*:44 (1987).

79. S. Terabe and T. Isemura, *Anal. Chem.*, *62*:650 (1990).
80. S. Terabe, M. Shibata, and Y. Miyashita, *J. Chromatogr.*, *480*:403 (1989).
81. A. S. Cohen, S. Terabe, J. A. Smith, and B. Karger, *Anal. Chem.*, *59*:1021 (1987).
82. S. Fanali, *J. Chromatogr.*, *474*:441 (1989).
83. J. Szejtli, "Cyclodextrins and Their Inclusion Complexes," Akademiai Kiado, Budapest, (1982).
84. H. J. Issaq, *J. Liq. Chromatogr.*, *11*:2131 (1988).
85. D. W. Armstrong, M. Hilton, and L. Coffin, LC–GC 9:646 (1991), and references therein.
86. D. W. Armstrong, W. Li, C. D. Chang, and J. Pitha, *Anal. Chem.*, *62*:914 (1990).
87. J. Snopek, I. Jelinek, and E. Smolokova-Keulemansova, *J. Chromatogr.*, *411*:153 (1987).
88. A. Gutman, A. Paulus, A. Cohen, N. Grinberg, and B. Karger, *J. Chromatogr.*, *448*:41 (1988).
89. S. Terabe, H. Ozaki, K. Otsuka, and T. Ando, *J. Chromatogr.*, *332*:211 (1985).
90. S. Terabe, Y. Miyashita, O. Shibata, E. R. Benhart, L. R. Alexander, D. B. Patterson, B. Karger, K. Hosoya, and N. Tanaka, *J. Chromatogr.*, *516*:23 (1990).
91. H. Nishi and M. Matsuo, *J. Liq. Chromatogr.*, *14*:973 (1991).

5

Organic Solvents in Capillary Electrophoresis

Ernst Kenndler

Institute of Analytical Chemistry, University of Vienna
Vienna, Austria

Organic solvents are of potential interest for capillary electrophoresis of small ions for several reasons. In the most simple case, they are used to enlarge the range of applications because of the enhanced solubility of analytes in these media, especially for organic solutes. In a more subtle way, greater flexibility for the adjustment of the separation selectivity of the system is gained when using organic or mixed aqueous–organic solvents. This flexibility is based on the influence of these solvents on the physicochemical parameters decisive for separation.

Before discussing in detail the particular effects and their individual importance, the advantage using organic solvents will be approached in a pragmatic way, namely, by treating the question of the electrophoretic similarity of buffer systems of different pHs, consisting either of pure aqueous solutions or of mixed aqueous–organic solvents. It is obvious that this discussion will be directed to small ions rather than to large molecules such as biopolymers.

SOLVENT SELECTION

In a general approach the similarity of such systems can be expressed (in the same way as in chromatography or spectrometry) by numerical values derived by procedures based on chemometry (cf. eg. [1]). The central point is to answer the question of how the particular systems relate to each other relative to the electrophoretic migration properties of the separands [2]. This interrelation must be calculated (when the commonly used correlation coefficient is only a measure for its linearity), and finally depicted in a form useable for interpretation of the similarity of the buffer systems. From these results, a selection of the appropriate combinations of different buffer systems can be rationally made.

Several algorithms are described for such evaluations. The discussion presented here will be focused on cluster analysis. This procedure includes several steps: selection of the data units, feature selection, definition of the measure of similarity, construction of the dissimilarity matrix, selection of a clustering algorithm for the agglomeration of the units, visualization, for example by a dendrogram.

A dendrogram resulting from such a hierarchical clustering procedure with a single linkage algorithm [2] is shown in Fig. 1. For the interpretation of this graph it must be pointed out that the Δ value shown is a measure of the similarity: identical systems are linked together with a Δ value of 1, totally dissimilar systems with a value of 0. The larger the Δ value is where the systems are linked together, the larger is their similarity.

It can be seen from Fig. 1 that the pure aqueous buffer systems with pH 6, 7, 8, and 9 are very similar, when the migration properties of the solutes are considered as the taxonomic attributes. These systems form a cluster with a high value of Δ. It can also be seen that the two buffers with methanol as solvent (see buffers 9* and 10*), which are found in the same subcluster, are linked to all other aqueous systems at the lowest Δ value given in the dendrogram (0.45): they have the lowest similarity compared with these buffers. This demonstrates clearly the different electrophoretic properties of the aqueous and the methanolic systems and supports the use of organic solvents to enhance the selectivity of buffers for the electrophoresis of small ions.

Classification of Solvents

Many solvents are described in the physicochemical literature as a medium for ion transport and acid–base phenomena, and several excellent reviews have been published on this topic [3–8]. The discussion presented here concerning the classification of solvents is closely related to the clear treatments of this problem given by Bates [3] and King [4], respectively.

The solvents can be classified according to several schemes, which are more or less arbitrary and depend on the particular subject of interest, such as solvation, ion association, acid–base equilibria, or other. One of the classification schemes is based primarily on the dielectric constant, ε, with a value of 30, for example, as the distinction between two main classes. The scheme of the most important solvents based on this concept is shown in Table 1. It leads to the one class with high ε (larger than 30), often termed *polar solvents*, and the other class with lower values for ε, called *nonpolar solvents*. In the former class ion pairing is negligible, and the strength of an acid in these solvents can be described by a numerical value that is independent of the base with which it reacts. In the latter class, ion pairing can

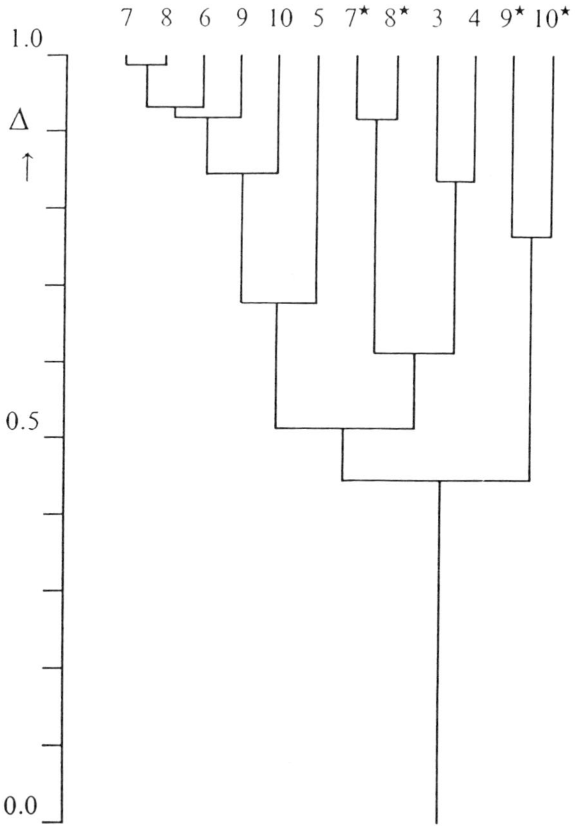

Figure 1 Dendogram demonstrating the similarity of buffer electrolytes with water or methanol as solvents. The pH of the buffers are indicated by the numbers on the top of the graphs. Methanolic buffers are marked with an asterisk. The dendrogram was calculated by cluster analysis from the relative mobilities of 55 anions. △, normalized euclidian distance. (From Ref. 2)

play a dominant role, and the acidity of the solute depends on the reacting base. The choice of the value 30 separates methanol from the other alcohols; they are found in two different classes. Each of these two classes can be divided further into two subclasses, namely into *hydrogen-bonded* (*protic*) and *non–hydrogen-bonded* (*aprotic*) solvents. The solvents with high ε, which are unable to donate H-bonds (although they can accept H-bonds) are called *aprotic, dipolar solvents*. Typical members of this

Table 1 Classification of Solvents

High ε		Low ε	
H-bonded	Non–H-bonded	H-bonded	Non–H-bonded
Water	Acetonitrile	Ethanol	Acetanhydrid
Deuterium oxide	Dimethylacetamide	*n*-Propanol	Acetone
Hydrogen peroxide	Dimethylformamide	*iso*-Propanol	Benzene
Methanol	Hexamethylphosphorotriamide	*n*-Butanol	Chloroform
Ethylene glycol	Dimethyl sulfoxide	*tert*-Butanol	Carbon tetrachloride
	Nitromethane	Methyl cellusolve	Chlorobenzene
Sulfuric acid	Nitrobenzene		Cyclohexane
Fluorosulfuric acid		Acetic acid	Dioxane
Hydrofluoric acid	Propylene carbonate	Trichloroacetic acid	Ethyl ether
Formic acid	Sulfolane		*n*-Hexane
Hydrocyanic acid		Ammonia	Pyridine
		Ethylenediamine	Sulfur dioxide
Ethanolamine			Tetrahydrofuran
Formamide			
Acetamide			
N-Methylformamide			
N-Methylacetamide			

Source: Ref. 4.

class are dimethyl sulfoxide (DMSO) and acetonitrile. Although acetone is not found in this class, it behaves similar to the aprotic, dipolar solvents and thus, is placed in this class. This example also demonstrates the somewhat arbitrary classification scheme.

Each of the subclasses can be further subdivided according to the acid–base properties of the solvents. These can be protogenic, protophilic, or amphiprotic. For a more extended discussion the reader is referred to the special literature [3–8].

Solvents Used in Capillary Electrophoresis

For the application of solvents (either pure or as constituents of an aqueous–organic mixture) several properties other than those mentioned in the foregoing are decisive in capillary electrophoresis. Because a low content, or even traces, of water can drastically influence the physicochemical properties of these solvents, it is often more favorable not to apply water-free solvents (the purification of is often rather complicated, or they are not easily kept pure, especially when they are hygroscopic), but to work with mixed organic–aqueous solvents with a defined and reproducible water content. Another advantage for use of such mixed solvents is the possibility to gradually change a certain property of interest (e.g., p*K*, or actual mobility).

It follows that, in general, the miscibility with water is one decisive factor for the selection of organic solvents for electrophoresis from the list given in Table 1. Other factors will be the chemical stability against water, a sufficient solubility to buffer substances, the ease of purification, chemical inertness against the separands, appropriate properties for handling (boiling point, viscosity), low toxicity (and low price). Given these criteria, the many solvents presented in Table 1 are reduced to only a few solvents for potential use in electrophoresis (Table 2).

These solvents are the same as those applied in high-performance liquid chromatography (HPLC), although for other reasons, namely, to adjust the selectivity and the retention range (the analysis time) in that chromatographic method by affecting the partition coefficients of the solutes. In capillar electrophoresis these solvents are used for other purposes, namely, to influence the effective mobilities of the separands, either by the actual mobilities or (and) the p*K* values (or other dissociation constants; e.g., for ion pairing) to achieve an appropriate electrophoretic resolution. The discussion presented here will focus on zone electrophoresis and isotachophoresis (displacement electrophoresis) because, for these methods, the effective mobility is the central property required for separation.

Isoelectric focusing and electrokinetic chromatography will not be a part of this contribution. Neither method applies effective mobility as a parameter for electromigration. Furthermore, in isoelectric focusing, the separands are almost exclusively biopolymers, for which the application of organic solvents is usually coun-

Table 2 Physical Properties of Solvents Applicable to Capillary Electrophoresis

Solvent	Molecular mass	b.p. (°C)	f.p (°C)	ρ (g/cm^3)	ε	η (cP)	μ (Debye)
Water	18.02	100.0	0.0	0.9971	78.30	0.8903	1.85
Methanol	32.04	64.6	−97.8	0.7866	32.70	0.5445	1.70
Ethanol	46.07	78.3	−114.5	0.7851	24.55	1.089	1.69
Acetonitrile	41.05	81.6	−44.9	0.7768	35.95	0.3409	3.92
N,*N*-Dimethylformamide	73.10	153	−61	0.9443	36.71	0.796	3.86
Dimethyl sulfoxide	78.13	189	18.5	1.0958	46.7	1.96	3.96
Acetone	58.08	56.2	−94.6	0.7850	20.70	0.3040	2.88
1,4-Dioxan	88.11	101.3	11.8	1.0269	2.209	1.196	0.45
Tetrahydrofurane	72.11	66	−108.5	0.8811	7.58	0.460	1.75

b.p., boiling point; f.p., melting point; ρ, density; ε, dielectric constant; η, viscosity coefficient; μ, dipole moment. Values refer to 25°C.

Source: Data from Ref. 5.

terproductive. The aspect of changing the electroosmotic flow by the solvent composition is, on the other hand, of substantial interest for electrokinetic chromatography.

When one considers the decisive factors for the selection of organic solvents for use in electrophoresis as generally discussed in the foregoing, it is remarkable that in the literature no solvents other than those given in Table 2 have been described, namely methanol [9–19], acetonitrile [19–22], DMSO [15–18], dimethylformamide [23], acetone [18], tetrahydrofurane [19], ethanol [19,20], propanol [19,20], and dioxan [20].

When electrophoresis is carried out under experimental conditions during which electroosmosis occurs, the organic solvents added to affect the separand properties can automatically influence the magnitude of the electroosmotic flow. Not only the electroosmotic mobility of the bulk liquid phase is changed, but also its pH dependence. A part of these effects can be explained by the shift in pK of the dissociable surface groups located on the capillary wall, which is one of the reasons for the existence of an electric double layer.

It can be concluded that the following properties can be expected to be influenced when applying organic solvents as constituents of the buffer solution:

Absolute or actual mobilities
pK values
Electroosmotic flow and zeta potential

The effect of organic solvents on these properties will be discussed in this chapter. All the occurring effects result from solvation or desolvation in the particular solvents. Therefore, a theoretical approach will be briefly discussed that offers a quite clear interpretation of these effects by describing a parameter for the relative stability of the particles in the different solvents. It is based on the concepts of the *transfer activity coefficient* and the *medium effect*.

Concept of Transfer Activity Coefficient and Medium Effect

The chemical potential, μ_i, of a species i in a solvent, S, and in water, W, respectively, can be expressed by [3–8]

$$ {}^{s}\mu_i = {}^{s}\mu_i^0 + RT \ln {}^{s}c_i + RT \ln {}^{s}\gamma_i \tag{1} $$

$$ {}^{w}\mu_i = {}^{w}\mu_1^0 + RT \ln {}^{w}c_i + RT \ln {}^{w}\gamma_i \tag{2} $$

When the standard states, μ_i^0, are taken for solutions of the hypothetical molality of 1 (c_i expressed on the molal scale), then the activity coefficients ${}^{s}\gamma_i$ and ${}^{w}\gamma_i$ are depending on only these standard states.

From Eqs. 1 and 2 the following expression is obtained:

$$\frac{{}^{s}\mu_i^0 - {}^{w}\mu_i^0}{RT} = \ln {}^{w}\gamma_i - \ln {}^{s}\gamma_i \tag{3}$$

The so-called medium effect of the solvents on the species i is then given by

$$\ln {}_{m}\gamma_i = \frac{{}^{s}\mu_i^0 - {}^{w}\mu_i^0}{RT} \tag{4}$$

where ${}_{m}\gamma_i = {}^{w}\gamma_i / {}^{s}\gamma_i$ is the transfer activity coefficient or medium activity coefficient (which is a "degenerate" activity coefficient, because it does not depend on the concentration).

The medium effect, $\ln {}_{m}\gamma_i$, (also expressed by $\log {}_{m}\gamma_i$) is proportional to the reversible work of transferring 1 mol of species i from infinite dilution in water to infinite dilution in solvent, S. When the species is more stable in S, then the medium effect has a negative value, and when it is more stable in water the value is positive. When applied to the proton, the medium effect is a measure of the basicity of the solvent S relative to water.

The transfer activity coefficients for the single species can be estimated by only approximation methods, because with thermodynamic measurements, only summation parameters can be obtained for all species involved in the equilibrium. This fact often leads to large variations of the estimated values, strongly depending on the kind of approximation, not only for the absolute number of $\log {}_{m}\gamma_i$, but even for its sign. Thus, very divergent values can be found in the literature. Examples will be given later.

The concept of the medium effect as present here can, however, be very useful in discussing the stabilization of ions in different solvents, and also can be used for a straightforward interpretation of the changes of the pK values dependent on the solvent composition. It can also serve to explain the changes of the electroosmotic flow caused by the pH of the buffer in different solvents.

INFLUENCE OF ORGANIC SOLVENTS ON MOBILITIES OF IONS

For clearness, the definitions of the different ionic mobilities, as used in this chapter, will be repeated here: the *absolute mobility*, m_{abs}, is the limiting mobility of an ionic species at infinite dilution; it is related to the limiting single ionic conductance, λ^0, by $m_{abs} = \lambda^0/F$, F being the Faraday constant.

The *actual mobility*, m_{act}, is the mobility of the ion at given concentration; for weak electrolytes it is the mobility of the totally dissociated or protonated particle and depends on the ionic strength of the solution. The *effective mobility*, m_{eff}, is the mean mobility of the ions of a weak electrolyte, consisting of different ionic species,

k, given by the actual mobilities and the respective degrees of protolysis (or complexation), according to $m_{eff} = \Sigma (m_{act,k} - m_{act,(k-1)}) \alpha_k$; α_k is the degree of dissociation of the particular species. For monovalent ions, where $k = 1$, the effective mobility is obviously given by $m_{eff} = m_{act} \alpha$.

The *apparent mobility*, m_{app}, which is relevant only in systems with electroosmotic flow, is the vector sum of the effective mobility of the solute and the *electroosmotic mobility*, m_{eo}, of the bulk liquid: $m_{app} = m_{eff} + m_{eo}$.

In this chapter the influence of some organic solvents on the absolute and the actual mobility, respectively, will be discussed. The variation of the effective mobility is caused by changes in the pK value of the separands, which is discussed later. It is based on the relation of the degree of dissociation α, on the pK and the pH of the buffer, given by $\alpha = 1/(1 + 10^{pK_a - pH})$ for anions, and by $\alpha = 1/(1 - 10^{pK_a - pH})$ for cations.

Absolute Mobilities

Only very limited data on absolute mobilities in nonaqueous solvents have been published in the literature, restricted to very few types of solvents, and to a few small ions of physicochemical, but little analytical interest.

It is obvious that the absolute mobility is determined (besides the charge) by the size and the shape of the solvated ion (whereby the radius of the solvated ion, the Stoke's radius, can differ significantly from the crystallographic radius of the particle) and by the viscosity of the solvent. In reality, a linear dependence of the absolute mobilities on the viscosity, as implied by Walden's rule (which assumes that, for a given ionic species, the product of the limiting ion conductivity, λ^0, and the viscosity coefficient, η, in different solvents is constant, given by $\lambda^0\eta$ = const) is observed in only few cases, because this oversimplified view does not take into account the microenvironment of the migrating ion, which is not a continuum, but consists of solvent particles of a size similar to that of the ions. Especially for solvents with a high dipole moment, the molecules are often oriented differently toward the ion during its migration. Furthermore, it does not account for the different sizes of the solvated ions in the respective solvents. Therefore, the variation of the solvent leads to quite complex changes in the ionic mobilities, and no model exists that enables a quantitative treatment of these variations that is in accordance with the experimental data.

In Table 3 the absolute mobilities of some ions are given in water, acetonitrile, DMSO, and methanol as solvents, together with the Walden products [24–28]. It can be seen that the mobilities are larger in acetonitrile relative to water, and are smaller in DMSO. In methanol the mobilities are smaller compared with water for small ions with large hydration shells, but are larger for large ions with low hydration numbers. All these ions under consideration are, however, very simple, and do not

Table 3 Absolute Ionic Mobilities, m_{abs}, and Walden Products, $\lambda^0\eta$, in Different Solvents

	Water		DMSO		Acetonitrile		Methanol	
Ion	m_{abs}	$\lambda^0\eta$	m_{abs}	$\lambda^0\eta$	m_{abs}	$\lambda^0\eta$	m_{abs}	$\lambda^0\eta$
Li^+	40.0	0.34	11.6	0.22	71.8	0.24	41.0	0.22
Na^+	51.9	0.46	15.0	0.29	79.7	0.27	46.8	0.25
K^+	76.2	0.65	15.6	0.30	86.6	0.29	54.5	0.29
$(CH_3)_4N^+$	46.5	0.40	19.4	0.37	98.0	0.33	71.2	0.33
$(C_2H_5)_4N^+$	33.8	0.29	17.8	0.34	87.9	0.29	62.7	0.33
$(C_3H_7)_4N^+$	24.2	0.21	14.1	0.27	72.8	0.24	47.8	0.25
$(C_4H_9)_4N^+$	20.2	0.17	12.2	0.24	63.6	0.21	40.4	0.21
$(C_5H_{11})_4N^+$	18.2	0.16	—	—	58.0	0.19	35.9	0.19
Cl^-	79.2	0.68	23.7	0.46	92.2	0.31	54.3	0.28
J^-	79.6	0.68	24.4	0.47	105.7	0.35	65.1	0.34
ClO_4^-	69.8	0.60	25.1	0.48	107.8	0.36	76.6	0.40
CH_3COO^-	42.4	0.36	—	—	110.9	0.37	40.8	0.21
$(C_6H_5)_4B^-$	20.4	0.18	11.5	0.22	60.4	0.20	—	—

λ^0, limiting single ion conductivity; η, viscosity coefficient. $T = 25°C$.
m_{abs} in (10^{-5} cm^2 V^{-1} s^{-1}); $\lambda^0\eta$ in (Poise cm^2 Ω^{-1} mol^{-1}).
Source: Values from Refs. 24–28.

permit one to derive general rules for the changes in the mobilities for more complex organic analytes.

It should be mentioned, that Walden's rule seems to be obeyed by some examples given in Table 3, not for water, but for some organic solvents, for which the product $m\ \eta$ has values in a quite narrow range for a series of ions. In general, this relation is not valid owing to the foregoing reasons.

More refined theories on ion migration take into account the dynamic aspect of the orientation of polar solvent molecules during the movement of the ions through the medium, as introduced and discussed, for example, by Fuoss, Boyd, or Zwanzig [see, e.g., 29], with the dipol moments of the solvent molecules as parameters. They also fail, however, in quantitatively describing the absolute mobilities as a function of the solvent.

Although some results have been published on the solvation of ions in different mixed media (e.g., for methanol–water [30–32], DMSO–water [33,34], and acetonitrile–water [35]), which are based on the transfer activity coefficients and the standard free energy of transfer (see later), only very few data are available on the variation of the absolute mobilities with the solvent composition of binary aqueous–organic mixtures [25,36]. The relations observed are also complex and do not allow any generalization to predict such effects for other analytes.

Actual Mobilities

The changes of the actual mobilities in mixed solvents are even more complex to describe and are discussed here only in a phenomenological way. The actual mobilities of some anions in binary methanol–water mixtures, and in a ternary mixture of water, methanol, and DMSO are given in Table 4. It can be seen that the mobilities decrease when organic liquid is a constituent of the solvent. A large effect is found for the small chloride ion, for which the mobility is reduced even by a factor of 3. The smallest effect is observed for the large trinitrobenzoate ion, with a reduction of only 20%. However, this reduction is not really related to the change of the viscosity because the Walden product also does not remain constant for the particular ions. A variation of these products by about 50% is observed, reflecting the complexity of the parameters involved, including solvation, structure of the solvent, and so on.

It can be seen also from Fig. 2 that the product $m\,\eta$ is far from being constant for the example given. This result was typical for the anions of small aliphatic acids.

For cations of the alkylammonium type (NR_4^+ = type compounds from monomethyl- to tetrabutyl-) the changes of the actual mobilities with solvent composition have been published [16] for the water–methanol and water–methanol–DMSO systems as mentioned earlier. It can be seen from Fig. 3 that the mobilities of the small ions (different methyl- and ethylammonium ions) decrease, which is most pronounced for methanol and DMSO, and the strongest effect was observed for the NH_4^+ ion. Dimethyl sulfoxide always reduces the mobilities. Methanol as a cosolvent of water had almost no influence on medium-sized ions, such as tetramethyl-,

Table 4 Actual Mobilities, m_{act}, of anions in different solvents

	Water		Methanol–Water[a]		DMSO–Methanol–Water[b]	
Ion	m_{act}	$m_{act}\,\eta$	m_{act}	$m_{act}\,\eta$	m_{act}	$m_{act}\,\eta$
2,6-Dihydroxybenzoate	34.4	0.31	27.5	0.27	23.8	0.38
2,4,6-Trihydroxybenzoate	31.6	0.28	24.2	0.24	21.0	0.34
2,4,6-Trinitrobenzoate	26.9	0.24	24.0	0.24	20.3	0.33
Chloride	74.2	0.66	37.0	0.37	25.5	0.41
Trichloroacetate	34.6	0.31	27.5	0.27	23.4	0.38

[a]70/30 mol%

[b]30/40/30 mol%

m_{act} in (10^{-5} cm^2 V^{-1} s^{-1}), $m_{act}\,\eta$ in (10^{-10} NV^{-1}).

The values refer to 25°C and an ionic strength of 0.01 (mol L^{-1}).

Source: Data from Ref. 15.

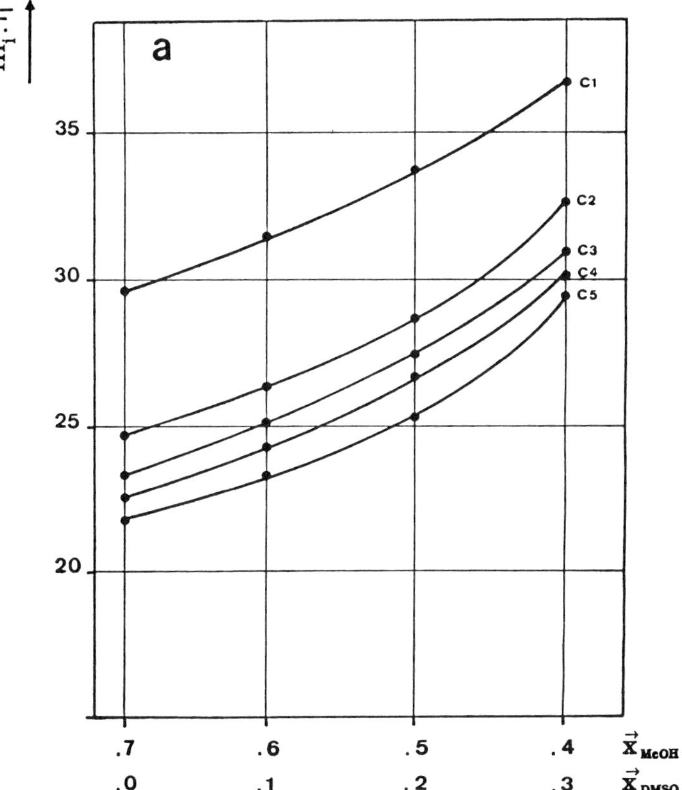

Figure 2 Product $m_i\,\eta$ (10^{-12} NV^{-1}) of the actual mobility, m_i, of nonsubstituted aliphatic carboxylic acids, and the viscosity coefficient for mixed aqueous–organic solvents. The mole fraction, x, of water is 0.3 in all solvents. $T = 30°C$. Code of solutes: C1, formate; C2, acetate; C3, propionate; C4, butyrate; C5, valeriate. (From Ref. 17)

tetraethyl-, and tetrapropylammonium, but increases the mobilities of the larger di-, tri-, and tetrabutylammonium.

A more interesting analytical aspect is that the organic cosolvent also sometimes changes the selectivity of the buffer systems by inverting the sequence of the mobilities. This effect was observed especially for the small alkylammonium ions, and again most pronounced for NH_4^+. This ion in water has the highest actual mobility of the compounds under consideration, when the mobilities are arranged in falling order, but exhibits only the sixth highest mobilitity in the ternary solvent system. From Fig. 3 it can be seen that the organic cosolvents also invert the order of mobilities for other pairs of ions.

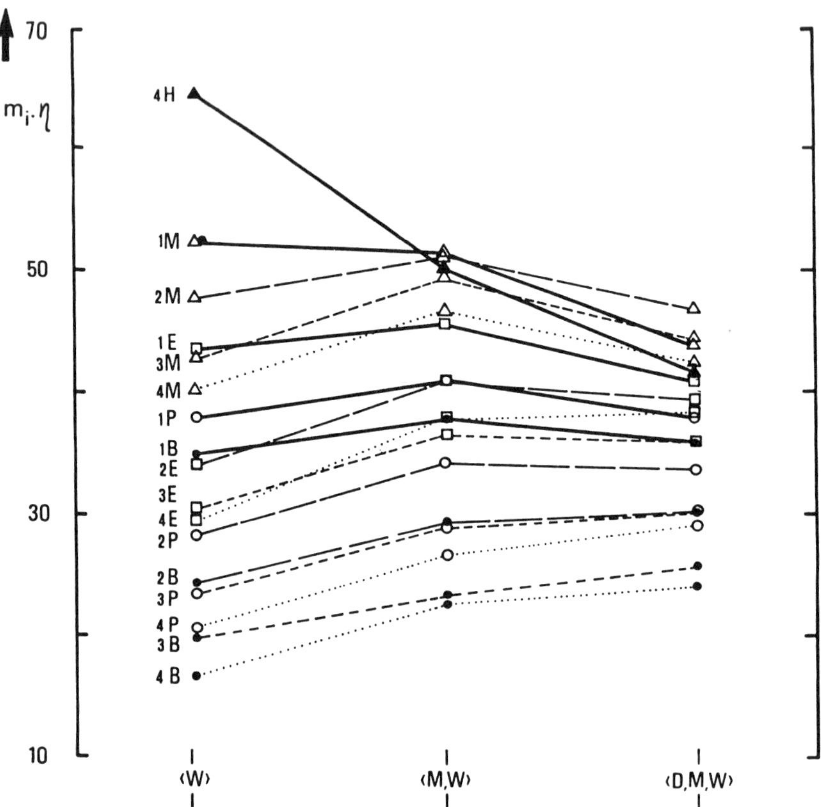

Figure 3 Influence of organic solvents on the electrophoretic selectivity: change of the migration sequence of ammonium ions. The actual mobility, m_i, of ammonium (4H) and aliphatic amines was corrected for the different viscosity of coefficient, η, of the solvents, given by $m_i\ \eta$ (10^{-12} NV^{-1}). Code of solvents: (W), water; (M,W), methanol–water (70:30 mol%); (D,M,W), DMSO–methanol–water (30:40:30 mol%). Code of solutes: 1M, methylammonium; 2E, diethylammonium; 3P, tripropylammonium; 4B, tetrabutylammonium; T = 25°C. (From Ref. 16)

For this quite simple type of component, an explanation of this effect can be given that is based on the solvation ability of methanol and DMSO: both are good solvators for cations (Table 5) and thus can compete with the water molecules for the solvation sites. Especially when the charge density of the ion is high and when the charge center is not sterically hindered, preferential solvation of these molecules can occur. Consequently, the different methyl- and the monoalkyl-substituted components, and to the largest extent NH_4^+, will be strongly influenced.

Table 5 Solvation Abilities of Solvents for Ions

Solvent	Anion solvation	Cation solvation
Water	+ +	+ +
Methanol	+ −	+ +
DMSO	−	+ +
Acetonitrile	− −	− −

INFLUENCE OF ORGANIC SOLVENTS ON p*K*

ΔpK_a and Standard Free Energy of Transfer

A theoretical approach to interpretation of the variation in p*K* with solvent or solvent composition is based on the concept of the transfer activity coefficient presented in the introductory chapter, whereby all species involved in the acid–base equilibrium are considered. Two cases can be distinguished, namely the ionization of a molecular or neutral acid, HA, given by

$$HA = H^+ + A^- \tag{5}$$

and the protolysis of a cation acid, HB^+, described by

$$HB^+ = H^+ + B \tag{6}$$

For HA, the reversible work done on the system in transferring 1 mol of H^+ and A^- from infinite dilution in water to infinite dilution in the particular solvent and transferring 1 mol of HA from solvent *S* to water, is equivalent to the standard free energy of transfer, ΔG_t^0, and is given by

$$\Delta G_t^0 = RT \ln \left[\frac{({}_m\gamma_{H^+})({}_m\gamma_{A^-})}{{}_m\gamma_{HA}} \right] \tag{7}$$

where ${}_m\gamma_{H^+}$, ${}_m\gamma_{A^-}$ and ${}_m\gamma_{HA}$ are the transfer activity coefficients of the particular species, as discussed earlier.

In the same way, the standard free energy of transfer for a cation acid is given by

$$\Delta G_t^0 = RT \ln \left[\frac{({}_m\gamma_{H^+})({}_m\gamma_{B})}{{}_m\gamma_{HB^+}} \right] \tag{8}$$

The change of the ionization constants, K_a, is related to ΔG_t^0, leading to

$$\Delta pK_{a,HA} = p^sK_{a,HA} - p^wK_{a,HA} = \log \left[\frac{({}_m\gamma_{H^+})({}_m\gamma_{A^-})}{{}_m\gamma_{HA}} \right] \tag{9}$$

for a neutral acid, and to

$$\Delta pK_{a,HB^+} = p^sK_{a,HB^+} - p^wK_{a,HB^+} = \log \left[\frac{({}_m\gamma_{H^+})({}_m\gamma_B)}{{}_m\gamma_{HB^+}} \right] \tag{10}$$

for cation acids. *W* and *S* indicate water and the solvent *S*.

It follows that ΔpK_a reflects the stabilization (or destabilization) of the particular species, occurring in the equilibrium, by water and solvent *S*.

It must be mentioned here that the simple interpretation of the p*K* shift, based on electrostatic effects of the dissociation equilibrium, as given by Born's treatment, is not adequate for an interpretation of the influence of the solvent. In a first assumption the contribution of the neutral molecules, HA or B, respectively, is neglected, because it is small compared with that of the ionic species. The contribution of H^+ to ΔpK_a, according to the medium effect on H^+, $\log {}_m\gamma_{H^+}$, depends on the basicity of the solvent relative to water. Positive values for $\log {}_m\gamma_{H^+}$ indicate that H^+ is more stable in water, compared with solvent *S*, and negative values implicate a higher basicity of solvent *S* than water.

Pure Solvents

The medium effect on the proton is summarized for different solvents in Table 6. It is seen that the values, which are taken from different literature sources vary over a wide range. This is because the single transfer activity coefficient of H^+ can be estimated by approximation only from a sum parameter, consisting of the contribution of all species of the electrolyte. This is the usual way for estimating single-ion properties. A unique exception is the possibility to determine single ionic mobilities.

The values of the medium effect, $\log {}_m\gamma_{H^+}$, for methanol are between +3.1 and −2.6 [4], a range of nearly six orders of magnitude for the transfer activity coefficient. These results hardly allow a definite description of the basicity of meth-

Table 6 Averaged Experimental Values for the Medium Effects, $\log {}_m\gamma_{H^+}$, on the Proton for Transfer From Water to Solvent

Solvent	$\log {}_m\gamma_{H^+}$
Methanol	0.0 ± 2.0
Ethanol	2.5 ± 1.8
n-Butanol	2.3 ± 2.0
DMSO	−3.6 ± 2.0
Acetonitrile	4.3 ± 1.5

The values are on the molal scale.
Source: Data from Ref. 4, are based on the different literature sources cited therein.

anol. Ethanol, on the other hand, is clearly less basic than water. Acetonitrile is (surprisingly) also less basic than water, because it exhibits large positive values for log ${}_{m}\gamma_{H^+}$. In contrast with these solvents, DMSO is a stronger base than water, according to the large and negative values for log ${}_{m}\gamma_{H^+}$.

From Eqs. 9 and 10 for ΔpK_a, it follows that not only log ${}_{m}\gamma_{H^+}$, the medium effect on the proton, is decisive for the shift of the pK values caused by the solvent composition, because the transfer activity coefficients of A^- and HB^+, respectively, are also involved. They are involved, however, in a different manner: in the logarithmic term of the equation concerning the neutral acid (Eq. 9), it is the product of the transfer activity coefficients of the ionic species (H^+ and A^-), whereas for cation acids the ratio of ${}_{m}\gamma_{H^+}$ and ${}_{m}\gamma_{HB^+}$ of the ionic species, H^+ and HB^+, occurs (Eq. 10). It should be mentioned once more that the transfer activity coefficient of the noncharged species is of minor importance on both cases.

Whereas in the expression for the pK_a shift for HA, ions with different sign occur, in the expression for the cation acid, only ions with the same sign are found. This leads to different consequences: for neutral acids the ability of the solvents to stabilize both types of ions, cations (H^+) as well as anions plays a role. For cationic acids, in contrast, the solvation ability for only cations (H^+ and HB^+) is important. As water stabilizes cations and anions, and because the ability for stabilizing both these species is often totally different for organic solvents (Table 5), namely, at a loss of the anions, the pK_a values of neutral acids increase mostly when going from pure aqueous to organic (or mixed aqueous–organic) solvents. Even a higher basicity of the organic solvent, as in DMSO, is overcompensated by the low solvation ability relative to anions. The net effect thus leads to a reduction of the acidity of the neutral acid. This effect can be seen for pure solvents from Table 7. The pK_a values increase by about 5 units in pure methanol, compared with water, and about 7

Table 7 Thermodynamic pK_a Values of Acids in Different Pure Solvents at 25°C

	pK_a			
Acid	Water	Methanol	DMSO	Acetonitrile
Hydrochloric	−7	1.2	2.0	8.9
Hydrobromic	−9	—	1.8	5.5
Acetic	4.76	9.52	12.6	22.4
Chloroacetic	2.86	7.81	8.9	18.8
Dichloroacetic	1.29	6.3	5.9	15.8
Benzoic	4.20	9.28	11.0	20.7
Phenol	9.98	14.4	16.4	27.2
Anilinium	4.6	—	3.8	10.6

Source: Data from Refs. 24, 37–41.

units higher values are observed in pure DMSO for the less acidic solutes (benzoic acid, phenol). A drastic increase is found in pure acetonitrile, by up to 17 units, but in the presence of even small amounts of water this large change is not reached.

For cationic acids, on the other hand, the ability of the solvents to stabilize H^+ often goes parallel with the ability to solvate other cations. Both ${}_m\gamma_{H^+}$ and ${}_m\gamma_{HB^+}$, vary to about the same extent, and the effects, reflected by the ratio ${}_m\gamma_{H^+}/{}_m\gamma_{HB^+}$, are thus counterbalanced. The net effect remains relatively small, and the pK_a values of cation acids show a less pronounced variation with the solvent composition, at least for those solvents that have a good ability to stabilize cations, such as methanol or DMSO. This can indeed be seen from Table 7: the anilinium acid has about the same pK_a in DMSO as in water, a behavior that is totally different from that of a neutral acid. Also in acetonitrile, in which, for instance, acetic acid (with a comparable pK_a in water) had a ΔpK_a of 17.5 units, the pK of anilinium increase by only a relatively small 6 units.

Mixed Aqueous–Organic Solvents

When comparing experimental values of the pK in mixed solvents with theoretical predictions, normally it is not the thermodynamic ionization constant, related to the activity of the solutes, that is determined [42]. Usually, not even the concentration ionization constant related to a pH scale in pure aqueous solutions is determined by potentiometric titration, because the measurements are carried out without taking into account changes in the liquid junction potentials between pure aqueous and mixed aqueous–organic phases in the measuring chain. The calibration of the

Table 8 Apparent pK_a' Values of Neutral Acids in Mixed Solvents

	pK_a'			
Acid	Water	Methanol–water (70:30 mol%)[a]	Methanol–water (Water <1 mol%)[b]	DMSO–methanol–water (30:40:30 mol%)[a]
Formic	3.75	5.25	7.06	6.70
Acetic	4.76	6.62	8.47	8.20
Propionic	4.87	6.78	8.78	8.44
Butyric	4.81	6.84	8.75	8.43
Glycolic	3.82	5.55	7.55	7.05
Lactic	3.86	5.73	7.75	7.09
Benzoic	4.20	6.42	8.30	7.40
2-Chlorobenzoic	2.92	5.12	7.18	6.52
3-Chlorobenzoic	3.82	5.84	7.78	6.85
4-Chlorobenzoic	3.97	5.99	7.98	7.20

T = 25°C.

Source: Values for water Refs. 15 and 17; [a]Refs. 15 and 17; [b]Ref. 14.

glass–calomel electrode is normally carried out with pure aqueous buffer systems, whereas the potentiometric titrations are carried out in mixed solvents. This leads to a systematic deviation of the "apparent" pK value (indicated by pK') determined in this way, from the pK values related to the aqueous scale. This pK' value, however, is an appropriate parameter for the adjustment of the effective mobility by the pH (or the apparent pH$'$, respectively) as long as the pH$'$ of the buffer is set under the same conditions, under which the pK' was measured. This is normally true. It can be assumed that the systematic error arising from applying pure aqueous solutions for calibration, on the one hand, and mixed solvents for electrophoresis, on the other hand, is counterbalanced and nearly eliminated in this way. The relation between pH$'$, pK', and the effective mobility is still valid under these circumstances. Indeed, for binary mixtures, the deviation of the apparent pK' value from the thermodynamic pK is in the range of only a few tenths of a pK unit or even less [15,43–45].

In Tables 8 and 9, pK'_a values of several solutes are given. In contrast with the actual mobilities, a general trend can be observed for neutral acids (see Table 8): the pK'_a values always increase (as with pure organic solvents) depending on the concentration of the organic cosolvent. For methanol, an increase of about 1–2 pK units is observed when 70 mol% of water is substituted by methanol. For a higher methanol content (water concentration only 1 mol% or lower) the pK'_a values are about 4 units higher than in water.

A similar increase is found when DMSO is a solvent constituent, but this increase is still observed at a much higher water concentration, compared with the 1%, namely, even at about 30 mol% water. This is in accordance with the literature, in which it was found that DMSO drastically raises the pK_a values of neutral acids [44].

Table 9 Apparent pK'_a Values of Aliphatic Cationic Acids in Mixed Solvents

	pK'_a		
Solute	Water	Methanol–water (70:30 mol%)	DMSO–methanol–water (30:40:30 mol%)
Ammonium	9.25	8.76	9.73
Methylammonium	10.66	9.89	10.69
Dimethylammonium	10.73	9.86	10.47
Trimethylammonium	9.80	8.38	8.79
Ethylammonium	10.70	9.81	10.59
Propylammonium	10.69	9.95	10.72
Tributylammonium	9.93	9.24	9.38

T = 25°C.

Source: Data from Ref. 16.

Again it is more interesting analytically that the extent of the pK_a shift is not identical for different solutes. This effect can be clearly seen from Fig. 4, showing the fluctuation of the pK_a' values about the regression line (the linear correlation coefficient, r, is 0.945). These deviations are much higher than the statistical error of the determination of the pK' values, which is only about 0.02–0.03 pK units. Thus a systematic deviation takes place, leading to a variation of the selectivity of

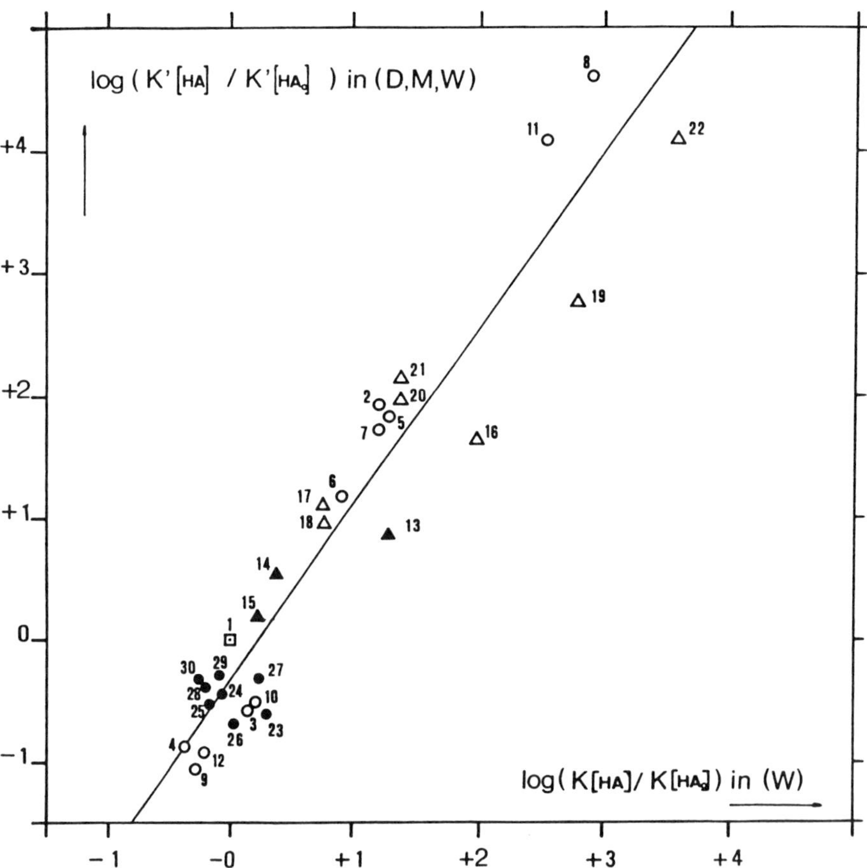

Figure 4 Selective change of the pK'_a values of 30 benzene carboxylic acids with solvent composition. The systematic variation is indicated by the fluctuation of the measured values about the regression line (linear correlation coefficient, r = 0.945), which is much higher than expected owing to the statistical error. The ionization constants, K, of the solutes HA are related to that of benzoic acid (HA_0). Solvent code as in Fig. 3. The numerical code of the solutes is not given here in detail. (From Ref. 15)

the solvent systems. It follows that when using the different solvent systems, a variation of the electrophoretic migration should be observed even at the same pH′. This is shown in Fig. 5 by the electropherograms of five test components (in the isotachophoretic or displacement electrophoretic mode) at the same pH′ of 5.7 in water and in a ternary mixture consisting of water, methanol, and DMSO: whereas in the pure aqueous buffer, the five components migrate in the same zone, all substances are separated in the ternary solvent because the pK' values are selectively shifted by the solvent composition to values that cause different effective mobilities of the separands.

Acetonitrile also shifts the pK_a of neutral acids to higher values. The pure solvent has the most pronounced effect on the pK_a, even compared with methanol or DMSO (see Table 7), owing to its lower solvation ability for cations than for anions (see Table 5), so that all ionic species occurring in the dissociation reaction are destabilized when transferred from water to pure acetonitrile (one expection is Ag^+, which forms stable complexes with acetonitrile). However, in mixed solvents, its influence lies between methanol and DMSO [45]. This effect was used to separate benzoic acid derivatives by capillary zone electrophoresis, as shown in Fig. 6. The separation of cationic analytes in pure acetonitrile (used rather for the enhancement of the solubility of the analytes) is demonstrated in Fig. 7.

It was pointed out earlier that the solvents have only a minor influence on the pK_a values of cationic acids, according to Eq. 10. Indeed it can be seen from Table 9 that the change with solvent composition is far away from the effects found for neutral acids: methanol decreases the pK'_a values slightly (to about 0.5–0.8 units), whereas, in the ternary solvent, about the same values as in water are observed (with the exception of the trialkylammonium acids, that show about the same value as in methanol–water).

In contrast with the mobilities, a rule can be given for an estimate of the influence of organic solvents on the ionization constants: the pK_a values of neutral acids increase with increasing content of organic solvent, and the effect increases in the order methanol < acetonitril < DMSO, when aqueous–organic mixtures are compared on the molar ratio scale. The increase can be several pK units. The pK_a values of cationic acids, on the other hand, are affected to a much lower extent, and the changes can be positive or negative, compared with water.

VARIATION OF ELECTROOSMOTIC FLOW AND ZETA POTENTIAL WITH SOLVENT COMPOSITION

When electrophoresis is carried out in capillaries made from fused silica, glass, polyfluorocarbon, such as Teflon, or polyethylene, and the inner surface remains untreated, an electroosmotic flow occurs, under the influence of an electric field, that is directed toward the cathode. This flow is due to an electric double layer formed by an immobile negative surface charge on the capillary wall, which is

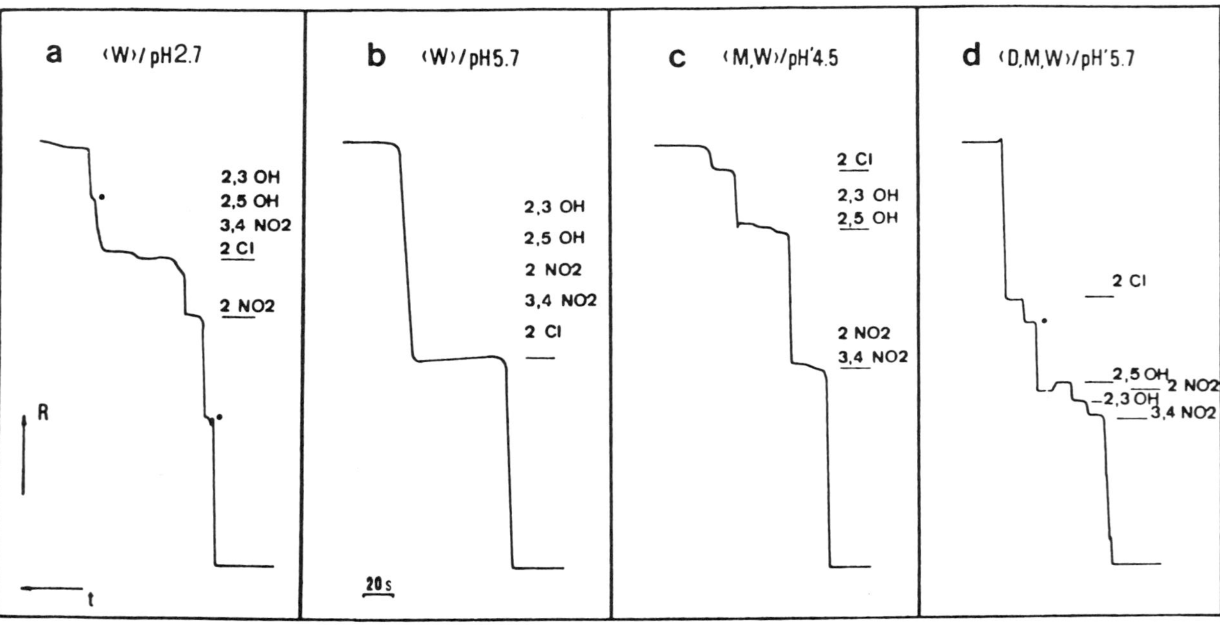

Figure 5 Electropherogram of substituted benzene carboxylic acids in different buffer systems. The isotachopherograms in b and d are obtained at the same (apparent) pH 5.7 of the buffer, but with different solvents. Solvent code as in Fig. 3. Code of solutes: 2,3 OH: 2,3-dihydroxybenzoate; 2,5 OH: 2,5-dihydroxybenzoate; 2 NO_2: 2-nitrobenzoate; 3,4 NO_2: 3,4-dinitrobenzoate; 2Cl: 2-chlorobenzoate. The asterisks indicate impurities. R = resistance measured with a conductivity detector, *t* time. (From Ref. 15)

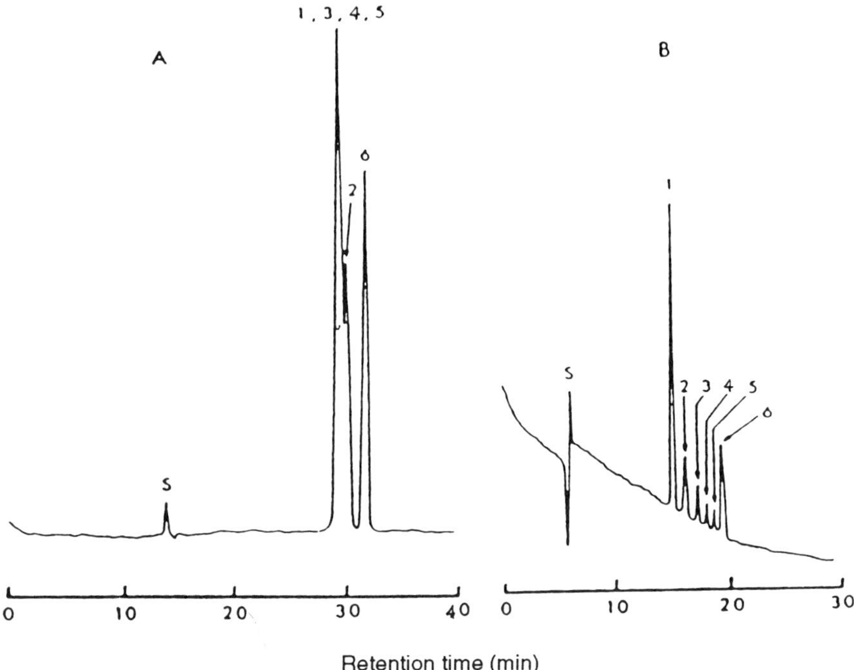

Figure 6 Effect of acetonitrile on the separation of anions (positional isomers of amino- and methylbenzoic acids) by capillary zone electrophoresis. (A) aqueous buffer, pH 7; (B) mixed solvent (acetonitrile–water, 50 vol% each), pH 7. Detection: UV absorbance at 254 nm. Peak assignment for the benzoates as follows: 1, *p*-amino-; 2, *m*-amino-; 3, *p*-methyl-; 4, *m*-methyl-; 5, *o*-methyl-; 6, *o*-amino-. (From Ref. 19)

counterbalanced by cations in the liquid buffer. Some of these cations migrate in an applied electric field and cause a net flow of solvent to the cathode side of the capillary by partially transferring their mechanic impulse onto the solvent molecules. The sources of the negative charge of the double layer are (besides adsorbed ions such as OH^- at high pH) acidic silanol groups, in the case of fused silica, forming silanolate anions. For organic polymers, such as polyfluorocarbon (a material that has been used in isotachophoresis for many years), polyethylene, polyvinylchloride, polyurethane, carboxyl groups, incorporated into the surface, seem to be responsible for the negative surface charge. This assumption is supported by the fact that the pH dependence of the electroosmotic velocity shows a shape similar to a titration curve, with inflection points at pH 5.3 for silica and 4.8 for polyfluorocarbon and polyethylene [46]. These points are interpreted as pK_a values of the acidic groups positioned on the surface.

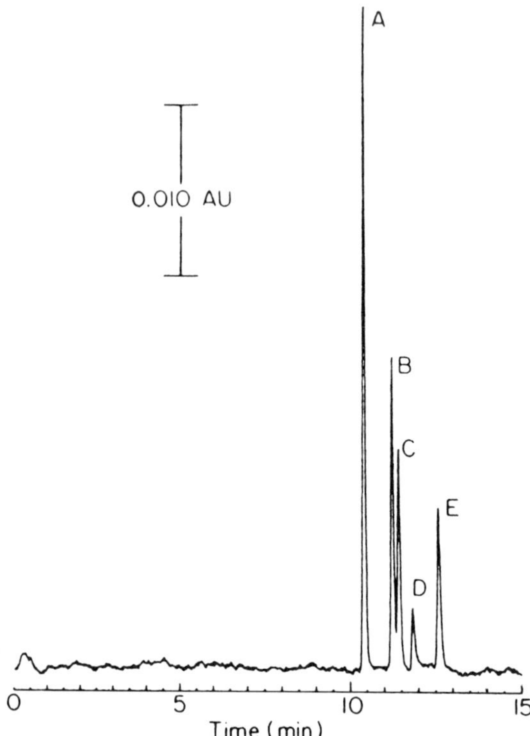

Figure 7 Separation of cations by capillary zone electrophoresis in an unbuffered electrolyte system with acetonitrile as solvent. The separands are quinoline-type components. Detection: UV absorbance at 229 nm. Assignment of solutes: A, isoquinoline; B, quinoline; C, quinaldine; D, 8-quinolol; E, acridine. (From Ref. 22)

From the foregoing discussion, it can be assumed that organic solvents should influence the electroosmotic velocity, which is related to the degree of dissociation by the pH of the buffer and the p*K* of the ionizable surface group, of a kind that is typical for weak neutral acids. Indeed a shift of the inflection points of the velocity vs pH curves to values about 1.5 units higher is observed in fused silica capillaries with 50 vol% methanol, and even larger increases for the same volume percentage are found with higher alcohols, acetonitrile, DMSO, and acetone, for which a value of 8.5 was reached for the pK_a' of the surface silanol group [18].

The effect of the solvent composition on the electroosmotic velocity in fused silica capillaries is shown in Fig. 8. The measurements were carried out at such high pH values that the surface silanol groups were fully ionized. It can be seen that the electroosmotic velocity is always reduced by an increasing percentage of

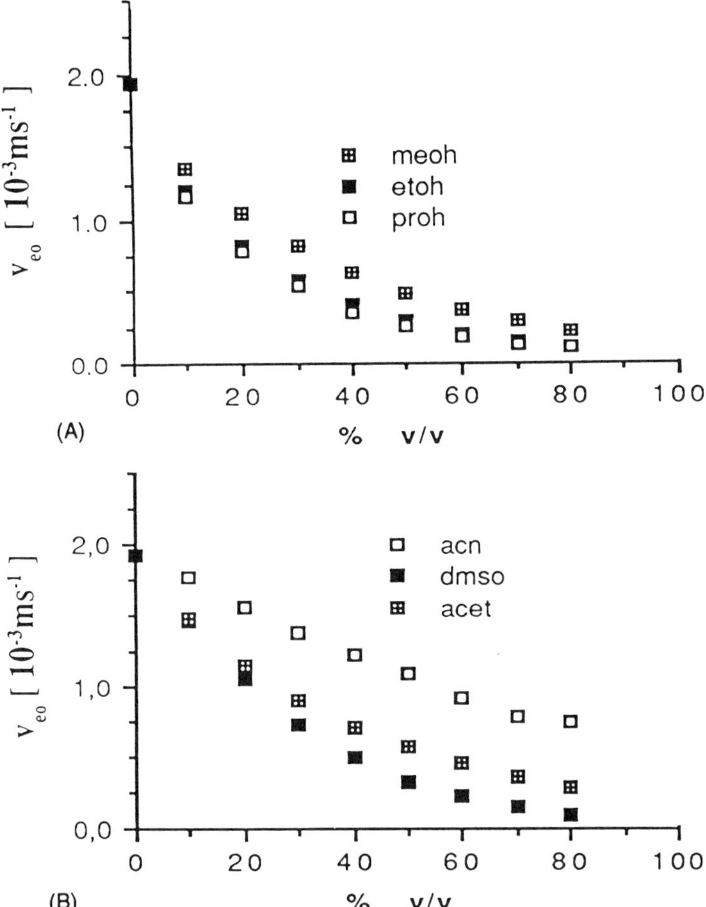

Figure 8 Effect of the composition of the mixed aqueous–organic solvent on the electroosmotic velocity, v_{eo}, in fused silica capillaries (T = 25°C). Solvent code: (A) meoh, methanol; etoh, ethanol; proh, 2-propanol; (B) acn, acetonitrile; dmso, dimethyl sulfoxide; acet, acetone. (From Ref. 18)

organic solvent in the solvent mixture. This effect cannot be related solely to a change in the viscosity, because the viscosity vs percentage organic solvent curves have a maximum (Fig. 9). If the change in the viscosity alone were solely responsible for the variation of the electroosmotic velocity, a minimum at a content between 40 and 70 vol% should occur, followed by an increase with increasing content of organic cosolvent. This is clearly not observed. The constant decrease of the elec-

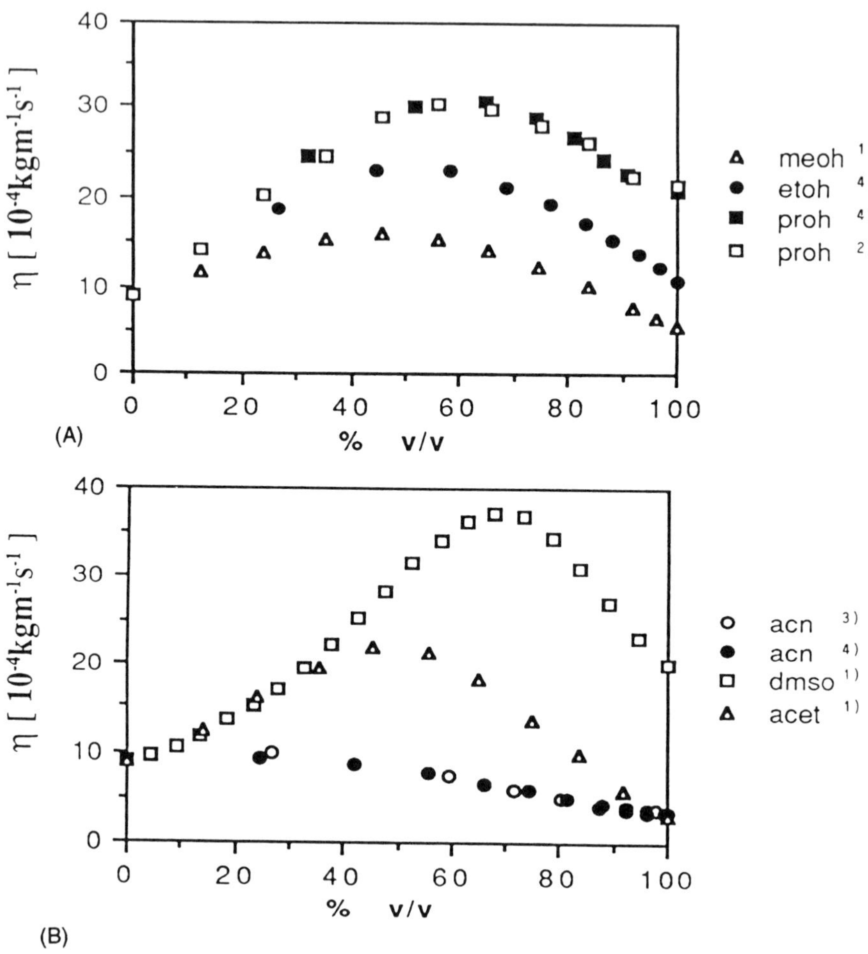

Figure 9 Viscosity coefficient, η, vs composition of mixed aqueous–organic solvents at 25°C. Solvent code: (A) meoh, methanol; etoh, ethanol; proh, 2-propanol; (B) acn, acetonitrile; dmso, dimethyl sulfoxide; acet, acetone. (From Ref. 18)

troosmotic velocity is, in fact, not only caused by the viscosity change, but also by a decrease of the zeta potential, which was found for fused silica with nearly all of the solvents investigated. This behavior of fused silica is in contrast with that found for organic synthetic polymers, for which the zeta potential (not the electroosmotic flow) remains about constant over a wide range of solvent composition, and decreases at a higher percentage of organic solvent to a much smaller extent than in fused silica. Although the zeta potential is nearly unaffected by the solvent, the electroosmotic flow varies [46], and the latter is a more important parameter in electrophoresis than the former because it determines the apparent ionic mobility and, thus, the migration properties of the separands.

To summarize, all organic solvents that are constituents of aqueous–organic mixtures decrease the electroosmotic flow, even under conditions in which the groups responsible for surface charge are fully ionized, namely, at sufficiently high pH. This decrease is directly related to the content of organic solvent. The strength of the effect in fused silica follows the sequence acetonitrile < acetone < methanol < ethanol, propanol < DMSO, for comparable volume percentages of the individual solvent. In capillaries made from organic polymers the same trend is found.

REFERENCES

1. D. L. Massart, B. G. M. Vandeginste, S. N. Deming, Y. Michotte, and L. Kaufmann, *Chemometrics: A Textbook*, Elsevier, Amsterdam (1988).
2. E. Kenndler and B. Gassner, *Anal. Chem.*, *62*:431 (1990).
3. R. G. Bates, Medium effect and pH in nonaqueous solvents. *Solute–Solvent Interactions* (J. F. Coetzee and C. D. Ritchie, eds.), Marcel Dekker, New York, pp. 45–96 (1969).
4. E. J. King, Acid–base behavior. *Physical Chemistry of Organic Solvent Systems* (A. K. Covington and T. Dickinson, eds.), Plenum Press, London, pp. 331–403 (1973).
5. A. K. Covington and T. Dickinson, Introduction and solvent properties. *Physical Chemistry of Organic Solvent Systems* (A. K. Covington and T. Dickinson, eds.), Plenum Press, London, pp. 1–23 (1973).
6. I. M. Kolthoff and M. K. Chantooni, General introduction to acid–base equilibria in nonaqueous organic solvents. *Treatise on Analytical Chemistry*, Part I, Theory and Practice, Vol. 2, Sect. D (I. M. Kolthoff and P. J. Elving, eds.), John Wiley & Sons, New York, pp. 239–301 (1979).
7. A. P. Popov and H. Caruso, Amphoprotic solvents. *Treatise on Analytical Chemistry*, Part I, Theory and Practice, Vol. 2, Sect. D (I. M. Kolthoff and P. J. Elving, eds.), John Wiley & Sons, New York, pp. 303–347 (1979).
8. I. M. Kolthoff and M. K. Chantooni, Dipolar aprotic solvents. *Treatise on Analytical Chemistry*, Part I, Theory and Practice, Vol. 2; Sect. D (I. M. Kolthoff and P. J. Elving, eds.), John Wiley & Sons, New York, pp. 349–384 (1979).
9. J. L. Beckers and F. M. Everaerts, *J. Chromatogr.*, *51*:339 (1970).
10. J. L. Beckers and F. M. Everaerts, *J. Chromatogr.*, *68*:207 (1972).
11. F. M. Everaerts, J. L. Beckers, and T. P. E. M. Verheggen, *Isotachophoresis, Theory, Instrumentation and Applications*, *J. Chromatogr. Libr. Ser.* Vol. 6, Elsevier, Amsterdam, (1976)
12. J. C. Reijenga, G. V. A. Aben, T. P. E. M. Verheggen, and F. M. Everaerts, *J. Chromatogr.*, *260*:241 (1983).
13. J. C. Reijenga, H. J. L. A. Slaats, and F. M. Everaerts, *J. Chromatogr.*, *267*:85 (1983).
14. T. Hirokawa, T. Tsuyoshi, and Y. Kiso, *J. Chromatogr.*, *408*:27 (1987).
15. E. Kenndler and P. Jenner, *J. Chromatogr.*, *390*:169 (1987).
16. E. Kenndler and P. Jenner, *J. Chromatogr.*, *390*:185 (1987).

17. E. Kenndler, C.Schwer, and P. Jenner, *J. Chromatogr.*, *470*:57 (1989).
18. C. Schwer and E. Kenndler, *Anal. Chem.*, *63*:1801 (1991).
19. S. Fujiwara and S. Honda, *Anal. Chem.*, *59*:487 (1987).
20. H. Yoshida and Y. Hirama, *J. Chromatogr.*, *298*:243 (1984).
21. S. Tanaka, T. Kaneta, and H. Yoshida, *J. Chromatogr.*, *472*:303 (1989).
22. Y. Walbroehl and J. W. Jorgenson, *J. Chromatogr.*, *315*:135 (1984).
23. Y. Hirama and H. Yoshida, *J. Chromatogr.*, *322*:139 (1985).
24. B. G. Cox, *Annu. Rep. Chem. Soc.* *70*:249 (1973).
25. Landolt-Börnstein, *Zahlenwerte und Funktionen*, Vol. 2, Part 7, 6th ed., Springer, Berlin (1969).
26. R. L. Kay and D. L.Evans, *J. Phys. Chem.*, *70*:2325 (1966).
27. R. L. Kay, D. L. Evans, and G. P. Cunningham, *J. Phys. Chem.*, *73*:3322 (1969).
28. D. L. Evans, T. Tominaga, J. B. Hubbard, and P. G. Wolynes, *J. Phys. Chem.*, *83*:2669 (1979).
29. M. Spiro, Conductance and transference numbers. *Physical Chemistry of Organic Solvent Systems* (A. K. Covington and T. Dickinson, eds.), Plenum Press, London, pp. 635ff: (1973).
30. C. F. Wells, *J. Chem. Soc. Faraday Trans. I*, *69*:984 (1973).
31. C. F. Wells, *J. Chem. Soc. Faraday Trans. I*, *70*:694 (1974).
32. C. F. Wells, *J. Chem. Soc. Faraday Trans. I*, *74*:636 (1978).
33. C. F. Wells, *J. Chem. Soc. Faraday Trans. I*, *77*:1515 (1981).
34. B. G. Cox, R. Natarajan, and W. E. Waghorne, *J. Chem. Soc. Faraday Trans. I*, *75*:1780 (1979).
35. R. Natarajan and W. E. Wagborne, *J. Chem. Soc. Faraday Trans. I*, *75*:86 (1979).
36. D. Feakins and D. A. O'Shaughnessy, *J. Chem. Soc. Faraday Trans. I*, *74*:380 (1978).
37. P. D. Bolton and C. H. Rochester, *Trans. Faraday Soc.*, *66*:1348 (1970).
38. C. S Leung and A. Grunwald, *J. Amr. Chem. Soc.* *74*:696 (1970).
39. O. M. Konovalov, *Zh. Fiz. Khim.*, *39*:693 (1965).
40. M. K. Chantooni, Jr. and I. M. Kolthoff, *J. Phys. Chem.*, *79*:1176 (1975).
41. M. K. Chantooni, Jr. and I. M. Kolthoff, *Anal. Chem.*, *51*:133 (1979).
42. A. Albert and E. P. Serjeant, *The Determination of Ionization Constants. A Laboratory Manual*, 3rd. ed., Chapman & Hall, London, (1984).
43. R. G.Bates and R. A. Robinson, Acid–base behaviour in methanol–water solvents. *Chemical Physics of Ionic Solutions* (B. E. Conway and R. G. Barradas, eds.), John Wiley & Sons, New York (1966).
44. J. Juillard and N. Simonet, *Bull. Soc. Chim. Fr.*, 1883 (1968).
45. A. Widhalm, *Graduation Report*, University Vienna (1988).
46. W. Schützner and E. Kenndler, *Anal. Chem.*, *64*:1991 (1992).

6

Controlling Migration Behavior in Capillary Electrophoresis: *Optimization Strategies for Method Development*

Morteza G. Khaledi and Changyu Quang

North Carolina State University, Raleigh, North Carolina

Rachhpal S. Sahota

Proctor & Gamble Pharmaceuticals, Norwich, New York

Joost K. Strasters

Sterling Winthrop Pharmaceuticals Research Division, Malvern, Pennsylvania

Scott C. Smith

Magellan Laboratories, Inc., Research Triangle Park, North Carolina

High-performance capillary electrophoresis (HPCE) has gained much popularity over the past decade, mainly because of the combination of certain advantages that resemble, and sometimes surpass, those of high-performance liquid chromatography (HPLC). Currently, HPCE seems to be in a situation similar to HPLC in the 1970s and beginning of the 1980s: commercial instruments and capillaries are readily available, and numerous descriptions of applications are appearing in the literature.

In spite of the tremendous growth and popularity of HPCE during the past decade, however, HPLC remains the main technique for method development in bioanalytical laboratories. An important advantage of HPLC is the feasibility of controlling solutes' partitioning (and, consequently, the chromatographic retention and selectivity) by changing the mobile phase composition. The quality of separation can be easily improved through careful manipulation of one or more of the several experimental parameters, such as concentration and the type of organic modifiers—pH, concentration, and the type of ion-pairing reagents. The influence of different variables on retention and the effective strategies for optimizing of these variables have been the focus of many intensive investigations [1–4].

Because of the fundamentally different migration mechanisms between the two techniques, however, the exact role of these factors on the HPCE migration behavior should be carefully examined.

Resolution in HPCE is quantitatively described by the following equation [5]:

$$R = \left(\frac{\sqrt{N}}{4}\right)\left(\frac{\mu_2 - \mu_1}{\mu_{avg} + \mu_{eo}}\right) \tag{1}$$

As a result, separation can be enhanced by increasing efficiency, N, selectivity in migration, $\Delta\mu$, and by reducing the electroosmotic mobility, μ_{eo}.

The enormous separation power of HPCE stems from the possibility of generating many theoretical plates in a short period. For separation of compounds with nearly identical electrophoretic mobilities, however, the high efficiency of the technique might be inadequate. For such compounds, manipulation of the buffer composition for selectivity enhancement should be examined. In addition, an extension of the list of selectivity parameters and development of optimization protocols results in a broader applicability of HPCE separations.

This chapter provides an overview of a systematic effort in this laboratory to achieve a better understanding of the influence of different solvent characteristics on migration behavior in capillary zone electrophoresis (CZE) and micellar electrokinetic capillary chromatography (MECC).

METHOD DEVELOPMENT

Strategies and General Considerations

Solvent composition affects electrophoretic mobility and electroosmotic flow velocity. Consequently, the quality of HPCE separations and analysis times depends on the selection of buffer composition. Method development in HPCE is still based on trial and error and on the researcher's intuition and experience. The number of controllable variables is considerable, thus the number of possible variations of these variables is too large to screen all combinations for a given problem. For example, pH, type of buffer, ionic strength, type and concentration of organic cosolvents, type and concentration of surfactants in electrokinetic chromatography, and inclusion of chemical equilibria, such as ion exchange and complexation, can have significant effects on migration behavior in HPCE [6–16]. Therefore, it is essential to determine if and how migration behavior (and separation selectivity) of solutes is influenced by these variables.

Although the usefulness of HPCE methods has been extensively discussed in the literature, most reports have focused on specific applications [6–16]. For HPCE to be accepted as a major analytical technique, both a better understanding of the factors that influence separation and a more general and systematic approach in method development are required. The method development strategy must provide an answer to which variables are the most appropriate to use and how to set up initial experiments to search the selected parameter space in an efficient way.

Ideally, one should be able to predict the behavior of a solute a priori from its structure. Thus, without any experimentation, one should be able to design a separation at the optimal experimental conditions. In HPLC, this approach has never gained widespread popularity owing to the complex partitioning behavior. For HPCE, this approach seems somewhat more promising because of the absence of a stationary phase. For "unknown" compounds in the mixture, or unknown migration behavior of the solutes, a certain amount of experimentation will be unavoidable. Consequently, the experimental design to collect this information will determine the efficiency of the development process.

Two general approaches can be used in the optimization of HPCE separations:

Sequential Techniques (Simplex)

Although initially propagated for use in HPLC, the use of sequential techniques, such as the simplex method, are currently less favored. In this approach, the optimum is reached through a series of experiments. Each experiment in the sequence is based on the observed response of the previous experiments (e.g., minimum resolution) and, in turn, influences the choice of conditions for the subsequent experiments. However, the complex behavior of solutes as a function of the experimental conditions often results in changes in peaks' elution order and, consequently, in multiple local minima. As a result, the current practice in HPLC is the use of this type of technique only if the more powerful interpretive strategies fail [17]. Since peak reversals are equally common in CZE and MECC when the buffer composition is varied, the same conclusions will apply for these separations. The sequential strategy is very problem-specific. One may have to repeat the process all over again as a single component is added to, or removed from the mixture. In addition, this approach does not provide any general information about the migration behavior of individual solutes within the parameter space, because no attempt is made to predict or observe the migration pattern of individual components. An electropherogram is viewed rather as an individual, single entity. No useful information is thus obtained about the role of a particular variable that can be generalized to other HPCE separations.

Simultaneous Experimental Designs

Simultaneous design strategies are based on a predefined set of experiments. The experimental parameter space is searched for the set of variable values that gives the optimum separation. The search usually involves tracing a multivariate grid. One obvious advantage in using the grid-search approach is that it provides a better understanding of the system. Adding new compounds to the mixture translates into understanding the migration patterns of the new compounds alone.

One way of determining the migration pattern of a mixture component is to experimentally obtain its migration at each experimental condition. However, for this, an exhaustive grid search may be required, which can be prohibitively expensive and time-consuming.

To reduce the number of required experiments, one can derive quantitative models to describe either the variations of the observed response of the system (e.g., minimum resolution) or the migration of solutes (e.g., mobility) as a function of variables.

Modeling the overall response of the system is useful when the individual compounds in a mixture cannot be distinguished. A Plackett–Burman statistical design (a reduced factorial design) was used to optimize the resolution–analysis time of the separation of a mixture of steroid esters using MECC, varying the buffer concentration, pH, acetonitrile concentration, micelle concentration, and addition of a second surfactant [18].

Preferably, however, one should try to apply an interpretive strategy; that is the migration behavior of the individual components in a mixture is quantitatively described. In this approach, an attempt is made to understand the migration behavior of each component in the mixture as a function of experimental variables. Once the migration pattern of the solutes of interest in a mixture is simulated, the prediction of any electropherogram for any given combination of the variables within a given parameter space becomes feasible. This can then be translated into a measure of quality of separation or a criterion. A computerized search of the parameter space will thus produce the location of a global optimum (within this space). As a result, several conditions can be optimized simultaneously on the basis of a few experiments. Simultaneous multivariable optimization is the most effective and efficient approach owing to the interactive nature of several selectivity characteristics in CE.

Modeling of Migration Behavior

A quantitative description of the migration in CZE or MECC is beneficial for two apparent reasons: First, in method development, the behavior of a solute under different experimental conditions can be predicted on the basis of a limited number of experiments, thereby facilitating the optimization of retention and selectivity in practical separations. Second, an in-depth understanding of the migration behavior in HPCE as well as that of the exact role of important variables that influence migration and selectivity can be achieved. As a result, one could take full advantage of any additional selectivity that the use of different variables might provide in different applications. Finally, since the behavior of each solute in a mixture is modeled, useful information about the relation between migration and chemical structure can be obtained. An interesting prospect will be to predict the migration behavior on the basis of structural properties. By using the mathematical models, HPCE can be used for the determination of physicochemical characteristics of solutes, such as dissociation constants and partition coefficients into micelles. Determination of other equilibrium constants can be easily included—one interesting example being drug–protein-binding constants.

Depending on the type of sample, the selected variables and the available theoretical background, one can choose three types of approaches for modeling the migration behavior:

Physical Models

Electrophoretic mobility can be directly related to important solvent and solute properties, such as dielectric constant, viscosity, zeta potential, and size.

$$\mu = \frac{q}{6\pi\alpha\eta}$$

where q is the charge, α is its radius, and η is the buffer viscosity. Both q and α are rather sensitive to the electrolyte buffer used.

The use of fundamental physical models is always preferred, since an in-depth understanding of the migration mechanism is achieved. Extensive use of these models for optimization of HPCE separation is not feasible, however, because the physical parameters are difficult to measure. This will require numerous experiments to determine the relevant constants in the equation. However, the predictions can be expected to be fairly accurate, provided the initial experiments have been designed such that they sufficiently cover the selected parameter space.

Semiempirical Models

One can directly relate the observed migration behavior and the solvent characteristics through the use of semiempirical models. The constants in these equations are physically meaningful and reflect the physicochemical properties of solutes or their interaction with the surrounding environment. One can then obtain useful information about the migration mechanism of solutes. This approach, however, is limited to a small number of variables.

Phenomenological models that describe the influence of pH and micelle concentration on migration behavior of charged and uncharged solutes in CZE and MECC are such examples. These models are generally useful to describe the chemical equilibria, such as acid–base, partitioning into a pseudophase, or complexation. Another example of semiempirical modeling is the use of linear solvation free energy relationships (LSER) to describe the influence of solute–solvent interactions on migration behavior [19].

Empirical Models

For certain variables or combination of variables (e.g., temperature, ϕ_{org}/pH) physically meaningful models are unavailable. Empirical models can be quite useful in minimizing the number of required experiments to predict the optimum conditions. An advantage of empirical modeling is that the models can include any combination of variables, whereas the semiempirical models are limited to specific ones. On the other hand, physically meaningful information about solutes or their interactions with the media is not obtained from the empirical approaches.

These models can be based on regression techniques (e.g., polynomial or linear) or factor-based methods (e.g., factor analysis). An example of the use of polynomial models with interaction terms is to optimize the resolution on the basis of an overlapping resolution map [20]. Alternatively, it is possible to apply a simple linear model and a limited number of initial experiments to get a first impression of the location of the optimal separation [21].

In the following sections the influence of different parameters and the use of phenomenological models, target factor analysis models, and iterative regression strategies for the prediction and optimization of CZE and MECC separations will be discussed.

PARAMETERS IN HIGH-PERFORMANCE CAPILLARY ELECTROPHORESIS SEPARATIONS

In free-solution CZE, migration behavior of solutes is a function of their charge, size, and shape. Several variables can change these solutes properties and, consequently, their migration pattern. The most notable variables include pH, organic solvents, temperature, type of buffer, ionic strength, and addition of complexing reagents. The most important factor that influences selectivity of large groups of compounds is pH, which determines the charge of solutes. Obviously, the type of variable that influences selectivity in migration is a function of sample type. For instance, pH has a great influence on migration behavior of ionizable compounds. The selective effect of other factors on migration is limited to certain groups of compounds. The use of organized media, such as micelles and cyclodextrins (CDs) will be treated separately under the context of electrokinetic chromatography.

It is also important that some variables such as pH and organic solvents change electroosmotic flow in addition to selectivity, thereby influencing resolution. One

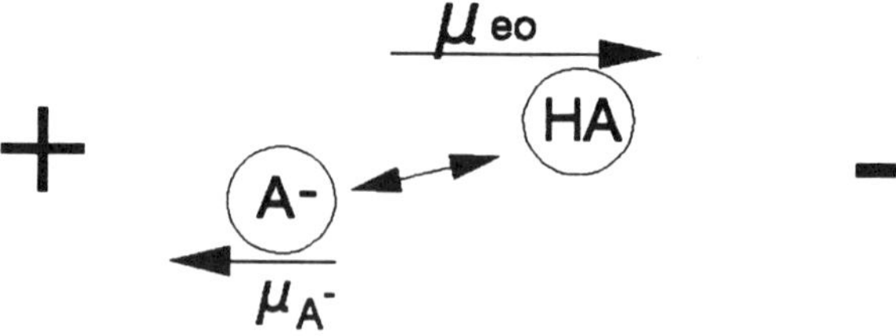

Figure 1 A schematic representation of the migration of an acidic compound in capillary zone electrophoresis.

can thus conclude that although it is important to identify the factors that affect migration selectivity, their influence on electroosomotic flow cannot be ignored. Therefore, it is important to differentiate those situations during which a better separation is achieved by simply reducing electroosmotic flow (i.e., better separations in longer analytic times) from those in which selectivity enhancement plays an important role.

Optimization of pH in Capillary Zone Electrophoresis

The migration behavior of a typical acid and a base in CZE are illustrated in Figs. 1 and 2. When the solute is uncharged (HA or B), it moves with the buffer at the electroosmotic velocity. A simple phenomenological model assumes that the electrophoretic mobility of an acid (or a base) is directly proportional to the fraction of dissociation of acid ($F_{A^-} = K_a/(K_a + [H^+])$ (or protonation of a base) and the mobility of the fully dissociated acid (or fully protonated base) is

$$\mu = \frac{\mu_{A^-}\,(K_a/[H^+])}{1 + (K_a/[H^+])} \tag{2}$$

where μ is the electrophoretic mobility of the partially dissociated solute, μ_{A^-}, is the mobility of the anionic form of the acid, and K_a is the acid dissociation constant [22].

This equation predicts a sigmoidal behavior for the variation of mobility as a function of pH. At low pH values ($pK_a - 2$) the acid is fully protonated and the electrophoretic mobility is zero. Conversely, at high pH values ($pK_a + 2$), the acid is fully deprotonated and the mobility is the limiting mobility μ_{A^-}. If the values of the two constants (pK_a and μ_{A^-}) for a given solute are known, mobility of the solute can be predicted over the entire pH range. A similar equation can be written for polyprotic acids as

$$\mu = \frac{\mu_0 + \mu_1(K_{a1}/[H^+]) + \mu_2\,(K_{a1}K_{a2}/[H^+]^2) + \ldots}{1 + (K_{a1}/[H^+]) + (K_{a1}K_{a2}/[H^+]^2) + \ldots} \tag{2a}$$

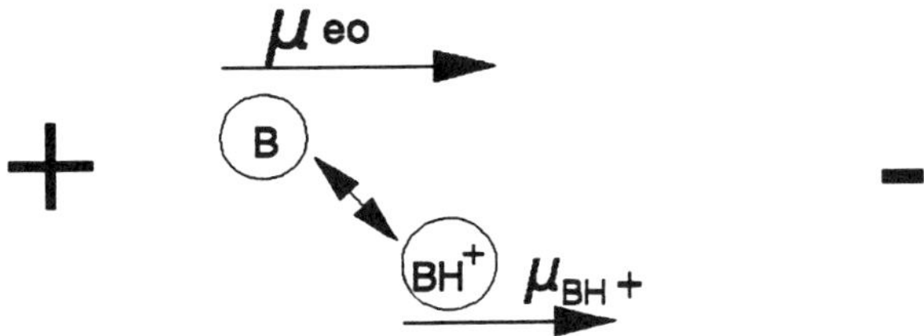

Figure 2 A schematic representation of the migration of a basic compound in capillary zone electrophoresis.

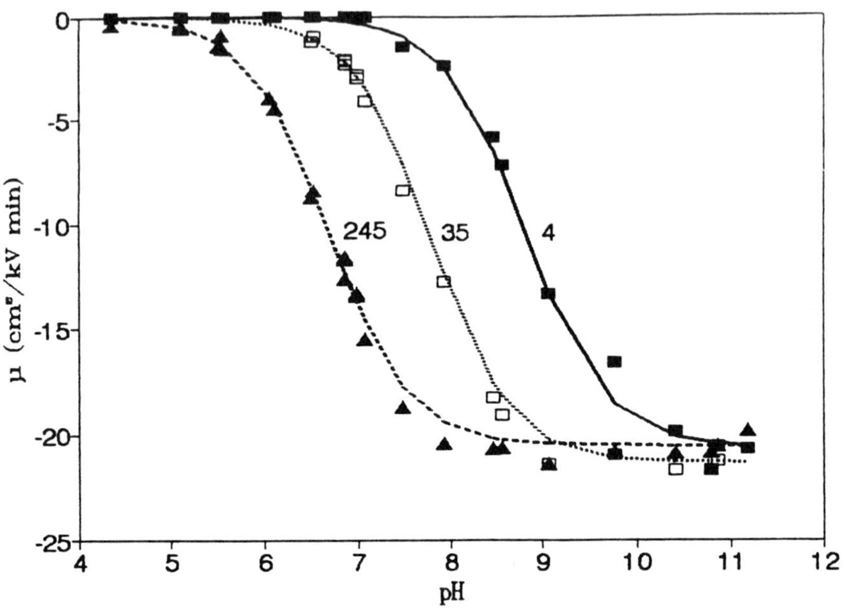

Figure 3 The dependence of mobility on pH in the absence of surfactant for three chlorophenols. (From Ref. 23)

Table 1 Estimates of μ_{A^-} and pK_a by Weighted Nonlinear Regression of Data From CZE Experiments[a]

	WNLIN	
Compound	μ_{A^-}(cm² kV⁻¹ min⁻¹)	pK_a
2-Chlorophenol	−23.92	8.20
	±0.58	±0.03
3-Chlorophenol	−22.40	8.83
	±.99	±0.04
4-Chlorophenol	−20.71	8.83
	±0.23	±0.05
2,4-Dichlorophenol	−21.94	7.56
	±0.21	±0.01
2,5-Dichlorophenol	−22.53	7.18
	±0.21	±0.01
3,5-Dichlorophenol	−21.72	7.79
	±0.51	±0.02
2,4,5-Trichlorophenol	−20.88	6.71
	±0.26	±0.02

[a]Using Eq. 2.
Source: Ref. 24.

An estimate of dissociation constants can be obtained by fitting the experimentally measured mobility data at different pH values to the model using weighted nonlinear regression (WNLIN) or weighted linear regression (WLIN) of the linearized model [23]. In other words, based on a few initial mobility measurements it is possible to simulate the variation of mobility over a wide range of pH values.

Figure 3 shows that the equation fits the experimental data quite nicely. Estimates of the μ_{A^-} and pK_a values for a group of chlorophenols are listed in Table 1. An excellent agreement is observed between the predicted and observed mobilities as illustrated in Fig. 4 in which the slope of the line is 1.007 and $r^2 = 0.996$.

The accuracy of predictions depends on the accuracy of the model (see Eq. 2), the reproducibility of data, the proper selection of the parameter space, and the number of initial experiments.

Number of Required Measurements

One important criterion to evaluate the effectiveness of an optimization procedure is the minimum number of initial experiments that is required to predict the optimum conditions for separation. The operating pH range should encompass the migration behavior of the fully protonated and dissociated forms. The accuracy of

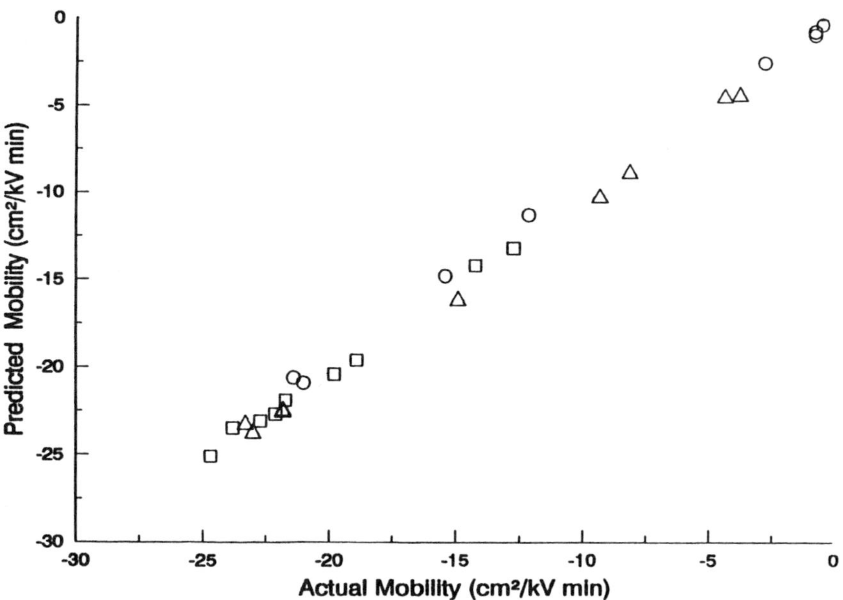

Figure 4 The correlation between predicted and actual mobility for phenol and eight substituted phenols at different pH values pH 8.2 (□), 9.5 (△), 10.5 (○). (From Ref. 24)

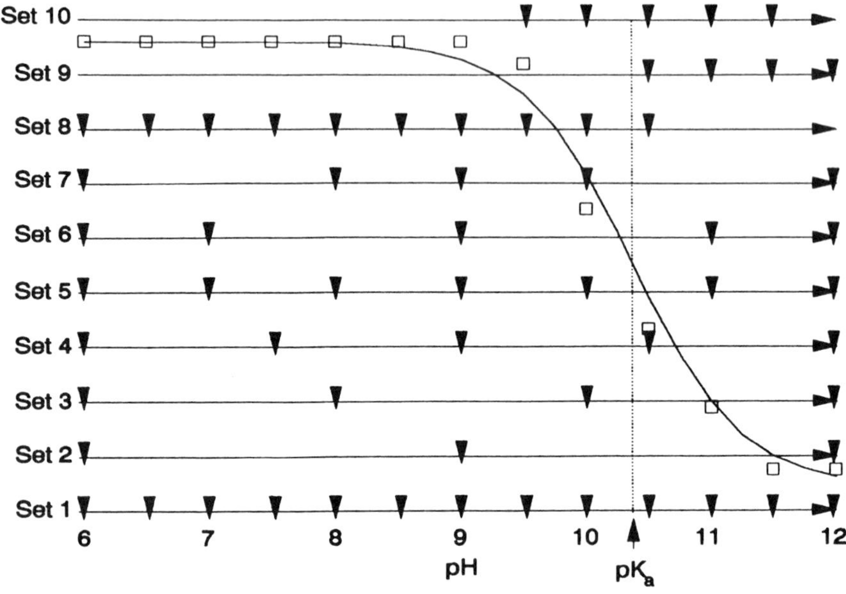

Figure 5 Data sets used in determining pK_a and m_{A^-} superimposed over the mobility of 4-isopropylphenol as a function of pH. Triangular tick marks along each pseudo-x-axis indicate the pHs included in the data set; □ indicates the average mobility at each pH. (From Ref. 24)

Table 2 Comparison of Anion Mobilities and Dissociation Constants Calculated from Reduced pH Data Sets for 4-Isopropylphenol

Set number	Number of pH data sets	m_{A^-} Estimate	m_{A^-} 95% CI[d]	pK^a Estimate	pK^a 95% CI
Evenly spaced sets[a]					
1	13 (0.5)[b]	−14.04	0.82	10.37	0.16
2	3 (3)[c]	−17.71	70.52	11.49	7.27
3	4 (2)	−13.71	0.28	10.20	0.09
4	5 (1.5)	−13.77	1.06	10.27	0.31
5	7 (1)	−13.75	0.65	10.27	0.21
Other sets					
6	5 (6, 7, 9, 11, 12)	−13.87	0.94	10.38	0.38
7	5 (6, 8, 9, 10, 12)	−13.75	1.05	10.27	0.33
8	pH pK_a	−31.87	50.01	10.89	0.91
9	pH/pK_a	−13.79	0.78	10.22	0.25
10	$pK_a - 1 < pH > pK_a + 1$	−14.82	4.29	10.42	0.46

[a]All pH data sets cover the range of pH 6–12. The increments between the evenly-spaced sets are indicated in parentheses.
[b]Represents the best estimate of the anion mobility and dissociation constant.
[c]The nonlinear regression iteration did not converge.
[d]Confidence interval
Source: Ref. 24.

estimated parameters by WNLIN depends on where along the mobility versus pH curve the data points are gathered. The effect of the number of initial data points and the pH region on the estimation of the parameters of Eq. 2 is shown in Fig. 5 and Table 2. Examination of the data in Table 2 reveals that as few as four evenly spaced pHs (set 3) are required to accurately estimate pK_a and μ_{A^-}. Unevenly spaced data can also provide good estimates of pK_a and μ_{A^-}, as seen with sets 6 and 7. Sets 8, 9, and 10 examine the quality of the estimate if only a small pH range (approximately 2 pH units) is considered.

Using Literature pKa

According to Eq. 2, if the value of pK_a is available, only one experiment would be needed to determine μ_{A^-} and, subsequently, predict the migration behavior of each solute over the entire pH range. This can be done by measuring mobility at high pH (pK_a + 2), which is a direct measurement of μ_{A^-}. The effect of using literature pK_a values on the goodness of the fit of Eq. 2, as compared with the WNLIN estimates is illustrated in Fig. 6. When the literature and WNLIN pK_a values are close (e.g., for phenol), the mobility–pH curve using Eq. 2 closely fits the actual data. If there is a significant error (as with 3,5-dichlorophenol), the estimation of mobility could be off by as much as 20% in the region about the pK_a.

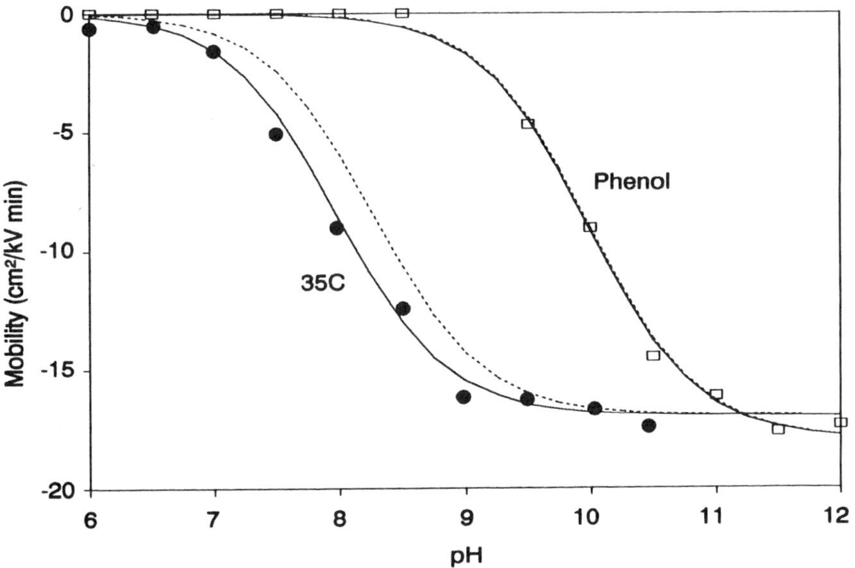

Figure 6 Comparison of the use of literature values of pK_a, (dashed line) and WNLIN estimate of pK_a (solid line) in Eq. 3 for phenol (□) and 3,5-dichlorophenol (●) relative to the actual data. (From Ref. 24)

Elution Order Prediction

Uncertainty in the mobilities of two solutes, with relatively close mobilities, can lead to uncertainty in their elution order if the mobilities are relatively close. It is possible to check the degree of overlap of the error envelopes as an initial screening to see if elution order reversal may be a possibility. The error envelopes of two closely eluting pairs of solutes are presented in Fig. 7. This figure illustrates the wide variation that can exist in the error envelope (compare 35C with the others) and the high degree of overlap of error envelopes between some closely eluting solutes (4I and 4M). This shows that, over the pH range, it is possible for the elution order of neighboring peaks to be reversed: 4I and 4M below pH 10.5, and 35C and 245C above pH 10.5. Note that the error in mobility is smallest at pH $< pK_a - 1$. However, operating in this pH range is impractical, because the solute has little or no mobility.

Prediction of Optimum

The migration behavior of substituted phenols were predicted, based on four initial experiments within the pH range of 6–11 using Eq. 2 (Fig. 8). Note that the pairs 4I and 4P, 4B and 4C, and 4E and 4M coelute over almost the entire pH range. The pK_as and μ_{A^-}s of these three pairs at 25C were nearly identical. In Fig. 9, the

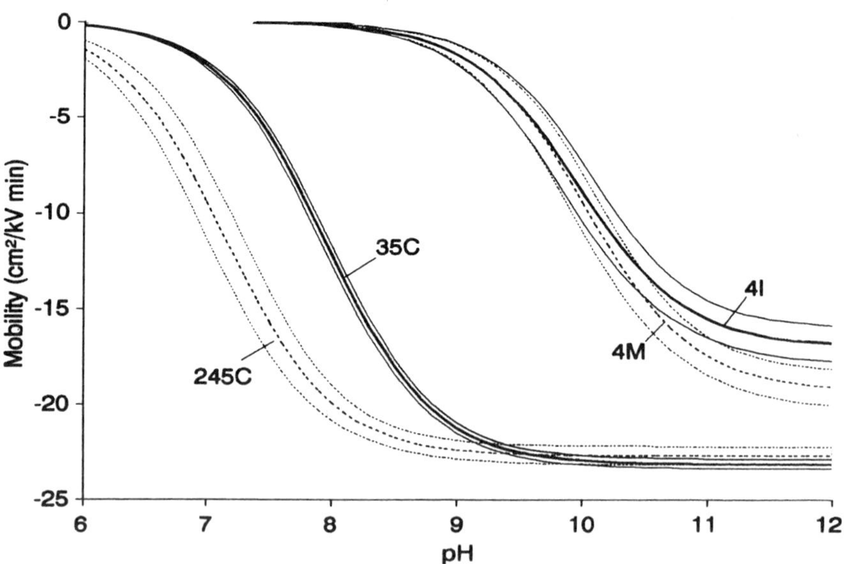

Figure 7 The variation in mobility (center lines) and standard deviation as a function of pH for two closely eluting peak pairs. (From Ref. 24)

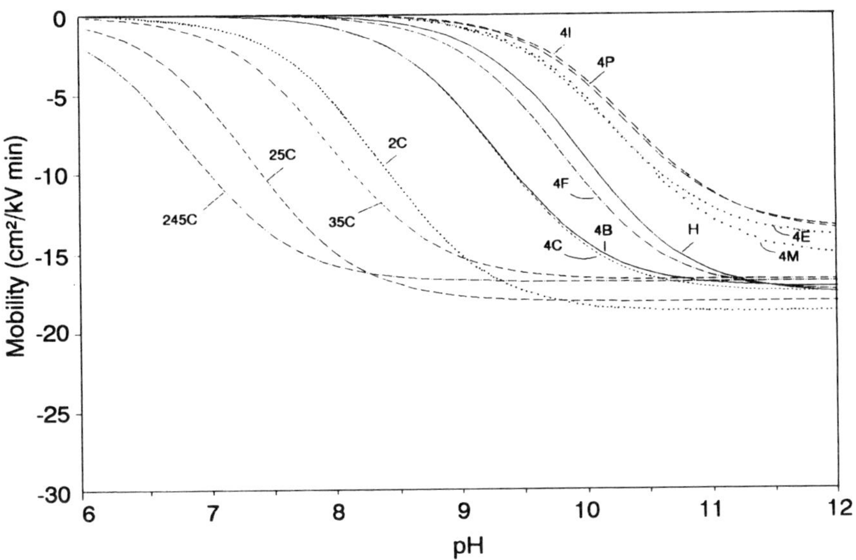

Figure 8 The mobility of 12 substituted phenols as a function of pH at 25°C. 2,4,5-Trichloro-(245C); 2,5-dichloro-(25C); 3,5-dichloro-(35C); 2-chloro-(2C); 4-chloro-(4C); 4-bromo-(4B); 4-floro-(4F); phenol(H); 4-methoxy-(4M); 4-ethoxy-(4E); 4-iodo-(4I); and 4-propoxy-(4P) phenols. (From Ref. 24)

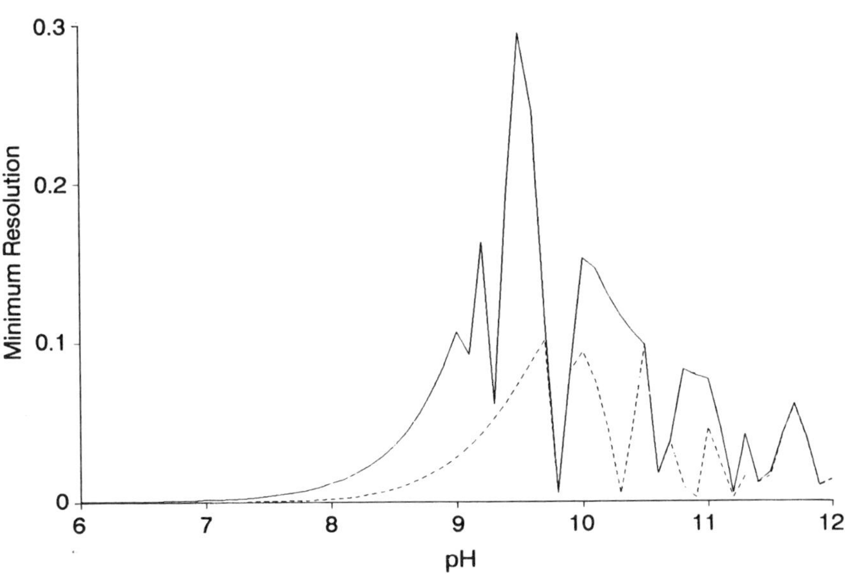

Figure 9 The window diagram of the minimum resolution of the worst-resolved peak pair. (From Ref. 24)

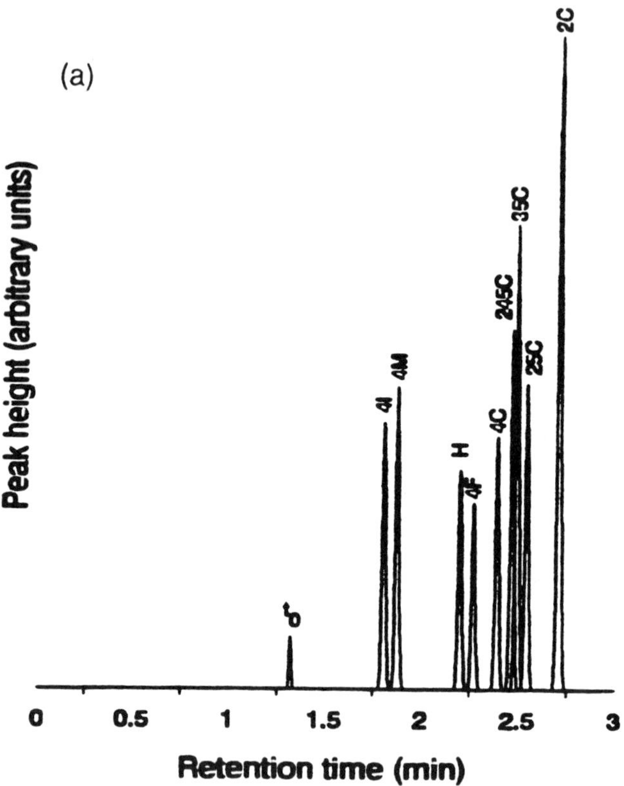

Figure 10 The (a) actual and (b) predicted electropherograms of the nine phenols. (From Ref. 24)

window diagram using all 12 substituted phenols is compared with the window diagram of one of the solutes in each of the coeluting pairs is removed. Removing the three solutes shifts the optimum pH and greatly improves the predicted minimum resolution. Figure 10 illustrates a good agreement between the predicted and observed electropherogram of a mixture of nine phenols at an optimum pH of 10.5. With examples for which coelution of solutes occurs over the entire pH range (see Figs. 8 and 9) the use of a second variable (e.g., micelles, organic solvents, temperature) to enhance the resolution is essential.

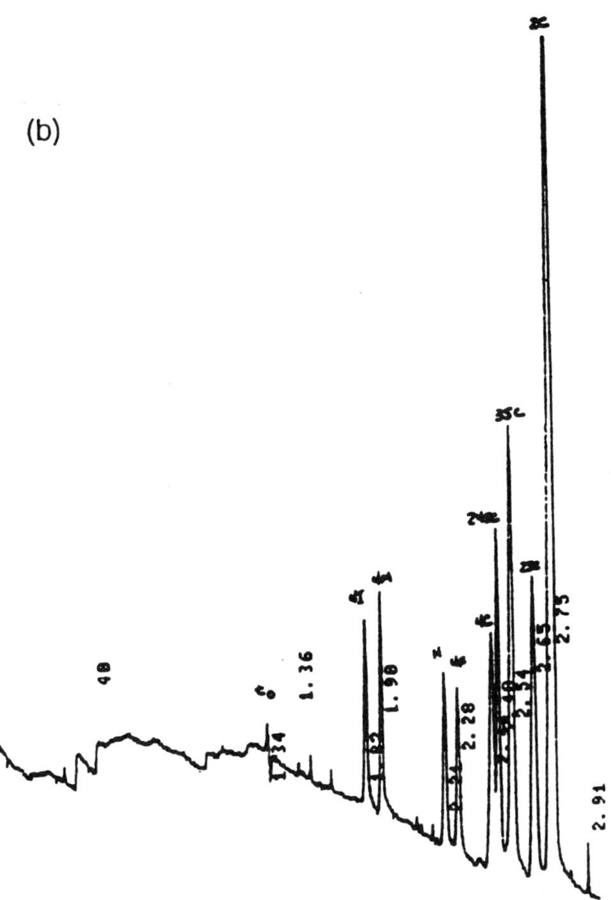

Temperature Effect in Capillary Zone Electrophoresis Separations

The data presented in Table 3 illustrate the considerable difference in both pK_a and μ_{A^-} as a function of temperature for all nine substituted phenols. The mobility–pH curves for phenol, 4-isopropylphenol, and 3,5-dichlorophenol presented in Fig. 11 illustrate the effect of temperature change on selectivity. The solutes' mobilities increase at higher temperature, probably because of a change in their diffusion coefficients, as described by the Einstein-Nernst equation. However, the increase in mobilities is different for various solutes, which affects the selectivity between solutes.

Table 3 Comparison of Dissociation Constants and Anion Mobilities of Nine Substituted Phenols at 25° and 40°C

	pk_a		Mobility ($cm^2\ kV^{-1}\ min^{-1}$)	
	25°C	40°C	25°C	40°C
4I[a]	10.37	9.95	−14.04	−16.92
4M	10.29	10.01	−15.54	−19.26
H	9.98	9.69	−17.96	−22.59
4F	9.81	9.59	−17.76	−22.93
4C	9.30	9.11	−17.81	−22.79
2C	8.38	8.29	−18.97	−25.26
25C	7.32	7.34	−18.34	−23.47
35C	7.98	7.95	−16.98	−23.18
245C	6.83	7.14	−17.08	−22.74
4P	10.33		−13.87	
4E	10.22		−14.36	
4B	9.29		−17.49	

[a]See Fig. 8 for abbreviations.
Source: Ref. 24.

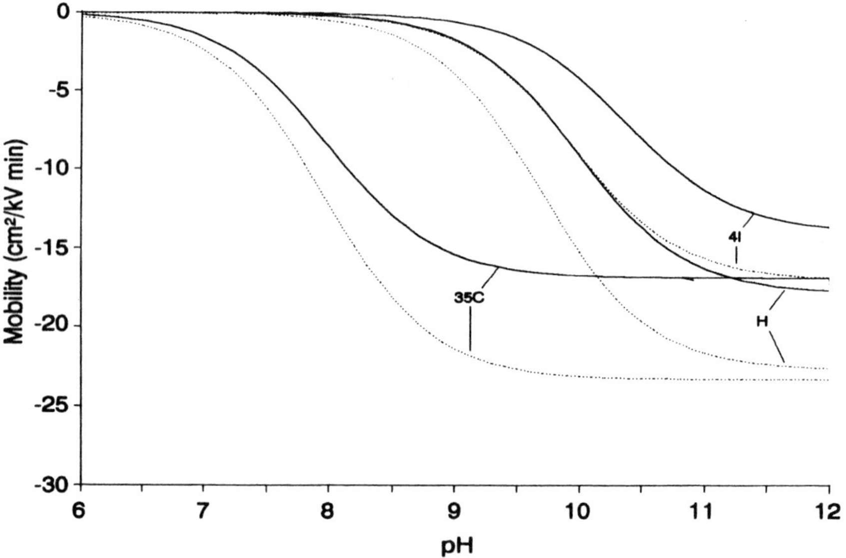

Figure 11 The mobility of phenol, 4-isopropylphenol, and 3,5-dichlorophenol as a function of pH at 25°C (solid line) and 40°C (dotted line). (From Ref. 24)

In addition, the downward shift in pK_a with increased temperature leads to higher mobility at lower pH. The combination of changes in pK_a and μ_{A^-} can substantially alter selectivity, as seen with the elution order reversal of 3,5-dichlorophenol and 4-isopropylphenol.

Capillary Electrophoresis in Nonaqueous Media

Organic solvents have been used in aqueous media of CZE and MECC separations for two primary reasons: enhancing solubility of solutes and reducing the electroosmotic flow. Performing electrophoretic separations in the presence of nonaqueous solvents can also be advantageous from the selectivity point of view, owing to possible preferential solvation as well as different acid-base properties of analytes in the presence of organic solvents. The use of totally nonaqueous media in CE is virtually

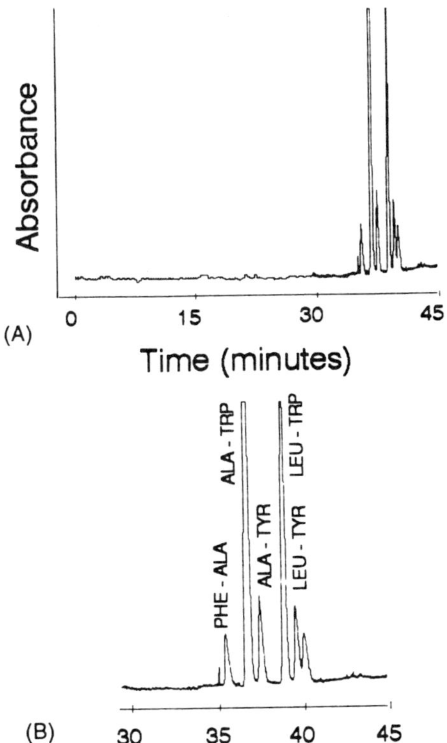

Figure 12 Separation of a mixture of peptides in CZE using formamide as solvent. [NaH_2PO_4], 0.25 M; apparent pH, 3.0; capillary length, 57 cm (50 cm up to the detector); id, 75 μm (A) the entire chromatogram; (B) enlarged segment of the electropherograms that contained the peptide peaks. (From Ref. 26)

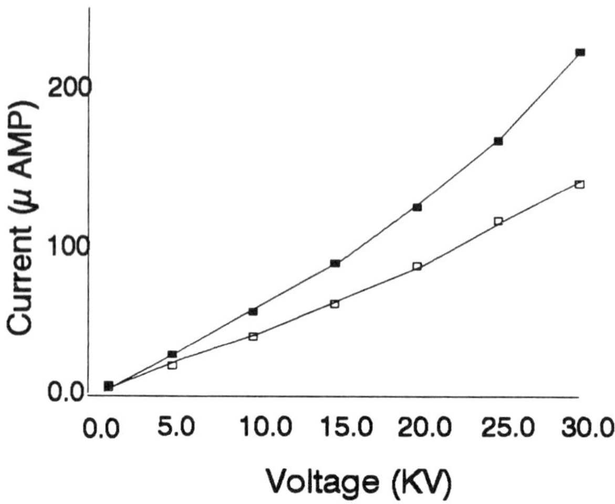

Figure 13 Current as a function of voltage in formamide (open squares) and aqueous buffer (solid squares). [K_2HPO_4], 0.1 M in formamide and 0.04 M in aqueous buffer. (From Ref. 26)

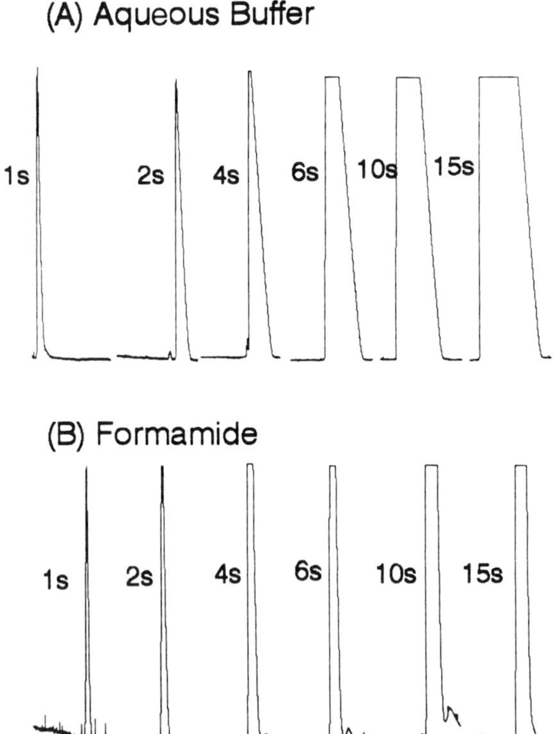

Figure 14 Effect of sample stacking in (A) aqueous buffer and (B) formamide. Injection time, *ns* = *n* (1, 2, 4, 6, 10, or 15) seconds. (From Ref. 26)

an unexplored area. To the best of our knowledge, there has been only one report of a CZE separation in acetonitrile [25].

Water is a suitable medium for electrophoresis because of its high dielectric constant, which facilitates the dissolution and dissociation of salts and, consequently, provides an electrically conducting medium. In addition, an enormous amount of available fundamental physicochemical information, such as acid–base properties, ion mobility, and conductance, for aqueous media facilitates their application. There are several organic solvents that have physical properties similar to water and, therefore, are suitable media for electrophoresis.

Figure 12 shows an electrophoretic separation of a group of dipeptides at low pH in formamide. Electroosmotic flow is slower in this solvent compared with that in water [26]. An interesting phenomenon is that ionic mobilities in formamide are much lower than that in water; thus, the electric current is much smaller in the former medium at a constant ionic strength and electric field strength (Fig. 13 i vs E). As a result, one can either use a higher salt concentration or higher electric field strength at a given operating current. The use of higher salt concentrations increases the amount of sample that can be injected without severe loss of efficiency (i.e., sample capacity is increased). This is illustrated in Fig. 14. In addition to extending the sample capacity, another implication of this property is that it makes the application of sample stacking more feasible in formamide. The possibility of using higher electric field strengths can result in reduction of analysis time or in efficiency enhancement. This is shown in Fig. 15 where the CZE experiments were performed at constant current and salt concentration in the aqueous and nonaqueous media. The field strength, therefore, is larger in formamide.

These results, along with different selectivity, indicate the potential of nonaqueous media in broadening the applicability of CE separations.

MICELLAR ELECTROKINETIC CAPILLARY CHROMATOGRAPHY SEPARATIONS: CHARGED AND UNCHARGED SOLUTES

First introduced by Terabe and co-workers, MECC has extended the enormous power of CZE to the separation of uncharged solutes [27,28]. The high efficiency of CZE is often adequate to separate charged compounds with very small differences in mobility; therefore, one must have compelling reasons for using micelles in a CZE system for the separation of charged compounds.

Micelles provides both ionic and hydrophobic sites of interaction simultaneously, making MECC preferable to CZE for the separation of mixtures of charged and uncharged solutes. Another application is the separation of charged solutes with identical electrophoretic mobilities. The question is whether micelles (or other forms

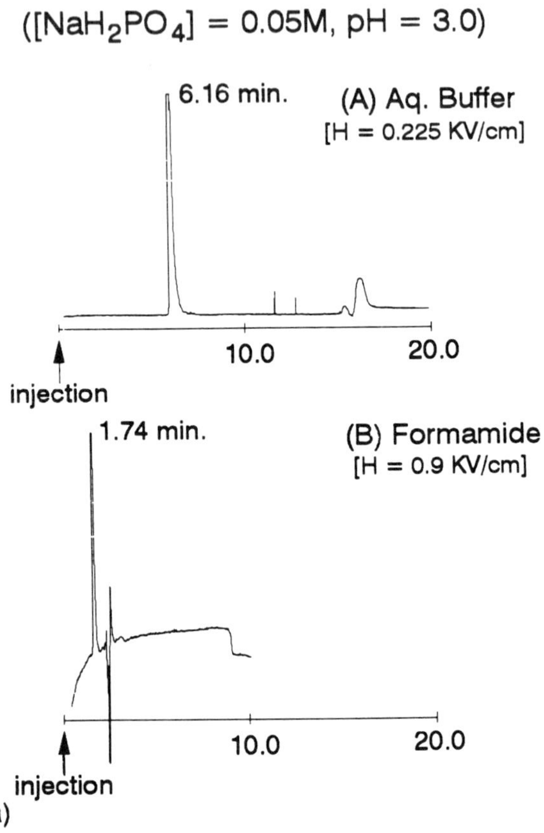

Figure 15 (a) Comparison of migration time of a cation at 100 μAMP in (A) aqueous buffer and (B) formamide. (b) Comparison of migration time of an anion at 85 μAMP in (A) formamide and (B) aqueous buffer. (From Ref. 26)

of organized media) would provide the additional selectivity needed for the separation of compounds of this nature.

Similar to chromatography, the primary mechanism for separation of uncharged solutes in MECC is their differential partitioning into micelles. A specific feature of MECC, however, is that the elution times of neutral compounds are encompassed within an elution window, defined by the elution time of an "unretained" solute, t_0, and the migration time of micelles, t_{mc}. Resolution in MECC separation of uncharged solutes is defined (28) as

$$Rs = \left(\frac{\sqrt{N}}{4}\right)\left(\frac{\alpha - 1}{\alpha}\right)\left(\frac{k'_2}{1 + k'_2}\right)\left(\frac{1 - t_{e0}/t_{mc}}{1 + (t_{e0}/t_{mc})k'_1}\right) \tag{3}$$

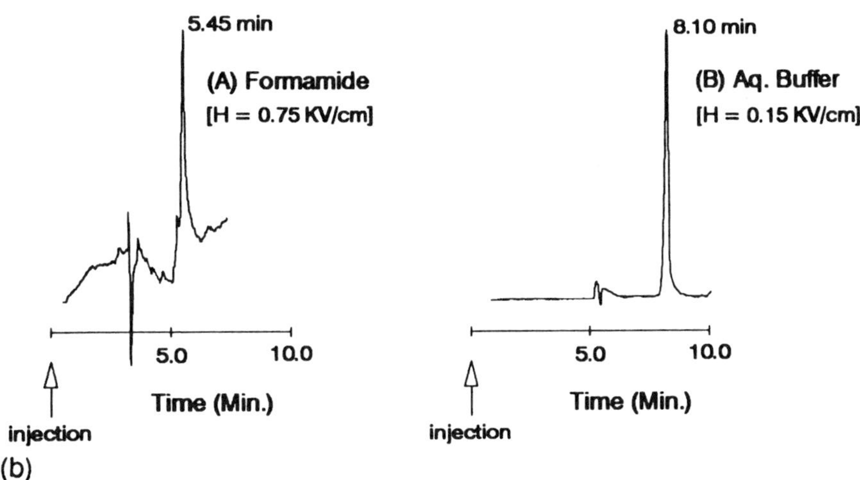

(b)

Therefore, in addition to efficiency, resolution in MECC can be enhanced by optimizing selectivity, α, migration factor, k', and increasing the elution window (reducing t_{e0}/t_{mc}). This can be accomplished by a careful selection of several solvent variables such as micelle concentration, addition of organic solvents, use of mixed micelles, and pH for ionizable compounds.

In this section, different models are presented to quantitatively describe the migration behavior of acidic and basic compounds in MECC systems with anionic micelles (e.g., sodium dodecyl sulfate; SDS). The use of these models for the optimization of separation will also be described.

Two important variables that influence the electrophoretic migration of ionizable solutes are pH and micelle concentration. The models derived in the following describe the effects of these two factors on migration factor, (k'), and electrophoretic mobility, μ.

For the derivation of the models, a phenomenological approach was used. This approach, which has been previously reported for HPLC as well as MECC [29–31], assumes that the net migration parameter of an ionizable solute (k', μ, or velocity) is the weighted sum of the migration parameter of the solute in the undissociated and dissociated forms, in the micelles and in the bulk solvent.

Migration Factor k' Versus [M]

In MECC, the extent of retention of uncharged solutes is directly proportional to their interaction with the micelles. As in chromatography, the term *capacity factor*,

k', has been used to describe the retention of solutes owing to the existence of a partitioning mechanism. In this chapter, the term *migration factor* is used instead of capacity factor for k', which seems to be more representative of the process occurring in MECC, especially for the ionizable compounds.

Migration factor is linearly related to micelle concentration (which is proportional to the phase ratio). This linear relation can be used for separation optimization of uncharged solutes in MECC (32).

$$k' = P_{mw}\, \nu\, ([S] - CMC) \tag{4}$$

where P_{mw} is the partition coefficient of solute into micelles, ν is the molar volume of the surfactant, and $[S]$ is the total surfactant concentration. This equation is valid for both uncharged and ionizable compounds at low micelle concentrations. The main difference between the uncharged and acidic solutes is that the P_{mw} for acidic solutes is a function of the pH of the medium as described in the following.

Uncharged Solutes

Terabe et al. have already reported the migration model for uncharged solutes [27,28]. Figure 16 presents the typical migration behavior of a neutral solute in MECC. The mobility of a neutral solute in MECC, μ_n, is proportional to the mobility of the micellar phase, μ_{mc}:

$$\mu_n = \left(\frac{k'}{1 + k'}\right) \mu_{mc} \tag{5}$$

Since *migration factor*, k', is defined as the ratio of the amount of solute associated with micelles to that in the aqueous phase, the term $k'/(k' + 1)$ represents the mole fraction of the solute in micellar pseudophase. Equation 5 can also be expressed as

$$k' = \frac{\mu_n}{\mu_{mc} - \mu_n} \tag{6}$$

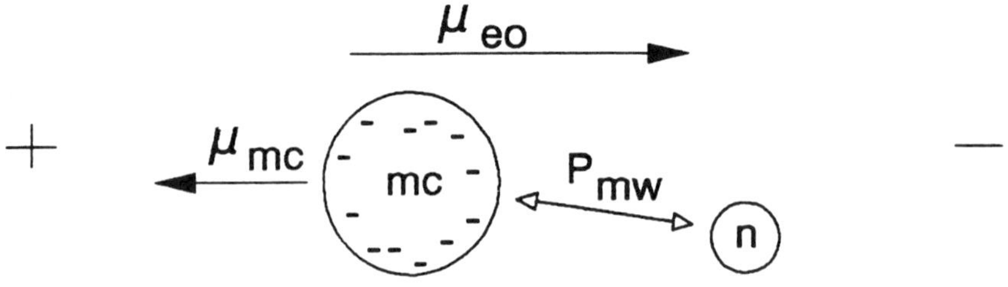

Figure 16 The interaction of a neutral solute (n) with a micelle (mc) in MECC. (From Ref. 23)

By substituting retention times for mobilities, one would arrive at the following equation, which was originally reported by Terabe et al. [27]:

$$k' = \frac{t_r - t_{e0}}{t_{e0}(1 - t_r/t_{mc})} \quad (7)$$

Figure 17 shows the linear relation between k' and total surfactant concentration for several neutral compounds. As shown, all lines almost pass through the same intercept (equal to the critical micelle concentration [CMC]; 0.004 M), and the slope of the line (P_{mw}) increases with the hydrophobicity of the compound. This value is smaller than the frequently reported literature CMC value of 8 mM (in pure water and at 25°C). This is due to the presence of buffer and, consequently, higher ionic strength as well as higher operating temperature in this study. A good agreement exists between the P_{mw} values measured by MECC and those reported in the literature. The data presented in Fig. 17 and Table 4 demonstrate that the relation in Eq. 4 is accurate for our test solutes.

The migration behavior of ionizable solutes in MECC is more complicated than those of the neutral compounds for two reasons. First, ionizable solutes have an extra migration mechanism that influences the separation process; that is, in addition to partitioning into (or association with) micelles and moving at the velocity of micelles, ionizable compounds possess their own characteristic electrophoretic mobilities. Second, the involvement of additional chemical equilibria (acid–base, ion-pairing) would further add to the complexity of the system. Accordingly, the number

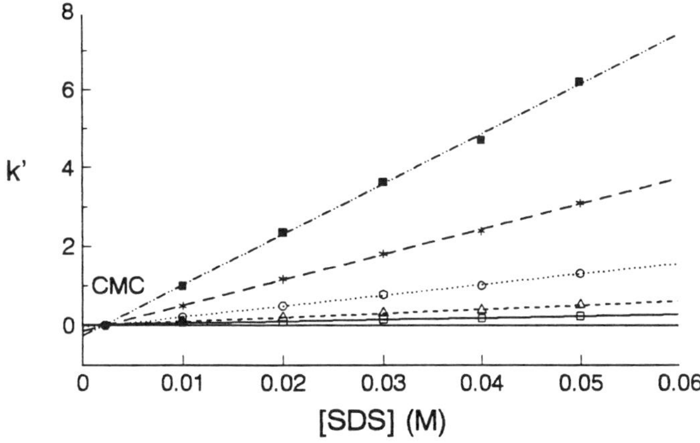

Figure 17 The relation between k' and [SDS] for the neutral solutes 2-naphthol (■), toluene (*), nitrobenzene (O), phenol (△), and resorcinol (□). (From Ref. 23)

Table 4 P_{mw} of Neutral Solutes from Slopes of k' vs [SDS] as Shown in Fig. 17

Solute	P_{mw} (lit. [27])	P_{mw} (Fig. 17)
Resorcinol	22	19
Phenol	52	49
Nitrobenzene	135	131
Toluene	318	305
2-Naphthol	656	720

Source: Ref. 23.

of experimental variables that influence the separation would be rather extensive. Some examples include pH, surfactant type and concentration, and concentration and type of the buffer. These additional equilibria, if optimized effectively and efficiently, would provide additional tools to manipulate selectivity and, thereby, enhance separation.

Anions and Acidic Solutes

The migration behavior of a weak acid (and that of an anion) is shown schematically in Fig. 18 [23]. Note that an anionic solute is a special case of an acid that is fully dissociated. In contrast with uncharged solutes, not all retention of anions is explained by interactions with the micellar pseudophase, since these compounds have a negative electrophoretic mobility in the aqueous phase. Therefore, in MECC the mobility of an anion would be the weighted sum of the mobility of the micellar

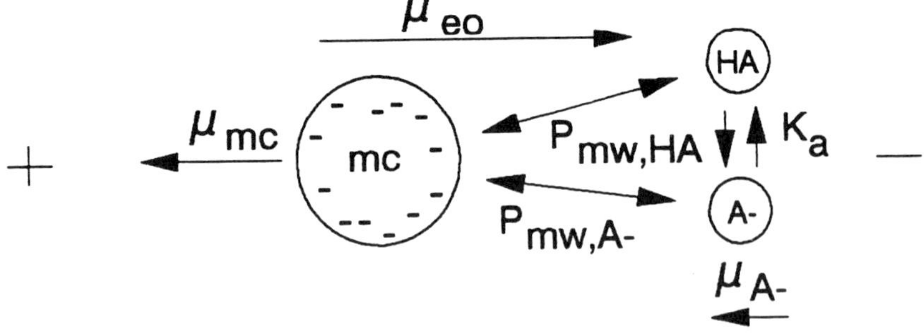

Figure 18 The interaction of a weak acid (HA) and its conjugated base (A^-) with a micelle (mc) in MECC. (From Ref. 23)

phase and its own mobility in the aqueous phase. This can be accounted for by the observed overall mobility in the absence of micelles, μ_0:

$$\mu_n = \left(\frac{k'}{1 + k'}\right) \mu_{mc} + \left(\frac{1}{1 + k'}\right) \mu_0 \tag{8}$$

As shown in Fig. 18, two migrating fractions can be discerned: the acid–conjugated base pairs that are associated with the micelles and migrate at a mobility of μ_{mc}, and the anionic (dissociated) form in the aqueous phase moving at a mobility of μ_0. Another important migration mechanism is through the electroosmotic flow. All equations, however, are expressed in terms of electrophoretic mobilities, because the electroosmotic mobility, μ_{e0}, is constant for all components in the system.

Note that in deriving these equations, several assumptions have been made: the mobility of micelles is not altered when associated with analyte, secondary chemical equilibria with buffer components do not occur, the possible effects of the ionizable analytes and their equilibria on the zeta potential and, hence, electroosmotic flow are insignificant.

From Eq. 8, one can derive the following equations:

$$k' = \frac{\mu - \mu_0}{\mu_{mc} - \mu} \tag{9}$$

or

$$k' = \frac{t_r - t_0}{t_0 (1 - t_r/t_{mc})} \tag{10}$$

where t_0 is the retention time in the absence of micelles. A similar equation has been reported in the literature in terms of velocity, rather than mobility; however, it was not verified experimentally [33].

The linear relation between k' and $[S]$ is also valid for ionic solutes (both anionic and cationic). The major difference between the ionic and nonionic compounds would be in the relation between k' and different mobilities and that for ionizable compounds P_{mw} is a function of the pH. Figure 19b shows the k' versus $[S]$ plot for a group of acidic compounds (chlorinated phenols), using the corrected model (see Eq. 9). As shown, all lines pass through the same origin (CMC). When the same plot was made using the uncorrected model, there were widespread intercept values for different compounds (see Fig. 19a). The P_{mw} values that were determined from the corrected and uncorrected models are listed in Table 5. For 2CP and 3CP, the two P_{mw} values were almost identical because, at pH 7, both of these compounds are nearly protonated in micellar solution and, therefore, there μ_0 is almost zero. The corrected P_{mw} values for other compounds were all smaller than the uncorrected values. This is expected, because the contribution of the electrophoretic mobilities of the ions have been removed.

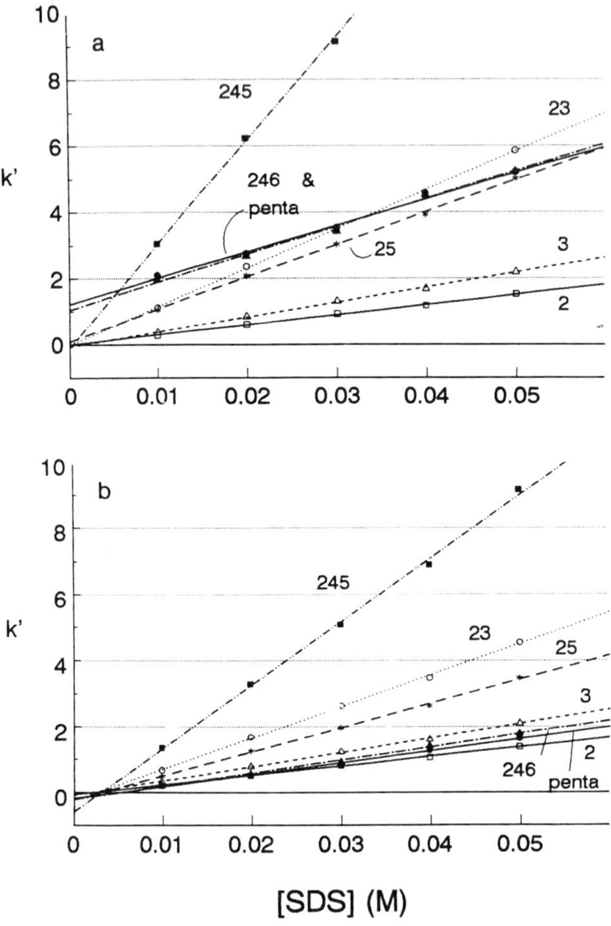

Figure 19 The relation between k' and [SDS] (a) without and (b) with correction for solute mobility in absence of micelles for seven chlorophenols. (From Ref. 23)

Table 5 P_{mw} of Chlorophenols From Slopes of k' vs [SDS] as Shown in Fig. 19

Solute	P_{mw} (uncorrected)	P_{mw} (corrected)
2CP	122	116
3CP	179	175
23CP	467	380
25CP	389	292
245CP	1241	766
246CP	334	159
PentaCP	316	145

Cationic and Basic Solutes

It can be expected that the observed migration of these compounds is more difficult to describe. This is due to the electrostatic interaction between the free surfactant ions in the mobile phase with the cations, which introduces an additional equilibrium (ion-pairing formation) into the equations.

The various equilibria playing a role in the migration of a cationic solute in MECC are depicted in Fig. 20. First, there will be a dissociation of the solute in the aqueous phase, represented by the equilibrium constant K_b, which also explains the major influence of pH on the observed migration [34]. Both the base, B, and its conjugate acid, BH^+, will partition into (or associate with) the micellar phase, each according to its own distribution coefficient, $P^{m}_{BH^+}$ or P^{m}_{B} respectively. Also, there will be an interaction between the free ionized surfactant and the positively charged cations. Finally, there exists the possibility of solute-wall interactions. This ion-pairing equilibrium is defined by the equilibrium constant K_{IP}. Since the concentration of free surfactant in micellar solutions is in the first approximation constant and equal to the critical micelle concentration (CMC), we can define the constant K_I = (CMC) K_{ip}. The observed mobility of a solute will be a combination of the mobilities of the various species of the solute, according to the fractions, f, of these species.

According to the phenomenological approach, the observed mobility can now be expressed as

$$\mu_s = f_{BH^+,aq}\,\mu_c + (f_{BH^+,mc} + f_{B,mc})\,\mu_{mc} \tag{11}$$

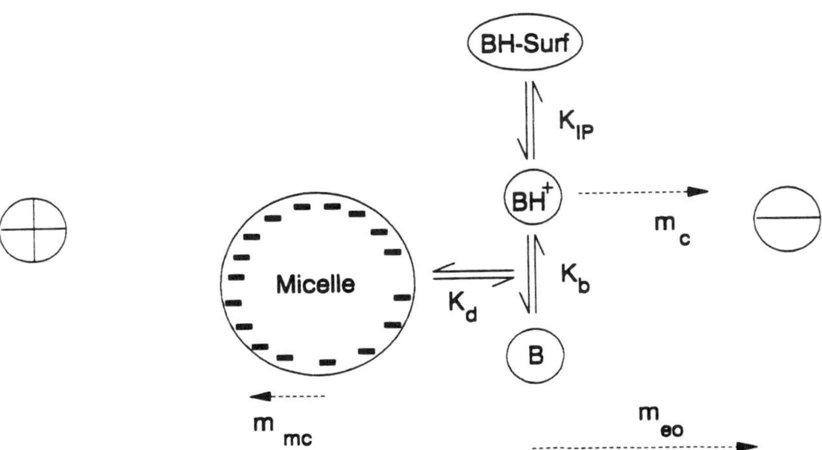

Figure 20 A schematic representation of the migration mechanism in the MECC of a cationic solute. (From Ref. 34)

where μ_c refers to the mobility of the cationic species in the aqueous phase. The following equation can be derived for cationic solutes:

$$k' = \frac{\mu - \left(\dfrac{1}{1 + K_a/[H^+] + K_I}\right)\mu_c}{\mu_{mc} - \mu} \tag{12}$$

From Eq. 12, several observations are immediately apparent:

1. As in all calculations for k' in MECC, this equation shows that k' will approach infinity if the observed mobility of the solute becomes equal to the mobility of the micelles (i.e., complete partitioning of all species of a solute into the micelles). This is directly linked with the limited migration range in MECC.

2. In complete complexation of the cations with the surfactants (i.e., K_I approaches infinity), Eq. 12 will reduce to

$$k' = \frac{\mu}{\mu_{mc} - \mu} \tag{13}$$

which is identical with the relation derived for uncharged solutes (see Eq. 6). This is in accordance with expectations, since the cation–surfactant complex will not have an electrophoretic mobility of its own.

3. In the absence of complexation (i.e., K_I approaches zero) Eq. 12 reduces to

$$k' = \frac{\mu - \mu_0}{\mu_{mc} - \mu} \tag{14}$$

which is identical with the relation derived for anionic solutes (see Eq. 9). The mobility is now completely defined by the mobility of the micellar phase and the mobility of the solute in the absence of micelles.

The plausibility of Eqs. 12–14 (and, therefore, the calculated k' values) can be demonstrated by examining the relation indicated in Eq. 4: a plot of the k' values calculated using the appropriate equation (Eqs. 12, 13, or 14) versus the surfactant concentration should result in a linear function for the relation to be a valid alternative. This is shown in Figs. 21, 22, and Table 6.

Indirect Determination of t_{mc}

Accurate determination of migration factor, k', is crucial in systematic optimization of MECC separations. Error analysis of k' revealed that the amount of relative error in k' depends on t_o/t_{mc} ratio as well as the range of k' (35). As shown in Fig. 23, an exponential increase in relative error is observed at "high" k' values. The range of acceptable k' would depend on the t_o/t_{mc} ratio. The best results are observed at

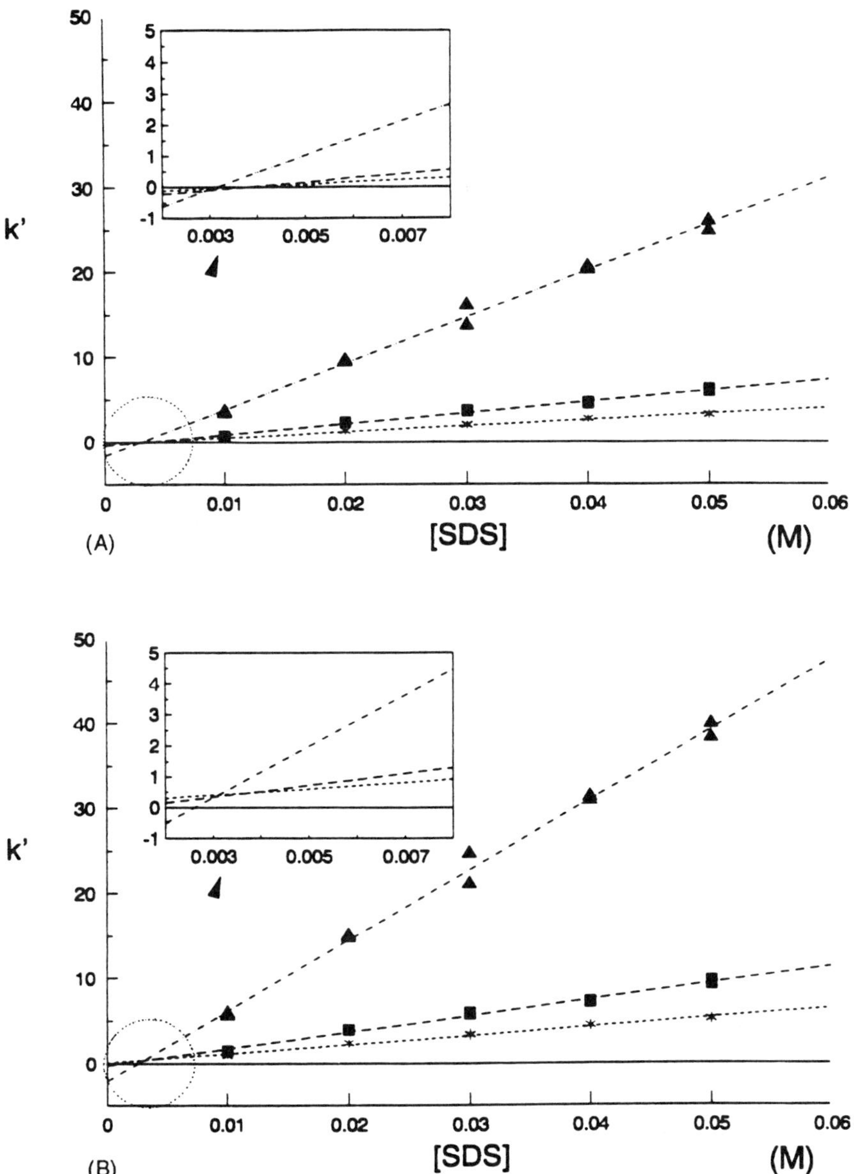

Figure 21 Migration factors k' calculated for dopamine (square), *p*-hydroxybenzylamine (asterix) and norephedrine (triangle) using Eq. 6 (A) or Eq. 9 (B) as a function of the surfactant concentration [SDS]. The area of intersection with the *x*-axis is shown further enlarged in the upper left corner of the frames. (From Ref. 34)

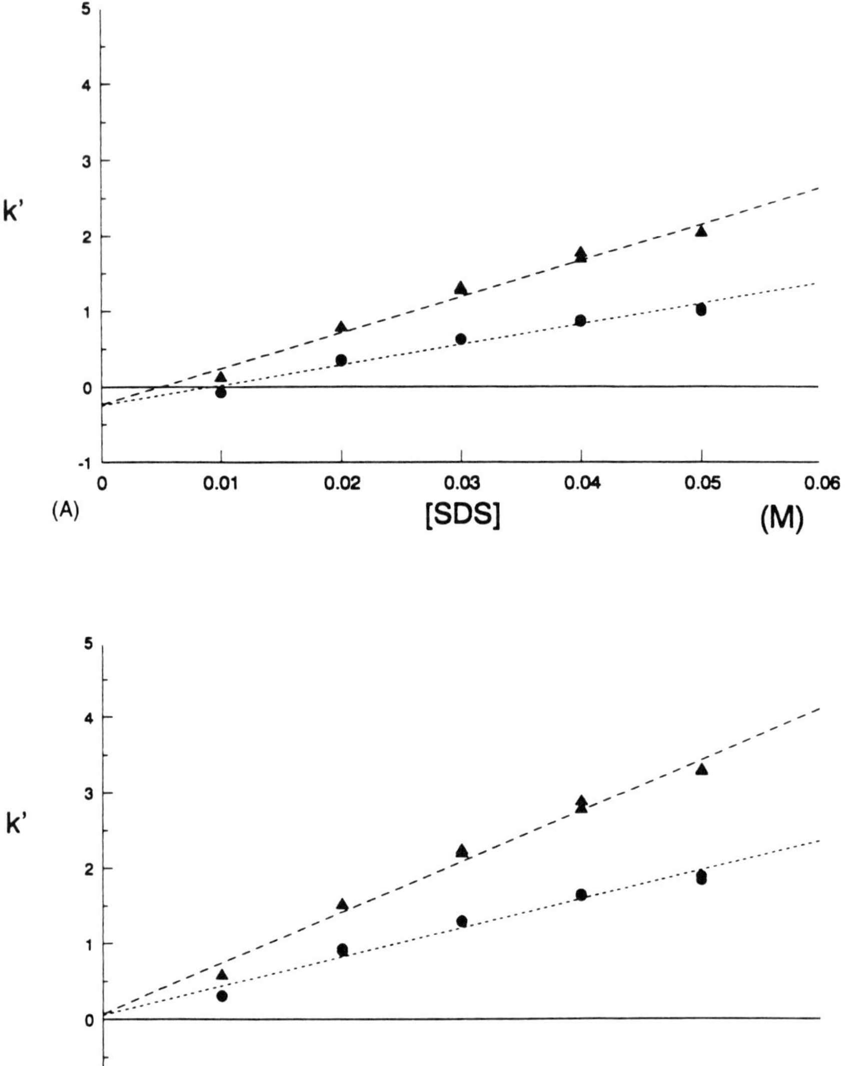

Figure 22 Migration factors k' calculated for norepinephrine (circle) and epinephrine (triangle) using Eq. 6 (A) or Eq. 9 (B) as a function of the surfactant concentration [SDS]. (From Ref. 34)

Table 6 The Critical Micelle Concentration and Distribution Ratio P_{mw} Values Derived From Eqs. 6 and 9

Compounds	CMC (mM)	P_{mw}	Eq.
Catechol	4.3	28	6
Methylcatechol	4.0	68	6
p-Hydroxyphenylalanine	4.2	9.0	9
Methyldopamine	3.8	11	9
Dopamine	3.7	520	6
p-Hydroxybenzamine	3.7	380	6
Norephedrine	3.1	2200	6
Ephedrine	1.3	2800	6
Norepinephrine	9.2	110	6
Norepinephrine	−1.6150	12	6
Epinephrine	5.1	180	6
Epinephrine	−1.7	260	9

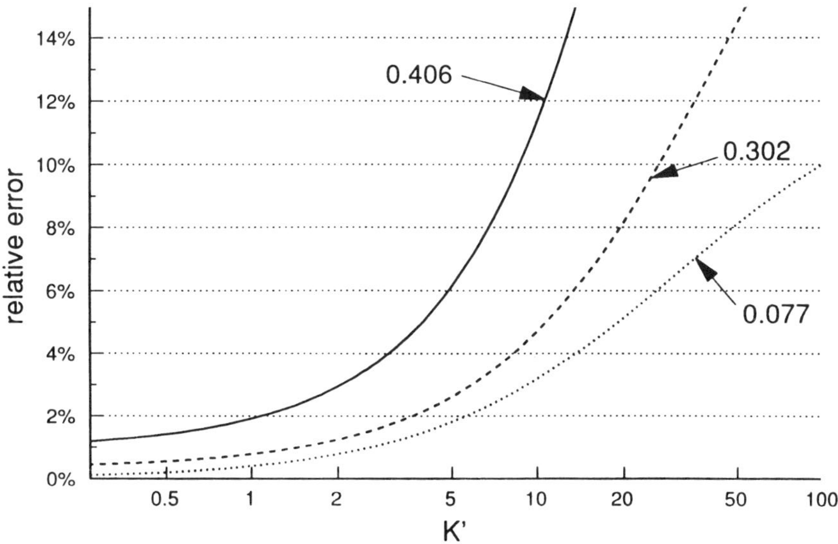

Figure 23 Relative error in k' in MECC as a function of k' at different t_o/t_{mc}. (From Ref. 35)

larger elution windows that can be achieved by the use of organic solvents or mixed micellar systems (Fig. 24).

This also emphasizes the necessity for accurate determination of t_o and t_{mc} in MECC. The direct measurement of neither of these two quantities is feasible.

The retention of an unretained compound in MECC (which has no electrophoretic mobility and would not interact with micelles) is determined through the measurement of retention time of methanol. This is problematic, as methanol does not absorb UV light. Therefore, its retention time should be measured from a baseline disturbance caused by the elution of methanol. It is often difficult to rationalize the correct position on the baseline disturbance as an accurate measure of t_o. This is especially true considering that the shape of the baseline disturbance is a function of buffer composition and is often irreproducible.

Hydrophobic compounds migrate at a velocity similar to that of the micelle owing to their strong association. As a result, the migration time of a highly hydrophobic compound (such as Sudan III) has been used as a measure of t_{mc}. The direct measurement of migration time of Sudan III (or other highly hydrophobic compounds) is not experimentally straightforward. The low aqueous solubility of these compounds, their broad peaks caused by long migration times, along with the low sensitivity of absorbance detectors in CE, create serious problems in detecting

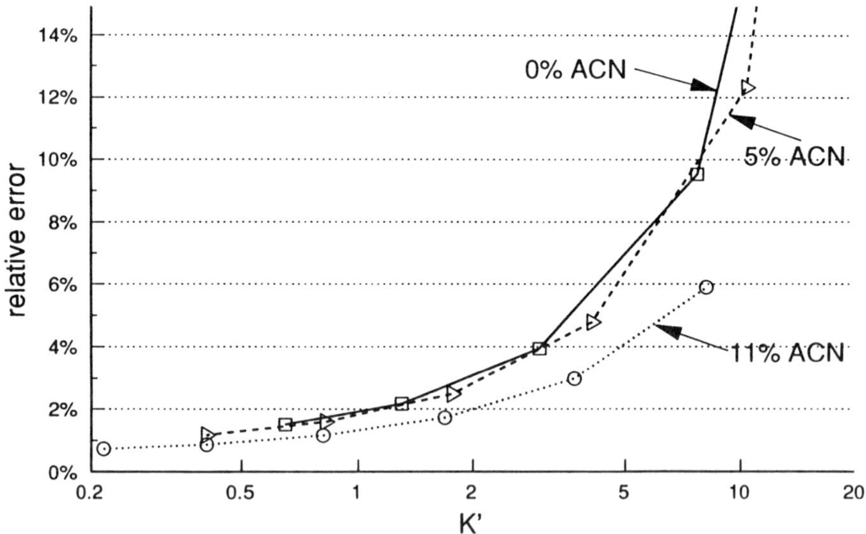

Figure 24 Relative error in k' in MECC as a function of k'. Separation buffer contained 0.02 M SDS and 0% (open squares), 5% (open triangles), or 11% (open circles) acetonitrile. (From Ref. 35)

them. In addition, the assumption that Sudan III completely associates with micelles has not been verified experimentally. This assumption is especially doubtful for an MECC system that includes organic cosolvents in the micellar solution, since solute–micelle binding constants generally decrease upon the addition of organic solvents to the aqueous medium. In certain surfactant systems, such as fluorocarbon micelles, it is even more difficult to detect a peak for Sudan III.

One can determine t_{mc} and t_o values indirectly from the migration behavior of homologous series. There is a linear relation between logarithm of k' and number of $-CH_2$ groups as $\ln k' = a N_c + b$.

The linearity of $\ln k'$ versus N_c depends on the accuracy of t_{mc} and t_o. We used an iterative process to search for the best t_{mc} value that provided the best linear relation between $\ln k'$ and N_c. There was a good agreement between the predicted t_{mc} results and the observed values [35].

pH EFFECT IN MICELLAR ELECTROKINETIC CAPILLARY CHROMATOGRAPHY

Acidic Solutes

The migration factor of an acid is the weighted sum of the migration factor of its undissociated (HA) and dissociated (A^-) forms [36] as

$$k' = F_{HA}\, k'_{HA} + F_{A^-}\, k'_{A^-} \qquad (15)$$

where k'_{HA} and k'_{A^-} are the limiting migration factors of the acid in the associated and the dissociated forms, respectively. The F values are the mole fractions of the acid in the associated and dissociated forms in aqueous solution. Migration factor of a weak acid in MECC can be described as

$$k' = \frac{k'_{HA} + k'_{A^-}\,(K_a/[H^+])}{1 + (K_a/[H^+])} \qquad (16)$$

Interestingly, Eq. 16 has exactly the same format as that of the HPLC techniques [29–31].

Since k' is directly related to the solutes' partition coefficients into micelles, P_{mw}, it is important to explore the dependence of P_{mw} on the pH.

For acidic solutes

$$P_{mw} = \frac{P_{HA} + P_{A^-}\,(K_a/[H^+])}{1 + (K_a/[H^+])} \qquad (17)$$

where P_{HA} and P_{A^-} are the limiting partition coefficients of the associated and dissociated form of the acid from aqueous phase into micelles.

For basic solutes

$$P_{mw} = \frac{P_{BH^+} + P_B(K_a/[H^+])}{1 + (K_a/[H^+]) + K_I} \tag{18}$$

The net electrophoretic mobility of an acidic solute in an MECC system can also be quantitatively described through the phenomenological approach as follows:

$$\mu = (F_{HA,mc} + F_{A^-,mc})\mu_{mc} + F_{A^-}\ \mu_{A^-,aq} \tag{19}$$

where μ is the net mobility of an acidic solute, μ_{mc} is the mobility of micelles, and μ_{A^-} is the moblity of A^- in the aqueous solution; F values with the subscripts or superscripts HA, A^-, mc, and aq represent the mole fraction of solute in the protonated and dissociated forms in micelles and water, respectively, and are related to physicochemical properties of solutes (such as binding constants to micelles and dissociation constants), pH, and micelle concentration as

$$F_{HA,mc} = \frac{K_{HA}\,[M]}{(1 + K_{HA}\,[M]) + (K_a/[H^+])\,(1 + K_{A^-}\,[M])} \tag{20}$$

Similar equations can be derived for $F_{A^-,mc}$ and $F_{A^-,aq}$:

$$F_{A,mc} = \frac{(K_{A^-}\,[M])\,(K_a/[H^+])}{T} \tag{21}$$

$$F_{A^-} = \frac{K_a/[H^+]}{T} \tag{22}$$

where T is the denominator of Eq. 20. (Note that K is solute–micelle-binding constant and is related to partition coefficient P_{mw} as $K = P_{mw}\upsilon$)

The following equation can then be written for mobility of acidic solutes in MECC by substituting Eqs. 20–22 into Eq. 19:

$$\mu_s = \frac{\left(K_{HA} + \dfrac{K_a}{[H^+]}K_{A^-}\right)[M]\,\mu_{mc} + \left(\dfrac{K_a}{[H^+]}\right)\mu_{aqA^-}}{1 + K_{HA}\,[M] + \dfrac{K_a}{[H^+]}(1 + K_{HA}\,[M])} \tag{23}$$

Equation 23 directly shows the relation between mobility and the two important variables, pH and $[M]$ that have a great influence on the migration behavior. It can be simplified by defining the two limiting mobility terms μ_{HA} and μ_{A^-} in micellar media as

$$\mu_{HA} = \frac{K_{HA}\,[M]\mu_{mc}}{1 + K_{HA}\,[M]} \tag{24}$$

$$\mu_{A^-} = \frac{K_{A^-}\,[M]\mu_{mc} + K_a/[H^+]\,\mu_0}{1 + K_{HA}\,[M]} \tag{25}$$

and the apparent dissociation constant in micellar solutions

$$K_{app} = K_a\left(\frac{K_{A^-}\,[M] + 1}{K_{HA}\,[M] + 1}\right) \tag{26}$$

then, Eq. 23 could be further simiplified to

$$\mu = \frac{\mu_{HA} + \mu_{A^-}\,(K_{a,app}/[H^+])}{1 + (K_{a,app}/[H^+])} \tag{27}$$

Equations 16 and 27 predict a sigmoidal relation between both mobility and migration factor with the pH. The fit for the mobility–pH model in MECC (see Eq. 27) is good for less hydrophobic compounds, such as monochlorophenols. The results deteriorate for the di- and trichlorophenols (Fig. 25; Table 7). The main

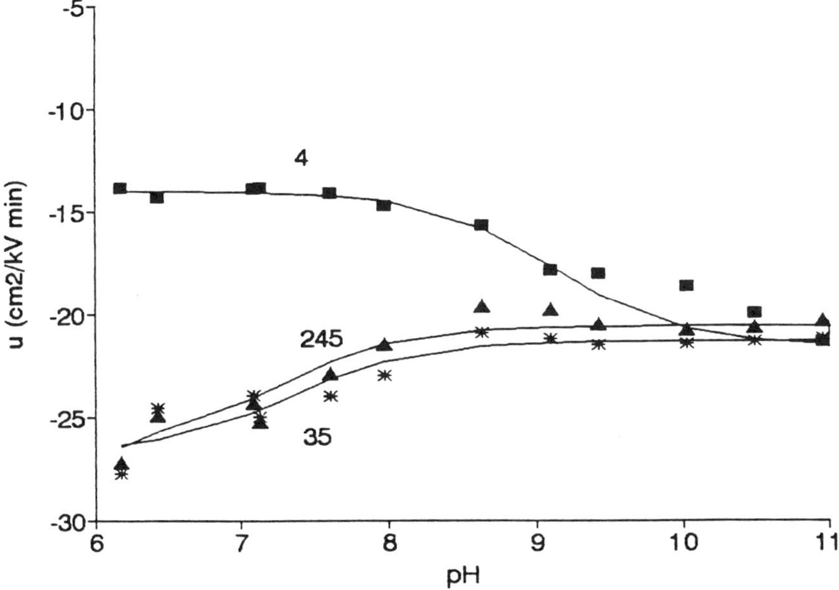

Figure 25 The dependence of mobility on pH in the presence of 20 mM SDS for 4CP, 35CP, and 245CP using weighted nonlinear regression. Lines represent the predicted behavior according to Eq. 29 and presented in Table 7. (From Ref. 23)

Table 7 Estimates of μ_{HA}, μ_{A^-}, and pK_a by Weighted Nonlinear Regression of Mobility Data From MECC Experiments at 20 mM SDS[a]

Compound	μ_{HA}	μ_{A^-}	p$K_{a,app}$
2-Chlorophenol	−11.82	−22.53	7.93
	±0.18	±???	±0.05
3CP	−13.97	−22.28	8.67
	±0.13	±0.14	±0.08
4CP	−13.98	−21.57	9.13
	±0.13	±0.13	±0.07
24CP	−22.55	−21.25	7.89
	±0.34	±0.18	±0.68
25CP[b]	−21.76	−22.06	9.12
	±0.36	±0.10	±3.84
35CP	−24.88	−21.21	9.83
	±0.41	±0.18	±0.21
245CP	−25.58	−20.53	7.43
	±0.67	±0.19	±0.21

[a]Using Eq. 27.
[b]WNLIN failed to converge.
Source: Ref. 23.

reason behind this is that the variation in the net mobility of these hydrophobic phenols with the pH in MECC is small.

In contrast with the mobility model, the migration factor model in MECC (see Eq. 16) gave reasonably good fits for the same set of compounds (Fig. 26) [23]. This is interesting since the k' values were basically calculated from the mobility data using Eq. 9.

By using these equations, the variations in mobility and migration factor as a function of pH and micellar concentration can be predicted on the basis of a few initial experiments. This is illustrated in Figs. 27 and 28. The predictions agree well with the experimental results as shown in Fig. 29 [37]. The disagreement between the predicted and the observed behavior for some of the halogenated phenols (see Fig. 29) is because the pK_as of these compounds lie outside the pH range that was used in this study. Similar to CZE, more accurate predictions are achieved if the selected operating pH range is wide enough to encompass both the protonated and the dissociated forms.

Cationic Solutes

Much like anionic solutes, similar expressions for quantitative description of migration behavior can be derived for cationic solutes. For this, the possible ion-pair

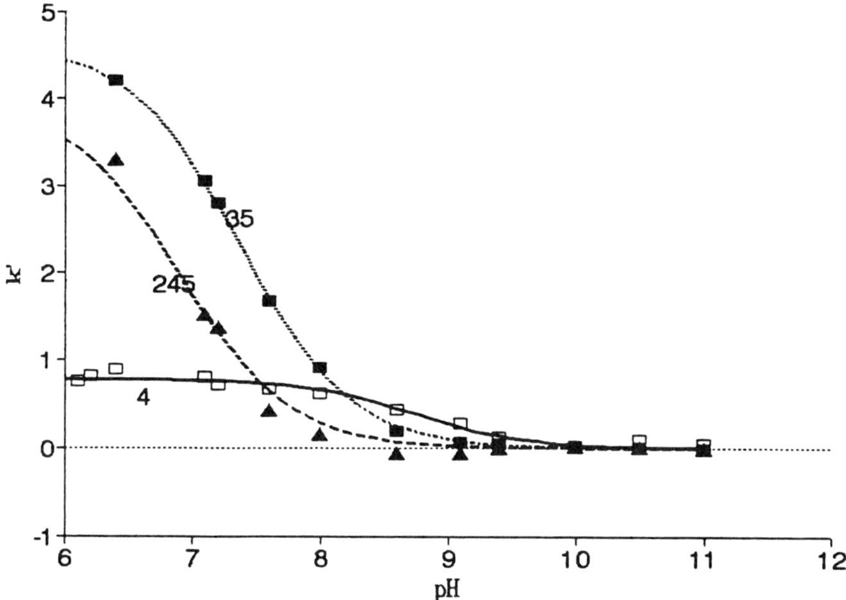

Figure 26 The dependence of k' on pH in the presence of 20 mM SDS for three chlorophenols using weighted nonlinear regression. (From Ref. 23)

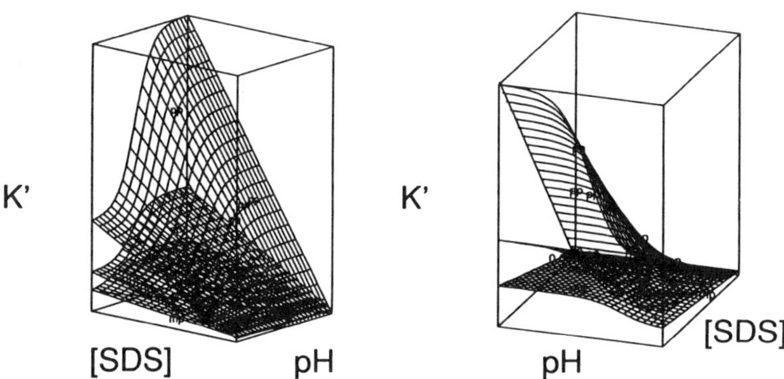

Figure 27 Migration surfaces of three alkoxyphenols in migration factor (k') as functions of pH and SDS concentration. (From Ref. 36)

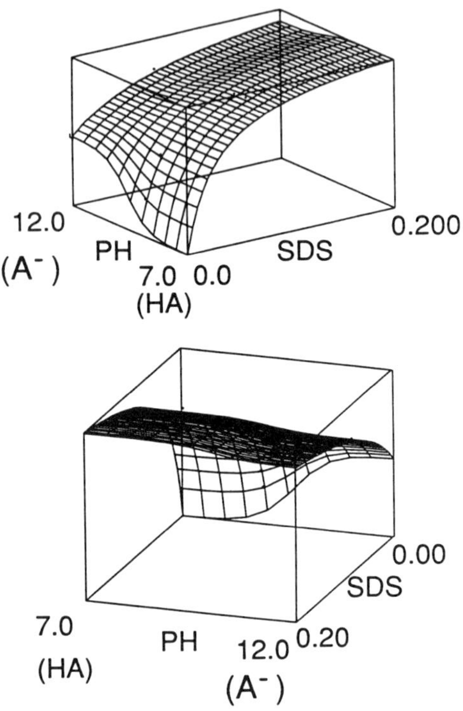

Figure 28 Migration surfaces of ethoxylphenol in mobility (m) as functions of pH and SDS concentration. (From Ref. 36)

interaction between positively charged solutes and surfactant monomers has to be taken into account. By using a similar phenomenological approach, the mobility of cationic solutes can be expressed as functions of pH and SDS concentration [36] as

$$\mu_s = \frac{\left(K_{BH^+}^m + \dfrac{K_a}{[H^+]} K_B^m\right)[M]\,\mu_{mc} + \mu_{aqBH^+}}{1 + K_{BH^+}^m [M] + K_{ip}\,\mathrm{CMC} + \dfrac{K_a}{[H^+]}(1 + K_B^m [M])} \tag{28}$$

where K_{ip}, and μ_{aqBH^+} are the ion-pair formation constant and ionic mobility of protonated form BH^+ in aqueous phase. Replacing K_as in Eq. 28 by

$$K_a = K_{app} \frac{1 + K_{BH^+}^m [M] + K_{lp}\,\mathrm{CMC}}{1 + K_B^m [M]} \tag{29}$$

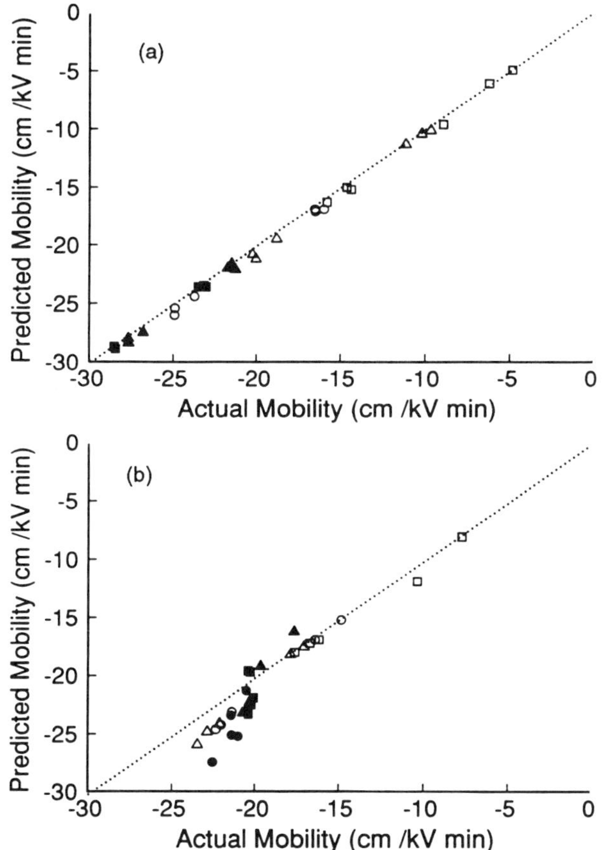

Figure 29 The correlation between predicted and observed mobilities of (a) alkylphenols and (b) halogenated phenols in MECC. (From Ref. 36)

and substituting limiting mobilities μ_{BH^+} and μ_B into the equation, a simplified mobility equation for cationic solutes is obtained.

$$\mu_s = \frac{\mu_{BH^+} + [\mu_B(K_{app}/[H^+])]}{1 + (K_{app}/[H^+])} \tag{30}$$

In a similar fashion, the migration factor for basic solutes can be described as a function of hydrogen ion concentration and micelle concentration:

$$k' = \frac{k'_{BH^+} + (k'_B K_a/[H^+])}{1 + (K_a/[H^+]) + K_{ip}\ \text{CMC}} \tag{31}$$

Since $k' = K([S] - \text{CMC})$:

$$k' = \frac{K^{\text{m}}_{\text{BH}^+} + (K^{\text{m}}_{\text{B}}K_{\text{a}}/[\text{H}^+])}{1 + (K_{\text{a}}/[\text{H}^+]) + K_{\text{ip}}\,\text{CMC}}([S] - \text{CMC}) \tag{32}$$

With a limited number of initial experiments, an estimate of the constants in Eqs. 28 and 32 can be obtained through a nonlinear fit of these equations. Consequently, the migration behavior of individual solutes can be predicted (Figs. 30 and 31). It is obvious that the effect of one factor, pH, on migration depends on the value of the other factor $[S]$, involved. Furthermore, with the parameters available for each compound, electropherograms at different buffer conditions (pH and SDS concentration) can be simulated.

A group of aromatic amines were chosen as test solutes (36). These organic amines have a $\text{p}K_{\text{a}}$ range of 8–11. The use of pH to manipulate HPLC separation of amines is not feasible owing to the chemical instability of *n*-alkylsilicas at high pH. Consequently, these solutes exist almost exclusively in their protonated, pos-

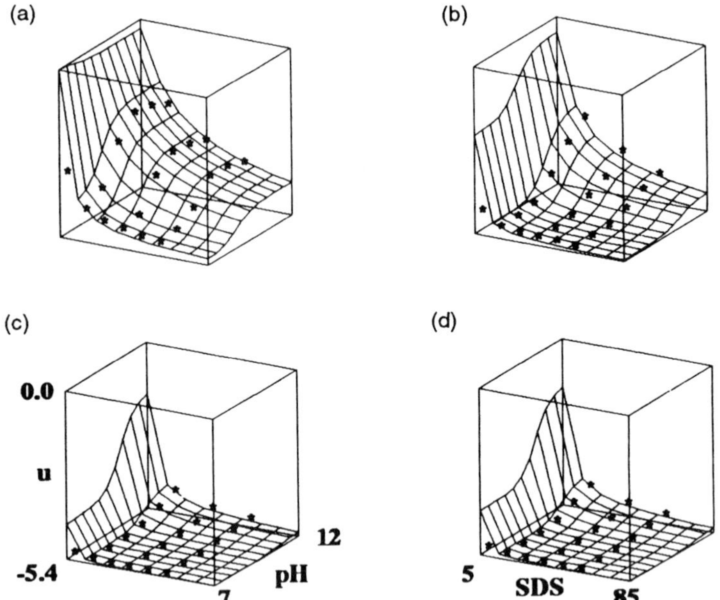

Figure 30 Migration surfaces of organic amines in mobility as function of pH and SDS concentration: (a) nicotine, (b) ephedrine, (c) 4-chlorophenethylamine, (d) 2-(4-tolyl)-phenethylamine. (From Ref. 36)

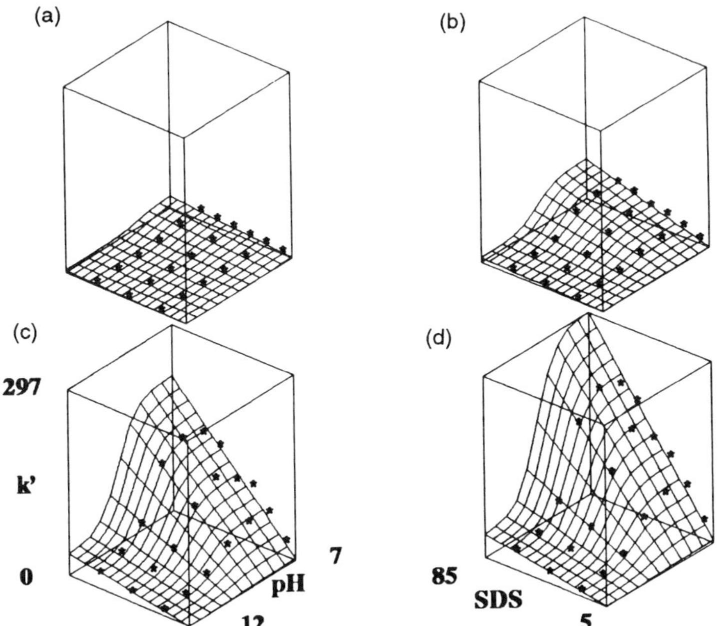

Figure 31 Migration surfaces of organic amines (a) nicotine, (b) ephedrine, (c) 4-chlorophenethylamine, (d) 2-(4-tolyl)-phenethylamine in migration factor as function of pH and SDS concentration. (From Ref. 36)

itively charged forms (alkylammonium ion) in the operating pH range of HPLC with silica-based stationary phases.

The CZE separation of amines in basic pH range is complicated by the silanophilic interactions of amines with the capillary wall. In addition, the electroosmotic flow is maximized in the basic pH range, which causes a fast elution of an unretained peak (i.e., short t_o). Since cationic solutes elute before t_o, the practical separation window (between time zero and t_o) is very short to achieve a reasonable separation of a complex mixture. Reducing the pH significantly improves the peak shape and reduces the electroosmotic flow; however, selectivity is sacrificed, as all amines have very similar charges. Figure 32a is intended to clarify the concept. Despite the high separation efficiency of CZE, the separation of a complex mixture of organic amines is still difficult to achieve because of the limited separation window and nearly identical mobilities of solutes at low pH.

In the presence of micelles, one can operate in the basic pH range, as the solute–micelle interactions compete effectively with the silanophilic interactions

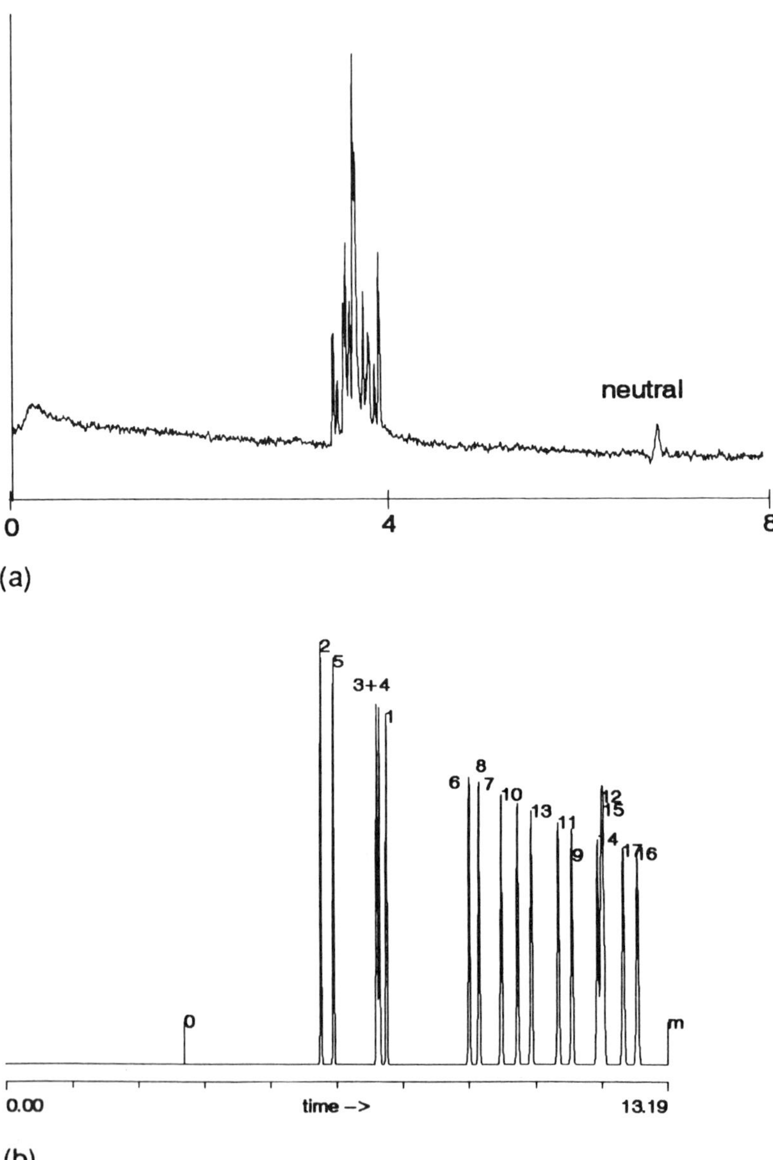

Figure 32 (a) Electrophoregram of 18 organic amines in 0.05 M phosphate buffer (pH 5.0). Capillary, 570 mm (500 mm to detector); electric field strength, 316 V cm^{-1}; detection wavelength, 214 nm. (b) Predicted electrophoregram of 17 organic amines based on mobility model at pH 11.5; SDS 60 mM. (c) Practical separation of 18 organic amines in 0.05 M phosphate and carbonate mixed buffer (pH 11.5, 0.06 M SDS). (d) Practical separation of 18 cationic amines in the same buffer as in Fig. 33b. modified with 10% MeCN. Solute

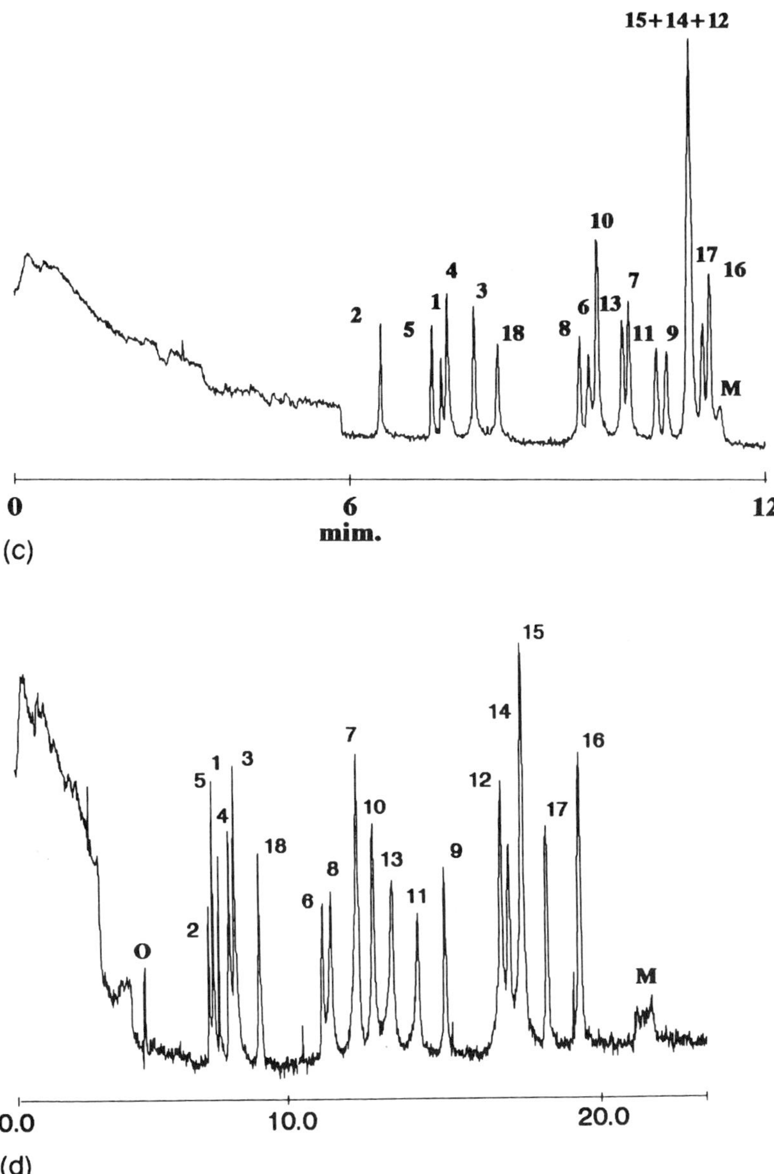

identification: 1, nicotine; 2, 4-nitrobenzylamine; 3, benzylamine; 4, 3-methyldopamine; 5, norephedrine; 6, ephedrine; 7, phenethylamine; 8, 4-nitrophenethylamine; 9, *N*-methylphenethylamine; 10, 4-chlorobenzylamine; 11, 2-methylpenethylamine; 12, phenylpropylamine; 13, 4-bromobenzylamine; 14, 4-chlorophenethylamine; 15, 2-(4-tolyl) phenethylamine; 16, phenylbutylamine; 17, methylphenylbutylamine; 18, 4-fluorobenzylamine. (From Ref. 36) and, consequently, the peak shape is improved, and pH can be manipulated to achieve an acceptable separation.

Figures 32b shows the predicted electropherogram of the separation of 18 organic amines, based on the mobility model, and Fig. 32c shows practical MECC separation of the same mixture (36). Good agreement is observed in terms of the separation pattern and solute elution order. The quality of the separation can be improved by including 10% acetonitrile that reduces t_o/t_{mc} ratio (see Fig. 32d).

Micellar–Induced pK_a Shift

One important factor that should be considered in a study of the pH effect on the migration behavior and separation of ionizable compounds by MECC is the micellar-induced pK_a shift. Consideration of micellar-induced shift of ionization constants is important when comparing the capabilities of MECC and CZE for the separation of ionizable compounds. The extent of ionization of acids in the MECC system depends on the apparent ionization constant in micellar solutions, which is different from the aqueous ionization constants. More importantly, the magnitude of the pK_a shift is a function of the solutes' structural properties.

This would create two interesting situations. The first is when two compounds with similar pK_a values in aqueous solutions would have different pK_a in micellar media (e.g., amino acids and small peptides) [37]. As a result of this phenomenon, an enhanced separation selectivity is observed.

Another possibility is that compounds with widely different pK_a values in an aqueous medium and different lipophilicities would have similar ionization constants in micellar solution. Figure 33 illustrates the titration curves for four chlorophenols in purely aqueous (Fig. 33a) and micellar (Fig. 33b) media. As shown, in the absence of micelles the pK_a values of these compounds are widely different, although the addition of micelles reduces the difference in pK_a values. Here, inclusion of micelles would reduce the variations in selectivity as a function of pH and might actually lead to a loss of selectivity from this leveling effect. However, one should be cautious about generalizing the latter conclusion, since the solute charge is not the only factor that affects selectivity in MECC. As shown in Fig. 34, changing the micellar concentration from 20 to 100 mM SDS has greatly improved the poor selectivity between the two monochlorophenols over the entire pH range. Micellar concentration influences the extent of pK_a shift as evidenced by Eq. 26. Again, the rate of variation of p$K_{a,app}$ with micellar concentration is an intrinsic property of the solute.

Migration Parameters in Micellar Electrokinetic Capillary Chromatography: μ Versus k'

Migration behavior of ionizable solutes in MECC can be expressed in terms of migration factor or mobility.

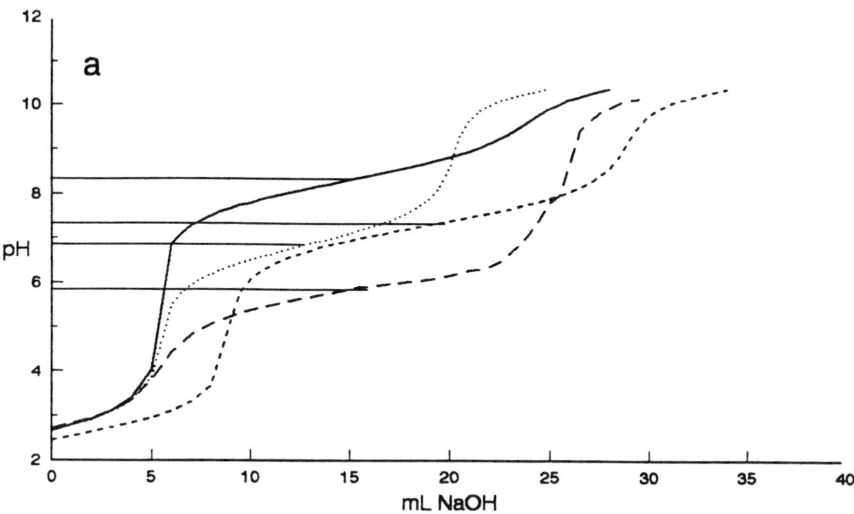

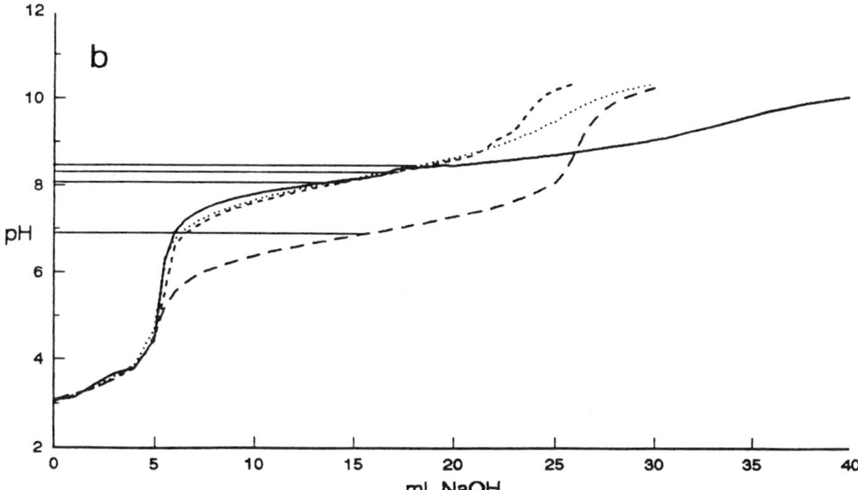

Figure 33 Titration of 2CP (—), 25CP (- - -), 245CP (· · ·), and PCP (— —) with 0.005 M NaOH in (a) aqueous and (b) 50 mM SDS solutions. Ionic strength without SDS, 50 mM using NaCl. (From Ref. 23)

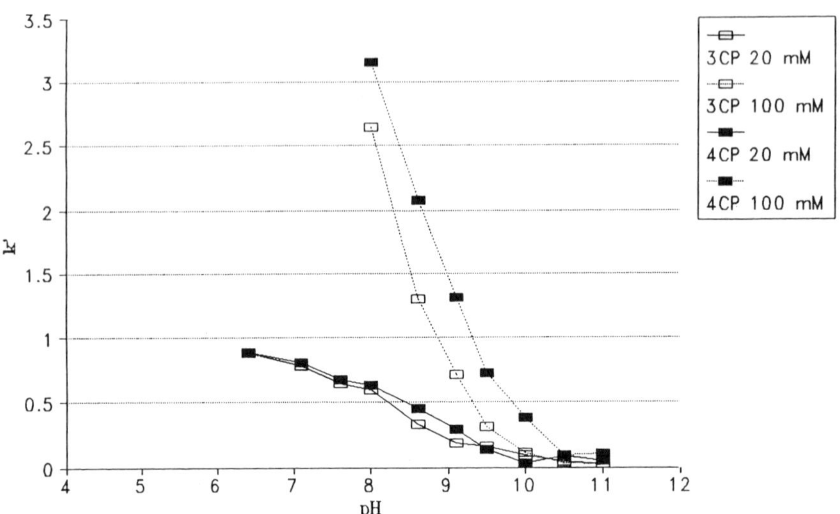

Figure 34 The dependence of k' on pH in the presence of 20 mM SDS (dotted line) and 100 mM SDS (solid line) for 3CP and 4CP. (From Ref. 23)

For all solutes, migration factor decreases with an increase in pH (i.e., $k'_{HA} > k'_{A^-}$) as shown in the sigmoidal curves (see Fig. 26). This is expected, since the protonated acid (HA), which is the dominant form at the lower pH values, interacts to a greater extent with micelles, compared with the dissociated form (A^-), which is electrostatically repelled from the micelles and, therefore, has a smaller partition coefficient into micelles than the protonated form.

Unlike k', the trend of variations of μ in micellar media with pH depends on the type of the solute (see Fig. 25). For example: for monochlorophenols, the mobility increases with an increase in pH (i.e., $\mu_{HA} < \mu_{A^-}$). For the tri- and pentachlorophenols the trend is reversed; that is, the mobility of the more hydrophobic compounds decreases with an increase in pH, $\mu_{HA} > \mu_{A^-}$. This is shown in Fig. 25 for 4CP, 35CP, and 245CP. For all three compounds, the trend for k' versus pH is the same, whereas the μ versus pH is different.

The different behavior in mobility can be explained by examining the relation shown in Eq. 19:

$$\mu = F_{HA}^{mc}\, \mu_{mc} + F_{A^-}^{aq}\, \mu_{A^-}^{aq} + F_{A^-}^{mc}\, \mu_{mc}$$

At the low pH values, the first term dominates and $\mu = \mu_{HA} = F_{HA}^{mc}\, \mu_{mc}$. The contribution of the second and the third terms to the overall mobility increase with the pH and at high pH values:

$$\mu = \mu_{A^-} = F_{A^-}^{aq}\, \mu_{A^-}^{aq} + F_{A^-}^{mc}\, \mu_{mc}$$

The contribution of the third term is probably not significant, owing to the electrostatic repulsion of the dissociated solute from the anionic micelles that results in small $F^{mc}_{A^-}$ values. Therefore, the trend of the μ–pH curve is determined by the terms $F^{mc}_{HA}\ \mu_{mc}$ and $F^{aq}_{A^-}\ \mu^{aq}_{A^-}$. The mobility of micelles is larger than any of the test solutes; however, the contribution of the protonated acid HA becomes larger than the dissociated form A^- only if it interacts strongly with micelles. This is true for trichlorophenols. Therefore, for these compounds $F^{mc}_{HA}\ \mu_{HA} > (F^{aq}_{A^-}\ \mu^{aq}_{A^-} + F^{mc}_{A^-}\ \mu_{mc})$. The interaction of the monochlorophenols is considerably less, to the extent that the solute mobility in the dissociated form is more than that of in the HA form (Fig. 35a,b).

Note that the fractions of solutes in micellar media are a function of solute–micelle interactions and micellar concentration. This is especially pronounced for protonated solutes. As a result, at higher micellar concentrations the difference between the mobilities of HA and A^- is reduced for more polar solutes, whereas it is increased for the more hydrophobic ones. This can be seen in the predicted mobility versus pH plot at two different micellar concentrations (see Fig. 28) as well as the three-dimensional plots of mobility versus pH/[SDS] (Fig. 36a,b).

In general, for less hydrophobic compounds and at low micellar concentrations, the difference between μ_{HA} and μ_{A^-} is much larger than that for the hydrophobic compounds. In other words, the changes in mobility with the pH variation is greater for the less hydrophobic monochlorophenols (see Fig. 36). The situation is different for the k' versus pH function. The difference in k'_{HA} and k'_{A^-} increases with the solute hydrophobicity.

These two situations seem to contradict one another as far as the effect of pH on selectivity is concerned. Given the mobility data, for more hydrophobic solutes, pH is not an effective variable for selectivity manipulation at low micellar concentrations, since there is not much difference in the mobilities of the protonated and the dissociated forms of an acid. At higher micellar concentrations, however, the changes in mobility of more polar compounds with pH is greatly reduced, although one observes larger variations in mobility versus pH for more hydrophobic compounds (see Fig. 36). On the other hand, one would observe drastic changes in k' as the degree of the acid dissociation is changed through pH variations. Note, however, that k' represents the degree of solute interactions with micelles. The degree of this interaction does not necessarily indicate the true migration behavior of ionizable solutes in MECC; therefore, mobility is probably a more representative variable for solute migration.

It is worth reiterating at this point, that these observations illustrate the usefulness of migration modeling for achieving an in-depth understanding of the migration behavior, in addition to predicting the behavior for optimization purposes.

Multiparameter Optimization in Micellar Electrokinetic Capillary Chromatography

Migration behavior of ionizable solutes in MECC is more complicated than in CZE. The electropherograms shown in Fig. 37 provide further support for this observation.

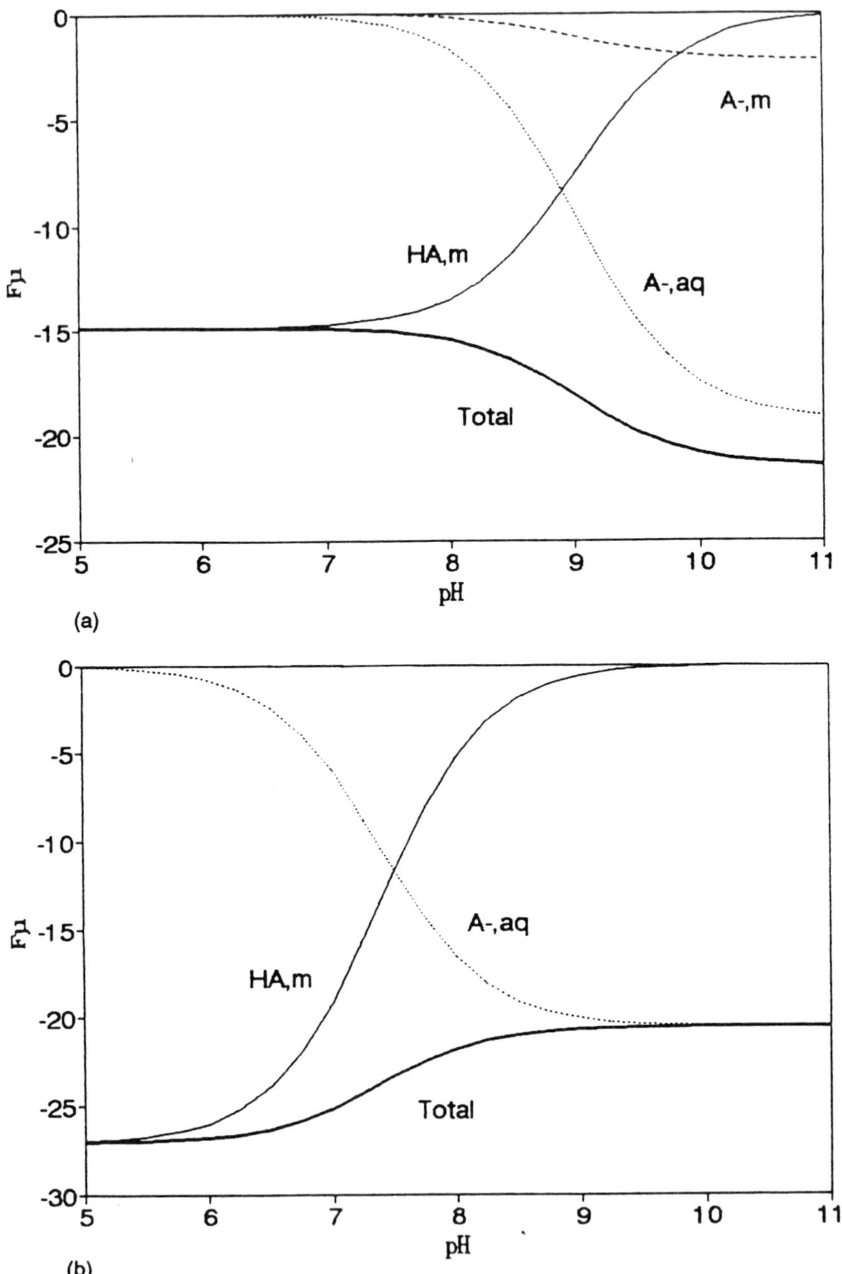

Figure 35 The dependence of the product of relative fraction × mobility of (a) 4CP and (b) 245CP on pH for $HA_m\mu_{mc}$, $A^-_m\mu_{mc}$, $A^-_{aq}\mu_o$, and total. (From Ref. 23)

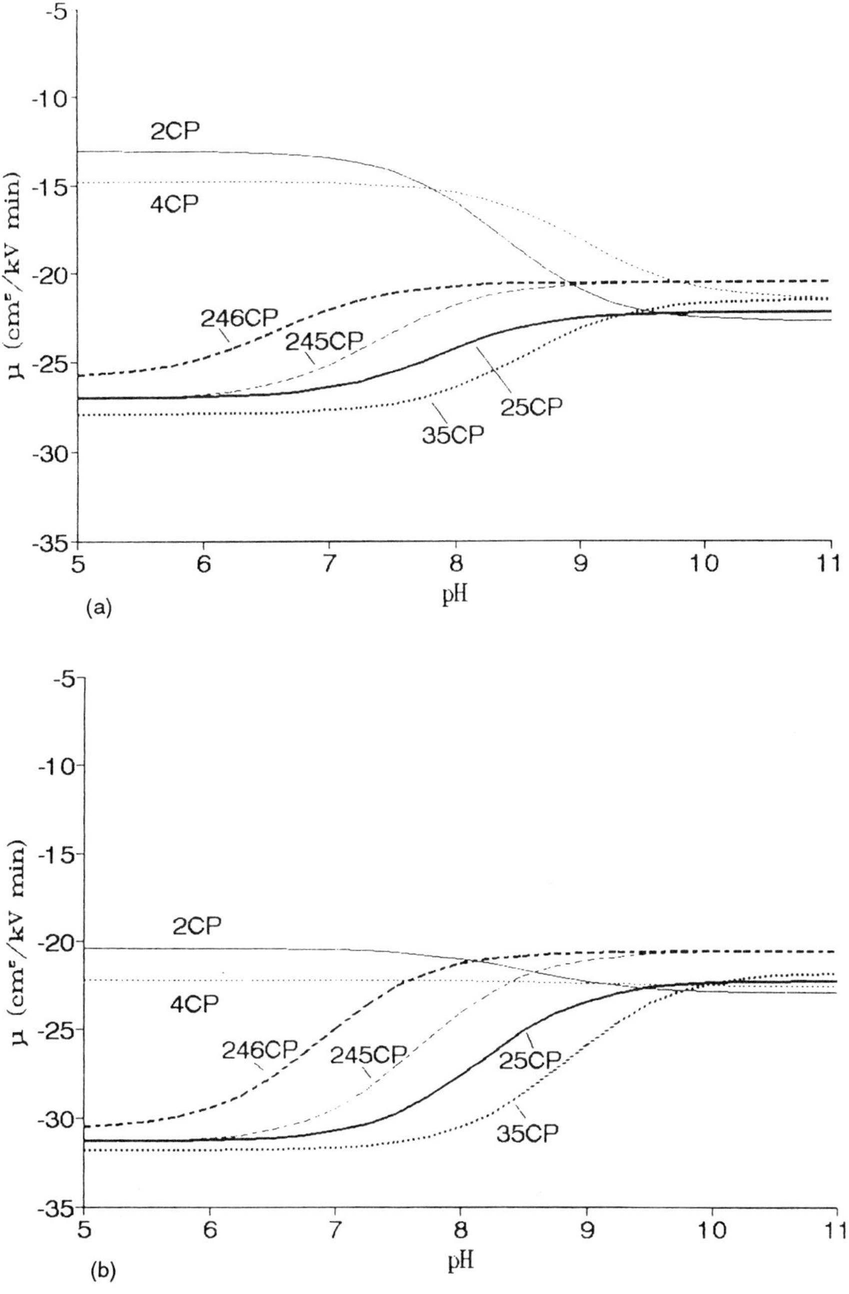

Figure 36 (a) The dependence of μ on pH at 20 mM SDS and (b) the predicted μ at 40 mM SDS for chlorinated phenols. (From Ref. 23)

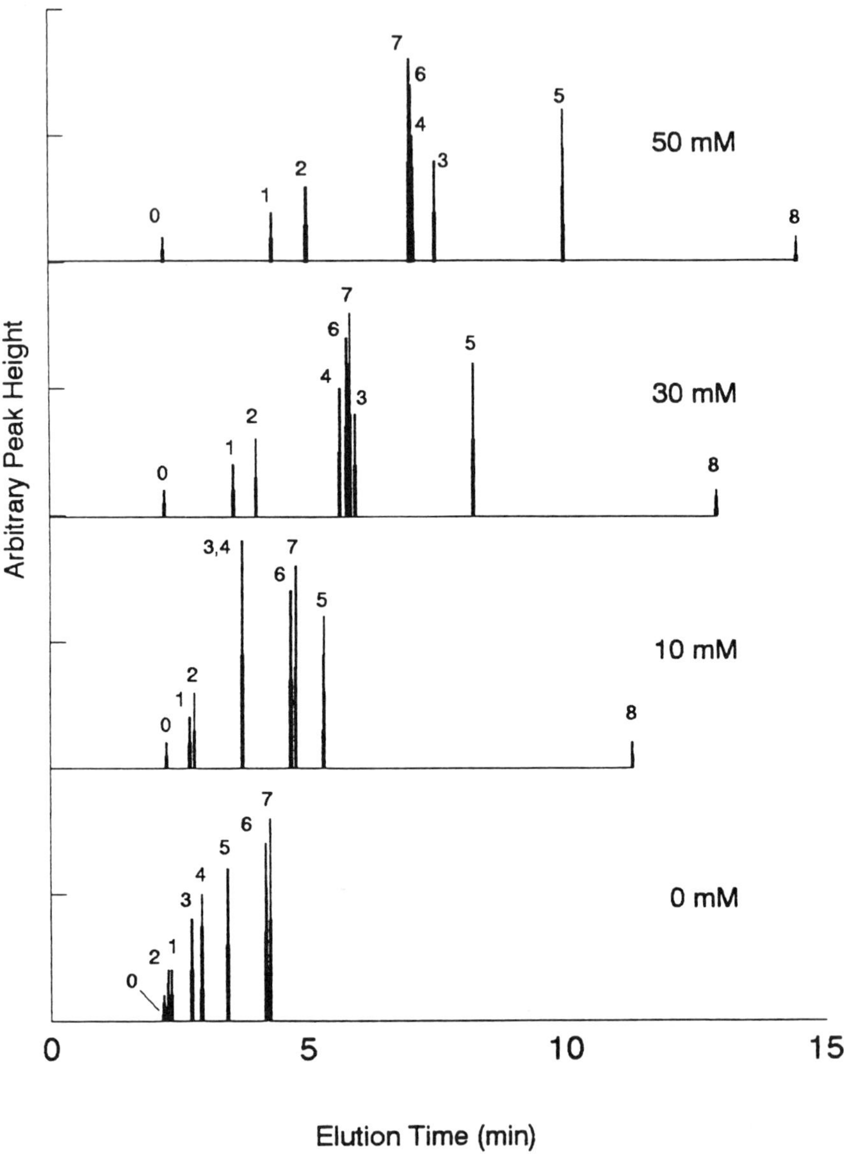

Figure 37 The elution times of (0) t_0, (1) 2CP, (2) 3CP, (3) 23CP, (4) 25CP, (5) 245CP, (6) 246CP, (7) PCP, and (8) t_{mc} at 0, 10, 30, and 50 mM SDS in 50 mM phosphate buffer. Electropherograms are reconstructed from individual experiments. Peak heights are arbitrary and vary to assist in peak identification. (From Ref. 23)

As shown, when micelles are not present, the elution order of the chlorophenols is according to their charge and size (or molecular mass). Upon addition of micelles and with changing micellar concentration at a constant pH, however, the elution order changes greatly. Consider an example in which a mixture of 20–30 solutes in a mixture are to be separated by MECC, which would require the optimization of at least two important variables: namely, pH and micellar concentration. One important issue that should be addressed is whether these two variables can be optimized independently of one another or is there a need for simultaneous optimization of the variables. From the preliminary results and the dependence of $pK_{a,app}$ on micellar concentration and solute type, one can conclude that the simultaneous multivariable optimization would be a more effective strategy. Interpretive methods of optimization that are based on the use of a retention model would be quite appropriate here [1].

With the foregoing described models, one can predict the migration behavior as a function of the two variables based on a limited number of experiments. The initial experiments should be designed such that one would get an estimate of the important physicochemical properties of an acid: P_{HA}, P_{A^-}, and $K_{a,app}$. If these values are known, one can predict the mobility and migration factor of a compound at different pH and micellar concentrations. In principle, five or six initial measurement at different pH or [SDS] values are adequate to predict migration behavior in MECC using nonlinear fitting of the models. It is important that the regression techniques require no knowledge of different physicochemical properties of solutes, such as ionization constants and partition coefficients. The predictions are generally good as illustrated. Once the migration surface for each solute is known, finding an optimum composition that provides the best separation would not be difficult. This can be achieved through a plot of criterion (e.g., minimum resolution) versus variables.

MIXED MICELLES IN ELECTROKINETIC CHROMATOGRAPHY

Resolution in MECC can be enhanced through a careful manipulation of a solute's-binding constant to micelles (which directly influences migration factor and selectivity) as well as through extending the elution window (i.e., reducing t_o/t_{mc}).

Mixed micelles can be used to manipulate retention, selectivity, and t_o/t_{mc} to enhance resolution. Several mixed micellar systems in aqueous media can be useful in MECC separations: (a) mixed surfactant micelles, (b) surfactant micelles–organic modifiers [39], and (c) surfactant micelles–cyclodextrins [40].

There are three types of mixed surfactant micelles [41]:

1. Mixed surfactants with similar headgroup charges: These are considered ideal mixed micelles the CMC of which can be predicted according to the ideal solution theory.
2. Mixtures of fluorocarbon (FC) and hydrocarbon (HC) surfactants that show a positive deviation from ideality: The CMC of the latter mixture is larger than

what is predicted according to an ideal solution. This has been attributed to a "phobicity" effect between —CH_2 and —CF_2 groups. It has been suggested in the literature that, at a specific composition of FC and HC mixtures, two distinct types of micelles form, one rich in FC surfactants and rich in HC surfactants [41].

3. Mixtures of ionic and nonionic surfactants: These mixtures exhibit a negative deviation from ideality; that is, the CMC of the mixture is smaller than that of the ideal solution. This is due to the reduction in electrostatic repulsion between the headgroups of ionic surfactants owing to the incorporation of the nonionic headgroups.

Figure 38 shows the typical migration of an uncharged solute in mixed micellar electrokinetic chromatography (MMEC). The upper frame corresponds to a situation for which only one type of mixed micelle exists, whereas the lower frame illustrates the existence of two types of micelles with their corresponding mobilities and partition coefficients.

We have investigated the use of FC–HC mixed micelles in electrokinetic chromatography [42]. A test mixture of aromatic compounds was selected. These compounds belong to several classes and represent a wide range of polarities and functional groups. The migration behavior of these compounds were studied in three surfactant systems: pure hydrocarbon (lithium dodecyl sulfate; LiDS) micelles, pure fluorocarbon (lithium perfluorooctane sulfonate; LiPFOS) micelles, and mixtures of LiDS and LiPFOS.

The migration behaviors of alkyl nitriles were studied. There is a linear relation between the logarithm of migration factor (ln k') and the number of —CH_2 groups of

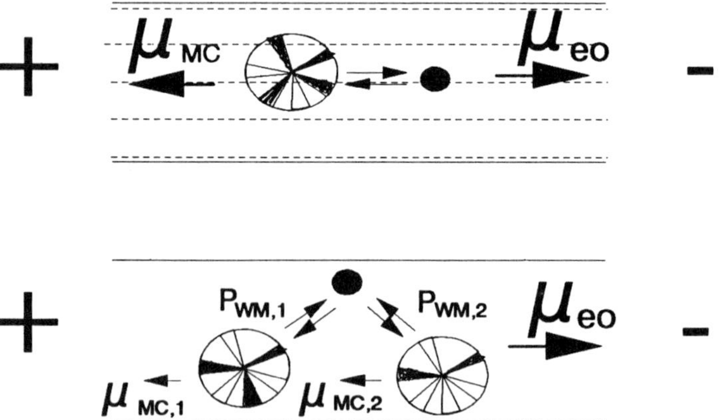

Figure 38 A schematic representation of mixed micellar electrokinetic chromatography. Extension of retention window (t_1/t_{mc}) and selective manipulation of Pwm.

homologues (N_c). This linear relation between the migration factor and a structural characteristic of solutes is of great use in migration mechanism studies. The linear behavior was observed for all three surfactant systems as shown in Fig. 39. The slopes of the ln $k'-N_c$ lines show the hydrophobic selectivity in the micellar system. As expected, the pure LiDS had the largest and the pure LiPFOS had the smallest slope indicating the larger interactions between the hydrocarbon solutes with hydrocarbon micelles versus fluorocarbon micelles. In both pure surfactant systems the intercept changed greatly with the micellar concentration; however, the intercept of ln $k'-N_c$ lines for the mixed micellar systems changed slightly with a change in mole fraction of the two surfactants at a constant total micellar concentration. The retention of this homologous series was also used for the determination of t_{mc} in FC and mixed micellar systems.

One of the factors that influences resolution is the size of the elution window, shown by the t_o/t_{mc} ratio. The t_o/t_{mc} values in FC systems are smaller than in HC systems. Therefore, in mixed systems one should consider the optimum compositions (both the total micellar concentration and mole fraction of individual surfactants) that would provide the optimum migration factor, selectivity, and t_o/t_{mc} values.

The effect of micellar concentration on migration behavior in MECC is quantitatively described by Eq. 4, which shows a linear relation between migration factor k' and surfactant concentration. The x-intercept of this line is equal to the CMC, and the slope is the binding constant of the solute to micelle. Figure 40 shows that

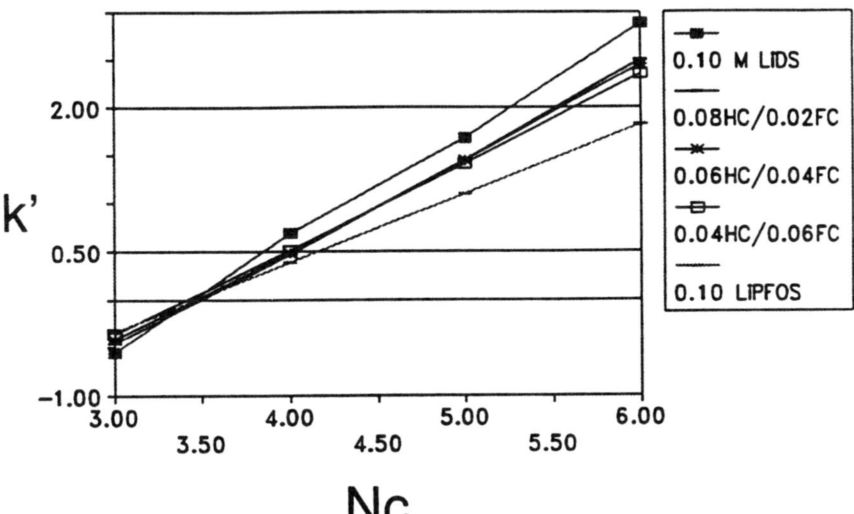

Figure 39 k' as a function of carbon number in fluorocarbons–hydrocarbons MMEC. Total surfactant concentration, 100 mM; at different mole fraction of fluorocarbons.

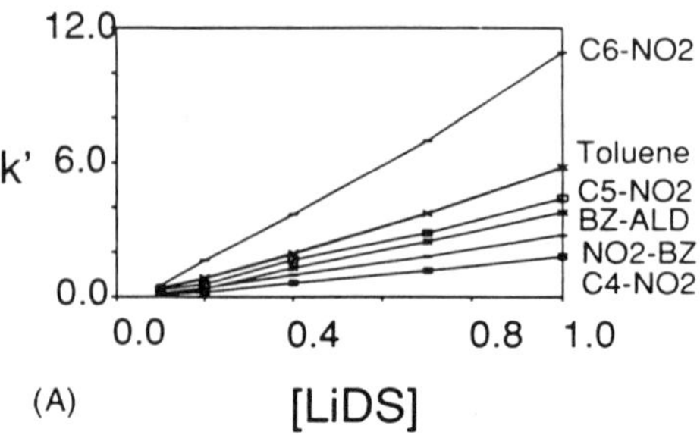

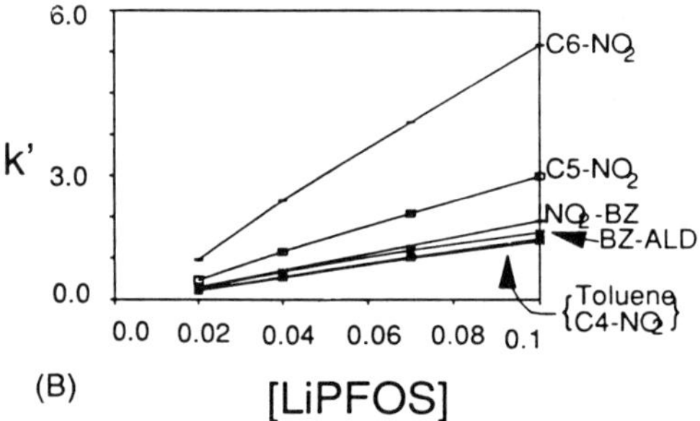

Figure 40 k' as a function of the surfactant concentration in MECC (A) hydrocarbon surfactants and (B) fluorocarbon micelles. (From Ref. 42)

an excellent linear relation was observed for pure HC and pure FC systems. An important difference between the FC and HC surfactant systems is the degree of interaction of different solutes, represented by the solute–micelle binding constants or the slopes in Fig. 40. For HC surfactant, the solute–micelle binding increases with the hydrophobicity of the solute, therefore k' (toluene) $>$ k' (benzaldehyde) $>$ k' (nitrobenzene). In FC micellar system, however, the order is reversed owing to the phobicity effect between hydrocarbons and fluorocarbons. As a result, toluene, which is the most hydrophobic compound, has the smallest binding constant to FC micelles in the group of the three compounds. The solute–micelle-binding constants for all aromatic compounds in the LiDS and LiPFOS are compared in Fig. 41. This clearly shows a great potential for manipulating the solutes' binding to micelles by using surfactant mixtures. Consequently, the chromatographic selectivity and resolution can be controlled. The chromatograms shown in Fig. 42 further illustrate this. As can be seen, by changing the mole fraction of HC and FC surfactants (at

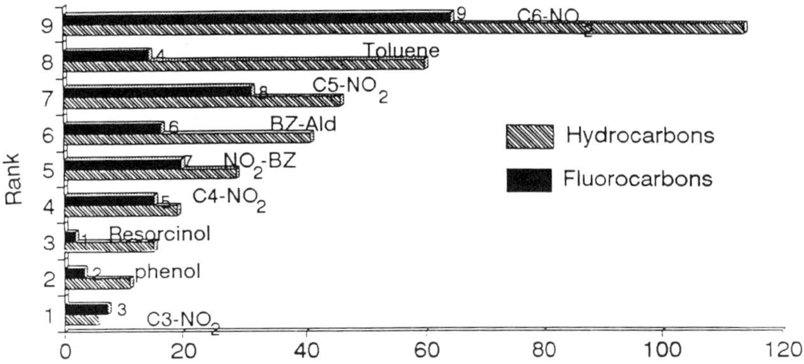

Figure 41 Solute–micelle-binding constants, K_{wm} for hydrocarbon and fluorocarbon surfactants. (From Ref. 42)

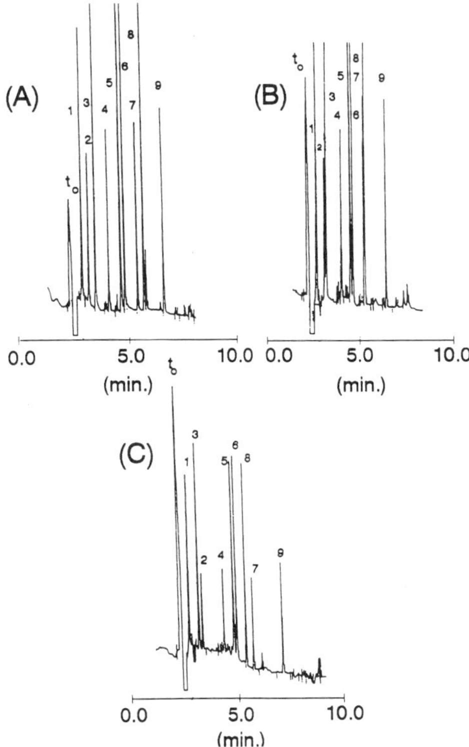

Figure 42 Effect of the surfactant ratio on selectivity. Total surfactant concentration, 100 mM; mole fraction of the fluorocarbon surfactant, 0.2 (A), 0.4 (B), and 0.6 (C). Solutes: 1, resorcinol; 2, C_3NO_2; 3, phenol; 4; C_4NO_2; 5, benzaldehyde; 6, nitrobenzene; 7, C_5NO_2; 8, toluene; 9, C_6NO_2. (From Ref. 42)

a constant total concentration), the elution order of several peaks has changed; most notably being the large variation of toluene upon increasing mole fraction of FC surfactant. Although t_o/t_{mc} ratio has also changed, the overall separation time was relatively constant. The separation shown in Fig. 42 resembles that in HPLC, in which selectivity can be manipulated at a constant solvent strength.

EMPIRICAL APPROACHES

Sometimes it is difficult to describe the migration behavior of solutes as a function of certain variables through the use of phenomenological or mechanistic models. For example, the influence of organic cosolvents (ϕ_{org}) can be quantitatively described in conjunction with variables such as temperature, pH, or micellar concentration, through the use of either a physical model (which will require many experiments) or a completely empirical approach.

The empirical approaches can be classified into two categories: (a) regression techniques that are based on linear or polynomial least square (curve fitting) and (b) factor-based methods such as target transformation factor analysis (TTFA). In the following sections the use of a linear interpolation-based method (iterative regression) and target transformation factor analysis for the optimization of variables in CE will be discussed.

Target Transformation Factor Analysis

Usually, polynomial equations with interaction terms can be used in the form of

$$Y = a_1X_1 + a_2X_2 + \ldots + a_nX_n + b_1X_1X_2 + b_2X_1X_3 + \ldots$$

where Y is a migration factor (μ or k') and $X_1 \ldots X_n$ are the parameters (e.g., pH, $[S]$, ϕ_{org}, or other) and $X_1X_2, \ldots$ are the interaction terms. Instead of modeling migration behavior of solutes, one can directly relate the quality of separation (e.g., Y = minimum resolution) to the variables of interest for optimization of a separation [20].

Figure 43 shows a typical data matrix ([Y] matrix of **mxn** dimensions) in CE separations. The **n** column vectors of this matrix are composed of migration value (k' or mobility, or even migration times) measured at n different variable values (e.g., pH, ϕ_{org}), whereas the **m** rows correspond to m individual solutes in a mixture.

The main problems with the use of polynomial modeling are the many experiments required and the colinearity between the columns of the data matrix, which indicates the existence of redundant information. The dimensionality of the matrix is larger than what is really needed (i.e., one can model migration behavior with fewer experiments). A typical example is illustrated in Fig. 44 in which the migration of solute has been measured in a three-dimensional parameter space, whereas only a surface (a two-dimensional space) is adequate to describe the behavior of the solutes.

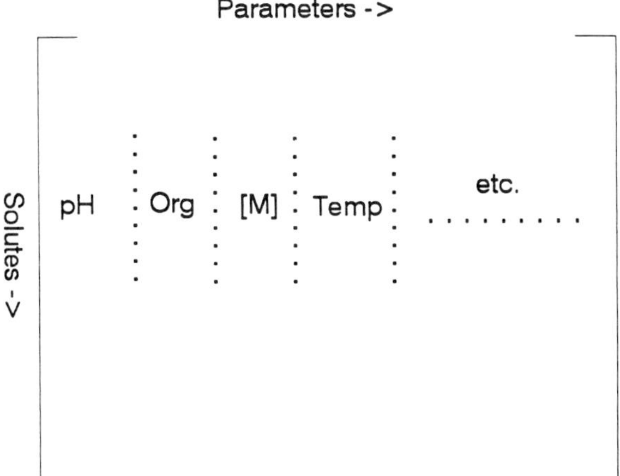

Figure 43 A schematic representation of a typical data matrix. Empirical description of migration behavior: migration factor = $ax_1 + bx_2 + cx_3 + \ldots$. Problems: Multicolinearity (redundant information) and large number of experiments. (From Ref. 44)

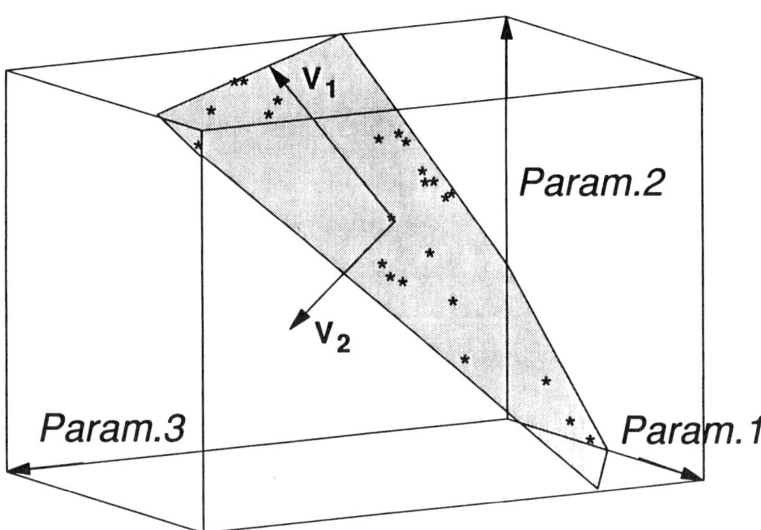

Figure 44 A schematic representation of dimensionality analysis of an apparently three-dimensional data set. (From Ref. 44)

This problem can be solved by performing factor analysis of the original data matrix of **mxn**. In factor analysis, k orthogonal, abstract factors are found that best describe the variations of the original data set in n dimensions, provided $k < n$. For example, the two abstract factors of V_1 and V_2 in Fig. 44 are the result of a factor analysis of the original three-dimensional vectors.

The basic premise of factor analysis modeling is that the variations of migration can be described by fewer factors. This would reduce the number of experiments for the prediction of migration behavior.

Target factor analysis is of particular help in grid search-based optimization strategies. It helps (a) in the recognition of the true dimensionality of the original data set, (b) in the recognition of the most appropriate experimental conditions for obtaining the necessary data for new compounds, and (c) in characterizing the response surface for the migration behavior of the given class of solutes.

The process of TTFA comprised two steps of factor analysis and target transformation. As a result of factor analysis, the original data matrix $[D]$ is decomposed into two matrices, one is the row matrix $[R^{\#}]$ that contains the k factors for m solutes and column matrix $[C^{\#}]$ that is called the loading, which is equivalent to the coefficients of the parameters in a regression equation. The product of $[R^{\#}]$ $[C^{\#}]$ will produce a matrix $[D^{\#}]$ as

$$[D^{\#}] = [R^{\#}]\,[C^{\#}] \tag{33}$$

The theory of target factor analysis is now well established [43]. The columns of the row matrix $[R^{\#}]$ are abstract factors and represent the coordinates of the data points in the space of the new set of orthogonal axes, given by the column matrix, $[C^{\#}]$. These abstract factors may not always be easy to interpret. A chemist is generally interested in the real or more meaningful solutions. Often, one can rotate the axes of $[C^{\#}]$, slightly away from orthogonality and can still work in the lower dimensional space (as compared with the original data matrix). The objective of target transformation is to rotate the set of k orthogonal axes of $[C^{\#}]$ such that some real descriptor of a compound can qualify to be the coordinates for that compound in the space of the rotated axes. In other words, target transformation replaces the row or the score matrix $[R^{\#}]$ by real factors matrix $[R^1]$. This is done by finding a suitable transformation vector $[\mathbf{T}]$ and postmultiplying Eq. 33 such that:

$$[D^{\#}] = [R^1][C^1] \tag{34}$$

In the case of CE optimization, $[R^1]$ can be column(s) of migration factors or physicochemical descriptors of solute structures.

In a successful factor analysis $[D^{\#}] = [D]$ within the range of experimental errors. In other words, one can reproduce the original matrix D within experimental errors by using orthogonal factors with a smaller dimensionality.

Usually knowledge about the chemical structure of the set of solutes becomes the basis for the selection of $[R^1]$. One initially tries a large number of migration factors or solute descriptors to select a set of k descriptors that can serve as acceptable coordinates in the space of the axes given by $[C^1]$. So each column of $[R^1]$ is correlated well with a different member in the set of the orthogonal columns of $[R^{\#}]$.

The overall process of target factor analysis can be described graphically from Fig. 45. Once the matrices $[R^1]$ and $[C^1]$ are available, one is prepared to make the predictions. The following two examples illustrate the usefulness of TTFA for the prediction and optimization of CE separations based on a few experiments [44].

Example 1: Optimization of pH

The data set $[D]$ consists of the electrophoretic mobilities for eight *para*-substituted phenols, obtained at 8 pH values between pH 8.0 and 12.0. Factor analysis showed that more than 99% of variation in the data can be explained by only two factors. Target transformation depicts that the columns of $[D]$ corresponding to the mobilities at pH 9.5 and 11.5 are the most suitable pair of targets. These two columns of $[D]$ thus represent $[R^1]$ of Eq. 34; $[C^1]$ of Eq. 34 represents the modified loadings obtained by premultiplying $[C^{\#}]$ by $[\mathrm{T}]^{-1}$. Figure 46 gives a plot of the two loadings of the rows of $[C^1]$ as a function of pH. The loading 1, corresponding to the core mobility at pH 9.5 is more important for prediction at lower end of the

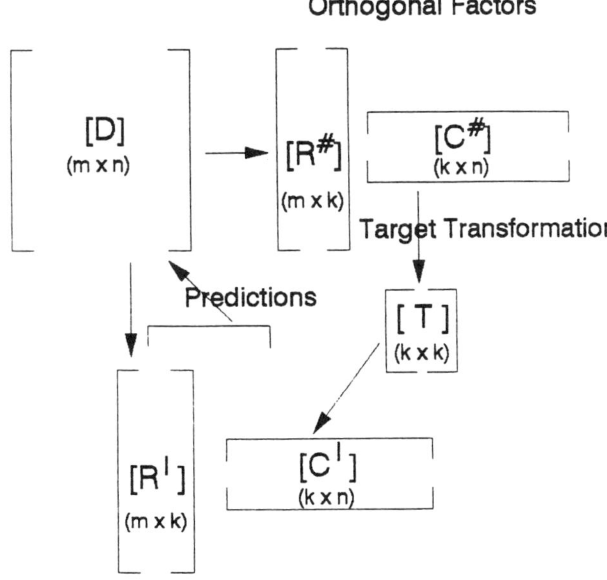

Figure 45 A graphic representation of the target factor analysis process. (From Ref. 44)

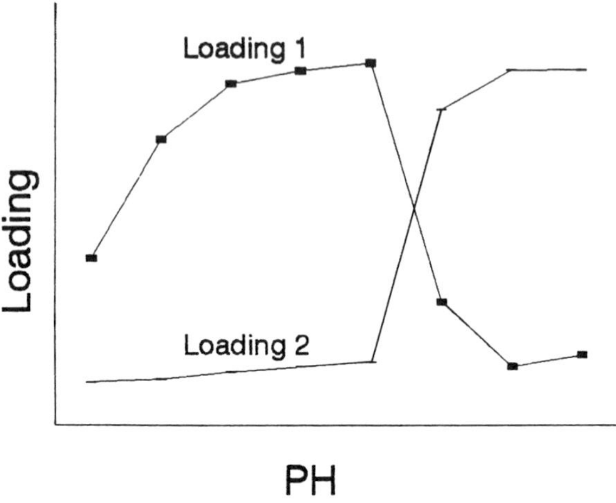

Figure 46 Loadings as a function of pH. (From Ref. 44)

pH range, whereas 2, the loading corresponding to the core mobility at pH 11.5 is more important at higher pH range. Because of their nonlinear variation, simple regression of these curves is not adequate. Spline cubic fits through both these loadings seem most appropriate to interpolate loadings between this pH range [44].

As a first test of the target factor model, we regenerated our experimental data using the two core mobilities for each phenol. Figure 47 shows good correlation between the experimental and the calculated mobilities. We then took a test set that included seven different *para*-substituted phenols. The core mobilities—the mobility at pH 9.5 and 11.5—were obtained experimentally for the test phenols. Mobilities were predicted for each phenol between the pH limits of 8.0 and 12.0. Figure 48 shows a procedure for TTFA prediction of mobilities that was adapted from the application of TTFA in LC [45]. As shown, the mobilities for each new solute over the entire pH range can be predicted by first measuring the "core" mobility of a solute at pH values of the target matrix and then multiplying the mobility by the previously calculated loading matrix. In other words, the prediction of migration behavior of new solutes over a wide pH range can be accomplished by measuring their behavior at only two pH values.

Maximization of minimum resolution was used as the optimization criterion. The pH 11.2 was found to be the optimum pH. A mixture of these phenols was run at this pH. Figure 49 shows good agreement between the experimental and the predicted electropherograms.

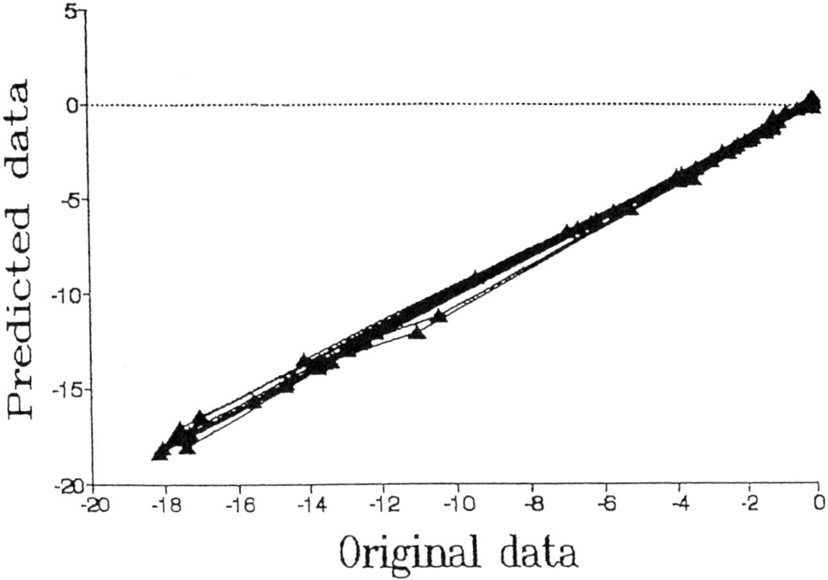

Figure 47 Comparison of the experimental data with that predicted using target factor modeling. (From Ref. 44)

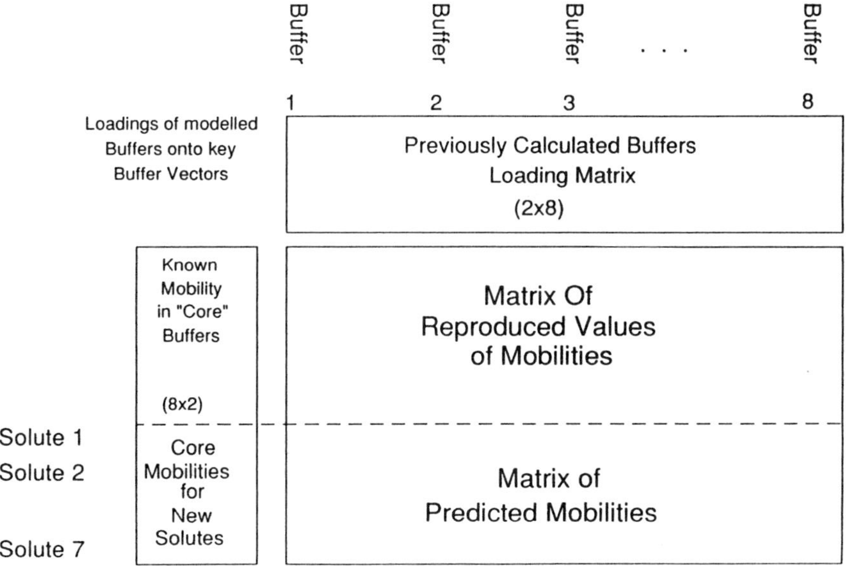

Figure 48 Procedure for prediction of mobilities of compounds using TTFA. (Adapted From Ref. 45)

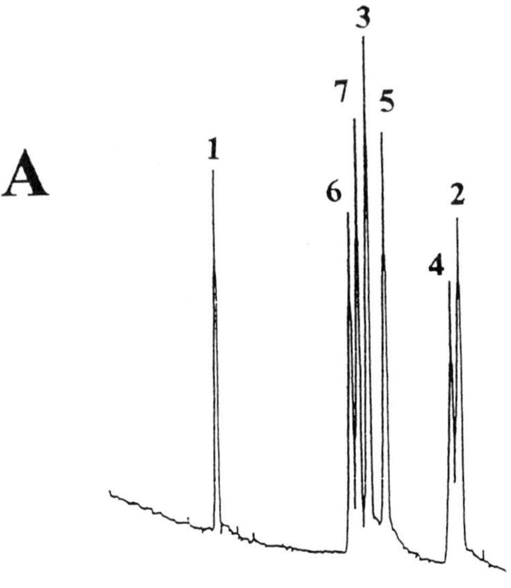

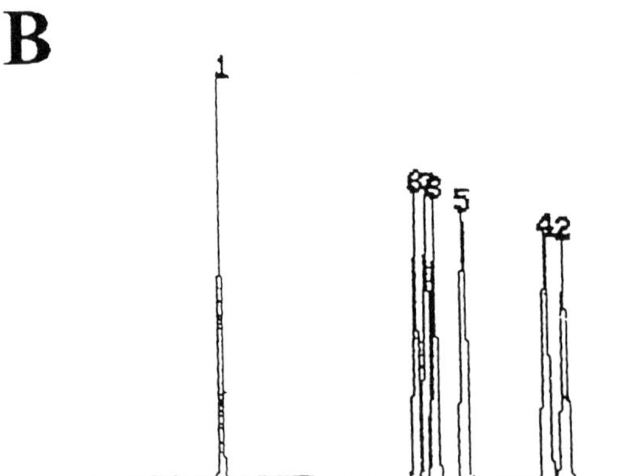

Figure 49 A comparison of (A) the experimentally observed and (B) the predicted electropherograms. pH is the only parameter optimized. (From Ref. 44)

Example 2: Optimization of pH/ϕ_{org}

Example 2 demonstrates the extension of this technique to optimizing more than one parameter. The data set consists of the mobilities for a set of 19 peptides (di- and tripeptides). The mobility data was obtained at different pHs (7, 7.75, 8.5, 9.25, and 10) and the concentrations (7.5, 15, 22.5, and 30%) of acetonitrile. The data matrix, thus, has a dimension of 19 × 20. The objective is to predict mobilities of peptides at any pH and acetonitrile concentration within the experimental range. Factor analysis shows that most of the variation in the data can be explained using first three principle factors. Target transformation showed that the set of mobility columns of [D], corresponding to acetonitrile [ACN] concentration of 22.5% at pH values of 7.0, 8.5, and 10 are the most suitable set of target vectors. Figure 50 shows the three loadings (rows of [C^1]) as a function of pH and percentage ACN. A cross-validation of the data was performed, and the results of predictions are shown in Fig. 51. There is a good correlation between the predicted and the experimental data. The loading surfaces shown in Fig. 50 are based on all 19 peptides; hence, they were not used in these predictions. In cross-validation, for every peptide, the remaining 18 peptides were used as the training set. The targets chosen in each iteration (which were not always the same for each peptide) along with their corresponding loading surfaces were used for making the predictions.

Iterative Regression Approach

It is possible to apply a simple linear model and a limited number of initial experiments to get a first impression of the location of the optimal separation. The initial model is then refined by additional experiments in the proximity of this optimum. This method, which is known as iterative regression optimization strategy, was first developed by Drouen et al. for reversed-phase LC separations [46]. It was also applied in ion-pair chromatography [47]. More recently, successful applications for two- and three-parameter optimization in micellar liquid chromatography were reported from this laboratory [48,49]. The following examples illustrate the preliminary results of the application of this technique in MECC, which was presented previously [21]. First, the migration of all individual compounds in a mixture is modeled as a linear function of the variables using a minimum number of initial experiments. Figure 52 shows the optimization of one variable, micellar concentration in MECC. For a one-variable situation, such as micelle concentration in MECC, a linear model of k' versus [S] is derived on the basis of two initial values of surfactant concentrations. The top frame of Fig. 52 shows the two initial chromatograms at 10 mM and 50 mM SDS concentrations. The parameter space for a given optimization is encompassed by the initial variable values. Then the migration factor of the solutes in the mixture at surfactant concentrations other than those used in the actual measurements within the parameter space were predicted through

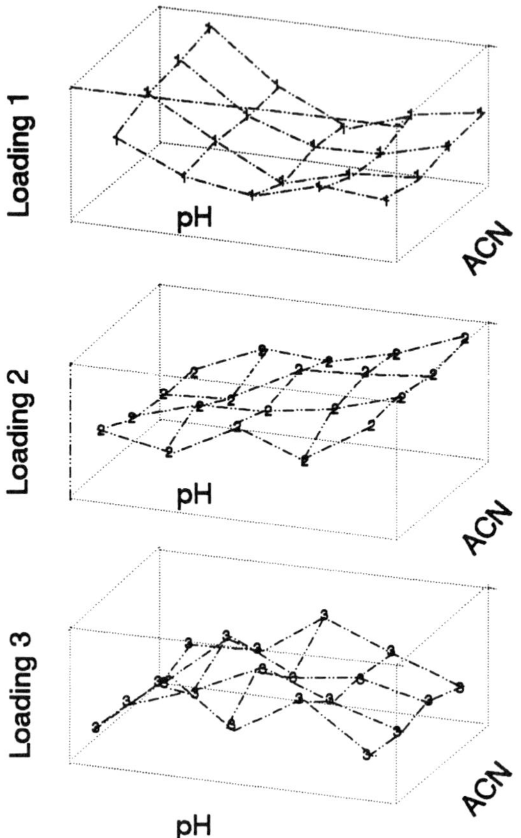

Figure 50 Loadings as a function of pH and acetonitrile concentration for the peptide data set. (From Ref. 44)

interpolation of the assumed linear model of k' versus [SDS] (shown in the second frame of Fig. 52). From the predicted migration values of all the compounds in the mixture, the quality of the separation at all surfactant concentrations within the parameter space is calculated and an optimum is predicted (the third frame in Fig. 52). By means of a computer, chromatograms can be constructed based on these models in each point in the parameter space. The lower frame of Fig. 52 shows the excellent agreement between the predicted and the measured separation.

The success of predicting the optimum based on a minimum number of experiments depends on the linearity of k' versus variable and to some extent on the

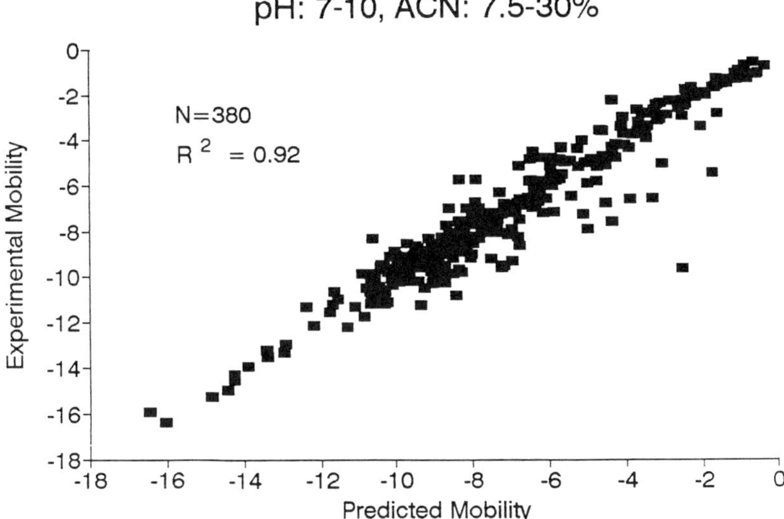

Figure 51 Correlation between the predicted and the experimentally observed mobilities for the peptide data set. (From Ref. 44)

reproducibility of migration behavior. If the linearity assumption is not accurate, additional experiments will be required to locate the optimum. Figure 53a shows the result when pH is used as a variable. Because of the inherent nonlinearity of the migration behavior as a function of this variable, an additional measurement is required to obtain a satisfactory separation and a good agreement between predicted and observed separation. As shown in Fig. 53a, the two initial measurements were performed at pH 8.0 and 11.0. An optimum was predicted at pH 10.0, which differed from the observed. The additional migration data at pH 10.0 was then incorporated into the model and a new optimum at pH 8.5 was predicted, which agreed better with the observed results (see Fig. 53b).

SUMMARY AND OTHER CONSIDERATIONS

In this chapter the usefulness of interpretive methods for the optimization of "selectivity" variables in CZE and MECC was examined. The influence of several buffer (or solvent) variables on electrophoretic and electrokinetic migration were quantitatively described. Migration patterns of solutes in a mixture can be predicted on the basis of a few initial experiments. This method is useful in several ways: (a) a fundamental knowledge is obtained about the migration mechanism of solutes, which

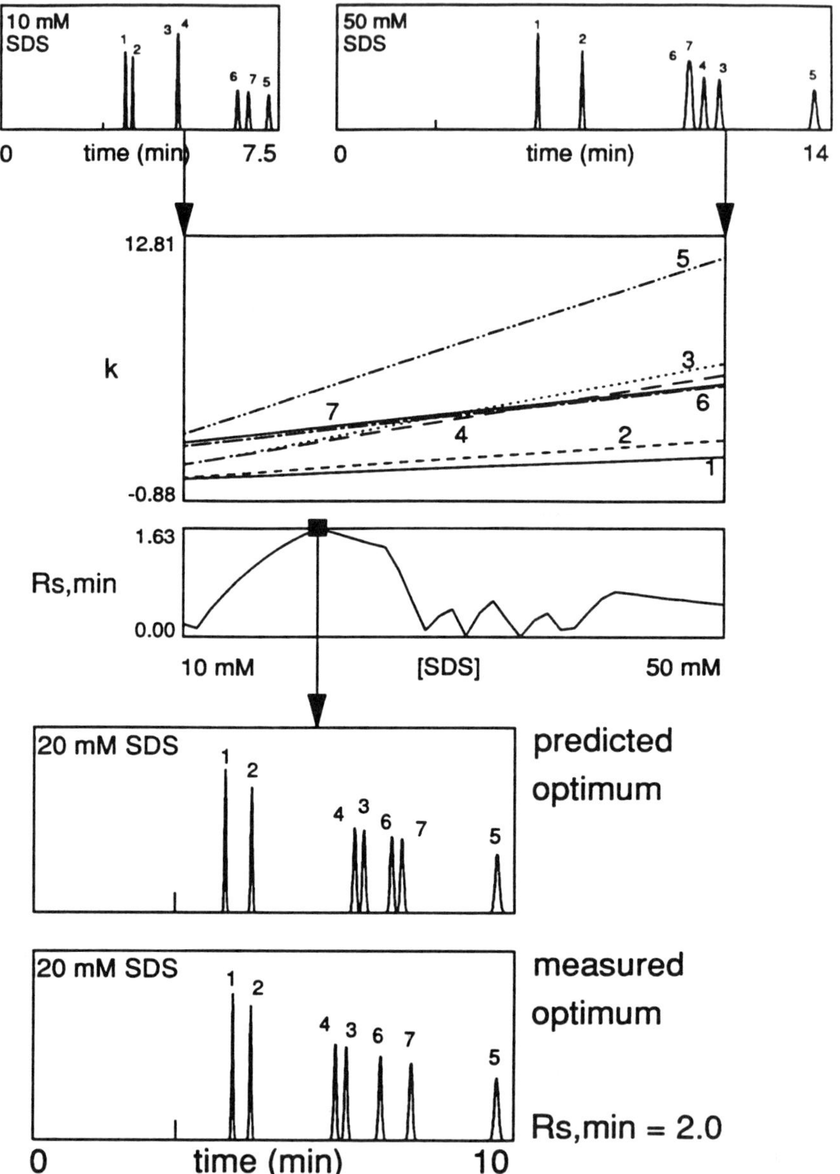

Figure 52 Optimization of the MECC-separation of a set of chlorophenols by varying the micellar concentration. Experimental conditions: capillary id 50 μm, od 375 μm; length 55 cm, separation length 38 cm; temperature 35°C; voltage 18 kV; 50 mM phosphate buffer, pH 7, SDS concentration varying between 10 and 50 mM. Solutes: 1, 2-chlorophenol; 2, 3-chlorophenol; 3, 2,3-dichlorophenol; 4, 2,5-dichlorophenol; 5, 2,4,5-trichlorophenol; 6, 2,4,6-trichlorophenol; and 7, pentachlorophenol. (From Ref. 21)

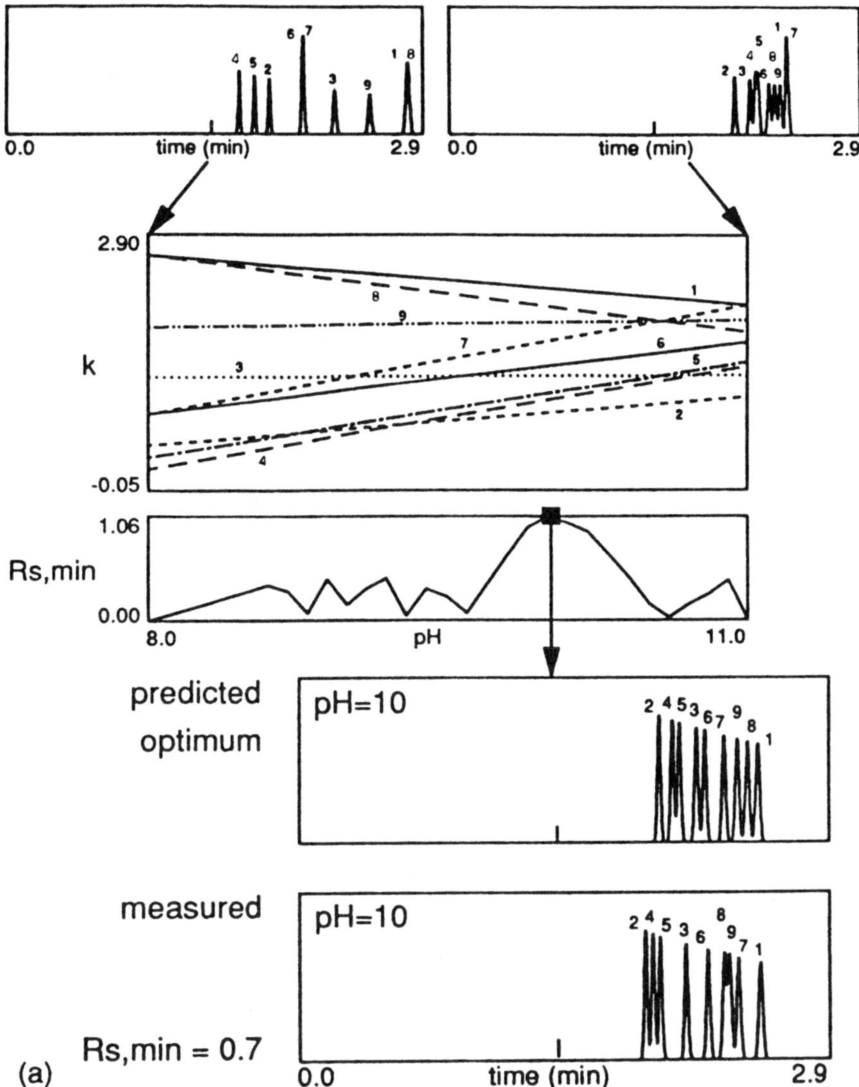

Figure 53 Optimization of the MECC separation of a set of chlorophenols by varying the pH. Experimental conditions: capillary id 50 μm, od 375 μm; length 43 cm, separation length 30 cm; temperature 40°C; voltage 20 kV; 50 mM phosphate buffer, 20 mM SDS concentration, pH varying between 8 and 11. Solutes: 1, 4-propylphenol; 2, 4- methylphenol; 3, 4-ethylphenol; 4, phenol; 5, fluorophenol; 6, 4-chlorophenol; 7, 2-chlorophenol; 8, 3,5-dichlorophenol; and 9, 2,5-dichlorophenol. (From Ref. 21)

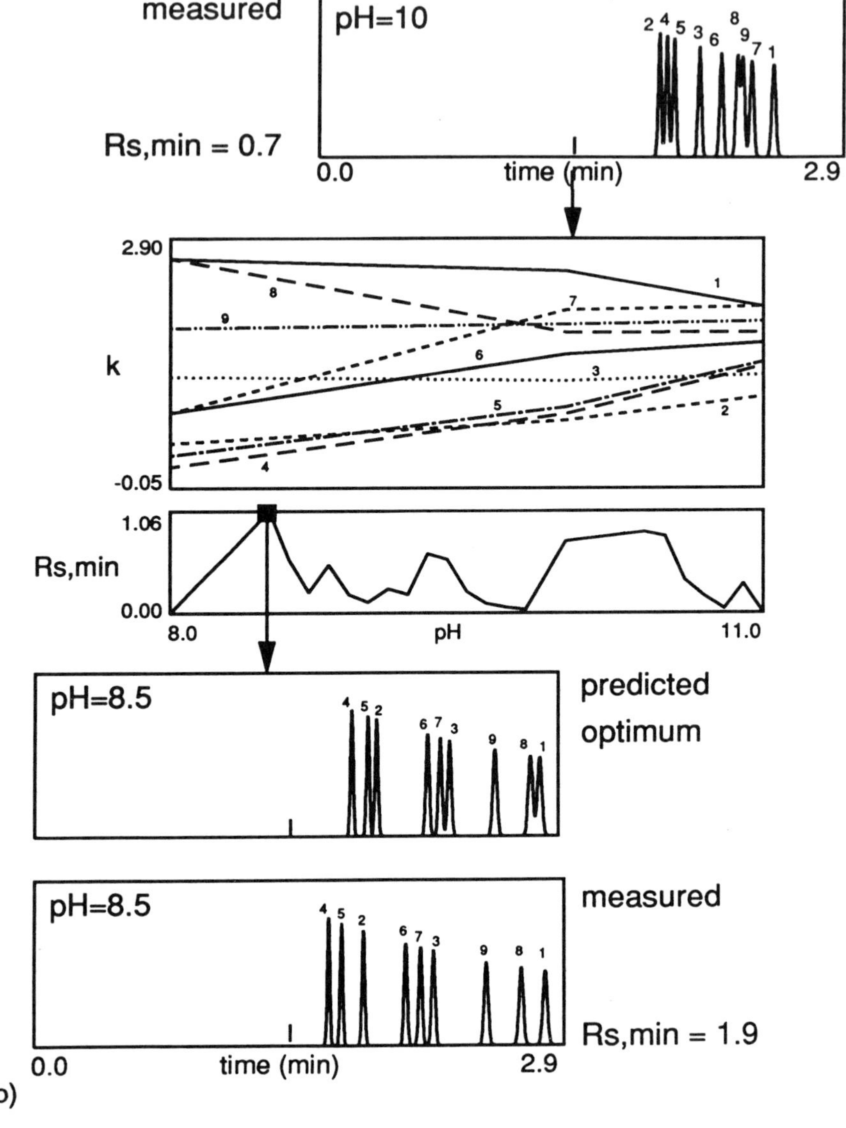

Figure 53 (continued)

can be related to their structural properties; (b) it would facilitate the identification of the important selectivity variables for a given separation; and (c) the optimum solvent composition can be predicted with a minimal experimental effort.

There are several other relevant and important issues that should be considered in optimization of CE separations, which are briefly mentioned in the following:

Reproducibility of Migration Behavior

Repeatability of migration behavior is a crucial factor for a successful method development. Two approaches can be considered to enhance the reproducibility of migration behavior in CE. One is a careful control of the operating variables, such as ionic strength of the buffer, age of the capillary, previous capillary treatment, applied voltage and operating current, temperature control, and rinsing protocols. The other approach is the use of appropriate standardization methods, such as migration indices, to enhance the transferability of the CE data [50]. Many attempts have been made to enhance column-to-column retention reproducibility in HPLC. These efforts have been frustrated mostly because of the existence of stationary phases with variable characteristics and the complex cooperative effects of mobile phase and stationary phase in LC. In a CE separation, the mobility of individual solutes is based on the characteristics of the system (such as buffer composition or ionic strength). These factors can be easily controlled and standardized, thereby making it easier to transfer CE migration data.

We have recently reported the results of an extensive study on the influence of different operating variables, such as the frequency and composition of rinsing solutions, ionic strength, the presence of Sudan III in MECC samples, and thermal equilibration times for the capillary [51]. The frequency of rinsing the capillary and the solutions used for rinsing had the greatest effect on migration reproducibility. In addition, the migration behavior of solutes that interact with micelles is not repeatable unless the proper rinse protocol is applied. A lack of rinsing can even result in a change of selectivity and elution order of the peaks (Fig. 54). A correlation between inconsistent migration behavior and fluctuations in electric current was observed, which might indicate the existence of nonequilibrium conditions between the buffer and the capillary wall.

Peak Tracking

To apply interpretive strategies, one should be able to track the peaks in the various chromatograms or electropherograms. When the composition of the sample is known, this can be done by individual injection of the various solutes. However, to save time and, more importantly, to analyze unknown samples, applying some form of specific detection is necessary to perform peak tracking. Multiwavelength detectors are especially powerful in this context [52]. Fortunately, interfacing of the CE instruments with these types of detectors is feasible.

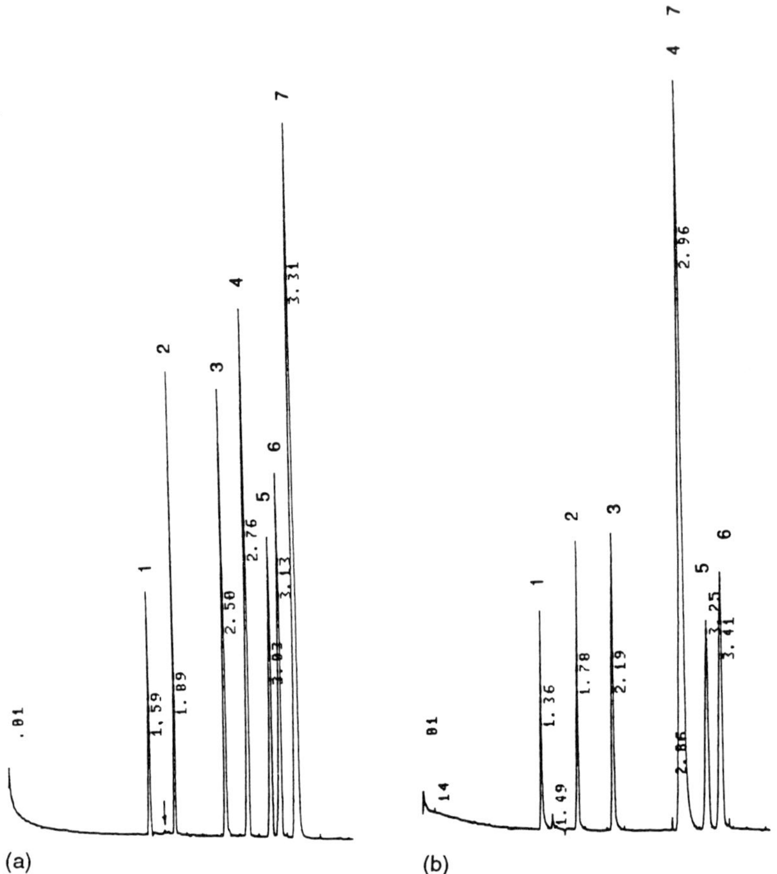

Figure 54 The effect of rinsing on reproducibility of a test mixture. Conditions: 50 mM phosphate buffer, pH 7; 10 mM SDS; 15 kV; L_t, 40 cm, L_s, 26.5 cm. (1) arterenol; (2) 2CP; (3) 23CP; (4) 2,4,6-trimethylbenzoic acid; (5) *p*-hydroxybenzoic acid; (6) *m*-methylbenzoic acid; (7) 245CP. (a) run 1, immediately after rinsing; (b) run 20; no rinsing was done before runs 2–20; (c) run 21, after rinsing before run; (d) run 30; capillary was rinsed before each run for 21–30. Retention times are in minutes. (From Ref. 23)

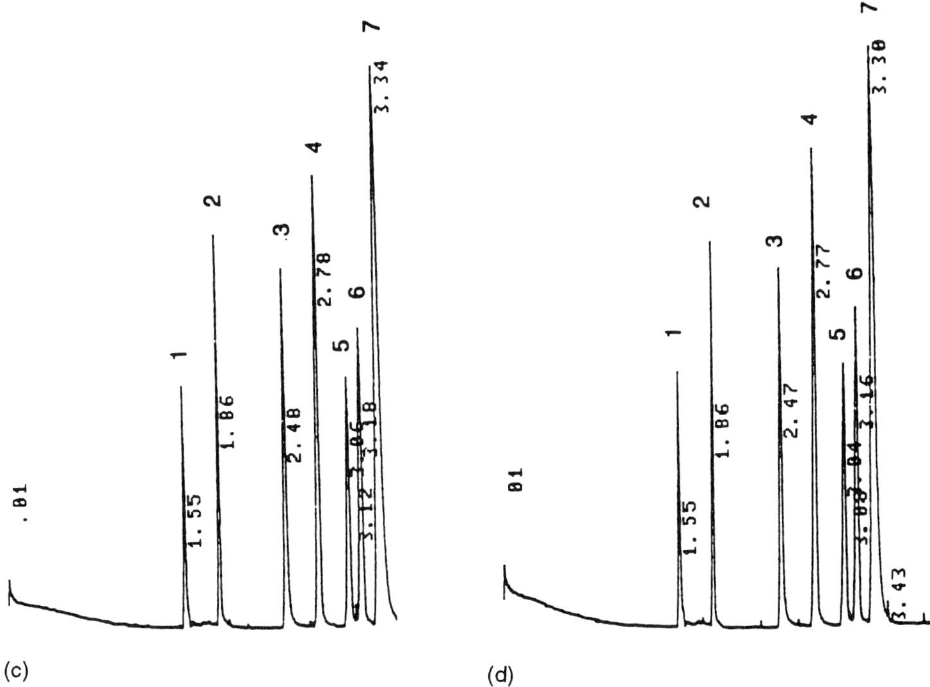

Control of Electroosmotic Flow

Electroosmotic flow has an important effect on resolution and analysis time in CE separations. In many routine analyses, achieving the best separation in a minimum time period is desired. There have been several reports in the literature about controlling electroosmotic flow through changing the buffer characteristics (e.g., addition of organic solvent, use of a cationic surfactant, and changing pH), modification of the capillary walls, and application of an external electric field [53,54].

System Optimization

A major reason for the popularity of CE techniques is the possibility of generating very high efficiencies. Several system variables such as capillary dimensions, electric field strength, and ionic strength can be optimized to maximize the number of theoretical plates. One should then consider how other factors, such as sample size, analysis time, and detectability, would be influenced by these variables.

In HPLC, optimizing column parameters (to maximize column efficiency) requires much time and effort; thus, it is not a primary remedy for improving separation. This is not true, however, for HPCE for which column efficiencies within the range of open tubular GC can easily be generated. As a result, the urgency for optimizing selectivity in HPCE has been lessened, compared with that in HPLC. This greatly facilitates the method development process by CE, as optimizing a few important variables may be adequate to separate even complex mixtures.

Undoubtedly, the popularity of CE technology will continue to grow in the future. The vast knowledge on the enhancement of separation selectivity in HPLC and the wealth of information that is available on the method for the optimization of the important variables are of great use in HPCE separation. It will be possible to advance the electrophoretic techniques very rapidly by implementing the knowledge obtained in the last decade for structured method development. This will result in a mature technology that can easily be applied to both routine and specialized problems in the analytical sciences.

ACKNOWLEDGMENT

We thank Mohammadreza Hadjmohammadi, David Smith, and Bing Ye for their invaluable contributions to different CE research projects in our group. Funding of our research by grants from the National Institutes of Health (First Award, GM 38783), NIH-BMRG, and Glaxo, Inc. in gratefully acknowledged.

REFERENCES

1. P. J. Schoenmakers, *Optimization of Chromatographic Selectivity—A Guide to Method Development*, *J. Chromatogr. Libr.*, Vol. 35; Elsevier, Amsterdam (1986).
2. S. Ahuja, *Selectivity and Detectability Optimizations in HPLC*, John Wiley & Sons, New York (1989).
3. J. C. Berridge, *Techniques for the Automated Optimization of HPLC Separations*, John Wiley & Sons, New York (1985).
4. L. R. Snyder, J. L. Glajch, and J. J. Kirkland, *Practical HPLC Method Development*, Wiley, New York (1988).
5. J. W. Jorgenson, and K. D. Lucacs, *Anal. Chem.*, *53*:1298 (1981).
6. S. Terabe, Y. Toshiyuki, T. Nobuo, and A. Mikio, *Anal. Chem.*, *60*, 1673 (1989).
7. Y. Walbroehl and J. W. Jorgenson, *Anal. Chem.*, *58*:479 (1986).
8. A. S. Cohen, S. Terabe, J. A. Smith, and B. L. Karger, *Anal. Chem.*, *59*:1021 (1987).
9. S. J. Fanali, *Chromatographia*, *474*:441 (1989).
10. D. E. Burton, M. J. Sepaniak, and M. P. J. Maskarinec, *Chromatogr. Sci.*, *24*:347 (1986).
11. A. Guttman, A. Paulus, A. S. Cohen, N. Grinberg, and B. L. J. Karger, *Chromatographia*, *448*:41 (1988).
12. I. Jelinek, J. Dohnal, J. Snopek, and E. J. Smolkova Keulemansova, *Chromatographia*, *435*:496 (1988).

13. R. A. Wallingford, and A. G. Ewing, *J. Chromatogr.* *441*:299 (1988).
14. W. G. Kuhr, *Anal. Chem.*, *62*:403R (1990).
15. F. E. Regnier, and J. K. Towns, *J. Chromatogr.*, *516*:69 (1990).
16. J. Liu, K. A. Cobb, and M. J. Novotny, *Chromatography*, *468*:55 (1988).
17. A. F. Fell, T. M. Bridge, and M. H. Williams, *J. Pharm. Biomed. Anal.*, 6:555 (1988).
18. J. Vindevogel, and P. Sandra, *Anal. Chem.*, *63*:1530 (1991).
19. B. Ye, and M. G. Khaledi, Unpublished results.
20. S. K. Yeo, C. P. Ong, and S. F. Y. Li, *Anal. Chem.*, *63*:2222 (1991).
21. J. K. Strasters, H. A. H. Billiet, S. C. Smith, and M. G. Khaledi, "15th International Symposium on Column Liquid Chromatography," Basel, Switzerland, June 3–7 (1991).
22. C. J. O. R. Morris, and P. Morris, *Separation Methods in Biochemistry*, Ch. 18, Interscience Publishers, New York (1963).
23. M. G. Khaledi, S. C. Smith, and J. K. Strasters, *Anal. Chem.*, *63*:1820 (1991).
24. S. C. Smith, M. G. Khaledi, *Anal. Chem.*, *65*:193 (1993).
25. Y. Walbroehl, and J. W. Jorgenson, *J. Chromatogr.*, *315*:135 (1984).
26. R. S. Sahota, and M. G. Khaledi, *Anal. Chem.*, (submitted for publication).
27. S. Tarabe, K. Otsuka, K. Ichikawa, A. Tsuchiya, and T. Ando, *Anal. Chem.*, *56*:113 (1984).
28. S. Tarabe, K. Otsuka, and T. Ando, *Anal. Chem.*, *57*:834 (1985).
29. C. S. Horvath, W. Melander, and I. Melnar, *Anal. Chem.*, *49*:142 (1977).
30. J. P. Foley, and W. E. May, *Anal. Chem.*, *59*:102 (1987).
31. A. H. Rodgers, J. K. Strasters, and M. G. Khaledi (in preparation).
32. J. P. Foley, *Anal. Chem.*, *62*:1302 (1990).
33. K. Otsuka, S. Terabe, and T. Ando, *J. Chromatogr.* *348*:39 (1985).
34. J. K. Strasters, and M. G. Khaledi, *Anal. Chem.*, *63*:2503 (1991).
35. D. M. Smith, J. K. Strasters, and M. G. Khaledi (manuscript in preparation).
36. C. Y. Quang, J. K. Strasters, and M. G. Khaledi, *Anal. Chem.* (submitted for publication).
37. S. C. Smith, and M. G. Khaledi, *J. Chromatogr.* (in press, 1993).
38. M. G. Khaledi, and A. H. Rodgers, *Anal. Chim. Acta*, *239*:121 (1990).
39. A. T. Balchunas, and M. J. Sepaniak, *Anal. Chem.*, *60*:617 (1988).
40. S. Terabe, Y. Miyashita, O. Shibata, E. R. Barnhart, L. R. Alexander, D. G. Patterson, B. L. Karger, K. Hosoya, and N. J.Tanaka, *Chromatographia*, *516*:23 (1990).
41. J. F. Scamehorn, (ed.), *Phenomena in Mixed Surfactant Systems, ACS Symp. Ser. 311*, American Chemical Society (1986).
42. M. G. Khaledi, and R. Hadjmohammadi, *Anal. Chem.* (submitted for publication).
43. E. R. Malinowski, and D. G. Howery, *Factor Analysis in Chemistry*, John Wiley & Sons, New York (1980).
44. R. S. Sahota, and M. G. Khaledi, *Anal. Chem.*, (submitted for publication).
45. C. H. Lochmuller, S. J. Briener, C. E. Reese, and M. N. Koel, *Anal. Chem.*, *61*:367 (1989).
46. A. C. J. H. Drouen, H. A. H. Billiet, P. J. Schoenmakers, and L. de Galan, *Chromatographia*, *16*:48 (1982).
47. H. A. H. Billiet, A. C. J. H. Drouen, and L. de Galan, *J. Chromatogr.*, *316*:231 1984).

48. J. K. Strasters, E. D. Breyer, A. H. Rodgers, and M.G. Khaledi, *J. Chromatogr.*, *511*:17 (1990).
49. J. K. Strasters, S. Kim, and M. G. Khaledi, *J. Chromatogr.*, *586*:221 (1991).
50. T. T. Lee, and E. S. Yueng, *Anal. Chem.*, *63*:2842 (1991).
51. S. C. Smith, J. K. Strasters, and M. G. Khaledi, *J. Chromatogr.*, *586*:221 (1991).
52. J. K. Strasters, H. A. H. Billiet, L. de Galan, B. G. M. Vandeginste, and G. J. Kateman, *Liq. Chromatogr.*, *12*:3 (1989).
53. C. S. Lee, W. C. Blanchard, and C. T. Wu, *Anal. Chem.*, *62*:1550 (1990).
54. C. T. Wu, T. Lopes, B. Patel, C. S. Lee, *Anal. Chem.*, *64*:886 (1992).

7

Dynamic Changes of Electrolyte Systems in Zone Electrophoresis

Petr Boček and Petr Gebauer

Institute of Analytical Chemistry
Czech Academy of Sciences
Brno, Czech

The basic requirement of a successful zone electrophoretic (ZE) analysis is that the sharp zones of the solutes to be separated pass the detection cell at sufficiently different times. In common zone electrophoretic systems, this is equivalent to the existence of sufficient differences in the migration velocities of the solutes, based on differences in their effective (net) electrophoretic mobilities. In a typical case, the solutes are weak acids or bases, and the most simple and usual way of affecting their effective mobilities is the selection of a proper pH for the background electrolyte. It generally holds that the most effective operational pH region is found at pH values near the pK_a values of the solutes, at which the effect of pH on their dissociation degree is the most pronounced. The mobility curve [1] (i.e., the dependence of the effective mobility ($\bar{\mu}$) versus the pH of the background electrolyte) is a most useful tool here. In simple cases, an explicit formula can be derived; for example, when looking for the optimum pH for the separation of two weak monohydric acids A and B, the following expression is obtained [2]:

$$pH_{opt} = \frac{pK_A + pK_B}{2} - \log \frac{\sqrt{\mu_A/\mu_B} - \sqrt{K_B/K_A}}{1 - \sqrt{\mu_B K_B/\mu_A K_A}}$$

where K_A, K_B, and μ_A, μ_B are the dissociation constants and ionic mobilities of acids A and B, respectively.

In practice, it is not a rare event that solutes of very close ionic mobilities differ substantially in their pK_a values. Then the problem may often arise that these solutes

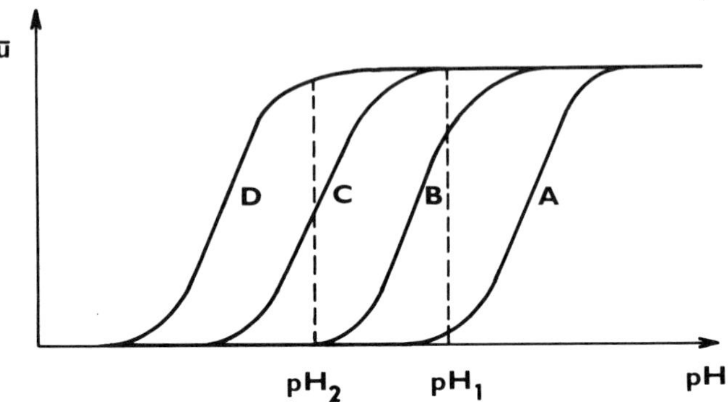

Figure 1 Mobility curves of four weak acids.

cannot be separated in one run at a constant background electrolyte composition and pH. Figure 1 illustrates this, by showing the mobility curves of the weak acids A, B, C, and D, which have the same ionic mobilities, but differ in their pK_a values. At pH_1, substances A and B are partially dissociated and can be separated easily from substances C and D; these, however, are both dissociated completely, and their resolution is impossible. At a lower pH value (pH_2), the resolution of C and D is sufficient owing to their reduced dissociation degree; however, here, the dissociation degree of A and B is negligible, which does not permit their separation. Apparently, to carry out two analyses at two different pH values is the only solution to this problem. Such a separation problem is, however, a typical candidate to be solved by using the dynamic changes of the electrolyte system to achieve the complete separation of all components within one run.

The principle of any type of dynamic change of an electrolyte system is that the composition of the background electrolyte is no longer constant both with time

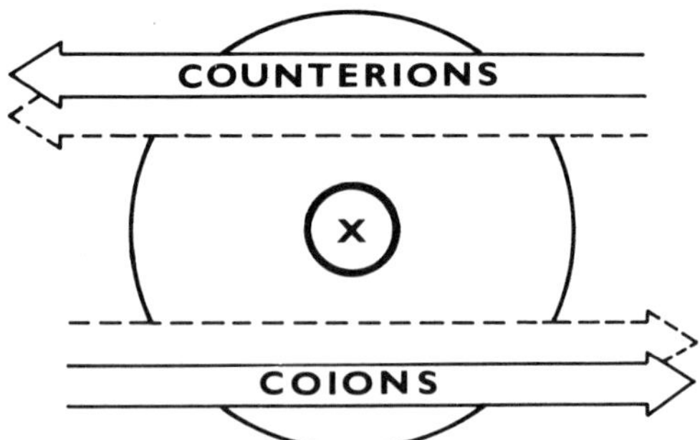

Figure 2 Scheme for flows of background electrolyte ions surrounding the solute ion X.

and along the separation path. The concept of the ionic matrix, which affects the effective mobility of a solute, is important here, reflecting that the mobility of a migrating ion is controlled by its surroundings. It should be emphasized that this surrounding forms a dynamic system in which the respective ions flow in and out. Figure 2 shows the scheme of the ionic matrix surrounding a solute X, formed by the coions and counterions of the background electrolyte. As long as this ionic matrix is constant, the electrophoretic migration of ion X proceeds in a constant way. Any (even temporal) change in the flow of any coion or counterion directly affects the migration of X. By changing or modulating the flows of these ions we can change or modulate the migration of the solute X and, thereby, achieve its separation from other solutes. One important principle should be mentioned here: species once separated under suitable conditions (in a suitable ionic matrix where they possess mutually different mobilities) are no longer mixed when again returning to conditions under which they have identical effective mobilities. Briefly said, the separation once obtained survives, and the solutes possessing the same effective mobilities migrate in parallel.

It is clear that in an electric field all ions are moving and, therefore, all changes in an electrolyte system have a dynamic character. This implies the method for

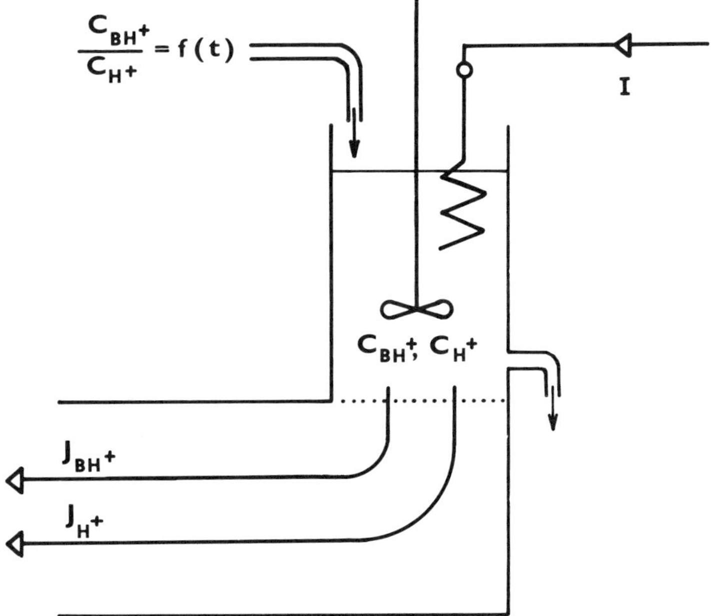

Figure 3 Instrumental principle for generation of dynamic changes by changing the contents of the electrode chamber.

generation of these changes, which consists in modifying the composition of the electrolyte at the inlet or outlet of the separation capillary.

HOW TO GENERATE DYNAMIC CHANGES

Two general ways are possible to generate dynamic changes in electrolyte systems:

The principle of the first one is shown in Fig. 3. It consists of a programmed change in the composition of the electrolyte in an electrode chamber that can be used for both the cathode and anode chambers. If, say, the electrode chamber is anodic and contains a solution of cations BH^+ and H^+, then the passage of electric current I through the system results in flows J_i of both cations out of the electrode chamber into the separation capillary. The ratio of these flows depends on the ratio of the concentrations of these cations in the electrode chamber. By changing this

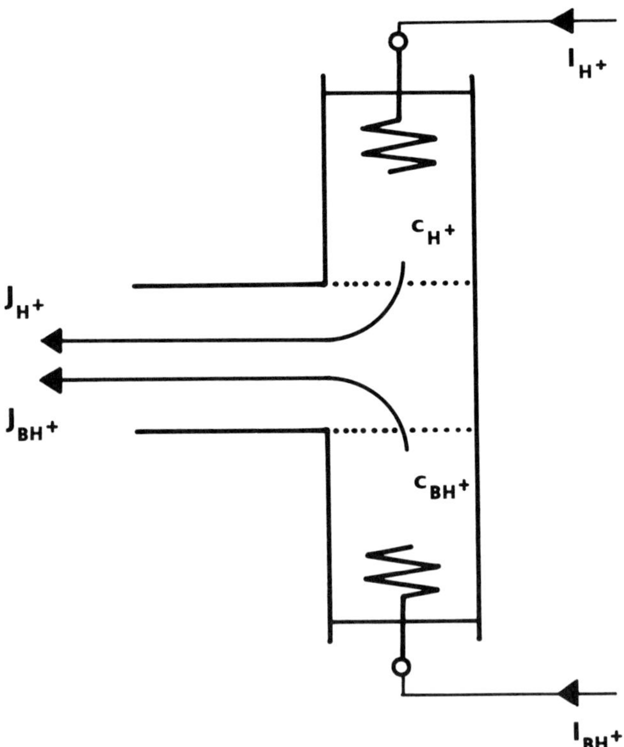

Figure 4 Instrumental principle for generation of dynamic changes by using a pair of electrode chambers of same polarity. (From Ref. 3)

ratio, a dynamic change (here, a pH change) is generated and sent into the separation system. The required nature of the dynamic change predetermines the instrumental method used for its generation; for example, for a steplike change the setup shown in Fig. 3 can be used, and the procedure is based on a brief interruption of the analysis during which the electrode chamber is refilled with the new (modifying) electrolyte.

Figure 4 depicts the principle of the second possibility, for which the separation capillary is connected to a pair of electrode chambers of the same polarity. In the simplest case (when using the example from Fig. 3) each electrode chamber contains the solution of a single ion, that is, BH^+ and H^+, respectively. By controlling the ratio of the electric currents flowing through the electrode chambers, the ratio of the flows of both ions into the separation system is controlled. The simplest steplike change can be generated here by only switching over from one to the other electrode chamber.

TYPES OF DYNAMIC CHANGES

There are three basic types of dynamic changes that can be generated in zone electrophoretic systems (Fig. 5). The most simple is a step (see Fig. 5a) that represents a sharp discontinuity in the composition of the background electrolyte migrating through the separation column. This causes each solute to migrate electrophoretically for some time in one background electrolyte and then for a time in a changed background electrolyte. Prerequisite to a proper function of such a step is its constant shape; this presumes that the step must have self-sharpening properties (discussed in the following section).

The gradient (see Fig. 5b) represents a continuous profile in the composition of the background electrolyte; usually, this profile changes when migrating through

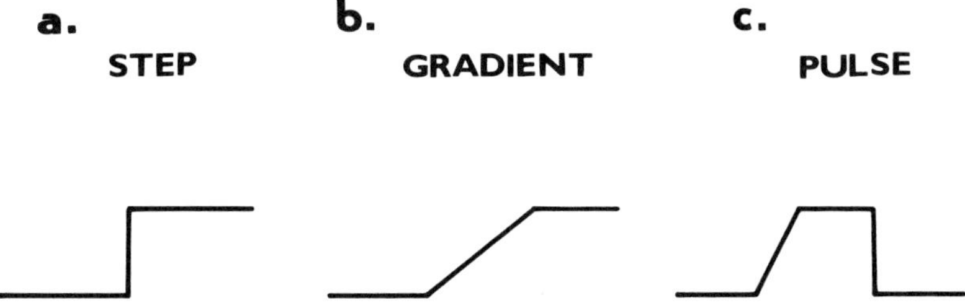

Figure 5 Typical shapes of dynamic changes of electrolyte systems.

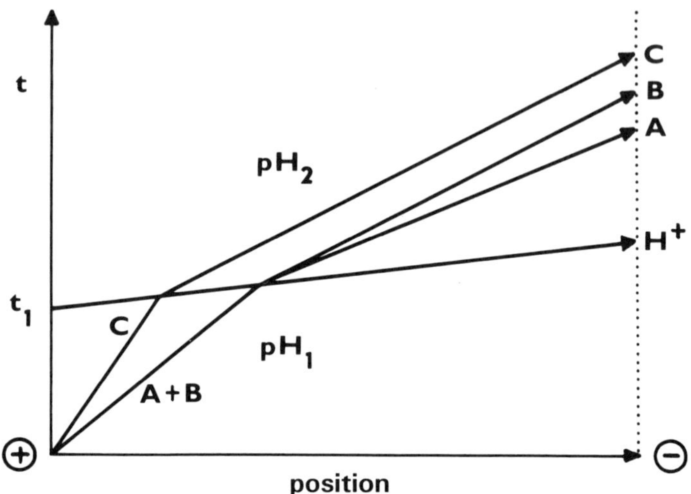

Figure 6 Migration trajectories of cations A, B, and C and of a comigrating step of H^+.

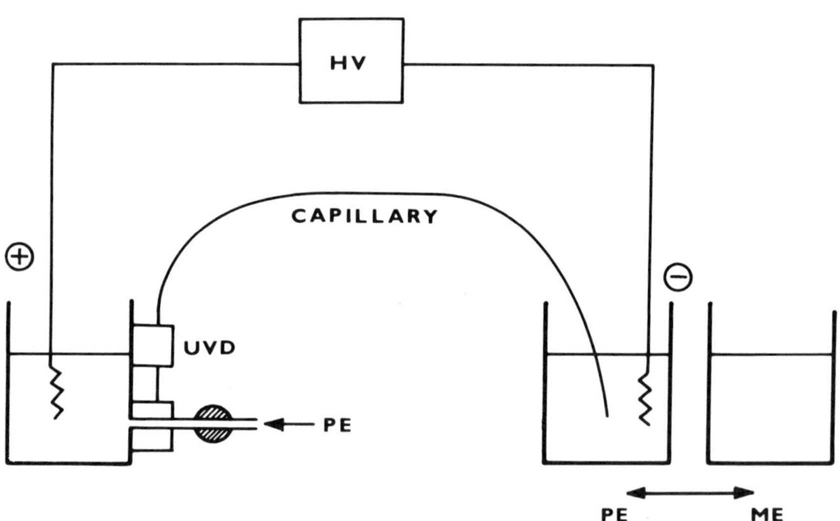

Figure 7 Instrumental setup for the generation of systems with a dynamic step. PE, primary electrolyte; ME, modifying electrolyte; HV, high-voltage power supply; UVD, UV absorption detector. (From Ref. 5)

the separation column. Therefore, the shape of a gradient may be different when observed at different positions of the column. The separation conditions thus continually change with time.

The pulse (see Fig. 5c) is a combination of both previous examples. It represents a temporal change of the background electrolyte migrating through the column. Here, the solutes migrate temporally under different separation conditions and are then detected at conditions of the original background electrolyte.

Dynamic Step

As indicated in the foregoing, the dynamic step represents a migrating sharp discontinuity in the composition of the background electrolyte. Its function can be explained by the example of the separation of three cathodically migrating weak bases using a dynamic step of H^+ as the coion. Figure 6 shows the trajectories (time vs coordinate plots) of all three solutes and of the H^+ step. It is seen that A and B are not separated at the pH of the original background electrolyte. At time t_1, the step starts to migrate through the system and modifies the background electrolyte. In the modified background electrolyte, A and B are separated. Although the migration velocities of B and C are now constant, these substances, having once been separated, no longer mix one with the other. The combination of two pH levels thus always gives the combination of their individual benefits.

Figure 7 shows a typical setup of instrumentation that can be used to create systems with dynamic steps. The only difference from simple devices for zone electrophoresis is that two electrolyte vials are operated at one side of the capillary. One vial is filled with the original background electrolyte (primary electrolyte) and the other with a modifying electrolyte, creating the step. In a manner similar to that currently used for electromigration sampling, the capillary (and the electrode) is dipped into one vial for a part of the analysis and into the other for the rest of the analysis.

An experimental example of how an H^+ step can be used for the separation of cathodically migrating proteins (Table 1) is shown in Figure 8 [4]. Figure 8a presents

Table 1 Proteins Separated Using an H^+ Step

Protein	Abbreviation	Molecular mass (M_r)	pI
Cytochrome C	Cyt	13,000	9.4
Lysozyme	Lys	14,000	11
Myoglobin	Myo	17,500	8.1
Carbonic anhydrase	CA	29,000	5.9
Trypsinogen	Try	24,000	9.3

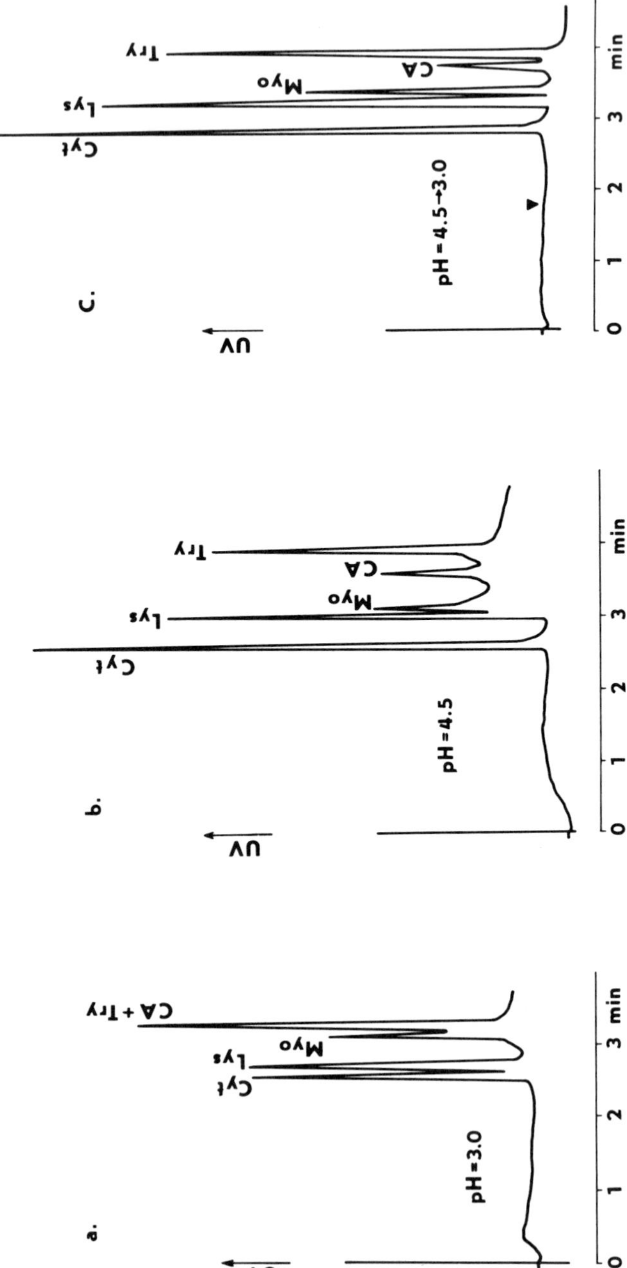

Figure 8 Cationic zone electrophoresis of proteins (see Table 1) in 0.05 M Tris–HCl buffer (a) at pH 3.0, (b) at pH 4.5, (c) using a 4.5–3.0 dynamic pH step. (From Ref. 4)

Table 2 Model Mixture Separated Using an OH^- Step

Substance	Abbreviation	Mobility (10^{-9} m^2 V^{-1} s^{-1})	pK_a
2,5-Dinitrophenol	2, 5-DNP	31.3	5.22
4-Nitrophenol	4-NP	33.4	7.15
3-Nitrophenol	3-NP	33.4	8.40
3-Chlorophenol	3-CP	33.4	9.02
4-Chlorophenol	4-CP	33.4	9.38
Phenol	P	34.4	9.99

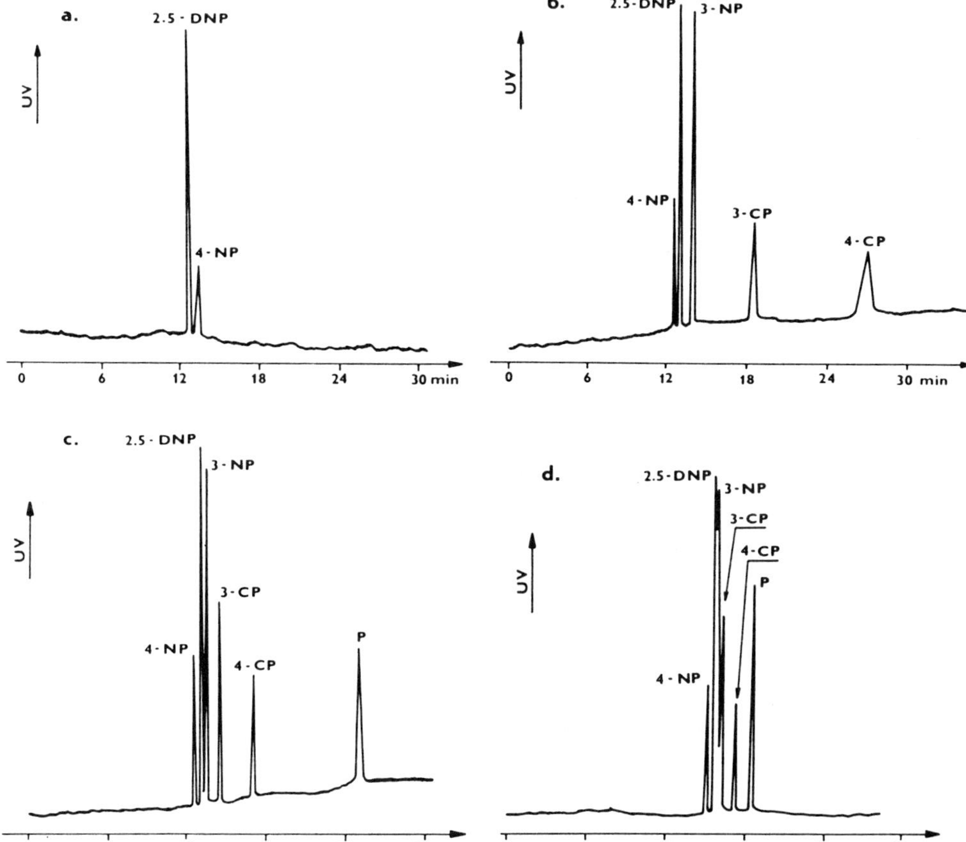

Figure 9 Anionic zone electrophoresis of substances from Table 2 in 0.01 M lithium oxalate at (a) pH 7.2, (b) pH 8.3, (c) pH 9.5, (d) pH 10.5. (From Ref. 5)

the zone electrophoretic separation at pH 3.0; only partial resolution of myoglobin, carbonic anhydrase, and trypsinogen is obtained. At pH 4.5 (see Fig. 8b), the separation is improved; however, extensive tailing of some protein peaks disturbs the separation result. The use of a dynamic 4.5–3.0 pH step solves the problem, as is seen from Fig. 8c, showing complete resolution and symmetric peaks.

The next example demonstrates the use of a step of the coion in anionic zone electrophoresis [5]. The characteristics of the model substances used are given in Table 2. In Fig. 9, the dependence of zone electrophoretic separation of some derivatives of phenol on the pH of the background electrolyte is shown. At low pH, some of the substances never migrate, owing to their low dissociation degree. At high pH, dissociation is sufficient; however, their mobilities are too close together to reach good separation. The trajectories diagram in Fig. 10 shows the action of a pH step on the separation, during which the step of the coion (carbonate) creates inherently a sharp step of OH^-. Here, the primary electrolyte separates all sample components; then a step of an increased pH is generated that mobilizes all the separated substances to bring them into the detector. The experimental record of a

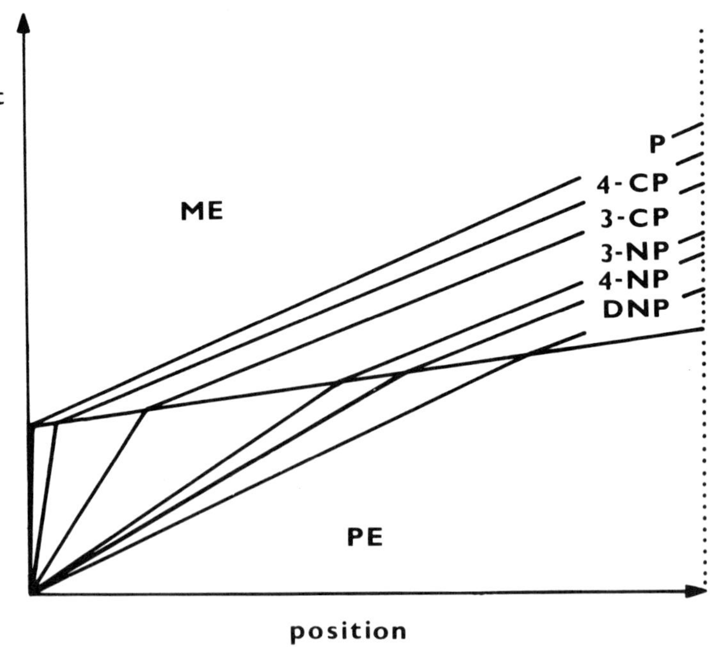

Figure 10 Migration trajectories of the analyzed substances and of a comigrating step of carbonate (OH^-). For abbreviations, see Table 2. (From Ref. 5)

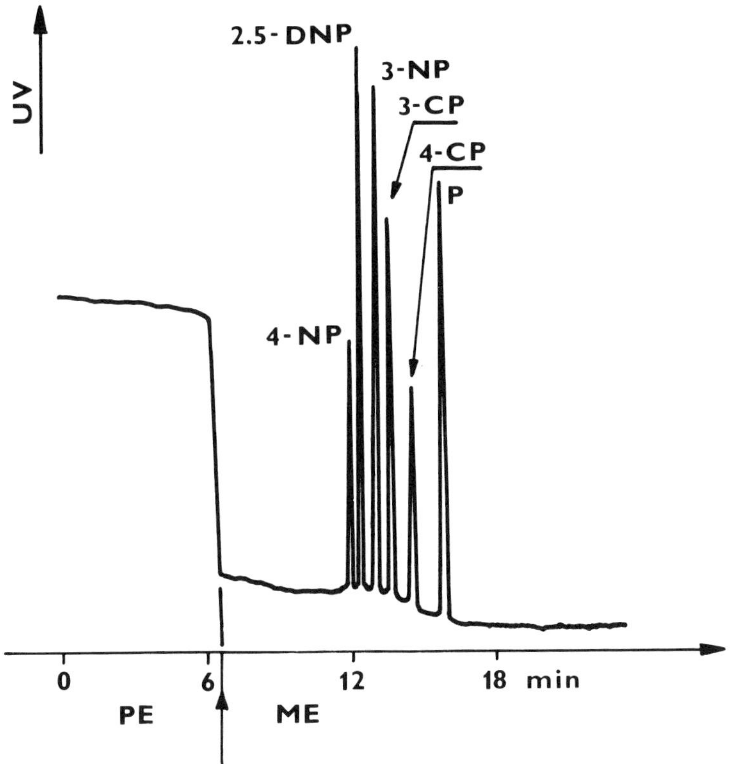

Figure 11 Separation of the substances from Fig. 9 using a step of coion (OH^-). Primary electrolyte: 0.01 M lithium oxalate, pH 7.2; modifying electrolyte: 0.01 M Na_2CO_3, pH 11.0. (From Ref. 5)

Table 3 Model Mixture Separated Using a Step of Counterion

Acid	Abbreviation	Mobility (10^{-9} m^2 V^{-1} s^{-1})	pK_a
Picric	PIC	31.5	0.71
3,5-Dinitrobenzoic	DNB	29.3	2.82
2,6-Dinitrophenol	2,6-DNP	31.3	3.71
2,4-Dinitrophenol	2,4-DNP	31.3	4.02
Nicotinic	Nic	34.6	4.82

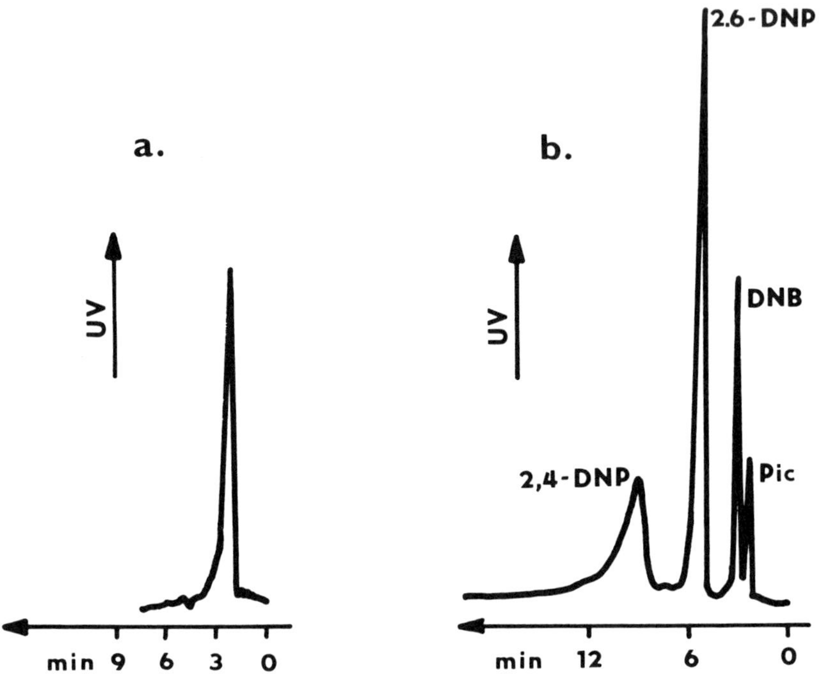

Figure 12 Anionic zone electrophoresis of a mixture of acids in 0.01 HCl adjusted with KOH to (a) pH 6.5, (b) pH 3.5. For abbreviations, see Table 2. (From Ref. 6)

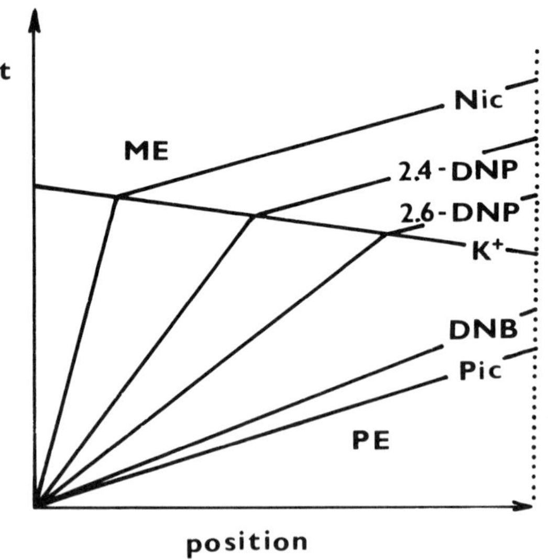

Figure 13 Migration trajectories of the analyzed substances and of a countermigrating step of K^+/H^+. For abbreviations. see Table 3. (From Ref. 6)

separation using such a step is shown in Fig. 11, from which both complete resolution and fast migration are seen.

In an analogous manner, a step of counterion can be generated and used. By way of example, the anionic separation of a mixture of substances (Table 3) may be described in which a step of the countercation is used [6]. Figure 12 shows the zone electrophoretic analysis of a mixture of these anions at two different pH values.

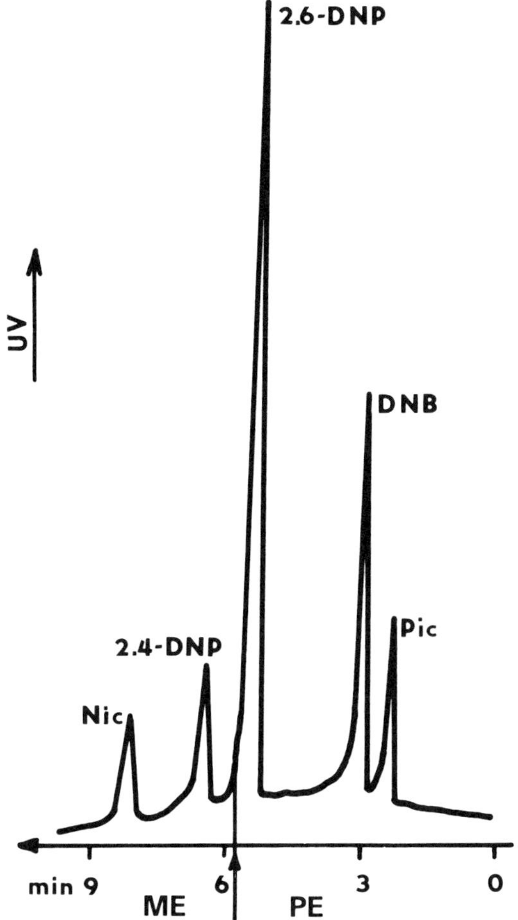

Figure 14 Separation of substances from Fig. 12 using a step of counterion (H^+/K^+). Primary electrolyte: 0.01 HCl adjusted with KOH to pH 3.5; modifying electrolyte: 0.01 HCl adjusted with KOH to pH 6.5. (From Ref. 6)

Depending on pH, one can see either fast migration and poor resolution (see Fig. 12a; at this high pH the substances are ionized enough, but their net mobilities are very close) or good resolution, but very slow migration, of some components (see Fig. 12b; nicotinic acid has not reached the detector).

Figure 13 shows the diagram of the trajectories explaining the mechanism of the separation process when using a pH step. Both the primary and the modifying electrolytes are mixtures of HCl + KCl at different pHs. In the primary electrolyte, at a low pH, sufficient resolution is reached for all components. Then the migrating step formed by an increased K^+/H^+ concentration ratio increases the pH in the system, thereby mobilizing the separated components. Figure 14 shows the corresponding separation using the step.

Dynamic Gradient

The principle of the dynamic gradient is very similar to that of the dynamic step. The difference is that the profile in the electrolyte composition migrating through the column is not steplike but forms a ramp; moreover, the profile of the gradient changes with time. In analogy with the step, the gradient may also be formed either to migrate against the migrating sample components (gradient of counterion composition) or to migrate in the same direction (usually with some delay, gradient of coion composition).

Figure 15 shows a possible instrumental setup for creating a gradient of coion. The primary electrolyte in the anodic electrode chamber is modified by adding the modifying solution in a defined way by the pump.

The experimental example of the dynamic gradient method comprises the separation of cations of purine and pyrimidine bases (Table 4), by using a dynamic

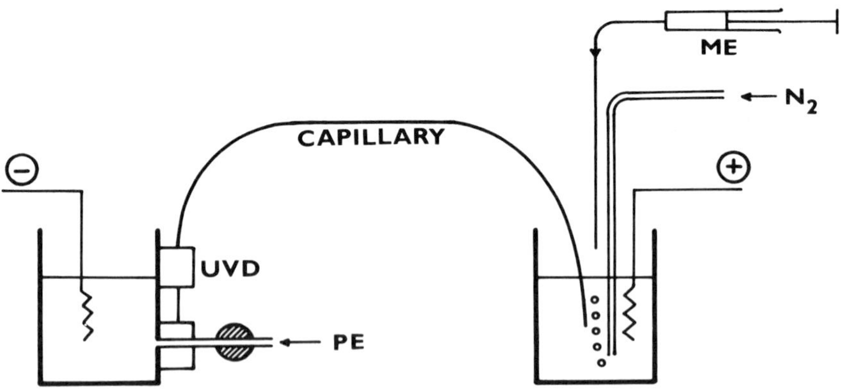

Figure 15 Instrumental setup for generation of systems with a dynamic gradient. PE, primary electrolyte; ME, modifying electrolyte; UVD, UV absorption detector. (From Ref. 6)

Table 4 Purine and Pyrimidine Derivatives Separated Using an H^+ Gradient

Substance	Abbreviation	pK_a
4,6-Diaminopyrimidine	4,6-DAP	6.0
Cytosine	C	4.6
5-Methylcytosine	5-MC	4.6
Adenine	A	4.1
2-Aminopurine	2-AP	3.7
6-Benzylaminopurine	6-BAP	>4
Guanine	G	3.3
5-Aminouracil	5-AU	3.2
5-Bromocytosine	5-BC	3.0
Hypoxanthine	HX	1.9
Guanosine	Guo	1.9

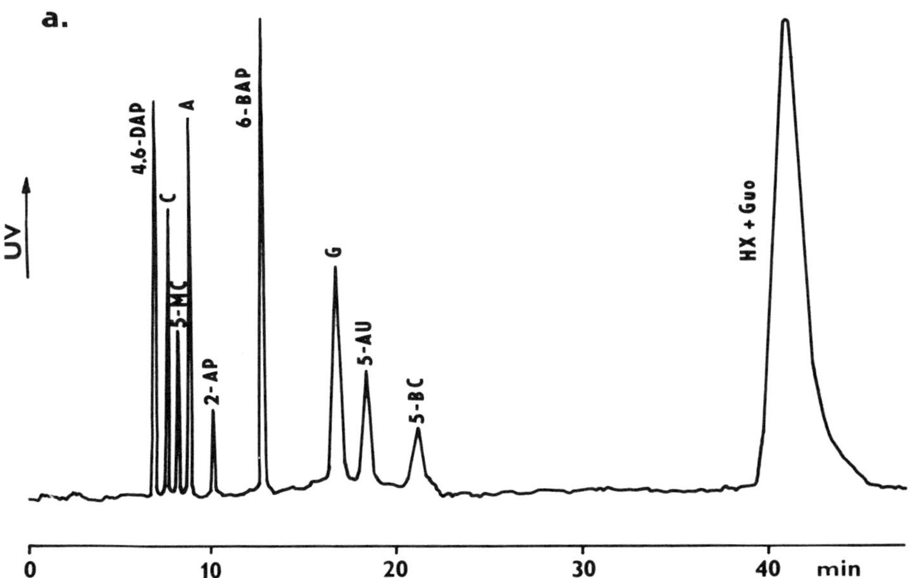

Figure 16 Cationic zone electrophoresis of a mixture of cationic substances in 0.01 M Tris–chloroacetate at (a) pH 3.5, (b) pH 2.2. For abbreviations, see Table 4. (From Ref. 7)

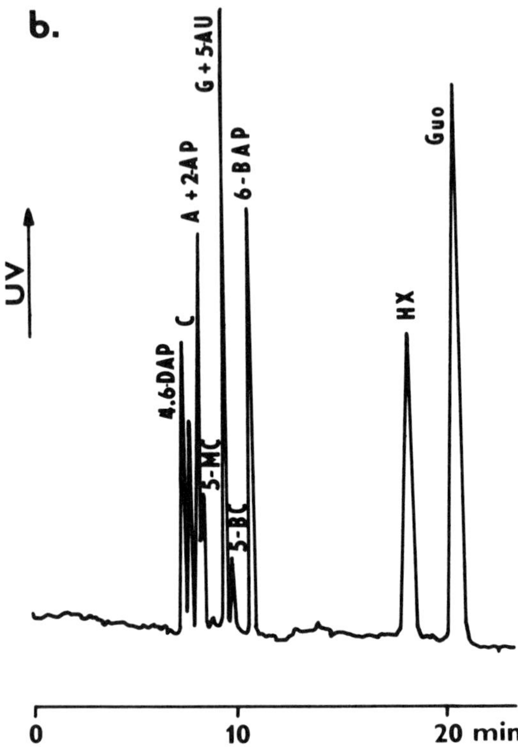

Figure 16 (continued)

gradient of H^+ as a coion [7]. Figure 16a,b shows zone electrophoretic analyses at pH 3.5 and 2.2, respectively. At the higher pH value, some components (HX + Guo) migrate very slowly and are not separated; at the lower pH value, they are much faster and are separated; however, resolution of other sample components is lost. The use of a dynamic pH gradient brings good resolution and fast migration of all sample components (Fig. 17).

Dynamic Pulse

The dynamic pulse is the third possibility for dynamically changing the background electrolyte. The advantage of the pulse is that after it has migrated through a certain point of the column, the original background electrolyte is restored, and the detection of the components can be performed under constant and known conditions [8].

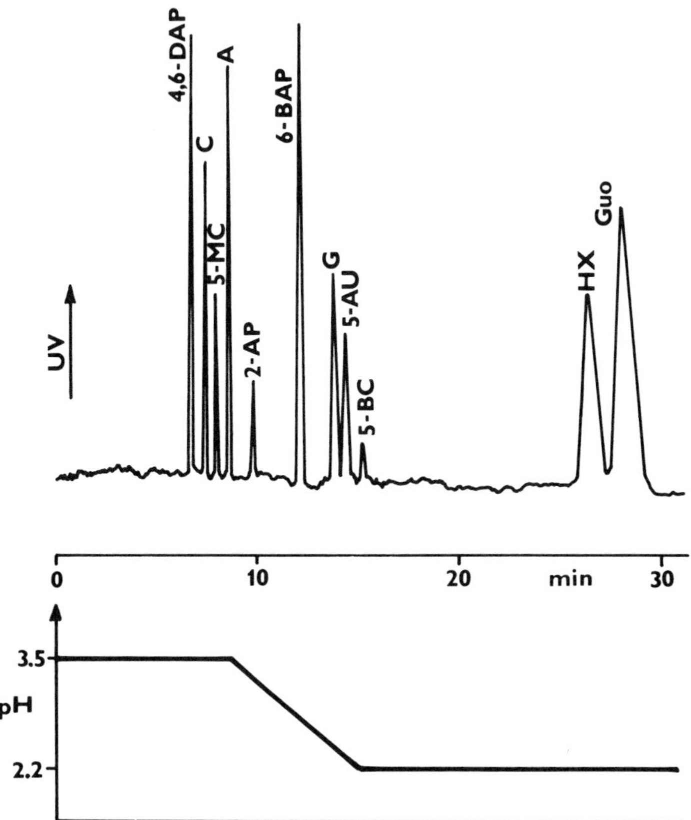

Figure 17 Separation of substances from Fig. 16 using a dynamic pH gradient. Primary electrolyte: 0.01 M Tris–chloroacetate, pH 3.5; modifying solution: 0.5 M trichloroacetic acid. Pumping was switched on between 9 and 14 minutes of the analysis time; the resulting pH in the anode chamber was 2.2 (see the lower panel). (From Ref. 7)

The principle of pulse action is explained by Fig. 18 showing the scheme of anionic separation of three acids A, B, and C using a pulse of H^+ (here counterion). The acids have close ionic mobilities and are fully dissociated in the background electrolyte. Their migration velocities, therefore, are equal, and they do not separate from each other. When migrating into the H^+ pulse, the acids are selectively retarded according to their pK_a values and thereby resolved. Then the acids migrate into the detector; they are separated, although they migrate again with equal velocities. This can be clearly seen from the trajectories diagram (Fig. 19).

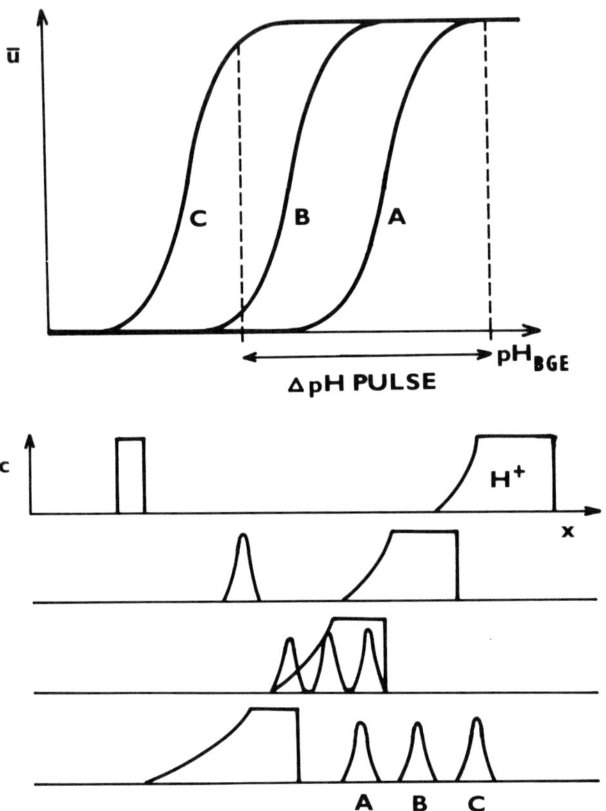

Figure 18 Principle of the separation of three anions, A, B, and C, using a countermigrating pulse of H^+. (From Ref. 9)

In Fig. 20, a possible instrumental setup for generating a pulse of counterion is shown. The system consists of a pair of anodic electrode chambers containing the primary and the modifying electrolyte. By switching over for some period from the primary to the modifying anode, the pulse is generated.

Figure 21 shows an experimental example of the application of the foregoing type of H^+ pulse on the separation of some acids [9]. The run in Fig. 21a was performed by zone electrophoresis at a high pH (5.6) at which no separation is seen. Because the pK_a values of the mixture (Table 5) differ by approximately 4 units, the separation problem cannot be solved by decreasing the pH of the background electrolyte (see also Fig. 1). The use of an H^+ pulse easily solves the problem, as can be seen from Fig. 21b.

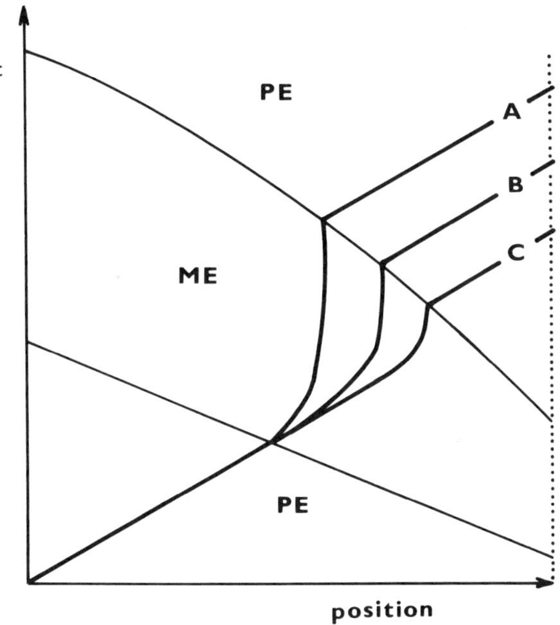

Figure 19 Migration trajectories of the anions from Fig. 18 and of a countermigrating pulse of H^+. (From Ref. 9)

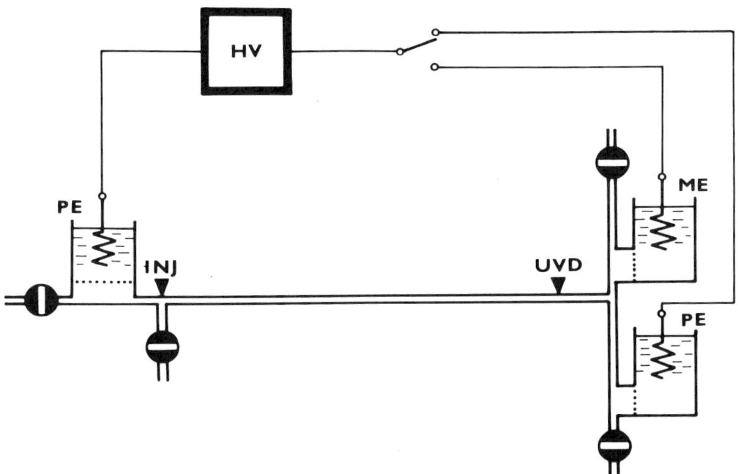

Figure 20 Instrumental setup for the generation of systems with a dynamic pulse. PE, primary electrolyte; ME, modifying electrolyte; UVD, UV absorption detector; HV, high-voltage power supply; INJ, sampling point. (From Ref. 6)

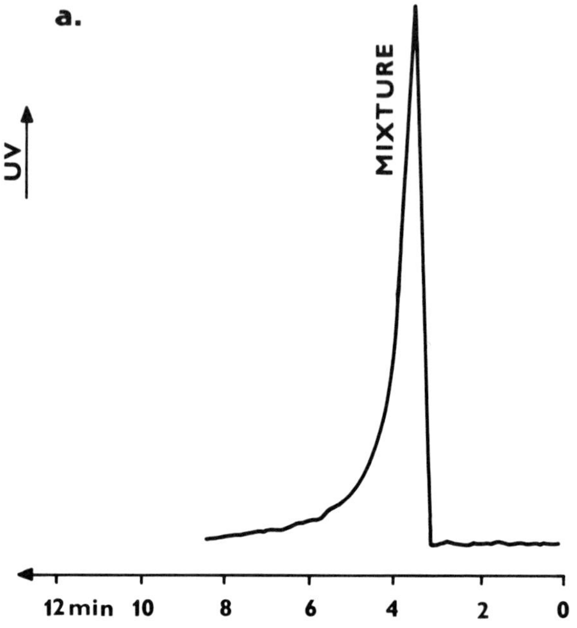

Figure 21 Anionic separation of acids (a) by zone electrophoresis in 0.01 M KCl, pH 5.6; (b) by using a 1-min long H^+ pulse (primary electrolyte: 0.01 M KCl, pH 5.6; modifying electrolyte: 0.01 M HCl). For abbreviations, see Table 5. (From Ref. 9)

Table 5 Model Mixture Separated Using an H^+ Pulse

Acid	Abbreviation	Mobility (10^{-9} m^2 V^{-1} s^{-1})	pK_a
Picric	Pic	31.5	0.71
3,5-Dinitrobenzoic	DNB	29.3	2.82
2,4-Dinitrophenol	2,4-DNP	31.3	4.11
Cinnamic	CIN	28.3	4.44
Sorbic	SOR	33.4	4.77

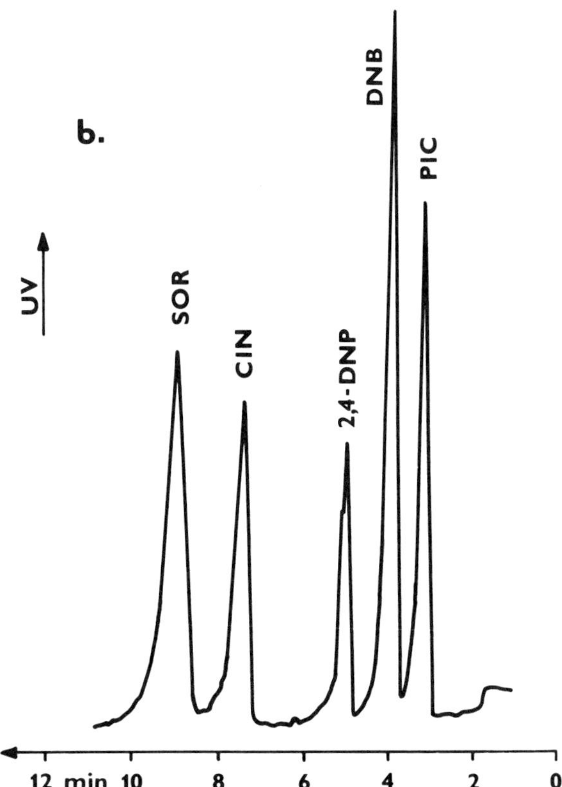

CONCLUSION

Dynamic changes of electrolyte systems offer significant benefits and extend the separation potential of electrophoresis (Fig. 22):

1. Selective resolution is achievable at different pH levels during one run, making possible a complete resolution of sample mixtures that cannot be separated in one background electrolyte.
2. Mobilization of too slowly moving species is possible by a suitable change of, for example, pH, so that reasonable analysis times may be reached for all solutes analyzed.
3. Selective retardation of some species is possible for a controlled period, which may bring about the required separation while keeping fast analysis and defined constant detection conditions.
4. Reduction of peak tailing is possible by the focusing effect of a pH gradient.

a. Selective separation

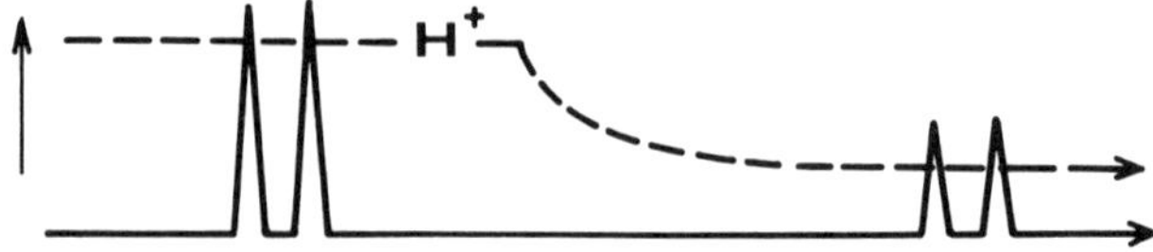

b. Mobilization

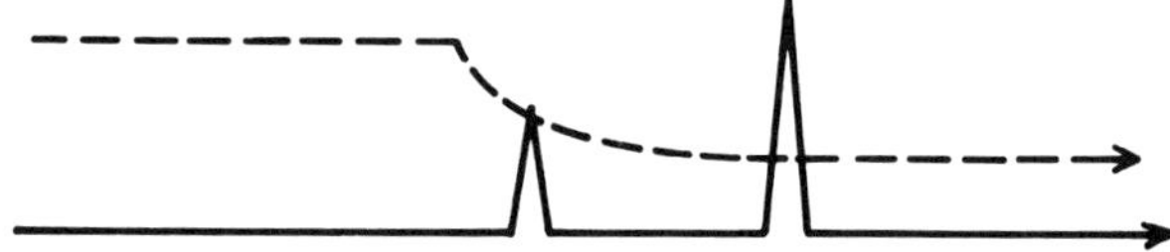

c. Retardation

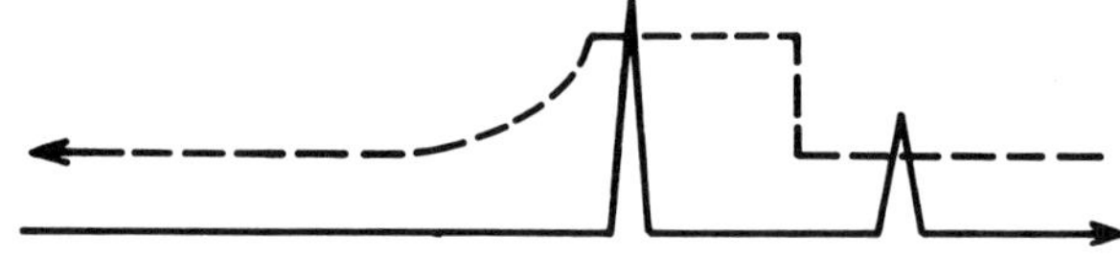

d. Focusing

Figure 22 Analytical aspects of dynamic changes of pH.

A major advantage of dynamic changes is the flexibility that allows one to generate various types of dynamic changes using the same instrumentation. It should be noted that the range of possible dynamic changes is not limited to changes of pH, but changes of other electrolyte properties, such as complex-forming equilibria, are also possible.

REFERENCES

1. K. Klepárník and P. Boček, *J. Chromatogr.*, 569:3 (1991).
2. F. Foret and P. Boček, *Advances in Electrophoresis*, vol. 3 (A. Chrambach, M. J. Dunn, and B. J. Radola, eds.), VCH Publishers, New York, pp. 273–347 (1989).

3. J. Pospíchal, M. Deml, P. Gebauer, and P. Boček, *J. Chromatogr.*, *470*:43 (1989).
4. F. Foret, S. Fanali, and P. Boček, *J. Chromatogr.*, *516*:219 (1990).
5. J. Sudor, J. Pospíchal, M. Deml, and P. Boček, *J. Chromatogr.*, *545*:331 (1991).
6. J. Sudor, Z. Stránský, J. Pospíchal, M. Deml, and P. Boček, *Electrophoresis*, *10*:802 (1989).
7. V. Šustáček, F. Foret, and P. Boček, *J. Chromatogr.*, *480*:271 (1989).
8. P. Gebauer, Deml, M., J. Pospíchal, and P. Boček, *Electrophoresis*, *11*:724 (1990).
9. P. Boček, M. Deml, and J. Pospíchal, *J. Chromatogr.*, *500*:673 (1990).

III
CAPILLARY COLUMN

8

Chemical Derivatization of Fused Silica Capillaries

Fred E. Regnier and Dan Wu

Purdue University
West Lafayette, Indiana

INTRODUCTION

Fused silica has the desirable features of being (a) easy to fabricate into capillaries, (b) optically transparent across the UV and visible spectrum, (c) available in internal diameters (id) ranging from 10 to 300 μm, (d) mechanically strong, (e) flexible when coated with polyimide, and (f) inexpensive. Unfortunately, fused silica also has undesirable features. The major limitation of fused silica is that the surface has approximately 8 $\mu m/m^2$ of silanol groups that behave as a weak acid, ionizing in water, with a broad titration curve from pH 3.5 to 9 [1]. The presence of these anionic groups at the surface of capillaries influence electrophoretic separations in several ways. One is to trigger liquid flow (i.e., electroendosmosis or electroosmotic flow; EOF) in capillaries during electrophoresis [2]. The EOF in fused silica capillaries as a function of pH is seen in Fig. 1 [3]. Electroosmosis results from migration of the ionic double layer adjacent to the capillary wall toward the negative electrode when a potential is applied across the electrophoretic system. According to theory [4], electroosmotic velocity is described by the equation

$$\mu_{eo} = \frac{\varepsilon \zeta E}{4\pi\eta}$$

where μ_{eo} is the electroosmotic mobility of a neutral species, ε is the dielectric constant, ζ is the zeta potential at the wall, and η is viscosity. Transport velocity in fused silica capillaries by electroosmosis is almost always greater than by elec-

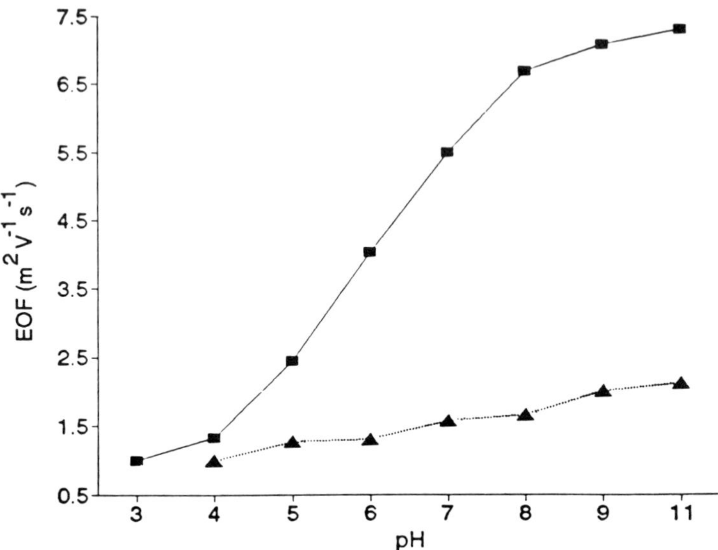

Figure 1 Dependence of electroosmotic flow (m^2 V^{-1} s^{-1}) $\times$ 10^8 on pH for uncoated (■) and Brij 35/alkylsilane-coated (▲) capillary (75-μm id). (From Ref. 3)

trophoresis at pH values greater than 6. This means that all substances will be swept to the negative electrode, regardless of their charge. The positive feature of this phenomenon is that a single detector placed near the negative electrode may be used to detect both anionic and cationic species. At pH values less than 3, silanol ionization is repressed to the extent that EOF approaches zero. This has been used effectively in the separation of peptides (Fig. 2) [5–7].

Electroosmosis also has several disadvantages. The first is that electroosmotic transport essentially shortens capillaries by eluting analytes from the column before electrophoretic resolution is complete [8]. Fractionation of a protein mixture on an uncoated capillary with high electroosmotic flow (Fig. 3) and a coated capillary with low flow (see Fig. 3b) shows that resolution is increased substantially by decreasing the EOF. This problem can be reduced by the use of longer capillaries [9].

Another disadvantage of electroosmosis is that, when coupled with the adsorption of cationic proteins, it can cause band-spreading and poor reproducibility of elution time [10]. Adsorption of cationic proteins at the capillary inlet produces axial heterogeneity of the zeta potential along the length of the capillary. In some situations, adsorbed protein can even reverse the surface charge at the head of the column. The net effect is that the rate, and even the direction, of electroosmotic pumping at the capillary wall can vary along the length of the column. To compensate for these axial differences in electroosmotic pumping, complex flow profiles develop

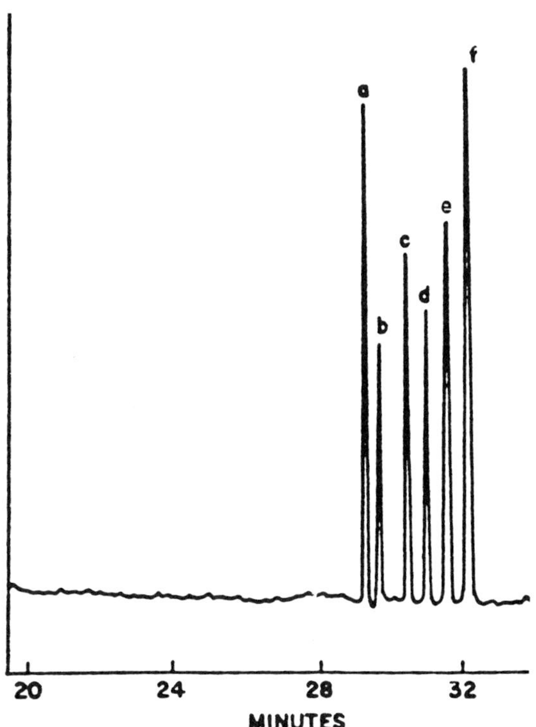

Figure 2 Separation of dipeptides in low-pH phosphate buffer. Conditions: 110 cm of 50-μm–id fused silica column, 75-cm separation length; buffer, 150 mM H_3PO_4, pH 1.5. Sample: tyrosyl-x dipeptides, for which x is (a) glycine, (b) alanine, (c) valine, (d) leucine, (e) glutamic acid, (f) tyrosine. (From Ref. 5)

within the column that compromise efficiency [11,12]. Adsorption of 10 ng or less of polyethylene imine caused a 50% reduction in theoretical plates [11]. The effect of adsorbed proteins on the transport velocity of analytes along the length of column is equally dramatic (Fig. 4). This figure illustrates that 10 ng of a positively charged protein, such as lysozyme, can cause the transport velocity of the neutral mesityl oxide marker to vary along the length of the column. These data establish that adsorption of matrix proteins has an instantaneous effect on both efficiency and transport velocity of other components in the mixture. In addition, it may be concluded that axial differences in either ionic strength or pH, which alter zeta potential, can diminish efficiency and influence analyte transport velocity.

Electrostatic adsorption of proteins to fused silica capillaries is a problem that is most serious at low ionic strength and intermediate pH (Fig. 5). Basic proteins

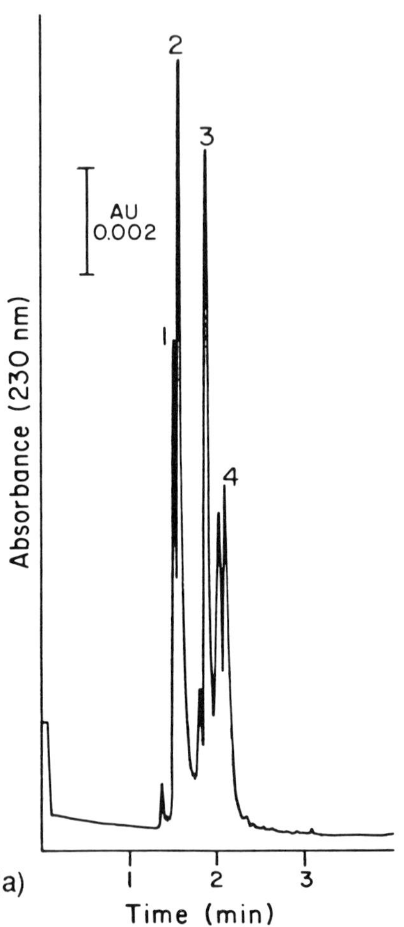

Figure 3 (a) Separation of proteins on uncoated capillaries. Conditions: capillary length 45 cm; separation length 25 cm; 75-μm id; separation potential, 15 kV, 333 V/cm, Tris-glycine buffer, pH 8.30. (1) Carbonic anhydrase (2) α-amylase, (3) ovalbumin, (4) trypsin inhibitor. (b) Separation of proteins on C18/PF108-coated capillary. (1) Trypsin inhibitor, (2) ovalbumin, (3) α-amylase, (4) carbonic anhydrase. Separation conditions are the same as (a). (From Ref. 9)

are the most likely to adsorb, because their charge is opposite that of fused silica. In contrast, acidic proteins are electrostatically repelled from the surface and show minimal adsorption. Although silanol ionization is greatest above pH 8–9, ionization of amino groups in polypeptides is repressed at basic pH. At extremes in pH, many proteins are denatured, deamidation of asparagine residues is greatest,

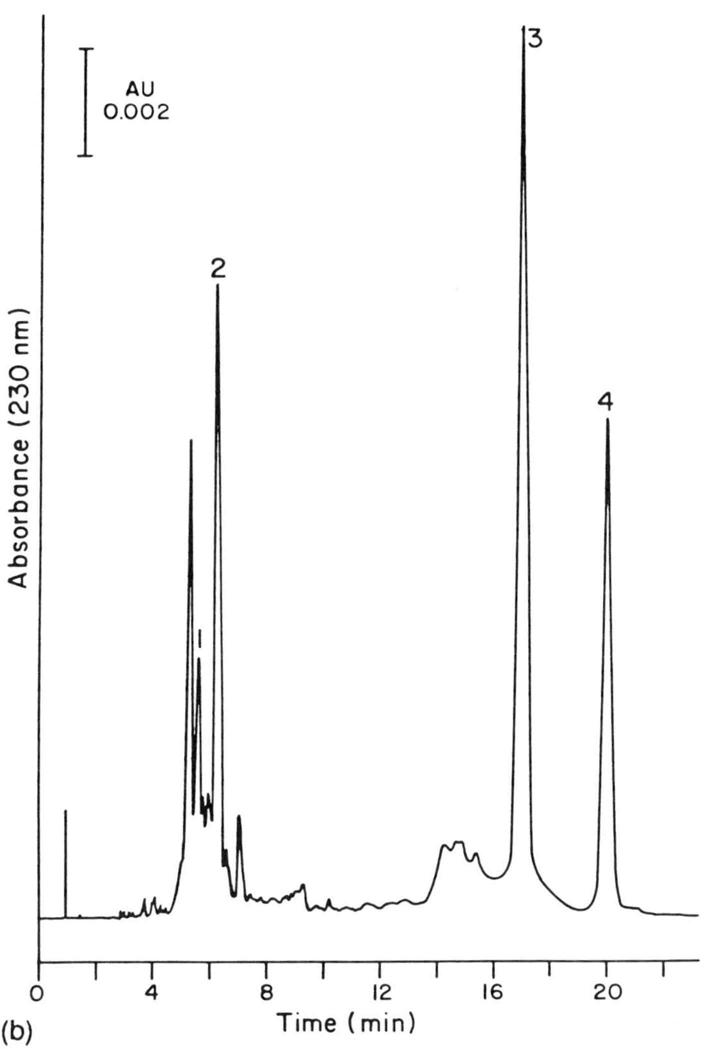

and charge differences between polypeptide species are often not as large as at intermediate pH. Minimum electrostatic interactions occur at acidic pH. Under these conditions silanol and carboxyl ionization is suppressed and amines are fully ionized.

It has been shown in the foregoing that silanol effects dominate the contribution of fused silica capillaries to electrophoretic separations. The purpose of this chapter is to examine the various methods that have been used to diminish or control these

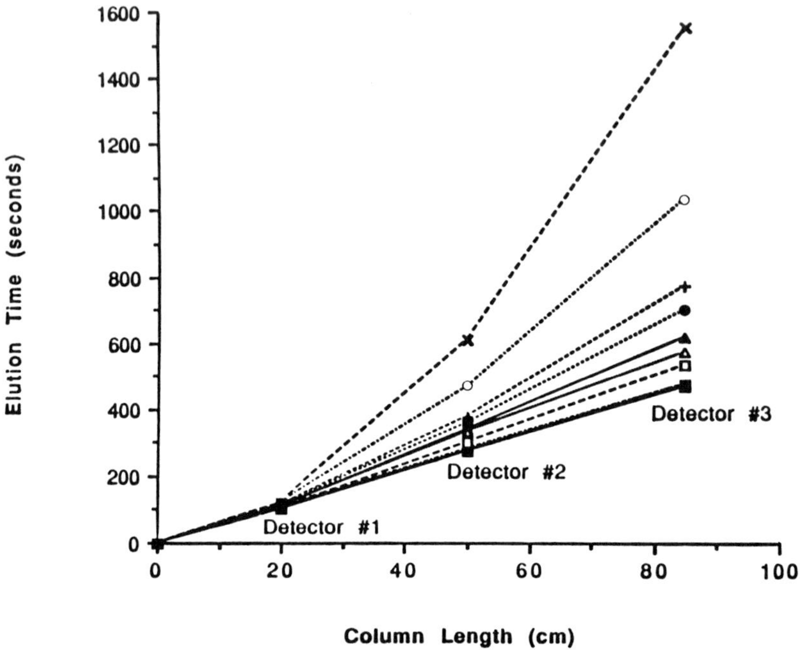

Figure 4 Effect of individual protein adsorption on neutral marker (mesityl oxide) elution time for proteins ranging in pI value from 3.3 to 11 on an uncoated capillary. Pepsin, β-lactoglobulin A, β-lactoglobulin B, and ovalbumin (those with pI < 5.2) all fall on the same line as that of neutral marker injected alone. Conditions: 0.01 M phosphate buffer, pH 7.0; 280 V/cm, 75-μm × 100-cm capillary; detectors (254 nm) at 20 cm, 50 cm, and 85 cm from injection. (×) Lysozyme, pI 11.0; (○) Cytochrome C, pI 10.2; (+) Chymotrysinogen, pI 9.2; (●) Ribonuclease A, pI 9.3; (▲) Myoglobin, pI 7.3; (△) Conalbumin, pI 6.3; (□) Carbonic anhydrase, pI 6.2; (■) Ovalbumin, pI 4.7; Lactoglobulin B, pI 5.2; Lactoglobulin A, pI 5.1; Pepain, pI 3.2; and Natural Marker. (From Ref. 11)

effects. The discussion will begin with the use of mobile-phase additives that suppress the charge properties of silanols and, then, progress on to coating techniques that either sequester or sterically block the interaction of analytes with surface silanols. The focus of the discussion will be on surface modification as it relates to proteins, because they present one of the most difficult classes of compounds to separate by capillary electrophoresis.

ADDITIVES

Dynamic modification of surface silanols through the use of mobile-phase additives has been used in chromatography for decades [13]. The value of this technique in capillary electrophoresis will be examined in the following sections.

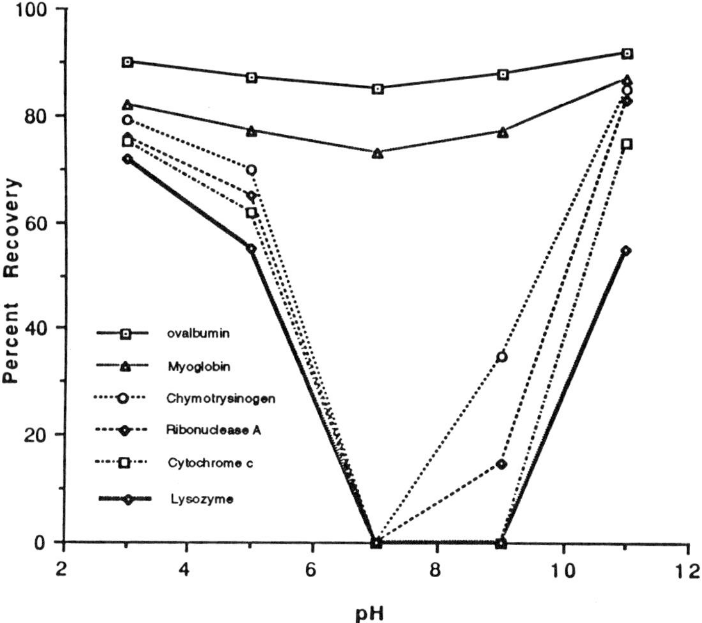

Figure 5 Protein recovery vs pH for six proteins ranging in pI value from 3.1 to 11.1 on an uncoated 75-μm × 100-cm capillary, using two detectors 50 cm apart. Conditions: pH 3 and 5, 0.01 M acetate; pH 7, 0.01 M phosphate buffer; pH 9 and 11, 0.01 M diazomethane; detection at 214 nm; 300 V/cm. (From Ref. 11)

Acidic pH

Low-pH buffers and acidic additives provide one of the simplest methods to suppress silanol ionization [14–18]. When the pH is less than 3, electroosmotic flow drops essentially to zero, and separations are based totally on electrophoretic mobility. Very high resolution of peptides in the tryptic digest of a protein can be achieved under acidic conditions (Fig. 6). Because only amino groups are ionized at acidic pH, it would seem counterintuitive that peptides would be well resolved by capillary zone electrophoresis (CZE). Other variable, such as solute size, ion-pair formation, and solvation, play an important role in CZE of peptides.

Suppression of silanol ionization (i.e., electrophoresis at low pH) is equally effective in protein separations [19]. There are, however, other variables to be considered with proteins. One is that charge differences among proteins are generally not maximum at acidic pH. Titration curves and electrophoretic titration maps show that the optimum pH for resolving proteins varies between pH 3 and pH 10 [20]. Selectivity is compromised by carrying out electrophoretic separations exclusively at acidic pH. Another consideration is protein structure. At low pH the

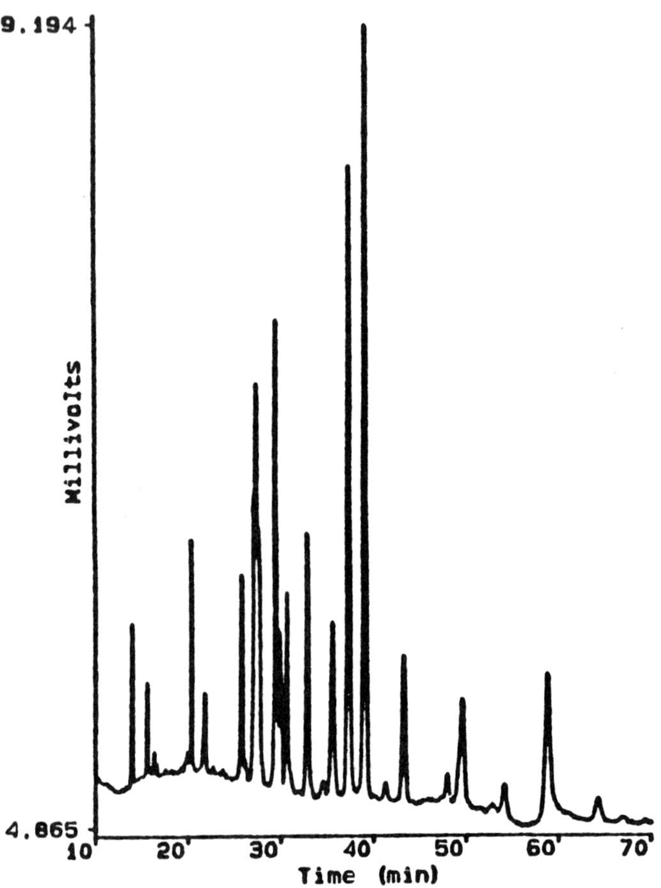

Figure 6 Capillary electrophoresis of the tryptic digest of human growth hormone (hGH) in a 50-cm × 50-μm–id capillary; pH 2.5 phosphate buffer; detected at 200 nm. (From Ref. 18)

structure of most proteins is substantially different than under physiological conditions (i.e., conformational changes have generally occurred). Denaturation can promote one or more of the following phenomena: dissociation of multimeric proteins into subunits, the transition of different conformational states to the same structure, loss of biological activity, and precipitation.

Basic pH

High pH has also been used in protein and peptide separations [21] as noted earlier (Fig. 7). At pH 10 or higher, surface silanols and solute carboxyl groups are

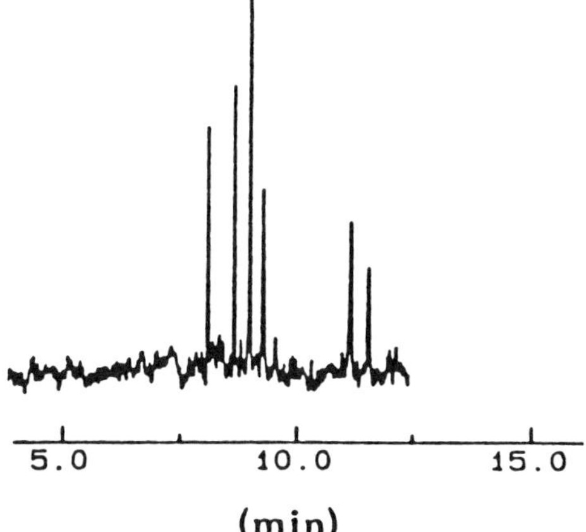

Figure 7 Separation of model proteins at pH 8.22 (tricine/KCl 10:20 mM). Conditions: capillary total length 101 cm, separation length 55 cm, 20 kV. Elution order: myoglobin (whale skeletal muscle), myoglobin (horse heart), conalbumin, carbonic anhydrase, β-lactoglobulin B, and β-lactoglobulin A. (From Ref. 21)

completely ionized. In contrast, amino groups in polypeptides either are nonionized or partially ionized. This means that most proteins will have a net negative charge at high pH and be electrostatically repelled from the negatively charged capillary wall. The problems associated with separation at basic pH are the same as those experienced at acidic pH.

Competing Ions

Still another approach is to add salts [22], amines [23], or zwitterions [24] that compete with cationic solutes for surface silanols. A salt concentration of 100 mM or greater also sharpens peaks and increases solute recovery by blocking electrostatic interactions with the surface of the capillary. The negative feature of the high-salt approach is that joule heating increases because heat production is proportional to ionic strength at constant potential. Operating potential must often be limited at high-ionic strength. Multiply charged amines such as putrescine, cadaverine, and morpholine, work in much the same way as salts, but at lower ionic strength [21,23,25]. Polyamines are more tenaciously bound to silanols than sodium ions. These additives, too, increase recovery and sharpen peaks.

COVALENT MODIFICATION WITH SIMPLE SILANES

The term *simple silanes*, as used here, refers to organosilanes in which an organic moiety is bounded to the surface of silica through a single silicon atom. Polymeric coatings that are attached to surfaces through multiple silicon atoms will be treated separately.

Organosilanes have been used for more than two decades to modify the surface of silica-based separation media [26]. Hydrophilic silylating agents that control protein adsorption in size-exclusion chromatography materials [27] have been found to be equally efficacious in capillary electrophoresis [28]. Protein recovery from surface-modified, silica-based size-exclusion columns is generally greater than 90%, even with cationic proteins, such as lysozyme. Although protein recovery from simple silane-derivatized capillaries has not been quantitated in capillary electrophoresis, it is apparently assumed to be similar to that seen in size-exclusion chromatography. The efficacy of surface modification procedures in capillary electrophoresis has been determined more on the basis of peak asymmetry and whether proteins elute in descending order of their isoelectric point than on actual recovery data. Symmetrical peaks and elution according to net charge have both been used as indicators that there is no interaction with the capillary wall.

From literature dealing with the preparation of chromatographic media, we know that both halide and methoxy silanes may be coupled to surface silanols through one or more siloxane (Si—O—Si) bonds [29] as seen in the reactions

$$\begin{array}{ccccc} | & & | & & |\quad | \\ (Y)\text{-}Si\text{-}(CH_2)_3\text{-}X & + & \text{-}Si\text{-}OH & \text{----->} & \text{-}Si\text{-}O\text{-}Si\text{-}(CH_2)_3\text{-}X \\ | & & | & & |\quad | \\ (CH_3)_2 & & & & (CH_3)_2 \end{array}$$

and

$$\begin{array}{ccccc} & & OH & & OH \\ & & | & & | \\ (Y)_3\text{-}Si\text{-}(CH_2)_3\text{-}X & + & \text{-}Si\text{-}OH & \text{----->} & \text{-}Si\text{-}O\text{-}Si\text{-}(CH_2)_3\text{-}X \\ & & | & & | \\ & & \text{-}Si\text{-}O & & \text{-}Si\text{-}O \end{array}$$

Y in these reactions is a halide or alkoxy group and X, for fused silica capillaries for electrophoresis, has generally been either a glycidoxy [30] or amino group [31]. Steric limitations generally allow only half of the silanols to be sequestered, leaving approximately 4 $\mu m/m^2$ of surface silanols [32]. These residual silanols account for the substantial electroosmotic flow seen in simple silane-derivatized capillaries [28].

Neither glycidoxy- nor amine-derivatized capillaries are used directly for electrophoresis. The amine group ion-pairs with both surface silanols and analytes and has been found to be of little use in CE. Glycidoxy groups, in contrast, are subject to nucleophilic addition and will either react with proteins or be hydrolyzed to a diol during operation of the electrophoresis system. Consequently, glycidoxy-derivatized capillaries have been either converted to the glyceryl derivative [33] or substituted with another hydrophilic substance [34] according to the reaction

$$\text{-O-CH}_2\overset{\text{O}}{\widehat{\text{CH-CH}_2}} + \text{HO-R} \xrightarrow{BF_3 \cdot Et_2O} \text{-O-CH}_2\text{CH-CHOH-CH}_2\text{-O-R}$$

R may be —H, —$CH_2CHOHCH_2OH$ (glycerol), or —$(OCH_2CH_2)_nOH$ (polyoxyethylene).

Although not discussed in the electrophoresis literature, there are some important aspects of these coatings that are worth mentioning. First, the yield of product in nucleophilic substitutions to oxiranes decreases with increasing molecular size of the R group [35]. This means that the polyoxyethylene (POE) coating, in particular, will be a mixed phase of glycerol and POE. Reaction conditions will be critical in determining the ratio of these groups. Second, studies with chromatography systems suggest that hydrogen-bonding between oxygen residues in the glyceryl and POE phases and residual surface silanol groups is probable [36]. This allows coatings with sufficient steric freedom, such as POE, to actively mask the surface. It would be interesting to know if there is enough POE at the surface to mask the silanols.

Several other simple silanes have also been examined. Maltose and pentafluorobenzoic acid were coupled to γ-aminopropylsilane derivatization capillaries [14,37]. Reducing sugars react readily with primary amines to form a Schiff base that can be reduced to a secondary amine by sodium cyanoborohydride. This synthetic route was used to prepare capillaries with the maltose-bonded phases [14]. Because efficiency is much higher on the maltose-derivatized column than with the propylamine derivative alone, it was concluded that the disaccharide moiety shields the surface (Fig. 8). From this conclusion, it is surprising that the smaller glycerylpropyl derivative provides higher separation efficiency than the maltose-bonded phases. It should also be noted that Schiff base coupling produces an amphoteric-bonded phase. Since the column has both secondary amines in the bonded phase and residual silanols, it will have a net positive charge, and electroosmotic flow will be toward the anode under acidic conditions. The opposite will be true at basic pH.

Pentafluorobenzoic acid was bonded to the surface of capillaries by treating an aminopropylsilane-derivatized column with a 0.2-M solution of pentafluorobenzoyl chloride in toluene [37]. When operated at neutral pH and moderate ionic strength, a 20-μm–id column so coated gave greater than 300,000 theoretical plates with a series of proteins. More than 600,000 theoretical plates were obtained with several proteins (Fig. 9). Few laboratories can exceed this efficiency.

Our experience from both chromatography and electrophoresis indicates that the principal limitation of the simple silane coatings is their limited hydrolytic stability, particularly when the organosilane has extensive water solubility. Above pH 8, the siloxane bonds of water-soluble organosilanes are hydrolyzed, and the

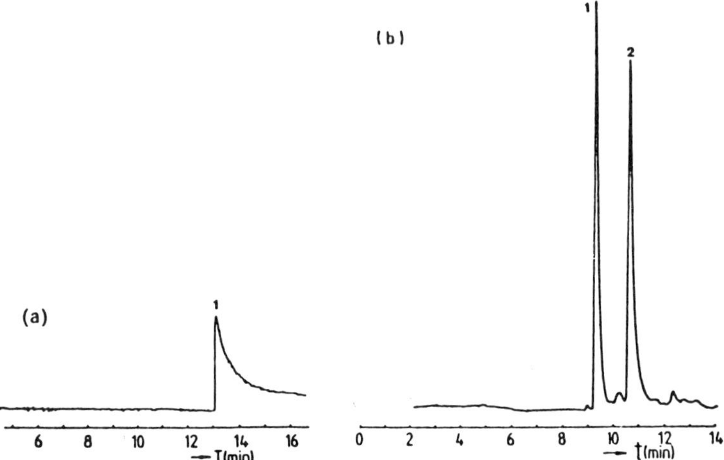

Figure 8 Electrophoregrams of (a) lysozyme on an aminopropyl-modified capillary and (b) of lysome and cytochrome *c* on maltose-modified capillary at pH 6.0. (a) 1, Lysozyme; (b) 1, lysozyme, 2, cytochrome *c*. (From Ref. 14)

bonded phase rapidly erodes from the surface [32]. When the organic portion of the coating is either of limited solubility of interacts with the surface, erosion of the organosilane is diminished and coating life is extended.

ADSORBED COATINGS

Adsorbed coatings provide another avenue for surface deactivation. It is well-known that a variety of substances adsorb to silica, either through electrostatic interactions or hydrogen bonding. When the surface of silica is derivatized with an alkyl silane, it is also possible to adsorb substances through hydrophobic interactions. Adsorbed coating can both sequester surface silanols and shield proteins from contact with the capillary wall. Questions that must be considered with adsorbed coatings are the strength with which the coating is held to the surface and whether the chemical and physical properties of the coating are compatible with the separation process being examined.

Adsorption to Surfaces

Adsorption of organic compounds or polymers to the surface of fused silica capillaries can play a major role in the separation process, as noted earlier; therefore, it is critical that the coating be stable. Either chemical degradation or erosion of the coating during operation will alter electroosmosis and perhaps analyte adsorption.

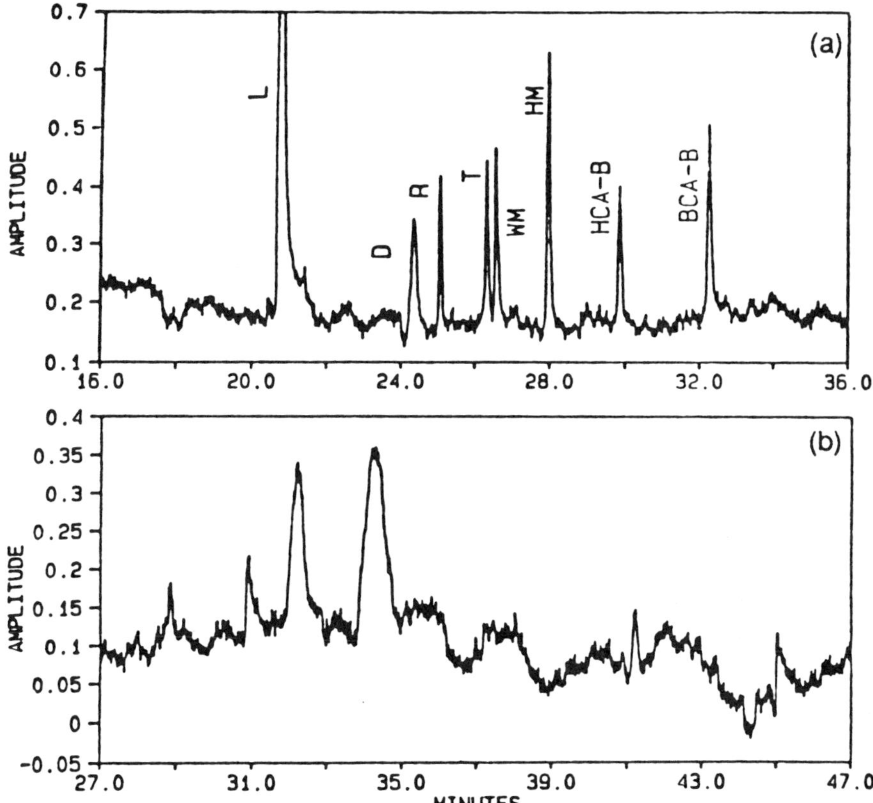

Figure 9 Elution profile of proteins and dimethyl sulfoxide (DMSO) on an aryl pentafluoro-treated capillary (a) and an untreated capillary (b). Conditions: 200:100 mM phosphate/KCl buffer, pH 7; capillary, 20-μm id; 250 V/cm; detected at 219 nm; separation length 100 cm. Elution order: L, hen egg white lysozyme; D, DMSO; R, bovine ribonuclease A; T, bovine pancreatic trypsinogen; WM, whale myoglobin; HM, horse myoglobin; HCA-B, human carbonic anhydrase B; BCA-B, bovine carbonic anhydrase B. (From Ref. 37)

The most critical question here is the strength with which the coating is adsorbed to the capillary. The coating process may be represented as

$$S + C \rightarrow CS$$

$$K = \frac{[CS]}{[S][C]}$$

where S is the silica surface, C is the substance that is coating the surface, CS is the coated surface, and K is the binding constant. When the binding constant is large, the coating is adsorbed with sufficient tenacity that there is little erosion of the coating during operation. Because the lifetime of the coating will be short when the binding constant is small, coating components must either be added to the separation buffers to displace the equilibrium toward adsorption, or techniques must be found to further stabilize the coating. Both of these approaches will be discussed in the following.

Physical and Chemical Properties of Adsorbed Coatings

Hjerten has noted that the general function of coatings in capillary electrophoresis is (a) to create a layer at the capillary surface that is of sufficiently high viscosity that electroosmotic flow is precluded and (b) to shield analytes from contact with the silica surface [10]. Although the low-molecular-weight additives discussed earlier might be thought of as an adsorbed coating, adsorbed coatings are generally oligomers or polymers. Random coil polymers typically adsorb at surfaces in a "loop-and-train" configuration. The thickness of the adsorbed layer is both a function of polymer molecular weight and the concentration of polymer in the coating solution. High concentration and high molecular weight produce thicker layers. Thicker coatings probably reduce both electroosmotic flow and shield the surface more effectively than thin coatings.

Neutral Polymers

Methylcellulose (MC) was one of the first adsorbed polymers to be used in capillary electrophoresis. The advantage of polymers such as methylcellulose is that they are neutral and hydrophilic. Methylation of cellulose disrupts intermolecular hydrogen bonding sufficiently that the polymer is soluble in water. Because it is not possible for MC to interact with a silica surface electrostatically, the bonding mechanism is apparently hydrogen bounding. Adsorbed MC coatings have been used effectively in both capillary zone electrophoresis and isoelectric focusing [10]. Further stabilization of MC coatings has been achieved by cross-linking with formaldehyde [10].

Adsorption of polymers at hydrophobic surfaces has also been used to create hydrophilic surface coatings. Derivatization of silica with octylsilane produces a surface that will adsorb polymers by hydrophobic interaction. Because hydrophobic surfaces do not adsorb hydrophilic polymers, it is necessary to use an amphipathic copolymer to be effective in this coating process. Hydrophobic portions of the polymer adsorb to the sorbent surface, whereas hydrophilic residues orient away from the surface toward the polar solution. Examples of species that have been used to create hydrophilic surfaces are surfactants from the Tween and Brij series (Fig. 10) and the BASF surfactant Pluronic F-108 [3,38]. In addition, various molecular weights of polyvinyl alcohol, polyvinyl pyrrolidone, and methylcellulose have been used to create adsorbed, hydrophilic coatings. A small amount of the surfactant or

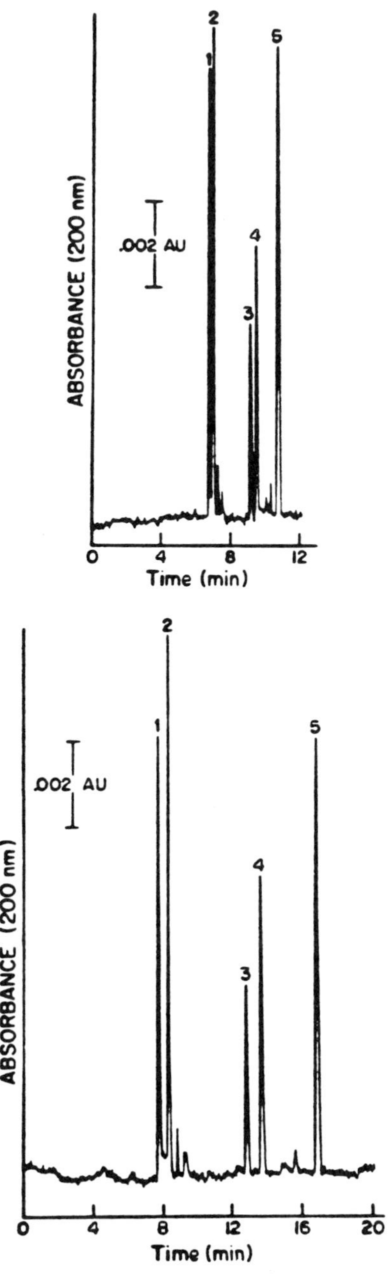

Figure 10 Electrophoregrams of five basic proteins: (1) lysozyme, (2) cytochrome *c*, (3) ribonuclease A, (4) α-chymotrypsinogen, and (5) myoglobin. Experimental conditions: 0.01 M phosphate buffer (pH 7.0) on a 75-μm × 50-cm (top) Tween 20/alkylsilane capillary and (bottom) Brij 35/alkylsilane capillary at 300 V/cm. (From Ref. 3)

polymer, ranging in concentration from 0.01 to 0.1% (w/w), is generally included in the buffer to replace polymer that might erode during column operation.

Studies with the hydrophobically adsorbed polymers show that the coating both shields proteins from contact with surface of the column and reduces EOF (Table 1). Significant electrostatic interaction of proteins with surface silanols is apparently precluded by the polymer layer. Recovery of proteins from Brij 35-coated capillaries was always greater than 95%. It remains to be seen whether this type of coating will be equally efficacious with more hydrophobic proteins, such as those found in membranes. In this example, very hydrophobic proteins might displace surfactants from the alkylsilane layer and be adsorbed. Other attractive features of these adsorbed coatings are that the EOF is relatively independent of pH, and it may be manipulated by using polymers of different size. Coatings that allow high EOF provide rapid, low-resolution analyses. In contrast, those coatings that give intermediate EOF provide higher resolution of complex mixtures. It would appear that coatings that reduce EOF to near zero would be the most desirable. Although it is true that zero EOF systems provide highest resolution, it should be recalled that anionic and cationic species move in opposite directions when there is no EOF. It is no longer possible to analyze all species in a single capillary when EOF is zero.

Polyamines

It is well known from chromatographic systems that polyamines readily adsorb to silica and may be used to modify the surface properties of the sorbent [39]. When the polyamine is applied at high concentration, the surface properties of silica are completely masked. Similar phenomena have been seen with fused silica capillaries. When polyamine is added to the running buffers and the solution is forced through

Table 1 Electroosmotic flow (EOF) of the C_{18} Coated Capillary (50-μm id) with Adsorption of Different Polymers

Adsorbed polymer	EOF ($m^2\ V^{-1}\ s^{-1}$) × 10^{-8}
TWEEN-20	2.030
BRIJ-35	1.650
MC-15	0.722
PVP	0.672
PVA-2000	0.645
MC-25	0.586
MC-400	0.574
MC-1500	0.428
PVA-124000	0.363
MC-4000	0.320
PF-108	0.200

a capillary, the adsorbed polyamine causes the surface of the column to acquire a positive charge, and the EOF is reversed [40]. This indicates that the surface of the capillary is coated with a sufficiently thick, or concentrated, layer of polyamine that the surface silanols are completely masked. Under these circumstances, negatively charged proteins move in the direction of the EOF toward the anode, and positively charged species migrate against the flow. Ion-pairing phenomena also play an important role in this system. At ion strengths less than 100 mM, negatively charged proteins would ion-pair with polyamine. Ion-pair formation with polyamine dissolved in the running buffer will have a major influence on the electrophoretic mobility of anionic proteins. The polyamine–protein complex will have a more positive charge than the protein itself. In fact, the electrophoretic mobility of the complex will be much more like that of a cationic rather than an anionic protein.

Polyamines adsorbed to a silica surface may also be cross-linked into a permanent coating [12]. This coating behaves as an anion exchanger, adsorbing negatively charged proteins. As a consequence, these columns may only be used for the separation of cationic proteins. Because polyamine-bonded–phase columns adsorb anions and the EOF is toward the anode, these columns are best for the separation of cationic proteins.

We have found that fused silica capillaries may also be coated with proteins, either by adsorption onto an octylsilane-derivatized surface, or by covalent attachment. Procedures described for immobilization of proteins on glass seem to be effective. Capillaries treated in this manner take on characteristics of the protein. Electroosmotic flow will be toward the cathode at high pH and the anode at low pH. The pH at which the EOF reverses will be near that of the isoelectric point of the protein. Immobilized enzyme and antibody columns will be seen more frequently in the future.

POLYMERIC COATINGS

The term *polymeric coatings*, as used here, refers to coatings that are composed of polymers covalently bonded at multiple sites to the fused silica surface of capillaries. The differentiation is made between single- and multiple-site binding of polymers, because coatings that are bound to surfaces at multiple sites will be difficult to erode and should have a longer life. There are theoretically two ways in which these coatings can be prepared. One would be to synthesize a linear copolymer from an organosilane monomer and some other hydrophilic monomer and then bond the preformed polymer to the surface. Because the copolymer has multiple methoxy- or chlorosilane residues, it could both bond to the surface at multiple sites and cross-link with adjacent polymer chains through siloxane bond formation. Although this approach has not been currently reported, it should provide a simple synthetic route to high-stability coatings.

A second approach is to couple the silanes to the surface first and then incorporate them into a polymer layer at the surface. Polyacrylamide, polyvinylpyrrolidone, and epoxy polymer coatings have been successfully prepared in this way. The organosilanes used in this process must have two different types of reactive groups; one that bonds the organosilane to the surface and a second that can be used to incorporate the organosilane into the polymer. Si—OCH_3 or Si—Cl groups are generally used to bond the organosilane to the silica surface. Three organosilanes have been most widely used to bond polymers to surfaces; γ-methacryloylpropyl, γ-glycidoxylpropyl, and vinyl silanes. The vinyl- and methacryloyl-containing silanes were used in free–radical-coupling reactions and the glycidoxy- or glycidyl-derivates were used to prepare epoxy coatings. Exact techniques by which the reactions were carried out will be described for each type of polymer.

Acrylates

From the combined literature on both chromatography [41] and electrophoresis [10], it is known that acrylate polymers may be grafted to the surface of silica in several ways. There are important differences in both the structure and stability of the polymeric coatings that will be derived from these various techniques. One of the first grafting approaches was to silylate the silica support with an organosilane, containing either a methacryl or vinyl group that was subsequently incorporated into an acrylate polymer. Incorporating monomers that are covalently bonded to a surface into a polymer is a widely used technique for coupling polymers to surfaces (Schemes 1 and 2). Free-radical catalysis is commonly used in this type of poly-

$$\equiv Si(CH_2)_3O\text{-}CO\overset{\overset{\displaystyle CH_3}{|}}{C}{=}CH_2 + \underset{\underset{\displaystyle CONH_2}{|}}{CH_2{=}CH} \text{ -----> } \equiv Si(CH_2)_3O\text{-}CO\overset{\overset{\displaystyle CH_3}{|}}{C}H\text{-}CH_2\ \underset{\underset{\displaystyle CONH_2}{|}}{(CH_2CH)_n}\text{-}$$

Scheme 1.

$$\underset{\underset{\displaystyle CONH_2}{|}}{CH_2{=}CH} \text{ ------> } \underset{\underset{\displaystyle CONH_2}{|}}{\text{-}(CH_2\text{-}CH)_n\text{-}}$$

Scheme 2.

merization. When polymerization is carried out by adding the silane-derivatized support to a solution containing acrylamide monomer, two types of polymers are formed. One is linear polyacrylamide in the solution, and the other is linear acrylamide coupled to the surface through the organosilane. Washing subsequent to polymerization removed nongrafted polymer. Although it might be concluded from the figure that a single silane is involved in the grafting and that polymerization is initiated at the surface, this probably is not true. It is likely that multiple silanes may be incorporated into the polymer, and that the sites of incorporation are random. In those instances for which polymer is coupled to the surface through multiple silanes, the polymer will be more difficult to leach from the surface.

The use of cerium salts to catalyze polymerization produces a polymer of different structure [42]. This catalyst initiates polymerization at hydroxyl groups. Through the use of a glycerylpropylsilane-bonded phase it is possible to prepare linear acrylamide polymers that are grafted to the surface at their termini (Scheme 3). Although not documented, this mode of grafting probably produces a less stabile coating than those grafted at multiple sites.

A recent study has focused on Si—C instead of Si—O—Si bonding as a means of increasing the stability of organic grafts to silica surfaces (Scheme 4) [19]. The concept on which this process is based is that silicon atoms at the surface of a support are bonded into the silica matrix through three siloxane bonds, instead of one or two, as with organosilane derivatives. It is seen in Scheme 4 that the process for attaching polymers by this process is almost identical with that used in Scheme 1, with the exception of bonding to the silica matrix.

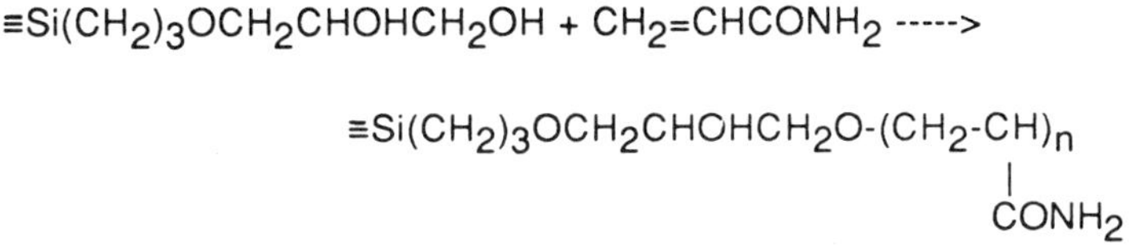

$$\equiv Si(CH_2)_3OCH_2CHOHCH_2OH + CH_2{=}CHCONH_2 \dashrightarrow$$

$$\equiv Si(CH_2)_3OCH_2CHOHCH_2O\text{-}(CH_2\text{-}\underset{\displaystyle CONH_2}{\underset{|}{CH}})_n$$

Scheme 3.

$$\equiv Si\text{-}CH{=}CH_2 \quad + \quad CH_2{=}CHCONH_2 \dashrightarrow \equiv Si\text{-}\underset{\displaystyle (CH_2\underset{\displaystyle CONH_2}{\underset{|}{C}}H)_n\text{-}}{\underset{|}{CH}}\text{-}CH_2\text{-}(CH_2\overset{\displaystyle CONH_2}{\overset{|}{C}}H)_n\text{-}$$

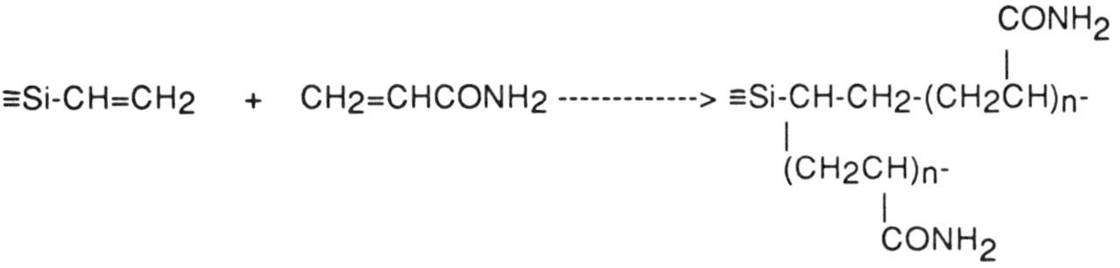

Scheme 4.

From our experience, the only limitation with the acrylate coatings is that the amide bond slowly hydrolyzes during extended use, with a concomitant increase in electroosmotic flow. This means that retention and resolution can vary over the course of weeks or months of use.

Polyvinylpyrrolidone

The process by which polyvinylpyrrolidone (PVP) phases are prepared and coupled to the surface of silica is essentially identical with those used in the preparation of acrylates. Tests with PVP coatings would indicate that they have no advantage over polyacrylamide coatings [5]. In fact, they appeared to be inferior.

Epoxy Coatings

Epoxy coatings have been known and used in liquid chromatography systems for almost two decades [37]. With slight modification, the same coatings have been used in electrophoresis [33,34]. Following physical deposition of a multifunctional oxirane onto the surface of a glycidoxypropyl- or glycerolpropylsilane-derivatized silica, polymerization is initiated with either a Lewis acid or tertiary amine catalyst, with subsequent formation of a highly cross-linked polymer (Scheme 5).

Hydroxyl groups of the organosilane are incorporated into this polymer as it is formed, anchoring it to the surface at many sites. Owing to the high degree of cross-linking and the many sites of attachment to the silica surface, this coating is potentially the most stable of any discussed in this review. The limitation is that small amounts of aldehyde can be formed in highly cross-linked epoxy polymers. These will oxidize during use and increase electroosmotic flow. Amine catalysis diminishes this problem [34].

SUMMARY

The need to control analyte adsorption in capillary electrophoresis (CE) has been recognized for more than a decade. Given the data presented in this chapter, it may be concluded that the issue of controlling adsorption in fused silica capillaries is essentially solved. The possible exception might be cationic or hydrophobic proteins that are present at less than 10^{-8} M. It has been shown that a wide variety of approaches ranging from adsorbed to covalently bonded coatings successfully control adsorption.

The issue in the future will be whether one of these coating strategies (i.e., adsorption or covalent bonding) will come to dominate capillary zone electrophoresis (CZE) and capillary isoelectric focusing (CIEF). It is interesting that chromatographers were confronted with the same dilemma. Here, the two techniques have coexisted for more than two decades. The advantage of dynamic coatings is that they present the user with an inexpensive method to fine-tune columns to one's

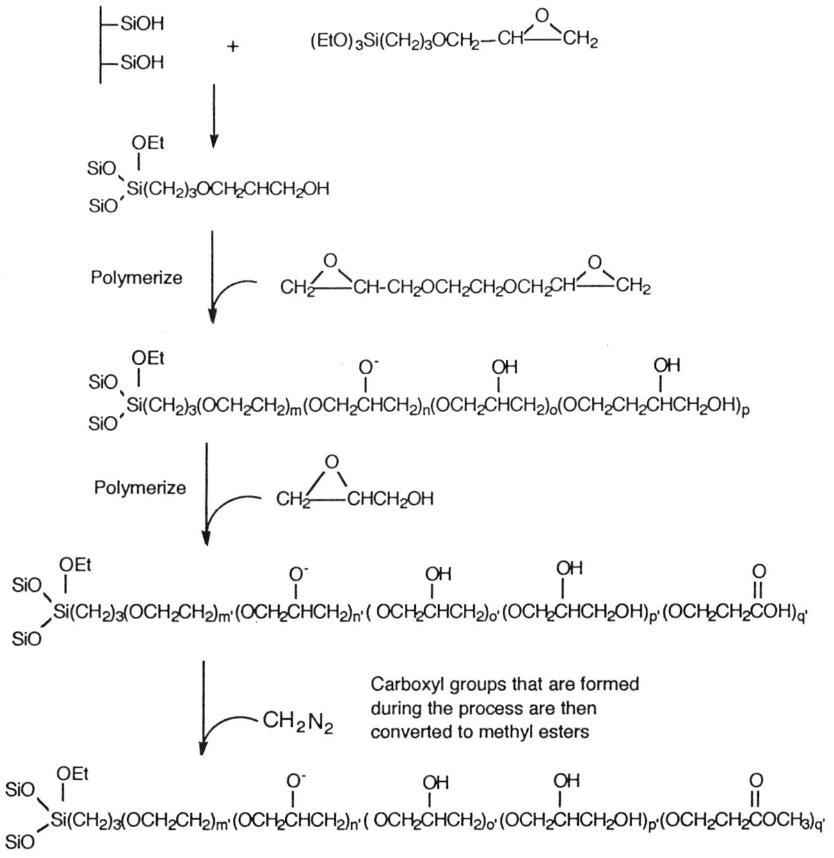

SiOH
SiOH
+
$(EtO)_3Si(CH_2)_3OCH_2-CH(O)CH_2$

SiO, SiO, OEt
$Si(CH_2)_3OCH_2CHCH_2OH$

Polymerize
$CH_2(O)CH\text{-}CH_2OCH_2CH_2OCH_2CH(O)CH_2$

$Si(CH_2)_3(OCH_2CH_2)_m(OCH_2CH(O^-)CH_2)_n(OCH_2CH(OH)CH_2)_o(OCH_2CH_2CH(OH)CH_2OH)_p$

Polymerize
$CH_2(O)CHCH_2OH$

$Si(CH_2)_3(OCH_2CH_2)_{m'}(OCH_2CH(O^-)CH_2)_{n'}(OCH_2CH(OH)CH_2)_{o'}(OCH_2CH(OH)CH_2OH)_{p'}(OCH_2CH_2C(=O)OH)_{q'}$

CH_2N_2
Carboxyl groups that are formed during the process are then converted to methyl esters

$Si(CH_2)_3(OCH_2CH_2)_{m'}(OCH_2CH(O^-)CH_2)_{n'}(OCH_2CH(OH)CH_2)_{o'}(OCH_2CH(OH)CH_2OH)_{p'}(OCH_2CH_2C(=O)OCH_3)_{q'}$

Scheme 5

specific needs through the addition of mobile-phase additives. Selection and control of EOF in CE is such an example. In contrast, covalently bonded phases have the advantage that additives are not required in those situations for which fine-tuning is unnecessary.

Given the fact that manipulation of EOF is an important variable in both CZE and CIEF, we conclude that multiple capillaries will be used in the future to select EOF. Perhaps dynamic coatings will be used more frequently in research, in which it is more important to alter EOF during optimization studies. In contrast, covalently bonded phases may be more useful in situations in which an instrument is dedicated to routine analysis.

REFERENCES

1. K. K. Unger, *Packings and Stationary Phases in Chromatographic Techniques*, Marcel Dekker, New York (1979).
2. J. W. Jorgenson, and K. D. Lukacs, *Science*, *222*:266 (1983).
3. J. K. Town, and F. E Regnier, *Anal. Chem.*, *63*:1126 (1991).
4. V. Pretorius, B. J. Hopkins, and J. K. Schieke, *J. Chromatogr.*, *99*:23 (1974).
5. R. M. McCormick, *Anal. Chem.*, *60*:2322 (1988).
6. M. A. Moseley, C. J. Deterding, K. B. Tomer, and J. W. Jorgenson, *Anal. Chem.*, *63*:109 (1991).
7. R. G. Nielsen, R. M. Riggin, and E. C. Rickard, *J. Chromatogr.*, *480*:393 (1989).
8. J. W. Jorgenson, and K. D. Lukacs, *Anal. Chem.*, *53*:1298 (1981).
9. X. W. Yao, D. Wu, and F. E. Regnier, *J. Chromatogr.* (submitted).
10. S. Hjerten, *J. Chromatogr.*, *347*:191 (1985).
11. J. K. Town, and F. E. Regnier, *Anal. Chem.*, (in press).
12. J. K. Town, and F. E. Regnier, *J. Chromatogr.*, *516*:69 (1990).
13. J. C. Giddings, *Dynamics of Chromatography; Principles and Theories*, Marcel Dekker, New York (1965).
14. G. J. M. Bruin, R. Huischen, J. C. Kraak, and H. Poppe, *J. Chromatogr.*, *480*:339 (1989).
15. P. D. Grossman, J. C. Colburn, J. J. Lauer, R. G. Nielsen, R. M. Riggin, B. S. Sittampalam, and E. C. Rickard, *Anal. Chem.*, *61*:1186 (1989).
16. J. S. Green, and J. W. Jorgenson, *J. Chromatogr.*, *478*:63 (1989).
17. A. Vinther, S. E. Bjorn, H. H. Sorensen, and H. Soeberg, *J. Chromatogr.*, *516*:175 (1990).
18. J. Frenz, S.-L. Wu, and W. S. Hancock, *J. Chromatogr.*, *480*:379 (1989).
19. K. A. Cobb. V. Dolnik, and M. Novotny, *Anal. Chem.*, *62*:2478 (1990).
20. W. Kopaciewicz, M. A. Rounds, J. Fausnaugh, and F. E. Regnier, *J. Chromatogr.*, *266*:3 (1983).
21. H. Lauer, and D. McManigill, *Anal. Chem.*, *58*:166 (1986).
22. Y. Walbroehl, and J. W. Jorgenson, *J. Microcolumn Separation*, *1*:41 (1989).
23. V. Rohicek, and Z. Deyl, *J. Chromatogr.*, *494*:87 (1989).
24. M. M. Bushey, and J. W. Jorgenson, *J. Chromatogr.*, *480*:301 (1989).
25. R. G. Nielsen, G. S. Sittampalam, and E. C. Rickard, *Anal. Biochem.*, *177*:20 (1989).
26. L. R. Snyder, and J. J. Kirkland, *Introduction to Modern Liquid Chromatography*, Wiley-Interscience, New York, 272 (1979).
27. W. W. Yao, J. J. Kirkland, and D. D. Bly, *Modern Size Exclusion Chromatography*, Wiley-Interscience, New York (1979).
28. T. Tsuda, K. Nomura, and G. Nakagawa, *J. Chromatogr.*, *248*:241 (1982).
29. J. J. Kirkland, J. C. Glajch, and R. D. Farlee, *Anal. Chem.*, *62*:2 (1989).
30. J. W. Jorgenson, and K. D. Lukacs, *Trends Anal. Chem.*, *3*:51 (1984).
31. B. J. Herrin, S. G. Shafer, S. V. Alstine, J. M. Harris, and R. S. Snyder, *J. Colloid Interfac. Sci.*, *115*:46 (1987).
32. K. K. Unger, N. Becker, and P. Roumeliotis, *J. Chromatogr.*, *125*:115 (1976).
33. J. K. Towns, J. M. Bao, and F. E. Regnier, *J. Chromatogr.*, *599*:227 (1992).

34. G. J. M. Bruin, J. P. Chang, R. H. Kuhlaman, K. Zegers, J. C. Kraak, and H. Poppe, *J. Chromatogr.*, *471*:429 (1989).
35. S. H. Chang, R. Noel, and F. E. Regnier, *Anal. Chem.*, *48*:1839 (1976).
36. S. H. Chang, K. M. Gooding, and F. E. Regnier, *J. Chromatogr.*, *120*:321 (1976).
37. S. A. Swedberg, *Anal. Biochem.*, *185*:51 (1990).
38. X. W. Yao, and F. E. Regnier, *J. Chromatogr.* (submitted; 1992).
39. T. G. Lawson, F. E. Regnier, and H. L. Weith, *Anal. Biochem.*, *133*:85 (1983).
40. J. E. Wiktorowicz, and J. C. Colburn, *Electrophoresis*, *11*:769 (1990).
41. T. Kremmer, and L. Boross, *Gel Chromatography*, John Wiley & Sons, New York (1979).
42. G. Mino, and S. Kaizerman, *J. Polym. Sci.*, *31*:242 (1958).

9

Technology of Separation Capillaries for Capillary Zone Electrophoresis and Capillary Gel Electrophoresis: *The Chemistry of Surface Modification and Formation of Gels*

Gerhard Schomburg

Max-Planck-Institut für Kohlenforschung
Mülheim an der Ruhr, Germany

Electrophoretic separations have been successfully performed and analytically applied since the very early work of Tiselius [1]. The versions that have been analytically most successfully applied were slab gel electrophoresis, isotachophoresis, and isoelectric focusing. Electrophoretic separations in capillaries were firstly performed by the groups of Mikkers et al. [2] and especially Jorgenson and Lukacs [3,4]. The latter introduced the mechanically flexible, polyimide-coated, thin-walled fused silica capillaries of narrow internal diameter (<100 μm) to the field of electrophoresis.

Systems of capillary electrophoresis as they have been realized for CZE (free capillary zone electrophoresis), CGE (capillary gel electrophoresis) [5], and MECC (micellar electrokinetic capillary chromatography) [6] must be considered as typical *miniaturized separation techniques* and are similar to the chromatographic methods of capillary gas or supercritical chromatography. Miniaturized separation techniques are performed in capillaries or columns of microdimensions. In high-performance liquid chromatography (HPLC), the aim of fast mass transfer between the mobile and stationary phase is also obtained either in narrow-bore capillaries or, at limited separation efficiency and without instrumental problems, in column packings of small nonporous or especially porous particles [7]. Miniaturization in chromatographic systems has been adopted mainly to increase the efficiency or the speed of separations, with the final aim of either higher resolution or higher signal/noise ratios in the detection of the separated species.

Miniaturization in CE is necessary for good dissipation of joule heat through the thin walls of fused silica capillaries. Large quantities of heat are formed at the high electric field strengths (<450 V cm^{-1}) that must be applied when fast electrophoretic separations are to be achieved. Without good heat dissipation and additional active cooling of the capillaries, too large of temperature gradients within the separation buffer along the capillary radius lead to viscosity and electromobility gradients. They impair the otherwise optimal flat profile of the electroosmotic flow and give rise to peak broadening and to a decrease in separation efficiency. Too strong of an increase in temperature within the capillary may also cause formation of bubbles and, in CGE, destruction of gel fillings.

The sample capacity of CE systems is very small, and the analytical application of such instrumentation strongly depends on the incorporation of highly sensitive, preferably, in-column detection methods that produce high enough signal/noise ratios in detector cells. The detection cell volumes have to correspond to the extraordinary small peak volumes of the separated species. Without sufficient detection sensitivity, the dynamic range of quantitative CE analyses is quite limited.

Adequate instrumentation for the different types of miniaturized electrophoretic separations has been developed commercially in the recent years by several companies, but currently, such instruments seem to be in the status of further optimization.

Problems that are related to the miniaturization feature and the corresponding low sample capacity also arise with the sample introduction and with the chemistry of the analyte molecules within the capillary itself during the separation by differential migration. The usually applied narrow-bore separation capillaries (<100 μm) are characterized by a large surface/volume ratio (i.e., the limited amounts of sample that can be separated come into contact and may undesirably interact with the large capillary surfaces). By adsorption, for example, of large molecules, such as proteins, at the surfaces of CE systems, efficiency and selectivity of separations are seriously impaired by lateral chromatographic equilibria of interaction of the analyte molecules with the walls; broad and even very flat peak profiles may be the consequence. With strong analyte surface interaction, a considerable portion of the generally small samples may not even reach the detector cell (i.e., the section of the capillary in which the in-column detection takes place).

The chemical status of the surfaces in fused silica capillaries determines, along with the pH of the running buffer, the electroosmotic flow (EOF) in electrophoretic systems. The EOF is a consequence of the electric double layers that are formed on the capillary surfaces by ionic interaction with the different types of ions contained in the buffer. The chemistry of the surface (i.e., the presence of functional groups or its adsorptivity) strongly influences the formation of the double layers and the zeta potential, which causes the electroosmotic flow (i.e., the shifting of the buffer against the surface in the electrical field).

As known from stationary-phase syntheses in gas chromatography (GC) and liquid chromatography (LC), the chemistry of silica surfaces is unfortunately complicated because of the type and variable concentration of silanol groups present. The concentration of silanols on the surfaces of fused silica capillaries depends strongly on the parameters of the drawing procedure applied, the purity of the initial silica tubing material, and the pretreatment of the surfaces by either acidic or basic etching. Systematic investigations on the concentration of the SiOH groups in fused silica capillaries have not yet been performed. The electroosmotic flow that evolves in such capillaries depends on the pH and type of the buffer. Additives or buffer components may undergo intermolecular interaction with the silanols. Therefore, and because of the disturbing interaction and adsorption of analyte molecules, the control of the chemistry of fused silica surfaces is of great importance for analytical applications of CZE and MECC. For different analytical reasons, it is desirable to be able to modify the electroosmotic flow deliberately with these modes of CE. The aim may either be a faster separation or the transfer of species with opposite charge against its direction of electromigration to the detection zone of the capillary. In CGE, electroosmotic flow should not arise because of extrusion of the gel filling from the capillary. Special surface pretreatments before the introduction of the actual gel into the capillary are necessary to achieve stable gel fillings.

In this chapter the different chemical methods of modification of the EOF that are known from the literature will be discussed. The EOF is strongly influenced by the dissociation of the silanols, which can be alterated by the pH of the buffer. It has been shown that the EOF can also be instrumentally modified by application of a perpendicular additional electrical field to the outside of the capillary [8–10].

The chemical methods of EOF modification involve derivatization of the silanols, coating by polymeric layers, or adsorption of special additives contained in the buffer. In MECC the EOF is determined by both the properties of the modified or unmodified surface and by the adsorption of different types of surfactants that are contained in the buffer to form the micelles in which uncharged molecules can also move by electromigration through the capillary. According to the observations of Terabe et al. [6,11,12], anionic surfactants do not influence the EOF in untreated fused silica capillaries under the conditions of MECC. This is different when cationic surfactants are added to the buffer [13,14]. A special chemical pretreatment of the fused silica surface must be applied in CGE. Within the narrow-bore capillaries, stable gel fillings have to be generated that may or may not be chemically bound to the surface. Chemical bonding was claimed to be necessary to avoid extrusion of the filling from the capillary [15] by zeta potentials that are effective in the double layer on the surface underneath the gel filling. The production of stable gel-filled capillaries for the separation of large charged molecules, such as oligonucleotides or proteins, requires a careful pretreatment of the fused silica surfaces. The chemistry and kinetics of the in situ formation of the polymers that are to form linear and crosslinked gels in the buffer are also complicated. The long-term stability of such gel-filled capillaries for repetitive analytical separations depends on the type of buffers, the temperature inside the capillary and, probably, also strongly on contamination of the gel filling by components of the sample matrix.

In the following, experimental results from my laboratory are discussed, which have been obtained with the modification of surfaces for CZE and CGE and the generation of different types of gels in fused silica capillaries. The surfaces were chemically pretreated by procedures of several steps. The results and the conclusions about the type and number of steps for the preparation of electrophoretic separation capillaries that are suitable for application in analytical practice are also based on the results of other workers, as far as they have been published or have become known from patents.

MODIFICATION OF FUSED SILICA SURFACES IN CAPILLARY ZONE ELECTROPHORESIS AND MICELLAR ELECTROKINETIC CAPILLARY CHROMATOGRAPHY

Etching the Native Fused Silica Surface

Fused silica (FS) capillaries are drawn at temperatures beyond 1400°C and probably contain only a few silanols on the inner surfaces, depending on the drawing conditions and the water content of the gaseous medium inside the initial tubing during the drawing.

The chemical pretreatment of FS capillaries with different SiOH concentrations for use in capillary electrophoresis could be an advantage from the analytical point of view. With such a procedure, surfaces with reproducible SiOH concentrations can be obtained. The EOF in CZE and the adsorption and fixation of surfactant molecules under the typical conditions of MECC may become more defined.

The pretreatment step that is most frequently mentioned by several authors in the field of CZE and MECC is etching with NaOH or KOH (0.1–1 M) at elevated temperature, in the range between 20° and about 180°C; in the work of some authors the sodium or potassium ions are removed after the etching by rinsing with dilute acid (0.1 M HCl), as described. Finally, the capillaries are thoroughly rinsed with water and occasionally, with methanol and more lipophilic organic solvents. If silanization of the silanols will subsequently be performed, water is removed at elevated temperatures under high vacuum, to avoid reaction with the silanization reagent. It is known from the thermal treatment of porous silicas (which are probably characterized by a much higher concentration of silanols on the surface) that, at temperatures of about 150°C and above, water cleavage from neighboring silanols takes place. At lower temperatures polar compounds, such as water, may be too strongly adsorbed to be removed, even under high vacuum. The complete removal of adsorbed water and other etching chemicals may not be easy, and it may be difficult to perform silane derivatization reactions with high reproducibility.

Unfortunately, reliable experimental data about the concentration of silanols on the silica surfaces in FS capillaries are difficult to obtain, and the effectiveness of the etching procedures can be concluded only from the change of EOF by measuring migration times of UV-absorbing neutral markers and the electrophoretic behavior in general. However, adsorption of the test markers at the surfaces may falsify such measurements. It is of great analytical interest if, in practice, FS capillaries can be easily produced that have defined EOF properties and adsorption of analyte molecules. Capillaries of reproducible performance should be obtainable by a simple procedure of etching and rinsing. The process of equilibration before a capillary is conditioned for analytical separations should not require too much time. Chemical derivatization or polymer coating should be applied, if this goal cannot be reached simply by etching. It is also desirable that contamination by adsorbed sample components, such as the different types of proteins, can be removed by acidic or basic rinsing, depending on the nature of the adsorbed compounds, to restore the former performance in a series of separations.

Silanol groups form negative charges on the surface by dissociation, dependent on the pH of the buffer medium; positively charged surfactants (as, for example, alkylammonium salts) effect the formation of lipophilic coatings. Such ionic coatings give rise to a high EOF and have been applied to fast CZE separations of inorganic anions [14].

The aforementioned etching procedures are time-consuming and require long equilibration times before stable conditions for practical analyses are achieved. Therefore, it seems to be of interest to modify the silica surfaces by procedures, such as silanol derivatization, ionic or nonionic adsorption, as well as polymer coating. The time for equilibration between the buffer and the surfaces may be shortened, and the chemical properties of the coatings may be suited to suppress interaction or adsorption of special analyte molecules. The control of the electroosmotic flow and the suppression of analyte adsorption are the major aims of surface modifications.

Chemical Modification by Silanol Derivatization

There are two different chemical reactions for the bonding of especially selected molecules to the FS surface that is to be modified. The preferably applied silane derivatization reaction is characterized by the formation of an SiOSiC bond. Chemical bonding can also be effected by formation of an SiC bond. The SiOH can be converted by reaction with $SOCl_2$ into SiCl groups that are then reacted with the Grignard reagent vinyl-MgBr to fix a double bond to the surface. This double bond can be copolymerized in the same way as that of the 3-methacryloxypropyltrialkoxysilane (MAPT) reagent. The Si atom, in the formed SiC bond, belongs to the SiO lattice of the capillary material and does not originate from the silanization reagent. This reaction for modification of fused silica surfaces has recently been used for the coating of CZE capillaries by Cobb and Novotny [16]. These authors claim that their coatings are characterized by a better stability in long-term use at high pH, and they have performed experiments that indicate this higher stability. Polyacrylamide chains that are fixed to the surface in this way may not be stable at higher pHs, however.

Another way of forming SiC bonds at the surface is to convert the SiCl bond into SiH and to perform a subsequent Pt-catalyzed reaction with olefins, as has been applied in HPLC [17,18].

Silanization with Monofunctional and Multifunctional Reagents*

$$R_1R_2R_3SiX + HO{-}\overset{|}{\underset{|}{Si}}{-} \rightarrow R_1R_2R_3Si{-}O{-}\overset{|}{\underset{|}{Si}}{-} + HX$$

$$R_1SiX_3 + HO{-}\overset{|}{\underset{|}{Si}}{-} \rightarrow R_1X_2Si{-}O{-}\overset{|}{\underset{|}{Si}}{-} + HX$$

(Fixation of R_1, R_2, R_3 by Si–O–Si)

or formation of $R_1XSi(-O-\overset{|}{\underset{|}{Si}}-)_2$

The silanization reagents that are mostly used in capillary electrophoresis are trichloro-, trimethoxy-, and triethoxysilanes. The same reagents have been extensively applied for the modification of porous silica in syntheses of stationary LC phases, especially also for the fixation of chiral selectors. Therefore, a large variety of reagents containing different substituents R_1 are commercially available (ABCR GmbH & Co. KG, Karlsruhe, Germany). It is obviously assumed that the fixation of the group R_1 is more stable if the Si atom of the reagent is bonded to the surface twofold by SiO bonds in a chelate-type arrangement. However, the formation of such an arrangement would require two closely neighboring SiOH groups.

It can be expected that in the presence of water on the fused silica surface during the silanization procedure the SiOH groups of two neighboring reagent molecules react by formation of SiOSi bonds with each other, forming a network as shown in Fig. 1. One of the reaction partners can also be bonded to the surface by a single SiO bond. Such a network may exhibit a higher stability of attachment and effect a better or different shielding of unreacted SiOH groups from the surface, but it may be difficult to generate coatings of reproducible thickness and chemical properties.

In the presence of water on the surfaces, or in the solvents, the multifunctional reagents can polymerize without any chemical bonding to the surface. By the same polymerization reaction stationary LC phases, which are highly selective for polyaromatic hydrocarbon (PAH) separations, have been reproducibly synthesized [19].

It is known from the chemical modification of porous silicas that the reactivity of the preferably applied alkoxy reagents $(RO)_3SiR_1$ is weak compared with that of certain monofunctional reagents containing the more reactive SiX groups such as

**Multifunctional* means that more than one group SiX can react with the SiOH.

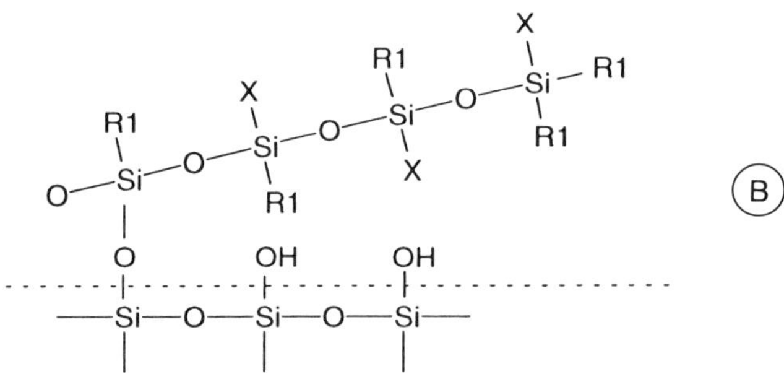

Figure 1 Silanization of SiOH silica surfaces using "multifunctional" reagents SiX_3 R_1.

SiHal or SiN. Reactions of monofunctional reagents with silanols are more defined and well known from the extensive work on syntheses of stationary LC phases, but they do not lead to a network, as shown in Fig. 1. Such networks can be considered as polymer coatings that are also chemically bonded. The monofunctional reagents lead to "brush-type" coatings.

Reagents of the SiX-type (X being halogen or amino) have the disadvantage that they form acidic or basic reaction products that may be strongly adsorbed on silica surfaces. For this reason we [20] have applied silanization reagents that contain SiO bonds to the enol of acetylacetone, which forms a less polar neutral reaction product with the active hydroxyl groups.

Silanization with Reagents That Contain a Functional Group in R_1 for Chemical Bonding After Derivatization

The reagent that has been extensively used to modify fused silica surfaces in CE is the γ-methacryloxypropyltrimethoxysilane (MAPT), which was first applied by Hjertén

[21,22]. The molecule, which is attached to the surface by the silanization reaction, contains a double bond (Fig. 2) by which this group can be copolymerized with acrylamide to form linear or, in the presence of bifunctional monomers, cross-linked polyacrylamide coatings. Such polar coatings have special chemical properties relative to suppression of EOF in CGE or to analyte adsorption as in CZE. It has been claimed that by the chemical bonding of cross-linked gels to the FS surface, the stability of polyacrylamide gel-filled capillaries would be improved [15]. It has been shown [23] that chemical bonding of the gel to the surface is unnecessary and is rather disturbing when stable and bubble-free gel-filled capillaries are to be achieved; the formation of bubbles in the polyacrylamide gels during the formation of the actual gel is prevented when the double bonds of the methacryloxypropyl group are used only for the preceding formation of a thin layer of linear polyacrylamide by copolymerization. The formation of bubbles is caused by the volume contraction during the polymerization, when the buffer solution of the acrylamide is prevented from moving freely along the capillary walls.

Several recent publications by different groups concern the production of linear polyacrylamide gel columns in which larger oligonucleotide molecules, especially DNA restriction fragments, can be separated. Such separations of larger DNA fragments can also be achieved with diluted solutions of different polymers that form polymer networks by "entangling" in the buffer [24–26].

The investigations on surface modification for CZE deal with the separation of different types of proteins according to hydrophobicity, isoelectric points, and other properties in acidic or basic buffers. McCormick has also used the MAPT reagent for silanization [27], but then generates a coating of poly(vinylpyrrolidinone) by radical copolymerization. The capillaries obtained could be applied for the separation of standard proteins at the low pH of 2.0; however, the peak shapes obtained in the CZE separations were tailed.

Another reagent that has been successfully used by Bruin et al. for the production of capillaries to be applied to CZE separations of proteins is the glycidoxypropyltrimethoxysilane [28]. The structure of this reagent is shown in Fig. 3. After the silanization, the surface is coated with a layer of polyethylene glycol (PEG 600), which is chemically bonded to the surface by reaction with the oxirane ring in the molecule attached to the surface. The oxirane ring is opened by in situ reaction on

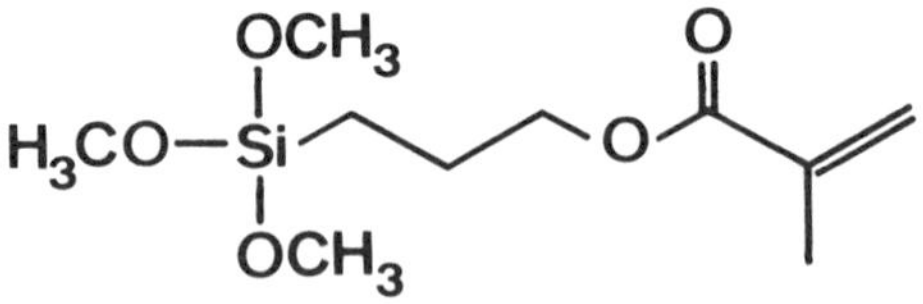

Figure 2 Structure of 3-*m*eth*a*cryloxy*p*ropyl*t*rimethoxysilane (MAPT).

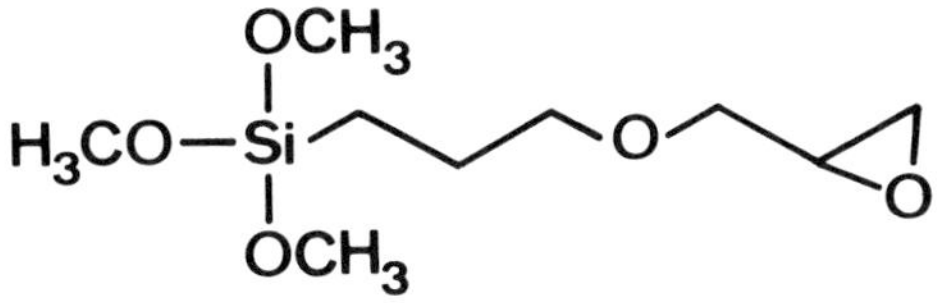

Figure 3 Structure of 3-*g*lycidoxy*p*ropyl*t*rimethoxysilane (GPT).

the surface catalyzed by $BF_3 \cdot Et_2O$. Good separations of standard proteins at a pH 4.1 were obtained, the peaks achieved were considerably tailed. The peak symmetries became worse at a pH 6.8. In Fig. 4, CZE separations of standard proteins are shown that have been obtained in my laboratory following the procedure described by Bruin et al. [28]. The reaction with the oxirane ring could also easily be performed with other hydroxylic or glycolic compounds.

The reagent aminopropyltrialkoxysilane was applied by Swedberg [29]. After the surface derivatization, the amino group was then reacted with pentafluorobenzoylchloride to give the structure that is shown in Fig. 5. Here, the hydrophobic interaction that disturbs the CZE separation of proteins is decreased by the repulsive effect of the fluoro-substitution in the derivatizing group. The separations shown were achieved at quite high ionic strength (250 mM) of the buffer.

The same reagent has been used by Bruin et al. [30] to bond maltosis by a special reaction with the amino group of the bonded moiety (Fig. 6). The CZE capillaries showed a strongly pH-dependent EOF from the influence of both the residual Si—OH and the NH_2 groups.

Summary of the Role of Silanization for Capillary Zone Electrophoresis, Micellar Electrokinetic Capillary Chromatography, and Capillary Gel Electrophoresis

The concentration of the SiOH groups on the surfaces is decreased by the described silane derivatization reactions. The residual SiOH strongly influences the EOF, depending on the pH of the buffer used and may cause the adsorption of basic molecules as in HPLC. By silanization, hydrophobic or polar molecules as substituents of the reagent are attached to the fused silica surface and modify the EOF and the adsorptivity for analyte molecules, depending on the chemical properties of the coating. Hydrophobic coatings obtained by silanization may be suited for the adsorptive bonding of ionic or nonionic surfactants by the hydrophobic part of such molecules. The groups that are attached to the surface by the silanization reagent may contain reactive groups by which further derivatization reactions may be performed on the capillary surface (in situ). The chemical properties and especially the charging of the coatings have a strong influence on the undesirable adsorption of large analyte molecules, such as proteins. For most proteins hydrophobic coatings are not favorable.

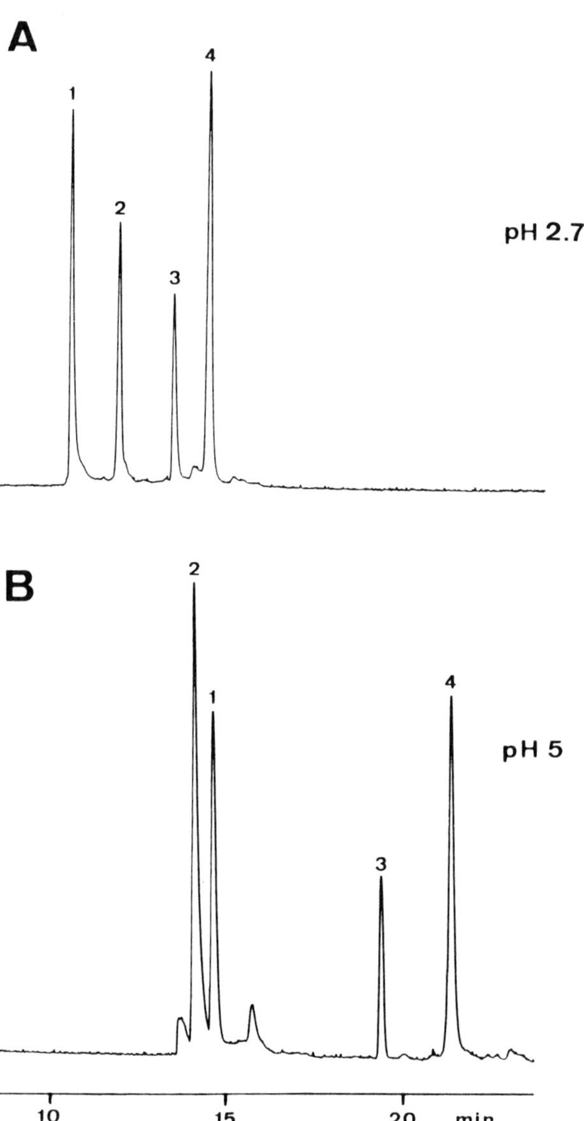

Figure 4 Protein separation with polyethylene glycol-modified capillaries at different pHs. Sample: 1, cytochrome *c*; 2, lysozyme; 3, ribonuclease A; 4, α-chymotrypsinogen A. Coating: PEG 550 bonded by γ-glycidoxypropyltrimethoxysilane. Capillary: 57-cm effective length, 70-cm total length; (A) 75-μm, (B) 50-μm id. Conditions: 20 kV (286 V cm^{-1}); (A) 64 μA, (B) 25 μA; 20°C. Buffer: 20 mM Na-phosphate, 30 mM NaCl; (A) pH 2.7, (B) pH 5.0. Injection: (A) 10 kV, 5 seconds; (B) 10 kV, 10 seconds. Detection: UV 214 nm.

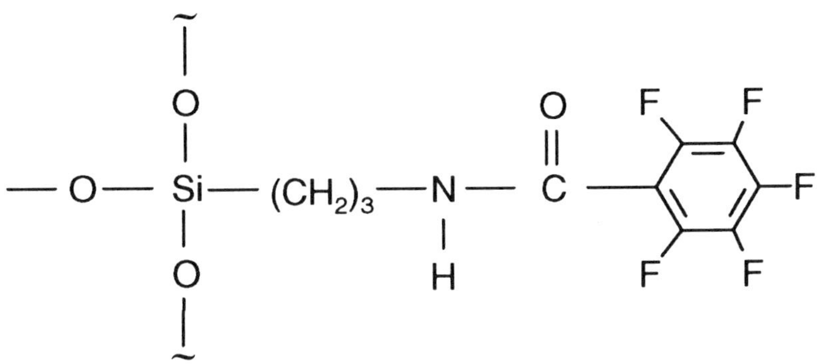

Figure 5 Structure resulting after reaction of the amino group of aminopropyltrioxysilane with pentafluorobenzoylchloride. (From Ref. 29)

Further investigations are necessary if the SiC bonds that are formed for the fixation of coating layers to the surface are more stable than the SiOSi bonds formed by the common silanization reactions.

The reproducible execution of silanization reactions employing a variety of mono- or multifunctional reagents of different reactivity is not without experimental problems that are related with the initial status of the surface after the preceding etching and rinsing procedures, which are mainly performed to increase the number of anchor groups on the surface for chemical bonding. Residual adsorbed water may cause formation of nonbonded reaction products or polymerization of the multifunctional reagents (such as X_3SiR_1). It would probably be of interest, in practice, if surface pretreatment steps that are too complicated could be omitted or replaced by simpler procedures as, for example, adsorptive coating.

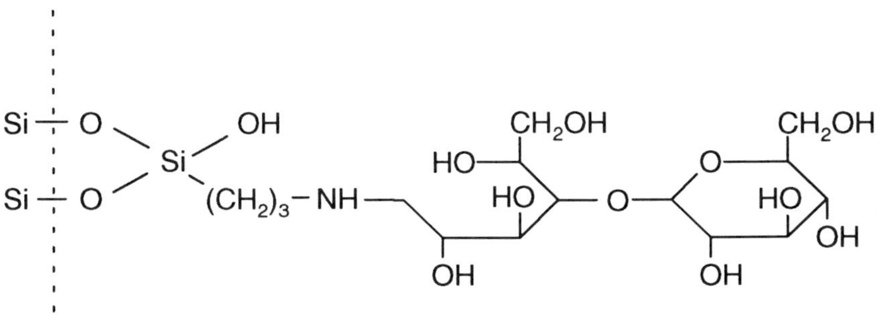

Figure 6 Bonding of maltose by aminopropyltrialkoxysilane. (From Ref. 30)

Modification of Fused Silica Surfaces by Adsorption of Modifiers Contained in the Buffer

In CZE and MECC the constituents of the buffers, or special additives contained in them, undergo a weaker or stronger intermolecular interaction with the FS surface. This interaction is also dependent on the pH of the buffer, the polarity of the buffer components, and the concentration of SiOH at the surface. On chemically modified (i.e., coated silica) surfaces, the concentration of the SiOH is reduced or residual SiOH may be shielded against such interactions by special coatings.

Influence of Surfactants on the Electroosmotic Flow in Micellar Electrokinetic Capillary Chromatography

By the addition of anionic surfactants to the buffers in MECC, the EOF which goes toward the cathode is only slightly changed [11]; however, the EOF can be reversed by addition of cationic surfactants. In CZE, the increase of the EOF may be of interest to accelerate analytical separations and, thereby, to increase the signal/noise ratio of the detection. This is true when the EOF is effective in the same direction as the migration of the charged species.

Influence of a Cationic Surfactant on the Electroosmotic Flow for Fast Separation of Inorganic Anions

Jones and Jandik have made use of a high EOF for the fast separation of inorganic anions [14]. For the anion separation, they added trimethyltetradecylammonium-bromide (0.50 mM) to the buffer, which also contains UV-absorbing anionic electrolyte for indirect UV detection.

Separation of Proteins in Capillary Zone Electrophoretic Systems

An important analytical application of CZE is the separation of proteins. The migration of different types of proteins in CE systems depends on their ionization (isoelectric point), size, and hydrophobicity at a given pH of the buffer. Figure 7 shows the influence of the ionic strength of the buffer on peak shape.

At a high buffer pH (e.g., pH 10) the dissociation of the SiOH on the surface is high, so that, because of the coulombic repulsion between the proteins and the walls, excellent CE separations of basic proteins are achieved with unpretreated fused silica capillaries [31]. A protein bears a net negative charge if the pH of its solution is higher than its isoelectric point, and silica gel has a negative charge at a pH higher than 2.

Jorgenson and Lukacs [4] found in 1983 that fused silica surfaces modified with glycol-containing groups decrease the adsorption of proteins, but in a CZE separation of standard proteins, the peak of cytochrome *c* exhibited only 70,000 theoretical plates, instead of the theoretically expected 2×10^6.

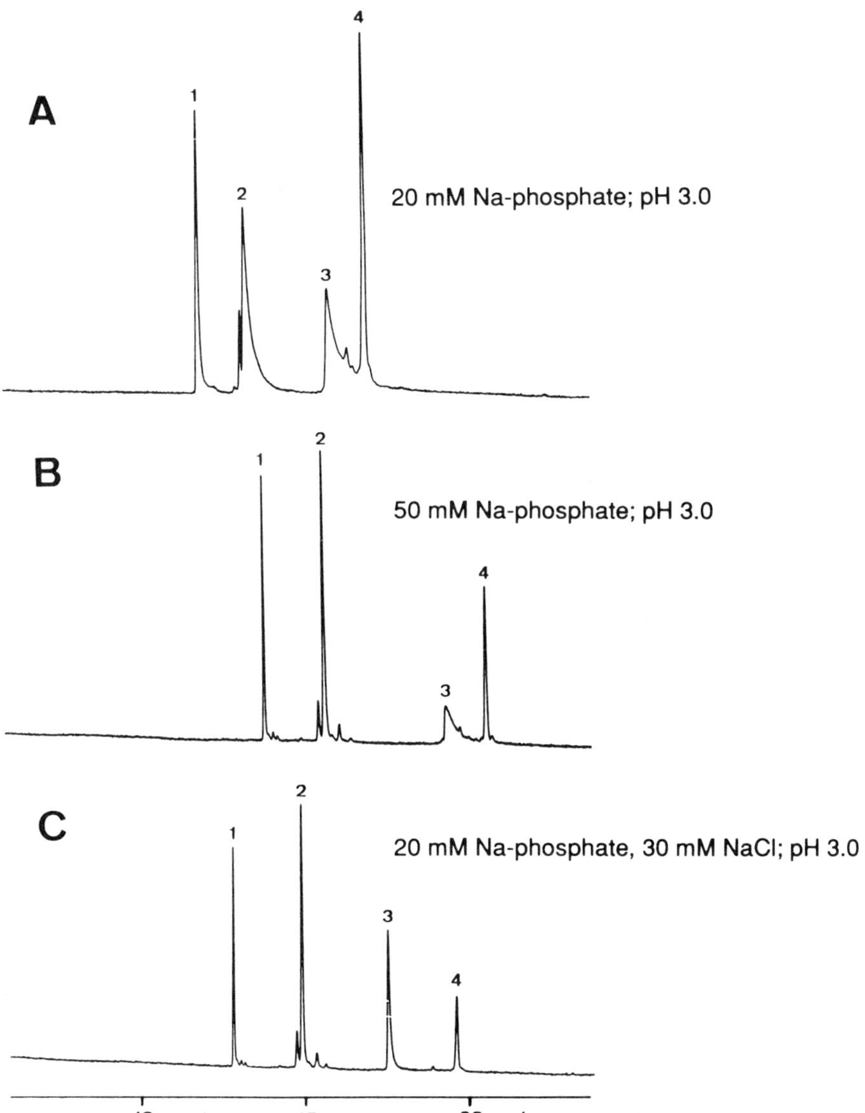

Figure 7 Influence of buffer composition on the separation of proteins. Sample: 1, cytochrome *c*; 2, lysozyme; 3, ribonuclease A; 4, α-chymotrypsinogen A. Capillary: 57-cm effective length; 70-cm total length; 75-μm id. Coating: linear polyacrylamide. Conditions: 20 kV (286 V cm^{-1}); (A) 17 μA, (B) 38 μA, (C) 54 μA; 20°C. Injection: (A) 10 kV, 10 seconds, (B) and (C) 5 kV, 7 seconds. Detection: UV 214 nm.

Many other procedures of surface modification for protein CZE separations have been described in the literature since then: Emmer et al. [32] applied a dynamic deactivation of fused silica surfaces by the cationic fluorosurfactant $CF_3(CF_2)_7SO_2NH(CH_2)_3N^+(CH_3)_3I^-$ for the separation of proteins. The capillaries were pretreated only by etching with sodium hydroxide. The surface charge becomes positive owing to the adhered bilayer of the cationic surfactant, and the EOF is directed toward the anode. The efficiencies and the peak symmetries in the electropherograms obtained were excellent. In spite of the strong interaction of the ammonium ion with the acidic surface, this method of dynamic surface modification is based on an equilibrium between the buffer and the surface. The adsorption of certain proteins is obviously reduced in a manner similar to the approach of chemical modification by silanization reported by Swedberg [29].

Recently, Bullock and Yuan [33] reported on the separation of basic proteins in uncoated fused silica capillary tubing using phosphate and formate buffers that contained the highly basic 1,3-diaminopropane, in concentrations of 30–60 mM, as a cationic surface modifier along with variable, but moderate, concentrations of alkali metal salts. The alkali salts are to suppress protein–capillary wall interaction, whereas the basic 1,3-diaminopropane probably undergoes a stronger interaction with the silanols than do the basic proteins. Good separations can be performed at pHs that are below the proteins' isoelectric points and in a pH range at which proteins are stable with an interesting flexibility of selectivity optimization by the pH (Fig. 8).

Gilges et al. [34] observed that untreated fused silica capillary tubing can be deactivated for the separation of basic proteins by operation with polyvinylalcohols (PVAs) contained in low pH phosphate buffers along with low concentrations of the common alkali salts (Fig. 9). After each or after a few separations, the capillary was rinsed with water and then refilled with fresh buffer containing PVA. By this method of deactivation by adsorption of a nonionic hydroxylic modifier, without preceding time-consuming etching or silanization procedures, separations with reproducible migration times and high efficiencies can be performed. The PVA can also be thermally immobilized before introduction of the buffer medium in a special procedure. Then the buffer does not need to contain PVA molecules as additives [75].

Excellent results in protein separations for efficiency and peak symmetry have been obtained by Towns and Regnier [35], who used nonionic surfactants as additives to the buffers (Fig. 10). These surfactants are adsorbed to the silica surfaces that were previously hydrophobized by C_{18}-silanization. The surfactants contained in the running buffers were the Tween 20, 40, or 80 and Brij 35 or 78 (the chemical structures are shown in Fig. 11). The hydroxylic or glycolic groups of these adsorbed surfactants generate a coating of soft interaction with the proteins, which obviously enables separations of the proteins with high efficiency and with very symmetric peaks. This system is based on the combination of surface derivatization and ad-

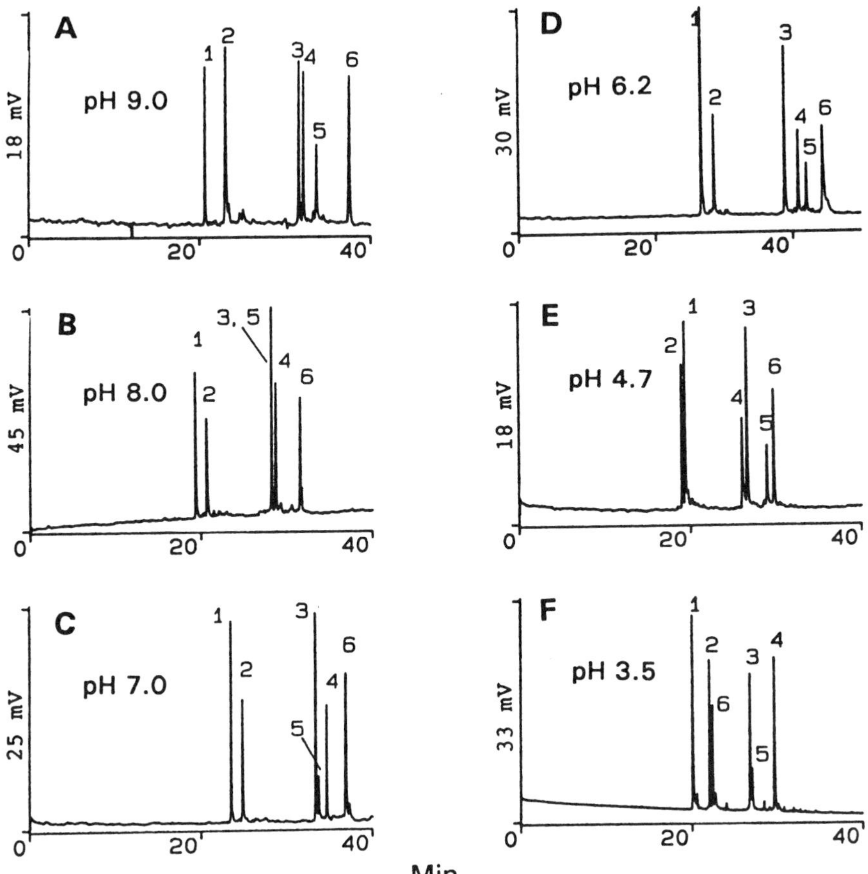

Figure 8 Separation of the basic proteins: 1, lysozyme; 2, cytochrome *c*; 3, ribonuclease A; 4, α-chymotrypsinogen A; 5, trypsinogen; 6, rhuIL-4, at pH 9.0, 8.0, 7.0, 6.2, 4.7, 3.5. Conditions: 75-cm × 50-μm id capillary column, 55-cm separation distance; 200 V cm^{-1} field strength at (A) pH 9.0 (37 μA), (B) pH 8.0 (40 μA), (C) pH 7.0 (43 μA), (D) pH 6.2 (46 μA), and 240 V cm^{-1} at (E) pH 4.7 (50 μA), pH 3.5 (50 μA). (From Ref. 33)

sorptive modification by a surfactant. The EOF in such systems of Brij 35-alkylsilane is very low and constant in a pH range between 4 and 11. Figure 12 compares the pH dependence of the EOF in uncoated and Brij-coated capillaries. The adsorbed surfactant in the systems proposed by Towns and Regnier [35] is leached gradually from coated surfaces; therefore, a small amount of the surfactant is added to the running buffer to maintain the coating.

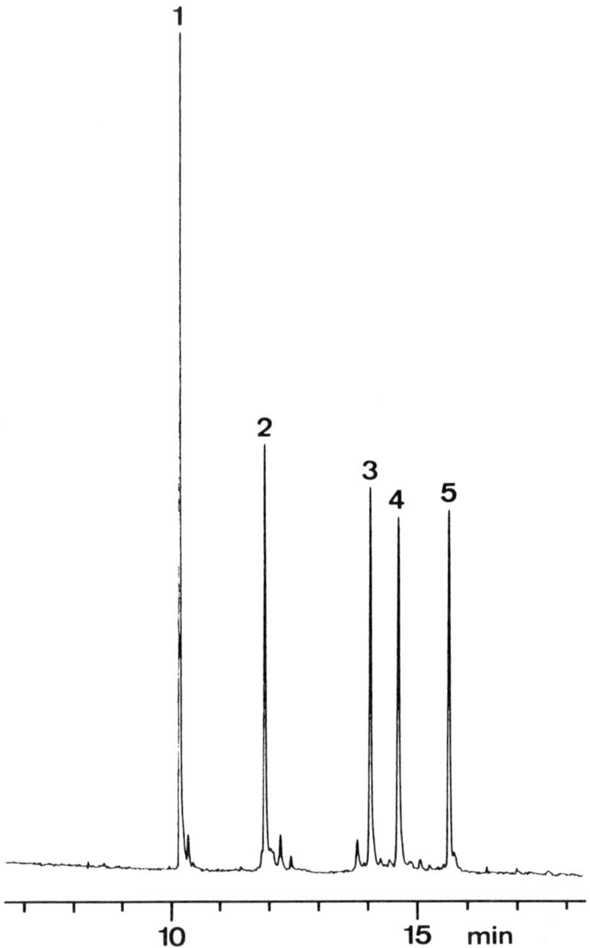

Figure 9 Separation of basic proteins with polyvinylalcohol as buffer additive. Sample: 1, cytochrome *c*; 2, lysozyme; 3 trypsin; 4, trypsinogen; 5, α-chymotrypsinogen A: 0.2 mg mL^{-1} of each. Capillary: 57-cm effective length, 70-cm total length; 75-μm id; untreated. Conditions: 25 kV (357 V cm^{-1}); 69 μA; 20°C. Buffer: 20 mM Na-phosphate, 30 mM NaCl pH 3.0; 0.05% (w/w) PVA average M_w 15000. Injection: 5 kV, 5 seconds. Detection: UV 214 nm.

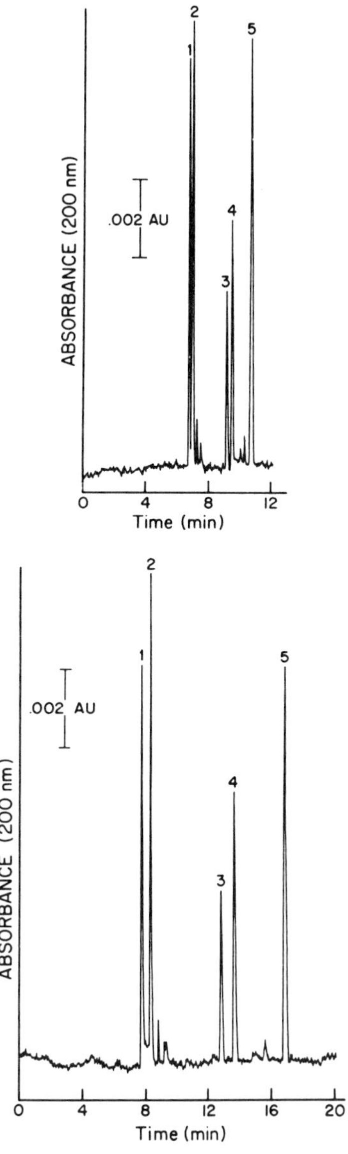

Figure 10 Electropherograms showing the separation of five basic proteins (1, lysozyme; 2, cytochrome *c*; 3, ribonuclease A; 4, α-chymotrypsinogen; and 5, myoglobin) using a 0.01 M phosphate buffer (pH 7.0) on a 75-μm × 50-cm (a) Tween 20/alkylsilane capillary and (b) Brij 35/alkylsilane capillary at 300 V cm^{-1}. (From Ref. 35)

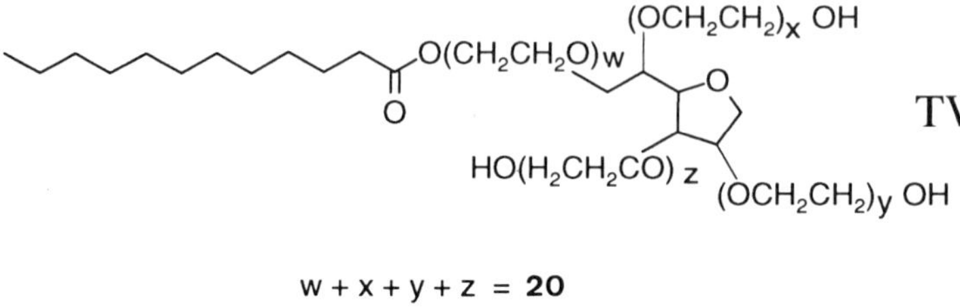

(OCH2CH2)x OH
O(CH2CH2O)w
O
O
HO(H2CH2CO) z
(OCH2CH2)y OH
w + x + y + z = 20

TWEEN® 20

(OCH2CH2)23OH

BRIJ® 35

Figure 11 Basic chemical structures of the Tween and Brij families of surfactants.

7.5
6.5
5.5
4.5
3.5
2.5
1.5
0.5
Electroosmotic Flow (m²/ V s) X 10⁸
Uncoated
BRIJ 35 / Alkylsilane
3
4
5
6
7
8
9
10
11
pH

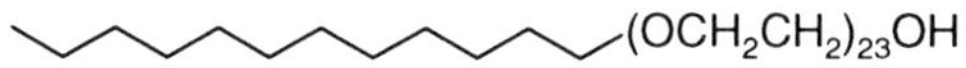

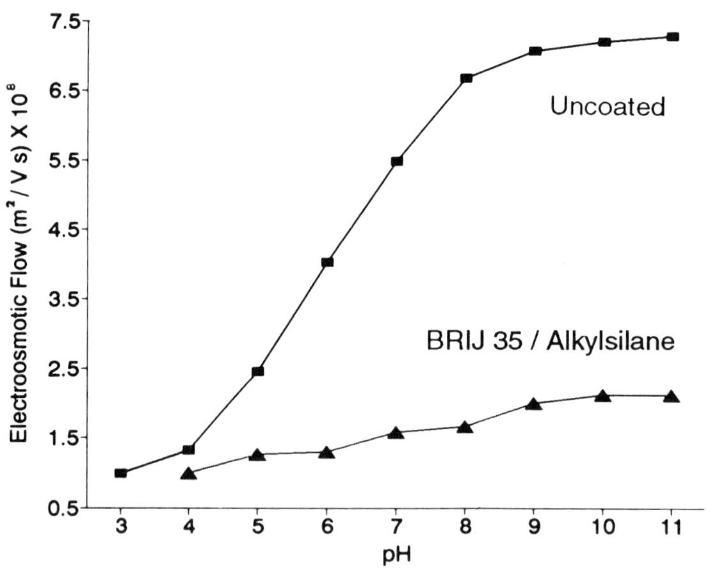

Figure 12 Dependence of electroosmotic flow ($m^2\ V\ s^{-1}$) $\times\ 10^8$ on pH for (■) uncoated and (▲) Brij 35/alkylsilane-coated capillary (75 μm × 50 cm). (From Ref. 35)

It still seems to be of general interest to produce chemically bonded polymer coatings that are well suited for protein separations. The surface modification procedure described by Bruin et al. [28], which was mentioned earlier, is of this type.

Towns and Regnier state in their excellent paper on this matter [35]: "Stable reproducible deactivation of the capillary is needed before CZE will be of general utility in protein separation." Such an approach is required if the same especially surface pretreated capillary is to be repetitively applied with the same buffer when analyzing a series of samples. If the samples are contaminated by compounds that may be adsorbed on the surface, restoration of the former surface properties by rinsing with solutions having extreme pHs becomes necessary. Therefore, permanent coatings must also be stable against the chemicals used in such rinsing procedures.

Modification of Fused Silica Surfaces by Polymer Coating

Different methods for coating fused silica surfaces by special polymers are well-known from the field of column technology for capillary GC and from syntheses of stationary phases for the different aqueous phase systems in LC. In capillary electrophoresis, the major objective is a weak interaction of the analyte molecules with the surface (i.e., the coating on it, lateral equilibria between the buffer and the capillary walls, or the coatings on them, may considerably impair the efficiency of electrophoretic separations).

Polymers can be immobilized on fused silica surfaces merely by adsorption, but in CE the polar components of the aqueous buffers have strong displacing properties and the polar water-soluble polymers will be leached from the surface. The solubility of polymers in the buffers depends on their polarity and molecular size and can be reduced by cross-linking.

Polymers can also be chemically bound to the surface, either by SiOH or by other functional groups. For chemically bound coatings in CZE and CGE, polymers such as polyacrylamides are mostly generated in situ in the capillaries by copolymerization with functional groups that have been fixed to the surface by previous derivatization steps. Prepolymerized compounds, such as the alkylpolysiloxanes, that contain functional groups such as SiH could also be chemically bonded by the reaction of this group with SiOH on the surface by forming a SiOSi bond.

Coatings Generated by Formation of Polymers at the Surface (In Situ)

POLYACRYLAMIDE COATINGS Coatings with linear polyacrylamide for the separation of proteins were first applied by Hjertén [22] and, recently, in a different sequence of surface reactions by Cobb et al. [16]. The in situ formation of the linear polymer is started from the double bond of the *methacryloxypropyl* group (of the

MATP silanization reagent) or from the vinyl group that has been attached to the surface by means of the Grignard reagent $CH_2CHMgBr$, respectively.

Kohr and Engelhardt [36] silanized KOH-etched capillaries with with trichlorovinylsilane and polymerized, within the rinsed capillary, solutions of different acrylamides in dichloromethane and methanol with 2,2′-azobisisobutyronitrile (AIBN) as radical starter at temperatures of about 120°C. The layer thickness was varied by the concentration of the monomers in the solvent. The authors also measured the electroosmotic flows in uncoated capillaries and observed hysteresis effects, depending on the direction of pH change from low to high or from high to low. Reproducible adjustment of electroosmotic flows was possible only at very low or very high pH of the buffers.

The performance of the protein separations achievable with capillaries produced by the three similar surface-coating methods was not yet optimal. We have made several polyacrylamide-coated capillaries for protein separations. The polyacrylamide coatings were achieved by in situ polymerization after fixation of double bonds to the surface, either by silanization with reagents that contain a copolymerizable double bond in the group R or by a vinyl-Grignard reaction, after conversion of the SiOH into SiCl bonds (Fig. 13).

Suitable separations in peak symmetry and efficiency have been achieved. The pH stability of the coatings probably mainly depends on the stability of the copolymerized polyacrylamide chain. The peak widths (efficiencies) and symmetries are obviously not as good as those expected from CE separations in capillaries. For the comparison of such protein separations with the good results that Towne and Regnier obtained with the dynamic (adsorbed) nonionic surfactant coatings, the pH of the buffer at which the separation has been performed must also be considered. Towne and Regnier have achieved good peak shapes at a somewhat higher pH, or even at neutral pH and low ionic buffer strength [35].

ALKYLPOLYSILOXANE COATINGS FORMED FROM MULTIFUNCTIONAL SILANIZATION REAGENTS IN THE PRESENCE OF WATER Silanizations with a trifunctional reagent such as X_3SiR, more or less, leads to chemically bound and cross-linked alkylpolysiloxane coatings, depending on the presence of low concentrations of water. It may be possible that the layer thickness can be controlled by the concentrations of the reagent and the water in the solvent [19] (see the reaction scheme of Fig. 1).

COATING BY POLYMERIZATION OF THE N–*CARBOXYANHYDRIDE OF* γ-*METHYLGLUTAMATE* Such coating leads to a moderate decrease of the EOF, only because the too thin coating layers are not suitable for hydrophobic proteins [37].

Coatings Generated by Deposition of Externally Synthesized Polymers

ALKYLPOLYSILOXANES AND POLYETHYLENE GLYCOLS In the well-developed technology of GC capillary columns, coatings of polymers can be generated

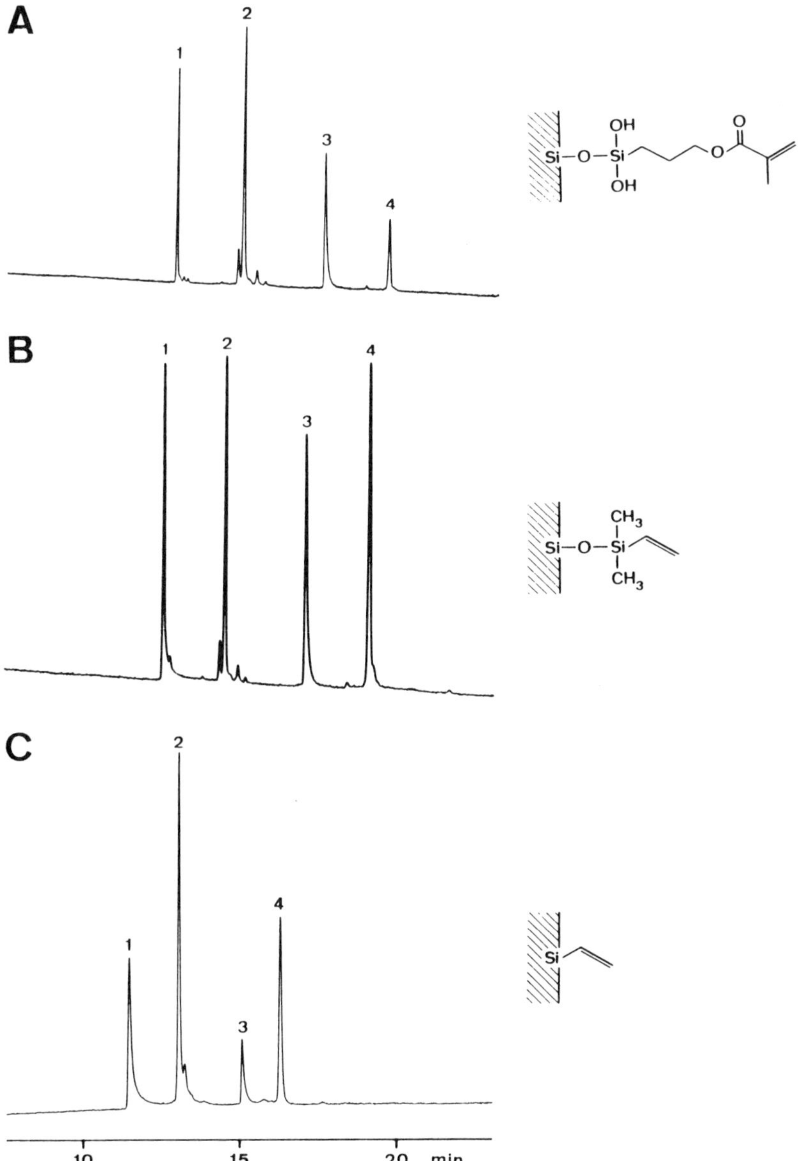

Figure 13 Comparison of polyacrylamide-coated capillaries. Samples: 1, cytochrome *c*; 2, lysozyme; 3, ribonuclease A; 4, α-chymotrypsinogen A. Coating: polyacrylamide bound to the capillary wall with different anchor groups; (A) silanization with γ-methacryloxypropyltrimethoxysilane, (B) silanization with vinyl dimethyl silylenolate of acetylacetone, (C) direct vinylation with vinyl-MgBr after chlorination with $SOCl_2$. Capillary: (A,B) 57-cm effective length, 70-cm total length; (C) 55/68 cm; 75-μm id. Conditions: 20 kV (A,B, 286 V cm^{-1}; (C) 291 V cm^{-1}); (A,B) 56 μA, (C) 60 μA; 20°C. Buffer: 20 mM Na-phosphate, 30 mM NaCl (A,B) pH 3.0, (C) pH 2.7. Injection: 10 kV, 5 seconds. Detection. UV 214 nm.

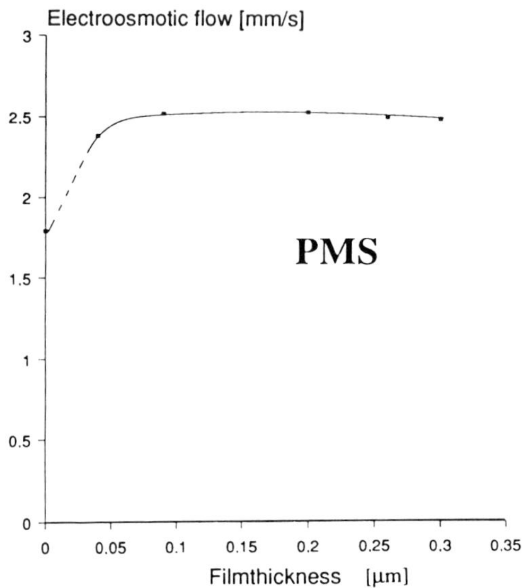

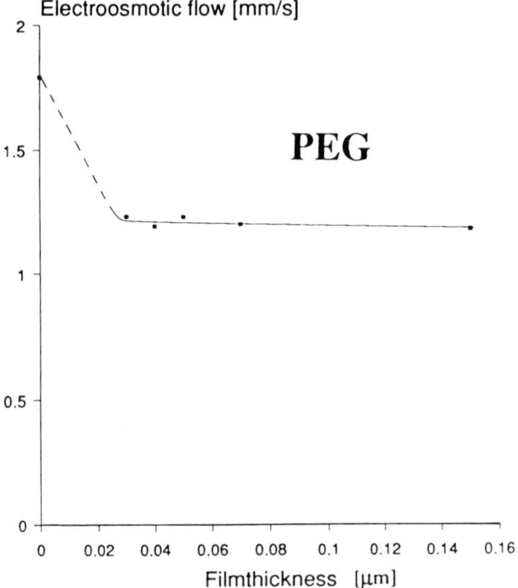

Figure 14 Influence of polymer coating of different thicknesses on electroosmotic flow.

on fused silica surfaces that are stable against rinsing with organic solvents. The immobilization of alkylpolysiloxane or polyethylene glycol (PEG) coatings can be achieved by cross-linking reactions. The immobilization of the polar PEGs for use with the aqueous buffers of CE may be more difficult.

The coatings, which can also be used for protein separations, have preferably been applied to avoid adsorption and only in the second instance for suppression of the EOF. Lipophilic coatings must be ruled out because of the disturbing and denaturing hydrophobic interactions with these large biomolecules. This may be different for the analytical applications of CZE and especially MECC to smaller molecules.

Terabe et al. [38] pointed out that fused silica capillaries coated by methylpolysiloxanes or by polyethylene glycols can be applied to either accelerate or retard separations through modification of the EOF by such coatings. Lux and associates [39] showed that a few layers of molecules (i.e., thicknesses of about 40 nm) were sufficient to achieve the maximum effect on the EOF. Such low film thicknesses could be determined by GC from the retention of a standard compound (*n*-nonane). The results of the EOF measurements and MECC separations are shown in Fig. 14 (EOF dependence on film thicknesses of methylpolysiloxane and polyethylene glycol coatings) and in Fig. 15 (MECC separation with different EOFs). With the slow EOF in a PEG-coated capillary, a much better resolution of the four purine derivatives could be achieved. The reverse behavior of the methylpolysiloxane-coated capillary is explained by the strong adsorption of the anionic surfactant SDS by the lipophilic alkyl group. This adsorption increases the negative charge of the surface. The more polar PEG coating, however, has a weak interaction with the lipophilic portion of the surfactant molecule. Adsorption of small and polar analyte molecules does not seem to be a problem relative to impairment of efficiency and peak symmetry.

CROSS-LINKED IONIZABLE POLYETHYLENEIMINE The coatings of ionic polyethyleneimine polymer for the CZE separation of proteins were obtained by cross-linking polyethyleneimine coatings by reaction with ethyleneglycol diglycidyl ether (Fig. 16). The EOF depends on the pH of the buffer and on the molecular size of polyethyleneimine because of the different layer thicknesses. Good protein separations have been obtained at pH 7 [35].

PREPARATION, CHARACTERIZATION, AND APPLICATION OF GEL-FILLED CAPILLARIES

The separation of large, charged biomolecules is an important area of application in modern CE because of the high efficiencies (several million theoretical plates in capillaries of less than 50 cm in length) that have been reported in recent publications [5,23,40]. Gel fillings in narrow-bore, fused silica capillaries produce friction to the migrating large molecules that is proportional to their molecular size as, for example, the number of base pairs (bp) in DNAs. Besides this molecular sieving

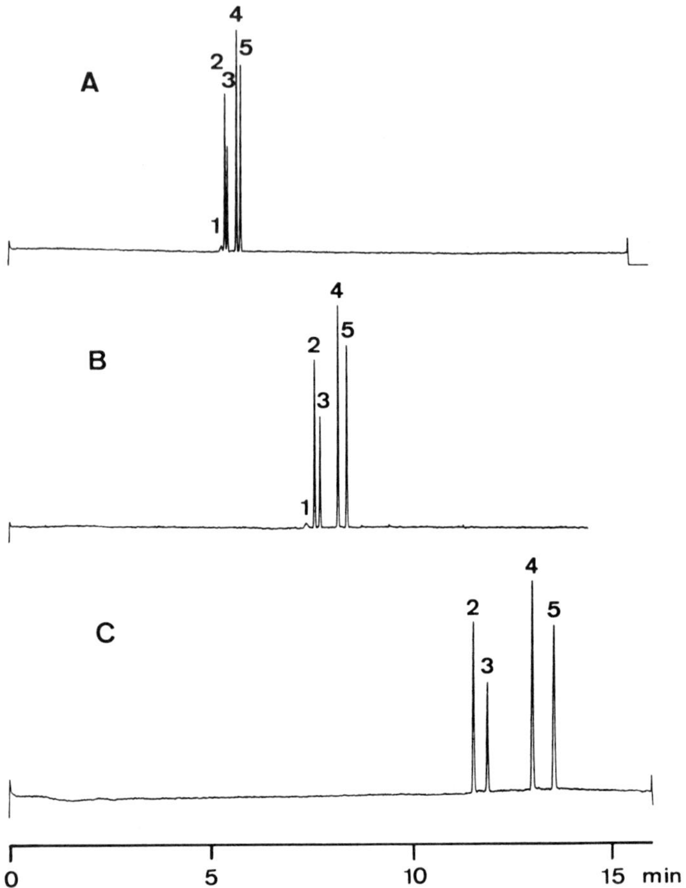

Figure 15 MECC separations of nucleobases using capillaries coated with polymers of different chemical structure. Sample: 0.2 mg ml^{-1} each; 1, water; 2, uridine; 3, cytidine; 4, guanosine; 5, adenosine. Column: 80-cm effective, 120-cm total length; (A) polymethylsiloxane OV−1 on FS, 50-μm id, film thickness 0.09 μm; (B) uncoated FS, 50-μm id; (C) polyethylene glycol CW20M on FS, 50-μm id, film thickness 0.04 μm. Buffer: 0.02 M phosphate buffer; 0.05 M SDS. Injection: electrokinetic; 7500 V; (A) 2 seconds, (B) 5 seconds, (C) 8 seconds. Separation voltage: 35 kV; Detection: UV 254 nm; Temperature: 298 K.

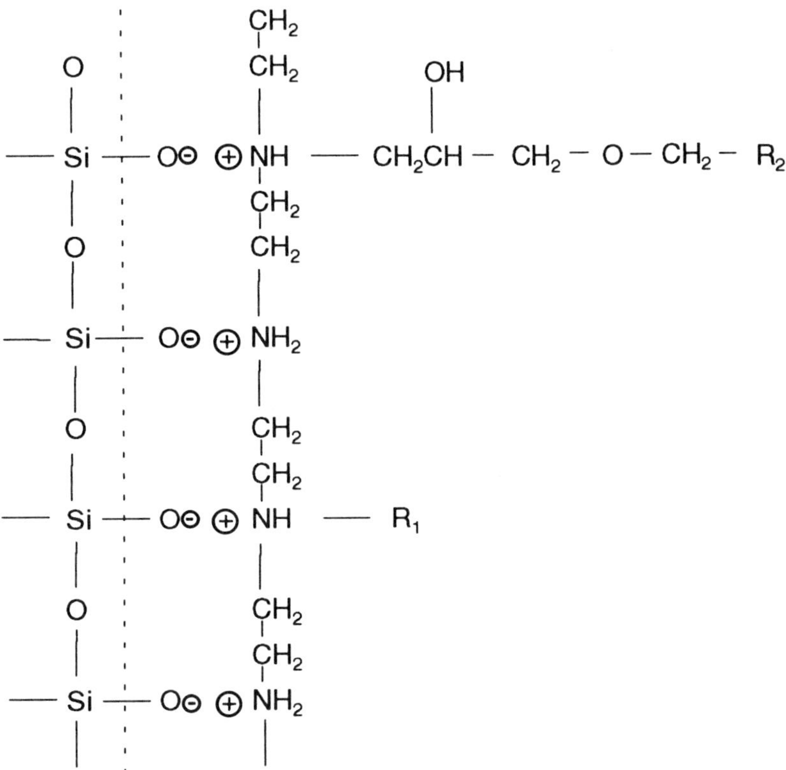

Figure 16 Formation of a polyethyleneimine capillary column coating by cross-linking. (From Ref. 35)

effect, gel fillings of polyacrylamides and agaroses exhibit an anticonvective effect that is the reason for the extremely high separation efficiencies [5,23,40–43].

In this area of the separation of large molecules, such as oligonucleotides or even DNA-fragments, CGE allows fast and highly efficient separations, together with a pronounced molecular size selectivity. Principally because of too slow a mass transfer of large biomolecules between the mobile and the stationary phase, HPLC separations are, even with the application of gradient elution, inferior to the separation efficiencies and resolution that are achievable in CGE [44–46]. For bioanalytical applications, the transfer of the well-known and highly developed slab gel separations into fused silica capillaries seems to be a promising approach. In the narrow-bore FS capillaries, much stronger electrical fields of up to 450 V cm^{-1} can be applied because of the effective heat dissipation.

Fast separations with high resolution of large molecules according to molecular size can be achieved. Meanwhile, speed and resolution of analogous slab gel separations could also be improved by the use of thin gels (10–100 μm) because of better heat dissipation [47–50].

Recent Progress in DNA Separations

Conventional slab gel electrophoresis is a time-consuming technique that is characterized by limitations for speed and resolution. Parallel separations of several samples are possible, however, but the slab gels can be used only once. The separation properties of the slab gels must be reproducible if the migration times obtained are to be used for identification of target or other components. Samples that contain components that migrate very slowly, or not at all, are accumulated at the origin of the separation on the slab. Such contaminants would disturb subsequent separations, but the slab gels are not designed for repetitive use. Slab gel separations can also be partially automated, but with modern CE equipment, entirely automated separations of series of samples are possible.

The following discussion on the work of different groups who have been and are still strongly involved in the further development of the new methods of CGE will primarily deal with the separation of nucleotides, especially DNA fragments. Nucleotide separations are, in principle, less demanding than the separations of other large charged biomolecules, especially of the ampholytic proteins, which are characterized by a wide range of molecular size and isoelectric points. Basic or acidic proteins require optimization of the buffer properties and, especially, an adequate modification of the surfaces that are suitable for universal application to different types of proteins.

Cross-linked Polyacrylamide Gels for the Separation of Oligonucleotides and DNA Sequencing Fragments With up to 500 Base Units

Several publications [5,51,52] have shown excellent gel electrophoretic separations (Figs. 17 and 18) mainly of oligonucleotides, but also of proteins. The oligonucleotide separations, which were performed at high field strength (up to 400 V cm^{-1}) and also using pulsed electrical fields [40], had such a high molecular size selectivity that at the high efficiencies of gel electrophoresis single-unit baseline resolution could be achieved for fragments up to 330 nucleotides in length. Fast separations with such a resolution are required for sequencing fragments that contain up to 400 base units. Because of the high interest in fast and efficient separations of the typical sequencing fragments, several laboratories became interested in the manufacture of such gel columns, with variable content of the gel in the buffer and with variable pore dimensions in differently cross-linked gels, as obtained by variation of the concentration of the bifunctional cross-linker in the initial solution of the monomer in the buffer [52–54]. The polymerization can be performed in situ

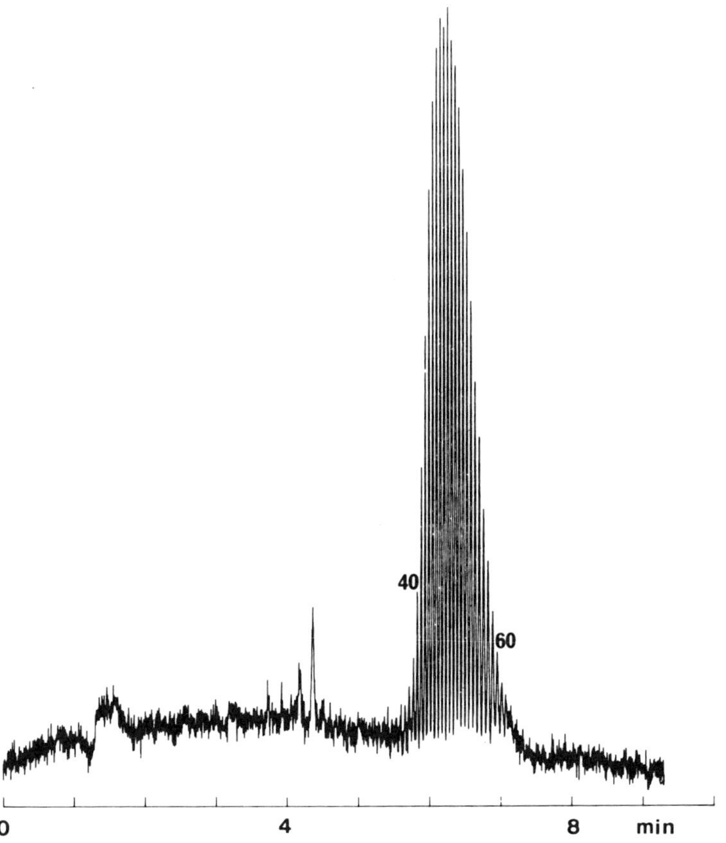

Figure 17 HPCE separation of polydeoxyadenylic acid mixture, $(dA)_{40-60}$. Capillary size, 270- × 0.075-mm id; running buffer, 0.1 M Tris/0.25 M borate/7 M urea. pH 8.3, and the gel contained 7.5% T and 3.3% C; applied field, 400 V cm^{-1}. (From Ref. 74)

(i.e., within the column) using ammonium peroxodisulfate as the radical starter in the presence of TEMED (*N*,*N*,*N*′,*N*′-tetramethylethylenediamine) as a catalyst. This polymerization and cross-linking process is accompanied by considerable volume contraction, which may be the reason for the formation of bubbles in the capillary that makes these capillaries unstable [23] (Fig. 19). This bubble formation has been observed in several laboratories; different measures for improvement of the procedures of gel formation have been proposed to avoid this problem [15,23,55]. This was tried by performing the polymerization under pressure or with addition of compounds such as polyethylene glycols to the initial solution of the acrylamide monomer. Doldnik and co-workers [56] proposed a method of smooth, isotachophoretic

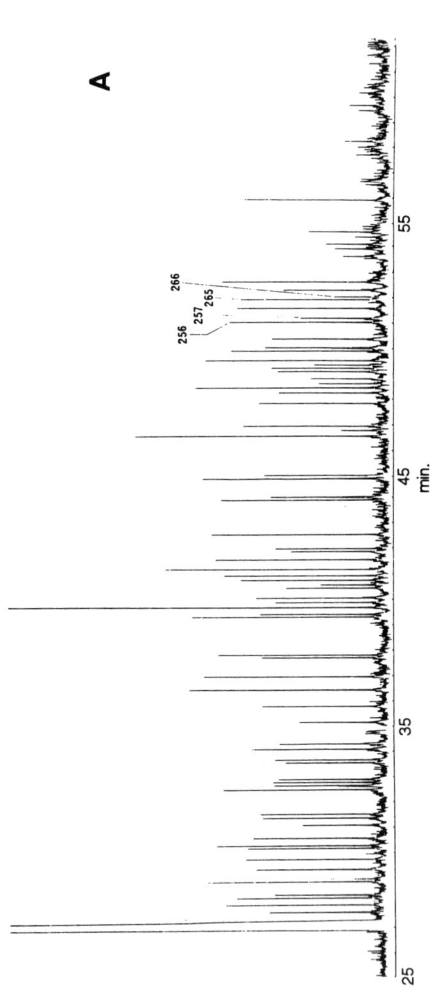
A
266
265
257
256
25
35
45
55
min.

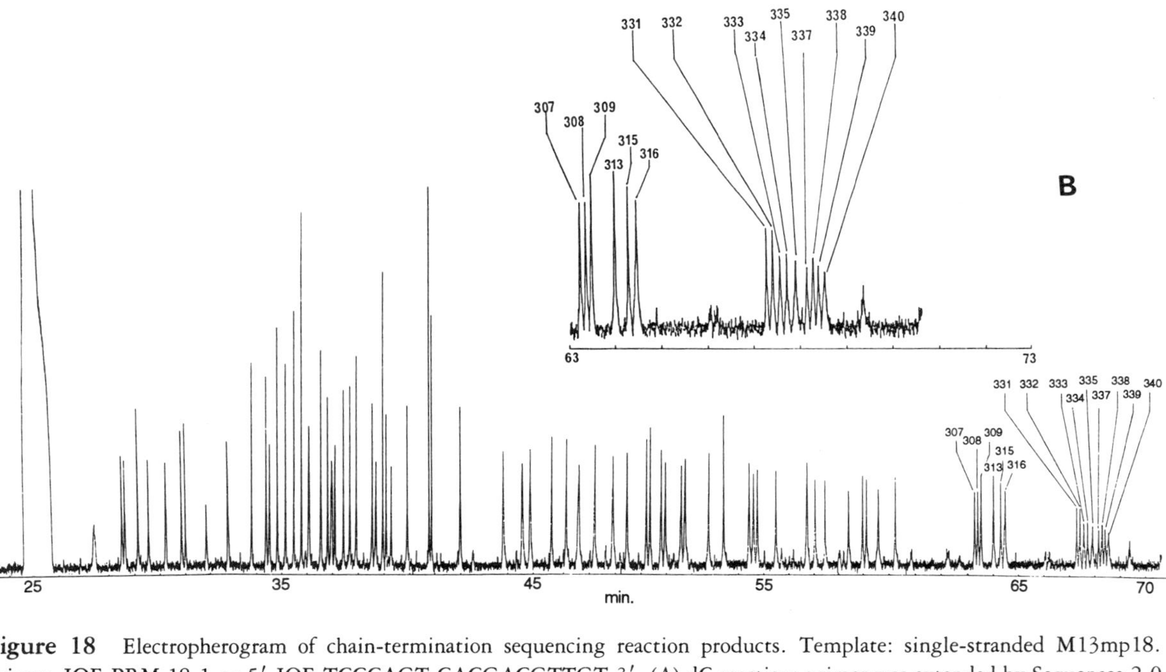

Figure 18 Electropherogram of chain-termination sequencing reaction products. Template: single-stranded M13mp18. Primer, JOE-PRM 18.1 = 5′-JOE-TCCCAGT-CACGACGTTGT-3′. (A) dC reaction: primer was extended by Sequenase 2.0 in the presence of ddCTP. (B) dT reaction: primer was extended by Sequenase 2.0 in the presence of ddTTP. Sequence reactions and electrophoretic conditions are described in the experimental section. Running conditions: (A) l = 500 mm, L = 650 mm, E = 350 V cm^{-1}, Buffer = 0.1 M Tris–borate, 2.5 mM EDTA, 7 M urea. pH = 8.0; (B) l = 750 mm, L = 920 mm, E = 310 V cm^{-1}, pH = 8.0 injection = 10 kV; 15 seconds. (From Ref. 52)

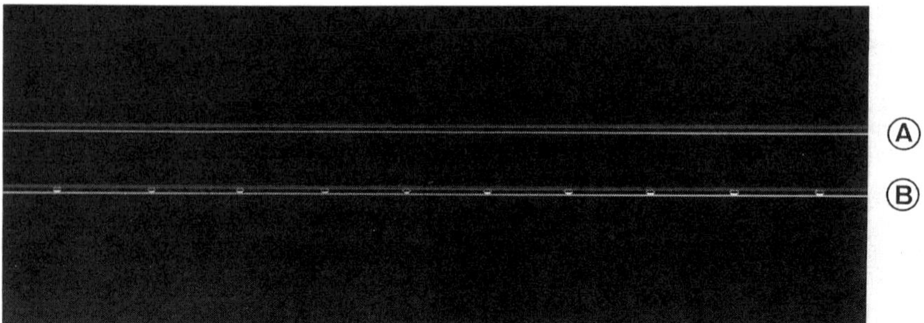

Figure 19 Polyacrylamide gel-filled capillaries: avoidance of bubble formation during production of capillaries filled with cross-linked gels. (A) Gels free of bubbles are obtained without surface pretreatment. (B) Equidistant bubbles are generated when surface is pretreated with 3-methacryloxypropyltrimethoxysilane.

polymerization. An electrical field is applied in which the polymerization initiator slowly migrates into the capillary, which is filled with the solution of the acrylamide monomer in buffer. In this manner bubble formation was not observed, according to these authors.

Even in recent publications, especially in that by Swerdlow et al. [54], on the present state of DNA sequencing using capillary gel electrophoresis, the following sentence is found: "An important issue is the performance of the gel filled capillaries under high electrical field strength; patents have been issued on the use of a bifunctional silane reagent to bind the acrylamide to the wall of the capillary to improve the gel stability" [15,55].

In the paper by Swerdlow et al. [54], which compares different DNA-sequencing methods, the following data are given, which characterize the future applicability of CGE in this field:

1. Sequencing rates up to 1000 bases per hour are obtained at electric field strength of 465 V cm^{-1}. This is roughly a 25 times higher sequencing rate than with conventional slab gel electrophoresis.
2. The sequencing rate for a single capillary is similar to that produced by a 24-lane slab gel.
3. At lower electric field strengths, capillary electrophoresis produces useful data for fragments larger than 550 nucleotides in length with two times better resolution than slab gel electrophoresis.
4. Minimizations of compressions (of subsequent peaks) when using "standard capillary sequencing gels" in special regions of the entire range of fragments

were achieved by slowing down the separation at modest electrical field strength and by addition of formamide.

The results obtained in my laboratory with PAG capillaries that were made by a different procedure [57] supply the same, if not better, resolution and peak symmetry in separations performed with the typical oligonucleotide or DNA test mixtures (Figs. 20 and 21).

The procedure for production of this type of polyacrylamide gel (PAG) capillaries chemical bonding of the separation gel to the walls does not suffer from the formation of bubbles at the polymerization step of acrylamide and bisacrylamide within the buffer. Gel-filled capillaries could be generated using this special surface pretreatment without any addition of chemicals (such as PEG or formamide) that are used to avoid the formation of bubbles.

The major characteristics of this procedure for making PAG capillaries (6% T, 5% C) are

1. The derivatization of the etched fused silica surface by the common bifunctional reagents, such as γ-methacryloxypropyltrimethoxysilane (MAPT), was not performed to effect a chemical bonding of the final gel to the capillary wall.
2. The derivatization was used to generate a thin coating of linear polyacrylamide, which reduces the electroosmotic pressure so that extrusion of the gel from the filled capillary does not occur.
3. The gels in the capillaries made according to this procedure are not fixed to the walls, they can be removed ("blowout"), for a refill with fresh gel but only if they are not too viscous.

The silanization reagent with a copolymerizable double bond in R is applied for the surface derivatization of the inner surface in the fused silica capillary and the chemical bonding of a thin polyacrylamide layer; both the reduction of the silanol concentration on the surface and the chemical bonding of the thin polymeric layer serve to suppress electroosmotic flow.

The stability of these capillaries was excellent also in other laboratories in series of more than 200 repetitive measurements at high electrical fields [58]. A very recent publication reports on investigations of the stability of PAG capillaries that had been made without any surface pretreatment in which the gel was not attached to the capillary walls [59]. The authors state that the chemically bonded capillaries (obtained according to the procedure of Karger and Cohen [15]), "although more stable, still suffer from problems with bubbles." Therefore, they investigated the disruption process by employing PAG capillaries in which the gel was not attached to the capillary wall and identified three sources or indications of gel instability:

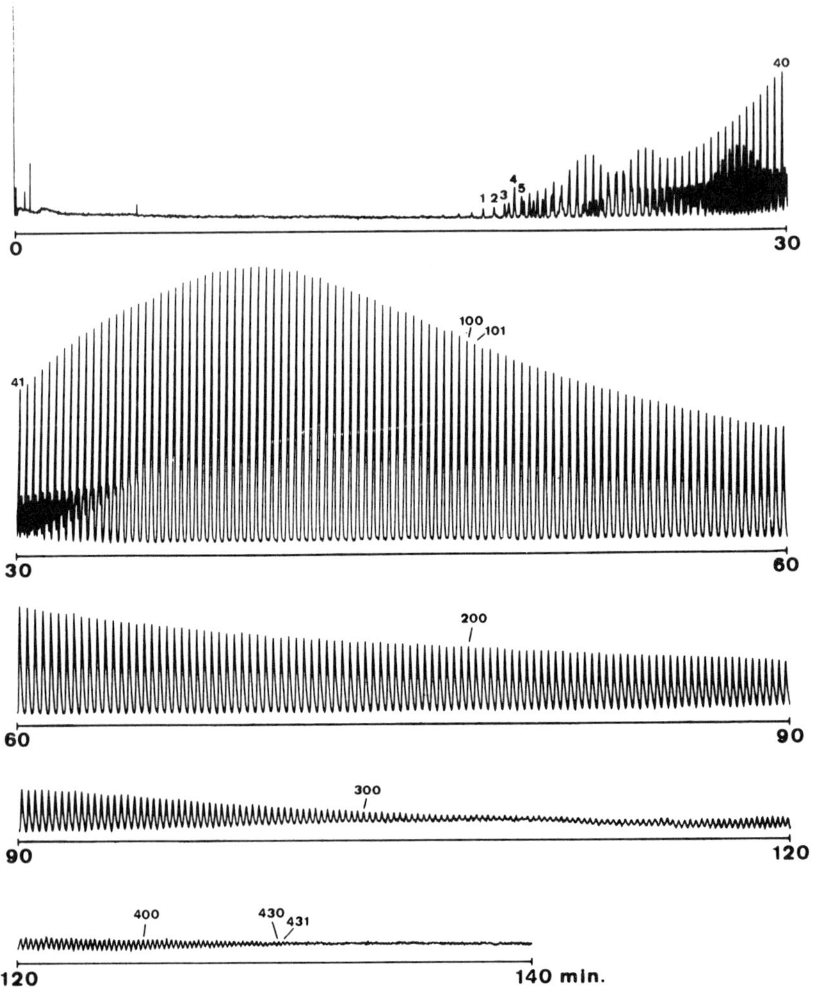

Figure 20 CGE separation of poly(uridine 5′-phosphate). Sample: poly(uridine 5′-phosphate); the numbers in the electropherogram indicate the number of nucleotide units. Capillary: 45-cm effective length, 100-μm id. Gel: polyacrylamide, 6% T, 5% C. Buffer: 0.1 M Tris, 0.25 M boric acid, 7 M urea. Separation conditions: 300 V cm^{-1}, 12 μA; injection: 5000 V, 2 seconds; detection: UV/260 nm.

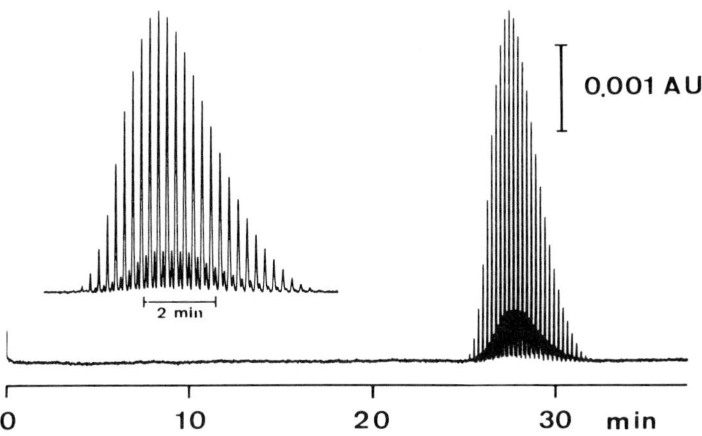

Figure 21 CGE separation of polydeoxyadenosine $Pd(A)_{40-60}$. Sample: oligonucleotides originating from Pharmacia. Capillary: 40-cm effective length, 100-μm id, surface pretreatment before filling with gel. Gel: polyacrylamide, 6% T, 5% C. Buffer: 0.1 M Tris, 0.25 M boric acid, 7 M urea. Separation conditions: 300 V cm^{-1}, 12 μA; injection: 5000 V, 2 seconds, detection: UV/260 nm.

1. Electroosmotic force (eliminated by formamide replacing urea in the buffer)
2. Decline of current because of ionic depletion at the gel–liquid interface (eliminated by adding denaturant and acrylamide monomer to the buffer in the reservoirs)
3. Sample-induced problems caused by the presence of template in DNA-sequencing samples (eliminated by improved enzymatic sample preparation)

Silanization with a X_3SiR reagent has been applied for the modification of the surface by adsorption of nonionic surfactants (Brij and Tween) on C_{18}-silanized FS surfaces in CZE of proteins [35]. The modifying surfactants are fixed by hydrophobic interaction, only. For the production of gel capillaries filled with low-viscosity solutions of weakly cross-linked or linear polyacrylamides, Heiger [60] executed a surface pretreatment by X_3Si-C_{18}-silanization and subsequent adsorption of alkylmaltose surfactants, before the solution of the polymers in the buffer was introduced into the capillary. This type of gel-filled capillaries was to be used in the separation of larger DNA (restriction) fragments [60] for a separation of a typical test mixture with restriction fragments ΦX174-RF DNA *Hae* III digest (Pharmacia LKB Biotechnology) (Fig. 22). The molecular sieving selectivity is dependent on temperature [61–63] (Fig. 23). It can be seen that the migration sequences of peaks 6 a, b, and c, as well as of peaks 8, 9, and 10 are changed in the narrow temperature range between 20° and 29°C. Similar observations have also been made [61,62]. For the

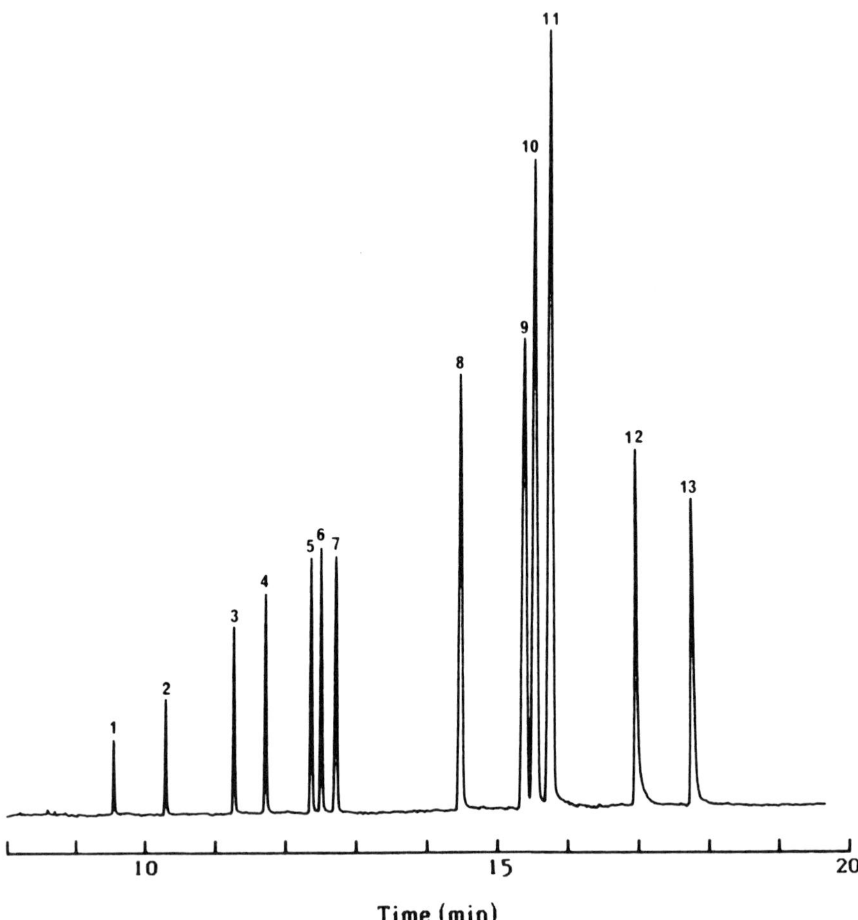

Figure 22 Separation of *Hae* III digest of ϕX174 DNA. *Eco*R I digest of pBR322 and *Eco*R I digest of M13mp18 on a 3% T, 0.5% C polyacrylamide capillary column. The 11 fragments of the ϕX174 digest have been identified as the first 11 peaks in the electropherogram, followed by pBR322 and M13mp18. Identification was made by spiking each peak with slab gel isolated species. Peaks: 1, 72; 2, 118; 3, 194; 4, 234; 5, 271; 6, 281; 7, 310; 8, 603; 9, 872; 10, 1078; 11, 1353; 12, 4363; and 13, 7253 base pairs. Conditions: effective length, 30 cm, total length, 40 cm, applied field, 250 V cm^{-1}, current generated, 12.5 μA, and sample was injected electrophoretically at 10 kV for 0.5 seconds. Buffer: 100 mM Tris–borate (pH 8.3), 2 mM EDTA. (From Ref. 40)

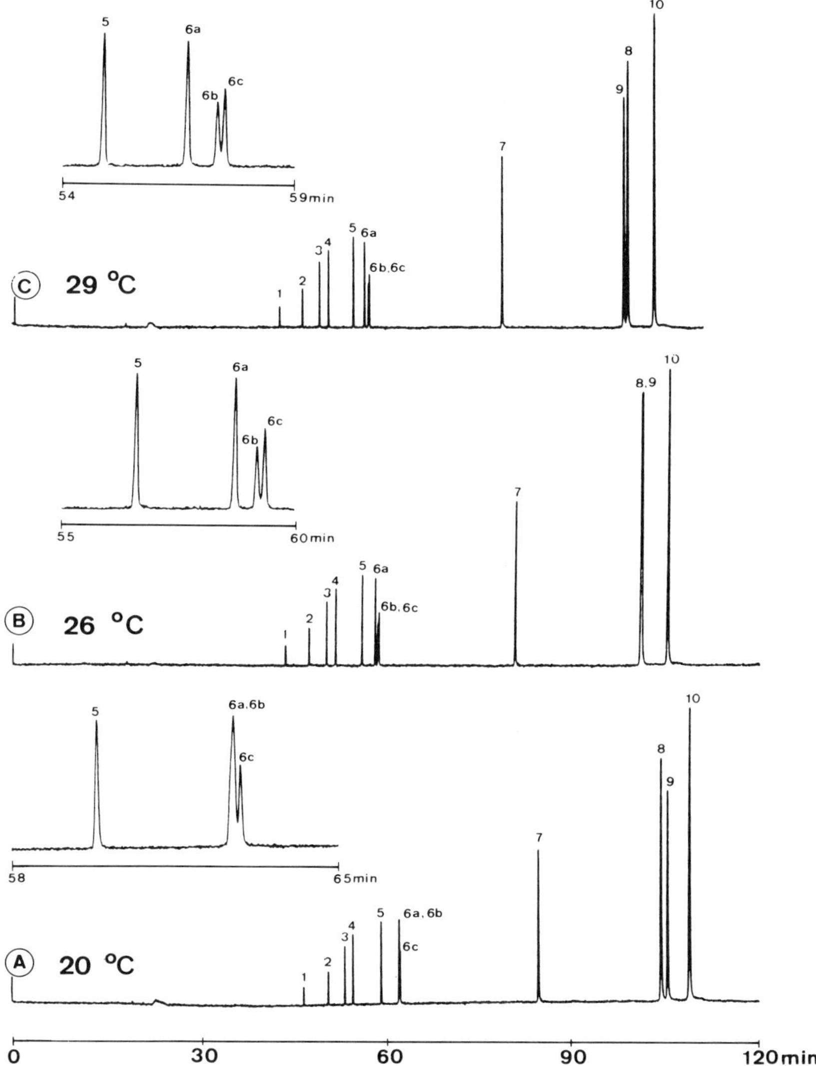

Figure 23 Influence of temperature on the CGE separation of DNA restriction fragments. Sample: ΦX-174-RF DNA *Hae* III digest. Capillary: 42-cm effective length, 100-μm id. Gel: linear polyacrylamide, 8% T. Buffer: 0.1 M Tris, 0.25 M boric acid. Temperature: (A) 20°C, (B) 26°C, (C) 29°C. Separation conditions: 160 V cm^{-1}; injection: 5000 V, 3 seconds; detection: UV/260 nm.

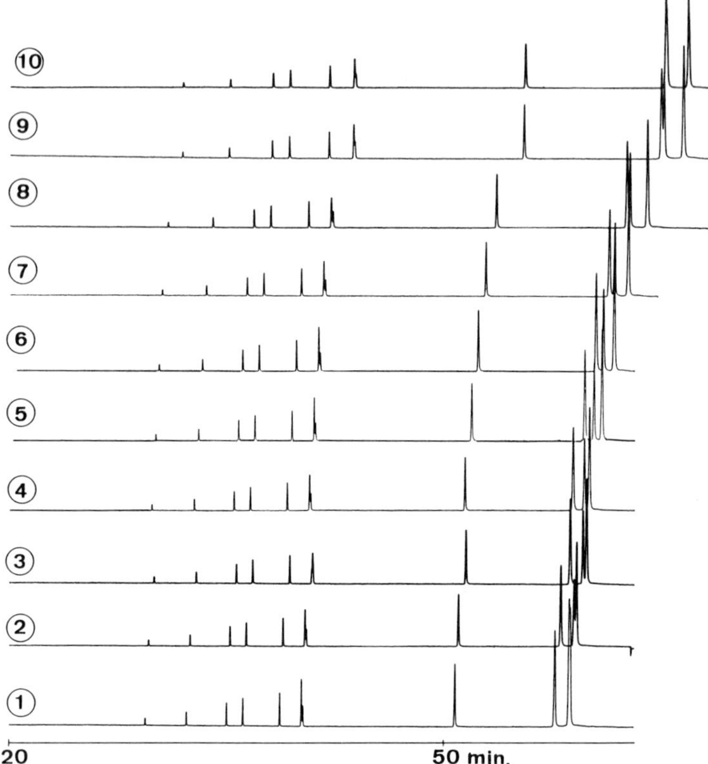

Figure 24 Migration times in linear polyacrylamide gel-filled capillaries during extended series of separation. Sample: ΦX-174-RF DNA *Hae* III digest. Capillary: 39.5-cm effective length; 100-μm id. Gel: linear polyacrylamide, 6% T. Buffer: 0.1 M Tris, 0.25 M boric acid. Temperature: 20°C. Separation conditions: 180 V cm^{-1}, 8 μA; injection: 5000 V, 3 seconds; detection: UV/260 nm.

separations of Fig. 24, Lux et al. [63] applied a capillary that had been obtained by polyacrylamide coating after silanization with MAPT. In such capillaries, the polyacrylamide chains are fixed to the surface. It was observed that the absolute and also the relative migration times of the constituents of the restriction fragment test mixture during a series of about ten separations (i.e., after use for about 15 hours) are increased or changed, respectively, probably owing to changes of the surface coating and the corresponding EOF. Therefore, it seems to be of advantage to apply fillings of gels or polymer network solutions for such separations that can be replaced by fresh fillings according to the "blowin–blowout" procedure. This has been demonstrated [61,64] using solutions of externally synthesized, weakly cross-linked polyacrylamides.

Separation of DNA Restriction Fragments in Buffers Containing Polymer Networks

The larger DNA restriction fragments with 500 up to 15,000 base units can no longer be separated in polyacrylamide gels of even minor cross-linking because the rigid pore structure in such gels is inadequate, considering the size and the geometry of the large DNA molecules, which is also influenced by the electrical field strength [65]. Other mechanisms of migration ("reptation") become effective with these large molecules [66,67]. In non–cross-linked (i.e., linear) polyacrylamide networks, which can be considered as "physical" gels or "entangled" polymer solutions, DNA restriction fragments can be separated with good efficiency and molecular sieving selectivity [40,43,68].

The absolute and relative migration times in buffers containing polymer networks depend on the

1. Chemical structure and the entangling of the polymer
2. Concentration of the polymer in the buffer
3. Viscosity of the polymer solution
4. Molecular size and molecular size distribution of the polymer
5. Type of buffer and type of special modifier in the buffer
6. Type of surface coating

For the DNA fragment separations, polymers with a chemical structure that is suited for moderately weak intermolecular interaction of the dipol–dipol type have successfully been applied. Water-soluble polymers, with a high amount of hydroxyl groups, which give rise to intra- and intermolecular hydrogen bonding in the polymer solution, proved to be well suited.

Excellent DNA fragment separations were achieved with the following polymers:

Polyacryamides in concentrations of approximately 3% T, 0.5% C (see Fig. 22)
(Hydroxyalkyl)cellulose in concentrations $< 1\%$ (Fig. 25)
Different types of agarose with special additives (Fig. 26)
Polyvinylalcohols (M_r 77,000, concentrations $< 4\%$) (Fig. 27)

All solutions of these polymers in buffers are of low viscosity and can easily be introduced and removed from the capillaries (blowin–blowout). The modification of the capillary walls before the introduction of such polymer solutions has been done by the methods that were first used for cross-linked and non–cross-linked polyacrylamides (i.e., by sequences of pretreatment steps comprising etching, silane derivatization, polymer coating, surfactant adsorption) [60,69]. It was found, however, that the polymers with the hydroxylic structures are suitable for modifying the surface by themselves, even without execution of any preceding surface pretreatment [26,69,70,71].

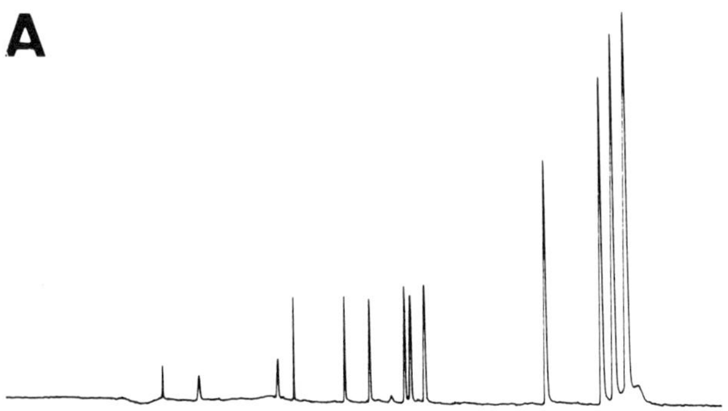

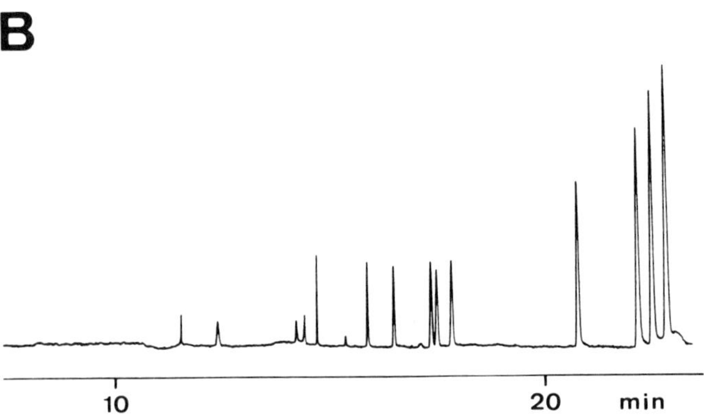

Figure 25 Separation of DNA restriction fragments with a solution of an entangled polymer. Sample: ΦX-174-RF DNA *Hae* III digest. Capillary: (A) 49.1-cm, (B) 47.0-cm effective length; 75-μm id. Coating: (A) bare fused silica capillary rinsed with a solution of polyvinylalcohol before filling with the hydroxyethyl-cellulose solution; (B) *N*,*N*-dimethylacrylamide copolymerized on a vinylized surface before filling with the hydroxyethyl-cellulose solution. Conditions: (A) 226 V cm^{-1}, 28 μA; (B) 220 V cm^{-1}, 28 μA; 20°C. Buffer: 20 mM Na-phosphate, pH 7; 1% hydroxyethyl-cellulose (middle viscosity). Injection: 10 kV, 4 seconds. Detection: 260 nm.

Criteria of such CE systems for the separation of restriction fragments are

1. Repetitive generation of a buffer and capillary fillings with reproducible absolute and relative migration times
2. Defined molecular mass and molecular mass distribution of the polymers used as a sieving matrix
3. Stability of the systems at higher pH

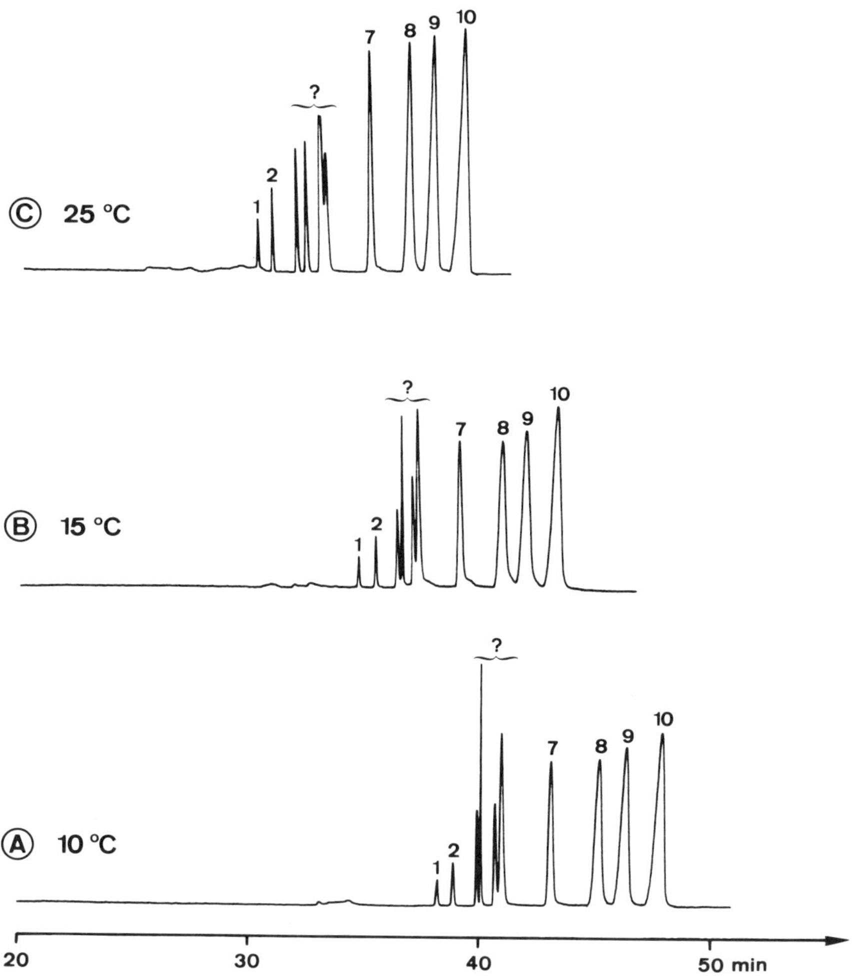

Figure 26 DNA restriction fragments separated on agarose gel at different temperatures. Sample: ΦX-174-RF DNA *Hae* III digest. Capillary: 72-cm effective length, 75-μm id (MOT-AG-39). Gel: agarose (Sigma), 2% in buffer. Buffer: 0.01 M Na_2HPO_4, 0.01 M NaH_2PO_4. Temperature: (A) 10°C, (B) 15°C, (C) 25°C. Separation conditions: 100 V cm^{-1}; injection: 20,000 V, 5 seconds; detection: UV/254 nm.

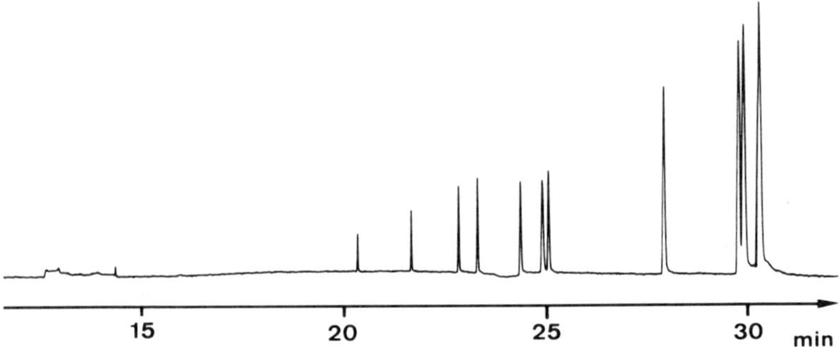

Figure 27 CE separation of DNA restriction fragments with networks of polyvinylalcohol in buffer. Sample: ΦX-174-RF DNA *Hae* III digest. Capillary: 48.2-cm effective length, 75-μm id. Coating: bare fused silica capillary. Buffer: 20 mM Na-phosphate, pH 7, 4% polyvinylalcohol (Aldrich, average MW 77,000–79,000; 100% hydrolyzed). Separation conditions: 223 V cm^{-1}, 25 μA, 20°C; injection: 3 kV, 4 seconds; detection: 260 nm.

MODIFICATION OF THE ANALYTE MOLECULES BY ADDITIVES TO THE BUFFER

The differences in the electromobility of analyte molecules depend on their chemical structure: size, polarity, charge, charge density, conformation, geometric structure and also on the environment of the analyte molecules in the buffer (i.e., on the composition of the buffers). Modifiers contained in the buffer may, besides the modification of the capillary surfaces, also undergo a special intermolecular interaction, ion-pair formation, or even complexation, with the analyte molecule, which gives rise to changes in the electromobility and, therefore, to differential migration (*selectivities*) for molecules of different molecular properties.

Several examples for such changes of CE selectivity can be found in the literature:

SDS–PAGE separations of proteins (CZE)
Large DNA fragments: improvement of resolution by the intercalation of ethidium bromide (CGE)
Cyclodextrins and other chiral selectors as additives in MECC: chiral recognition
Cyclodextrins as additives in CGE (in PAGE): chiral recognition
Urea or formamide as buffer components for CGE with polyacrylamides to avoid the alignment of single-stranded DNA fragments (CGE)
Ligand exchange of chiral analytes with copper–aspartame complexes (CZE)

The aforementioned changes in the composition of the migrating analyte molecule occur *in-capillary* (i.e., after introduction of the sample into the separation system).

Another important possibility for changing the migration behavior is derivatizations that are performed off-line or externally (precolumn). These changes of the analyte molecule within or outside the CE system are not subject of this review.

CONCLUSION

The analytical aims for optimization of the chemistry involved in the performance of the different CE systems can be summarized as follows:

1. *Avoidance or suppression of adsorption* of the different types of analyte molecules on the large surfaces of the fused silica capillaries. Weaker interactions cause decrease of efficiency and deterioration of peak symmetry; strong interaction prevents any elution of the analyte.
2. Adjustment or *control* (i.e., decrease, increase, or reversal) of *electroosmotic flow* (EOF).
3. *Generation of gels* within the capillary for *CGE* separations, preferably for large molecules.

In all three areas of investigations on the chemistry of CE, surface modification is the most important topic.

Silane Derivatization

The SiOH groups are distributed on the fused silica surfaces in variable concentrations, depending on the origin of the tubing. These silanols influence the charging of the surface and, thereby, the EOF, but also the adsorption of analyte molecules, especially those with basic properties. The silanols can also act as anchor groups for the derivatization of the surfaces. Small and oligomeric or polymeric molecules can be attached to the surface and form layers of special chemical properties relative to EOF or adsorption of especially large biomolecules. Another method of modification is the adsorption of polar small or large molecules and, especially, of surfactants. The type of molecules that can be adsorbed depends on the status of the surface, which can also be previously derivatized, for example, by silanization. By derivatization, the concentration of silanols on the surfaces is decreased; the residual silanols are shielded against intermolecular interaction by the coating layers obtained.

Difficulties with the reproducible execution of silanizations and of the subsequent in situ reactions are related to the following topics:

1. The concentrations of SiOH on the surfaces are not known and depend on the previously applied etching, rinsing, and drying procedures.
2. The silane derivatization reaction depends on the type of silanization reagent. Especially with trifunctional reagents, problems arise with the presence of water, which causes polymerization.

3. Polar reaction products of the silanization, such as HX especially HCl, may be adsorbed on the surface and may catalyze the polymerization of the multifunctional reagents.
4. Polar (e.g., hydrophilic) coatings are more difficult to generate; secondary reactions after silanization must be applied.

Adsorption of Modifiers

The adsorption of molecules that contain highly polar or charged functional groups seems to be a method of surface modification that is, in comparison with the sophisticated silanization procedures, easier to perform reproducibly either by simple rinsing or by addition of the "modifier" to the buffer. Adsorption of surfactants on lipophilic coatings seems to be useful in CZE (proteins) and also in MECC (variation of EOF).

Surface modification by adsorption can be restored simply by rinsing with fresh buffer or a solution of the modifier after a single or a few separations, by which time the properties of the coating are impaired. Similar methods to reestablish the separation systems are employed in CGE, when the not-too-viscous gel is replaced by fresh after a few separations to achieve reproducible separation conditions.

Stability of Modifying Coating Layers

The stability of any kind of modification layers on FS surfaces and also of gels, which may or may not be covalently bonded to the walls, is an important feature of analytical CE, especially when the blowin–blowout procedure cannot be applied. Then rinsing at extreme, especially at high pH may become necessary. Very few coatings are suited to withstand such procedures.

Modification of the Electrophoretic Selectivity by Additives to the Buffer

The absolute and relative electromobilities of charged molecules can be influenced by pH changes. The addition of surfactants may lead to the formation of micelles by which uncharged analyte molecules can also migrate in the electric field. Enantioselectivity is achievable by addition of chiral modifiers (e.g., cyclodextrins) or surfactants. Chiral additives can also be contained in polyacrylamide gels [72]. Any kind of complexation of the analyte molecules can be used to change the electromobilities.

In capillary gel electrophoresis of double-stranded DNA, the addition of ethidium bromide to the buffer improves the separation efficiency considerably [68].

Generation of Different Types of Gels or Buffers with Polymer Networks ("Entangled Polymers")

Cross-linked PAGs are generated in situ (i.e., within the capillary) after certain surface pretreatment procedures. The degree of cross-linking is determined by the concentration of cross-linker and, probably, by the applied procedure and conditions of the radical polymerization. It is an important criterion of analytical application of such cross-linked gels in capillaries, in comparison with slab gels, that the covalent bonding of the gels is achieved by copolymerization between the acrylamides and double bonds that have been attached to the surface by the preceding derivatization.

Agarose gels can be obtained by temperature-controlled gelation of the agarose buffer solution [42].

A promising area of future CGE may be the formation of gels containing selectors that also undergo special intermolecular (e.g., chiral) interactions with small molecules.

OUTLOOK

The major aim of application of CE methods is to make use of the extraordinary separation efficiencies of up to several million plates and the special electrophoretic selectivities that can also be achieved for large, but charged, molecules. Despite the high grade of miniaturization of the modern CE methods and the related detection problems, CE will become a method of high analytical usefulness; further investigations on the chemistry involved seem to be useful and necessary, however.

REFERENCES

1. A. Tiselius, *Trans. Faraday Soc.*, *33*:524 (1937).
2. F. E. P. Mikkers, F. M. Everaerts, and T. P. E. M. Verheggen, *J. Chromatogr.*, *169*:11 (1979).
3. J. W. Jorgensen, and K. D. Lukacs, *High Resolut. Chromatogr.*, *4*:230 (1981).
4. J. W. Jorgensen, and K. D. Lukacs, *Science*, *222*:266 (1983).
5. A. S. Cohen, and B. L. Karger, *J. Chromatogr.*, *397*:409 (1987).
6. S. Terabe, K. Otsuka, K. Ichikawa, A. Tsuchiya, and T. Ando, *Anal. Chem*, *56*:111 (1984).
7. M. V. Novotny, and D. Ishi (eds.), *Microcolumn Separations*, Elsevier Science Publishers, Amsterdam (*J. Chromatogr. Lib.*, *30*) (1985).
8. C. S. Lee, W. C. Blanchard, and C.-T. Wu, *Anal. Chem.*, *62*:1550 (1990).
9. C. S. Lee, D. McManigill, C. Wu, and B. Patel, *Anal. Chem.*, *63*:1519 (1991).
10. J. A. Lux, C. A. Keely, D. McManigill, and J. E. Young, Poster presented at the Fourth Int. Symp. High Performance Capillary Electrophoresis, Amsterdam, abstract book, p. 103 (1992).
11. S. Terabe, K. Otsuka, and T. Ando, *Anal. Chem.*, *57*:834 (1985).
12. S. Terabe, *Trends Anal. Chem.*, *8*:129 (1989).

13. T. Tsuda, *High Resolut. Chromatogr.*, *10*:622 (1987).
14. W. R. Jones, and P. Jandik, *J. Chromatogr.*, *546*:445 (1991).
15. B. L. Karger, and A. S. Cohen, Northeastern University, US Patents 485,706 and 485,707; Eur. Patent EP 324.539 A2 (1986, 1989).
16. K. A. Cobb, V. Dolnik, and M. Novotny, *Anal. Chem.*, *62*:2478 (1990).
17. J. E. Sandoval, and J. J. Pesek, *Anal. Chem.*, *61*:2067 (1989).
18. J. E. Sandoval, and J. J. Pesek, *Anal. Chem.*, *63*:2641 (1991).
19. L. C. Sander, and S. Wise, *Anal. Chem.*, *56*:504 (1984).
20. G. Schomburg, A. Deege, J. Köhler, and U. Bien-Vogelsang, *J. Chromatogr.*, *282*:27 (1983).
21. S. Hjertén, *J. Chromatogr.*, *270*:1 (1983).
22. S. Hjertén, *J. Chromatogr.*, *347*:191 (1985).
23. H.-F. Yin, J. A. Lux, and G. Schomburg, *High Resolut. Chromatogr.*, *13*:624 (1990).
24. H.-J. Bode, *Anal. Biochem.*, *92*:99 (1979).
25. P. D. Grossman, and D. S. Soane, *Biopolymers*, *31*:1221 (1991).
26. H. E. Schwartz, K. Ulfelder, F. J. Sunzeri, M. P. Busch, and R. G. Brownlee, *J. Chromatogr.*, *559*:267 (1991).
27. R. M. McCormick, *Anal. Chem.*, *60*:2322 (1988).
28. G. M. Bruin, J. P. Chang, R. H. Kuhlmann, K. Zegers, J. C. Kraak, and H. Poppe, *J. Chromatogr.*, *471*:429 (1989).
29. S. A. Swedberg, *Anal. Biochem.*, *185*:51 (1990).
30. G. M. Bruin, R, Huisden, J. C. Kraak, and H. Poppe, *J. Chromatogr.*, *480*:339 (1989).
31. H. H. Lauer, and D. McManigill, *Anal. Chem.*, *58*:166 (1986).
32. Å. Emmer, M. Jansson, and J. Roeraade, *J. Chromatogr.*, *547*:544 (1991).
33. J. A Bullock, and L.-C. Yuan, *J. Microcolumn Separations*, *3*:241 (1991).
34. M. Gilges, H. Husmann, M. H. Kleemiss, St. R. Motsch, and G. Schomburg, *High Resolut. Chromatogr.*, *15*:452 (1992).
35. J. K. Towns, and R. E. Regnier, *Anal. Chem.*, *63*:1126 (1991).
36. J. Kohr, and H. Engelhardt, *J. Microcolumn Separations, 3*:491 (1991).
37. D. Bentrop, J. Kohr, and H. Engelhardt, *Chromatographia*, *32*:171 (1991).
38. S. Terabe, H. Utsumi, K. Otsuka, T. Ando, K. Inomata, S. Kuze, and Y. Hanaoka, *High Resolut. Chromatogr.*, *9*:667 (1986).
39. J. A. Lux, H.-F. Yin, and G. Schomburg, *High Resolut. Chromatogr.*, *13*:145 (1990).
40. D. N. Heiger, A. S. Cohen, and B. L. Karger, *J. Chromatogr.*, *516*:33 (1990).
41. P. Bocek, and A. Chrambach, *Electrophoresis*, *12*:1059 (1991).
42. St. R. Motsch, M. H. Kleemiss, and G. Schomburg, *High Resolut. Chromatogr.*, *14*:629 (1991).
43. J. Sudor, F. Foret, and P. Bocek, *Electrophoresis*, *12*:1056 (1991).
44. Y. Baba, T. Matsuura, K. Wakamoto, and M. Tsuhako, *J. Chromatogr.*, *558*:273 (1991).
45. Y. Kato, Y. Yamasaki, A. Onaka, T. Kitamura, and T. Hashimoto, *J. Chromatogr.*, *478*:264 (1989).
46. M. A. Strege, and A. L. Lagu, *J. Chromatogr.*, *555*:109 (1991).
47. F. Sanger, and A. R. Coulson, *FEBS Lett.*, *87*:107 (1978).

48. L. M. Smith, R. L. Brumley, E. Buxton, H. Drossman, A. J. Kostichka, A. J. Luckey, and D. A. Mead, In *Abstracts, Genome Mapping and Sequencing*, Cold Spring Harbor Laboratory, Cold Spring Harbor, New York, p. 250 (1991).
49. J. Stegemann, C. Schwager, H. Erfle, N. Hewitt, H. Voss, J. Zimmermann, and W. Ansorge, *Nucleic Acids Res.*, *19*:675 (1991).
50. H. Garoff, and W. Ansorge, *Anal. Biochem.*, *115*:450 (1981).
51. A. Guttman, A. S. Cohen, D. N. Heiger, and B. L. Karger, *Anal. Chem.*, *62*:137 (1990).
52. A. S. Cohen, D. R. Najarian, and B. L. Karger, *J. Chromatogr.*, *516*:49 (1990).
53. J. A. Luckey, H. Drossman, A. J. Kostichka, D. A. Mead, J. D. Cunha, T. B. Norris, and L. M. Smith, *Nucleic Acids Res.*, *18*:4417 (1990).
54. H. Swerdlow, J. Z. Zhang, D. Y. Chen, H. R. Harke, R. Grey, S. Wu, N. J. Dovichi, and C. Fuller, *Anal. Chem.*, *63*:2835 (1991).
55. P. F. Bente, and J. Myerson, US Patent 4,810,456 (1989).
56. V. Doldnik, K. A. Cobb, and M. Novotny, *J. Microcolumn Separations*, *3*:155 (1991).
57. G. Schomburg, J. A. Lux, and H.-F. Yin, US Patent 626,603 (1991).
58. E. Uhlmann, F. M. Everaerts, and H.-F. Yin, Personal communication (1990).
59. H. Swerdlow, K. E. Dew-Jager, K. Brady, R. Grey, N. J. Dovichi, and R. Gesteland, *Electrophoresis* (in press; 1992).
60. D. N. Heiger, Thesis, Northeastern University, Boston (1991).
61. B. L. Karger, A. S. Cohen, D. N. Heiger, and K. Ganzler, Lecture and poster presented at the Third Int. Symp. High Performance Capillary Electrophoresis, San Diego (1991).
62. A. Guttman, and N. Cooke, *J. Chromatogr.*, *559*:285 (1991).
63. J. A. Lux, H.-F. Yin, and G. Schomburg, Poster presented at the Third Int. Symp. High Performance Capillary Electrophoresis, San Diego (1991).
64. A. Guttman, J. I. Ohms, and N. Cooke, Lecture and poster presented at the Third Int. Symp. High Performance Capillary Electrophoresis, San Diego (1991).
65. P. D. Grossman, and D. S. Soane, *Anal. Chem.*, *62*:1592 (1990).
66. C. R. Cantor, and C. L. Smith, *Annu. Rev. Biophys. Chem.*, *17*:287 (1988).
67. G. W. Slater, and J. Noolandi, *Biopolymers*, *25*:431 (1986).
68. A. Guttman, and N. Cooke, *Anal. Chem.*, *63*:2038 (1991).
69. M. H. Kleemiss, M. Gilges, and G. Schomburg, Unpublished results (1992).
70. S. Hjertén, L. Valtechava, K. Ehlenbring, and D. Eaker, *J. Liq. Chromatogr.*, *12*:2471 (1989).
71. M.-D. Zhu, D. L. Hansen, S. Burd, and F. Gannon, *J. Chromatogr.*, *480*:311 (1989).
72. A. Guttman, A. Paulus, A. S. Cohen, N. Grinberg, and B. L. Karger, *J. Chromatogr.*, *448*:41 (1988).
73. A. S. Cohen, A. Paulus, and B. L. Karger, *Chromatographia*, *24*:15 (1987).
74. A. S. Cohen, D. R. Najarian, A. Paulus, A. Guttman, J. A. Smith, and B. L. Karger, *Proc. Natl. Acad. Sci. USA*, *85*:9660 (1988).
75. Abstract M311, Fifth Int. Symposium on High Performance Capillary Electrophoresis, Orlando, Florida.

10

Capillary Electrophoresis with Coated Capillaries

Jörg Kohr and Heinz Engelhardt

University of Saarlandes
Saarbrücken, Germany

Capillary electrophoresis (CE) is one of the most exciting separation techniques in analytical chemistry developed in the last decade. With the appearance of commercially available instruments, CE made a step out of the scientist's laboratories in the direction of industrial routine analysis. In comparison with the development of gas chromatography (GC) and high-performance liquid chromatography (HPLC), CE is still in a developmental stage: both, GC and HPLC were taken as acknowledged analytical techniques when stable and reproducible stationary phases had been developed. In CE, the motivation for coating a capillary is quite different. There is no need for a stationary phase in capillary electrophoresis—it is even not desirable (except for electrochromatography). Especially for the separation of biomolecules (proteins, peptides, DNA) there are still many serious problems to solve, such as loss in efficiency, bad reproducibility of migration time, and electroosmotic flow because of sample–wall interactions. Several approaches in this field have been made by using the right pH, zwitterionic additives, and high ionic strength buffers. However, the chemical modification of the capillary wall is one of the most promising ways to solve these problems; that is, to reduce sample adsorption on the silica surface and to guarantee a controlled and reproducible electroosmosis.

THE CAPILLARY WALL

Electroosmosis

Electroosmosis is one of the important factors influencing the quality and reproducibility of CE separations. The total mobility of a sample molecule is a result of the vectorial addition of the electroosmotic mobility and the electrophoretic mobility.

$$\mu_{tot} = \mu_{electrophoretic} + \mu_{electroosmotic} \tag{1}$$

As can easily be seen from Eq. 1 and as is well-known from chromatography, a constant flow during analysis, from run to run and day to day is necessary to obtain reliable analytical results.

However, the velocity of the electroosmotic flow is strongly dependent on many parameters, such as pH, ionic strength, the buffer composition, and the chemical nature of the capillary wall. In fused silica capillaries, linear velocities of more than 4 mm s^{-1} can be achieved, which is about the same as in HPLC. Figure 1 shows the dependence of the electroosmotic mobility on the pH of the buffer. The sigmoidal curve can be explained by additional dissociation of silanol groups at higher pH values, resulting in higher electroosmotic mobilities. Also in this figure, the error bars indicate the reproducibility of electroosmosis at a certain pH value. Especially in the region of the turning point of the curve, error levels of more than 10% relative standard deviation (RSD) may occur. Some authors report a hysteresis of the electroosmotic flow when cycling from alkaline to acidic buffers and back. One reason for this behavior might be the slow equilibration of the silica surface with the buffer medium. Additionally, adsorption of the test solutes on the capillary wall also changes the zeta (ζ) potential and, consequently, varies the electroosmotic flow (see later).

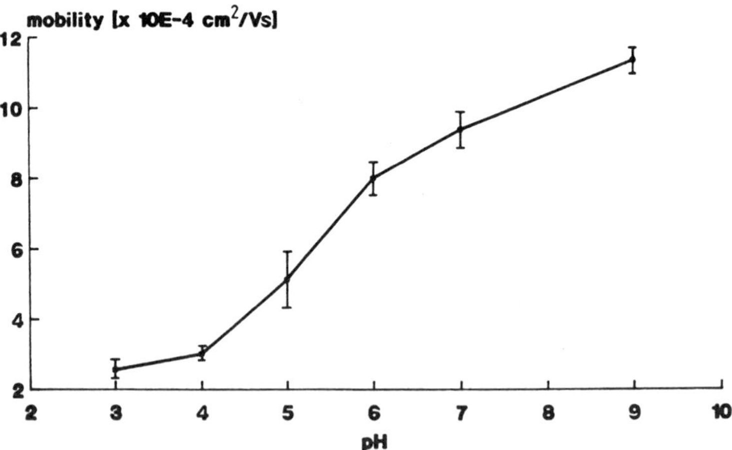

Figure 1 Dependence of the electroosmotic flow on pH. Capillary, untreated fused silica, 75-μm i.d.; buffer, 10 mM sodium phosphate; neutral marker, benzyl alcohol; voltage, 25 kV. (From Ref. 20)

Chemical modification of the inner capillary wall can cover the silanol groups and, accordingly, the electroosmotic flow will change. Sorption and desorption steps of buffer and sample molecules are also influenced by the chemical nature of the capillary surface and thus by the nature of the bonded functionalities. By chemical modification of the capillary wall, therefore, it should be possible to control the magnitude as well as the direction of, or even to eliminate, the electroosmotic flow (as an alternative to changing the pH of the buffer). Moreover, the use of a coating may help enhance the reproducibility of the separation in terms of a more constant flow rate and a controlled and reversible surface interaction with solute molecules.

Interactions Between Sample and Fused Silica Wall

In chromatography, the main mechanism in the separation process is the interaction between the sample and the stationary phase. In CE, however, no stationary phase exists, nor is it usually desirable. Different test solutes are separated in open tubular capillaries according to their different electrophoretic mobilities (i.e., different migration times in an electrical field).

The separation efficiency in terms of the total number of theoretical plates, N is given by Eq. 2.

$$N = \frac{\mu V}{2\,D} \tag{2}$$

where D is the solute's diffusion coefficient and V the applied voltage. For molecules with a higher relative molecular mass, Eq. 2 predicts efficiencies of several million theoretical plates. Therefore, proteins should be ideal samples fulfilling all the criteria just mentioned. However, as already reported by Jorgenson and Lukacs [1] at the beginning of the 1980s, the efficiencies for proteins observed in CE are very poor: biomolecules (especially proteins) tend to undergo serious interactions with glass and fused silica surfaces, from reversible binding to irreversible adsorption.

For reversible interactions, another model to describe efficiency in CE was proposed by Swedberg [2]. This equation resembles the well-known Golay equation in capillary GC. The first term in GC can be neglected because the injected volume is small, compared with the peak volume at the end of the separation and, therefore, it does not contribute significantly to peak dispersion.

$$H = \frac{l^2}{12\,L} + \frac{2\,D}{\mu} + \frac{k'\,\mu}{(1 + k')^2}\left[\frac{r^2 k'}{4\,D} + \frac{2}{k_d}\right] \tag{3}$$

where l = injected plug length; L = length of column to detection; D = diffusion coefficient of solute; k' = capacity factor; μ = linear velocity of the electroosmotic flow; r = internal radius of the capillary; and k_d = first-order dissociation constant.

From Eq. 3, it can be seen that these reversible interactions must be fairly small ($k' < 0.07$) for efficiencies to be fairly high (> 300,000 theoretical plates).

In irreversible adsorption, the whole sample or parts of the sample sticks to the capillary wall. A nonuniform adsorption of solutes to the charged fused silica wall may lead to a nonuniform charge distribution on it, resulting in locally different electroendosmotic migration velocities, causing asymmetric zones (peaks).

The adhesive forces between the sample and the capillary wall probably are electrostatic, resulting from the negatively charged silanol groups and positively charged functionalities of the samples. Additional nonspecific interactions, such as hydrogene bonding or van der Waals bonding may also occur. By modifying the wall chemically with a neutral coating, ion-exchange interactions with dissociated silanol groups are reduced. Moreover, steric hindrance can also prevent the solvents from being adsorbed. In this way, any interactions, other than electrostatic, are suppressed.

Types of Coating for Fused Silica Capillaries in Capillary Electrophoresis

In capillary electrophoresis, elements from chromatography as well as elements from electrophoresis can be found. In both methods, much experience in surface modification is available. Capillary electrophoresis takes advantage of this experience by transfer of this knowledge to the coating of fused silica capillaries. Two different types of chemical surface modification can be distinguished: Initial efforts to coat a capillary involved the use of traditional silane-coupling chemistry, using mono-, bi-, and trifunctional silane reagents, which attached to the silica silanols by siloxane bonds. The immobilized silane reagent can then be reacted in a second step to complete the coating procedure. A list of these conventional types of coatings, as now applied in capillary electrophoresis, is given in Table 1. The major limitation

Table 1 Conventional Coatings

Functionality	pH range[a]	Applications	Ref.
Trimethylsilyl-	7	Small molecules	1
Trimethylsilyl-	9	MECC	3
RP-C_8	9	Proteins	4
RP-C_{18}	9	Proteins	4,5
Polyethylenglycol	3–5	Proteins	6
Diol	3–5	Proteins	1,7
Glyceroglycidoxypropyl-	5	Proteins	8
Maltose	3–7	Proteins	7
Arylpentafluoro-	7	Proteins	2
α-Lactalbumin	8	Proteins	9

[a]Indicated range of stability or maximum of pH shown in the reference.

of this technique is the poor hydrolytic stability at basic pHs, owing to the siloxane bond used for immobilization.

This phenomenon is also known from the preparation of stationary phases in HPLC. One approach to solve it was coating the surface with polymer, which represents the second type of chemical surface modification. Polymer-coating procedures also can be classified in different ways.

1. Pretreatment of the surface using conventional silane chemistry to introduce anchor groups for copolymerization with appropriate monomers or oligomers
2. Adsorption of prepolymers at the silica surface, followed by in situ polymerizing and crosslinking

Polymer-coated surfaces exhibit an improved pH stability, at least up to a pH 9–10. A compilation of polymer coatings used in CE so far is listed in Table 2.

Surface Pretreatment of the Fused Silica Columns

The capillaries used in CE are produced by GC companys as nondeactivated open tubular materials, or they have their origin from optic fiber production. The quality of the inner surface of the "raw" fused silica columns might be quite different in their roughness and their ability to adsorb metal ions. Therefore, some authors propose a column pretreatment before chemical modification of the capillary wall. Most authors tend to wash the capillary with alkaline solutions (leaching) to enhance wetability of the surface. Another advantage is the generation of further free silanol groups to create additional binding sites for the chemical modification. Poppe and co-workers [6,7] used 1 M potassium hydroxide for 3 hours, whereas Regnier and Towns [15], McCormick [8], Cobb et al. [12], and Strege and Lagu [13] used sodium or potassium hydroxide (1 M) for 15–30 minutes. A few authors also report a wash with hydrochloric acid. Poppe and associates recommended an exchange of the introduced sodium ions (from the alkaline wash) by hydrogen ions, using 0.1 M

Table 2 Polymer Coatings

Functionality	pH range[a]	Applications	Ref.
Acrylamide	2–8	Proteins	10–12
	2–10	IEF	13,14
		DNA fragments	
Polyethyleneimine	3–11	Proteins	15
1-Vinyl-2-pyrrolidone	2–6	Proteins	8
Methylglutamate	2–9	Proteins	16
Polymethylsiloxane (OV-1)	7	MEC	17,18
Polyethylene glycol (Carbowax)	7	MEC	17,18

[a]Indicated range of stability or maximum of pH shown in the reference.

HCl. Balchumas and Sepaniak [3] removed adsorbed metal ions by washing with hydrochloric acid (0.1 M) as the only pretreatment step.

At the end of the procedures, the activated capillary was dried in a nitrogen stream at 80°–200°C and was ready for chemical modification. Since many authors do not mention any activating procedures, it seems not to be absolutely necessary for column pretreatment. However, from our experience, for reasons of reproducibility, it is highly recommended. Strege and Lagu [13] show the different efficiencies of the same separation in a pretreated and a nonpretreated fused silica capillary.

Characterization of the Coatings

Stationary phases in GC and LC are tested by numerous standard procedures and test mixtures. Moreover, many independent physical methods, such as elementary (CHN) analysis, FTIR, or NMR are available in LC to characterize these phases. However, the characterization of the chemically modified fused silica wall in CE is a major problem because of the small surface areas (in the range of square centimeters).

Most authors tend to use the change of electroosmotic flow caused by the coating procedure to demonstrate the modification of the capillary wall. Additionally, most types of coatings are developed to reduce protein–wall interactions. Thus, the efficiency of (model) protein separation is another test for the quality of the coating.

Besides these commonly used techniques, Regnier and Towns [15] reported the adsorption of picric acid to estimate the layer thickness of a positively charged polyethyleneimine coating. However, this method seems applicable (with reasonable results) only for charged multilayer coatings. Lux et al. [18] determined the film thickness by GC retention measurements, using *n*-nonane as a standard. This might be a useful technique for the characterization of nonpolar coatings with excellent surface shielding.

CONVENTIONAL COATINGS

Alkylsilane Modification

Preparation

The use of mono-, bi-, and trifunctional alkylsilanes for surface modification is well known from GC and HPLC. The silanol groups on the solid's surface are reacted with alkoxy- or chlorosilanes. The resulting modified surface has a more or less hydrophobic character (reversed phase), depending on the chain length of the aliphatic remnant of the coupling silane. To avoid precondensation of the single silane molecules, the reaction is normally carried out in dry solvents. Because of their reduced reactivity, alkoxysilanes might also be coupled in aqueous solution owing to their reduced reactivity. Balchumus and Sepaniak [3], for example, uses a 10%

solution of trimethylchlorosilane in dry toluene for 3 hours at 95°C. After the silanation, the column was rinsed with toluene, methylene chloride, and methanol, and was ready for use afterward.

Electroosmosis

Electroosmotic flow (EOF) is reduced when the silanol groups are modified with neutral (uncharged) alkylsilanes. Maa et al. [9] also report an influence of solvent used in the reaction: silylation in aqueous solution has a smaller impact on EOF than commonly used nonaqueous conditions. In Fig. 2 the EOF/pH characteristics of a C_8- and a C_{18}-coated capillary is shown in comparison with an untreated capillary [4]. Electroosmosis is reduced by 40% of the bare silica column. However, these capillaries have minimal variation of electroosmotic flow when changing pH. The slightly lower electroosmosis of the C_8 capillary might be due to a better coverage of the surface because of lower steric hindrance of the octyl groups.

Stability

It is well known in LC that silane-based stationary phases and the silica are stable up to pH 7.5. At higher pH values hydrolysis occurs. In Fig. 3, the stability in migration time of a neutral marker is demonstrated for $C_{8/18}$ and unmodified capillaries at pH 7 [4]. The bonded capillaries show a clear "breakin" period, during which the migration time for the neutral marker steadily decreases. Possible expla-

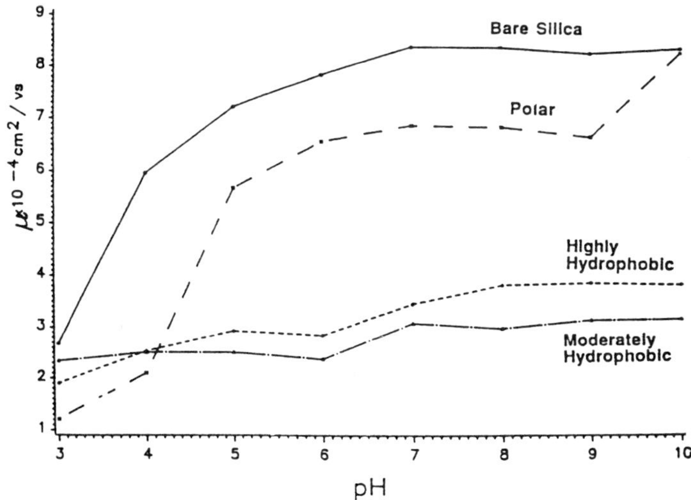

Figure 2 The EOF/pH characteristic of a highly (C_{18}) and a moderately (C_8) hydrophobic coated capillary in comparison with bare silica. Capillary, l(eff), 65 cm; buffer, 25 mM sodium phosphate; neutral marker, 0.01% benzyl alcohol; field, 200 V cm^{-1}. (From Ref. 4)

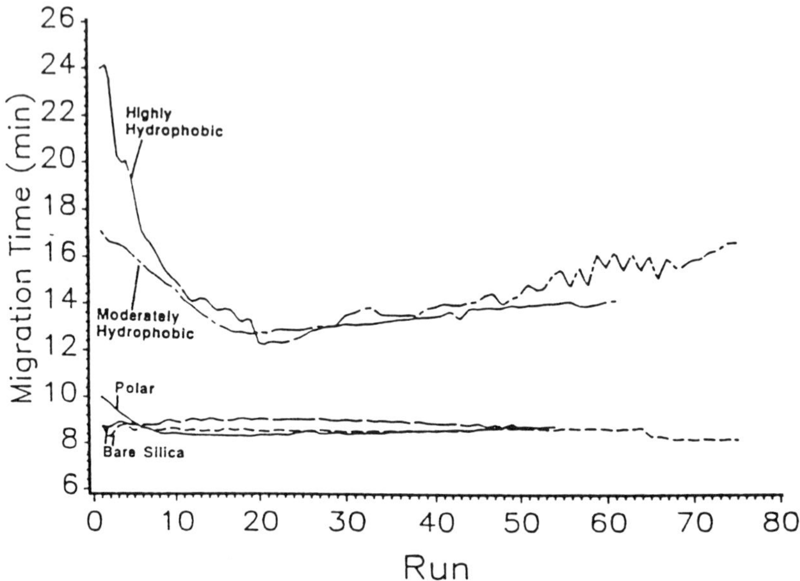

Figure 3 Stability of alkylsilane-modified fused silica capillaries: Migration time of a neutral marker as a function of run number. Conditions as in Fig. 2. (From Ref. 4)

nations are loss of unbound material or hydrolysis of the phase. Standard deviations of the electroosmotic mobility at pH 7, calculated by the authors, were worse in the bonded capillaries (15% for C_{18}, 8.5% for C_8, and 2.4% for untreated materials). A special equilibration procedure for the C_{18} coating improves reproducibility to 5.5%. Despite the possible hydrolysis, separations in buffers up to pH 9 are reported in the literature [3–5]. Dougherty et al. [4] even report a 1-min wash with 0.1 N NaOH between single runs.

Applications

The influence of reduced electroosmotic flow at pH 7 on resolution and separation time has been investigated by Jorgenson and Lukas [1], using trimethylchlorosilane-treated capillaries. A mixture of seven dansylated amino acids were separated, showing better resolution, but also a longer analysis time.

Balchumas and Sepaniak [3] used the same coating in micellar EC (MEC) for the separation of small amines. The reduced electroosmotic flow leads to increased peak capacity and resolution.

Protein separations with C_8 and C_{18} coatings were also reported by Sepaniak and associates [4,5]. A test mixture of myoglobin, conalbumin and β-lactoglobulin (A and B) was used to compare bonded and unbonded fused silica capillaries.

However, for conalbumin, a few injections were necessary before a reproducible separation could be achieved. On one hand, this may be due to hydrophobic properties of the alkylsilane coating, on the other hand, the surface silanols do not seem to be completely shielded.

Arylpentafluoro Coating

Preparation

Originally, the arylpentafluoro group was used to prepare highly hydrophobic stationary phases in LC. Surprisingly, a transfer to CE fused silica capillaries and to protein separations seems to be possible. In a two-step reaction, the fused silica wall was first treated with γ-aminopropyltrimethoxysilane by using standard procedures (see foregoing). After a drying step, the amino group was coupled with pentafluorobenzoylchloride dissolved in toluene [2].

Electroosmosis and Stability

Fused silica capillaries modified with the arylpentafluoro (AFP) coating were prepared such that a significant EOF (0.5 mm s^{-1}) resulted as neutral pH and moderate ionic strength [2]. The author gives no further information of EOF/pH characteristics or long-term stability of these capillaries. However, studies of the day-to-day reproducibility (over a 5-day period), with proteins as test solutes, is reported. Standard deviations (also dependent on the sample used) were less than 4.2% within a day and 7.6% from day to day. The coating test was performed at pH 7.

Applications

The AFP capillaries are used for the separation of proteins in open tubular CE; pH 7 is standard for this type of coating. Swedberg [2] shows a separation of seven standard proteins with pI values ranging from 11 to 6.9, yielding excellent efficiencies of several hundred thousand theoretical plates per meter (up to 600,000; Fig. 4). A special feature is that, in this system, a considerable EOF is retained and allows the transport of both positively and negatively charged proteins past the point of detection. By the application of small inner diameters (i.d. = 20 μm), higher buffer concentrations, or higher ionic strength (500 mM phosphate, or addition of 100 mM KCl) can be used to improve efficiency and suppress sample–wall interactions, as discussed earlier [19].

In conclusion, AFP coating in combination with high ionic strength buffers seem to be a good approach for CE separation of proteins.

Hydrophilic Hydroxyl and Polyether Coatings

Preparation

Hydrophilic functionalities, such as diol, polyethylene glycol (PEG), or polysaccharides, are used for the preparation of stationary phases in LC of proteins. Poppe

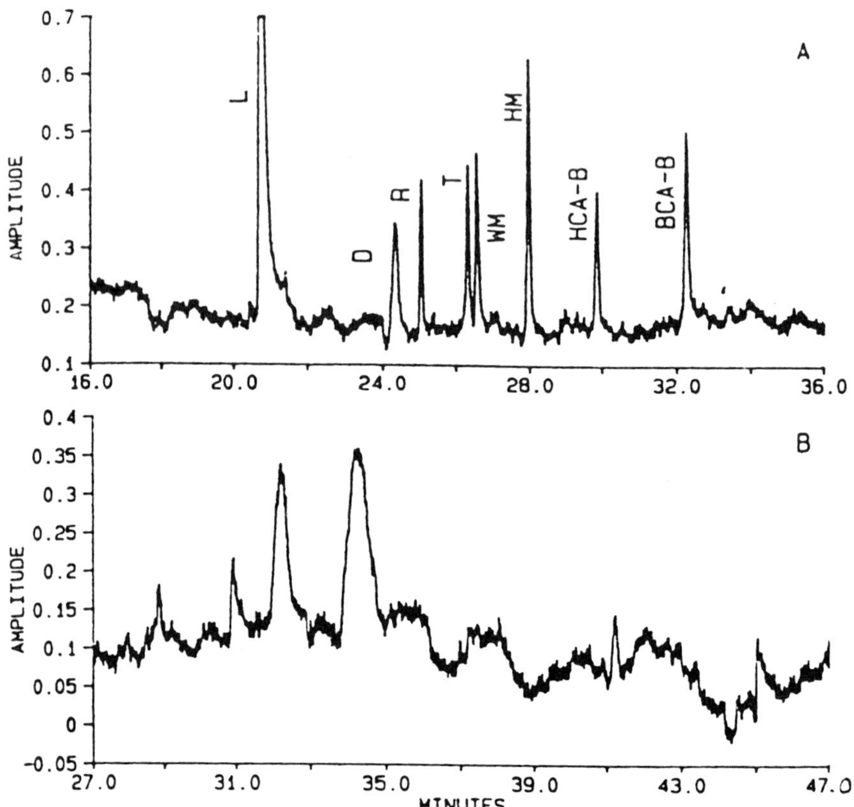

Figure 4 Elution profile of seven protein markers and DMSO on an APF-treated column (A) vs untreated fused silica (B). Run conditions: 200/100 mM phosphate/KCl solution pH 7, in 20 μm i.d. capillary. Applied voltage, 250 V cm^{-1}. Detection at 219 nm, l(eff) = 1 m. Samples: L, hen egg white lysozyme; D, DMSO; R, bovine ribonuclease A; T, bovine pancreatic trypsinogen; WM, whale myoglobin; HM, horse myoglobin; HCA-B, human carbonic anhydrase B; BCA-B bovine carbonic anhydrase B. (From Ref. 2)

et al. describe the modification of fused silica capillaries with these functionalities [6,7].

A dual-step procedure is used for the preparation. In the first step, the fused silica wall is reacted with γ-glycidoxipropyltrimethoxysilane. Afterward the reactive glycid function can be either hydrolyzed to the diol (using diluted HCl) or coupled with polyethylene glycol 600 (BF_3 catalysis). McCormick described a similar procedure to prepare glyceroglycidoxypropylsilane-modified capillaries [8]. He used (3-

glycidoxypropyl)diisopropylethoxisilane for the surface silylation and glycerol to open the epoxide function (BF_3 catalysis).

For carbohydrate coupling, the fused silica surface is first modified with γ-aminopropyltrimethoxysilane [7]. In the second step, the amino group is reacted with a dilute solution of maltose in methanol (sodium cyanoborhydride catalysis).

Electroosmosis and Stability

As expected, diol and polyethylene glycol coatings also show a reduced electroosmotic flow (diol: $\mu_{eo} = 10^{-4}$ cm^2 V^{-1} s^{-1} at pH 6, 50 mM phosphate buffer). For the separation of proteins, the working range for these capillaries is from pH 3 to pH 5. Capillaries were stable in this range over several months, and they even withstood treatment with diluted (0.01 M) HCl. Further information concerning stability and electroosmosis at higher pH values is not available.

For the maltose coating, a useful range from pH 3 to pH 7 is reported by the authors [7]. A reversal of electroosmotic flow occurs in the EOF/pH characteristics: After coupling of maltose, some free amino groups remain at the surface. At low pH values, these aminopropyl groups can be protonated and, therefore, the wall will have a positive charge directing electroosmosis to the anode. Maltose-based coatings are stable for protein separations at pH 4 with a relative standard deviation of 5% in migration time (over 1 week). Additionally, an antimicrobial agent had to be added to the buffer to avoid bacterial degradation of the coating.

Applications

The hydrophilic diol-, polyethylene glycol-, and maltose coatings were specially designed for the separation of proteins in CE. The first experiments in this field with diol modification were reported by Jorgenson [1]. Poppe et al., [4,7] got their best results on PEG-modified fused silica capillaries (Fig. 5). A mixture of seven basic proteins could be separated in a 30-mM phosphate buffer at pH 3.8. Plate numbers (per meter) between 80,000 and 150,000 are reported for these samples. McCormick [8] showed a similar separation at pH 4.5 with the glyceroglycidoxypropyl coating, but without calculating efficiencies.

For diol-based coatings, lower plate numbers up to 50,000 m^{-1} were found using the same samples. For the promising carbohydrate coating the efficiency was even lower ($N = 25,000$) when using lysozyme and cytochrome c as test solutes. According to the author, protein adsorption still occurs for this coating type, but to a smaller extent than with bare silica.

Protein-Bonded Coating

Preparation

Similarly to carbohydrate coatings, other bimolecules can also be attached to the inner capillary surface. Maa et al. [9] describe the use of α-lactalbumin as a coating

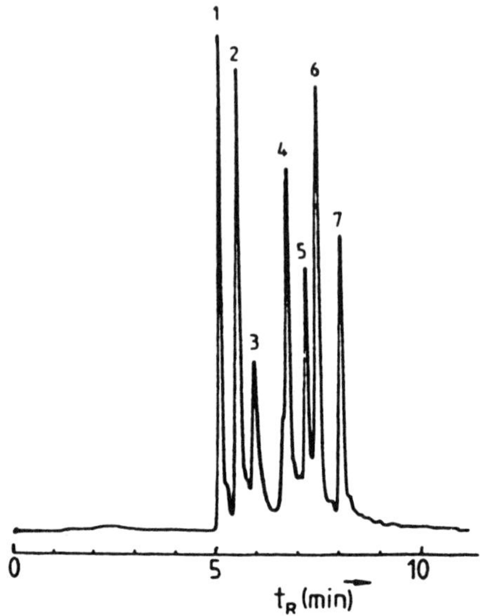

Figure 5 Separation of model proteins on a PEG-modified fused silica capillary. Conditions: injection 10 s; 10 kV; buffer, KH_2PO_4 30 mM pH 3.8. Samples: 1, cytochrome C; 2, lysozyme; 3, myoglobin; 4, trypsin; 5, ribonuclease; 6, trypsinogen; 7, chymotrypsinogen. (From Ref. 6)

for protein separation in CE. In a three-step reaction, the wall is first treated with aminosilane [2]. Then glutardialdehyde is reacted with the amino group to introduce a free aldehyde function as the protein-binding site. Finally, an aqueous solution of α-lactalbumin is pumped through the capillary to couple with the aldehyde.

Electroosmosis and Stability

Protein-bonded coatings show interesting EOF/pH characteristics. Because proteins are amphoteric molecules, they are positively charged at pH levels lower than pI and negatively charged at higher pH values. When pH = pI, the protein is uncharged. Consequently, electroosmotic flow of such coatings is directed to the anode at pH < pI and directed to the cathode when pH > pI. No electroosmosis occurs when pH = pI. An additional aspect of such capillaries is that, in principle, at any pH the magnitude and direction of the EOF can be adjusted by selecting the proper bonded amphoteric phase. However, the absolute value of electroosmosis is reduced by the chemical modification in comparison with bare silica. Data about EOF were available from pH 2 to pH 8, which is assumed to be the stability range of this type of coating. The authors [9] do not mention any influence of nonreacted amino groups similar to Poppe et al. [4,7] (see earlier section). Nevertheless, this effect could be another explanation for the reversal of the electroosmotic flow at low pH. The application of the coating procedure with basic proteins, resulting in a wider range of reversed electroosmotic flow, could prove the authors hypothesis of the coating structure.

Applications

Protein-bonded capillaries have been used for the separation of proteins in CE. Maa et al. [9] introduced this coating as a low-interaction phase yielding efficiencies in the range of 100,000 theoretical plates (proteins with pI values between 7 and 11 in a phosphate buffer, pH 7).

POLYMER COATINGS

Linear Polyacrylamide Coatings

Crosslinked polyacrylamide gels have proved to be an excellent matrix for electrophoretic separations of proteins such as in sodium dodecylsulfate–polyacrylamide gel electrophoresis SDS–PAGE, discelectrophoresis, or isoelectric focusing. This demonstrates the hydrophilic character of such gels coupled with practically no interaction with biomolecules.

In 1985, linear (i.e., noncrosslinked) polyacrylamide coatings were introduced in CE by Hjerten [10]. This type of polymer coating is as yet the only one commercially available in CE.

Preparation

Since the original work of Hjerten, in 1985, many publications have appeared dealing with preparation and application of linear polyacrylamide-coated capillaries. However, the original procedure for the preparation [10] has not significantly changed:

In a two-step reaction the silica wall is first silylated with γ-methacryloxypropyltrimethoxysilane, adjusted to pH 3.5 with acetic acid. Then, the vinyl groups in the second step are copolymerized with acrylamide in aqueous solution. *N*,*N*,*N*′,*N*′-Tetramethylethylenediamine (TEMED) is used as a catalyst for the polymerization and potassium persulfate as initiator. An improvement of the hydrolytic stability of the silane-based coating was reported by Cobb et al. [12]. According to their procedure, the silica wall was first chlorinated with thionyl chloride, followed by a Grignard reaction with vinyl magnesium bromide. Novotny and co-workers [12] claim that this direct Si—C bond to the silica is much more stable than the Si—O—Si—C bond. The polymerization step is identical with that described by Hjerten. Kohr and Engelhardt [20] transferred their method of polymer-encapsulated silica to capillaries. They used trichlorovinylsilane for silylation of the inner capillary wall. In the second step, acrylamide and α,α'-azoisobutyronitrile (AIBN) as initiators in dichloromethane were introduced into the capillary. After evaporation of the solvent, polymerization was induced by heat.

The linear acrylamide coating is still a monolayer type (i.e., the layer thickness is in the range of the molecular dimensions).

Electroosmosis and Stability

According to Hjerten [10] and Cobb et al. [12], electroosmotic flow is completely eliminated when their coating procedure is used. The stability of these modified capillaries is excellent at pH values below 3, but less stable at pH values above 8. This is indicated by an increase in electroosmotic flow and inconsistent migration behavior of the analytes. Cobb et al. [9] claimed an improved stability of the linear polyacrylamide coating up to pH 10.5 by modifying the vinylation step (see foregoing). This coating remained intact for 4 weeks a pH 9 (more than 150 injections) and for at least 5 days at pH 10.5.

With use of well-defined concentrations of acrylamide, it was possible [20] to prepare capillaries with a reduced, but not totally suppressed, electroosmotic flow. The reproducibility of the electroosmotic mobilities were much better than with uncoated capillaries, especially at pH 5 (turning point of the EOF/pH curve).

Applications

Linear polyacrylamide-coated capillaries are easy to prepare and, additionally, they are commercially available. Consequently, they have already found a wide field of applications.

Protein separation is an area of high interest in connection with coated capillaries. Successful application of this coating for protein separations was reported at high [12,14] and low [12,14,20] pH values. Since electroosmosis is totally eliminated, the analytes must be either all positively or negatively charged to reach the detector window (using the appropriate polarity of the power supply, adjusted accordingly). Therefore, the pH values must be chosen properly to ensure that all proteins will be of the same charge sign and migrate in the same direction. Figure 6 shows a separation of seven proteins at pH 9.5. The relative molecular mass of the proteins ranged from 2000 to 70,000 and the pI values from 4.3 to 7.6. Efficiencies of more than 200,000 theoretical plates per meter were achieved, and the RSD of the migration time was below 1% ($N = 5$).

With the use of linear polyacrylamide coatings, it was possible to transfer isoelectric focusing (IEF) of proteins to capillary electrophoresis [10,11,14,21]. In IEF a pH gradient is generated by an electrical field, and the proteins are focused at their appropriate pI values. In fused silica capillaries, it was necessary to reduce protein wall interactions as well as electroosmotic flow because EOF cannot guarantee a stable pH gradient with stable focused zones. Linear polyacrylamide-coated capillaries fulfill these requirements. A very nice separation of a model protein mixture

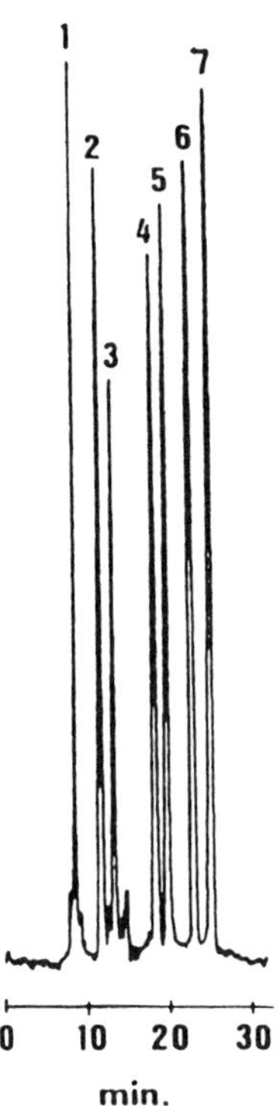

Figure 6 Capillary electrophoretic separation of a model protein mixture using a linear polyacrylamide-coated capillary (A). Conditions: buffer, 50 mM glutamine, triethylamine (pH 9.5); capillary, 50 μm i.d. × 60 cm (45 cm to detector); hydrodynamic injection, 5 s with 20-cm height differential; applied field, (A) 20 kV and (B) 10 kV. Samples: 1, insulin chain A; 2, bovine serum albumin; 3, chicken egg ovalbumin; 4, insulin; 5, α-lactalbumin; 6, β-casein; 7, insulin chain B. (From Ref. 12)

with pI values from 8.6 to 5.1 is illustrated in Fig. 7. Further applications (e.g., focusing of transferrin isoforms, hemoglobin variants, and so on) are described in the literature by Hjerten [10,11] and Bolger et al. [14].

Various techniques associated with CE, such as MEC, but especially polyacrylamide–gel capillary electrophoresis (CGE) have been successfully used to separate nucleosides, nucleotides, oligonucleotides, and DNA restriction fragments. Strege and Lagu [13] found the use of polyacrylamide-coated capillaries, in conjunction with buffers containing 0.5% methylcellulose, to be well suited for the separation of a wide size range of DNA restriction fragments (Fig. 8), with a resolution comparable with gel-filled capillaries. Coated capillaries in this field of application

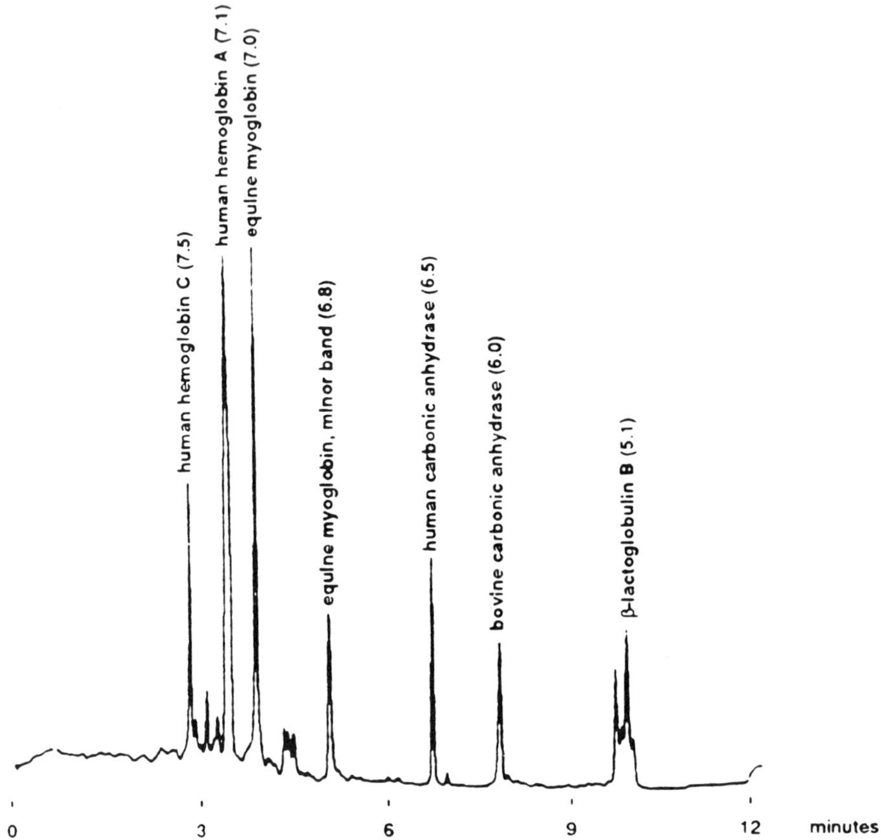

Figure 7 Isoelectric focusing of a model protein mixture using a linear polyacrylamide-coated capillary. (Reprinted courtesy of BioRad Company; see also Ref. 21)

can avoid many difficulties of CGE associated with poor gel-to-gel reproducibility, bubble generation, and gel matrix collapse and, therefore, may be a good alternative to the use of gel-filled capillaries.

Finally, linear polyacrylamide coatings were also used to prepare polyacrylamide gel-filled capillaries [22]. Since the electroosmotic flow is eliminated in these capillaries, no gel extrusion during analysis takes place, even at an electric field strength of 400 V cm^{-1}. Furthermore, gel destruction during the shrinking process is avoided because no bonding to the silica wall occurs. When compared with uncoated capillaries, an improvement of the peak shape was also observed for the separation of oligonucleotides.

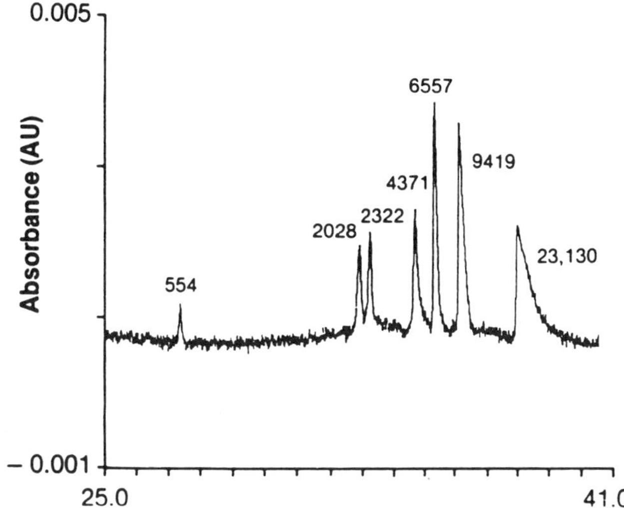

Figure 8 Electropherogram of the λ DNA *Hind* III digest separated at 10 kV in a polyacrylamide-coated capillary. Conditions: buffer, 50 mM Tris/borate, pH 8.0, 2.5 mM EDTA, and 0.5% methylcellulose (high viscosity). Resolved fragments are labeled by their size in base pairs. (From Ref. 13)

Poly(vinylpyrrolidone) Coating

Poly(vinylpyrrolidone) (PVP)-modified silica has been successfully used for the separation of proteins by both size-exclusion and hydrophobic-interaction chromatography. The use of PVP in CE has been proposed by Hjerten [10].

Preparation

For the polymer PVP-coating McCormick [8] used a two-step reaction similar to the procedure of Hjerten (see foregoing). The capillary wall is first treated with [(3-methacryloyloxy)propyl]trimethoxysilane in diluted acetic acid. After purging the column with water, 1-vinyl-2-pyrrolidone in an aqueous solution (containing TEMED and ammonium persulfate) is copolymerized with the vinyl groups.

Properties and Applications

The EOF/pH characteristics were not given in McCormick's paper [8]. However, it can be assumed that electroosmotic flow is suppressed in a way similar to that with the linear acrylamide coating. Moreover, no long-term studies of the stability of PVP capillaries have been demonstrated, but the author reports no observed deterioration in performance during use at low pH over weeks. Protein separations performed at pH 2 (60-mM phosphate buffer) are possible with PVP-coated fused silica capillaries, as illustrated in Fig. 9. Fifteen proteins, with M_r from 12 to 77 kDa and pI values from 4.5 to 11, were separated in less than 25 minutes with efficiencies up to 700,000 plates per meter. However, some proteins (such as insulin, ovalbumin, chymotrypsinogen) could not be separated with these capillaries because of irreversible adsorption to the capillary wall.

Nevertheless, PVP-modified fused silica capillaries seem to be a good alternative to linear acrylamide coatings for protein separations at low pH.

Polyethyleneimine-Bonded Coating

Preparation

Regnier and Towns described a positively charged polymer coating that is electrostatically adsorbed to the inner silica wall [15]. The synthesis of this coating material was similar to the preparation of silica-based anion-exchange packing materials for LC. Polyethyleneimine (PEI) of different molecular mass and in different concentrations was crosslinked on the capillary surface.

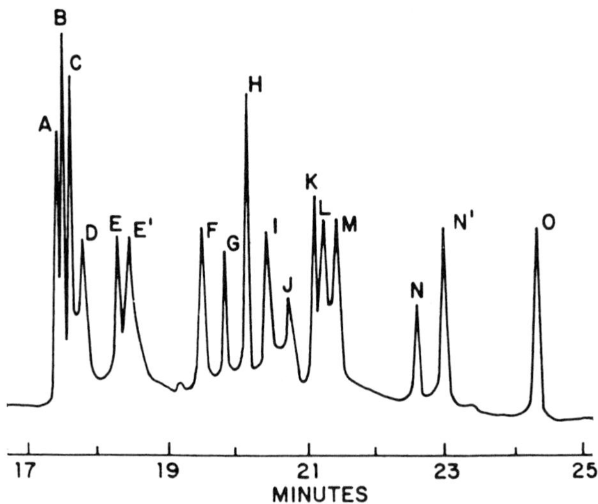

Figure 9 Separation of protein mixture on a PVP-modified capillary. Conditions: capillary, 52 μm i.d. l(tot) = 110 cm, l(eff) = 75 cm; buffer, 38.5 mM H_3PO_4, 20 mM NaH_2PO_4 pH 2.0; separation, 5–25 kV linear program in 150 s. Samples: A, β-lactoglobulin B; B, β-lactoglobulin A; C, lysozyme; D, human serum albumin; E, bovine serum albumin; F, cytochrome c; G, trypsinogen; H, whale myoglobin; I, transferrin; J, conalbumin; K, horse myoglobin; L, carbonic anhydrase B; M, carbonic anhydrase A; N, hemoglobin; O, parvalbumin. (From Ref. 8)

The procedure was described as follows: In the first step, PEI is adsorbed from methanolic solution to the silica surface (Fig. 10). After drying, the oligomers were crosslinked into a continuous film using ethylene glycol diglycidyl ether (EDGE) as the bifunctional reagent. Polyethyleneimines are known to be branched polymers with a ratio of primary, secondary, and tertiary amino groups of 1:2:1. These functionalities allow electrostatic interaction of the coating with the capillary wall as well as offering binding sites for crosslinking. The coating thickness was estimated, by ion-pairing experiments with picric acid, to be in the range of 1–7 nm.

Electroosmosis

Because of the high concentration of basic amino groups in the polymer, the coating is positively charged over a wide pH range, resulting in a reversed electroosmotic flow. The absolute value of the electroosmotic mobility in these capillaries at low pH is comparable with that of uncoated capillaries at high pH. The influence of molecular mass and polymer concentration on the electroosmotic flow was examined. It was found that EOF was directed to the anode for polymers with higher molecular masses ($M_r > 1800$) and for pH values between 3 and 11. For lower molecular mass polymers a sharp reversal of electroosmotic flow at basic pH values occurs from deprotonation of the amino groups and increased dissociation of the surface silanols. The influence of the polymer concentration on the EOF/pH characteristics is shown in Fig. 11. As expected, electroosmotic flow decreases with increasing pH owing to the gradual deprotonation of PEI. Electroosmotic mobility is always maximum with 5% PEI. A further increase in the polymer concentration (and coating thickness) decreases the flow rate again, which is not well understood.

Applications

Regnier and Towns showed that this polymer coating is suitable for protein separations in CE [15]. A separation of a test mixture (five proteins; pI 7–11) at pH 7 is demonstrated (Fig. 12). At this pH, all proteins are positively charged and migrate in the direction opposite the electroosmotic flow (elution after the neutral electroosmotic marker). Additionally, the positively charged wall repels the sample and minimizes protein adsorption.

Poly(methylglutamate)-Coated Capillaries

Bentrop and associates [16] reported a new type of polymer coating based on polypeptides adsorbed on the bare silica surface. The authors used the *N*-carboxy anhydride of γ-methylglutamate as monomer for the preparation of the polypeptide. After adsorption on the capillary wall, polymerization was accomplished by heat with the loss of carbon dioxide.

Electroosmosis and Stability

The EOF/pH characteristics showed that electroosmosis was only slightly reduced by poly(methylglutamate) coatings compared with uncoated capillaries. This implies

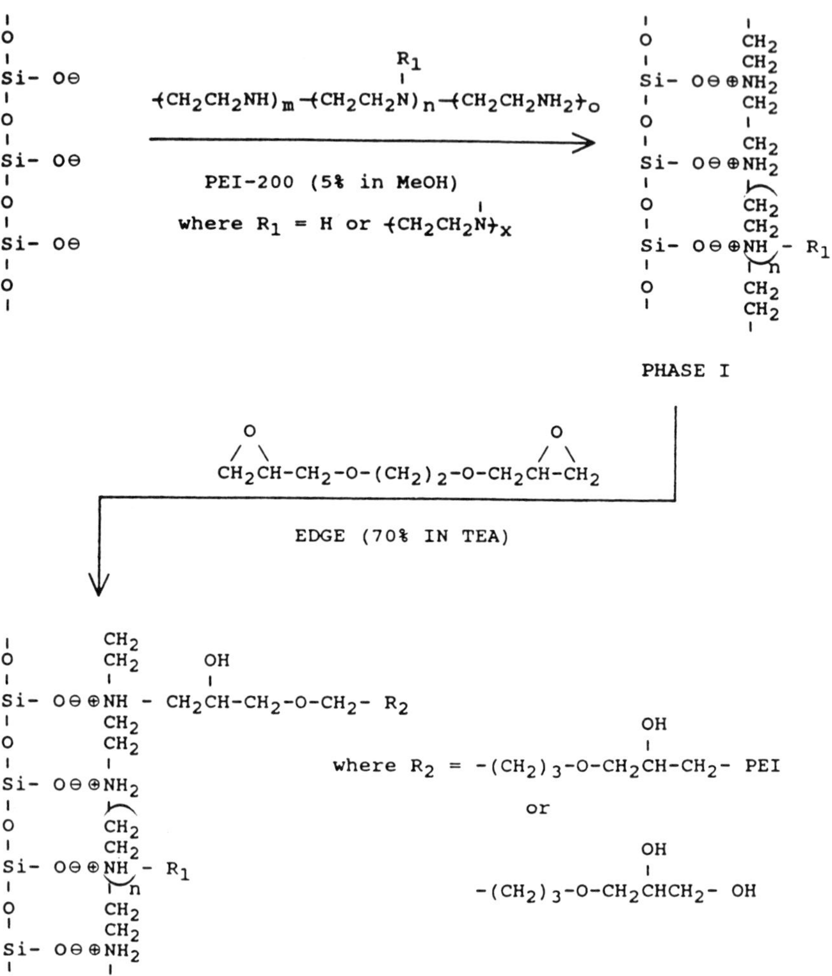

Figure 10 Synthetic route to an adsorbed PEI-bonded phase. (1) Adsorption of PEI 200 onto the fused silica capillary; (2) crosslinking of the PEI 200 polymer to stabilize the coating. (From Ref. 15)

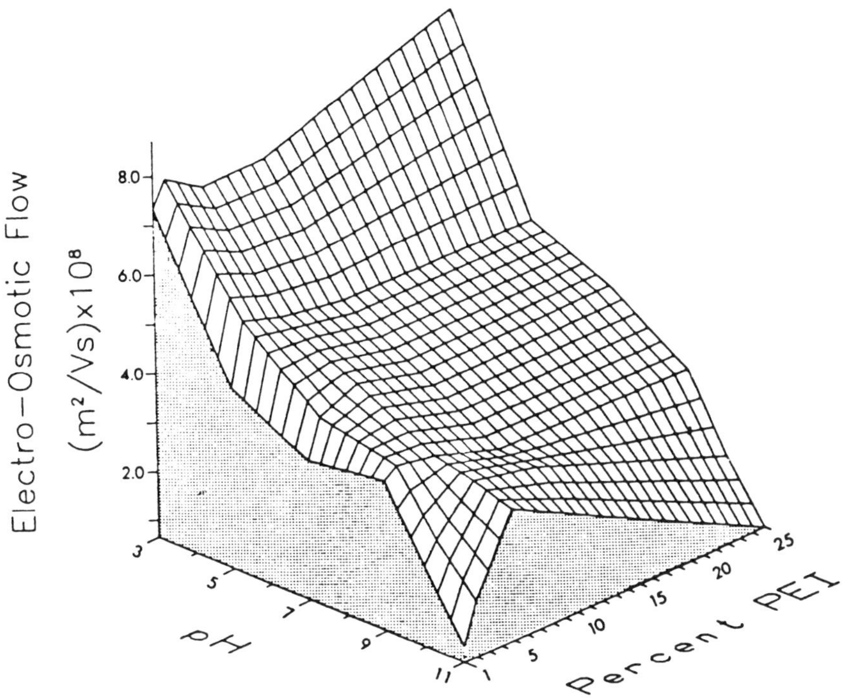

Figure 11 The EOF/pH characteristics of PEI-coated capillaries. Three-dimensional plot of electroosmotic flow as a function of polymer concentration (1, 2, 5, 10, and 25% in methanol) at a given pH. (From Ref. 15)

a very thin polymer layer and a silica surface that is not totally shielded. No long-term stability studies were reported. However, the coating was used for routine analysis of proteins at pH 7 over several weeks.

Applications

The applicability of poly(methylglutamate)-coated capillaries for protein separations in CE was demonstrated: a separation of model proteins was shown (pI range: 4.5–7), with efficiencies of more than 100,000 theoretical plates. Nevertheless, the broad peak of bovine serum albumin also indicates that this type of coating is not suitable for all types of protein separations. Further applications for the separation of complex enzyme mixtures (such as rat pancreas secretion proteins and cellulase) have also been reported (Fig. 13) [16].

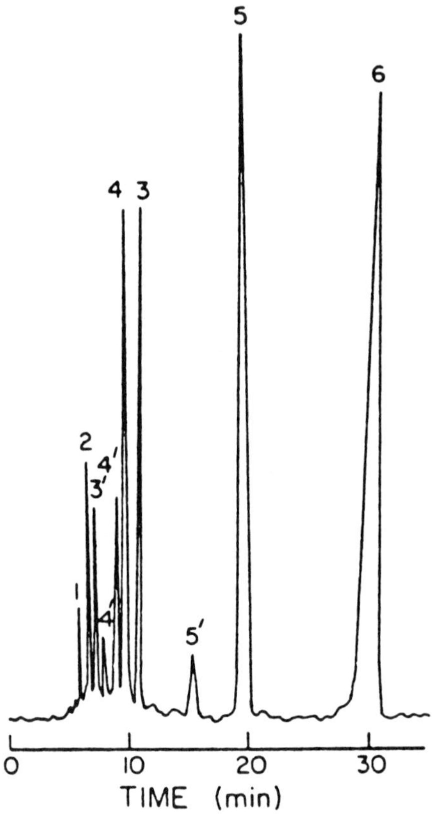

Figure 12 Capillary electrophoretic separation of model proteins on a PEI-200-EDGE-coated capillary. Conditions: capillary length, 50 cm; separation length, 35 cm; i.d., 75 μm; buffer, 0.02 M hydroxylamine/HCl at pH 7.0; separation potential 12.5 kV. Peaks: 1, mesityl oxide; 2, horse heart myoglobin; 3, bovine ribonuclease A; 4, bovine chymotrypsinogen A; 5, horse heart cytochrome c; 6, hen egg lysozyme. Peaks 3′, 4′, and 5′ are impurities in 3, 4, and 5 respectively. (From Ref. 15)

Polymer-Coated Gas Chromatography Columns in Capillary Electrophoresis

During the early developmental stage of CE, Terabe et al. suggested the introduction of conventional polymer-coated GC columns in CE [17]. Their aim was to control electroosmotic flow in MEC with commercially available polymethylsiloxane and polyethylene glycol-modified capillaries. Lux and associates [18] did systematic studies in this field. They examined the influence of the coating's layer thickness

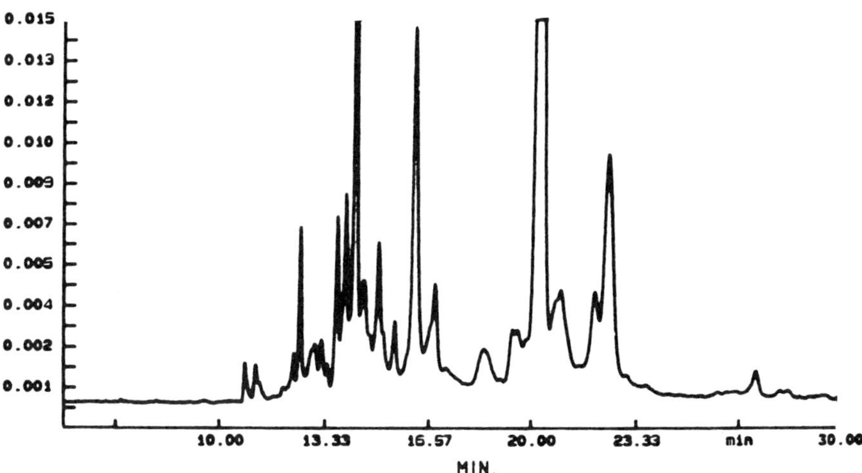

Figure 13 Separation of cellulase (enzyme complex) on a PMG-coated fused silica capillary. Conditions: capillary, 75 μm i.d. l(eff) = 34 cm; 30 mM phosphate buffer, pH 7.0; applied voltage, 8 kV (42 μA); detection, UV λ = 200 nm. (From Ref. 16)

on EOF. By using a nonpolar polymethylsiloxane coating in MEC (SDS was used as surfactant) the electroosmotic flow was increased, resulting in shorter analysis time and lower peak capacity. The increased flow rate, in comparison with bare silica, can be explained by the absorption of a higher amount of SDS on the coating, leading to a higher zeta potential. With polyethylene glycol coatings a decrease in electroosmosis occurred combined with longer analysis time and higher resolution. With this type of coating, silanol groups were probably better shielded, but significant adsorption of the surfactant was not noticed. Lux et al. found that these effects were independent of the layer thickness of the coating (for a film thickness above 40 nm) [18].

The separation of phenols and nitrobenzene derivatives [17], some purine derivatives [18], and nucleobases [18] by MEC indicates areas of application for polymer-coated GC columns in CE.

CONCLUSIONS

Chemical modifications of the inner capillary surface have become an excellent tool to control electroosmotic flow and to reduce irreversible sample wall adsorption (especially of proteins).

As known in chromatography, in CE polymeric coatings also offer many advantages in pH stability and complete surface shielding compared with the traditional silane-based–coating techniques. Nevertheless, numerous problems must be solved (e.g., reproducibility of coating procedures and long-term stability of the capillary columns). For the separation of proteins no universal coating has yet been found showing excellent properties for both acidic and basic proteins with higher molecular masses ($M_r > 100,000$).

An alternative to the chemical modification of the capillary wall is the dynamic coating performed by buffer additives. Jones and Jandik [23] controlled the electroosmotic flow by adding a cationic surfactant to the buffer to achieve fast separations of anions. Bushey and Jorgenson [19] used zwitterionic buffers of high ionic strength to separate proteins. Another successful example was given by Emmer and coauthors, who used a positively charged surfactant [24] to suppress interactions of basic proteins with the wall [for further information see Ref. 21].

The simplicity of the coating procedure, the renewal from run to run, and the flexibility are the main advantages of this method. Future developments will recommend one of these methods. Nevertheless, chemically modified and dynamically coated capillaries are and will be an important contribution to the progress of CE in a modern analytical laboratory.

REFERENCES

1. J. Jorgenson and K. Lukacs, *Science*, *222*:266 (1983).
2. S. Swedberg, *Anal. Biochem.*, *185*:51 (1990).
3. A. Balchumas and M. Sepaniak, *Anal. Chem.*, *59*:1466 (1987).
4. A. Dougherty, C. Woolley, D. Williams, D. Swaile, R. Cole, and M. Sepaniak, *J. Liquid Chromatogr.*, *14*:907 (1991).
5. M. Sepaniak, D. Swaile, A. Powell, and R. Cole, *J. High Resolut. Chromatogr.*, *13*:679 (1990).
6. G. Bruin, J. Chang, R. Kuhlman, K. Zegers, J. Kraak, and H. Poppe, *J. Chromatogr.*, *471*:429 (1989).
7. G. Bruin, R. Huisden, J. Kraak, and H. Poppe, *J. Chromatogr.*, *480*:339 (1989).
8. R. McCormick, *Anal. Chem.*, *60*:2322 (1988).
9. Y.-F. Maa, K. Hyver, and S. Swedberg, *J. High Resolut. Chromatogr.*, *14*:65 (1991).
10. S. Hjerten, *J. Chromatogr.*, *347*:191 (1985).
11. S. Hjerten, K. Ellenbring, F. Kilar, and J. Liao, *J. Chromatogr.*, *403*:47 (1987).
12. K. Cobb, V. Dolnik, and M. Novotny, *Anal. Chem.*, *62*:2478 (1990).
13. M. Strege and A. Lagu, *Anal. Chem.*, *63*:1233 (1991).
14. C. Bolger, M. Zhu, R. Rodriguez, and T. Wehr, *J. Liquid Chromatogr.*, *14*:895 (1991).
15. F. Regnier and J. Towns, *J. Chromatogr.*, *516*:69 (1990).
16. D. Bentrop, J. Kohr, and H. Engelhardt, *Chromatographia*, *32*:171 (1991).
17. S. Terabe, H. Utsumi, K. Otsuka, T. Ando, T. Inonata, S. Kuze, and Y. Hanaoka, *J. High Resolut. Chromatogr.*, *9*:666 (1986).
18. J. Lux, H. Yin, and G. Schomburg, *J. High Resolut. Chromatogr.*, *13*:145 (1990).
19. M. Bushey and J. Jorgenson, *J. Chromatogr.*, *480*:301 (1989).

20. J. Kohr and H. Engelhardt, *J. Microcol. Sep.*, *3*:491 (1991).
21. J. Mazzeo and I. Krull, *BioTechniques*, *10*:638 (1991).
22. H. Yin, J. Lux, and G. Schomburg, *J. High Resolut. Chromatogr.*, *13*:624 (1990).
23. W. Jones and P. Jandik, *J. Chromatogr.*, *546*:445 (1991).
24. A. Emmer, M. Jansson, and J. Roerade, *J. Chromatogr.*, *547*:544 (1991).

11

Capillary Zone Electrophoresis of Biopolymers with Hydrophilic Fused-Silica Capillaries

Ziad El Rassi and Wassim Nashabeh*

Oklahoma State University
Stillwater, Oklahoma

Soon after its introduction by Jorgenson and Lukacs [1], high-performance capillary zone electrophoresis (CZE), has become a major theme of research and development, and its potentialities are being increasingly exploited in many areas of the life sciences and biotechnology. This is mainly due to its unique selectivity, high resolution, high efficiency, and small sample requirements, as well as to its advanced instrumentation.

Free-zone electrophoresis was originally demonstrated by Hjerten [2] using high electric field strength and rotating glass tubes of 3-mm inner diameter (i.d.). Following this seminal work, Virtanen [3] reported the advantages of using glass columns with smaller inner diameter (i.d., 200–500 μm). In a move toward higher separation efficiencies, Mikkers et al. [4] demonstrated separations with plate heights less than 10 μm using Teflon tubes of 200-μm i.d. The major breakthrough in capillary zone electrophoresis in terms of resolution and separation efficiency was realized when Jorgenson and Lukacs [1,5] performed the separations of amino acids and peptides in open tubes of 75-μm i.d., with on-column fluorescence detection. Three years later, the applicability of CZE was extended to the analysis of neutral species by Terabe et al. [6], who introduced micellar electrokinetic capillary chromatography (MECC, or, as they prefer MEKC). This important modification of CZE allowed the separation of nonionic compounds by their differential partitioning

**Current affiliation*: Northeastern University, Boston, Massachusetts.

between a micellar pseudostationary phase and an aqueous phase. Other electrokinetic capillary chromatography (ECC) methods were reported including cyclodextrin [7], ligand-exchange [8], and ion-exchange [9] ECC. These methods further extended the usefulness of capillary zone electrophoresis to the analysis of various types of species. Another major progress was accomplished by Cohen and Karger [10] who introduced the first sodium dodecyl sulfate (SDS)–polyacrylamide gel-filled capillaries and made significant contribution to the utilization of these capillaries in the separation of large biomolecules and, in particular, oligonucleotides and DNA fragments [11–13].

With the aforementioned several variants, capillary zone electrophoresis has evolved into a powerful instrumental technique. Recently, Braun and Nagydiosi-Rozsa [14] assessed the growth rate of the technique over the last decade using literature-based countable items (e.g., citations to seminal papers and key words on CE and CZE registered in literature data bases). An exponential growth rate was obtained for all counted items, reflecting that the technique is in its growth interval and projecting an extremely fast growth rate in the near future. In addition, the significance of the technique is reflected by the large number of recent review articles on the subject covering the basic principles, the instrumental aspects, and the various applications of the technique.

Concerning the basic concepts in CZE, early reviews by Jorgenson and Lukacs [5,15] and Jorgenson [16,17] provided a simple theory of dispersion in open tubes and discussed the influences of many operational parameters on the quality of separations in CZE. More recently, an excellent study on the effects of diffusion, joule heating, and wall adsorption, as well as the difference in conductivity and pH between the solute zone and the surrounding medium on zone broadening in capillary electrophoresis was published by Hjerten [18].

Many comprehensive review articles have recently appeared. Wallingford and Ewing [19] discussed all aspects of capillary electrophoresis (e.g., theory, sample injection, detection, and separation methodology). Karger et al. [20] reviewed the basic principles of separation and instrumentation of capillary electrophoresis and illustrated some of the typical applications in the biological sciences with both open tubular and gel-filled capillaries. A more extensive review on capillary electrophoresis as an analytical tool was published by Kuhr [21]. More recently, Deyl and Struzinsky [22] provided an up-to-date review on the applicability and potential of capillary electrophoresis in the biochemical analyses.

More specific review articles focusing on either the instrumentation or the various applications of the technique have also appeared. A review on the instrumental aspects of CZE was published by Gordon et al. [23]. Ewing et al. reviewed the various detection modes in capillary electrophoresis [24]. This article also emphasized the advances of CZE in the area of separations-based sensors for the analysis of microenvironments in biological systems. Indirect detection techniques for capillary separations including capillary electrophoresis were covered by Yeung and Kuhr [25]. The applications of CZE in the area of biotechnology were covered by

Compton and Brownlee [26]. Novotny et al. discussed the recent advances in CE of peptides, amino acids and proteins [27]. More recently, Mazzeo and Krull reviewed the strategies of using coated capillaries and buffer additives in the separation of proteins by capillary zone electrophoresis and capillary isoelectric focusing [28]. Quantitative aspects of CE in the areas of pharmaceuticals, peptides, proteins, nucleic acids, clinical analyses, and organic and inorganic ions were reviewed by Goodall et al. [29].

In the last decade, extensive basic and applied research was directed toward improving the various aspects of capillary electrophoresis. An area of unique challenge has been the separations of biopolymers and, in particular, proteins. The major problem has been the pronounced affinity of the positively charged sites of macromolecules to the negatively charged fused silica surface as a result of the ionization of the surface silanol groups. This strong adsorption often gives rise to distorted peaks, poor resolution, low recovery of the separated analytes, and nonreproducible separations owing to the unpredictable changes in the magnitude of the electroosmotic flow.

To realize the full benefit of the high-resolving power of CZE, several attempts have been made to alleviate the problem of solute–wall interactions. These approaches can be classified into two main categories: One category comprises all the methods that allowed the use of untreated fused silica capillaries by providing certain electrophoretic conditions that minimize electrostatic interactions between the analytes and the capillary wall, and another category combines all the approaches dealing with the permanent modification of the capillary wall to produce an inert nonadsorptive surface. These efforts have enabled CZE to play a major role in the separations of proteins and other biopolymers.

The aim of this chapter is to provide a critical review of the recent advances in capillary zone electrophoresis of biopolymers. First, a brief account of the approaches employed with untreated capillaries, (e.g., extreme pH, high salt concentration, and buffer additives) is provided. Second, an in-depth treatment is devoted to the various surface modifications of fused silica capillaries with the hydrophilic coatings introduced thus far. Third, separation methodologies concerning proteins, peptides, nucleic acids, and carbohydrates are illustrated, and the advantages of performing these separations on fused silica capillaries with hydrophilic coatings are discussed. Finally, multidimensional electrophoretic systems (i.e., tandem operations) will be briefly outlined using two different coupled tandem systems, enzymophoresis and on-line preconcentration.

APPROACHES WITH UNTREATED FUSED SILICA CAPILLARIES

Untreated fused silica capillaries with their ionizable surface silanols function as cation exchangers toward positively charged species. Because of the low phase ratio (i.e., low k' values), solute–wall interactions may not be problematic for the separations of small and positively charged species by CZE, particularly when buffers of moderate ionic strengths are used. However, owing to the polyionic nature of large biomacromolecules (e.g., proteins), multipoint attachment of the analyte to the charged surface may cause band broadening or no elution of the analytes. In fact, capacity factors of only 0.05 can result in a 20-fold reduction in the separation efficiency of proteins [30]. Therefore, the separation of proteins with untreated fused silica are possible only under electrophoretic conditions whereby electrostatic interactions can be minimized. The strategies used so far to achieve this goal are (a) inducing coulombic repulsion between the analyte and the capillary wall by raising the pH of the buffer solution above the isoelectric points (pI) of the solutes; (b) operating with buffers of low pH, whereby the surface silanols are not ionized; (c) using relatively high salt concentration in the running buffer so that the counterions would compete with the solute for the available binding sites in a manner similar to ion-exchange chromatography; and (d) the inclusion of various additives in the running electrolyte that can either associate with the solutes or the silica surface, thereby reducing the extent of solute–wall interaction. These various ways of reducing solute adsorption to the capillary surface are briefly described in the following sections. The advantages and disadvantages of each approach are delineated.

Electrolyte pH

Lauer and McManigill were the first to report the separation of proteins at pH higher than their isoelectric points with untreated fused silica capillaries [31]. In this approach, negatively charged proteins are forced away from the capillary surface by coulombic repulsive forces. Under these conditions, proteins having different isoelectric points were separated with an average of half million theoretical plates. However, even at very high pH (i.e., pH 11.0), very basic proteins, such as lysozyme, still exhibited some interactions with the silica surface, and peak tailing was observed. More recently, Lee and Heo demonstrated the separation of red blood cell lysate at alkaline pH using bare fused silica capillaries [32]. Since some proteins may still adsorb to the surface, rinsing the capillary surface with 1.0 M sodium hydroxide solution between electrophoretic runs was recommended for reproducible separations.

McCormick performed the separation of a mixture of basic and acidic proteins using phosphate solutions, pH 1.5, with untreated fused silica capillaries [33]. Also, this approach was successful in the separation of closely related angiotensin II octapeptides as well as the separation of cytochrome c proteins from various sources that differ among each other by only three amino acids in a total sequence of 104 residues. The choice of a pH lower than the isoelectric point of the protein analytes in the mixture was aimed at (a) the reduction of the electroosmotic flow, so that more reproducible analysis can be obtained; (b) ensuring that all analytes will possess a net positive charge, so that cathodal migration can be achieved; and (c) most importantly, the formation of bound phosphate moieties on the capillary surface [34], which may result in reducing the magnitude of the negative charge on the capillary wall. Although protein bands were fairly symmetric, extremely low peak capacity was obtained. This was because the solutes were fully protonated at acidic pH, which greatly diminished charge differences between the analytes.

It follows then from the foregoing discussion that the use of extreme pH values for the analysis of biopolymers limits the selectivity of the system. This one would expect, since the differential migration of charged solutes in CZE is largely influenced by their degree of ionization, which can be conveniently manipulated by varying the pH of the electrolyte. It is highly desirable, therefore, to have the pH as a freely adjustable parameter.

High Salt Concentration

The magnitude of electrostatic interactions between basic proteins and the negatively charged surface of fused silica capillaries can be decreased by increasing the ionic strength of the running electrolyte owing to increased competition between the cationic counterions of the salt and proteins for the cation-exchange sites on the silica surface. Accordingly, Lauer and McManigill [35] demonstrated the separation of five standard proteins (e.g., lysozyme, trypsinogen, horse heart myoglobin, and β-lactoglobulin A and B) at pH 7.0 with an average plate number of 95,000 using a running electrolyte containing 0.25 M K_2SO_4. More recently, Green and Jorgenson [30] performed a detailed study on the effects of the concentration and nature of salts in the running electrolyte on the separation efficiency of proteins. Although useful, high ionic strength in the running electrolyte leads to an increase in the medium conductivity and, consequently, to system overheating. This limits the potential drop that can be applied to as low as 5 kV, and results in longer analysis time [30]. Even though efficient heat dissipation can be obtained through the use of capillaries with inner diameter of 25 μm or less, such small detection path length would lower the detection sensitivity and would necessitate the use of concentrated samples, which may deteriorate the separation efficiency. In fact, the increase in the sample zone concentration leads to significant changes in the local conductivity of the medium, which results in distortions in the local electric field around the sample zone. This phenomenon is generally manifested by band broadening and peak tailing [4].

In an attempt to overcome the conductivity problems associated with the use of high salt concentrations, Bushey and Jorgenson proposed the use of zwitterionic, instead of ionic, salts [36]. These zwitterionic salts do not contribute to the medium conductivity, but would likely associate with both the proteins and the negatively charged surface of the capillary, thereby minimizing solute–wall interactions. However, by using optimum conditions (i.e., 2 M betaine solution in 40 mM phosphate), the separation efficiencies were relatively low, about 3700 and 18,800 theoretical plates for lysozyme and α-chymotrypsinogen A, respectively. According to this study, when zwitterionic salts were combined with moderate concentration of alkali metal salts (e.g., 0.1 M K_2SO_4), the separation efficiencies were comparable with those obtained with high salt concentration (e.g., 0.25 M K_2SO_4). The use of relatively high zwitterionic salt concentrations yielded buffer solutions of higher viscosity, which decreased the medium conductivity. This viscosity effect added to the use of lower salt concentration, permitted higher potential drop to be applied, without greatly increasing the medium conductivity. For instance, the electrolyte containing 2 M betaine, 40 mM phosphate, and 0.1 M K_2SO_4, when used at 20 kV gave a lower current than 0.25 M K_2SO_4, and 40 mM phosphate operated at only 10 kV. As a result, the analysis time was reduced by almost 50%, without a substantial drop in efficiency. However, the major drawback of this method is the increased UV-absorbance background of most zwitterionic salts at low wavelength, which would either exclude the use of this detection mode or does not allow high-detection sensitivity to be achieved. One can envision from this study that viscosity agents such as sugars or ethylene glycol, which are transparent in the UV, could be substituted for zwitterions and may allow the use of high salt concentration without running into problems of high conductivity.

Electrolyte Additives

Several buffer additives have been tested in the separations of proteins with untreated fused silica capillaries. Ethylene glycol was reported [37] to be effective in enhancing the run-to-run reproducibility in the quantitative analysis of gamma globulins with naked fused silica capillaries at high pH. The effects of five different buffer additives, such as methylcellulose, ethylene glycol, Triton X-100R, and multicomponent mixtures of zwitterionic species (e.g., Z-60 and Z-4A) on the extent of hemoglobin adsorption to fused silica surfaces were recently investigated [38]. Among these additives, only Z-60 zwitterions yielded a substantial reduction of protein adsorption.

Recently, a cationic fluorosurfactant, the fluorad FC 134, was introduced as a new buffer additive [39]. Such surfactant, when used at concentrations higher than its critical micellar concentration (CMC), forms a micellar bilayer on the capillary wall owing to the hydrophobic interactions between its nonpolar chains. Although the first cationic layer neutralizes the negatively charged surface of the fused silica capillary, the second layer pointing toward to the buffer solution would produce a positively charged surface, thereby inverting the direction of the electroosmotic flow as illustrated in Fig. 1. This idea mimics the charge reversal coating introduced by Wiktorowicz and Colburn [40] (see next section). Four model basic proteins (Fig. 2) could be separated, with an average plate number of about 330,000 theoretical plates. The elution order of the separated proteins in the negative polarity mode was in the reversed order, with the less basic solute eluting first. The fluorad FC

A

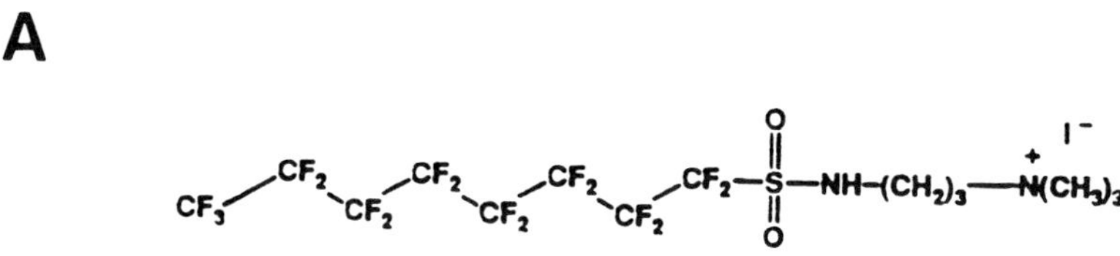

B

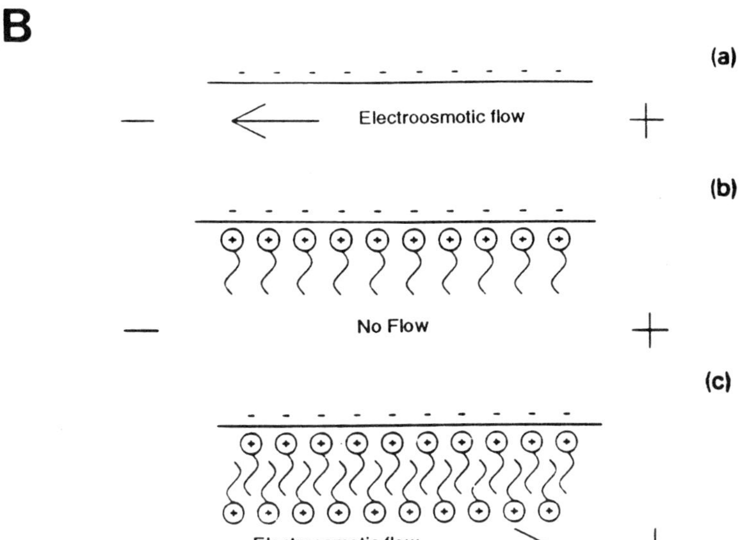

Figure 1 (A) General structure of Fluorad FC 134. (B) Schematic illustration of the capillary surface when (a) no surfactant is added, electroosmotic flow in normal direction; (b) in the presence of the surfactant, electrostatic adsorption of the positively charged surfactant head-groups to the negative silanol groups on the silica surface of the capillary inner wall; (c) in the presence of excess surfactant, admicellar bilayer formation by hydrophobic interaction between the apolar chains, resulting in a reversal of the electroosmotic flow. (From Ref. 39)

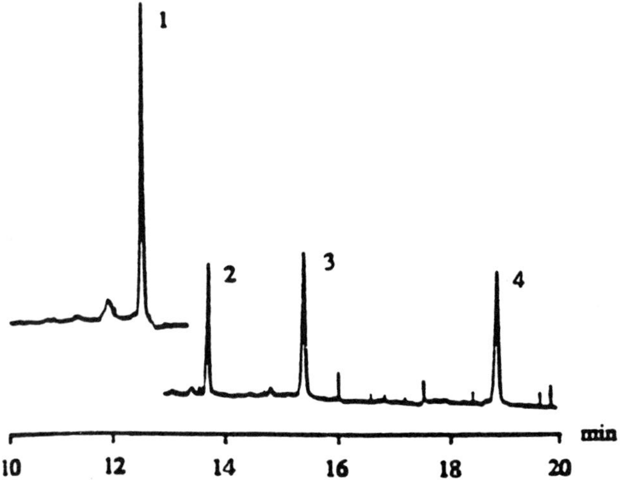

Figure 2 Electropherogram of four model proteins (myoglobin run separately). Capillary, 90 cm (to the detection point), 100 cm total length × 50 μm i.d.; electrolyte, 0.01 M phosphate solution, pH 7.0 containing 50 μg/mL FC 134; running voltage, −30 kV. Proteins: 1, myoglobin; 2 RNase A; 3, cytochrome c; 4, lysozyme. (From Ref. 39)

134 yielded a stable surfactant bilayer on the capillary wall and, as a result, the reproducibility of the migration time of the separated proteins was satisfactory. However, this approach requires the use of a pH that is lower than the isoelectric point of the analytes and, therefore, is useful in the analysis of basic proteins only in the physiological pH range.

Although untreated fused silica capillaries can be used in the analysis of proteins under certain conditions, various limitations do exist, such as narrow pH range, high conductivity, low detection sensitivity, and the ability to analyze only one class of proteins. All together this limits the range of applicability of untreated fused silica capillaries and does not provide a universal solution for the solute–wall interaction problem.

SURFACE MODIFICATION

In their pioneering work on electrophoresis in glass capillary tubes, Jorgenson and Lukacs concluded that the analysis of proteins and other large macromolecules would require more inert and nonadsorptive surfaces than untreated glass [1]. These early studies prompted the development of various types of hydrophilic coatings.

An ideal surface modification procedure should, in principle, meet the following criteria: (a) The hydrophilic coating should shroud sensitive biopolymers from direct contact with the surface proper of the capillary and should not interact with the biopolymer solutes over a wide range of operating conditions. (b) The hydrophilic coating should be chemically stable and resist hydrolytic degradation when in contact with aqueous solutions customarily used in capillary electrophoresis. (c) The surface modification should not completely inhibit the electroosmotic flow, which is essential for the separation of oppositely charged solutes in CZE. (d) The coating procedure should yield reproducible capillaries so that constancy in migration time of the analytes from run-to-run, day-to-day, as well as from column-to-column could be obtained. (e) The surface modification should be simple, involving only a limited amount of time and labor.

Several hydrophilic coatings covalently bonded to the capillary wall have been reported over the last few years and have showed significant improvements in the elution profile and separation efficiencies of biopolymers. These coatings can be classified into two types: neutral and charged. Although the neutral-type coatings have been useful for a wide range of biopolymers, the charged coatings have found limited applications.

Neutral Coatings

As early as 1983, Jorgenson and Lukacs [5] reported the modification of the surface of fused silica tubes with γ-glycidoxypropyltrimethoxysilane following an approach that was developed earlier [41] for the preparation of silica-based stationary phases for biopolymers high-performance liquid chromatography (HPLC). These surface-modified capillaries yielded sharp peaks and predictable elution order of the separated proteins (i.e., proteins of similar size eluted in the order of decreasing isoelectric point). Standard proteins as well as human serum blood proteins could be separated without significant adsorption to the capillary walls. However, according to these early attempts, the hydrophilic capillaries exhibited short-term stability and showed inconsistency in performance from column-to-column.

Two years later, glass or quartz tubes with noncrosslinked polyacrylamide coatings were reported by Hjerten [42]. The noncrosslinked polyacrylamide coating was obtained by first reacting the silica surface with a bifunctional silane, such as γ-methacryloxypropyltrimethoxysilane, in which one or two of the methoxy functions can react with the silanol groups on the capillary surface, while the acryl end groups provided the reactive sites for the attachment of the noncrosslinked polyacrylamide layer. This procedure led to the formation of a thin monomolecular layer, covalently attached to the glass surface (Fig. 3) as opposed to the methylcellulose coating described earlier by the same investigator [2] that produced a thicker hydrophilic layer. Also, the methylcellulose coating was time-consuming, entailing the treatment at low pH in the presence of formaldehyde and elevated temperatures. It should be noted that very thin polymeric layers (< 0.05 μm) were found efficient enough in covering the capillary surface silanols for analytical purposes [43]. How-

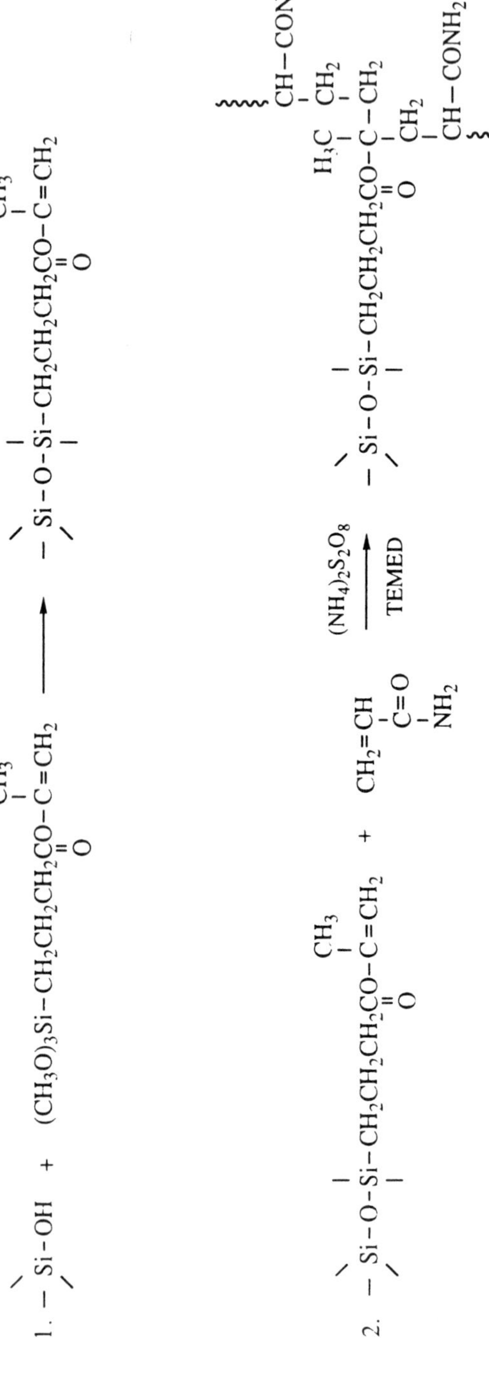

Figure 3 Reaction scheme for the preparation of polyacrylamide-coated capillaries.

ever, the polyacrylamide coating showed poor hydrolytic stability when operated at high pH. Recently, a more stable version of the polyacrylamide coating has been manufactured by Bio-Rad on fused silica capillaries and has found several useful biomedical applications [Bio-Rad, bulletin 1641].

In an attempt to enhance the coating stability over a wide range of pH, the polyacrylamide coating was attached to the silica wall by an Si—C bond [44], as opposed to the siloxane linkage (i.e., Si—O—Si, described by Hjerten [42]). The silica surface was first reacted with thionylchloride. Following, the chlorinated surface was converted to a vinyl-activated surface through a Grignard reaction with vinylmagnesium bromide. Finally, the bound vinyl functions were reacted with an acrylamide monomer in the presence of polymerizing agents, such as N,N,N',N'-tetramethylethylenediamine (TEMED) and ammonium persulfate to yield a linear noncrosslinked polyacrylamide coating. The procedure employed for the preparation of fused silica capillaries with vinyl groups was previously described for the modification of silica surface [45,46]. The polyacrylamide coating with Si—C bond yielded capillaries with excellent hydrolytic stability over the pH range 2–10.5. Enhanced run-to-run reproducibility was also observed with a relative standard deviation (RSD) of 0.21% for the cytochrome c solute as opposed to an RSD of 1.84% on untreated fused silica capillaries under otherwise identical conditions. Also, this coating increased the plate counts for acidic and basic proteins by a factor of approximately 4 when compared with untreated capillaries. However, this surface treatment substantially reduced the electroosmotic flow and, consequently, necessitated the use of extreme pH buffers to allow various proteins to travel in the same direction by their own electrophoretic mobilities. In fact, no protein analysis was carried out in the physiological pH range.

Following the procedure described by Hjerten [42], McCormick [33] coated the inner surface of the capillary with a linear polymeric layer of polyvinyl pyrrolidone (PVP) using γ-methacryloxypropyltrimethoxysilane as an organosilane spacer. As shown in Fig. 4, the PVP capillary allowed the separation of 15 proteins with relative molecular masses (M_r) ranging from 12 to 72 kDa and isoelectric points of 4.5–11 in less than 25 minutes. However, two of the analyzed proteins, namely bovine serum albumin and human hemoglobin, showed multiple peaks, a phenomenon that was attributed to the denaturation of these proteins to constituent subunits under the low pH buffer condition used in this study. This once more demonstrates the disadvantage of operating at extreme pH values. Also, that at low pH proteins, such as α-chymotrypsinogen A and ovalbumin, could not be eluted may be due to strong hydrophobic interactions with the coated surface.

Other types of hydrophilic polymeric coatings were also developed. Polyethylene glycol (PEG) 600 was bonded to the capillary wall [47] using γ-glycidoxypropyltrimethoxysilane as a coupling reagent. The PEG capillaries were stable for several months of use and showed a substantial decrease in the adsorption activity of the wall toward proteins when the capillary was operated at pH 3–5. However, at

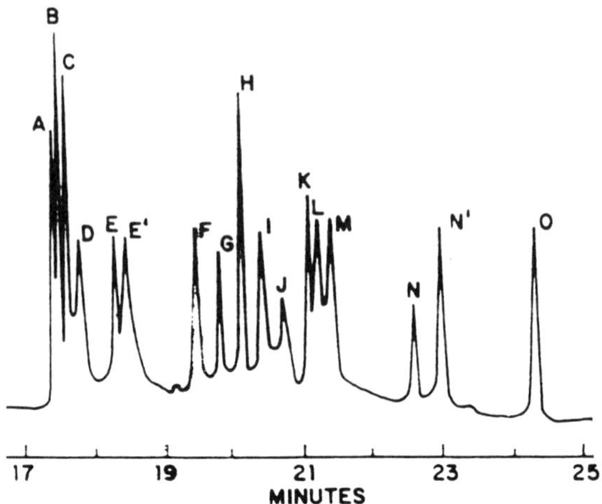

Figure 4 Electropherogram of 15 model proteins obtained on a PVP-modified capillary. Capillary, 75 cm (to the detection point), 110 cm total length × 52 μm i.d.; electrolyte, 38.5 mM H_3PO_4, 20 mM NaH_2PO_4, pH 2.0; running voltage, 5–25 kV linear program in 150 s. Proteins: A, β-lactoglobulin B; B, β-lactoglobulin A; C, lysozyme; D, human serum albumin; E, bovine serum albumin; F, cytochrome c; G, trypsinogen; H, whale myoglobin; I, transferrin; J, conalbumin; K, horse myoglobin; L, carbonic anhydrase B; M, carbonic anhydrase A; N, hemoglobin; O, parvalbumin. (From Ref. 33)

higher pH the peak shape was largely distorted, and a sharp decrease in the resolution was observed. This is probably the result of strong interactions between the analytes and residual silanols that became more ionized at higher pH values. In another approach, the same laboratory [48] reported a diol coating, which was prepared by binding γ-glycidoxypropyltrimethoxysilane to the capillary wall, followed by acid hydrolysis. The diol coating exhibited lower efficiency than that reported on the PEG capillary in the pH range 3–5, and similarly showed severe peak distortion at higher pH. In a third approach, the same investigators modified the capillary wall with carbohydrate moieties [48]. The capillary surface was first reacted with γ-aminopropyltriethoxysilane and thereafter with a maltose solution in the presence of sodium cyanoborohydride by a reductive amination reaction. This coating exhibited a significantly lower plate number (about 25,000) when compared with both the diol and PEG coatings. Because of its sugar content, the coating was susceptible to bacterial degradation that may develop in the aqueous buffer solutions with time. A good stability over the pH range 4–7 was observed only when an antimicrobial agent (e.g., sodium azide) was added to the running buffers.

Swedberg introduced an arylpentafluoro coating by first reacting the capillary surface with γ-aminopropyltriethoxysilane and then with pentafluorobenzoyl chloride [49]. These capillaries proved quite effective in minimizing solute adsorption when operated at high salt concentration. This report does not discuss the effect of unreacted amino groups, and no indication is made of whether these residual amino groups were scavenged or not. The separation of seven model proteins in the pI range 5.9–11.0 was shown, with an average plate number of 550,000 using a 200-mM phosphate buffer containing 100 mM KCl, pH 7.0. However, when operated at lower ionic strength (i.e., 125 mM phosphate), a substantial drop in the system efficiency was observed, and ribonuclease, trypsinogen, and whale myoglobin were not resolved. Furthermore, the lysozyme peak was not eluted, and that was attributed to sample aggregation at low ionic strength. Such a phenomenon is unlikely to occur, since lysozyme was efficiently analyzed by other researchers using buffers having the same or lower ionic strength [5,50]. It seems that the high ionic strength reduced hydrophobic interactions with the surface by a salting-in effect (i.e., reduction in retention due to salt binding to the protein analyte). The high salt dictated the use of capillaries with small inner diameter (20 μm) and, therefore, suffered from low detectability and limited sample capacity. More recently, Maa et al. deactivated the capillary surface with an amphoteric protein such as α-lactalbumin [51]. This surface-modification procedure was carried out by first converting the fused silica capillary to aminopropylsilyl-silica, and then the protein was immobilized through glutaraldehyde attachment. This coating showed satisfactory separation for basic proteins at neutral pH, with an average plate count that exceeded 100,000. As expected, owing to the presence of an immobilized protein on the capillary wall, the zeta potential of the capillary changed from positive to negative as the pH of the electrolyte was increased and, consequently, the direction of the electroosmotic flow was inverted. This behavior is shown in Fig. 5, whereby the flow is anodal at pH $<$ pI and cathodal at pH $>$ pI of the immobilized protein. Thus, with capillaries having surface-bound proteins, the direction of the electroosmotic flow can be switched by changing the pH.

Recently, besides polyacrylamide-coated capillaries from Bio-Rad, other coatings have become available from Supelco Inc. [The Supelco Reporter (1991). Vol. 10, No. 4]. This manufacturer has introduced a series of CElect-bonded phase columns for use in capillary electrophoresis. These columns are denoted as moderately hydrophobic C_8 (CElect-H^1), highly hydrophobic C_{18} (CElect-H^2), and polar hydrophilic phases (CElect-P). The hydrophobic columns reduced the electroosmotic flow by almost 40% when compared with untreated fused silica capillaries. More importantly, a relatively steady flow was obtained over the pH range 3–9. Unexpectedly, the reproducibility of the migration time was worse than that on untreated capillaries. Contrary to other observations [52] that a hydrophobic octadecylsilyl capillary phase readily adsorbed proteins, successful analysis of conalbumin and myoglobin was reported using the CElect-H^2 column. In addition, the CElect-P column proved efficient in the analysis of several proteins at near neutral pH and under low ionic strength conditions.

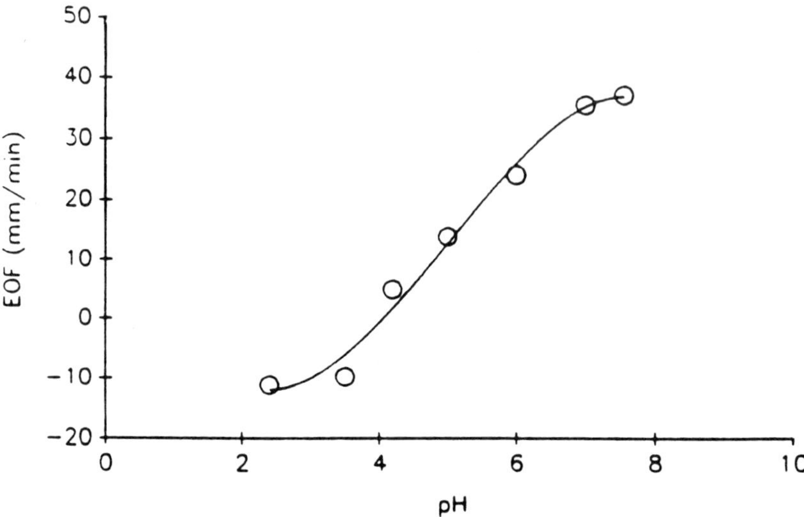

Figure 5 Plot of electroosmotic flow rate obtained on an α-lactalbumin-bonded capillary versus pH of the running electrolyte. (From Ref. 51)

Towns and Regnier introduced hydrophobic C_{18} capillary columns for the separation of acidic and basic proteins in the presence of nonionic surfactants [52]. The capillary surface was first reacted with octadecyltrichlorosilane in methylene chloride as the solvent to produce a C_{18} surface that is similar to that used in reversed-phase chromatography. Thereafter, micellar solutions of nonionic surfactants were hydrophobically adsorbed onto the C_{18} phase to yield a thick hydrophilic layer. According to this report, the nature of the nonionic surfactant, as well as its concentration, played important roles in the coating performance. Among five water-soluble surfactants investigated (e.g., Tween 20, 40, 80, and Brij 35 and 78), the surfactant Brij 35 gave the most satisfactory results in terms of plate height and peak capacity. This may be explained by the larger size of the polar headgroups of Brij 35 and its balanced hydrophile–lipophile character relative to the other surfactants. Also, for the first time, this study introduced a method for the determination of the percentage of protein recovery during electrophoretic runs. The percentage recovery was calculated by subtracting the solute peak areas obtained from two detectors set at 45 cm apart. Figure 6 illustrates the high protein recovery measured with the two-detector system placed at 20 and 65 cm from the injection end, using the C_{18} capillaries coated dynamically with Brij 35. It should be noted that the percentage recovery of proteins in the presence of Brij 35 was greater than 90%. In general, efficient and reproducible protein separations were achieved over a wide pH range.

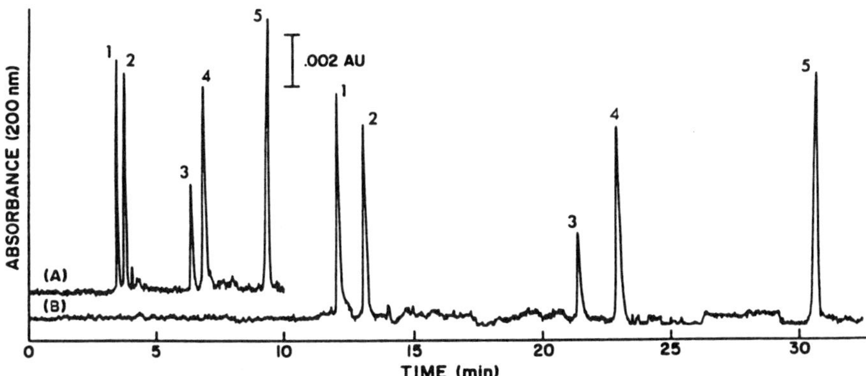

Figure 6 Electropherograms of model proteins on a Brij 35-coated capillary. Capillary, 80 cm total length × 75 μm i.d., with detectors at (A) 20 cm and (B) 65 cm from the injection end; electrolyte, 0.01 M phosphate, pH 7.0; running voltage 300 V cm^{-1}. Proteins: 1, lysozyme; 2, cytochrome c; 3, RNase A; 4, α-chymotrypsinogen A; 5, myoglobin. (From Ref. 52)

In a recent report from our laboratory [50], we have introduced two surface modification schemes that yielded stable and inert capillaries having moderate electroosmotic flow. The coating procedures involved the covalent attachment of polyether chains of various length to the inner surface of fused silica capillaries by siloxane bonds, either directly, using polyether chains with trimethoxysilane groups at both ends, or by attaching a top polyether layer to a polysiloxane sublayer already anchored to the capillary inner surface. The first type of coatings is referred to as *interlocked coatings* and the second as *fuzzy coatings*.

The idealized structure of the fuzzy coating is shown in Fig. 7. This hydrophilic layer falls into the category of stratified coatings. The first stratum is formed by covalent attachment of γ-glycidoxypropyltrimethoxysilane to the silanol groups on the inner surface of fused silica capillaries. Since the reaction is made in water, a glyceropropylpolysiloxane sublayer is formed in the first stratum. Thereafter, the capillary is reacted with an equimolar mixture of polyethylene glycol diglycidyl ether (M_r 600) and PEG 2000 or PEG 600 in dioxane leading to the formation of polyether chains of different lengths, with either one end or two ends covalently attached to the sublayer. The capillary codings F-600 and F-2000 are to denote the fuzzy coatings having PEG 600 or PEG 2000, respectively. In contrast with the coating reported by Bruin et al. [48], in which the PEG hydrophilic layer was not further reacted to enhance its shielding effect and stability, the crosslinked polyether top layer in F-600 and F-2000 produced chemically stable capillaries, and also

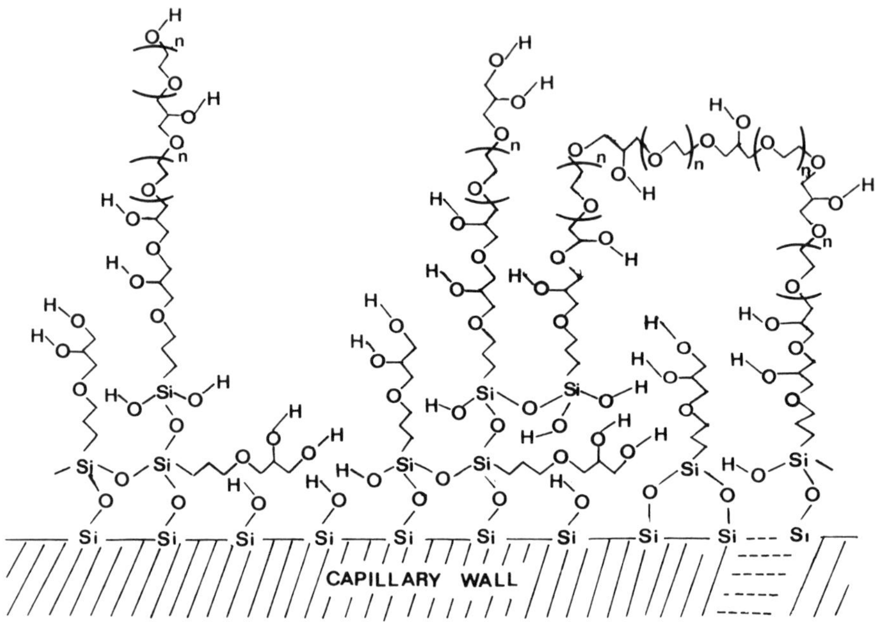

Figure 7 Schematic illustration of the idealized structures of fuzzy polyether coatings of fused silica capillaries. (From Ref. 50)

provided the hydrophilicity and protection required for biopolymer separation in open tubular fused silica capillaries.

Figure 8 depicts the idealized structure of the interlocked hydrophilic coatings. They consisted of polyether polysiloxane chains of which the monomeric units at both ends are covalently attached to the surface with possible interconnection. These coatings were prepared by allowing the capillary surface to react at 100°C with PEG 200 having trimethoxysilane at both ends in dimethyformamide (DMF) solution. The silicone functional groups at both ends of the polyethylene glycol molecules, reacted readily with the silanol groups on the inner walls of fused silica capillaries. It is also believed that the silicone functional groups at both ends of the polyethylene glycol molecules connect with each other, forming a homogeneously covered surface with polyether polysiloxane network. These capillaries were denoted as I-200.

The objective of the various modification procedures described in this work was to provide a series of hydrophilic capillaries with various levels of electroosmotic flow that suit the analysis of different biopolymers with high separation efficiencies (i.e., minimum solute–wall interactions). In fact, the foregoing surface modification procedures yielded capillaries with different magnitudes of electroosmotic flow, as illustrated in Fig. 9. The fuzzy coatings exhibited lower electroosmotic flow than

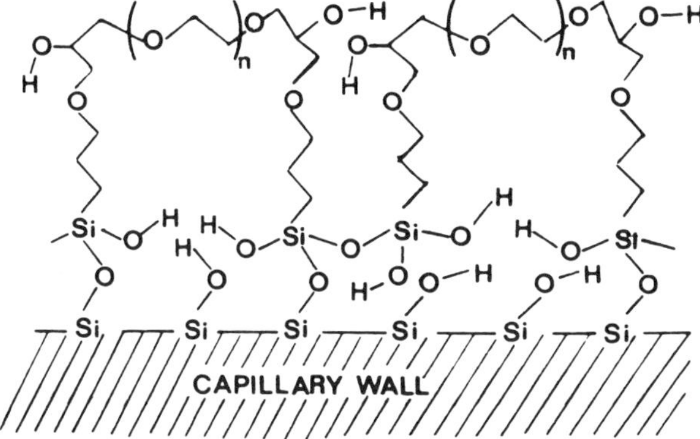

Figure 8 Schematic illustration of the idealized structures of the interlocked polyether coatings of fused silica capillaries. (From Ref. 50)

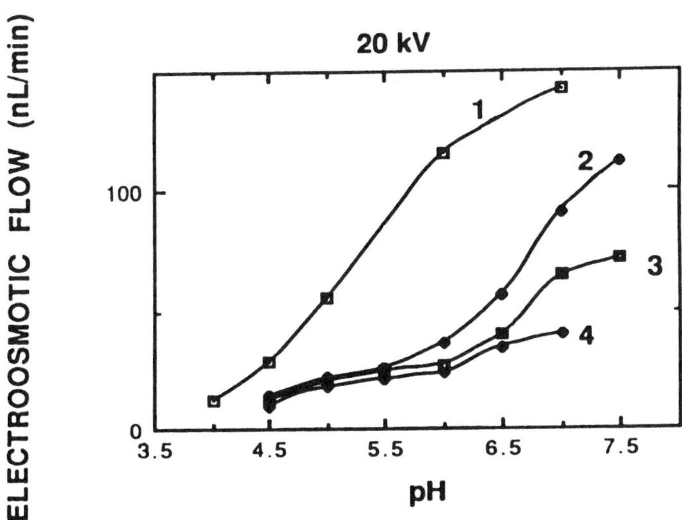

Figure 9 Plots of electroosmotic flow rates obtained on coated and uncoated capillaries versus pH of the electrolyte at 20 kV. Fused silica capillaries: 1, untreated; 2, I-200; 3, F-600; 4, F-2000. All tubes were 50 cm (to the detection point), 80 cm total length × 50 μm i.d.; electrolyte, 0.1-M phosphate solutions at different pH. Inert tracer, phenol. (From Ref. 50)

the interlocked ones. This is because the fuzzy coatings have an additional crosslinked polysiloxane sublayer that the interlocked coatings do not have (see Figs. 7 and 8). On the other hand, the difference in the flow rate between the two fuzzy coatings (i.e., F-600 and F-2000) may be related to the size of the polar top layer.

Furthermore, the various polyether capillaries yielded satisfactory separations of proteins by CZE. As shown in Fig. 10, relatively high separation efficiencies were obtained on the fuzzy 2000 capillary, with an average plate number of 270,000 for basic proteins. That the interlocked coatings exhibited, on the average, lower separation efficiencies may be due to significantly higher electroosmotic flow. The relatively high electroosmotic flow velocity obtained with I-200 capillaries permitted the analysis of both positively and negatively charged proteins in a single electrophoretic run in the positive polarity mode. Figure 11 illustrates the separation of eight acidic and basic standard proteins spanning a wide range of molecular masses.

All coatings exhibited constant performance in terms of efficiency and resolution for several weeks when operated at pH 6.0 or below, and for more than 80 hours

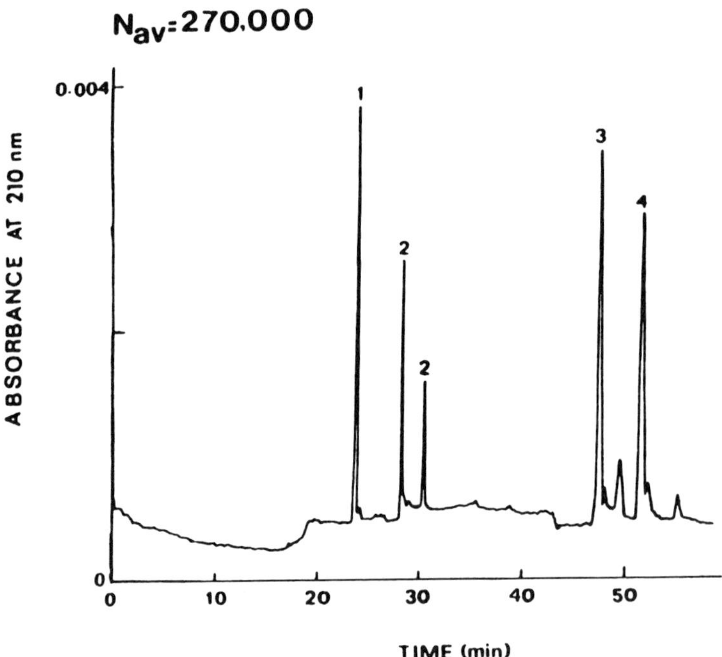

Figure 10 Electropherogram of basic proteins obtained on the fuzzy 2000-coated capillary. Capillary, 50 cm (to the detection point), 80 cm total length × 50 μm i.d.; electrolyte, 0.1-M phosphate solution at pH 7.0; running voltage, 17 kV. Proteins: 1, lysozyme; 2, cytochrome c; 3, RNase A; 4, α-chymotrypsinogen A. (From Ref. 50)

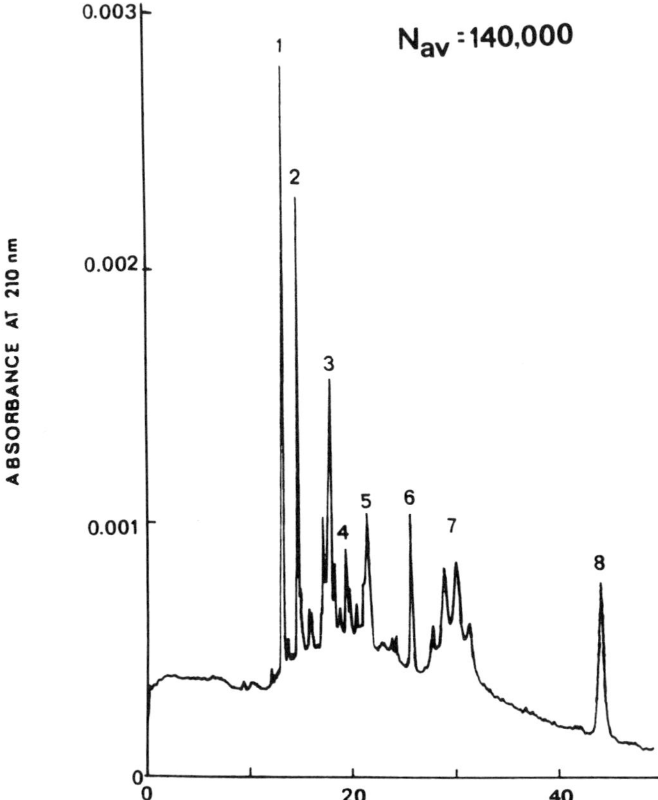

Figure 11 Electropherogram of basic and acidic proteins obtained on the I-200 capillary. Electrolyte, 0.1-M phosphate solution, pH 7.0; running voltage, 17 kV. All other conditions as in Fig. 10. Proteins: 1, lysozyme; 2, cytochrome c; 3, RNase A; 4, α-chymotrypsinogen A; 5, myoglobin; 6, carbonic anhydrase; 7, transferrin; 8, α-lactalbumin. N_{av} was calculated on the well-defined peaks (i.e., peaks 1, 2, 4, 6, and 8). (From Ref. 50)

when used in the pH range 6.5–7.0. In addition, the various coatings exhibited reproducible separations. For instance, the RSDs of the migration time of lysozyme on the interlocked polyether coating from run-to-run, day-to-day, and column-to-column were 0.9, 1.6, and 1.5%, respectively.

Charged Coatings

The idea behind charged coatings is to induce coulombic repulsive forces between the proteins and the capillary surface, both having the same net charge. It mimics the approach of using untreated fused silica capillaries at high pH. The charged coatings have the advantages over untreated capillaries of allowing a wider pH range to be used for the analysis of basic proteins. However, under most experimental conditions, these coatings do not permit the analysis of acidic proteins.

Towns and Regnier [53] developed hydrophilic capillaries having positively charged polyethyleneimine (PEI) layers by crosslinking PEI with ethylene glycol diglycidyl ether in the presence of triethylamine. The thick layer (30 Å) thus produced was effective in masking the silanol groups from interacting with the separated proteins. However, this coating was useful only for the analysis of positively charged proteins, which were repulsed from the surface, and showed pronounced solute–coating interactions with acidic proteins.

In another approach, Wiktorowicz and associates [40,54] introduced the concepts of electrocoating and dynamic surface charge reversal for the analysis of basic proteins at a pH lower than their isoelectric points. Electrocoating was defined by the authors as the process by which a coating agent is propelled through the capillary by virtue of electroosmotic flow. In this method, a positively charged reagent (not specified, presumably an amine polymer) was used to coat the capillary surface. As more reagent was introduced, the electroosmotic flow decreased proportionally until the flow was completely inhibited. Semilogarithmic plots of the electroosmotic flow velocity versus the run time at two different initial coating concentrations revealed a first-order reaction rate. Thus, by adjusting the run time and the concentration of the coating reagent, various levels of electroosmotic flow can be attained. By applying an excess of the coating reagent, charge reversal of the capillary surface would result (i.e., from negative to positive). This procedure resulted in a stable coating, such that the inclusion of the polymeric compound in the running buffer is not necessary, which diminished any potential interaction between the polymer and the analytes. By using negative polarity mode, the separations of the isozymes of lactate dehydrogenase and multiacetylated forms of H4 histones could be carried out. However, as with the PEI coating, this approach was useful only for the analysis of positively charged species and cannot be applied for a wider range of proteins.

SELECTED SEPARATION METHODOLOGIES WITH HYDROPHILIC CAPILLARIES

Capillary zone electrophoresis is increasingly used in biopolymer separation and is becoming an important microanalytical tool. This is partly due to the recent advances in column-coating technologies and to the development of automated instruments by several instrumentation companies. Besides minimizing solute–wall interactions, hydrophilic capillary coatings have greatly improved the reproducibility of the electrophoretic system, a parameter that is important for the quantitative and qualitative determination of the analytes.

This section is not intended to exhaustively review biopolymer CZE separations reported in the literature. Instead, selected applications will be illustrated including (a) separations of proteins in complex biological media, (b) elucidation of glycoprotein microheterogeneity, (c) tryptic peptide mapping, (d) analysis of DNA restriction fragments, and (e) separations of oligosaccharides. A special emphasis will be placed on the advantages of performing these separations on fused silica capillaries with hydrophilic coatings.

Proteins

Protein separations are the most challenging and yet interesting applications of capillary electrophoresis. Capillary zone electrophoresis, combining both the resolution of traditional electrophoresis and the advanced instrumentation of HPLC, should permit protein analysis with high-separation efficiencies. It is anticipated that, with the advent of commercial capillary electrophoresis systems and the improvements in the column technology, the usefulness of CZE in the proteins area will expand.

Jorgenson and Lukacs were the first to report the CZE of human blood serum proteins with surface-modified fused silica capillaries [5]. This early work demonstrated the potential of CZE in the protein area and pointed out the effects of the inherent properties of some proteins such as microheterogeneity and conformational changes on the quality of separations. Recently, Zhu et al. analyzed human red blood lysate proteins in less than 8 minutes using a Bio-Rad-coated capillary tube [38]. Excellent reproducibility in terms of peak height, peak shape, and migration time of human serum proteins were obtained for over 49 consecutive electrophoretic runs. The good precision was mainly the result of using coated capillaries, with buffers containing zwitterionic species, and purging the columns between runs with acidic solutions.

Very recently, the usefulness of CZE with coated capillaries has been extended to the monitoring of proteins constituents of biological fluids for the purpose of diagnosing disorders. Gurley et al [55] demonstrated the potential of CZE in the analysis of lung fluid proteins for the assessment of lung diseases that are associated with exposure to perfluoroisobutylene (PFIB). In this study, the CZE analysis of lung fluid proteins obtained from a rat that had been exposed to PFIB revealed the disappearance of an unidentified positively charged species as well as the increase by 40- and 30-fold in the response signal of the albumin and transferrin peaks, respectively, when compared with the electropherogram of a sample obtained from a normal rat. In addition, capillary electrophoresis provided a different selectivity than HPLC in this application. The CZE analysis of the transferrin peak isolated by reversed-phase chromatography (RPC) fractionation revealed the presence of two unresolved solutes in the HPLC fraction, namely, a guanidine contaminant and an IgG-like protein. Given these studies, it was concluded that the pulmonary edema resulting from PFIB exposure is accompanied by a massive translocation of blood compartment proteins (e.g., albumin, transferrin, and IgG) into the lung's alveolar compartment. This application clearly demonstrates the high resolving power of

CZE with hydrophilic coatings as a useful and complementary analytical tool to HPLC for monitoring protein changes induced by the inhalation of toxic materials. In another application from the same laboratory [56], various histone proteins (basic proteins) extracted from Chinese hamster cells were successfully analyzed by CZE with coated capillaries, as shown in Fig. 12.

The use of modified capillaries with moderate electroosmotic flow was also successful in the fingerprinting of crude protein mixtures in which the pI values of all the analyzed components were unknown. Figure 13 portrays an electropherogram of commercially available crude soybean trypsin inhibitor performed on a fused silica capillary with interlocked polyether coating (I-200) using 0.1 M phosphate, pH 6.5, as the running electrolyte [50]. It can be seen that the mixture contains at least 23 minor components besides the major analyte. Thus, this system can be used for assessment of purity and in quality control of protein extracts. This example illustrates the high efficiency that can be obtained in the analyses of complex protein mixtures with simple buffer systems.

Wenisch et al. [57] applied CZE with hydrophilic capillaries to monitoring the progress and purity level of proteins produced by recombinant DNA (rDNA) techniques, such as monoclonal antibodies against the gp-41 of human immuno-

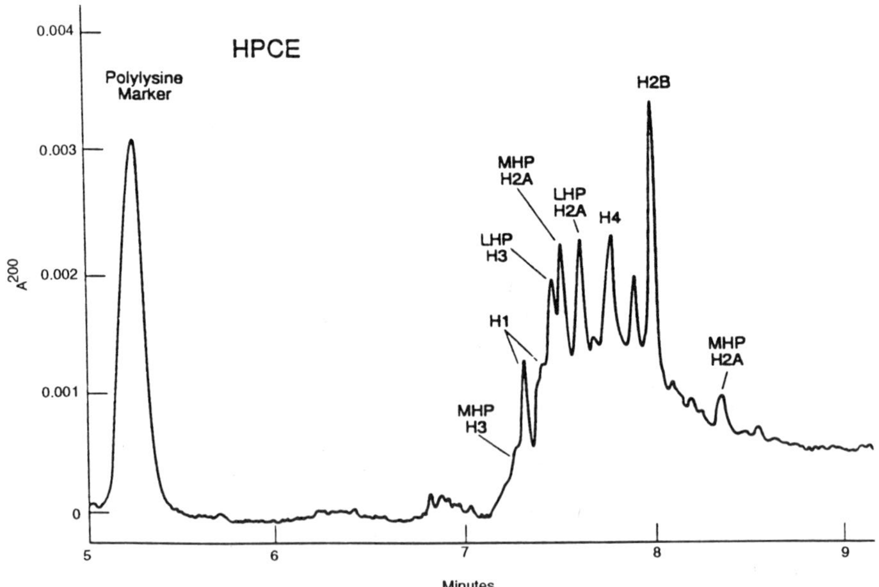

Figure 12 Electropherogram of histone proteins extracted from Chinese hamster cells. Capillary, 35 cm (to the detection point), × 50 μm i.d., with polyacrylamide coating; electrolyte, 0.1 M phosphate, pH 2.5; running voltage, 10 kV. (From Ref. 56)

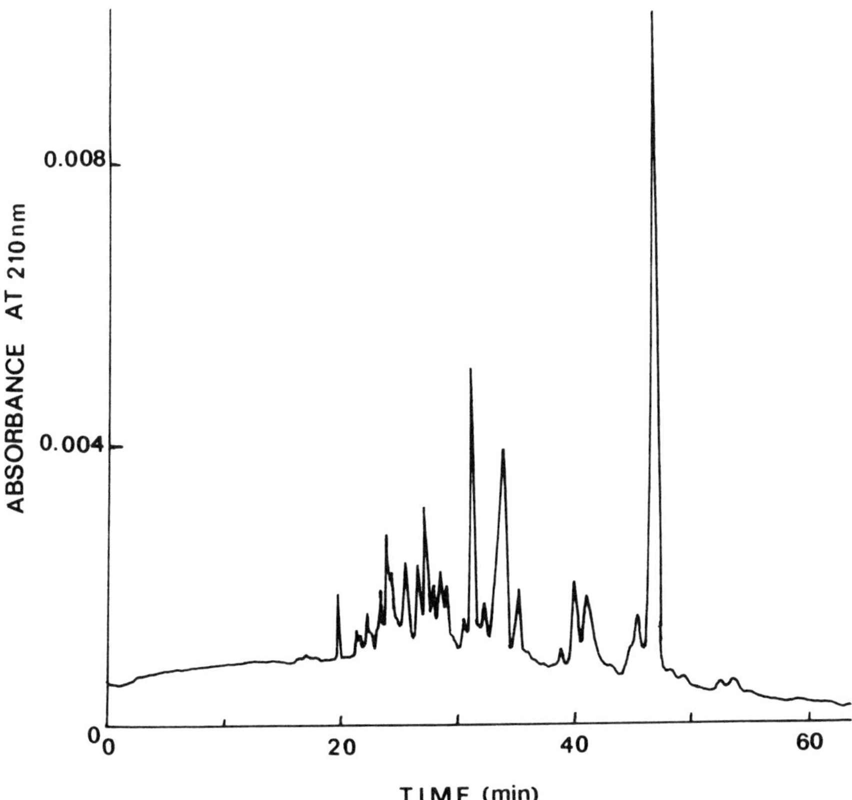

Figure 13 Electropherogram of crude soybean trypsin inhibitor. Capillary, I-200; electrolyte, 0.1-M phosphate solution, pH 6.5; running voltage, 17 kV. All other conditions as in Figs. 10 and 11. (From Ref. 50)

deficiency virus (HIV) and against superoxide dismutase. Even though the resolving power of CZE fell short from that of isoelectric focusing in immobilized pH gradients (IPG), it offered several advantages over IPG, mainly in terms of (a) speed of analysis, which can be reduced from as much as 10 hours in narrow-range IPG to only 20 minutes in CZE; (b) enhanced detection sensitivity; and (c) the possibility of accurate quantitation of the different protein zones, a very important piece of information in assessing the level of impurities in rDNA. This report has also demonstrated the usefulness of coated capillaries in the analysis of basic proteins, such as monoclonal antibodies, having slightly different pI values in the range 9.1–9.6.

Another interesting application of CZE with coated capillaries in the biotechnology area has been the comparison of the level of purity of biosynthetic proteins produced by rDNA to that of protein extracts from human tissue [58]. The CZE analysis of biosynthetic human growth hormone (met-hGH) yielded a single peak, whereas several peaks were obtained for growth hormone extracted from human pituitaries under otherwise identical conditions. The same study has demonstrated the usefulness of CZE in the rapid quantitative separation of native hGH from its deamidated variant. The deamidation of hGH occurs upon storage and corresponds to the conversion of asparagine residues 149 and 152 to aspartic acid. Because of differences in the charge/mass ratios, the additional negative charges of the carboxylic side chains of the aspartyl residue in the deamidated form permitted its separation from the native hGH. The amount of deamidated hGH variant determined by CZE was similar to that obtained by HPLC. In addition, as shown in Fig. 14, CZE yielded different selectivity than that obtained by HPLC, as the more acidic deamidated variant eluted before the native protein in the electrophoretic run.

Recent developments in biotechnology and biomedical research has engendered the need for powerful separation techniques for the characterization of proteins having closely related glycoforms. This is because the protein glycosylation pattern plays an important role in directing glycoproteins metabolic fate and their therapeutic functions. An important application of CZE in the area of glycoproteins has been the illustration of the glycoforms of recombinant human tissue plasminogen activator (rtPA), a fibrin-specific protein that has been approved for the treatment of myocardial infarctions [59]. The CZE analysis of rtPA using Bio-Rad hydrophilic capillaries elucidated the microheterogeneity of the glycoprotein, as manifested by the partial resolution of almost 15 glycoforms in a protein that has only four possible *N*-glycosylation sites. This report compared the CZE profile of an rtPA sample to that of a desialylated rtPA obtained through neuraminidase treatment. As shown in Fig. 15, the desialylated rtPA exhibited a much simpler CZE profile, indicating that the glycoprotein microheterogeneity is mostly the result of different levels of sialylation. Thus capillary zone electrophoresis can be efficiently employed in the analysis of glycoproteins with complex carbohydrate structures. Untreated fused silica capillaries can be also used in the separation of glycoproteins, provided that solute–wall interactions can be minimized (see earlier section on untreated capillaries). A typical example is the successful separation of ribonucleases A, B_1, and B_2 using 20 mM (cyclohexyl)aminopropanesulfonic acid, pH 11.0, as the running electrolyte [60]. These ribonucleases (RNases), having the same polypeptide chains, are either nonglycosylated (RNase A) or glycosylated at various level (RNases B_1 and B_2). At this high pH, the glycosylated proteins acquire more negative charges as the carbohydrate moieties become ionized. This ionization affects their electrophoretic mobilities and, consequently, permits their separation.

An area of challenge for most separation techniques is the analysis of membrane proteins. Capillary zone electrophoresis with hydrophilic coatings was not only successful in the analysis of water-soluble proteins, but also permitted the separation of hydrophobic membrane proteins. Josic et al. [61] investigated the potential of CZE in the analysis of extrinsic calcium-binding membrane proteins, as well as a

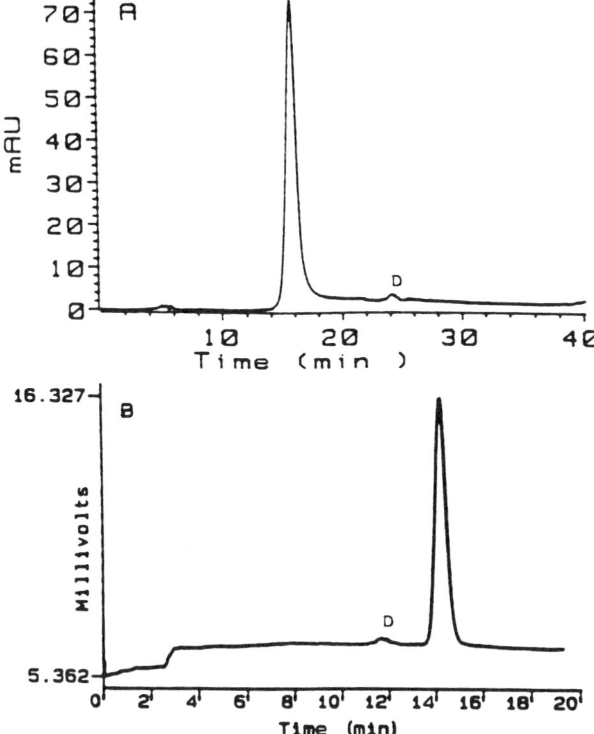

Figure 14 (A) Ion-exchange chromatography of hGH stored in solution for 2 weeks. The separation was performed on a silica-based DEAE column with an acetate gradient. The more acidic peak labeled *D*, corresponds to deamidated GH, as determined by sequencing and mass spectrometry of tryptic fragments. (B) Capillary electrophoresis at high pH of the sample shown in (A). Capillary, 20 cm (to the detection point), × 25 μm i.d., with polyacrylamide coating; electrolyte, phosphate solution, pH 8.0; running voltage, 8 kV. UV detection at 200 nm at 1 V/AU. (From Ref. 58)

mixture of three different intrinsic membrane glycoproteins, namely dipeptidyl peptidase IV (DPPIV), cell-CAM, and a third solute with an M_r of 95,000 Da. When the analysis was performed on an uncoated fused silica capillary, solute–wall interactions led to irreversible binding of the membrane proteins to the capillary wall. With the use of hydrophilic coated capillaries, a modest improvement in terms of the solute's peak area was obtained. Increased reproducibility and satisfactory analysis of these proteins were possible only after the addition of a high concentration of denaturing agent (7 M urea) to the running buffer. In our laboratories, the separation of human erythrocyte glycophorins was carried out successfully using

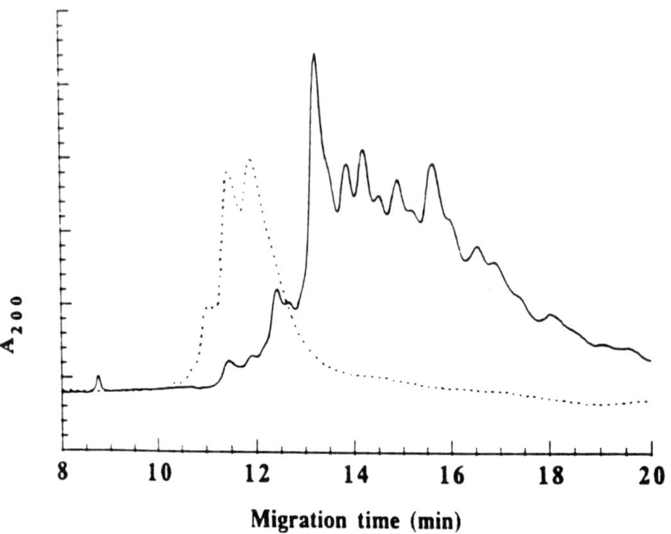

Figure 15 Overlay of CZE patterns of rtPA (———) and desialylated rtPA (········) performed on a polyacrylamide-coated capillary. Capillary, 20 cm (to the detection point), × 25 μm i.d.; electrolyte, 0.1 M ammonium phosphate pH 4.6 with 0.01% Triton X-100 and 0.2 M ε-aminocaproic acid; running voltage, 6 kV. (From Ref. 59)

polyether-interlocked capillaries with simple electrophoretic buffers. Figure 16 illustrates the baseline resolution of both sialo- and asialoglycophorin A in less than 10 minutes using 50 mM sodium phosphate at 10% (v/v) acetonitrile, pH 7.0, as the running electrolyte. The shorter migration time of the asialo forms is due to decreased net negative charge as a result of desialylation. This application adds another dimension to the usefulness of CZE with capillaries having hydrophilic coatings in protein analysis.

Besides the application of CZE to solving separation problems of biological significance, the technique has been useful in the determination of physicochemical properties of proteins. Walbroehl and Jorgenson determined the diffusion coefficient and effective charge values of model proteins using untreated fused silica capillaries at high salt concentration [62]. More recently, a semiempirical model for describing proteins electrophoretic mobility in capillary zone electrophoresis, was developed by Compton [63]. The model predicted that the molecular mass (M) dependency of mobility should be a continuous function of $M^{-1/3}$ to $M^{-2/3}$, depending on the magnitude of the protein molecular mass and buffer ionic strength. The model was also applied for the calculation of protein mobility at any given pH, and the titration curve of calculated mobility or charge versus pH was used to predict the appropriate choice of buffer pH for optimum separation conditions. This paper illustrated the

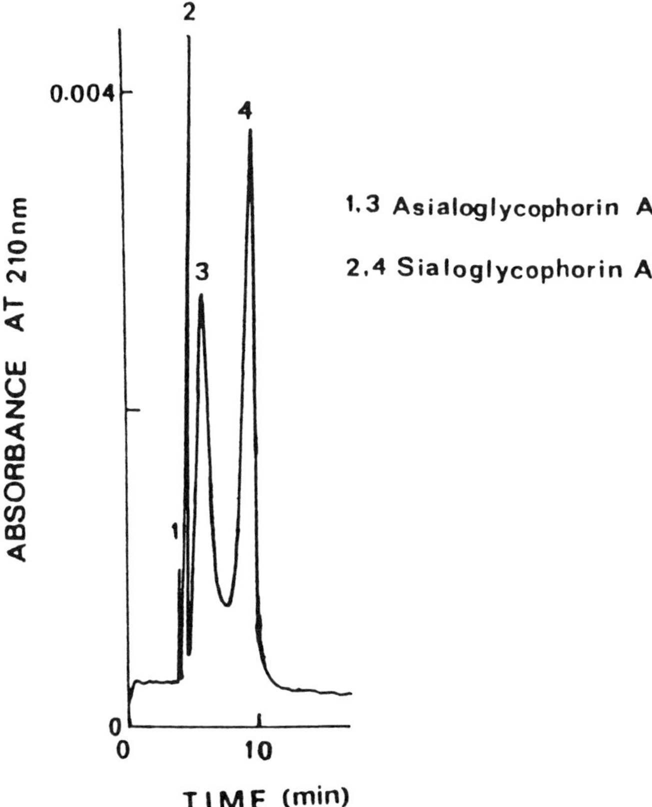

Figure 16 Electropherogram of human erythrocyte sialo- and asialoglycophorin A. Capillary, I-200, 34 cm (to the detection point), 64 cm total length × 50 μm; electrolyte, 50 mM phosphate, pH 7.0 containing 10% (v/v) acetonitrile; running voltage 20 kV.

validity of this approach by its application to the high-resolution of IgG isoelectrotypes using Bio-Rad-coated capillaries.

Peptides

Unlike proteins, short oligopeptides, with their limited number of charged groups, exhibit much less adsorption and, under most experimental conditions, can be separated by CZE without significant band broadening using untreated fused silica capillary.

Most of the research on CZE of peptides has focused on deriving mobility–structure relations to facilitate the prediction of their migration and the design of their separation [60,64,65]. Also, substantial effort has been devoted to improve peptide separations by investigating various electrolyte systems [27,66–68]. On the other hand, since the major thrust of peptide CZE is to characterize proteins, researchers have worked out various ways to couple CZE with MS for the detection of peptides and subsequent on-line determination of the amino acid sequence of proteins [69–74]. In this section, only peptide separations performed on coated capillaries are discussed, and the interested reader may consult the literature concerning peptide separations in untreated fused silica capillaries [1,25,64–67,75–77].

Tryptic peptide mapping necessitates highly reproducible systems to generate consistent maps that can be used in purity assessment of proteinaceous drugs, identification and localization of genetic variants, as well as in monitoring protein structural changes. Hydrophilic capillaries are well suited for this type of applications. Zhu et al. reported [38] the rapid and efficient separation of a tryptic digest of bovine serum albumin using surface-modified capillaries, prepared according to the procedure developed by Hjerten [42], with slight modification to enhance the coating stability at alkaline pH. When the peptide mapping was performed on untreated fused silica capillaries using the same conditions as with coated capillaries, most of the peptide components were strongly adsorbed to the capillary surface (Fig. 17). Frenz et al. carried out the CZE tryptic mapping of recombinant human growth hormone (hGH) [58]. Most of the peptide fragments were resolved in less than 12 minutes on a 20-cm coated capillary from Bio-Rad, as opposed to almost 2 hours in reversed-phase chromatography. Since capillary electrophoresis separation is based on different solute properties than used in RPC, it can provide a rapid and reliable method to check the purity of collected RPC fractions. As shown in Fig. 18, one of the peaks in the HPLC tryptic mapping yielded two well-resolved peaks when subjected to CZE.

Besides its important role in providing a reliable approach for checking the purity level of HPLC peptide fractions, CZE, when combined to HPLC, can be used to map a specific group of peptide fragments that was selectively enriched by HPLC. Nashabeh and El Rassi [78] demonstrated the usefulness of the combination of CZE with high-performance lectin affinity chromatography in the separation and characterization of peptide and glycopeptide fragments from the tryptic digest of human α_1-acid glycoprotein. This combination permitted CZE submapping of glycosylated and nonglycosylated fragments that were isolated on a concanavalin A silica-based stationary phases (Con A) before the electrophoretic run. Figure 19 illustrates the CZE submapping of the Con A nonreactive, Con A slightly reactive, and Con A strongly reactive peptides of human α_1-acid glycoprotein. The CZE submapping of Con A-reactive peptides produced peaks that are missing from the submaps of all other collected fractions (i.e., 0, 0′, and 1), but the components of which are found in the whole map (see area C_1, Fig. 19). This approach allows the monitoring of a group of peptides as well as the assessment of glycosylated fragments

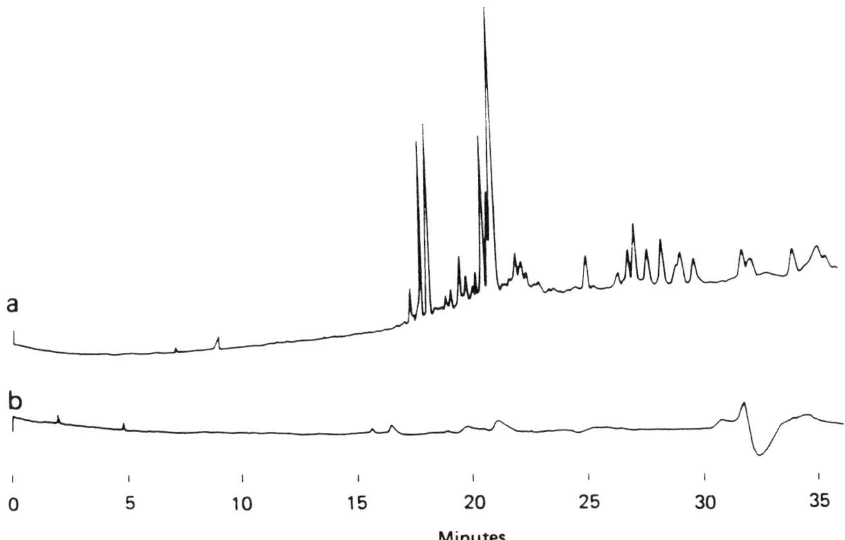

Figure 17 Separation of a tryptic digest of bovine serum albumin on (a) polyacrylamide-coated capillary and (b) uncoated fused silica capillary. Capillary, 50 cm (to the detection point), × 50 μm i.d.; electrolyte, 0.1 M potassium borate, pH 8.5 containing 0.2% methylcellulose-Z60; running voltage, 8 kV. (a) Negative-to-positive polarity; (b) positive-to-negative polarity. (From Ref. 38)

in the whole map. The methodology tested and developed in this report is expected to work also with other glycoproteins, and the CZE submapping of all the glycosylated tryptic fragments with different types of glycans may require the use of more than one lectin column in the prefractionation step.

Nucleic Acids

In addition to its high resolution, high separation efficiency, and precision instrumentation, capillary electrophoresis, with its small sample requirements, is becoming an important tool for analyses of nucleic acids, which most often are available in only minute amounts. Whereas small nucleic acid constituents and short oligonucleotides have been readily separated in various modes of capillary electrophoresis, large oligonucleotides and DNA fragments required the use of special medium that would bring about their separation according to their size. Indeed, the separation of nucleic acid constituents (i.e., bases, nucleosides, and small oligonucleotides) has been demonstrated with both capillary zone electrophoresis and micellar electrokinetic capillary chromatography. Cohen et al. [79] achieved the separation of uncharged bases and nucleosides at neutral pH employing the advantage of differential

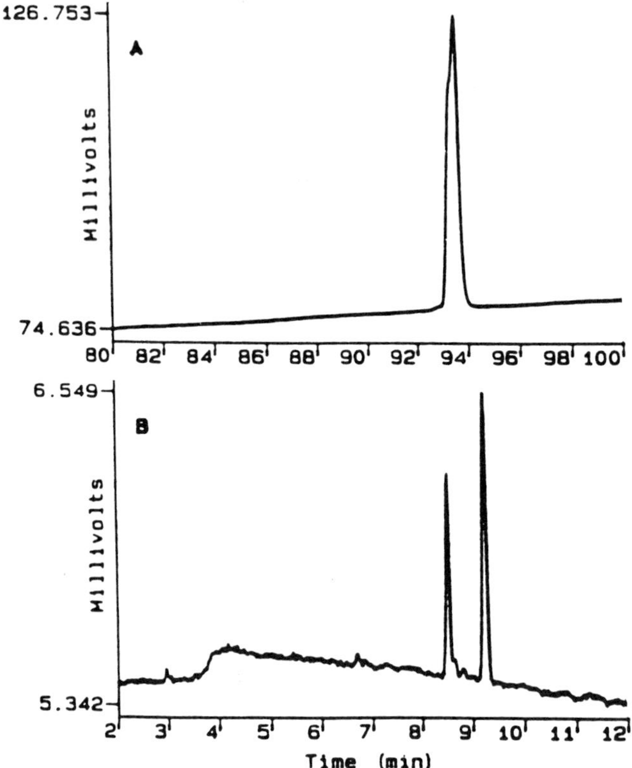

Figure 18 (A) HPLC and (B) CZE analyses of one of the fractions of the RPC tryptic map of hGH. Capillary, 20 cm (to the detection point), × 25 μm i.d., with polyacrylamide coating; electrolyte, phosphate buffer, pH 2.5; running voltage, 8 kV. (From Ref. 58)

solute partitioning into the interior of SDS micelles. In addition, this report showed that the combination of low concentration of divalent metal ions and SDS micelles facilitated the separation of closely related small oligonucleotides. The mechanism of separation was believed to occur through the differential complexation of the oligonucleotides with the metal ions, which are likely to adsorb on the surface of the micelles.

Realizing the benefit of using coated capillaries for the sake of reproducible separations, even though no interaction between solutes and capillary wall may occur, Takigku and Schneider evaluated the potential of CZE in the separation and quantitation of ribonucleoside triphosphates and deoxyribonucleoside triphosphates with polyacrylamide-coated capillaries [80]. Full resolution of the eight common NTPs and dNTPs, namely, ATP, dATP, CTP, dCTP, GTP, dGTP, dTTP, and

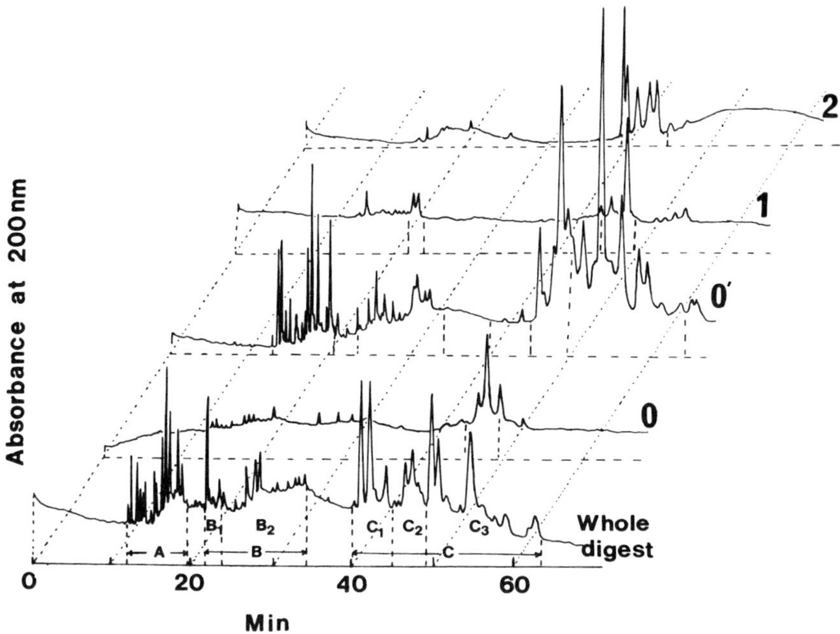

Figure 19 Capillary zone electrophoresis tryptic mapping and submapping of human α_1-acid glycoprotein. Capillary, fused silica tube with hydrophilic coating on the inner walls, 45 cm (to the detection point), 80 cm total length × 50 μm i.d.; electrolyte, 0.1-M phosphate solution, pH 5.0; running voltage, 22.5 kV. Symbols: fraction 0, Con A non-reactive (excluded from the column); fraction 0′, Con A nonreactive (unretained by the column); fraction 1, Con A slightly reactive (eluted with buffer); fraction 2, Con A strongly reactive (eluted with the haptenic sugar). (From Ref. 78)

UTP, was obtained in less than 18 minutes using 0.05 M phosphate, pH 2.7, containing 0.002 M EDTA as the operating buffer. Excellent reproducibility of the migration time of the nucleotides was observed with an RSD of less than 0.6%. Dolnik et al. carried out the separation of a mixture of polycytidine oligonucleotides with polyacrylamide-coated capillaries [81]. The addition of spermine, a cationic tetraamine, to the running electrolyte improved the resolution of the polycytidine mixture. This was the result of the binding of spermine to oligonucleotides, a process by which the effective mobilities of large oligonucleotides were considerably reduced, thereby favoring the separation of the mixture components in the order of increasing size. In a recent study from our laboratory on CZE of small nucleic acids [82] we have found that the addition of spermine to the background electrolyte at low concentration changes the zeta potential of the inner surface of a fused silica capillary from negative to positive, which resulted in the inversion of the direction of the electroosmotic flow. This phenomenon enabled the migration of the analytes to be in the same direction as the electroosmotic flow, which in turn yielded shorter analysis time.

Although the foregoing attempts were quite successful for short nucleic acid fragments, the major problem is that large oligonucleotides (e.g., double-stranded DNA fragments) are difficult to separate in open tabular CZE because of their identical mass/charge ratio, irrespective of their base pair numbers [83]. Various attempts have been made to overcome this inherent property of nucleic acid fragments and to bring about their separation by CZE with open tubes. A Tris-borate buffer containing 7 M urea at 0.1% SDS was reported [84] for the separation of 18 deoxyoligonucleotides ranging in size from 72 to 23,000 base pair (bp) using untreated open tubular capillaries. The mechanism of separation was explained by the binding of borate, urea, and SDS to the DNA molecules, which subsequently altered their conformation. Chin and Colburn [85] attempted the separation of DNA fragments in untreated fused silica capillaries with the Sepragene 5000 buffer, a tris-borate base medium containing hydrophilic methylcellulose derivatives. The DNA fragments eluted in the order of decreasing size, as the larger fragments associated more strongly with the cellulose derivatives in the buffer.

A more efficient approach for the separation of large DNA fragments has been the use of gel-filled capillaries. With these capillaries, the separation mechanism is based on gel-sieving effects. Cohen and Karger [10] were the first to demonstrate the use of SDS–polyacrylamide gel capillaries in biopolymer separations, and extended the usefulness of gel capillary electrophoresis to large DNA fragments. More recently, Cohen et al. [12] illustrated the use of gel-filled capillaries with laser-induced fluorescence for the separation and detection of DNA sequence reaction products at the subattomole level. Baseline resolution of fragments more than 330 bases long and differing in length by a single nucleotide was achieved (Fig. 20). Heiger et al. reported [13] the high-efficiency separation of double-stranded DNA fragments up to 12,000 bp using capillaries with low crosslinked polyacrylamide gel. Swerdlow et al. used polyacrylamide gel capillaries for the separation of deoxycytidine-terminated DNA fragments [86]. Although impressive separations can be obtained, the use of gel-filled capillaries is accompanied with many difficulties, such as poor gel-to-gel reproducibility, bubble formation under electrophoretic conditions, and gel matrix collapse.

A more convenient approach has been the use of coated capillaries, with buffers containing polymeric species acting as sieving additives. Such methods have combined the high efficiency and reproducibility of hydrophilic capillaries with the molecular-sieving effects of soluble polymers that induced nucleic acid separations. Heiger et al. reported [13] the separation of single-stranded oligonucleotides using linear polyacrylamide (i.e., noncrosslinked; liquidlike) capillary columns. Contrary to gel-filled capillaries that were only useable for multiple injections, the linear

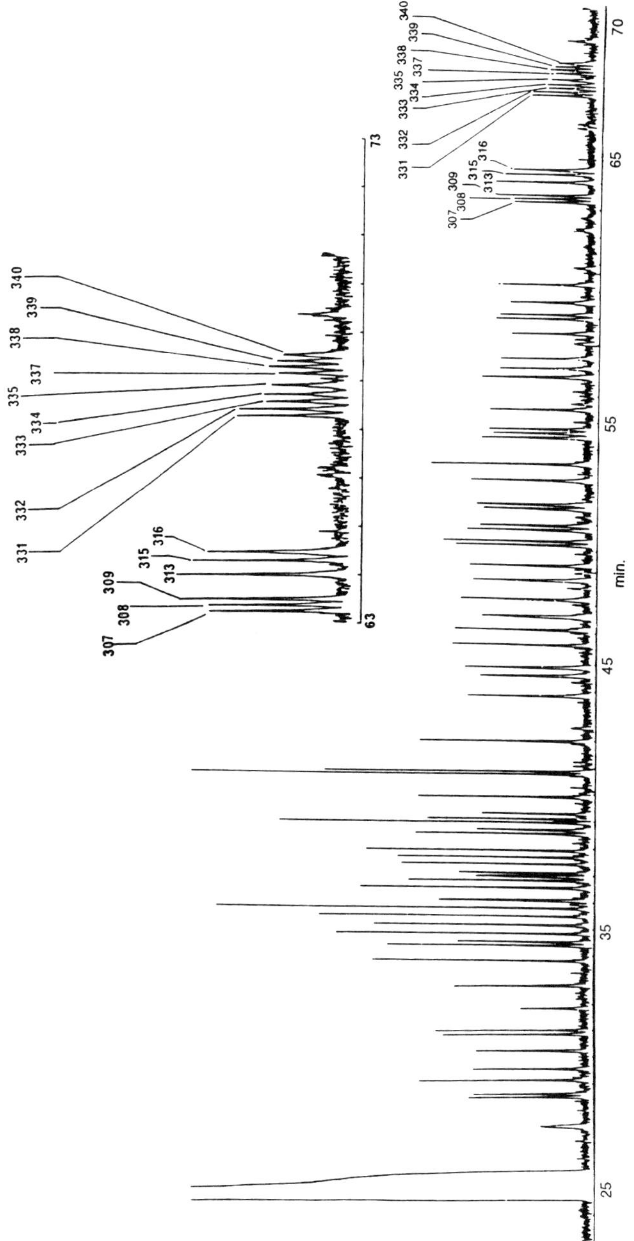

Figure 20 Electropherogram of chain-termination-sequencing reaction products. Template: single-stranded M13mp18. Primer, JOE-PRM18.1 = 5′-JOE-TCCCAGTCACGACGTTGT-3′; dT reaction: primer was extended by Sequenase 2.0 in the presence of dideoxythymidinetriphosphate. Running conditions: capillary, polyacrylamide gel-filled, 75 cm (to the detection point), 92 cm total length × 75 μm i.d.; electrolyte, 0.1 M Tris-borate containing 2.5 mM EDTA and 7 M urea, pH 8.0; running voltage, 310 V cm^{-1}. (From Ref. 12)

polyacrylamide capillaries were stable for several weeks of operation without degradation of performance. Zhu et al. demonstrated the separation of DNA 123-bp ladder fragments with a Bio-Rad-coated capillary [87]. Electrophoretic sieving of the DNA fragments was achieved by the inclusion of 0.5% (w/w) hydroxymethylcellulose in the background electrolyte as shown in Fig. 21. The use of diethylaminoethyl-dextran (DEAE-dextran) as a buffer additive was reported by Dolnik et al. [81]. Although DEAE-dextran showed promise for short oligonucleotide separations, excessive ionic interactions with large fragments hindered their separation. It seems that the polymeric additive should not undergo ionic interactions with the solutes, but rather act only as a sieving agent.

Schwartz et al. investigated the potential of polysiloxane-coated capillaries in conjunction with polymeric buffer additives for the optimum separations of the *Hae* III restriction fragment of phage ΦX 174 DNA [88]. The plots of mobility versus the logarithm of molecular mass (or bp number) of nucleic acids gave an S-shaped

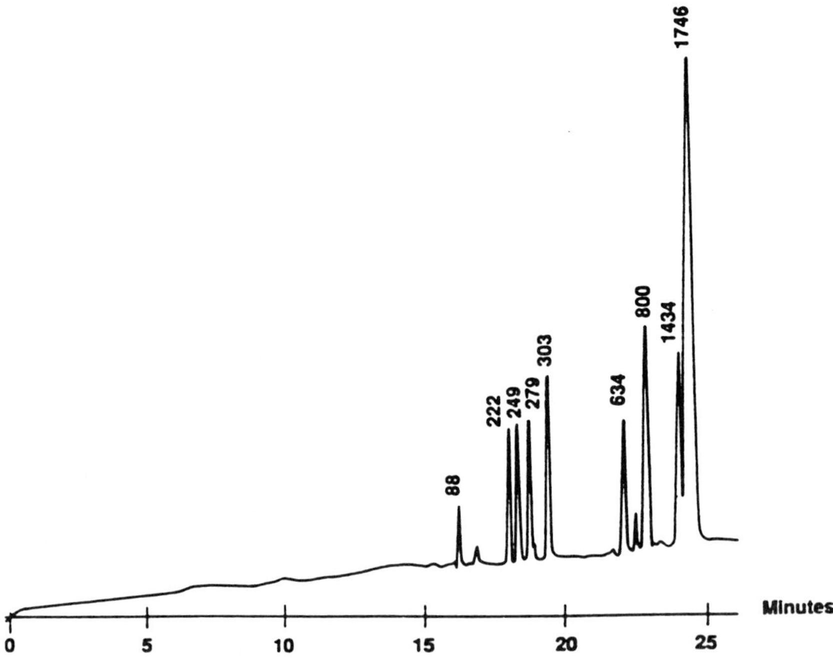

Figure 21 Molecular sieving of DNA standard, 88–1746 bp (as indicated at individual peaks). Capillary, fused silica tube with polyacrylamide coating on the inner walls, 50 cm (to the detection point), × 50 μm i.d.; electrolyte, 0.089 M Tris-borate buffer containing 2.5 mM EDTA, 7 M urea, and 5% hydroxymethylcellulose, pH 8.0; running voltage, 8 kV. (From Ref. 87)

curve, the middle portion of which was quasi-linear. The shallowness of the slope of the curves was used to evaluate the sieving efficacy of three polymeric buffer additives, namely, hydroxypropylmethylcellulose 4000 cP at 25°C (HPMC-4000), hydroxypropylmethylcellulose 100 cP at 25°C (HPMC-100), and polyethylene glycol (PEG 35,000). The steeper the slope of the S-shaped curve, the less effective is the sieving for a certain buffer additive and concentration. From these results, PEG was the least effective in separating DNA fragments. In addition, greater mobilities were obtained with the lower M_r HPMC, a result that is in agreement with previous findings [89] on the effects of the chain length of linear polyacrylamide. Furthermore, the addition of 10 mM ethidium bromide, an intercalating agent that binds to DNA by inserting itself between the base pairs of the double helix, was most effective in bringing about the baseline resolution of the 271 and 281 bp fragments. These fragments were not resolved with polymeric buffer additives under otherwise identical conditions. Ethidium bromide was also incorporated within the polyacrylamide gel network as a means of manipulating the selectivity of capillary gel electrophoresis separations [90].

Recently, Strege and Lagu compared the separation of 1-kbp DNA ladder in three different capillary electrophoretic systems [91]. These systems are (a) a buffer composed of 0.1-M Tris-borate pH 8.1, containing 7 M urea, 2.5 mM EDTA, and 0.1% SDS in untreated capillaries; (b) a Sepragene 5000 buffer using uncoated capillaries; and (c) a 50-mM Tris-borate buffer, pH 8.0 containing 2.5 mM EDTA and 0.5% methylcellulose in polyacrylamide-coated capillaries. As shown in Fig. 22, the full resolution of all the 1-kbp DNA ladder fragments was possible only by using a coated capillary and a methylcellulose sieving medium. In addition, this electrophoretic system proved efficient in the separation of the large oligonucleotide fragments in a λ DNA *Hind* III digest.

Carbohydrates

The analysis of carbohydrates, and in particular those associated with glycoproteins (i.e., glycans), is of primary importance in the biomedical and biotechnology areas. More than any other type of species, the separation of complex carbohydrates requires separation techniques of high-resolving power. Although CZE is well suited to play an important role in this area, its application to carbohydrate analysis has lagged well behind that of proteins and nucleic acids. This may be because most carbohydrates are neutral and lack chromophores in their structure, which make their separation and detection at low level rather difficult.

To overcome these difficulties, precolumn derivatization with tags that carry both the charge and the chromophore required for the electrophoresis and sensitive detection, respectively, has been the most widely used approach. Another alternative to precolumn derivatization of carbohydrates is the use of indirect fluorescence detection, a technique that has already been applied to several different liquid

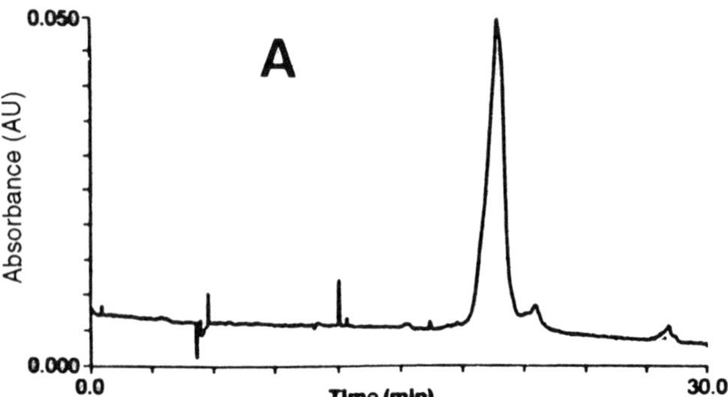

Figure 22 Electropherograms of the 1-kbp DNA ladder separated at (A) 20 kV in 0.1 M Tris-borate containing 2.5 mM EDTA, 7 M urea, and 0.1% SDS, pH 8.1; (B) 10 kV in Sepragene 5000 buffer; (C) 10 kV in a polyacrylamide-coated capillary in 50 mM Tris-borate containing 2.5 mM EDTA, and 0.5% methylcellulose, pH 8.0. (From Ref. 91)

chromatographic systems [92,93]. The detection in this approach is mainly based on charge displacement of the fluorescing buffer ions by the ionized analyte molecules. This creates a negative peak because the solute zone is deficient in the fluorescing ions, and the signal thus obtained is independent of any absorption or emission properties of the analyte. Typical separations of monosaccharides were demonstrated by Garner and Yeung [94] using 1 mM coumarin 343 as the background fluorophore in the running electrolyte. However, the ionization of sugars necessitated the use of a relatively high pH (i.e., pH 11.5), whereby the concentration of the hydroxide ions is no longer negligible relative to the concentration of the fluorescing ions. Under these conditions, OH^- competed with the fluorophore for displacement by the sugar solute. This problem, added to the small extent of ionization of most sugars, restricted the detection limit under optimized conditions to the femtomole range. For more details concerning the indirect detection methods for capillary separations, the interested reader is advised to consult the recent review on the subject by Yeung and Kuhr [25].

Following previous observations that carbohydrates can be electrophoresed in the form of anionic borate complexes [95], Honda et al. [96] were the first to report the CZE of monosaccharides tagged with 2-aminopyridine by reductive amination at their reducing termini using an alkaline borate buffer as the running electrolyte. With a similar buffer system, Nashabeh and El Rassi reported [78] the successful separations of 2-pyridylamino derivatives of monosaccharide constituents of glycoproteins.

More recently, Liu et al. [97] demonstrated that various mono- and oligosaccharides can be reacted with an ammonia solution to yield derivatives with primary

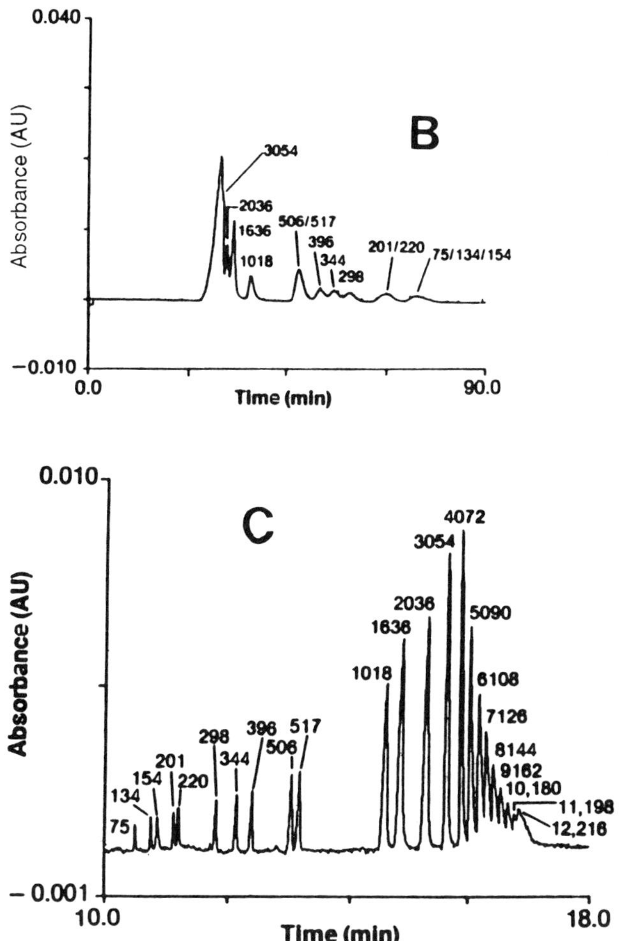

amine functions, which can then be derivatized with 3-(4-carboxybenzoyl)-2-quinolinecarboxaldehyde (CBQCA) to produce highly fluorescent isoindole derivatives, according to the reaction scheme depicted in Fig. 23. This enabled the detection of monosaccharides at the attomole level using laser-induced fluorescence detection, which is far below what can be achieved by indirect detection. In addition, the usefulness of this derivatization scheme was demonstrated by the mapping of complex oligosaccharides isolated from bovine fetuin by hydrazinolysis. Another report from the same group [98] has shown the separation of CBQCA-derivatives of partially hydrolyzed dextrin 15 using both open tubular and polyacrylamide gel-filled capillaries. Although longer analysis time was obtained with gel-filled capillaries, oligosaccharide fragments having high degree of polymerization (d.p.) showed better

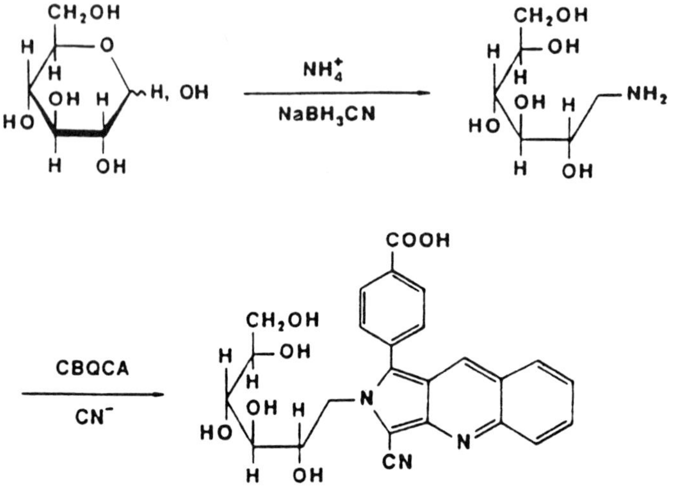

Figure 23 Reaction scheme of the derivatization of sugar with 3-(4-carboxybenzoyl)-2-quinolinecarboxaldehyde to yield highly fluorescent isoindole derivatives. For experimental details, see Ref. 97. (From Ref. 97)

resolution compared with open tubular analysis. (Fig. 24). Also, the separation of acidic oligosaccharides obtained by enzymatic digestion of chondroitin sulfate A and hyaluronic acid were reported with gel-filled capillaries [98].

Although efficient carbohydrate analysis can be achieved using both uncoated and gel-filled capillaries, we believe that the use of capillaries with hydrophilic coating for CZE analysis of carbohydrates can lead to more reproducible separations than untreated fused silica tubes and, in turn, more reliable quantitative analysis. In recent reports from our laboratory [78,99] we have demonstrated the potential of CZE with fused silica capillaries having hydrophilic coatings in the mapping of 2-pyridylamino derivatives of complex type of glycans cleaved from both human and bovine α_1-acid glycoprotein as well as that of high-mannose glycans from RNase B. Figure 25 illustrates the electropherogram of the oligosaccharides that were cleaved by endoglycosidase digestion and one of the peaks was identified as (mannose)$_5$(GlcNAc)$_2$ using an oligosaccharide standard. In all these studies, we found that the separations were best achieved by the inclusion of tetrabutylammonium bromide (Bu_4N^+) in the background electrolyte [99,100].

Figure 26 illustrates the high-separation efficiency of 2-pyridylamino derivatives of acetylated xyloglucan oligosaccharides from cotton cell walls (2-PA-XG) generated by cellulase digestion. It was performed on a fused silica capillary with a polyether-interlocked coating, using 0.1 M phosphate containing 50 mM Bu_4N^+, pH 4.75. In this study, the peaks were identified using fractions collected from reversed-phase chromatography of the xyloglucan, the structure of which was assessed by liquid secondary ion–mass spectrometry [101]. The peak numbering on the electropherogram (see Fig. 26) reflects the elution order obtained in RPC. As expected the elution pattern in CZE was different from that in RPC. In CZE, the elution order was governed mainly by the number of sugar residues and the degree of branching, whereas in RPC the elution order was influenced mainly by the size of the oligosaccharide and the hydrophobic character of the sugar residues. For instance, fragment 4, which is smaller than fragment 3, was more retarded on the RPC column. This may be attributed to the presence of a fucosyl residue in structure 3, which is more hydrophobic than any other sugar residues in the molecule. The same reasoning can explain the elution order for fragments 6 and 7. Referring to Fig. 26, the average plate count per meter was about 225,000, as calculated from the two major fragments, 5 and 8. In addition, structures 8 and 8a, which are nonasaccharides with the same extent of branching, could be resolved. The only difference is that fragment 8a has an acetyl group at the galactosyl residue that is not present in structure 8. This, once again, shows the high resolving power of CZE in recognizing small differences between the various xyloglucan fragments. The peaks labeled with asterisks may be attributed to other oligosaccharides present in the mixture.

To interpret the electrophoretic behavior of the various 2-PA-XG and to quantitatively describe the effects of the various sugar residues on their electrophoretic mobility, we have introduced a mobility-indexing system of the branched xyloglucan oligosaccharides relative to the linear 2-PA-GlcNAc$_n$ homologous series. The mobility indices, *MI*, of the 2-PA-XG fragments were calculated using the following equation:

$$MI = 100_n + 100\left(\frac{\log \mu_s - \log \mu_{n+1}}{\log \mu_n - \log \mu_{n+1}}\right)$$

where μ_s is the electrophoretic mobility of the 2-PA-XG solute, μ_n and μ_{n+1} are the electrophoretic mobilities for the two homologues with n and $n+1$ repetitive units that eluted before and after the xyloglucan fragment, respectively. The difference between the mobility indices of an adjacent pair of 2-PA-XG fragments, ΔMI, referred to as the group mobility index decrement contributed by a sugar residue added to a parent oligosaccharide molecule, was estimated for selected 2-PA-XG fragments. The ΔMI values obtained for the adjacent fragment pairs (1,2), (2,3), and (3,5) were almost identical (i.e., 52.5, 55.4, and 54.0, respectively [see Fig. 26]). Thus, the addition of a glucosyl residue to the linear core chain of the oligosaccharide showed a change in the mobility index decrement similar to the addition of a xylosyl residue at the glucose loci and behaved as one-half a GlcNAc residue in terms of its contribution to the electrophoretic mobility of the 2-PA-XG [see structures in Fig. 26]. However, the addition of a galactosyl residue to an already branched xylosyl residue exhibited less retardation (i.e., ΔMI of 38.4 for

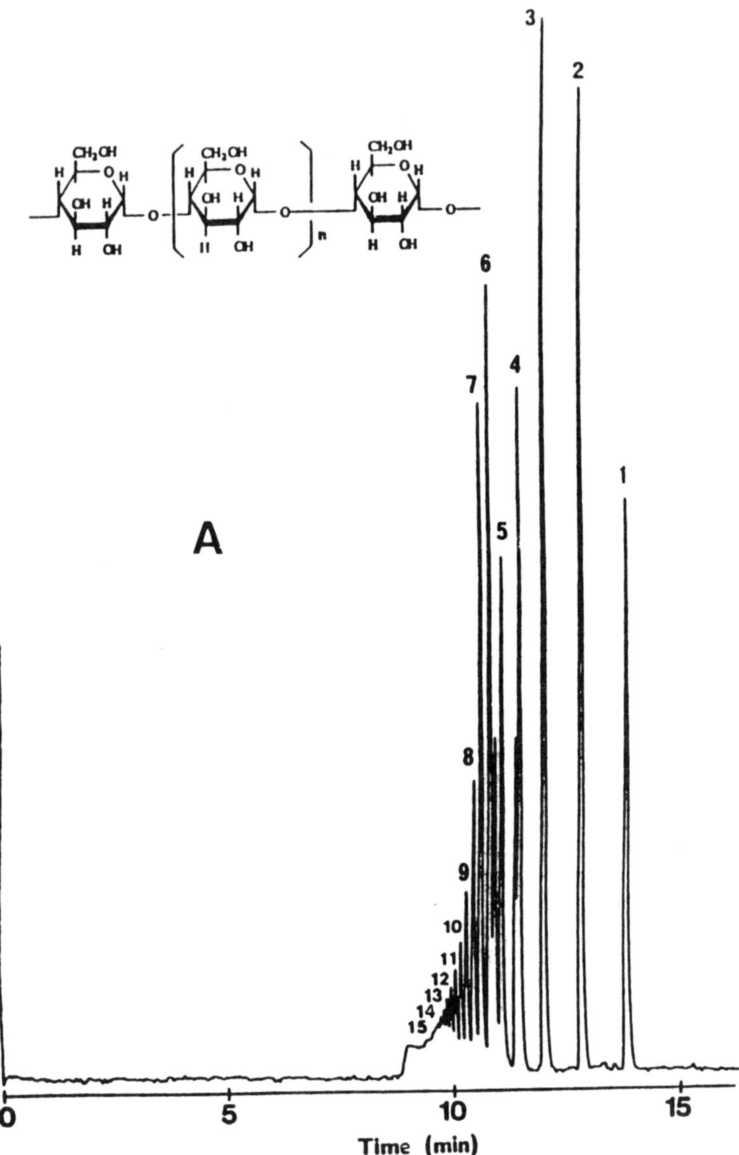

Figure 24 Electrophoretic separation of CBQCA-derivatives of a partially hydrolyzed polysaccharide (dextrin 15) in (A) open-tubular system and (B) polyacrylamide gel-filled column. (A) capillary, 58 cm (to the detection point), 88 cm total length × 50 μm i.d.;

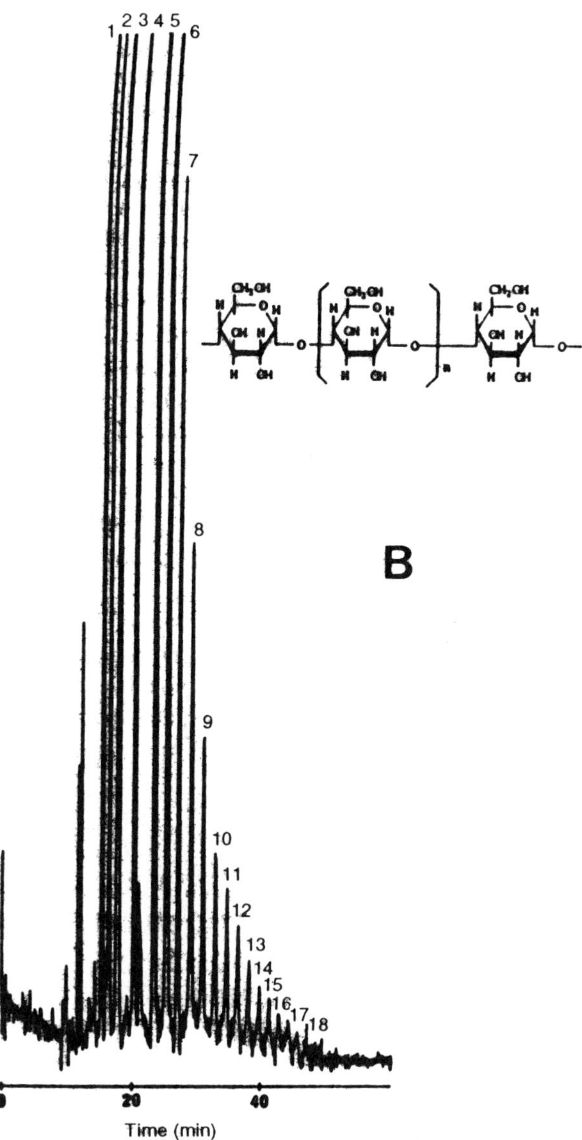

electrolyte, 0.01 M disodium phosphate solution–0.01 M sodium tetraborate decahydrate, pH 9.4; running voltage, 20 kV. (B) Capillary, 19 cm (to the detection point), 26 cm total length × 50 μm i.d.; electrolyte, 0.1 M Tris–0.25 M borate–7 M urea, pH 8.33; applied voltage 269 V cm^{-1}. (From Ref. 98)

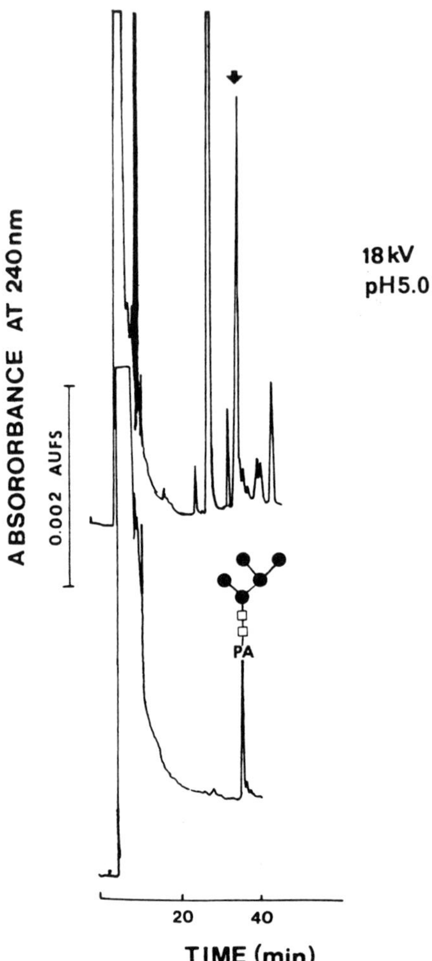

Figure 25 Capillary zone electrophoresis mapping of 2-pyridylamino derivatives of high-mannose oligosaccharides cleaved from bovine ribonuclease B (top electropherogram), and of 2-PA-(GlcNAc)$_2$-Man$_5$ standard (lower electropherogram). Capillary, fused silica tube with polyether-interlocked coating on the inner walls, 50 cm (to the detection point), 80 cm total length × 50 μm i.d.; electrolyte, 0.1-M phosphate solution containing 50 mM tetrabutylammonium bromide, pH 5.0; running voltage, 18 kV. Symbols: 2-AP, 2-aminopyridine; □, GlcNAc; ●, mannose. (From Ref. 99)

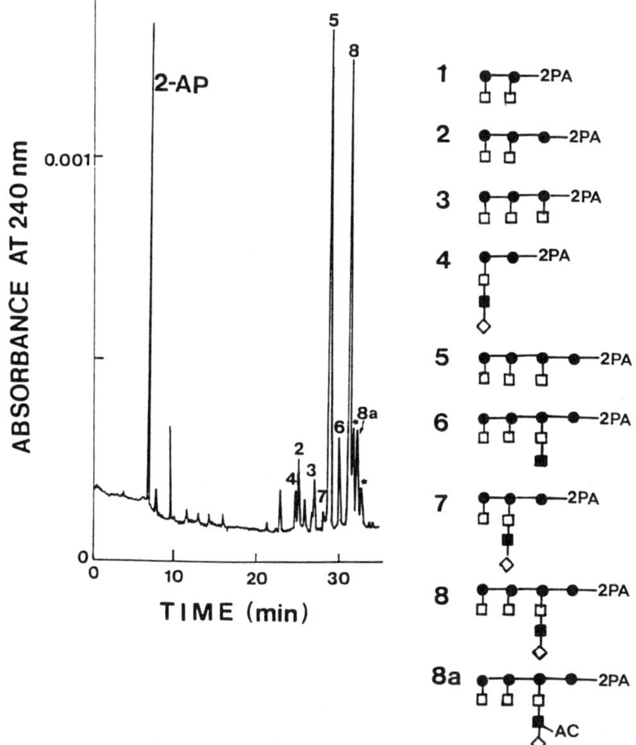

Figure 26 Capillary zone electrophoresis mapping of 2-pyridylamino derivatives of acetylated xyloglucan oligosaccharides from cotton cell walls. Capillary, fused-silica tube with polyether interlocked coating on the inner walls, 50 cm (to the detection point), 80 cm total length × 50 μm i.d.; electrolyte, 0.1-M sodium phosphate solution containing 50 mM tetrabutylammonium bromide, pH 4.75; running voltage, 20 kV. Symbols: 2-AP, 2-aminopyridine; ●, glucose, □, xylose; ■, galactose; ◇, fucose; Ac, acetyl group. (From Ref. 99)

the xyloglucan pair (5,6)) than the addition of a glucosyl or xylosyl unit to the backbone of the xyloglucan oligosaccharide. The same observation was made about adding a fucosyl residue to a branched galactosyl residue. Thus, as the molecule becomes more branched, the addition of a sugar residue does impart a slightly smaller decrease in its mobility. This approach may prove valuable in correlating and predicting the effects of several parameters, such as the nature, position, and number of sugar residues, on the mobilities of complex carbohydrates.

Although the derivatization with 2-aminopyridine was quite reproducible and allowed the electrophoresis of carbohydrates by CZE, the 2-pyridylamino derivatives

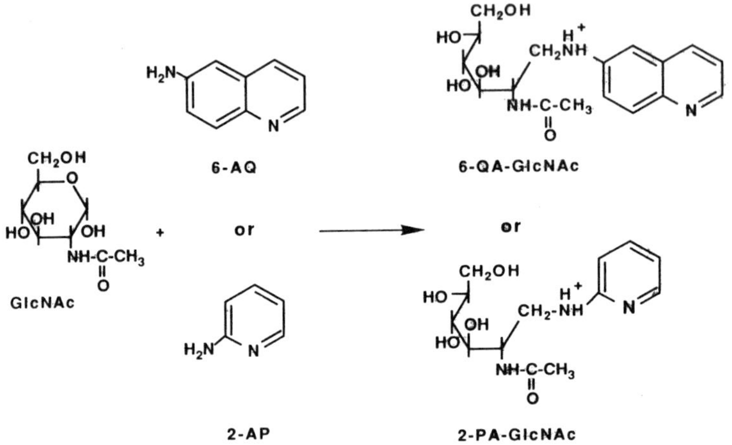

Figure 27 Reaction schemes for the derivatization of carbohydrates with 2-aminopyridine (2-AP) and 6-aminoquinoline (6-AQ). (From Ref. 99)

of mono- and oligosaccharides exhibited limited detection sensitivity by UV [96]. In search of a more sensitive tagging agent, we have investigated the performance of 6-aminoquinoline (6-AQ), and compared it with that of 2-AP using *N*-acetylchitooligosaccharides as model sugar solutes. Both tagging reagents were attached to the oligosaccharides at their reducing termini by reductive amination, according to the reaction scheme portrayed in Fig. 27, where *N*-acetylglucosamine is taken as a typical example. Spectral analysis of the sugar derivatives of both tagging agents revealed that the detectability of 6-QA-GlcNAc$_n$ was eight times more sensitive than that of 2-PA-GlcNAc$_n$ in UV analysis. The derivatization procedure described in this report [99] may prove useful for fluorescence detection. In fact, 6-aminoquinoline possess ideal fluorescence properties [102], in the sense that its excitation wavelength (355 nm) is far removed from its emission wavelength (550 nm) and, therefore, is expected to give low detection limits with fluorescence detection.

TANDEM OPERATIONS

This section briefly describes novel multidimensional electrophoretic systems in which short capillary tubes having immobilized specific functionalities are coupled in series with a hydrophilic separation capillary. The operational aspects and advantages of this approach will be illustrated using two different coupled tandem systems, enzymophoresis and on-line preconcentration. These two tandem operations have provided enhanced detectability and resolution and expanded the applicability of CZE.

Enzymophoresis

Enzymophoresis is a tandem format in which a short-fused capillary with immobilized enzyme on the inner wall is coupled in series to capillary zone electrophoresis. It has been introduced very recently by our laboratory [82] and is currently in progress. Enzymophoresis involves first the introduction of the substrate(s) into the enzyme reactor as a thin plug so that the enzymatic reaction will take place as the substrate(s) migrate through the open tubular reactor either by hydrodynamic flow or electromigration. The product(s) and the unreacted substrate(s) then enter the separation capillary, where they undergo differential migration under the influence of the applied field. The enzymic reaction and subsequent separation can be carried out either with the same or with a different background electrolyte.

Various enzymophoretic systems involving enzymes specific for RNAs were investigated and characterized over a wide range of operating conditions in both the qualitative and quantitative aspects of CZE of different nucleic acids. These systems, which comprised capillary enzyme reactors with immobilized ribonuclease T_1 (RNase T_1), hexokinase (HK) or adenosine deaminase (ADA), served as peak locator on the electropherogram, enhanced CZE selectivity by converting the analytes to more readily separable products, allowed the digestion and mapping of tRNAs, and facilitated the quantitative determination of unresolved analytes with good accuracy.

To illustrate the usefulness of enzymophoresis, some typical examples are briefly discussed in this section. Tandem RNase T_1–capillary enzyme reactor → CZE was useful for the identification and quantitative determination of various dinucleotides. As shown in Fig. 28a, a mixture of three dinucleotides, namely, GpU, GpA, and GpC, were not resolved by CZE alone using a buffer system containing 25 mM histidine, 25 mM MES, and 5mM spermine, pH 5.0. This is because these dinucleotides have approximately the same charge/mass ratio. As shown in Fig. 28b, with the on-line RNase T_1–capillary enzyme reactor, the dinucleotide mixture was converted to more readily separated products. Each of the three dinucleotides, GpA, GpU, and GpC, yielded the guanosine-2′:3′-cyclic monophosphate and its corresponding nucleoside (i.e., adenosine, uridine, and cytidine). Hence, the three dinucleotides that coeluted as a broad peak with CZE alone were transformed into four well-resolved peaks (i.e., G > p, U, A, and C) after a single pass through a 17-cm long RNase T_1–capillary enzyme reactor. The enzyme reactor could be readily used for the quantitative determination of the dinucleotides.

In addition, the RNase T_1–capillary enzyme reactor → CZE system proved very efficient in the on-line digestion and mapping of minute amounts of tRNAs and the simultaneous synthesis and separation of nanogram quantities of dinucleotides. Such a format employing simple CZE instrumentation can be easily automated and may become useful on the micropreparative scale for oligonucleotides of defined sequence that are produced at low level by chemical synthetic methods. Such oligonucleotides are important in the study of mechanisms involving protein synthesis.

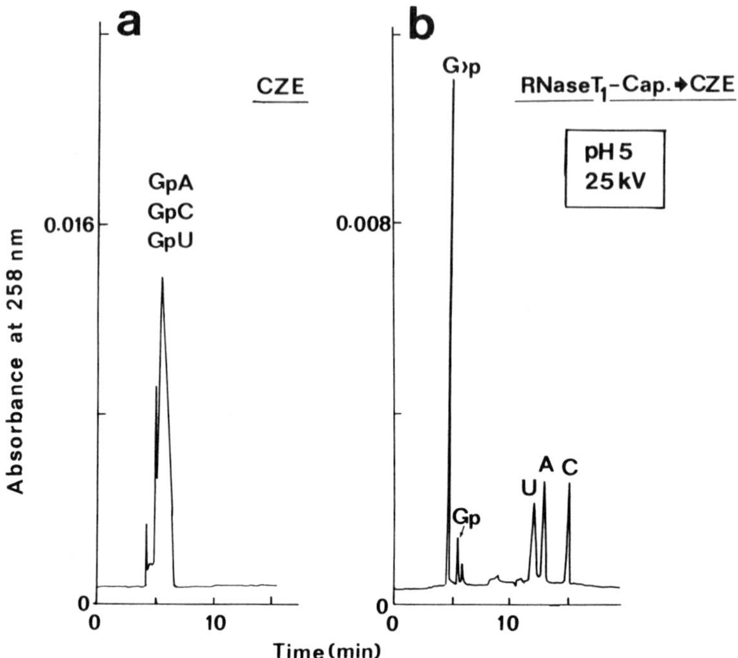

Figure 28 Typical electropherograms of dinucleotides and their RNase T_1 digest obtained by CZE alone (a) or by tandem RNase T_1–capillary enzyme reactor → CZE (b), respectively. Capillary enzyme reactor, 17 cm × 50 μm i.d.; separation capillary, interlocked polyether coating, 30 cm (to the detection point), 64 cm (total length) × 50 μm i.d.; substrate introduction, hydrodynamic mode; enzymic reaction was carried out using gravity-driven flow; separation step, as the plug of the reaction mixture entered the separation capillary, the enzyme reactor was disconnected and the voltage was turned on to 25 kV; background electrolyte 25 mM His, 25 mM MES, 5 mM spermine, pH 5.0. Symbols: GpU, guanylyl-(3′ → 5′)-uridine; GpA, guanylyl-(3′ → 5′)-adenosine; GpC, guanylyl-(3′ → 5′)-cytidine; G > p, guanosine-2′:3′-cyclic monophosphate; Gp guanosine-3′-phosphate; U, uridine; A, adenosine; C, cytidine. (From Ref. 82)

The tandem hexokinase–capillary enzyme reactor → CZE was useful in the partial conversion of adenosine triphosphate (ATP) to adenosine diphosphate (ADP). Such format, therefore, can be conveniently used to confirm the presence or absence of ATP in a complex biological matrix.

Another interesting capillary enzyme reactor was the one with immobilized adenosine deaminase, a highly specific enzyme that catalyzes the deamination of adenosine to inosine with the liberation of NH_3. By using the tandem ADA–capillary enzyme reactor → CZE system, the conversion of adenosine to inosine was complete, even when the ADA–capillary enzyme reactor was shortened to 3 cm, which corresponded to a contact time of about 2 minutes with the enzyme reactor. From this finding, and provided that the enzymic reactions are kinetically favored, various short capillary enzyme reactors having different immobilized enzymes can be connected in series at the inlet of a separation capillary and may become useful in the specific analysis of different solutes in a complex biological mixture.

In summary, enzymophoresis has proved suitable in the area of nucleic acids. This concept can be transposed to electrophoretic systems involving other types of species of relevance to many areas of the life sciences and biotechnology. In addition, the enzymophoresis systems have allowed repeated use and long-term stability of enzymes.

On-Line Preconcentration

In capillary zone electrophoresis, the injection volume must stay small (i.e., 1–5 nL) to achieve high-separation efficiencies. Consequently, dilute samples cannot be determined by CZE with concentration-sensitive detectors. They must be first concentrated so that a thin plug of the sample would contain a detectable amount of the analytes. Few attempts have been made to extend the use of CZE to the analysis of dilute samples of biological species. In one approach, on-line preconcentration by isotachophoresis (ITP) was investigated for proteins [103], peptides [104], and fluorescent amino acids [105]. These coupled ITP–CZE systems allowed the analysis of larger volumes of dilute samples by CZE and, in some instances, peptide solutions having 20 times lower concentration than possible by CZE alone could be easily determined by ITP–CZE [104]. The major drawback with ITP–CZE is that this preconcentration mode is difficult to automate and is limited by the choice of the electrophoretic buffer. In addition, positive and negative analytes cannot be determined at the same time. In another approach, Aebersold and Morrison [106] introduced electrophoretic preconcentration systems for the analysis of dilute peptide solutions by CZE. This method, which is based on stacking (i.e., an electrophoretic concentration step) before separation, permitted the analysis of peptide samples having five times lower concentration than that can be achieved by CZE alone. The advantage of electrophoretic over isotachophoretic preconcentration is that the former permits the determination of dilute samples of oppositely charged analytes. However, the limitation of the electrophoretic technique is the lower loadability in terms of volume of dilute samples that can be analyzed.

Very recently, in an attempt to increase the sample-loading capacity and improve the collection capability of the capillary electrophoretic system, an analyte concentrator containing an immobilized antibody was developed [107]. Capillary devices containing short segments packed with controlled-pore glass beads having surface-bound antibodies directed against methamphetamine were useful in concentrating urinary constituents separated by capillary electrophoresis. In fact, by using one analyte concentrator, it was possible to collect about 200 ng of methamphetamine in a single electrophoretic run. In addition, when using two analyte concentrators in series, more than 350 ng of purified material was generated in one injection. The purity of the collected fraction was further confirmed using off-line mass spectrometer analysis. This approach has the merit of increasing the capabilities of capillary zone electrophoresis as a semipreparative technique.

For the purpose of concentrating selectively dilute protein samples before their analysis by CZE, we have developed in our laboratory [108, 109] short capillaries having surface-bound iminodiacetic acid functions chelated with zinc. The metal chelate capillaries were coupled to interlocked polyether capillaries. The preconcentration with metal chelate capillaries is based on the affinity between proteins and the immobilized metal chelate on the capillary walls. Since this type of interaction is selective, only proteins having affinity for the chelated metal can be concentrated, thus allowing the detection of a protein or a group of proteins at a low level. This two-dimensional format employing tandem metal chelate affinity open tubular electrochromatography–capillary zone electrophoresis proved effective in the selective accumulation of detectable amounts of proteins from dilute samples, and it allowed the introduction of large volumes without affecting the separation efficiency. The minimum detectable concentration from dilute samples of carbonic anhydrase was approximately 25 times lower than with CZE alone. This on-line preconcentration system involves two steps: the accumulation followed by the desorption of the solutes, in which binding and debinding electrolytes are used successively. Since the accumulated solutes in the metal chelate capillary should be stripped off the capillary walls and introduced into the separation capillary as a thin plug, the debinding electrolyte should contain a strong competing agent that desorbs the metal and the protein from the binding sites on the surface of the metal chelate capillary. EDTA was a good debinding agent and allowed the desorption of proteins from the capillary surface, which were analyzed in the separation capillary downstream the preconcentration capillary tube.

ACKNOWLEDGMENTS

This work was supported in part by grant HN9-004 from the Oklahoma Center for the Advancement of Sciences and Technology, Oklahoma Health Research Program; from the University Center for Water Research; from the University Center for Energy Research; from the Agriculture Station, grant CSRS OKL 02109; from the College of Arts and Sciences, Dean Incentive Grant Program at Oklahoma State University; and from the Oklahoma Water Resources Research Institute.

REFERENCES

1. J. W. Jorgenson and K. D. Lukacs, *Anal. Chem.*, *53*:1298 (1981).
2. S. Hjerten, *Chromatogr. Rev.*, *9*:122 (1967).
3. R. Virtanen, *Acta Polytechn. Scand.*, *123*:1 (1974).

4. F. E. P. Mikkers, F. M. Everaetes, and T. P. E. M. Verheggen, *J. Chromatogr.*, *169*:11 (1979).
5. J. W. Jorgenson and K. D. Lukacs, *Science*, *222*:266 (1983).
6. S. Terabe, K. Otsuka, K. Ichikawa, A. Tsuchiya, and T. Ando, *Anal. Chem.*, *56*:111 (1984).
7. S. Terabe, H. Ozaki, K. Otsuka, and T. Ando, *J. Chromatogr.*, *332*:211 (1985).
8. P. Gozel, E. Gassman, H. Michelsen, and R. N. Zare, *Anal. Chem.*, *59*:44 (1987).
9. S. Terabe and T. Isemura, *Anal. Chem.*, *62*:650 (1990).
10. A. S. Cohen and B. L. Karger, *J. Chromatogr.*, *397*:409 (1987).
11. A. S. Cohen, D. R. Najarian, A. Paulus, A. Guttman, J. A. Smith, and B. L. Karger, *Proc. Natl. Acad. Sci. USA*, *85*:9660 (1988).
12. A. S. Cohen, D. R. Najarian, and B. L. Karger, *J. Chromatogr.*, *516*:49 (1990).
13. D. N. Heiger, A. S. Cohen, and B. L. Karger, *J. Chromatogr.*, *516*:33 (1990).
14. T. Braun and S. Nagydiosi-Rozsa, *Trends Anal. Chem.*, *10(9)*:266 (1991).
15. J. W. Jorgenson and K. D. Lukacs, *J. Chromatogr. Lib.*, *30*:121 (1985).
16. J. W. Jorgenson, *Anal. Chem.*, *58*:743A (1986).
17. J. W. Jorgenson, *ACS Symp. Ser.*, *335*:182 (1987).
18. S. Hjerten, *Electrophoresis*, *11*:665 (1990).
19. R. A. Wallingford and A. G. Ewing, *Adv. Chromatogr.*, *29*:1 (1989).
20. B. L. Karger, A. S. Cohen, and A. Guttman, *J. Chromatogr.*, *492*:585 (1989).
21. W. G. Kuhr, *Anal. Chem.*, *62*:403R (1990).
22. Z. Deyl and R. Struzinsky, *J. Chromatogr.*, *569*:63 (1991).
23. M. J. Gordon, X. Huang, S. L. J. Pentoney, and R. N. Zare, *Science*, *242*:224 (1988).
24. A. G. Ewing, R. A. Wallingford, and T. M. Olefirowicz, *Anal. Chem.*, *61(4)*:292A (1989).
25. E. S. Yeung and W. G. Kuhr, *Anal. Chem.*, *63*:275A (1991).
26. S. Compton and R. Brownlee, *Biotechniques*, 6:432 (1988).
27. M. Novotny, K. A. Cobb, and J. Liu, *Electrophoresis*, *11*:735 (1990).
28. J. R. Mazzeo and I. S. Krull, *BioChromatography*, *10*:638 (1991).
29. D. M. Goodall, S. J. Williams, and D. K. Lloyd, *Trends Anal. Chem.*, *10(9)*:272 (1991).
30. J. S. Green and J. W. Jorgenson, *J. Chromatogr.*, *478*:63 (1989).
31. H. H. Lauer and D. McManigill, *Anal. Chem.*, *58*:166 (1986).
32. K. J. Lee and G. S. Heo, *J. Chromatogr.*, *559*:317 (1991).
33. R. M. McCormick, *Anal. Chem.*, 60:2322 (1988).
34. B. M. Mitsyuk, *Russ. J. Inorg. Chem.*, *17*:471 (1972).
35. H. H. Lauer and D. McManigill, *Trends Anal. Chem.*, *5*:11 (1986).
36. M. M. Bushey and J. W. Jorgenson, *J. Chromatogr.*, *480*:301 (1989).
37. M. J. Gordon, K. J. Lee, A. A. Anias, and R. N. Zare, *Anal. Chem.*, *63*:69 (1991).
38. M. Zhu, R. Rodriguez, D. Hansen, and T. Wehr, *J. Chromatogr.*, *516*:123 (1990).
39. A. Emmer, M. Jansson, and J. Roeraade, *J. Chromatogr.*, *547*:544 (1991).
40. J. E. Wiktorowicz and J. C. Colburn, *Electrophoresis*, *11*:769 (1990).
41. S. H. Chang, K. M. Gooding, and F. E. Regnier, *J. Chromatogr.*, *120*:321 (1976).
42. S. Hjerten, *J. Chromatogr.*, *347*:191 (1985).

43. J. A. Lux, H. F. Yin, and G. Schomburg, *J. High Resolut. Chromatogr. Chromatogr. Commun.*, *13*:145 (1990).
44. K. A. Cobb, V. Dolnik, and M. Novotny, *Anal. Chem.*, *62*:2478 (1990).
45. H. Deuel, J. Hutschneker, U. Schobinger, and C. Gudel, *Helv. Chim. Acta*, *119*:1160 (1959).
46. J. Pesek and S. A. Swedberg, *J. Chromatogr.*, *361*:83 (1986).
47. G. J. M. Bruin, J. P. Chang, R. H. Kuhlman, K. Zegers, J. C. Kraak, and H. Poppe, *J. Chromatogr.*, *471*:429 (1989).
48. G. J. M. Bruin, R. Huisden, J. C. Kraak, and H. Poppe, *J. Chromatogr.*, *480*:339 (1989).
49. S. A. Swedberg, *Anal. Biochem.*, *185*:51 (1990).
50. W. Nashabeh and Z. El Rassi, *J. Chromatogr.*, *559*:367 (1991).
51. Y.-F. Maa, K. J. Hyver, and S. A. Swedberg, *J. High Resolut. Chromatogr.*, *14*:65 (1991).
52. K. J. Towns and F. E. Regnier, *Anal. Chem.*, *63*:1126 (1991).
53. K. J. Towns and F. E. Regnier, *J. Chromatogr.*, *516*:69 (1990).
54. J. E. Wiktorowicz, J. C. Colburn, and H. H. Lauer, "Ninth International Symposium on HPLC of Proteins, Peptides and Polynucleotides," Philadelphia, Abstr. 704 (1989).
55. L. R. Gurley, J. S. Buchanan, J. E. London, D. M. Stavert, and B. E. Lehnert, *J. Chromatogr.*, *559*:411 (1991).
56. L. R. Gurley, J. E. London, and J. G. Valdez, *J. Chromatogr.*, *559*:431 (1991).
57. E. Wenisch, C. Taver, A. Jungbauer, H. Katinger, M. Faupel, and P. G. Righetti, *J. Chromatogr.*, *516*:133 (1990).
58. J. Frenz, S.-L. Wu, and W. S. Hancook, *J. Chromatogr.*, *480*:379 (1989).
59. K. W. Yim, *J. Chromatogr.*, *559*:401 (1991).
60. P. D. Grossman, J. C. Colburn, H. H. Lauer, R. G. Nielsen, R. M. Riggin, G. S. Sihampalam, and E. C. Rickard, *Anal. Chem.*, *61*:1186 (1989).
61. D. Josic, K. Zeilinger, W. Reutter, A. Bottcher, and G. Schmitz, *J. Chromatogr.*, *516*:89 (1990).
62. Y. Walbroehl and J. W. Jorgenson, *J. Microcol. Sep.*, *1*:41 (1989).
63. B. J. Compton, *J. Chromatogr.*, *559*:357 (1991).
64. E. C. Rickard, M. M. Strohl, and R. G. Nielsen, *Anal. Biochem.*, *197*:197 (1991).
65. J. R. Florance, D. Konteatis, M. J. Macielag, L. A. Lessor, and A. Galdes, *J. Chromatogr.*, *559*:391 (1991).
66. F. S. Stover, B. L. Hajmore, and R. J. Mcbeath, *J. Chromatogr.*, *470*:241 (1989).
67. M. Novotny, K. A. Cobb, and J. Liu, *Electrophoresis*, *11*:241 (1989).
68. R. A. Mosher, *Electrophoresis*, *11*:765 (1990).
69. R. D. Smith, J. A. Olivares, N. T. Nguyen, and H. R. Udseth, *Anal. Chem.*, *60*:436 (1988).
70. R. D. Smith, H. R. Udseth, C. J. Barmages, and C. G. Edmond, *J. Chromatogr.*, *559*:197 (1991).
71. E. D. Lee, W. Mueck, J. D. Henion, and T. R. Covey, *J. Chromatogr.*, *458*:313 (1988).
72. R. Capricoli, C. Shumate, W. Siems, T. Tsuda, and H. Hill, *J. Chromatogr.*, *480*:247 (1989).

73. M. A. Mosely, L. J. Deterding, K. B. Tomer, and J. W. Jorgenson, *J. Chromatogr.*, *480*:197 (1989).
74. M. A. Mosely, L. J. Deterding, K. B. Tomer, and J. W. Jorgenson, *Rapid Commun. Mass Spectrom.*, *3*:87 (1989).
75. K. A. Cobb and M. Novotny, *Anal. Chem.*, *61*:2226 (1989).
76. R. G. Nielsen, G. S. Sihampalam, and E. C. Rickard, *Anal. Biochem.*, *177*:20 (1989).
77. R. G. Nielsen, R. M. Riggin, and E. C. Rickard, *J. Chromatogr.*, *480*:393 (1989).
78. W. Nashabeh and Z. El Rassi, *J. Chromatogr.*, *536*:31 (1991).
79. A. S. Cohen, S. Terabe, J. A. Smith, and B. L. Karger, *Anal. Chem.*, *59*:1021 (1987).
80. R. Takigku and R. E. Schneider, *J. Chromatogr.*, *559*:247 (1991).
81. V. Dolnik, J. Liu, J. F. Banks, Jr., and M. Novotny, *J. Chromatogr.*, *480*:321 (1989).
82. W. Nashabeh and Z. El Rassi, *J. Chromatogr.*, *596*:25 (1992).
83. J. J. Herman, *J. Polymer Sci.*, *18*:257 (1953).
84. A. S. Cohen, D. R. Najarian, J. A. Smith, and B. L. Karger, *J. Chromatogr.*, *458*:323 (1988).
85. M. A. Chin and C. J. Colburn, *Am. Biotechnol. Lab.*, 7(10A):16 (1989).
86. H. Swerdlow, S.-L. Wu, H. Harke, and N. J. Dovichi, *J. Chromatogr.*, *516*:61 (1990).
87. M. Zhu, D. L. Hansen, S. Burd, and F. Gannon *J. Chromatogr.*, *480*:311 (1989).
88. H. E. Schwartz, K. Ulfelder, F. J. Sunzeri, M. P. Busch, and R. G. Brownlee, *J. Chromatogr.*, *559*:267 (1991).
89. H. J. Bode, *Anal. Biochem.*, *83*:364 (1977).
90. A. Guttman and N. Cooke *Anal. Chem.*, *63*:2038 (1991).
91. M. Strege and A. Lagu, *Anal. Chem.*, *63*:1233 (1991).
92. S. Mho and E. S. Yeung, *Anal. Chem.*, *57*:2253 (1985).
93. W. D. Pfeffer, T. Takeuchi, and E. S. Yeung, *Chromatographia*, *24*:123 (1987).
94. T. W. Garner and E. S. Yeung, *J. Chromatogr.*, *515*:639 (1990).
95. E. M. Lees and H. Weigel, *J. Chromatogr.*, *16*:360 (1964).
96. S. Honda, S. Iwase, A. Makino, and S. Fujiwara, *Anal. Biochem.*, *176*:72 (1989).
97. J. Liu, O. Shirota, D. Wiesler, and M. Novotny, *Proc. Natl. Acad. Sci. USA*, *88*:2302 (1991).
98. J. Liu, O. Shirota, and M. Novotny, *J. Chromatogr.*, *559*:223 (1991).
99. W. Nashabeh and Z. El Rassi, *J. Chromatogr.*, *600*:279 (1992).
100. W. Nashabeh and Z. El Rassi, *J. Chromatogr.*, *514*:57 (1990).
101. Z. El Rassi, D. Tedford, J. An, and A. Mort, *Carbohydr. Res.*, *215*:25 (1991).
102. P. J. Brynes, P. Bevilacqua, and A. Green, *Anal. Biochem.*, *116*:408 (1981).
103. S. Hjerten, K. Elenbring, F. Kilar, J. L. Liao, A. J. C. Chen, C. J. Siebert, and M. D. Zhu, *J. Chromatogr.*, *403*:47 (1987).
104. V. Dolnik, K. A. Cobb, and M. Novotny, *J. Microcol. Sep.*, *2*:127 (1990).
105. D. S. Stegehuis, H. Irth, U. R. Tjaden, and J. Van Der Greef, *J. Chromatogr.*, *538*:393 (1991).
106. R. Aebersold and H. D. Morrison, *J. Chromatogr.*, *516*:79 (1990).

107. N. A. Guzman, M. A. Trebilcock, and J. P. Advis, *J. Liquid Chromatogr.*, *14*:997 (1991).
108. Z. El Rassi and J. Cai, "Third International Symposium on High Performance Capillary Electrophoresis," poster PM-51, San Diego, California, Feb. 3–6 (1991).
109. J. Cai and Z. El Rassi, *J. Liquid Chromatogr.* (in press; 1993).

12

Covalent Surface Modification for Capillary Electrophoresis: *Characterization and Effect of Nonionic Bondings on Separations in Capillary Electrophoresis*

Ann M. Dougherty*

Supelco, Incorporated
Bellefonte, Pennsylvania

Mark R. Schure

Rohm and Haas Company
Spring House, Pennsylvania

Although offering great potential as a new analytical technique, capillary electrophoresis (CE) also presents several problems, owing to the unique surface chemistry of silica capillary tubing. Theory, based solely on axial diffusion zone broadening, predicts efficiencies in the millions of theoretical plates. However, the theory predicting such high efficiencies assumes that solute–wall interactions are negligible [1–4]. Reported efficiencies, however, have been well below those predicted by this theory, particularly for samples containing large-molecular-weight solutes [5–10].

To reduce the solute–surface interaction between large-molecule solutes and silica surfaces, several different strategies have involved modification of the solvent buffer composition. For instance, solvent pH variation has been used [5,11,12] to either protonate the surface silanols or to alter the charge of the solute, such that electrostatic interactions are reduced. High ionic strength buffers or addition of alkali salts [8,13–15] have been used to reduce solute–wall interactions by electrostatic shielding of both the solute and the wall. Other additives, such as long-

**Current affiliation*: Beckman Instruments, Inc., Somerset, New Jersey.

chain alkyl amines [16,17], cationic fluorosurfactant [18], and polyethylene glycol [19], have been shown to reduce solute–wall interactions. However, high ionic strength and addition of salt to the buffer can lead to increased zone broadening, caused by joule heating in the column. Use of cationic surfactants usually entails regeneration of the surfactant coating on the column surface between analyses.

To circumvent some of these limitations, several approaches to reducing solute–wall interactions have involved covalently bonding compounds to the column wall, through the surface silanols. One of the earliest chemical surface modifications was prepared by Jorgenson and Lukacs [3], who used trichlorosilane to modify the column surface. Hjerten [20] was able to obtain near-zero electroosmotic flow with a bonded acrylamide phase, allowing the development of isoelectric focusing in capillaries, because of its low electroosmotic flow properties. However, this phase exhibited long-term stability problems [21]. Since then poly(vinylpyrrolidone) [12], polyethylene glycol [11,22], polyethyleneimine [23], maltose and epoxydiol [10,11] have been used to chemically modify the surface of silica capillaries.

All of the latter phases require a pH less than 5 to maintain either column stability or optimum separation conditions. An arylpentofluoro phase, prepared by Swedberg [8], was used in conjunction with high ionic strength buffers to obtain protein separations. Trimethylsilane [24] and hydroxylated polyether functionalities [6] have been used to covalently modify the capillary surface. These polyether-modified columns allowed the separation of a range of proteins at neutral pHs. Novotny et al. [25] produced a chemically modified surface that is an improvement on the original acrylamide bonding. Because of its negligible electroosmotic flow, low or high pH was required for separating proteins, as all proteins in the sample mixture had to be either anionic or cationic to travel past the detector window. All of the previously mentioned modifications generally result in altered (usually decreased) electroosmotic flow and a reduction of dependence of the electroosmotic flow velocity on pH. Both of these phenomena are due to chemical modification or shielding of the surface silanols.

The phase thickness [6,23,26] and degree of surface silanol deactivation (usually through covalent chemical modification) [6] have a direct effect on the electroosmotic flow properties. The nature of the chemical group bound to the column surface, the degree of surface deactivation, and the thickness of this surface chemistry determine the dependence of the electroosmotic flow on pH, as well as the direction of the flow. Thus, quantification of the electroosmotic flow can be used to measure the degree of surface deactivation. Use of a stable, bonded phase also reduces variation caused by changes in column surface from run to run, a result of solute binding to the wall of the column.

In this chapter, we will report on several developments from our laboratories concerning the production and understanding of bonded phase capillary columns suitable for large-molecule CE separations. Specifically, we will first examine theoretical considerations for zone broadening caused by solute–wall interactions. Next, aspects of the production of bonded phases will be given relative to window formation, characterization, and column stability. Finally, several experiments are shown that demonstrate various solvent conditions employed for optimal separations with bonded-phase capillary columns.

THEORETICAL TREATMENT OF ZONE BROADENING EFFECTS FROM THE USE OF COATED CAPILLARIES IN CAPILLARY ELECTROPHORESIS

Introduction

The analysis of zone broadening owing to retentive capillary walls is well known in gas, liquid, and supercritical chromatography in which the carrier fluid, driven by pressure, takes on a parabolic distribution of velocities because of viscous drag at the capillary wall. In these cases, there are three main contributions to zone broadening [27]: (a) axial molecular diffusion independent of the fluid velocity distribution; (b) Taylor dispersion [28], also known in chromatographic terms [27] as "nonequilibrium" zone broadening; and (c) resistance to mass transfer in the stationary phase due to kinetic effects such as pore diffusion. The first of these is well known in chromatography and considers a zone to broaden axially as a linear function of time. Taylor dispersion has its origins in the different convection paths and, hence, velocities that solutes can attain by radial diffusion in the flow–field velocity gradient. The third effect is characteristic of any chromatographic system in which solute reenters the flow stream after delay in the surface phase.

In CE, the velocity of the solute is approximately uniform in the radial direction, with the exception of the immediate wall area, where drag retards the molecule at the wall surface and causes the velocity to asymptotically approach zero. Hence, Taylor dispersion is greatly reduced in CE. Previous treatment of zone broadening in CE [4] has recognized that the main source of band broadening is axial molecular diffusion. Little theoretical consideration, however, has been given to the effects that finite solute–wall interactions may pose in practical separations by CE. It is the purpose of this section to consider the decrease in resolution relative to a "wall-less" capillary column when solute–wall interactions are taken into account in CE.

Solute Velocity Field in Capillary Electrophoresis

During electrophoresis in a capillary column with net wall charge, small regions of nonuniform fluid flow [29–31] exist in the vicinity of the electrical double layer at the wall because of electroosmotic flow. These different flow profiles are summarized in Fig. 1 for the cases of a pressure-driven fluid (A) and electroosmotic flow (B and C). We will assume, for the treatment here, that the solute molecule is very

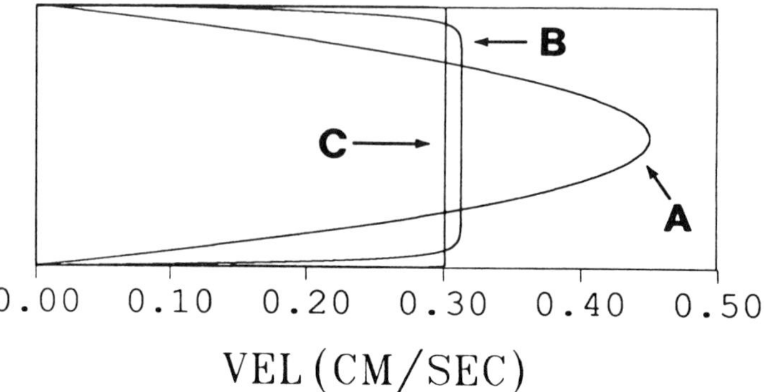

Figure 1 Comparison of various radial velocity distributions from chromatography and CE, with $v = 0.3$ cm s^{-1}, y-axis in the column cross section. (A) Normal parabolic flow exhibited by chromatography; (B,C) electroosmotic flow with $1/\kappa = 1.0$ μm, and $1/\kappa = 500.0$ Å, respectively. Radial velocity distribution for solute electrophoresis should resemble profile C.

small relative to the capillary column diameter, and that

$$\text{Vep} = \mathbf{V}_{eo} + \mathbf{V}_{ep} \tag{1}$$

where Vep is the net solute velocity vector, $\mathbf{V}_{eo}$ is the electroosmotic velocity vector of the fluid, and $\mathbf{V}_{ep}$ is the electrophoretic velocity vector of the solute. Note that vectors are denoted in boldface. The coordinate system used here considers x to be the nondimensional radial distance from the capillary center to the wall, and z is directed parallel to the electric field lines; hence, only the z component of these velocity vectors will be considered further. This summation of velocities will be used in the calculation of plate height, given in the following section.

Nondimensional Analysis

In chromatographic separation systems that use wall-coated capillary columns, such as capillary gas chromatography (capillary GC) and capillary liquid chromatography (capillary LC), a goal is to establish near-equilibrium conditions relative to solute partitioning or adsorption in the stationary phase. This is easily accomplished in gas chromatography because the inherent diffusivity of the solute in the gas phase allows very fast sampling of the wall region and allows the establishment of this dynamic equilibrium as the zone translates through the length elements of the column. However, this is not true in capillary LC because of the sluggishness of solute diffusion in the liquid mobile phase. A simple calculation based on one-dimensional random-walk diffusion can demonstrate this fact. Consider the one-dimensional Einstein equation which gives the average time, t', for a molecule to diffuse a distance d:

$$t' = \frac{d^2}{2D} \tag{2}$$

where D is the diffusion coefficient. Given some length of time, t, the average number of excursions, n, of length d, is given as $n = t/t'$. Considering that for any elution or migration technique $t = L_d/\bar{v}$, where L_d is the effective length of the column (from point of injection to detector) and $\bar{v}$ is the elution (or migration in CE) average cross-sectional velocity, we obtain through simple algebra:

$$n = \frac{2Dt}{d^2} = \frac{2L_dD}{\bar{v}d^2} \tag{3}$$

If d is identified as the capillary column diameter, then n is the number of average excursions or wall contacts that a nonretained solute of diffusion coefficient D makes when migrating through a column of length L_d with average velocity $\bar{v}$. In the case of capillary GC, where D is of the order 1 cm s^{-2} $\bar{v}$ is 1 cm s^{-1}, and for a column of length 10 m and diameter 50 μm, the average number of wall collisions is 80 million. As given in Table 1, for solutes and conditions typically found in capillary LC and CE, the number of average excursions or wall contacts is drastically reduced. This indicates that capillary LC cannot generate efficient separations at the same rate as GC; however, it also shows explicitly why capillary LC must be run at such low velocities.

Table 1 Number of Wall Contacts and Peclet Numbers for Large and Small Solutes in Columns of Differing Internal Diameters

Solute	mw	D (cm²/s)	n (d = 25 μm)	Pe (d = 25 μm)	n (d = 50 μm)	Pe (d = 50 μm)	n (d = 75 μm)	Pe (d = 75 μm)
Small molecules[a]								
Glycine	75	1.07×10^{-5}	2043	14	511	27	227	41
α-Alanine	89	9.10×10^{-6}	1747	16	437	32	194	48
p-Aminobenzoic acid	131	8.43×10^{-6}	1619	17	405	35	180	52
Citric acid	192	6.61×10^{-6}	1269	22	317	44	141	66
Proteins[b]								
Lysozyme	14,100	10.4×10^{-7}	200	140	50	280	22	241
β-Lactoglobulin	35,000	7.82×10^{-7}	150	186	38	373	17	559
Serum albumin	65,000	5.94×10^{-7}	114	246	29	491	13	737
Tropomysin	93,000	2.24×10^{-7}	43	651	11	1302	5	1954
Collagen	345,000	6.9×10^{-8}	13	2114	3	4227	1.5	6340

t = 10 minutes, L_d = 35 cm

Source: Diffusion coefficients: [a]Ref. 47; [b]Ref. 41.

To further characterize the transport of solute in CE, we examine the nondimensional Peclet number, *Pe*, which expresses the ratio of solute convection or migration to diffusion:

$$Pe = \frac{\bar{v}d}{D} \tag{4}$$

As shown in Table 1, separation of large solutes in CE typically occurs in the high-Peclet–number region, indicating that the axial movement of the solute is much larger than the diffusion of the solute. The consequences of this are as follows. For high-performance chromatographic elution systems, a large sampling of the stationary phase by the solute must take place for high-resolution separation of neighboring component zones. If the elution or migration speed of the solute is large relative to the diffusivity of the solute, sampling of the stationary phase will be reduced, and separation, based on solute–stationary phase interaction, will be of relatively low quality. On the other hand, in CE we want to minimize the amount of wall contact. The foregoing analysis indicates that because of the small diffusivity and large migration speed (high-*Pe* region), the number of wall contacts will be relatively small. This suggests that for CE, high *Pe*, and consequently low n conditions, will lower the risk for interactions. However, any time delay due to finite solute–wall interactions will cause the zones to broaden and subsequently even a small number of interactions must be avoided. In the case of larger diameter capillaries, however, temperature effects leading to zone broadening [32,33] become increasingly important and a balance must be struck between these two effects in choosing an optional capillary diameter.

For cases where some form of solute–wall interaction will invariably take place, for instance with solutes that form many strong hydrogen bonds with surfaces, the analogy with capillary chromatography is made and using previously developed theory [4,5], we can get a semi-quantitative view of resolution and plate height effects in CE due to solute–wall interactions.

Plate Height

Although the theory presented here was originally conceived for capillary LC with electroosmotic flow [29,30], for which solute convection is provided by electroosmotic flow and solutes are assumed to be uncharged (i.e., no electrophoresis takes place), a simple change of the convective frame of reference can be used to apply this theory to CE in coated capillary columns.

The three processes contributing to the zone broadening of solutes in capillary chromatography (axial molecular diffusion, Taylor dispersion, and the resistance to mass transfer in the stationary phase) can be written, respectively, as terms in a plate height equation:

$$H_c = \frac{2D}{\bar{v}} + C_m\left(\frac{a^2\bar{v}}{D}\right) + C_s\bar{v} \tag{5}$$

where C_m is the nondimensional coefficient that determines the extent of Taylor dispersion and C_s is the dimensional coefficient, with units of time, that determines the amount of resistance to mass transfer in the stationary phase. The quantity a is the column radius, equal to $d/2$. For the remainder of the treatment given here, it will be assumed that the stationary phase is some coating material that is immobilized on the capillary column wall, with the intention of minimizing the solute–wall surface interaction; terms pertinent to this interaction are given by the subscript c.

By using generalized dispersion theory [29], it has been shown for capillary LC using electroosmotic flow as the fluid driving force, that

$$C_m = \frac{[A_1 - 2A_2 + (1/4)] + 2(A_1 - A_2)k' + A_1k'^2}{(1 + k')^2} \tag{6}$$

where k' has its usual meaning in chromatography and

$$A_1 = \int_0^1 \frac{P^2(x)}{xdx} \tag{7}$$

$$A_2 = \int_0^1 xP(x)dx \tag{8}$$

$$P(x) = \int_0^x 2x'Q(x')dx' \tag{9}$$

$$Q(x) = \frac{v(x)}{\bar{v}} \tag{10}$$

where r is the dimensional distance from the capillary column center to the wall, so that $x = r/a$ and $v(x)$ is, as was mentioned previously, the fluid velocity as a function of radial position. The primed variables in Eq. 9 indicate dummy variables. The integrals in Eqs. 7–9 are conveniently evaluated by using a small computer and employing Simpson's three-point formula for numerical integration with typically 10001 grid points in r. The $P(x)$ integral given in Eq. 9 is calculated and stored in computer memory for 5001 values of x over the range of 0–1. In the case of uniform $v(x)$, the C_m term reduces to a particularly simple form [29]:

$$C_m = \frac{k'^2}{4(1 + k')^2} \tag{11}$$

For capillary LC with electroosmotic flow, the flow field is computed by considering $Q(x)$, which can be derived from fundamental considerations of the electrical double layer with thickness much smaller than the capillary column radius [29]:

$$Q(x) = \frac{\left[1 - \dfrac{I_0(\kappa x a)}{I_0(\kappa a)}\right]}{\left[1 - \dfrac{2I_1(\kappa a)}{\kappa a I_0(\kappa a)}\right]} \tag{12}$$

where I_0 and I_1 are zeroeth and first-order modified Bessel functions of the first kind and κ is the reciprocal double-layer thickness.

For the C_s term, both solution in the stationary phase (partitioning) and adsorption yield different functional forms. For partitioning [30]

$$C_s = \frac{\beta a^2}{6D_s} \tag{13}$$

and for adsorption [30]

$$C_s = 2\beta t_d \tag{14}$$

where

$$\beta = \frac{k'}{(k' + 1)^2} \tag{15}$$

and D_s and t_d are the diffusion coefficient of solute in the stationary phase and the mean desorption time of solute on an adsorptive phase. Since we consider the wall coating to be nonporous we only consider D_s and t_d to be zero for the remainder of the treatments here.

Having provided the mathematical detail necessary for evaluating H_c, we cast an additional equation as the ratio of the plate height in a coated capillary column, H_c, as given by Eq. 5, to the plate height in a "wall-less" capillary column, H_∞, given by the first term in Eq. 5. This ratio is

$$\frac{H_c}{H_\infty} = 1 + \frac{\bar{v}^2}{2D}\left[\frac{C_m a^2}{D} + C_s\right] \tag{16}$$

and expresses the severity of plate height increase with which solute–wall-coating interactions may occur, as a function of k' and other physicochemical, capillary, and operating parameters.

The consequences of zone broadening, through solute–wall-coating interactions, is best viewed through the reduction in resolution that may occur. Resolution, defined as $R_s = \Delta t/4\sigma$, where Δt is the difference in migration times between peaks and σ is the standard deviation of Gaussian peaks, is obviously reduced when σ is large. By using the general relationship [29],

$$H = \frac{L_d \sigma^2}{t^2} \tag{17}$$

and noting that t in Eq. 17 is equal to L_d/v for the case of no wall retention and to $(k' + 1)L_d/v$ for the case of a retained solute, we can cast Eq. 16 into the ratio of σ terms:

$$\frac{\sigma_c}{\sigma_\infty} = (k' + 1) \cdot \sqrt{\frac{H_c}{H_\infty}} \tag{18}$$

where σ_c is the standard deviation of a zone retained by the wall-coated capillary column and σ_∞ is the standard deviation of a zone with no wall interaction. Relative to two unretained zones with differences in migration time, Δt, migration times will be multiplied by the factor $k' + 1$ when retention occurs, assuming equal interaction energetics with the wall coating. Hence,

$$\frac{R_{s_c}}{R_{s\infty}} = \sqrt{\frac{H_\infty}{H_c}} \tag{19}$$

giving the ratio of resolution with retention to resolution expected for the idealized case of the wall-less column; we will refer to this ratio as *relative resolution*. A compact form for Eq. 19 can be obtained by using the approximate C_m term given by Eq. 11:

$$\frac{R_{s_c}}{R_{s\infty}} = \left[1 + \frac{\bar{v}^2 k'}{D(1 + k')^2} \cdot \left[\frac{k'a^2}{8D} + t_d\right]\right]^{-1/2} \tag{20}$$

and v is noted to be an average migration velocity when considering two zones of approximately equal migration velocities.

As used here k' is the ratio of the time that the solute spends in the wall coating (stationary phase), t_c, to the time the solute spends in free solution, t_s. One method for estimating t_c is to have an independent source of electrophoretic mobility data, gathered in an apparatus free of wall effects and at constant temperature. The difference in mobility between this wall-free data and that gathered by CE, at the same field strength, temperature, and buffer composition, would then be indicative of the solute residency time in the coating. This approach is impractical and difficult because commercially available instrumentation which can measure electrophoretic mobilities rarely can achieve the field strengths easily obtained in CE and because exact temperature control in CE is difficult. An alternative approach to estimating k' is to use a pump-driven flow with the capillary column and compare the elution time of the zone of a "target" molecule with either the elution time for a small reference molecule, assumed to be unretained, or an estimate of ideal t_s, obtained by measuring the flow rate. In this way t_c is the difference measurement between large- and small-molecule elution experiments, and this allows the estimation of

k'. Once k' is known, further modeling can be attempted. For CE calculations, we use Eqs. 5–10 and consider the flow field to be that given by Eq. 1. The net migration velocity is now biased by the electroosmotic flow, for which previous theoretical treatments have considered $\mathbf{V}_{eo}$ to be zero. The essential results are, however, the same with the exception of the choice of variables that govern the capillary column dimensions and the magnitude of the flow profile.

Figure 2 illustrates the calculation of relative resolution under a variety of conditions pertinent to CE. These results indicate that even a small amount of solute retention causes a great decrease in the relative resolution; on the order of ten times less resolution for a small solute in a 75-μm–diameter capillary column at k' values near unity.

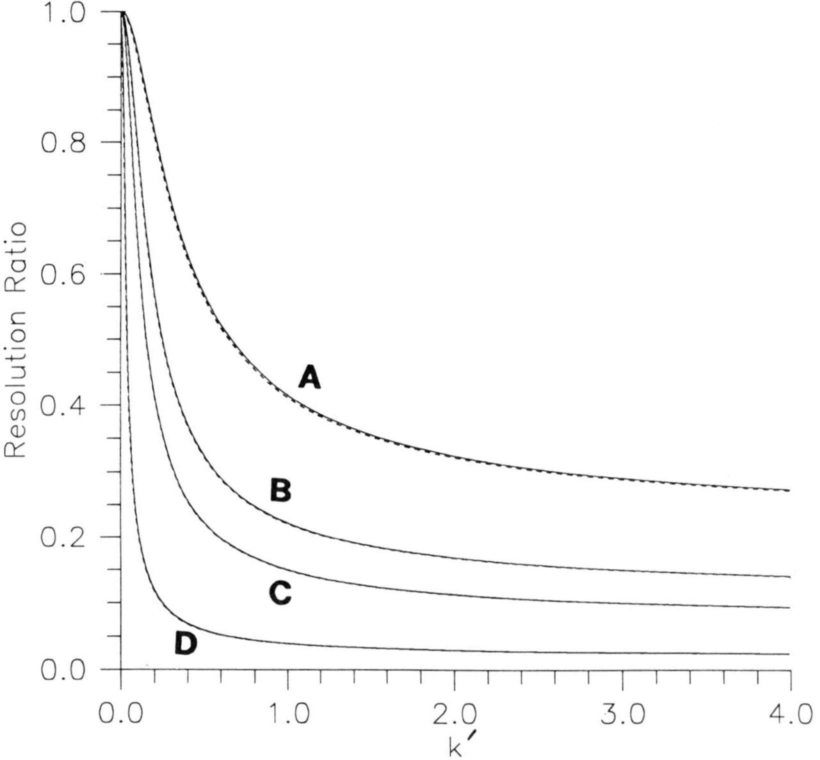

Figure 2 The quantity, $R_{sc}/R_{s\infty}$, as a function of k'. Conditions: V_{ep} = 0.14 cm s^{-1}; V_{eo} = −0.08 cm s^{-1}; 1/κ = 500 A; t_d = 0; L_d = 35 cm. (A) d = 25 μm; $D = 6 \times 10^{-6}$ cm^2 s^{-1}; (B) d = 50 μm, $D = 6 \times 10^{-6}$ cm^2 s^{-1}; (C) d = 75 μm, $D = 6 \times 10^{-6}$ cm^2 s^{-1}; (D) d = 50 μm, $D = 1.04 \times 10^{-6}$ cm^2 s^{-1}.

Furthermore, the results of Fig. 2 suggest that, once retention occurs, up to k' of unity, further retention causes little more loss of resolution. In other words, the most significant reduction in potential resolution by the CE method is associated with small levels of wall retention. In addition, it can be seen from Fig. 2 that the use of Eq. 11 (solid line) for approximating the C_m term is valid under the conditions used in this figure (the dashed line is the full treatment).

From the quantities given in Table 1, it is clear that plate height calculations may well approximate the results for small molecule solutes, which tend to sample the wall region from hundreds to thousands of times during their passage down the column. The results shown in Fig. 2 suggest that if some form of retention is inevitable, the use of narrower capillary diameters is highly desirable for this class of solutes.

Tailing

As mentioned previously, the equilibrium assumption used in plate height theory can be misleading, owing to the low n, high Pe region at which the CE separation of large molecules, as given in Table 1, may be performed. To increase accuracy and insight about this region of operation, preliminary experiments using digital simulation techniques have been employed to study the solute–wall interactions. These techniques are known to work well for systems with low n and high Pe, and do not suffer from the near-equilibrium assumption needed for plate height theory [35,36]. In addition, computer codes used previously for studies of field-flow fractionation [35,36] were easily modified to include a stochastic (poissonian) desorption term for solute release from the wall. The results of these simulations indicate that solutes with diffusion coefficients typically smaller than approximately 5×10^{-7} cm^2 s^{-1} give zone shapes with tailing peaks under a variety of conditions. This will lead to a further loss of resolution and has been observed in a variety of experiments.

We now consider the experimental aspects of providing an inert surface to prevent or reduce the solute–wall interaction maximum resolution in CE.

EFFECT OF SURFACE MODIFICATION ON SOLUTE BAND BROADENING

Experimental

Instrumentation

Capillary electrophoretic analyses were performed on an Applied Biosystems Model 270A and Model 270 HT automated system (Applied Biosystems, Inc., Foster City, California) and a Beckman P/ACE Model 2050 automated system (Beckman Instruments, Inc., Palo Alto, California). Data was acquired with an IBM Model PS/2 (IBM, Armonk, New York) using Beckman Gold Software and a Beckman 406

Analog Interface module. The GC analyses were performed on a Hewlett Packard Model 5890 gas chromatograph (Hewlett Packard, Avondale, Pennsylvania). A home-built apparatus for creating an on-column detection window by acid treatment consists of a soldering iron, a 25-mL buret with Teflon stopcock, a rheostat, a 150 × 75-mm petri dish, a buret clamp, extension clamps, and a ring stand. A high-temperature stripping device for creating detection windows using a heated coil is from Polymicro Technologies (Phoenix, Arizona).

Columns and Reagents

All CE columns used were 363-μm od, 50-μm id CElect series capillary electrophoresis columns and were manufactured by Supelco. Bare fused silica CE columns were purchased from Polymicro Technologies. The CElect series consists of a polymeric C_{18} phase (CElect-H250), a polymeric C_8 phase (CElect-H150), a polymeric C_1 phase (CElect-H50), and a nonionic polymeric polar phase (CElect-P150) that are chemically bound to the capillary surface. Proteins, surfactants, chiral compounds, and reagent-grade buffer components were purchased from Sigma Chemical Company (St. Louis, Missouri). Lysozyme for the experiments with high ionic strength was purchased from Calbiochem (San Diego, California). Reagent-grade sodium borate, acetonitrile, methanol, and sulfuric acid were purchased from Fisher Scientific (Pittsburg, Pennsylvania). All water used was purified with a Millipore RO6 and a five-bowl system (Millipore Corp, Bedford, Massachusetts).

Electrophoresis

Protein solution concentrations were approximately equal to or less than 1 mg mL^{-1} in purified water. Benzyl alcohol (0.01% w/w) was used as a neutral marker. In the stability studies reported, all electrophoretic analyses were performed using a 25-mM sodium phosphate buffer, pH 7.0. Analysis times ranged from 20 minutes for the bare (untreated) silica and polar-bonded columns to 30 minutes for the hydrophobic-bonded column series. In studies of pH effects, the ionic strength of the buffers was kept constant throughout the pH range. With the exception of the surfactant studies, all separations were performed using a 1-minute wash with 0.1 N NaOH, followed by a 4-minute rinse with running buffer. In the surfactant study, each column was rinsed for 30 to 60 minutes with buffer containing 0.5% (w/v) surfactant, followed by a 30-minute rinse with running buffer, containing 0.001% surfactant. After this initial coating of the bonded phase with surfactant, a 4-minute rinse with running buffer containing 0.001% surfactant was performed. This rinse was repeated between electrophoretic runs. All injections were done pneumatically. Experiments were performed using a field strength of 200 V cm^{-1}, except for the surfactant study for which a field strength of 350 V cm^{-1} was used. Plots of operating current versus applied voltage were inspected for nonlinear behavior, indicative of solution resistance changes from buffer heating. The nonlinear current versus voltage range was observed for field strengths in excess of 350 V cm^{-1}.

Formation of Detection Window

Chemical bonding was performed in a continuous 20-m coil, which is defined as a production batch, then cut into 1-m sections or columns. These columns were tested for surface inertness (as viewed by peak capacity and symmetry) and capacity factor. Each was subjected to one of the three window-burning techniques, then retested for inertness and capacity. The polyimide coating was removed from a 1-cm section of the capillary column by using fuming sulfuric acid, low-flame, or high-temperature stripping. To remove the polyimide coating with sulfuric acid, a marked section of column was rotated in a drop of fuming (330°C) sulfuric acid suspended from the tip of a heated soldering iron. Polyimide removal occurred in 3–5 seconds when using 98% sulfuric acid. Low-flame window burning was achieved by holding the marked section of the column in an open flame for 2–3 seconds. High-temperature stripping was done by holding the marked section of column in a high-temperature stripper source for 5–7 seconds, then wiping the charred polyimide from the surface of the fused silica. All GC analyses were performed on a 50-μm × 1-m C_{18}-bonded column at 40°, with a linear velocity of 31 cm s^{-1}, a split flow of 300 mL min^{-1}, and a 6000:1 split ratio. The sample was a C_{10}–C_{18} test mixture of *n*-alkanes, which was detected by FID at 300°C.

RESULTS

Column Characterization

EFFECT OF WINDOW FORMATION TECHNIQUE ON THE INTERNAL BONDING Since the column detection window is created after internal bonding of the phase, the effect of the different window-burning techniques was studied. Figure 3 illustrates the effect of the different-burning methods on the resulting gas chromatograms. Before window burning, no peak tailing is evident, as shown in Fig. 3. After processing a window with sulfuric acid, the k' of the column has increased, but no peak tailing, indicative of incomplete silanol surface deactivation, is evident. The GC analysis of columns in which windows were made by either high-temperature stripping and low-flame methods show peak tailing of alkanes, as demonstrated in Fig. 3. This suggests that some of the bonded phase has been lost or has degraded during these latter window-burning processes [37]. Although the window-processing methods affect only a small (1- to 4-cm) section of the column, this is sufficient to alter the surface character. Given the results of this study, subsequent columns were processed using the sulfuric acid method.

STABILITY To test the stability of the chemical surface modification, a series of consecutive analyses were performed on each column. A conditioning period is evident for all of the bonded columns, during which the electrophoretic mobility of the neutral marker (benzyl alcohol) increases. The duration of this conditioning period increases with increasing hydrophobicity and chain length. Figure 4 shows the electrophoretic mobility of 0.01% benzyl alcohol over a series of consecutive

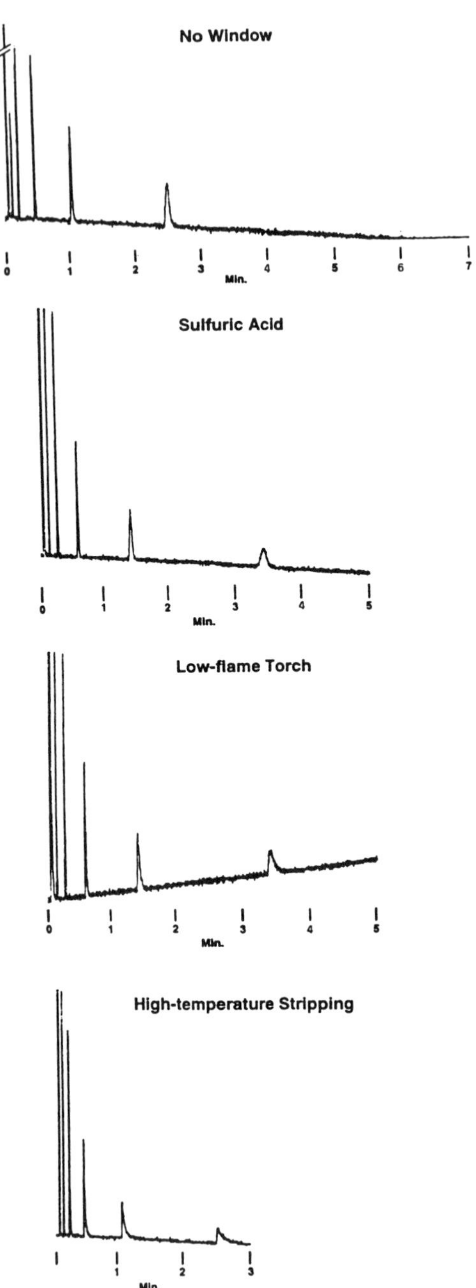

Figure 3 Effect of the window-processing method on the deactivation of C_{18}-bonded (CElect-H250) column surface, as evaluated by GC. Conditions listed in the text.

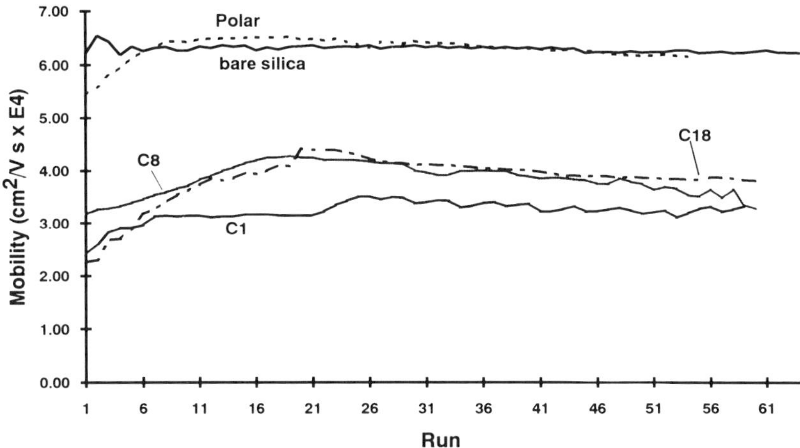

Figure 4 Electrophoretic mobility of a neutral marker over 75 runs using bare silica columns and covalently bonded columns. Conditions: column dimensions: 50-μm id, 100-cm total length, 65-cm effective length column. Buffer: sodium phosphate, pH 7.0, 200 V cm^{-1}; 3 nL vacuum injection of 0.01% (w/w) benzyl alcohol; 1-minute rinse with 0.1 N NaOH, followed by 4-minute buffer rinse before each run. (From Ref. 46)

analyses on a bare silica column and on four bonded columns, each bonded with one of the different phase types (C_{18}, C_8, C_1, and polar). Figure 5 shows conditioning of four C_8-bonded columns from the same 20-m production batch, as defined earlier. All exhibit the same characteristic conditioning pattern. This conditioning trend is somewhat, but not completely, reversible. After rinsing columns with 25–50% methanol, the electroosmotic flow in the column decreases to a level seen earlier in the conditioning period and, on further use, again slowly increases until it reaches the "fully conditioned" state. In fully conditioned columns, run-to-run variation of the electroosmotic flow is minimal for a given set of electrophoretic conditions. When columns were stored in water or electrophoresis buffer, either overnight or long-term, no changes in day-to-day electroosmotic flow was observed. However, after being evacuated and stored with no liquid in the column, the columns exhibited a brief (one to two runs) conditioning period, as measured by electroosmotic flow, before returning to their fully conditioned state.

There are several possible explanations for the change in electroosmotic flow during the conditioning period. First, the surface phase is bonded under totally nonpolar conditions, in which the surface silanols are in their uncharged protonated (SiOH) form. During rinsing with 0.1 N NaOH before the electrophoretic runs, a population of these silanols can deprotonate to the silanoate anion ($SiOH \rightarrow SiO^-$), increasing the net negative charge on the wall, resulting in increased electroosmotic

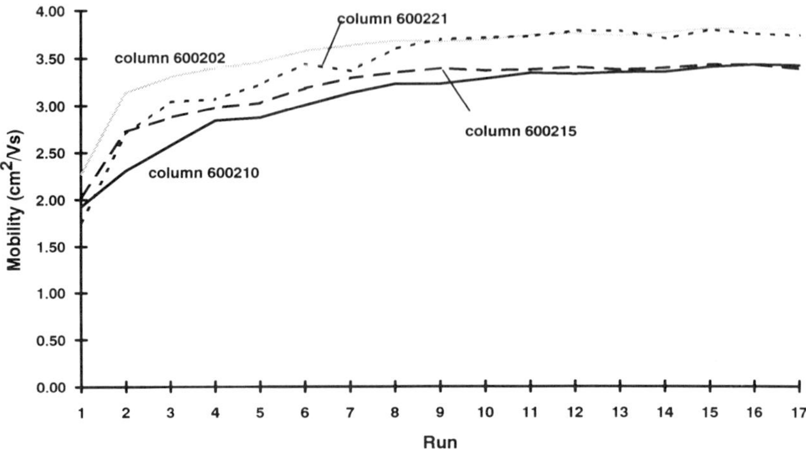

Figure 5 Stability and electrophoretic flow properties of four C_8-bonded (CElect-H150) columns from the same production lot. Conditions as those for Fig. 3.

flow. Since it is difficult to regain the completely protonated (SiOH) state of the column in aqueous solution, the initial surface conditions are never reached. Therefore, the slower electroosmotic flow observed for the first several runs on a new column cannot be reproduced. Second, loss of unbound surface phase during the initial use of the column may account for some of the change in electroosmotic flow, although there is little evidence of a loss of phase, as baseline perturbations during detection at 200 nm are not observed. The third explanation is that impurities in the column, as a result of phase synthesis, may slowly diffuse and desorb from the wall region. This also would explain the correlation of an increasing conditioning period duration with increasing alkyl chain length of the phase.

ELECTROOSMOTIC FLOW CHARACTERISTICS Electroosmotic (eo) flow velocities for untreated and bonded columns are shown in Table 2. Conditions were as for Fig. 4. Relative standard deviations were calculated using only postconditioning data points. The higher postconditioning relative standard deviations for the C_{18}- and C_1-bonded columns were due to a gradual decrease in the electrophoretic mobility of the neutral marker as run number increased. This could be due to adsorption of benzyl alcohol on the hydrophobic surface of the column over several runs, further shielding the surface silanols. This variability was not seen with polar-bonded columns, in which hydrophobic adsorption would not be expected. Contrary to expected outcome, the electroosmotic flow velocity decreased with shorter-chain length. This suggests better surface coverage with decreasing chain length and may be due to less steric hindrance during surface bonding with the shorter-chain compounds, allowing better coverage of the wall surface silanols.

Table 2 Relative Standard Deviation of Electroosmotic Flow Velocity in Untreated and Covalently Bonded Capillary Columns

Column	Phase type	Number of runs	Eo flow velocity (cm min^{-1})	Overall RSD (%)	Post-conditioning RSD (%)
Untreated no. 1	None	50	7.28	1.2	na[a]
Untreated no. 2	None	129	7.63	2.4	na
CElect-P150	Neutral polar	55	7.57	3.7	1.7
CElect-H250	C_{18}	109	4.84	15.2	5.5
CElect-H150	C_8	79	4.34	8.6	8.6
CElect-H50	C_1	58	3.87	16.2	3.1

[a]No conditioning required.
Source: Ref. 46.

The correlation between electroosmotic flow and phase thickness has been well documented [22,23]. Although the reduction of electroosmotic flow by the polar bonding is statistically insignificant, this phase does show improvements over bare silica for the separation of proteins; again, steric shielding may possibly explain this observation. Table 3 illustrates a comparison of electroosmotic flow velocities for four C_8-bonded columns from the same production batch, as defined earlier. Conditions were the same as for Fig. 5. The relative standard deviation for electroosmotic flow was determined for the first run on each column as well as for the postconditioning flow average. The relative standard deviation for the first runs is twice that for the calculated flow averages. Therefore, to determine the working variation in electroosmotic flow, the flow must be measured after the column is conditioned.

Table 3 Comparison of Electrophoretic Mobility and Electroosmotic Flow in C_8-bonded (CElect-H150) Columns From a Single Production Batch

Column	Electroosmotic flow preconditioning	Velocity postconditioning (cm/min)	Postconditioning RSD (%)
600202	2.730	4.498	1.7
600210	2.307	3.994	2.9
600215	2.434	4.044	0.6
600221	2.107	4.407	1.9
RSD of velocities	10.9%	6.0%	

Conditions same as for Fig. 7.

The advantages of reduced electroosmotic flow are most obvious in applications during which the solute is migrating in a direction opposite that of the electroosmotic flow. In these cases if the magnitude of the electroosmotic velocity is larger than the magnitude of the electrophoretic velocity, the solute will never migrate past the detector. This becomes critical when solute electrophoretic velocity is small and may require pH adjustment so that the electroosmotic and electrophoretic flows are in the same direction.

EFFECT OF pH ON ELECTROOSMOTIC FLOW Figure 6 describes the change in electrophoretic mobility of benzyl alcohol with changes in pH, using 25-mM sodium phosphate buffer at the indicated pHs. Above pH 3, both the polar-bonded and bare silica columns show the characteristic deprotonation of the silanol groups [38]. For all three of the hydrophobic-bonded phases, electroosmotic flow displays less dependence of pH, whereas flow velocity is maintained at one-third to one-half that of bare silica. This phenomenon has been described for other surface modifications [6,23]. This offers the advantage of optimizing the separation by using variations in pH, without having to compensate for large changes in electroosmotic flow. However, one still has the problem of having to include electroosmotic flow in optimization considerations. It is notable that there are no data on the C_1-bonded column, owing to irreproducible results at pH 4 and column blockage at pH 3. It is, therefore, recommended that this column be used at pHs higher than 4.0.

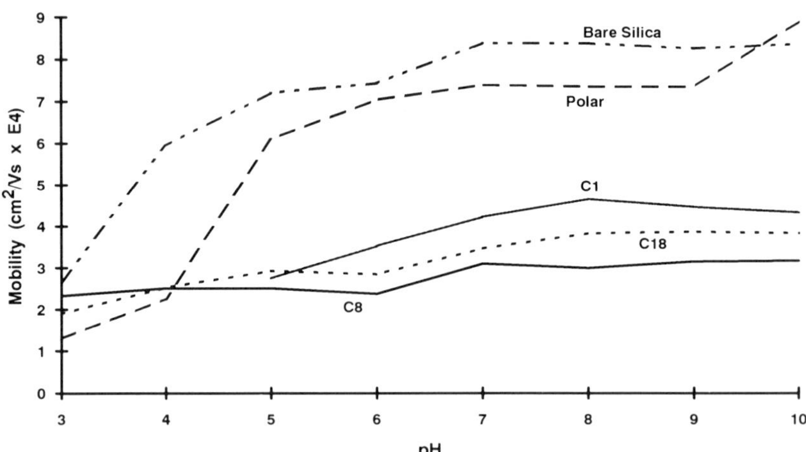

Figure 6 Dependence of electrophoretic mobility of a neutral marker on pH for each of the bonded columns and an unbonded fused silica column. Conditions: 25 mM sodium phosphate, pH as indicated. Other conditions as for Fig. 3. (From Ref. 46)

Protein Separations

PROTEIN SEPARATIONS WITHOUT ADDITIVES Proteins were separated at approximately pH 7 on each of the bonded-phase columns and on untreated fused silica columns, to assess the silanol-shielding properties of the phases in protein separations. Proteins used in these separations (and their pIs) are listed as follows, along with the corresponding numbers used as peak labels in Figs. 7–10:

Protein	pI
1. Myoglobin (horse heart)	7.4
2. Conalbumin (iron-free)	6.6
3. β-Lactoglobulin B	5.3
4. β-Lactoglobulin A	5.1

Figure 7 shows a series of separations of the protein mixture at three pHs on an untreated bare silica column, using 10-mM sodium phosphate/6-mM sodium borate buffer at three pH values. Even for these moderately acidic proteins, a high pH was needed to obtain separation of all four compounds. However, at pH 9, reduced peak height for β-lactoglobulin B, when compared with all other components, suggests irreversible adsorption or solute degradation.

Figure 8 shows the separation of the protein mixture on a polar-bonded column under the same conditions as Fig. 7. Although pH 9 again gave the best separation, usable separations were also obtained at pH 7 and pH 8. As discussed earlier, this column retains substantial silanol character. However, the phase shields the surface sufficiently to improve the separations, when compared with untreated bare silica. More evidence of this phenomenon will be given later in the discussion of ionic strength effects.

The results of CE separation using the C_1 phase at pH 7.0, 7.5, and 8.0 are given in Fig. 9. All proteins were separated in each analysis, with pH 7.5 offering the optimum resolution. Figure 10 shows the separation of the four proteins on a C_8-bonded column, using conditions similar to Fig. 9. Unlike the separation on the C_1-bonded column, resolution on the C_8-phase column is maximal at pH 6.2. The effect of lowering the dependence of electroosmotic flow on pH is evident in the series of separations shown in Figs. 7–11. When performing the foregoing separations on the hydrophobic-bonded columns, all protein elution times increased as the pH is increased (see Figs. 9 and 10). Because of increases in electroosmotic flow, on the polar-bonded column the peaks eluted more rapidly as the pH was increased (see Fig. 8). However, the longer elution times, at lower pH, on the polar-bonded column may have been partly due to retention of the protein molecules. This solute retention could be a result of the protonation of amine groups on the protein at pHs less than 8, giving the protein molecule a higher net positive charge,

pH 7.0

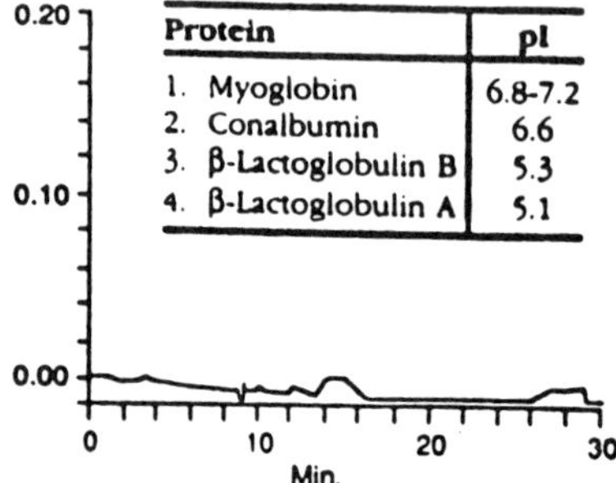

Protein	pI
1. Myoglobin	6.8-7.2
2. Conalbumin	6.6
3. β-Lactoglobulin B	5.3
4. β-Lactoglobulin A	5.1

pH 8.0

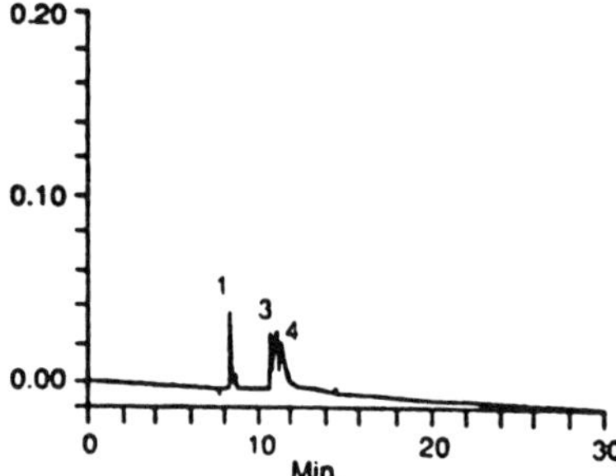

pH 9.0

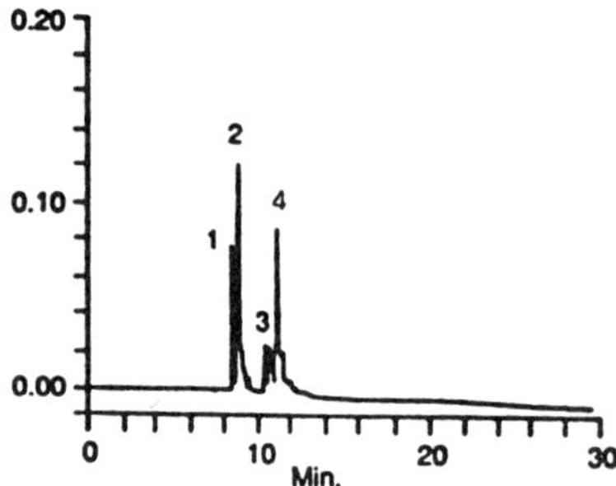

Figure 7 Protein separations on an untreated silica column. Conditions: 10 mM sodium phosphate/6 mM sodium borate, pH as indicated; 3 nL vacuum injection; 200 V cm^{-1}; 50-μm id, 100-cm total, 65-cm effective length column; all proteins are 1 mg mL^{-1} each in water, numbered as in the text; Detection; 200 nm, UV. (From Ref. 46)

pH 7.0

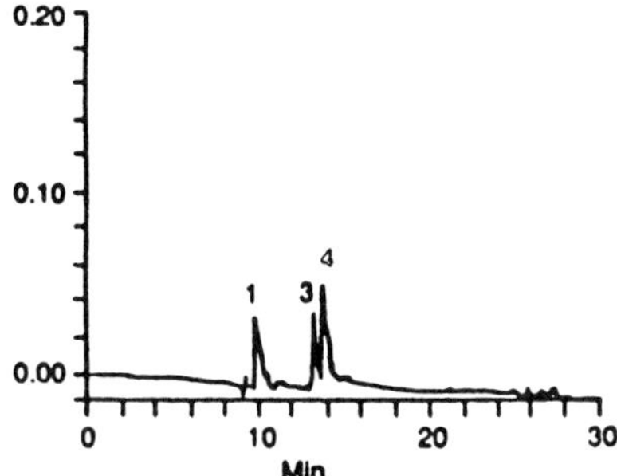

pH 8.0

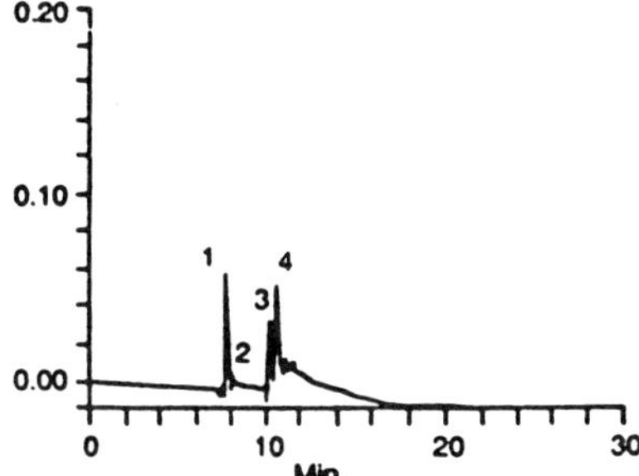

pH 9.0

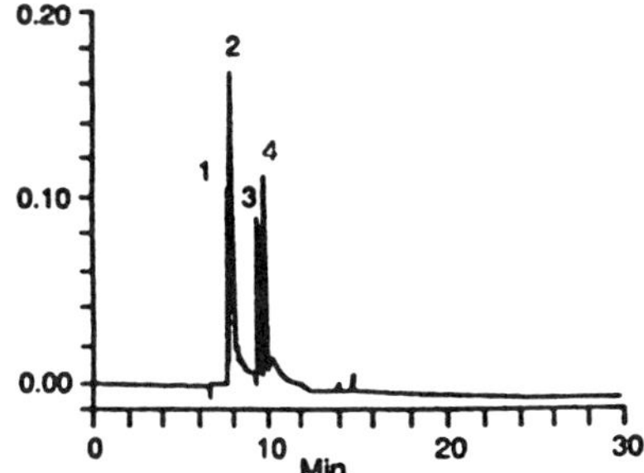

Figure 8 Protein separations on a bonded polar-phase column (CElect-P150). Conditions as for Fig. 7. (From Ref. 46)

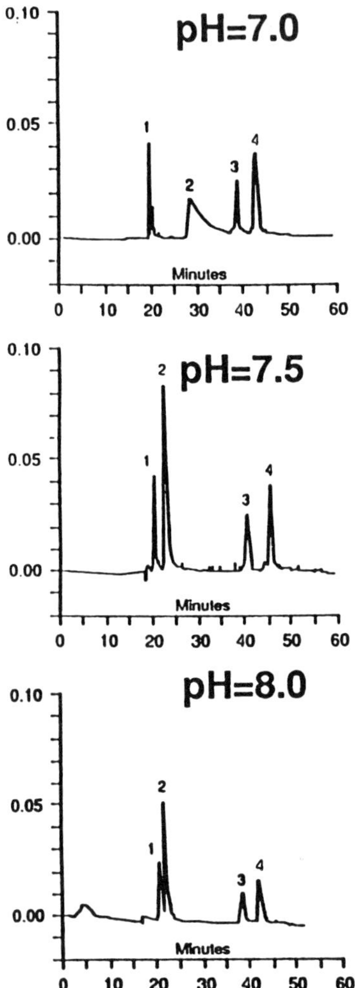

Figure 9 Effect of pH on optimization of a protein separation on the bonded C_1 phase (CElect-H50) column. Conditions as for Fig. 7.

which increases the possibility of ionic interaction with any unshielded silanoate anions on the surface of the capillary column. Assuming no solute–wall interactions and given the increasingly negative charge of the proteins with increasing pH, one would expect the proteins to elute later with increasing pH under normal polarity conditions. Under the conditions used for these separations, the magnitude of the electroosmotic velocity is larger than the magnitude of the electrophoretic velocity;

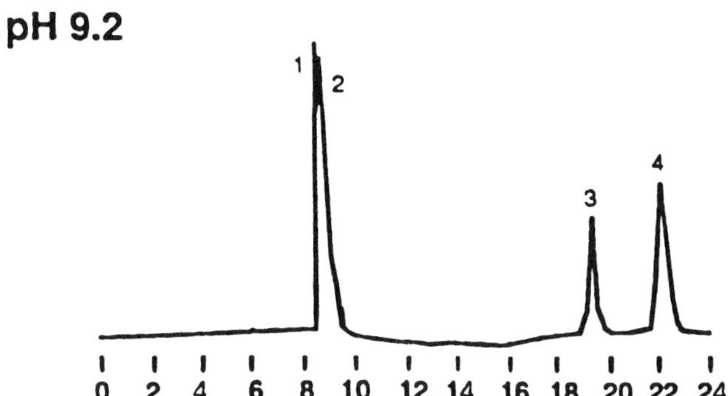

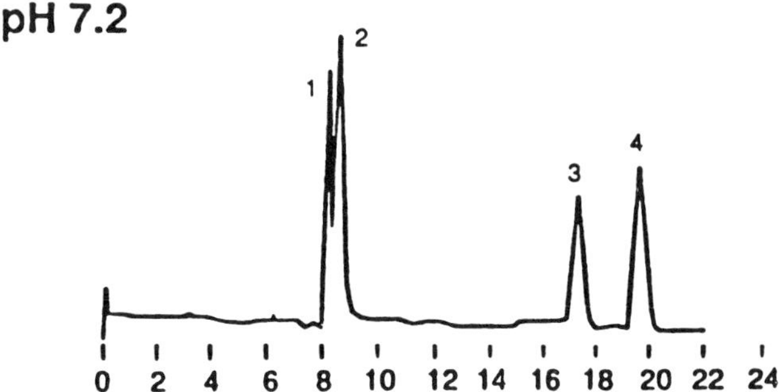

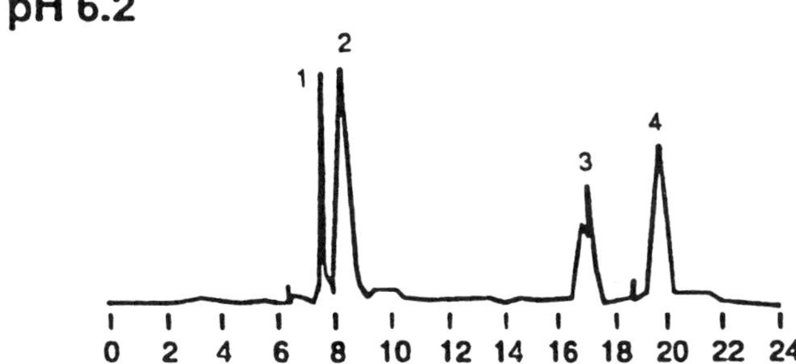

Figure 10 Effect of pH on optimization of a protein separation on the C_8-bonded phase (CElect-H-150) column. Conditions as for Fig. 7. (From Ref. 46)

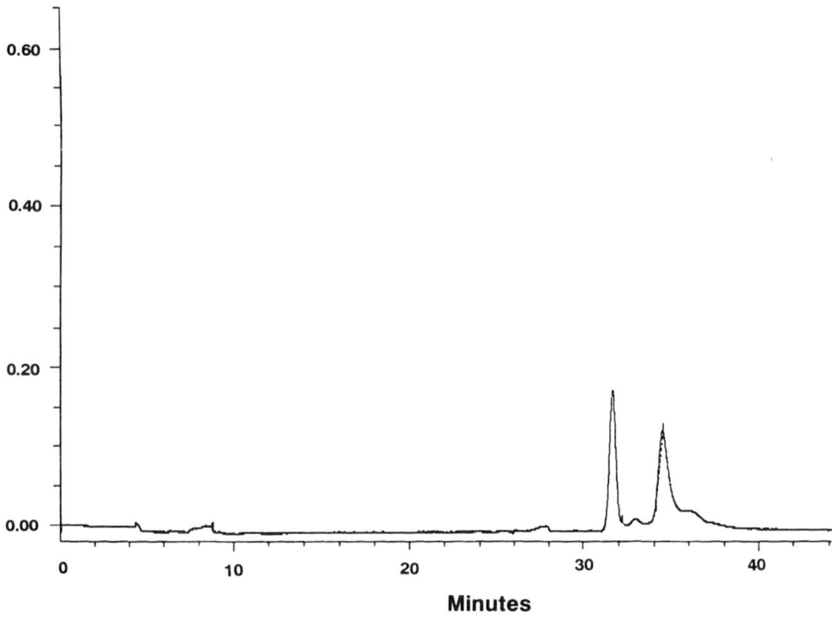

Figure 11 Separation of neutral proteins on a C_{18}-bonded phase (CElect-H250) column at pH 8. Conditions as for Fig. 7.

hence, the solute moves from injection at the anode to detection at the cathode. Figure 11 shows the separation of the four protein mixture at pH 8.0 on a C_{18}-bonded column. The separation did not improve with changes in pH. The broad-peak shape and long-elution times seen in this separation suggest solute–phase or solute–wall interactions might be occurring during the separation.

Figure 12 shows the separation of the casein peptides on an untreated fused silica column and a C_1-bonded column [39]. At high pH, separations are similar, but the peaks are compressed on the untreated silica column. This compression is due to the higher electroosmotic flow inherent in the untreated column. However, there is a marked difference in the separations at low pH, at which many of the peptides are positively charged. The coverage of the surface by a bonded-phase coating appears to reduce interaction of these peptides with the surface silanols and allows elution of these peptides at low pH.

Electrophoresis of other protein matrices, such as egg white proteins (lysozyme, conalbumin, ovalbumin) and a mixture of basic proteins (lysozyme, cytochrome *c*, ribonuclease), under the conditions used in the foregoing, did not yield acceptable separations on either the untreated or bonded columns. Changes to the buffer and additives that improved these separations will be discussed in the following paragraph.

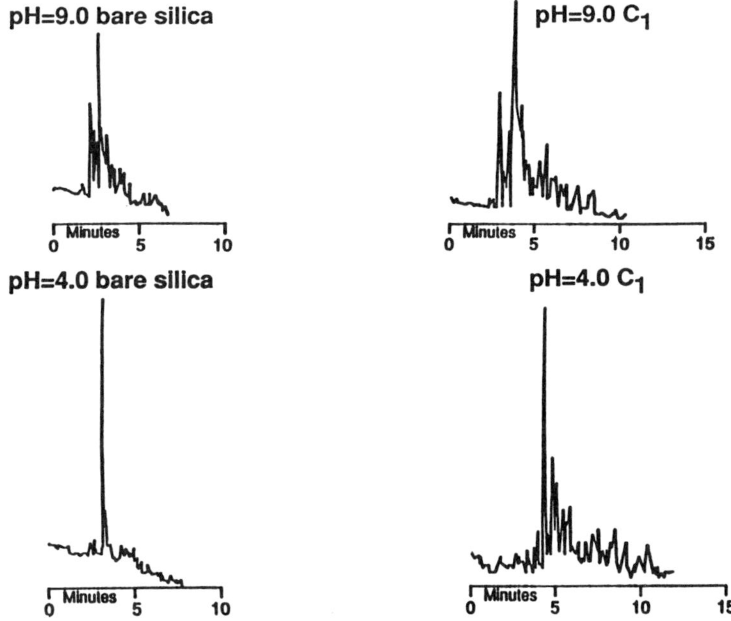

Figure 12 Separation of peptides resulting from the tryptic digestion of casein on untreated silica and a bonded C_1(CElect-H50) column at high and low pH. Conditions as for Fig. 7, except column length is 35-cm effective, 45-cm total.

Table 4 gives the number of theoretical plates for the aforementioned separations; as seen from this table, the best-achieved efficiency (myoglobin) is 223,000 plates per meter or 144,490 plates ($H = 4.50$ μm) for a column length of 65 cm. The injection volume used here is 3 nL, which gives a length-based variance owing to finite injection width of 1.93×10^{-3} cm^2 using the formula $\sigma_i^2 = l^2/12$, where l is the length of the injection plug [40]. In addition, the length-based variance of the zone for axial diffusion is given as $\sigma_a^2 = 2Dt = 1.08 \times 10^{-3}$ cm^2 for $D = 11.3 \times 10^{-7}$ cm^2 s^{-1} [41] and t equal to 8 minutes. Given $N = L^2/\sigma^2$, where N is the number of theoretical plates and σ^2 is the sum of injection and diffusion variances, the number of calculated theoretical plates is 1,403,000 ($H = 0.46$ μm), or approximately nine times larger efficiency than that observed from experiment. Most of this efficiency loss may be due to finite solute–wall interaction and, as suggested by Eq. 19, will lead to roughly one-third the potential resolution when compared with an ideal wall-less capillary column. This suggests that assumptions of negligible solute–wall and solute–bonded phase interactions could be misleading, and that methods used to reduce or eliminate these interactions must be evaluated. Use of other covalent surface-modified columns has also produced efficiencies much

Table 4 Efficiencies for Protein Separations on Bonded CE Columns

Column	Protein/buffer pH	Efficiency (plates per meter)
Polar (CElect-P)	Conalbumin/8.0	7,350
	Conalbumin/8.5	18,600
	Conalbumin/9.0	43,300
	Myoglobin/7.5	23,600
	Myoglobin/8.0	145,000
	Myoglobin/8.5	223,000
	Myoglobin/9.0	148,000
C_8 (CElect-H1)	Myoglobin/7.0	31,900
	Conalbumin/7.0	12,800
	β-Lactoglobulin B/7.0	27,900
	β-Lactoglobulin A/7.0	26,600
C_1 (CElect-H)	Myoglobin/7.5	30,600
	Conalbumin/7.5	64,700
	β-Lactoglobulin B/7.5	81,800
	β-Lactoglobulin A/7.5	62,000

lower than predicted by theory, which considers axial diffusion as the sole contributor to zone broadening [6,8,9,18,21,26].

EFFECT OF INCREASING IONIC STRENGTH Addition of salts to the buffer, with concomitant increase in ionic strength, has been used to reduce ionic interaction between a protein and activated silanol groups on the tubing wall [15]. Use of high-ionic strength is limited by the increased heating in the capillary owing to the larger conductivity of the buffer solution. Reduction of heating can be accomplished by reducing the radius of the capillary or reducing the field strength. Since the polar-bonded column demonstrates considerable silanol character, this column responds well to use of higher-ionic strength to improve separations. Higher salt concentrations used in conjunction with the hydrophobic phases usually result in a suboptimal separation, compared with those obtained at low-ionic strength. Use of the polar-phase column is suggested for separation of proteins that exhibit considerable hydrophobic character and may interact with the hydrophobic phases. For example, the separation of egg white proteins (Fig. 13), which is unsuccessful on the hydrophobic phases at both low- and high-ionic strength, was successful on a polar column when using a 100-mM sodium phosphate/60-mM sodium borate, pH 8, buffer [48]. This buffer, described in the foregoing, is ten times the concentration of the standard buffer used for protein separations on the hydrophobic columns.

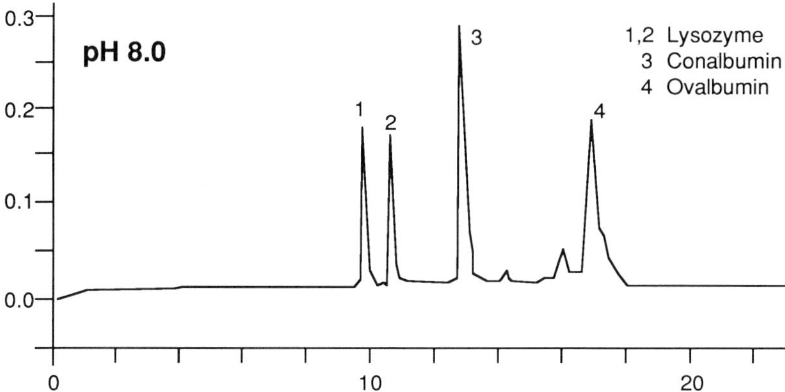

Figure 13 Separation of egg white proteins on a polar (CElect-P150) column. Conditions: buffer, 100 mM sodium phosphate/60 mM sodium borate, pH 8.0; detection, 200 nm; field strength, 200 V cm^{-1}; column, 50-μm id, 50-cm effective length/85-cm total length; sample, 4.5 nL, vacuum injection.

The increase in the buffer strength to 100 mM sodium phosphate/100 mM sodium borate, pH 8, resulted in an unstable baseline, possibly because of joule heating, with subsequent changes in refractive index at the detector. Use of this higher-ionic strength buffer in conjunction with a hydrophobic bonding resulted in a degradation of the separation. It is possible that sufficient joule heating in the column can begin to denature these proteins, exposing more hydrophobic areas of the protein, thus enhancing hydrophobic interaction of the protein with the hydrophobic phase. Increased ionic strength did not improve separations of the four protein mixtures described earlier, the basic protein mixture, or egg white proteins on any of the hydrophobic phases.

EFFECT OF SURFACTANTS AS ADDITIVES Utilization of the hydrophobic phases for micellar electrokinetic capillary chromatography (MECC) has extended the elution range of the technique when compared with bare fused silica capillaries [39]. Separations of 4-ethylamino-7-nitrobenzofurazan (NBD)-derivatized octyl amine and dodecyl amine were performed on both an untreated fused silica column and a C_8-bonded column at varying methanol concentrations. A buffer consisting of 10 mM sodium phosphate/6 mM sodium borate, pH 9.2, containing 50 mM sodium dodecyl sulfate (SDS) was used with a field strength of 300 V cm^{-1} in a 50-μm id, 60-cm (injection to detection length)/65-cm (total length) column. Figure 14 demonstrates the difference in selectivity on each column at the various methanol concentrations. The selectivity, as defined by k_1'/k_2', was greater for the C_8-bonded column at a lower methanol concentration than for the untreated silica column. The

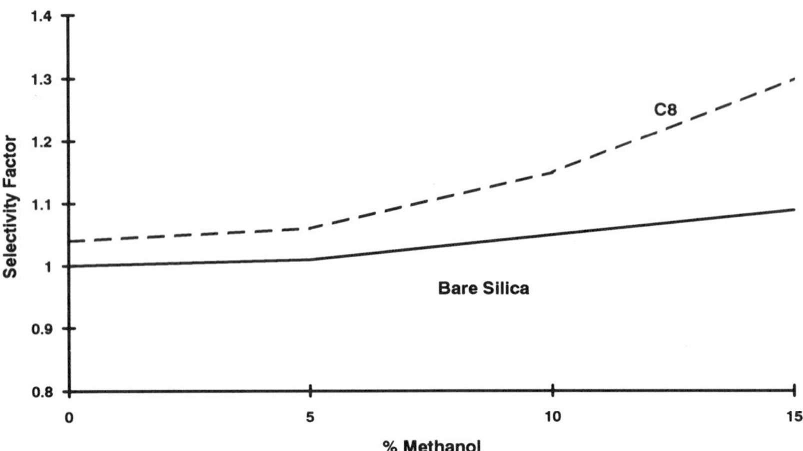

Figure 14 Column selectivity at varying methanol concentrations in an untreated fused silica column and C_8-bonded phase (CElect-H150) columns. Conditions listed in text.

peak capacity, n, and the elution range, t_0/t_m, are related by

$$n = 1 + \frac{\sqrt{N}}{4} (\ln t_m/t_0) \tag{21}$$

where t_0 is the migration time of a solute unretained by the micelle and t_m is the migration time of a micelle in solution. Figure 15 illustrates the difference in the elution ranges of each of the columns at the various methanol concentrations. Elution range for the C_8-bonded column is extended at lower concentrations of methanol, compared with the elution range on the bare silica column under the same electrophoretic conditions. The C_8-bonded column also displayed a larger peak capacity, as shown by Fig. 16. Data for t_0/t_m from Fig. 15 and $N = 200{,}000$ plates were used to calculate data points for Fig. 16. Extended elution range and increased peak capacities for MECC separations using a trimethylchlorosilane-modified column, has been reported by Balchunas and Sepaniak [24]. This extension of the elution range was presumably due to the reduction in electroosmotic flow, which decreases the net flow of the micelles in solution. However, the possibility of interaction of both the micelles and the solutes with the hydrophobic stationary phase can also contribute to decreased apparent mobility of the solutes. Chiral separations of 1,1′-bi(2-naphthol) were performed using a bile salt, sodium deoxycholate, as the surfactant. Chiral separations using bile salts as surfactants have been performed previously by Nishi et al. [42,43] and Cole et al. [44]. By using 5% methanol in the running buffer, a chiral separation was obtained on the C_8-bonded column, whereas no

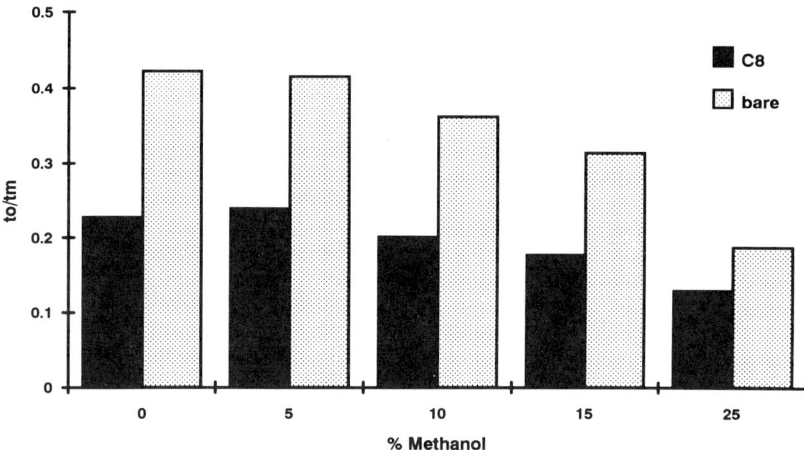

Figure 15 Elution ranges at varying methanol concentration in an untreated fused silica column and a C_8-bonded (CElect-H150) columns. Conditions as for Fig. 14.

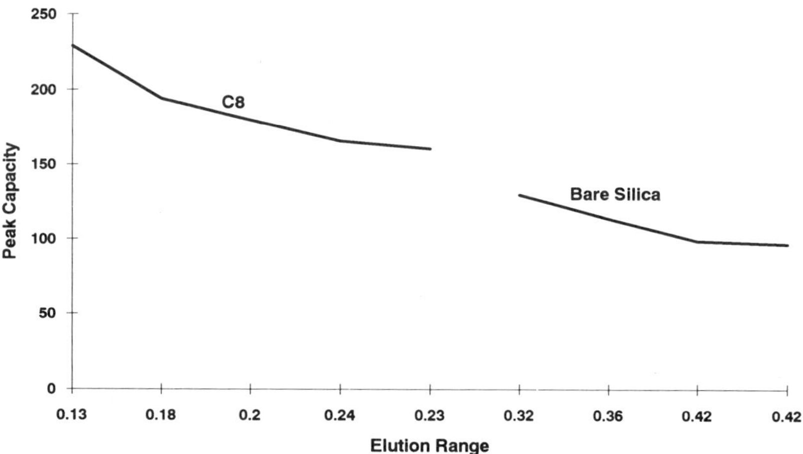

Figure 16 Peak capacity on a bare silica column and C_8-bonded column at different elution ranges for separation of NBD-derivatized amines in MECC. Conditions as for Fig. 14.

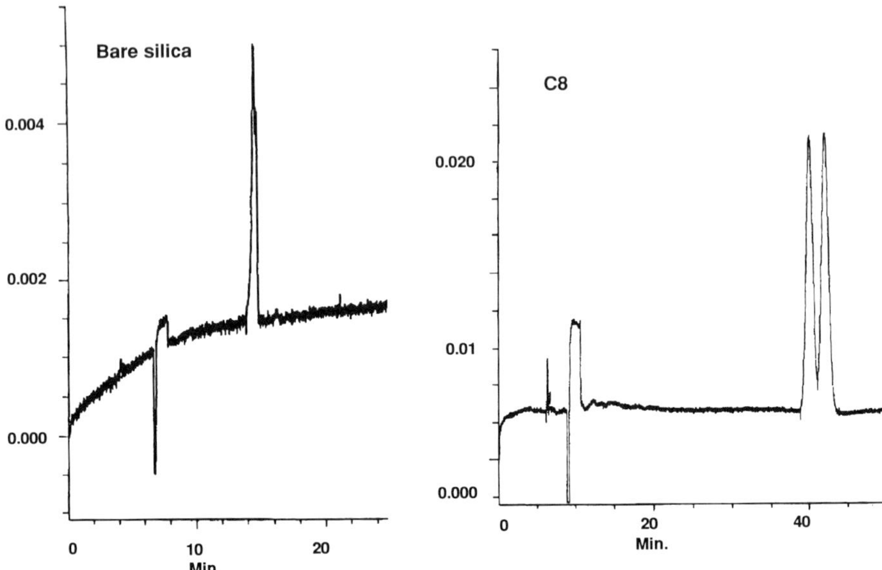

Figure 17 Chiral separations of 1,1′-bi-2-naphthol on a bare fused silica column and a C_8-bonded (CElect-H150) columns. Conditions: 10 mM sodium phosphate/6 mM sodium borate, pH 8.0/50 mM sodium deoxycholate/5% methanol; 200-V cm^{-1} field strength; 5-second pressure injection. Column: 50-mm id, 60-cm effective, 67-cm total length; detection, 214 nm

separation was seen on the untreated silica column under the same conditions (Fig. 17). No chiral separation was observed on the bare silica column until the concentration of methanol in the electrophoresis buffer reached 15%. At 15 and 20% methanol, chiral separations could be performed on both columns; however, analysis times on the C_8-bonded column became very long (>60 minutes) at these higher methanol concentrations. Also, increases in the critical micelle concentration of detergents become significant at higher levels of organic modifiers, and disruption of the micelles can occur [45]. Similar results were obtained for the separation of 1,1′-binaphthalene-2,2′-diyl hydrogen phosphate on both the untreated silica column and the C_8-bonded column under the same conditions.

One method of reducing protein–wall interactions involves adsorbing surfactants to the surface of the capillary column before electrophoretic analysis of solutes. Wiktorowicz and Colburn [16] rinsed a bare silica column with long chain cationic surfactants, which ionically bond to the surface of a bare silica column, before electrophoresis to reduce solute–wall interactions during protein separations. This technique of binding surfactant to the column surface before electrophoresis was later done by Towns and Regnier [23] on the surface of hydrophobic-bonded capillary columns.

Since the separations of a mixture of basic, relatively hydrophobic proteins was unsuccessful on the hydrophobic phases at both low and high ionic strength, attempts were made to adsorb surfactants to these hydrophobic bondings. Coating of the columns with these surfactants was described in the experimental section. Figure 18 shows a comparison of separations of three basic proteins on the hydrophobic-bonded columns, to which Brij 35 has been absorbed. Brij 35 is a nonionic surfactant possessing the structure $CH_3(CH_2)_{11}(OCH_2CH_2)_{23}H$. Once Brij 35 was adsorbed to the column surface, each of the columns appeared to give the same separation of the three proteins, suggesting the coverage with Brij 35 is equivalent for each of the bonded columns. This indicates that the chain length of the bonded phase does not determine the extent of surface binding. As seen by the results of the electroosmotic flow comparison given in Table 2, the columns with shorter (C_1 and C_8)-bonded alkyl chains provide better coverage of the silanol surface and possibly a higher level of overall surface hydrophobicity than the columns with longer (C_{18}) alkyl chains. Association of Brij 35 to the hydrophobic surface was expected to improve the separation by providing a more hydrophilic surface, a result of the hydrophilic headgroup of Brij 35. Brij 35-coated columns gave better efficiencies for separations (Table 5) of the neutral proteins (conalbumin, myoglobin, and the lactoglobulins), compared with the bonded columns alone (Fig. 19). Analysis times for separations on the Brij-coated columns were also significantly longer than for a bonded column alone, owing to slower electroosmotic flow in the surfactant-coated columns. Table 6 summarizes the electrophoretic mobility of a neutral marker before and after coating the columns with the various detergents.

The surfactant coating, however, did not improve separations of all protein mixtures. Resolution of the egg protein mixture, consisting of lysozyme, conalbumin, and ovalbumin, was nonexistent on Brij 35-coated columns. Coating the columns with SDS did not improve separation of either the egg proteins or the basic proteins, as expected, because of the negative charge imparted to the column surface by the SDS coating. The SDS was adsorbed to the column surface before the electrophoretic run by rinsing the column with a 0.5% solution of SDS in the electrophoresis buffer, followed by a 30-minute rinse with electrophoresis buffer containing no surfactant. It should be noted that, because of the ability of SDS to bind proteins, SDS was not added to the electrophoresis buffer. However, the possibility exists that SDS that has desorbed from the column surface can bind with proteins.

Zwittergent 3-14, a zwitterionic detergent, was used as the zwitterionic surfactant for surfactant coating of the C_{18} column. This coating gave results similar

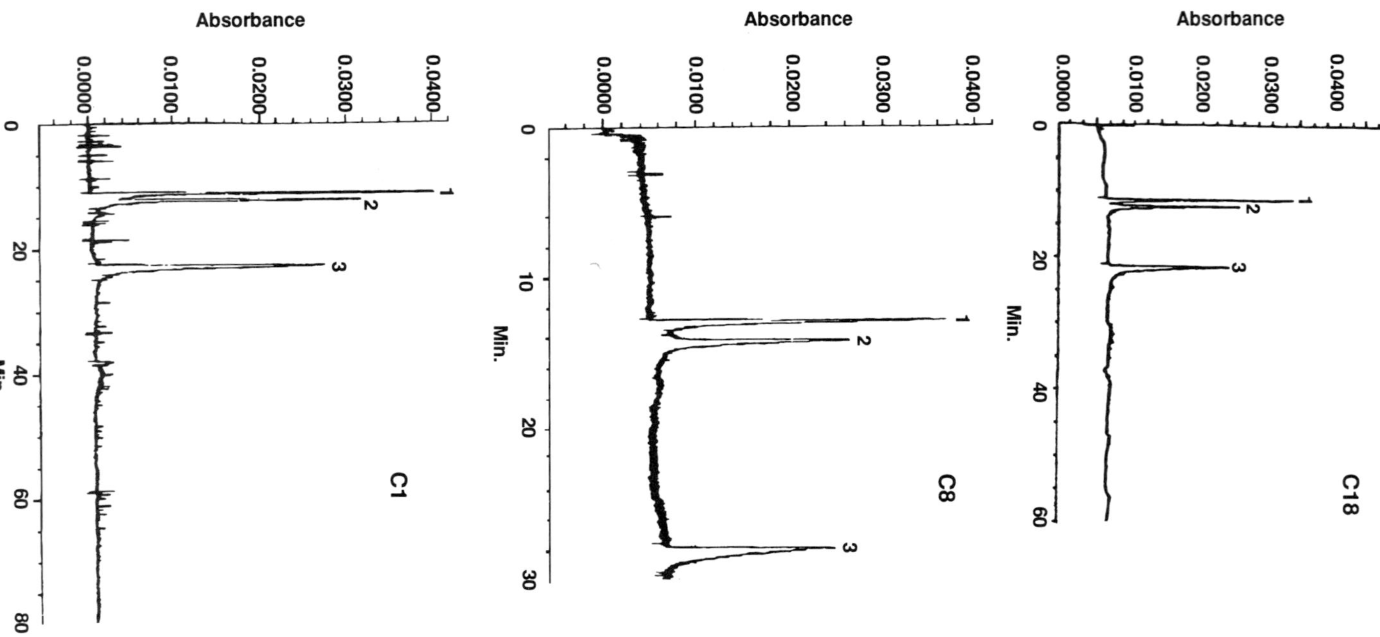

Figure 18 Separation of three basic proteins on hydrophobic-bonded columns dynamically coated with Brij 35. Conditions: 10 mM sodium phosphate, pH 7.0/0.001% Brij 35; 350 V/cm^{-1}; 5-second pressure injection; column, 50-μm id, 60-cm effective, 67-cm total length; detection, 214 nm UV. Sample: all proteins 1 mg/mL each. (1) Lysozyme, (2) cytochrome c, (3) ribonuclease A.

Table 5 Efficiencies for Protein Separations Performed Using Dynamic Surfactant Coatings on Hydrophobic-Bonded Columns

Bonding/surfactant	Protein	Efficiency (plates per meter)
C_{18}/SDS	Myoglobin	3,090,000
	Conalbumin	618,000
	β-Lactoglobulin B	1,270,000
	β-Lactoglobulin A	1,070,000
C_{18}/Brij 35	Myoglobin	2,920,000
	Conalbumin	343,000
	β-Lactoglobulin B	658,000
	β-Lactoglobulin A	150,000
C_1/Brij 35	Lysozyme	269,000
	Cytochrome *c*	124,000
	Ribonuclease A	129,000
C_{18}/Brij 35	Lysozyme	245,000
	Cytochrome *c*	176,000
	Ribonuclease A	225,000
C_8/Brij 35	Lysozyme	360,000
	Cytochrome *c*	228,000
	Ribonuclease A	500,000

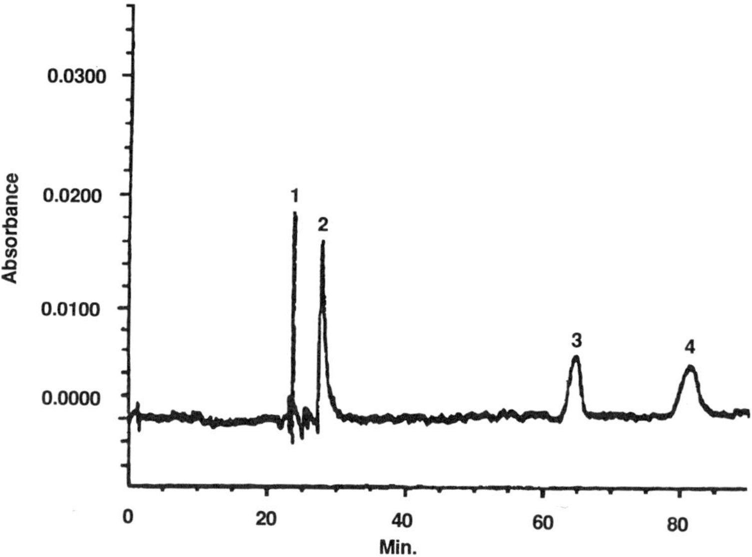

Figure 19 Separation of neutral proteins on an SDS-modified C_{18}-bonded column. Conditions as for Fig. 17. Proteins numbered as listed in the text.

Table 6 Comparison of the Electrophoretic Mobility of a Neutral Marker (Benzyl Alcohol) Before and After Coating Hydrophobic CElect Phases with Surfactants

		Mobility ($cm^2\ V^{-1}\ s^{-1}$)	
Bonding (column)	Surfactant	Precoating	Postcoating
C_{18} (CElect-H250)	Zwittergent 3-14	3.29×10^{-4}	1.095×10^{-4}
C_{18} (CElect-H250)	Brij 35	3.36×10^{-4}	0.400×10^{-4}
C_8 (CElect-H150)	Brij 35	2.23×10^{-4}	0.371×10^{-4}
C_1 (CElect-H50)	Brij 35	3.513×10^{-4}	0.318×10^{-4}
C_{18} (CElect-H250)	SDS	3.657×10^{-4}	1.446×10^{-4}

to those seen for the SDS-coated C_{18} columns. The structure of Zwittergent 3-14 ($C_{14}H_{29}N(CH_3)_2(CH_2)_3SO_3^-$) gives each molecule a set orientation for the anionic and cationic groups. It is this orientation that could give the surface more anionic character, as the polar headgroup is likely to be oriented toward the aqueous phase, whereas the alkyl chain would associate with the hydrophobic bonding. However, this coating did not give reproducible separations.

The addition of cetyl dimethylethyl ammonium bromide (CTAB) to the surface of a C_{18} phase initially reversed the electroosmotic flow, but even with 0.001% CTAB in the running buffer, the flow changed direction after 15–20 analyses. Basic proteins and egg proteins were separated on the CTAB-coated columns, but these separations degraded after several analyses. Higher levels of CTAB in the running buffer may help maintain the coated surface and give more reproducible separations.

These results suggest that adsorbing surfactants to the surface of a hydrophobic-bonded column reduces the hydrophobicity of the effective surface and improves efficiencies, possibly by decreasing hydrophobic interactions of the proteins with the bonded hydrophobic phase. These surfactant coatings may also serve to create a thicker surface layer that further shields solutes from the active surface silanol groups.

SUMMARY

The results of theoretical treatment indicates that solutes similar to those given in Table 1 are susceptible to a large loss of resolution if even a small amount of wall interaction is present; hence, the utmost care must be given to the design and implementation of coatings that will minimize this interaction. Although these results have been qualitatively confirmed by several experiments in which zone quality was greatly improved by using coated capillary columns, the magnitude of this effect on resolution is quantitatively seen to be very large, even at small levels of retention. Small, charged molecules, although susceptible to resolution degradation, do not have the potential loss of resolution that polymers may have. Furthermore, theory implies that small-molecule separations may profit from capillaries with smaller diameters when solute–wall interactions can not be eliminated.

Results of the characterization of the bonded-phase columns suggest these bonded phases possess long-term stability, even under alkaline rinse conditions. However, columns must be conditioned, as defined earlier, to assure reproducible electroosmotic flow in each column. The hydrophobic-bonded phases reduced the electroosmotic flow velocity, whereas the neutral polar-bonded phase exhibited electroosmotic flow velocity similar to that of bare fused silica. For the hydrophobic-bonded phases, electroosmotic flow velocity was inversely proportional to the alkyl chain length of the phase, possibly owing to more complete coverage of smaller alkyl chains on the silica surface. The maximum reduction in electroosmotic flow was 50% of the flow seen in bare fused silica under the same electrophoretic conditions. Although the polar-bonded phase exhibited the same pH dependence as bare fused silica column, the hydrophobic phases significantly reduced the dependence of electroosmotic flow on pH.

Protein separations were achieved from pH 6 to 9 on these bonded columns, but the resulting efficiencies were lower than predicted by axial diffusion theory. Increasing the ionic strength when performing separations on the polar-bonded phase reduced solute–wall interactions and improved protein separations. However, use of high-ionic strength is limited by joule heat produced in the capillary column. Surfactants, adsorbed to the hydrophobic bondings and added to the electrophoresis buffer at below critical micelle concentrations, seemed to be the most effective in reducing protein–wall interactions. However, efficiencies for these separations were still below those predicted by theory. Therefore, not all solute–wall interaction has been eliminated by the use of surfactant-coated capillary columns.

The C_8-bonded phase extended the elution range and increased peak capacity when used for separations in micellar electrokinetic capillary chromatography. Use of this bonded column permitted separation of compounds at lower methanol concentrations than for a bare silica column, which reduced the possibility of micelle dispersion. However, this separation capability was at the expense of longer analysis times.

Use of covalently modified columns for capillary electrophoresis can minimize solute–wall interactions during separation of proteins at neutral pHs. Modifiers of both the buffer and the bonded surface help to further reduce solute–surface interactions and improve protein separations. However, comparison of the predicted and experimental efficiencies suggest that solute–wall interactions are still occurring.

ACKNOWLEDGMENTS

The authors wish to thank Mary Swan, Keith Duff, George Lein, Julie McCulloch, and Binyamin Feibush of Supelco for their valuable discussions and support. The

authors also wish to thank Dave Swaile, Rod Cole, and Michael Sepaniak for their cooperative efforts in this work.

APPENDIX: SYMBOLS

a	Capillary radius
A_1, A_2	Plate height theory constants
C_m	Nondimensional Taylor dispersion coefficient
C_s	Resistance to mass transfer dimensional coefficient
D	Free solution solute diffusion coefficient
D_s	Solute diffusion coefficient in stationary phase
d	Capillary diameter
H	Generalized plate height
H_c	Plate height for retentive wall-coated capillary column
H_∞	Plate height for retentionless capillary column
I_0, I_1	Modified Bessel functions of the first kind
k'	t_c/t_s
L_d	Capillary length from point of injection to detector
n	Average number of wall contacts or excursions through distance d or peak capacity
Pe	Nondimensional Peclet number
$P(x)$	Integral used in plate height theory
$Q(x)$	Nondimensional radial velocity profile
r	Dimensional distance from center of capillary to wall
R_s	Resolution
R_{s_c}	Resolution of zones for a retentive capillary
$R_{s\infty}$	Resolution of zones for a wall-less capillary
t	Time
t'	Time for diffusive excursion over distance d
t_c	Total time of solute association with coating
t_d	Average desorption time for adsorbed solute
t_m	Time for micelle migration
t_o	Time for solute migration by electroosmotic flow
t_s	Total time of solute in the solution phase
Δt	Difference in migration time between adjacent peaks
$\mathbf{v}$	Net migration velocity vector of the solute
$\mathbf{v}_{eo}$	Electroosmotic velocity vector

$\mathbf{v}_{ep}$	Electrophoretic velocity vector of the solute
$\bar{v}$	Average cross-sectional velocity
$v(x)$	Velocity of solute as a function of radial coordinate
x	Nondimensional distance from capillary center to wall, r/a.
z	Elution or migration coordinate
β	$k'(1 + k')$
κ	Reciprocal double-layer thickness
σ	Standard deviation of a Gaussian peak
σ_c	Standard deviation of a Gaussian peak with retention
σ_∞	Standard deviation of a Gaussian peak with no retention

REFERENCES

1. F. Foret, M. Deml, and P. Bocek, *J. Chromatog.*, *452*:601 (1988).
2. J. W. Jorgenson and K. D. Lukacs, *Science*, *222*:266 (1983).
3. J. W. Jorgenson and K. D. Lukacs, *Anal. Chem.*, *53*:1298 (1981).
4. X. Huang, W. F. Coleman, and R. N. Zare, *J. Chromatog.*, *480*:95 (1989).
5. H. H. Lauer and D. McManigill, *Anal. Chem.*, *58*:166 (1986).
6. W. Nashabeh and Z. E. Rassi, *J. Chromatog.*, *559*:367 (1991).
7. C. Schwer and E. Kenndler, *Anal. Chem.*, *63*:1801 (1991).
8. S. A. Swedberg, *Anal. Bioch.*, *185*:51 (1990).
9. J. K. Towns and F. E. Regnier, *Anal. Chem.*, *63*:1126 (1991).
10. G. J. M. Bruin, J. C. Huisden, J. C. Kraak, and H. Poppe *J. Chromatog.*, *480*:339 (1989).
11. G. J. M. Bruin, J. P. Chang, R. H. Kuhlman, K. Zegers, J. C. Kraak, and H. Poppe, *J. Chromatog.*, *471*:429 (1989).
12. R. M. McCormick, *Anal. Chem.*, *60*:2322 (1988).
13. J. A. Bullock and L. Yuan, *J. Microcol. Sep.*, *3*:241–248 (1991).
14. M. M. Bushey and J. W. Jorgenson, *J. Chromatog.*, *480*:301 (1989).
15. J. S. Green and J. W. Jorgenson, *J. Chromatogr.*, *478*:63 (1989).
16. J. E. Wiktorowicz and J. C. Colburn, *Electrophoresis*, *11*:769 (1991).
17. T. Kaneta, T. Shunitz, and H. Yoshida, *J. Chromatogr.*, *538*:385 (1991).
18. A. Emmer, M. Jansson, and J. Roeraade, *J. Chromatog.*, *547*:544 (1991).
19. M. J. Gordon, K. Lee, A. A. Aias, and R. Zare, *Anal. Chem.*, *63*:69 (1991).
20. S. Hjerten, *J. Chromatog.*, *347*:191 (1985).
21. K. A. Cobb, V. Dolnick, and M. Novotny, *Anal. Chem.*, *62*:2478 (1990).
22. J. A. Lux, H. Yin, and G. Schomberg, *J. High Resolut. Chromatog.*, *13*:145 (1990).
23. J. K. Towns and F. E. Regnier, *J. Chromatogr.*, *516*:69 (1990).
24. A. T. Balchunas and M. J. Sepaniak, *Anal. Chem.*, *59*:1467 (1987).
25. M. Novotny, K. A. Cobb, and J. Lui, *Electrophoresis*, *11*:735 (1990).
26. J. Kohr and H. Engelhardt, *J. Microcolumn Separation*, *3*:491 (1991).
27. J. C. Giddings, *Unified Separation Science*, John Wiley & Sons, New York (1991).
28. V. Ananthakrishnan, W. N. Gill, and A. J. Barduhn, *AIChE. J.*, *11*:1063 (1965).

29. M. Martin and G. Guiochon, *Anal. Chem.*, *56*:614 (1984).
30. M. Martin, G. Guiochon, Y. Walbroehl, and J. W. Jorgenson, *Anal. Chem.*, *57*:561 (1985).
31. G. J. M. Bruin, P. P. H. Tock, J. C. Kraak, and H. Poppe, *J. Chromatog.*, *517*:557 (1990).
32. J. H. Knox, *Chromatographia*, *26*:329 (1988).
33. J. M. Davis, *J. Chromatog.*, *517*:521 (1990).
34. J. C. Giddings, *Dynamics of Chromatography*, Marcel Dekker, New York (1965).
35. M. R. Schure, *Anal. Chem.*, *60*:1109 (1988).
36. M. R. Schure and S. K. Weeratunga, *Anal. Chem.*, *63*:2611 (1991).
37. T. J. Stark, R. D. Dandeneau, and L. Mering, 1980 Pittsburg Conference, Atlantic City, N.J., Abstr. 002 (1980).
38. Lambert and Middleton, *Anal. Chem.*, *62*:1585–1587 (1990).
39. A. M. Dougherty, C. L. Woolley, D. L. Williams, D. F. Swaile, R. O. Cole, and M. J. Sepaniak, (1991) "Stable Bonded Phases for Capillary Electrophoresis," Second International Symposium on High Performance Capillary Electrophoresis, Lecture T-2, Feb. 5, (1991).
40. E. Grushka and R. M. McCormick, *J. Chromatog.*, *471*:421 (1989).
41. A. L. Lehninger, *Biochemistry*, Worth Publishers, New York, p. 176 (1979).
42. H. Nishi, T. Fukuyama, M. Matsuo, and S. Terabe, *J. Microcolumn Separations*, *1*:234 (1989).
43. H. Nishi, T. Fukuyama, M. Matsuo, and S. Terabe, *J. Chromatog.*, *498*:313 (1990).
44. R. O. Cole, M. J. Sepaniak, and W. L. Hinze, *J. High Resolut. Chromatog.*, *13*:579 (1990).
45. J. Neugegebauer, "A Guide to the Properties and Uses of Detergents in Biology and Biochemistry," Calbiochem Corp., La Jolla, Calif., pp. 4–9 (1988).
46. A. M. Dougherty, C. L. Woolley, D. L. Williams, D. F. Swaile, R. O. Cole, and M. J. Sepaniak, *J. Liq. Chromatogr.*, *14*:907 (1991).
47. CRC Handbook of Chemistry and Physics, 52nd ed. The Chemical Rubber Co., Cleveland, Ohio, (1971).
48. J. McCulloch, personal communication, (1991).

IV
INSTRUMENTATION

13

Direct Control of Electroosmotic Flow in Capillary Electrophoresis by Using an External Electric Field

Pei Tsai and Cheng S. Lee

University of Maryland
Baltimore, Maryland

GENERAL

In a typical aqueous solution with small binary electrolytes within a capillary column, the silica surface has an excess of anionic charge resulting from the ionization of surface silanol groups. The cationic counterions to these anions are in the electrostatic diffuse layer adjacent to the capillary wall. The electrostatic potential across the diffuse layer is called the zeta (ζ) potential. In the presence of an electric field, these hydrated cations migrate toward the cathode and drag solvent with them. Therefore, *electroosmosis*, the flow of solution in an applied electric field, is dependent on the polarity and magnitude of the zeta potential at the capillary wall [1,2].

In capillary zone electrophoresis (CZE), the electroosmotic flow adds the same velocity component to all solutes, regardless of their ionic status. The electroosmotic flow is often strong enough to cause all solutes to elute at one end of the capillary. The electroosmotic flow affects the residence time of a solute in the capillary. The effect of rapid electroosmotic flow is to sweep all solutes quickly through the capillary, leaving little time for zones to be separated. Thus, the resolution of two zones in the presence of electroosmotic flow can be written as

$$R = 0.117(\mu_{ep,1} - \mu_{ep,2}) \left[\frac{V}{D(\mu_{ep,avg} + \mu_{eo})} \right]^{1/2} \tag{1}$$

where $\mu_{ep,1}$ and $\mu_{ep,2}$ are the electrophoretic mobilities for the two solutes, V is the applied voltage, D is the solute's diffusion coefficient, μ_{eo} is the electroosmotic mobility, and $\mu_{ep,avg}$ is the average electrophoretic mobility [3]. On the basis of this equation, excellent resolution of substances having very similar mobilities can be achieved by balancing electroosmotic flow against electrophoretic migration. This resolution will be obtained, however, at a large expense in analytic time.

In addition, the peak area in capillary electrophoresis is a function of the amount of sample present, its extinction coefficient, and the velocity of the solute peak. Unlike high-performance liquid chromatography (HPLC), the peaks in capillary electrophoresis are traveling past the detector at different speeds. A slower-moving peak will have a greater-integrated area than a fast-moving peak, even if the extinction coefficient and amount of material are identical for each peak. Thus, a small change in the electroosmotic flow may have dramatic effects on the retention time and peak area for both qualitative and quantitative applications.

Application of capillary electrophoresis to the separation of proteins is particularly challenging because of the tendency of macromolecules to adsorb onto the capillary wall. The slightest extent of adsorption causes appreciable band broadening and tailing, with greatly reduced separation efficiency, and moderate adsorption can lead to completely destroyed peak shapes and poor reproducibility. In addition, the adsorbed proteins modify the capillary surface, usually decreasing the electroosmotic flow substantially. This leads to a completely unpredictable migration for all protein zones. Protein adsorption is believed to be primarily an electrostatic interaction between basic amine functionalities on proteins and the silanoate on the capillary wall [3]. Such protein–wall electrostatic interactions can be manipulated with the control of the surface charges at the capillary–aqueous interface. Thus, the polarity and magnitude of the zeta potential at the capillary–aqueous interface not only controls the electroosmotic flow, but also affects the adsorption of protein onto the capillary wall. Reported attempts to affect the zeta potential involve the deactivation of silica capillary by physically coating with surfactant [4] or by silane derivatization [3,5,6]. Additional approaches include the addition of chemicals to the separation buffer [3,7], or manipulation of the charges on the proteins and silica capillary wall to prevent the adsorption by coulombic repulsion [8,9].

Direct Control of the Zeta Potential and Electroosmosis

To enhance separation resolution and to prevent protein adsorption, it is clearly of interest to be able to directly and dynamically control the polarity and magnitude of the boundary condition at the aqueous–capillary wall, the zeta potential. A physical method involving the use of an additional electric field applied from outside of the capillary for the direct control of the zeta potential and, thereby, the electroosmotic flow in capillary electrophoresis has been proposed. The changes in the direction and speed of the electroosmotic flow in the presence of an additional electric field were recently observed and demonstrated in our laboratory [10,11] by using the current-monitoring method [12] and the UV-marker method [13]. A complete capillary electrophoresis system with a UV detector for delivering electric fields both externally, through the annulus between the inner and outer capillaries, and internally, across the buffer solution in the inner capillary, is shown in Fig. 1 and is described by Lee et al. [11]. This novel approach vectorally couples the externally applied electric potential with the internal electric potential in the inner capillary. The potential gradient between the external and internal electric potentials is perpendicular across the inner capillary wall and controls the polarity and magnitude of the zeta potential on the interior surface of the inner capillary. Because the direction and flow rate of the electroosmosis is dependent on the polarity and magnitude of the zeta potential, the electroosmotic flow can, accordingly, be directly manipulated by simply varying the external electric potential. There are several advantages in this instrumentational approach over the surface coatings and chemical methods for the direct control of electroosmosis:

1. Only one type of capillary is required.
2. The direct and dynamic manipulation of electroosmosis during the separation is obtained.

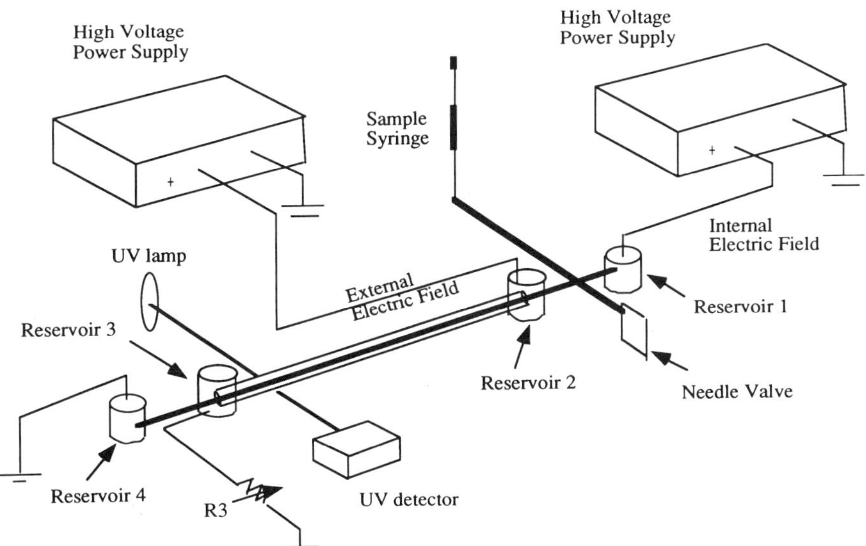

Figure 1 A complete capillary electrophoresis system with a UV detector delivers the electric fields both externally, through the annulus between the inner and outer capillaries, and internally, across the buffer solution in the inner capillary. (From Ref. 11)

3. A high degree of optimization for the separation efficiency and resolution is achieved.
4. A higher degree of automation is established.

Capacitor Model

To investigate the fundamentals of this instrumentational approach for directly controlling electroosmosis, a capacitor model, as shown in Fig. 2, was proposed and used by Lee et al. [14]. The capacitor of the electrostatic diffuse layer at the inner capillary–outer solution interface is in series with the capacitor of the inner capillary and the capacitor of the electrostatic diffuse layer at the inner capillary–inner solution interface. The change in the zeta potential at the inner capillary–inner aqueous interface, $\Delta\zeta$, owing to the applied potential gradient across the inner capillary wall, ΔV, is given by

$$\Delta\zeta = \frac{\Delta V}{C_{ei}/C_c} \tag{2}$$

where C_{ei} is the capacitance of the electrostatic diffuse layer at the inner capillary–inner aqueous interface and C_c is the capacitance of the inner capillary tubing [14]. The predicted value of the zeta potential is then converted to the controlled elec-

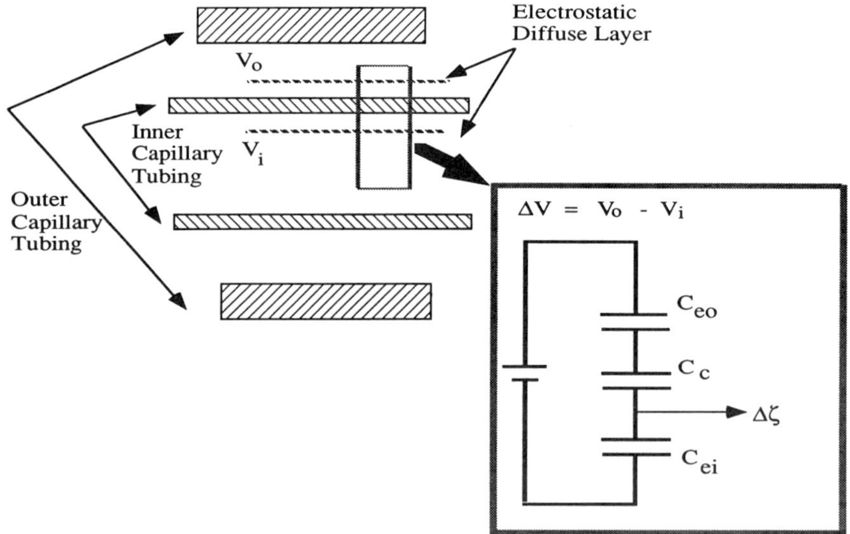

Figure 2 The proposed capacitor model for predicting the change in the zeta potential caused by the application of an external electric field. (From Ref. 14)

troosmotic mobility, μ_{eo}, by the equation

$$\mu_{eo} = (\varepsilon/\eta)\,\zeta \tag{3}$$

where ε and η are the permittivity and viscosity of the solution, respectively [1,2]. The predictions obtained from the capacitor theory were then compared with the experimental results and are summarized in Table 1. One typical plot of the comparison between the theoretical and experimental values of electroosmotic mobility against various applied potential gradients is shown in Fig. 3 [14]. The results indicate that the capacitor theory predicts reasonably well the trends of the experimental measurements at various solution pHs, electrolyte concentrations, and capillary dimensions. For the discussion of various capillary dimensions, the capacitance of the electrostatic diffuse layer decreases with the decrease in the inner diameter of inner capillary tubing. Therefore, the effectiveness of our approach for controlling the zeta potential can be enhanced by using a capillary tubing with a smaller inner diameter (see Eq. 2). For a capillary tubing with a fixed inner diameter (id), the capacitance of capillary tubing increases with the decrease in the outer diameter

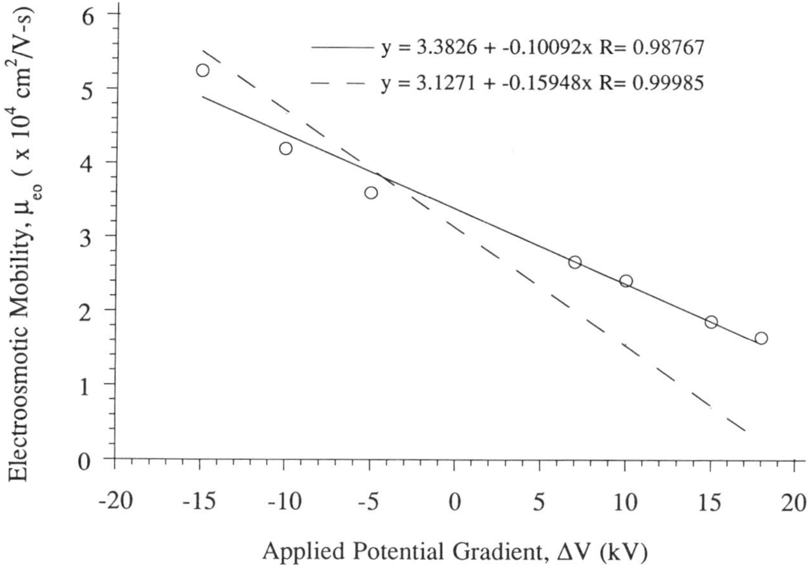

Figure 3 Plot of the electroosmotic mobility, μ_{eo}, against the applied potential gradient, ΔV. The experimental data at 45-mM/55-mM KCl solution (pH 7) with the use of a 25-μm id and 150-μm od inner capillary in the current monitoring method are shown as open circle and the solid line. The predictions obtained from the capacitor theory are shown as the dashed line. (From Ref. 14)

Table 1 The Effect of External Electric Field on Electroosmosis

Dimension of capillary tubing (id/od in μm)	Electrolyte concentration (M)	Solution pH	Slope of μ_{eo} vs ΔV[a] ($\times 10^7$ cm^2 V^{-1} s^{-1})	Slope of μ_{eo} vs ΔV[b] ($\times 10^7$ cm^2 V^{-1} s^{-1})	% of the difference[c]
75/375	1 mM/2 mM	5	−0.37	−0.28	24
50/192	1 mM/2 mM	6	−0.47	−0.29	38
50/150	1 mM/2 mM	6	−0.61	−0.31	49
25/150	1 mM/2 mM	6	−0.73	−0.44	40
25/150	10 mM/20 mM	7	−0.30	−0.26	13
25/150	45 mM/55 mM	7	−0.16	−0.10	38
50/192	45 mM/55 mM	7	−0.10	−0.041	59
50/192	45 mM/55 mM	6	−0.11	−0.045	59
50/192	45 mM/55 mM	5	−0.13	−0.067	48

[a] The slope was obtained from the capacitor theory.
[b] The slope was obtained from the experimental results.
[c] The percentage of the difference = (the slope predicted by the capacitor theory − the slope obtained from the experimental results)/(the slope predicted by the capacitor theory).
Source: Ref. 14.

(od). As shown in Eq. 2, the effectiveness of our approach, therefore, can increase with the use of thinner capillary tubing.

To investigate the separation efficiency of capillary electrophoresis with the direct control of electroosmosis, frontal analysis of dimethyl sulfoxide as an UV marker was examined [14]. The number of theoretical plates measured from the experiment was compared with the prediction based on molecular diffusion alone. The comparison clearly indicated that there was no measurable additional dispersion and band broadening induced by the direct control of electroosmosis.

APPLICATIONS OF DIRECT CONTROL OF THE ZETA POTENTIAL AND ELECTROOSMOSIS

Direct Control of the Zeta Potential and Electroosmosis on Coated Surfaces

To further investigate the applicability of our approach for directly controlling electroosmosis, silica capillaries (50-μm id and 150-μm od), coated with amino and butyl phases were studied. Capillaries coated with amino and butyl phases were provided by Dr. Swedberg of Hewlett Packard Laboratories. The procedures for preparing the capillaries are described elsewhere [15]. The electroosmotic mobility, μ_{eo}, as a function of pH in the absence of an external electric field is shown in Fig. 4 for indicating the characteristics of capillaries coated with amino and butyl phases. The flow rate of electroosmosis is assigned as positive when the direction of flow is toward the cathode end. The direction of electroosmosis, however, is toward the anode end in the capillary coated with amino phase, when the solution pH is below 6; the zeta potential at the amino phase is changed from negative to positive at low solution pHs.

As shown in Fig. 5, the electroosmotic mobility of coated capillaries is changed with the application of an external electric field. For instance, the electroosmotic mobility (toward the anode end) of amino phase at pH 4 and 10-mM phosphate buffer is increased in absolute value from $-2.93 \pm 0.02 \times 10^{-4}$ cm^2 V^{-1} s^{-1}, in the absence of an external electric field, to $-3.64 \pm 0.04 \times 10^{-4}$ cm^2 V^{-1} s^{-1}, in the presence of a +6-kV potential gradient. In contrast, the electroosmotic mobility is decreased to -2.57 ± 0.01 cm^2 V^{-1} s^{-1} with the application of a −7.8-kV potential gradient. For the capillary tubing coated with butyl phase, the direction of electroosmosis is reversed from $0.82 \pm 0.02 \times 10^{-4}$ cm^2 V^{-1} s^{-1}, in the absence of an external electric field, to $-0.74 \pm 0.04 \times 10^{-4}$ cm^2 V^{-1} s^{-1}, in the presence of a +5-kV potential gradient at pH 4 and 10-mM phosphate buffer. The electroosmotic flow is then enhanced to $2.36 \pm 0.04 \times 10^{-4}$ cm^2 V^{-1} s^{-1} in the presence of a − 10.8-kV potential gradient.

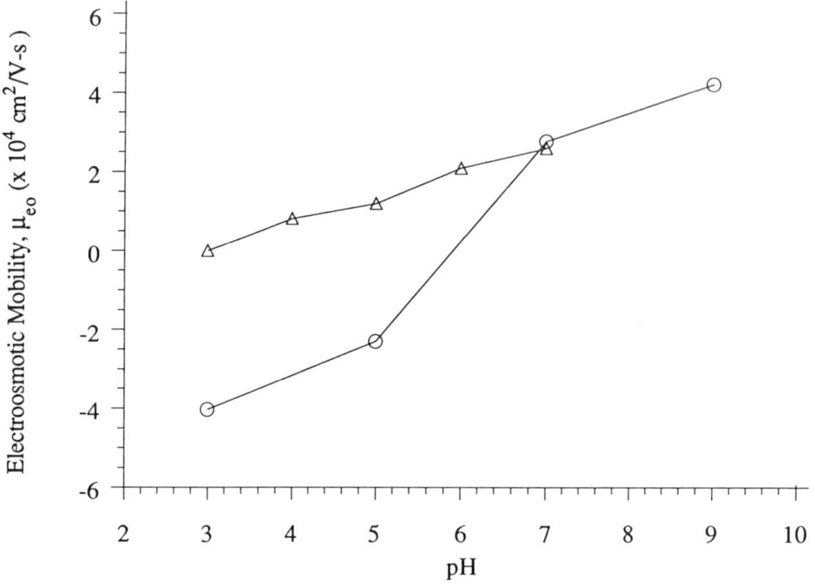

Figure 4 The electroosmotic mobility, μ_{eo}, of coated capillaries against the solution pH in 10-mM phosphate buffer. The mobility data for amino and butyl phases are shown as open circle and triangle, respectively.

Direct Control of Electroosmosis for Protein Separations

Low pH phosphate buffers have been used by McCormick [9] for reducing the negative charge of fused silica surface as well as for inducing some protective screening of the silica surface by phosphate groups. Operation at low pH values [pH less than the isoelectric point (pI) of all proteins] also ensures that the proteins in a sample will have a net positive charge and, thus, will migrate in the same direction (toward the cathode). The major limitation of performing CZE separations at such a low pH is that the peak capacity of the separations is low because the acidic buffer fully protonates the proteins, thereby diminishing charge differences between the species.

As shown in Fig. 6, the protein mixture in a solution of 10-mM phosphate buffer at pH 2.5 is not completely separated and resolved in the absence of an external electric field. With the application of a +3-kV potential gradient across the inner capillary wall, significant improvement on the protein separation is clearly observed in Fig. 7. The direction of electroosmosis is reversed from $0.65 \pm 0.02 \times 10^{-4}$ cm^2 V^{-1} s^{-1}, in the absence of an external electric field, to $-1.94 \pm 0.03 \times 10^{-4}$ cm^2 V^{-1} s^{-1}, in the presence of a +3-kV potential gradient [16].

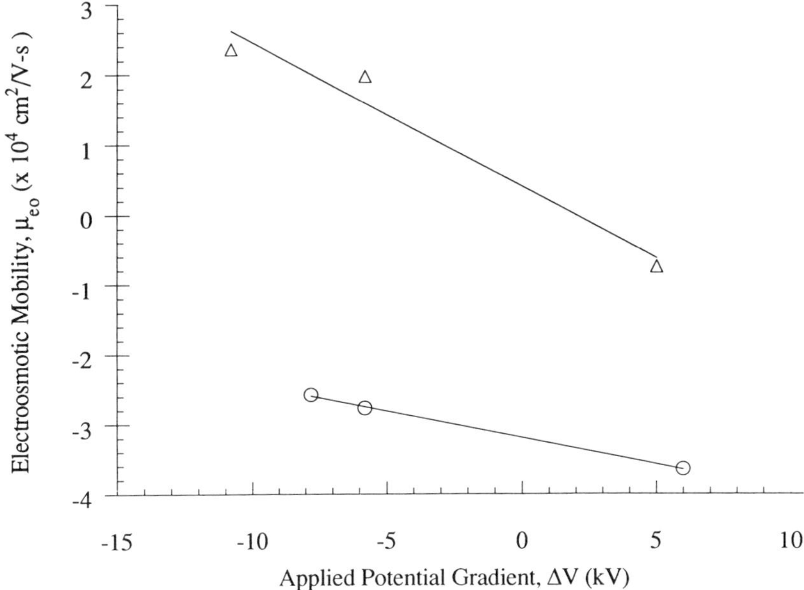

Figure 5 The electroosmotic mobility, μ_{eo}, of coated capillaries with the application of various potential gradients in 10-mM phosphate buffer at pH 4.

The controlled electroosmotic flow affects the migration time and zone resolution of protein mixtures. In addition, the changes in the magnitude and polarity of the zeta potential causes the various degrees of protein adsorption onto the capillary through the electrostatic interactions. The influence of external electric field on protein adsorption, therefore, is evaluated in terms of the separation efficiency. The values of total spatial variances for proteins in a solution of 10-mM phosphate buffer at pH 5 are measured from their peak profiles at various applied potential gradients. The results, summarized in Table 2, clearly indicate the strong effect of external electric field on protein adsorption at the capillary wall. The lowest separation efficiency and serious band tailing of proteins appear in the presence of a −3.7-kV potential gradient. In contrast, the adsorption of proteins onto the capillary wall is significantly reduced with the application of +4-kV potential gradient for obtaining the highest separation efficiency. The reduction of protein adsorption also contributes to the great improvements on the migration reproducibility of proteins. Thus, the zone resolution of proteins with the application of positive potential gradients is enhanced by the increase in both the separation efficiency and the migration time [16].

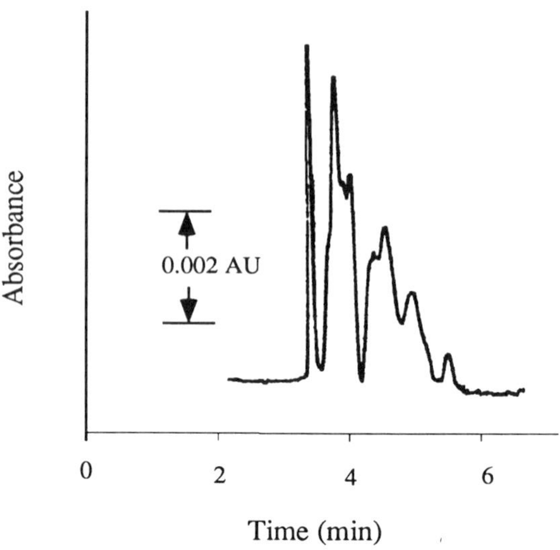

Figure 6 Zone electrophoretic separation of proteins in the absence of an external electric field: buffer, 10-mM phosphate at pH 2.5; capillary, 50-μm id and 150-μm od, length to detector 16 cm and overall length 25 cm; voltages, 0.5 kV and 5 seconds for injection, 4 kV for electrophoresis. (From Ref. 16)

Direct Control of Electroosmosis for Phenylthiohydantoin–Amino Acid Separations

Micellar electrokinetic capillary chromatography (MECC) was introduced by Terabe [17] as one of the separation modes of capillary electrophoresis for the separation of neutral compounds. The retention time of a solute that interacts with the micelles will fall in a "retention window" between the elution time (t_0) of a solute that has little or no interaction with the micelle (i.e., dimethyl sulfoxide used in this study) and the elution time (t_{mc}) of a solute that is 100% solubilized (i.e., Sudan III used in this study) by the micelles. One key obstacle in the application of MECC to the separation of a complex sample is its limited retention window. The *peak capacity*, which is defined as the maximum number of resolved peaks, is clearly dependent on the specific range of the retention window.

A neutral solute with no interaction with the micelle will elute at a time, t_0, determined solely by the magnitude of the electroosmotic flow. Alternatively, a solute that is completely micelle-solubilized will elute together with the micelle. The net flow velocity and the elution time of the micelle, t_{mc}, are determined by the summation of its electrophoretic velocity and the electroosmotic flow. Thus,

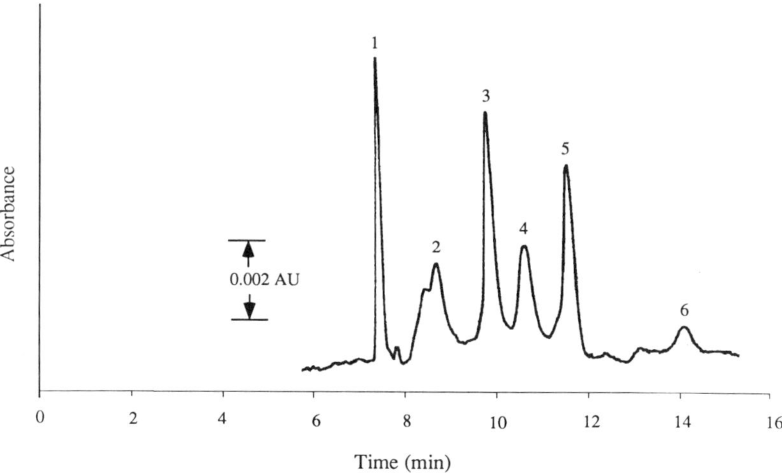

Figure 7 Zone electrophoretic separation of proteins in the presence of a +3-kV potential gradient. Elution order: 1, lysozyme; 2, bovine serum albumin; 3, ribonuclease A; 4, α-chymotrypsin; 5, α-chymotrypsinogen A; 6, hemoglobin A. Other conditions are the same as in Fig. 6. (From Ref. 16)

Table 2 The Values of Total Spatial Variances for Separation Peaks at Various Applied Potential Gradients

	Value of total variances at various potential gradients (cm^2)[a]		
Proteins	−3.7 kV	No external electric field	+4 kV
Lysozyme	0.256	0.085	0.0142
Cytochrome *c*	0.051	0.037	0.0102
Ribonuclease A	0.0233	0.0183	0.0085
α-Chymotrypsinogen A	0.0233	0.0171	0.0051

[a]The experimental error in measuring the value of total spatial variances was about 5−10% for various potential gradients for over five runs.
Source: Ref. 16.

the range of the retention window and the separation resolution of MECC are clearly dependent on the direction and flow rate of electroosmosis.

Phenylthiohydantoin (PTH)–amino acids, which are amino acid derivatives resulting from the Edman degradation of peptides and proteins, are important materials for determining amino acid sequences [18]. Otsuka et al. [19,20] have applied MECC for the separation of 18 PTH-amino acids. Three pairs of peaks were poorly resolved: Thr–Ser, Val–Met, and Leu–Trp without manipulation of the sodium dodecyl sulfate (SDS) concentration, the applied voltage, and the length of the capillary tubing [19,20]. The change in SDS concentration was to affect the capacity factor between the micelle and amino acid. The increase in the applied voltage and the length of the capillary tubing were used to increase the separation efficiency.

In this study, the neutral PTH–amino acids including Phe, Trp, Leu, Met, and Val were selected as the model mixture for investigating the enhancement of separation resolution in MECC with the direct control of electroosmosis. As shown in Table 3, the electroosmotic mobility, μ_{eo}, and the mobility of the SDS micelle, μ_{mc} (the summation of micelle's electrophoretic velocity and the electroosmotic flow), in a solution of 10-mM phosphate/50-mM SDS, at pH 3.5 are changed with the application of various potential gradients. At pH 3.5, the electrophoretic velocity of the micelle toward the anode is greater than the electroosmotic flow. Therefore, the UV detector is located at the cathode end for measuring μ_{eo} with dimethyl sulfoxide, and then switched to the anode end for determining μ_{mc} with Sudan III. The mobility and the retention time in a 50-μm id and 150-μm od capillary, as summarized in Table 3, are assigned as positive when the direction of movement is toward the cathode.

As summarized in Table 3, the electroosmotic mobility and the ratio of $-t_{mc}$ to t_0 as the range of the retention window all increase with the application of negative potential gradients across the capillary wall. The increase in the retention window greatly improves the separation resolution of PTH–amino acids as demonstrated in Fig. 8. The PTH–amino acid mixture is injected at the cathode end and detected near the anode end. The elution order is Phe, Trp, Leu, Met, and Val, with the more hydrophobic solutes being eluted earlier with the micelle phase. The capacity

Table 3 The Effect of External Electric Field on Retention Window

	μ_{eo} ($\times$ 10^4 cm^2 V^{-1} s^{-1})	μ_{mc} ($\times$ 10^4 cm^2 V^{-1} s^{-1})	t_0 (min)	t_{mc} (min)	$-t_{mc}/t_0$
No external field	+0.78	−4.20	+16.82	−3.17	+0.19
$\Delta V = -4.8$ kV	+1.47	−3.53	+9.09	−3.77	+0.41
$\Delta V = -7.8$ kV	+2.38	−2.62	+5.61	−5.08	+0.91

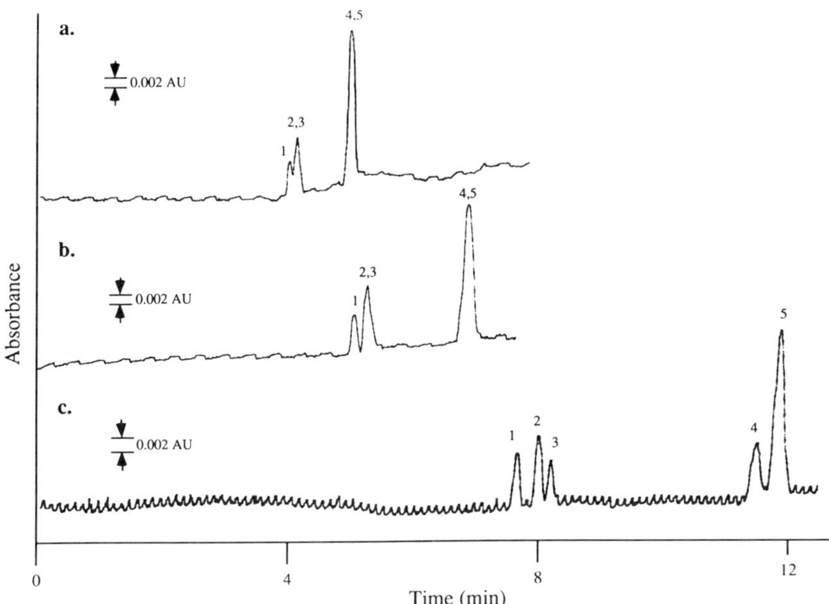

Figure 8 MECC separation of PTH–amino acids (a) in the absence of an external electric field; (b) in the presence of a −4.8-kV potential gradient; (c) in the presence of a −7.8-kV potential gradient. Elution order: 1, PTH-Phe; 2, PTH-Trp; 3, PTH-Leu; 4, PTH-Met; 5, PTH-Val. Buffer, 10-mM phosphate/50-mM SDS, at pH 3.5; capillary, 50-μm id and 150-μm od, length to detector 16 cm and overall length 25 cm; voltages, 1.0 kV and 5 seconds for injection, 5 kV for electrophoresis.

factors measured from the experiment are 4.88 for Phe, 4.37 for Trp, 4.14 for Leu, 2.45 for Met, and 2.37 for Val. With the application of a −7.8-kV potential gradient, Met and Val are baseline-resolved with the difference of 0.08 in the capacity factor.

SUMMARY

A physical method involving the use of an additional electric field applied from outside of the capillary for the direct control of the zeta potential and the electroosmotic flow is proposed and demonstrated. Factors, such as capillary dimensions and solution conditions, affecting the direct control of electroosmosis are analyzed both experimentally and theoretically with the capacitor theory. This method is applicable to both bare silica capillary and capillary with various surface coatings. Significant improvements on the separation efficiency and resolution of proteins in CZE and

PTH–amino acids in MECC are established by simply controlling the zeta potential and the electroosmotic flow with the application of an external electric field.

ACKNOWLEDGMENT

We are appreciative to Dr. Swedberg of Hewlett Packard Laboratories for providing the capillaries coated with amino and butyl phases. We thank C. T. Wu for his help in the preparation of this manuscript. Support for this work by Petroleum Research Fund Grant administrated by the American Chemical Society, National Science Foundation (CTS-9108875 and DIR-9105016), and Hewlett Packard Laboratories is gratefully acknowledged.

REFERENCES

1. C. L. Rice and R. Whitehead, *J. Phys. Chem.*, *11*:4017 (1965).
2. R. J. Hunter, *Zeta Potential in Colloid Science: Principles and Applications*, Academic Press, New York (1981).
3. J. W. Jorgenson and K. D. Lukacs, *Science*, *222*:226 (1983).
4. A. Emmer, M. Jansson, and J. Roeraade, *J. Chromatogr.*, *547*:544 (1991).
5. B. J. Herren, S. G. Shafer, J. V. Alstine, J. M. Harris, and R. S. Snyder *J. Colloid Interface Sci.*, *115*:46 (1987).
6. S. Hjerten, *J. Chromatogr.*, *347*:191 (1985).
7. M. J. Gordon, K.-J. Lee, A. A. Arias, and R. N. Zare, *Anal. Chem.*, *63*:69 (1991).
8. H. H. Lauer and D. McManigill, *Anal. Chem.*, *58*:166 (1986).
9. R. M. McCormick, *Anal. Chem.*, *60*:2322 (1988).
10. C. S. Lee, W. C. Blanchard, and C.-T. Wu, *Anal. Chem.*, *62*:1550 (1990).
11. C. S. Lee, C.-T. Wu, T. Lopes, and B. Patel, *J. Chromatogr.*, *559*:133 (1991).
12. X. Huang, M. J. Gordon, and R. N. Zare, *Anal. Chem.*, *60*:1837 (1988).
13. T. Tsuda, K. Nomura, and G. Nakagawa, *J. Chromatogr.*, *248*:241 (1982).
14. C. S. Lee, D. McManigill, C.-T. Wu, and B. Patel, *Anal. Chem.*, *63*:1519 (1991).
15. S. Swedberg, *Anal. Biochem.*, *185*:51 (1990).
16. C.-T. Wu, T. Lopes, B. Patel, and C. S. Lee, *Anal. Chem.*, *64*:886 (1992).
17. S. Terabe, K. Otsuka, K. Ichikawa, A. Tsuchiya, and T. Ando, *Anal. Chem.*, *56*:111 (1984).
18. L. Stryer, *Biochemistry*, 2nd ed., W. H. Freeman & Co., New York, p. 53 (1981).
19. K. Otsuka, S. Terabe, and T. Ando, *J. Chromatogr.*, *332*:219 (1985).
20. K. Otsuka and S. Terabe, *J. Microcolumn Separations*, *1*:150 (1989).

14

Semipreparative Capillary Electrophoresis and Its Advantages

Takao Tsuda

Nagoya Institute of Technology
Nagoya, Japan

The qualitative information obtained by capillary electrophoresis (CE) separation is derived from migration time, which depends on the mobility of a solute and, sometimes, on the partition between a medium and a micelle or the inner wall surface of the capillary. Mobility is dependent on the relative molecular mass, number of charges, and solvation, among others. Although these physical factors are not difficult to determine, compared with the factors affecting the partition, still it is not possible to estimate the mobility of a solute from only the available physical data. The lack of good qualitative information is one of the disadvantages in capillary zone electrophoresis (CZE) and micellar electrokinetic chromatography (MECC). One of the ways to overcome this problem is to connect the CZE equipment with spectral instruments, such as a mass spectrometer. But the on-line combination with CZE and a spectral instrument is as yet limited to a few examples.

Another way is to obtain a fraction of the desired solute and then to measure the spectra of that solute. In this way, we can obtain any spectra desired if the amount of the solute obtained is enough for analysis. Therefore, semipreparative instrumentation or micropreparative operation in CZE is very important.

The injection amount in conventional CZE is generally about 10 nL. This amount is 1/100–1/1000 that required in conventional liquid chromatography; that is, the sample amount that is necessary for spectral measurement or structural determination is not usually obtained in a single CZE run. Therefore, it is necessary to develop a semipreparative CZE method.

Semipreparative CZE provides another advantage. Because of the large sample injected in semipreparative CZE, the detection of solute is easier, with a gain in concentration sensitivity. In conventional CZE, the sensitivity of UV detection is about 10^{-5} mol L^{-1}, although the sensitivity in absolute amount is quite high, as much as 10 pg with a molecular mass of 100 and a 10 nL injection volume in this concentration. Sometimes a high sensitivity for the absolute amount is meaningless, as the sample volume is usually about 1 μL or more, and it is often difficult to obtain a sample in which the concentration is more than 10^{-5} mol L^{-1}. In this case, if possible, we prefer to inject a large amount of sample into the capillary. Both amperometric detection and laser fluorescence detection have quite high concentration sensitivities and may be comparable with the sensitivity of conventional liquid chromatography. The geometry of the microdetection in these detectors would reduce the noise level, but in UV detection, there is no advantage from the reduced size of a cell path length. Fluorescence detection with a xenon lamp also may not have the advantage of a microcell design compared with the design of the ordinary cell used for liquid chromatography.

Although the injection amount for CZE is 10 nL, we need about 10 μL or more of sample for one injection, because we do not yet have a method to handle very small volumes of liquid. Accordingly, we could not do the following operation: pick up 10 nL of liquid sample, carry it to the inlet of the capillary column, and inject it into the column. Here, we simply use a sample of 1/1000 for separation, and the rest remains as unused sample.

Semipreparative capillary electrophoresis may help overcome the disadvantage of conventional CZE for the sample amount used and the sensitivity of concentration detection. Two studies are relevant: a column of multiple bundled capillaries and a rectangular capillary column with a large cross-sectional area.

For spectral measurements or sequence analysis, it is necessary to obtain a large enough sample. If we use a semipreparative CZE, we may obtain enough sample with a single separation run.

One of the alternative methods to solve this problem, when using a conventional capillary column, is to accumulate the sample fraction by multiple runs. The application of multiple runs will be described later for the sequence analysis of a peptide.

Another alternative method is the use of an on-line analyte concentrator in the capillary column head. This method uses the maximum-loading capacity of a conventional capillary column. It is useful for obtaining the fraction of solute that would have a relatively high concentration, and it is easy to use for spectral measurement or sequence analysis. This method also has the advantage for detecting a very dilute sample, because the concentration of the sample zone becomes higher after it has been desorbed from the concentrator.

These approaches for semipreparative collection of samples using CE, for example, the use of a bundle of multiple capillaries, the use of a concentrator in the column head, and the collection of fractions with multiple runs will overcome the temporary disadvantages of CE, namely, its low concentration sensitivity and an amount of the separated sample that is too small for structural determinations.

There are several methods to achieve the semipreparative mode in capillary electrophoresis. Each method has its own advantages and disadvantages. In any event, if the proposed device requires a complex process, it may still be used in some special cases even though its experimental conditions are restricted. As there are only a few studies on semipreparative capillary electrophoresis, the proposed methods described in this chapter are still in the developmental stage.

THE SAMPLE-LOADING CAPACITY

Sample-loading capacity may be proportional to the cross-sectional area of the capillary column (Table 1). The injection amount for a 50-μm id capillary column is usually 10 nL. As the cross-sectional areas for the rectangular capillary with 50 × 1000 μm is 25 times larger than that of the 50-μm round capillary, the injection amount for the former column would be 250 nL. The sample size is nearly comparable with the injection capacity of gas chromatography using a large-bore capillary column. We need a capillary column that has a larger cross-sectional area and is able to attain good heat dissipation; the latter is the key to obtaining a good separation.

Use of Isotachophoretic Preconcentration in One or Tandem Capillary Columns

In isotachophoresis, each sample component has its own zone during the separation process. The concentration of the sample component in its own zone depends on

Table 1 Comparison of Total Injection Volume by Several Methods

Method	Maximum volume of sample employed
Isotachophoretic preconcentration	250 nL
Use of on-line analyte concentrator	500 nL[a]
A bundle of multiple capillaries	3–7 times
Sample micropreparation by multiple runs	5–10 times
Use of rectangular capillary column	
50 × 1000 μm	25 times[b]
100 × 2000 μm	100 times[b]

The number of injections is one except with the sample for micropreparation with multiple runs.
[a]This treated injection volume is quite large. It corresponds to more than 30 times the conventional injection volume.
[b]This number was obtained by comparison with the cross-sectional surface area of 50-μm id circular capillary columns.

the concentrations of electrolyte in the leading solution or in the terminal solution. The running medium for capillary zone electrophoresis has a higher concentration of electrolyte than that for the sample solution.

The process of isotachophoretic preconcentration in CZE is shown in Fig. 1, which assumes that the process will proceed with a quasi-constant current state.

After filling the capillary with a running medium (R), a large amount of sample solution(s) is injected at step I in Fig. 1. The conductivity of the sample solution remains less than that of the running medium. The charged species a, b, and c are in the sample solution; r is in the running medium. The electrolyte in the running medium is expressed as $r^+ r^-$. Their mobilities are assumed to be $r > a > b > c$. After a certain period, solutes a and b would form plug zones, as shown in step III. (see Fig. 1). The solutes a and b are more concentrated in plug zones compared with their original concentration because of the requirement in isotachophoresis that the concentrations (C) of a and b are determined by the concentration of the running electrolyte $r^+ r^-$ and depend on the mobility (U) of the ionic species concerned, as follows [1,2]:

$$C_{a+} = C_{r+} \left(\frac{U_{a+}}{U_{r+}}\right)\left(\frac{U_{r+} + U_{r-}}{U_{r+} + U_{r-}}\right)$$

In the plug zone, there is no distribution of solute; that is, every local volume concentration in the small plug zone is equal to that of the solute. Then, the running medium overtakes plug zones a and b during step IV to V, and these plug zones begin to form zones in which the concentrations have a gaussian distribution in step V. The solute, c, is also beginning to show a gaussian zone profile after step III.

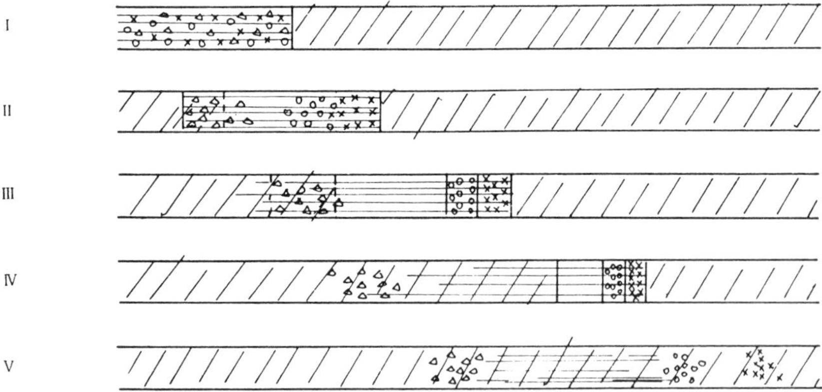

Figure 1 The process of isotachophoretic preconcentration in capillary zone electrophoresis. Sample components a, b, and c are expressed as X, ○, and △, respectively. A running medium and a solvent for sample components are expressed as ///// and ≡, respectively.

This isotachophoretic preconcentration has been studied by Dolnik et al. [3] and Hjerten [4]. Dolnik et al. used two capillaries [3]. The first capillary was used for isotachophoretic preconcentration; then it was connected to a second capillary for the separation by CZE. The first capillary was combined with the second capillary by a small Teflon tubing tip.

Although Dolnik et al. [3] used a tandem isotachophoresis–CZE system, it could be performed in one capillary column. For this we would use a sample solution that has a lower conductivity than the running medium and inject a relatively larger volume than 10 nL. The inlet of the capillary column is then set in the reservoir of the running medium. The preconcentration will occur first, followed by the separation step.

Aebersold and Morrison proposed a most interesting method to concentrate an ampholyte [5]. They used the condition that a ampholyte in the sample solution has an opposite charge to it in the running medium; that is, if the running medium is moving toward the negative electrode and the ampholyte has a negative charge in the sample solution, the ampholyte first delays in the zone of sample solution and meets the running medium coming backward. After the ampholyte is solvated with the running medium, the charge becomes positive, and it moves to the positive electrode with high mobility. Consequently, the ampholyte is concentrated in a very narrow zone. Aebersold and Morrison [5] applied this method to a dilute peptide sample and found that the sensitivity became five times greater than that of the ordinary injection method.

Use of an On-Line Analyte Concentrator for Increasing the Concentration of the Sample

In liquid chromatography (LC) we commonly use a small cartridge or a precolumn packed with LC supports. These concentrators are very powerful, especially when the sample solution is extremely dilute. We need these devices for CZE. They enlarge the total amount of the original injected sample solution. Therefore, the concentration detection limit of the sample solution becomes lower than without use of the device.

Merion et al. [6] and Guzman et al. [7] developed a device that placed an on-line analyte concentrator at the head of the capillary column. The concentrator is composed of LC support materials or chemically coated controlled porous glass, which were kept in a narrow space at the head of the capillary column by using two frits. The typical sample injection sequence is shown in Fig. 2. After the running medium is equilibrated with the capillary column that has the on-line analyte concentrator at its head, a sample solution is introduced into this column. The maximum amount of sample volume injected is about 1 μL. After the sample components adsorb on the LC supports or form a complex with the affinity materials,

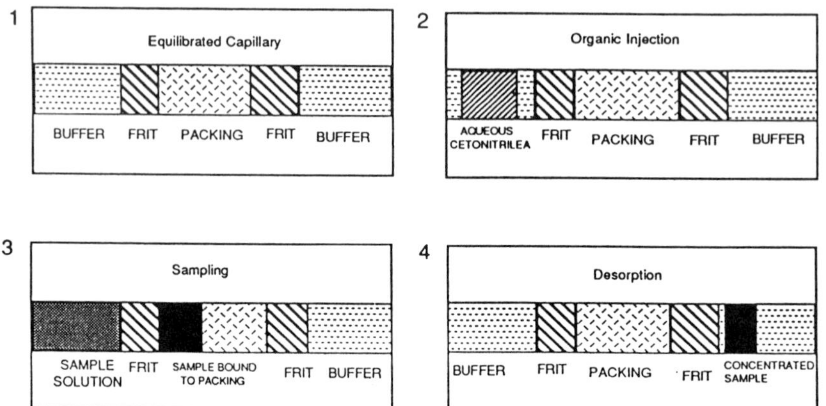

Figure 2 Injection sequence at the analyte concentrator. The capillary column with an analyte concentrator is Waters AccuSep C/PRP. [Courtesy of Sakai, Milipore Japan, Limited (Waters Chromatography), Tokyo, Ref. 6]

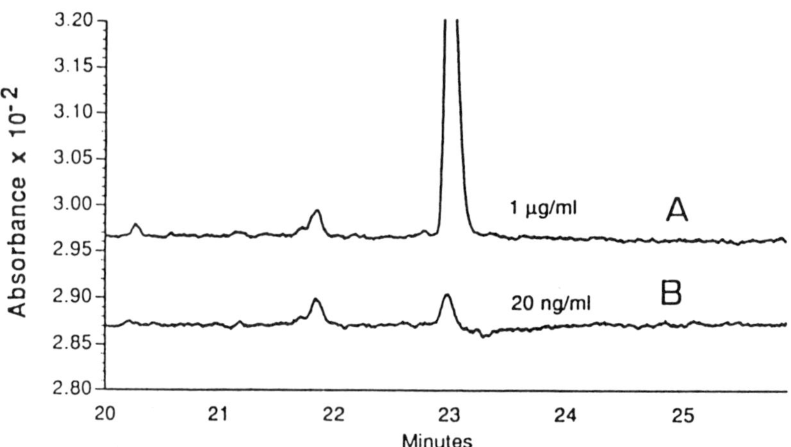

Figure 3 Detection limit of angiotensin I at pH 4.0 using a capillary with an analyte concentrator. Capillary, AccuSep C/PRP; buffer, 25 mM Na Citrate, pH 4.0; eluent, 25 mM Na Citrate, pH 4.0/acetonitrile 5/95; sample, angiotensin I in buffer + 1 M AccuPure Z1-methyl; injection, 999 s at 15 kV; elution, 10 s at 7.5 kV; run, 15 kV; detection, 214 nm. [Courtesy of Sakai, Milipore Japan (Waters Chromatography), Tokyo, Ref. 6].

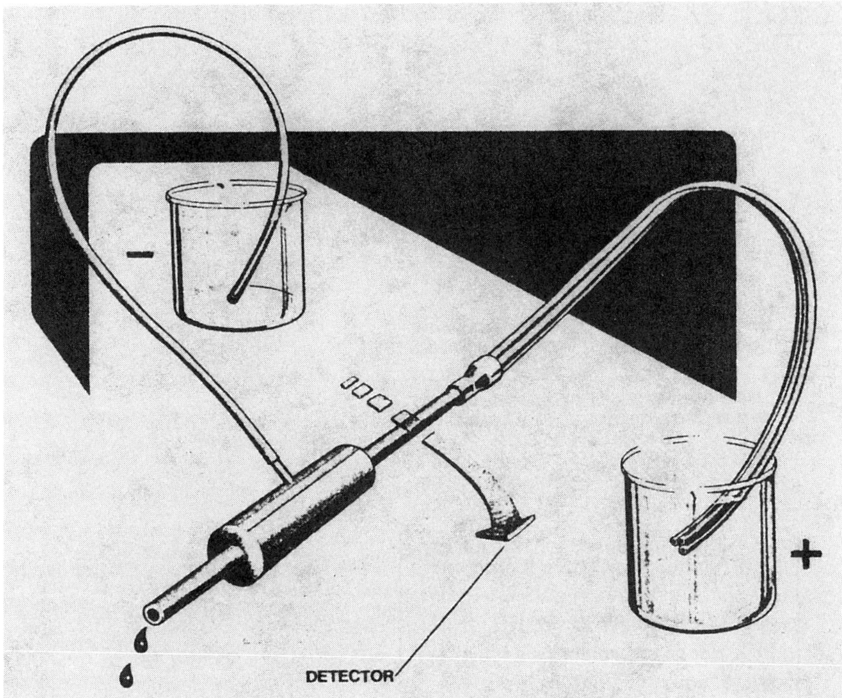

Figure 4 Schematic diagram of a continuous fraction collection for semipreparative capillary electrophoresis. The system includes both the microgrounding device (see Fig. 5a) and the link tube (see Fig. 5b). (From Ref. 9)

then they are eluted by the organic solvent (e.g., aqueous acetonitrile), as shown in Fig. 2.

In this method, the loading capacity of the column itself is not changed, but we are able use the maximum-loading capacity of this column because it is possible to concentrate each the sample component up to the maximum-loading capacity. This is the great advantage of this method. The disadvantage is that the media used for loading, elution, and running are very restricted; that is, these three media should form a good admixture for the injection and separation sequences. Other disadvantages include that such a column with a head on-line concentrator may be expensive, and the column itself is fragile. The handling of this special column is not as simple as that of the sample cartridge used in conventional liquid chromatography.

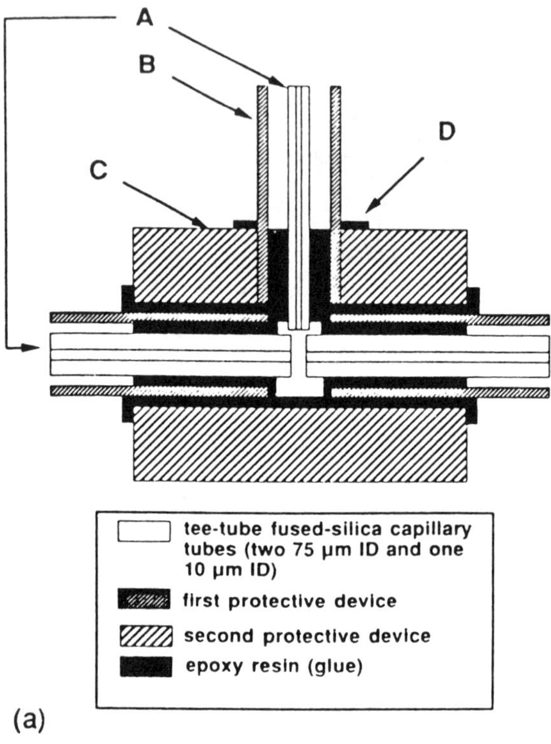

Figure 5 (a) Schematic diagram of the tee assembly unit used as a grounding device. (A) Three fused-silica capillary tubes were used to form the tee assembly unit. Two of these tubes were generated by breakage of a single capillary column (75-μm id, 375-μm od fused silica capillary) in which the separation process occurred. The third capillary tube (10-μm id 150-μm od fused silica capillary) was placed at the narrow interconnection point of the other two capillaries. At this point, the electrical-grounding process was initiated. (B) The first protective device was composed of a 530-μm id, 660-μm od fused silica capillary, covering and protecting both the 75- and the 10-μm columns. (C) The second protective device was composed of Teflon tubing acting as a supportive structure to give strength to the connecting components of the tee assembly unit. (D) Epoxy resin, to glue all spaces and produce a leakage-free system. (b) Schematic representation of the link tube. The diagram depicts three inlet capillaries fused into one outlet capillary. A minimum distance between inlet and outlet capillaries was maintained to avoid distortion of the electropherographic peaks. (From Ref. 9)

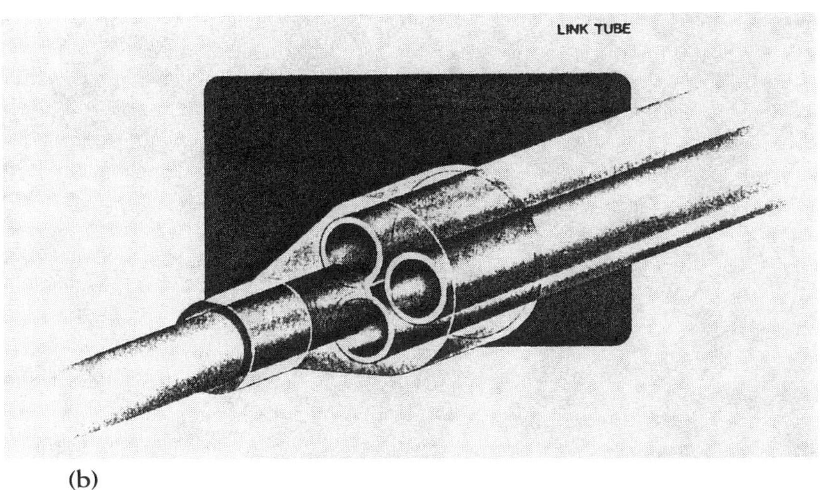

(b)

One typical example for its use is shown in Fig. 3. The injection amount is 500 nL and the concentrations of angiotensin I in the original sample solutions A and B in Fig. 3 are 1 μg mL^{-1} and 20 ng mL^{-1}, respectively. The detection limit of angiotension I without a concentrator is 1 μg mL^{-1}. This example clearly demonstrates the advantages of using the on-line concentrator in a capillary column; namely, a high gain in the detection limit and the possibility of using a very dilute sample for analysis.

A Bundle of Multiple Capillaries

Guzman et al. [8,9], Fujimoto et al. [10], and Anbe et al. [11] studied a bundle of capillaries in which several capillaries are bound into an analytical column. This design increases the total cross-sectional area of the analytical column; that is, the cross-sectional area of the column increases in proportion to the number of capillaries bound and to the sample-loading capacity for one injection. The schematic design is shown in Figs. 4 and 5. One end of the bundle is connected to a capillary (see Fig. 5-2), followed by a detector, a tee (see Fig. 5-1), and a fraction collector. From one end of the tee, a narrow fused silica capillary (10-μm id) leads to a terminal reservoir. The face-to-face design at the connector point inside the tee (see Fig. 5-1) and between the bundle and a capillary (see Fig. 5-2) are most important to avoid band broadening.

Guzman et al. [8,9] attempted to bundle up to five capillaries. As the sample amount injected is proportional to the number of capillaries bound, the peak height of a solute is also proportional to it (Fig. 6). The concentration detection limit for

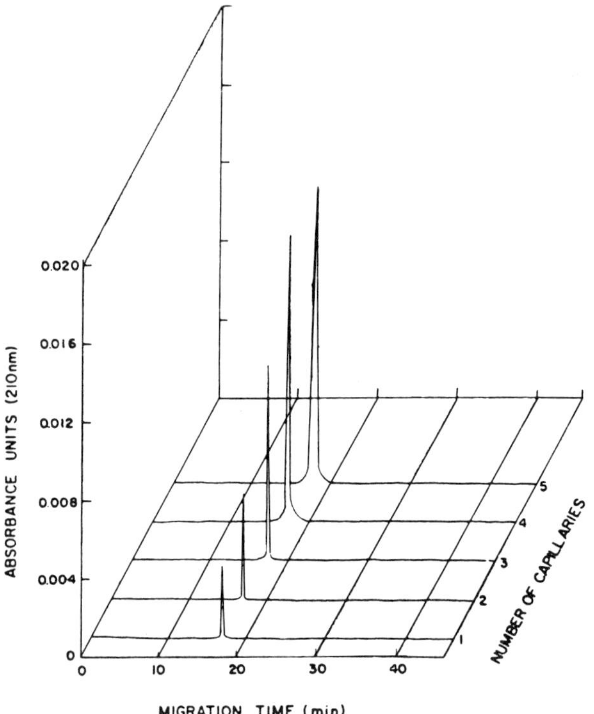

Figure 6 Use of multiple capillaries as a micropreparative collection system. Multiple capillaries were used to increase the sample-loading capacity to increase both the detector response and the collection capability of the system when using capillary electrophoresis. For this experiment 50 μg mL^{-1} antibody solution was used. (From Ref. 9)

an antibody sample is about 500 ng mL^{-1} using a column bound with five capillaries [9].

Electropherograms presented by Guzman et al. [8,9] show adequate resolution, although those of Fujimoto et al. [10] do not. Fujimoto et al. thought this low efficiency might come from band broadening by the tee-connection. The construction of the tee requires a very fine technique to keep the volume as small as possible.

The studies of Guzman et al. [8,9] and Fujimoto [10] were the first to attempt on-line fraction collection on a semipreparative scale. Although the bound column does not have a large sample capacity, the capacity is about five times larger than that of an ordinary capillary column in which the optimum sample-loading volume capacity is about 10 nL.

Sample Micropreparation by Multiple Runs

Another approach to collecting a large amount of a solute is achieved by use of multiple runs in a conventional capillary column [12,13]. The solute is collected in the same vial during the multiple runs. The reproducibility of the solute's elution time should be enough to avoid contamination by impurities coming from the original solution. When the sample component reaches the end of the column, the applied voltage is stopped, and the vial at the end of the column is transferred to the collecting vial. The electrovoltage is then reapplied for collecting the fraction. This process may cause a problem in fluctuation of the elution time. Therefore, use of an automatic HPCE system is recommended. Another problem is the electrochemical reaction at the surface of an electrode that has been placed into the collection vial during the elution of a target solute. It occasionally causes the degradation or reduction–oxidation of the solute. Third, to obtain a highly concentrated solution, the solvent in the collection vial should be as small as possible, and the solvent level should be adequate and constant during the operation. Akiyama and Omori recommend maintaining the humidity of the environment surrounding the vials by putting an extra vial filled with water around the collecting vial [12].

Determination of the amino acid sequence of a peptide or a protein requires relatively pure materials, preferably present at 100–200 pmol levels or 5–10 pmol of a carefully purified peptide. One example is shown in Figs. 7 and 8 [12]. The sample used is a synthesized peptide (16 mers), which has the following structure.

H-Asn-Thr-Asp-Gly-Ser-Thr-Asp-Thr-Gry-lle-leu-Gln-lle-Asn-Ser-Ars-OH

The original sample contained impurities (see Fig. 7), but the collected sample does not (see Fig. 8). In Fig. 8 the collected synthesized peptide eluted at 5.7 minutes, and other elution peaks seen between 4.3 and 4.7 minutes may be impurities from the running buffer. The operation of five micropreparative runs has been accomplished without contamination owing to the fractionation elution time of the solute. Akiyama and Omori used 300 μg mL^{-1} of the original peptide solution and injected 42 nL per run (This amount is rather large.) They used a vial containing 20 μL of pure water for collecting. After five runs, the peptide concentration is 67 ng (37 pmol) in 20 μL of water (3.3 μg mL^{-1}). They then performed its sequence analysis, and determined up to 15 complete cycles. The last amino acid (Arg) was obtained as the sign of its existence.

Rectangular Capillary Columns With a Large Cross-Sectional Area

Preparative columns in gas or liquid chromatography have several times larger cross-sectional areas compared with those of conventional columns for analytic use. From

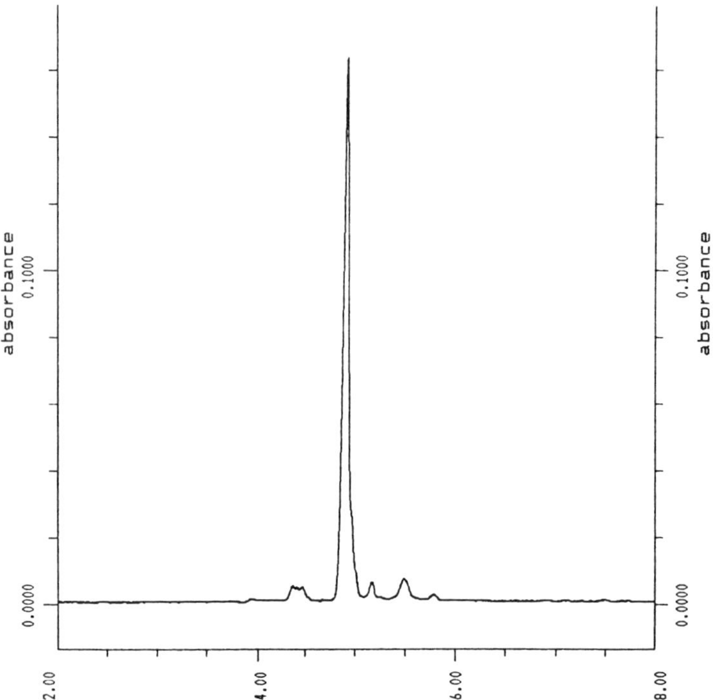

Figure 7 Electropherogram of a synthesized peptide of 16 mers. Medium, 100-mM borate buffer, pH 9.0; injection, 42 nL of 300 μg mL^{-1} sample solution; column, 75-μm id × 57 cm (50 cm from inlet to detector); applied voltage, 20 kV; detection, UV 200 nm; temp, 20°C. (Courtesy of Akiyama and Omori, From Ref. 12)

this analogy, it would appear favorable for the preparative columns in CZE to have a large cross-sectional area. Although the use of a bundle of multiple capillaries is one of the solutions, it may be better to use single capillary columns for this purpose. In CZE the column should have a good ability to dissipate heat. For this requirement, a rectangular capillary would be a suitable choice. Because of the large height/width ratio of a rectangular capillary and, hence, an extremely high surface area/volume ratio, rectangular capillaries are extremely efficient at dissipating heat, compared with conventional circular capillaries [14]. Thus, larger-volume rectangular capillaries can be used for CZE. Rectangular capillaries, which are commercially available from Wilmad Glass Co. (Buena, New Jersey) are borosilicate glass and are not available with any protective coating (e.g., polyimide). A significant disadvantage of the Wilmad Glass Co. capillaries is their thin walls; in all cases the available wall thickness is equal to the width of the narrowest dimension—for example,

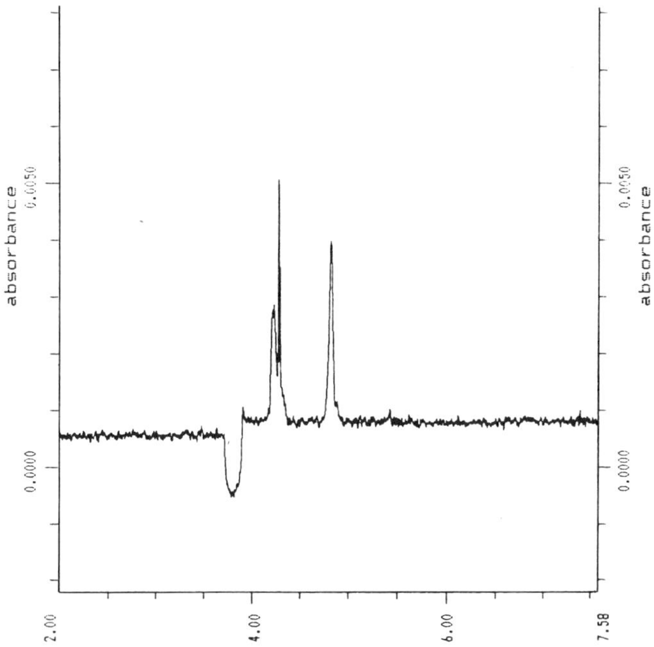

Figure 8 Electropherogram of a collected peptide after five micropreparative runs. Experimental conditions are same as Fig. 7. (Courtesy of Akiyama and Omori, From Ref. 12)

50 × 1000-μm capillaries have 50-μm walls, shown in Fig. 9. As these capillaries are extremely fragile, it is necessary to handle them very carefully and to keep them as straight as possible. Because of the fragility of these rectangular capillaries, special instrumentation should be designed that is different from what we would ordinarily use for circular fused silica capillaries (e.g., injection process and detection device) [15,16].

Another problem with use of rectangular capillaries as columns is that the flow resistance in the rectangular capillaries with a large cross-sectional area is very low. Thus, a small difference in levels of the reservoirs at both ends of column causes pressurized flow of the running medium. One should also use gravity injection in the same manner used for circular capillary columns. The schematic design of the instrumentation used for rectangular capillaries is shown in Figs. 10 and 11.

A split injection system in Figs. 10 and 11 is made under continuous flow, using a pump and a rotary valve. The sample is introduced into the capillary by electrokinetic injection. The injection procedure does not interrupt the applied voltage, and is a unique feature that may prove useful for automatic operation of CZE. The concentration of the analyte carried from the injector to the inlet of the

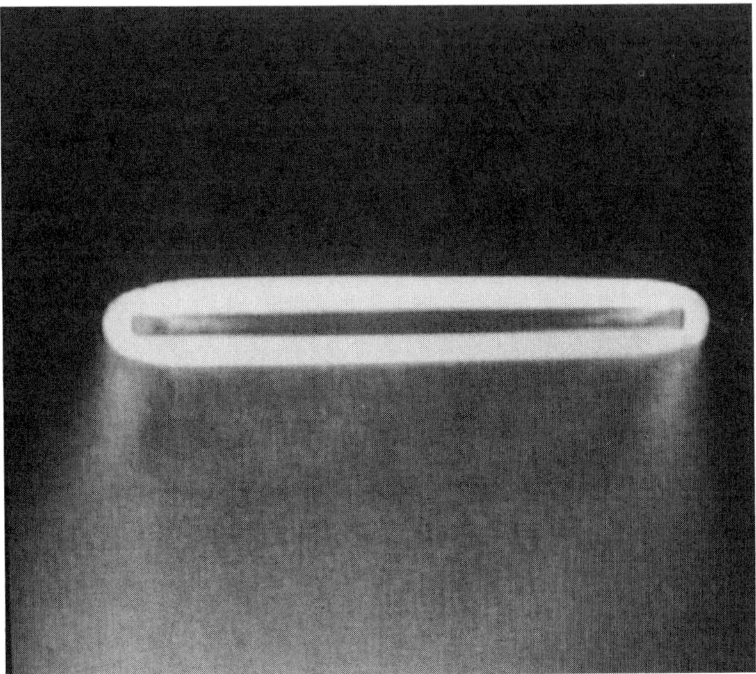

Figure 9 Photomicrograph showing the cross section of the 50 × 1000-μm id rectangular capillary with 50-μm walls.

column is lower at the column inlet, by a factor of approximately eight, than the concentration introduced into the injector. The relative standard deviations of reproducibility of split injection is about 4.2%.

If the flow pattern in CZE is a poiseuille flow profile, the flow speed between the center and the corner of the rectangular capillary is quite large, and it decreases considerably in resolution. This phenomenon is found in open-tubular liquid chromatography with rectangular capillaries [17]. But the flow pattern in CZE is similar to the plug flow pattern or the mixture of the poiseuille and plug flow pattern. (In the latter the contribution of poiseuille flow is a very small percentage) [17–21]. Therefore, the rectangular capillary as a column in CZE would be used with a fine resolution.

Table 2 and Fig. 12 show the theoretical plate number for a solute obtained by using round, square, and rectangular capillary columns. The theoretical plate numbers obtained are independent of the geometric differences of columns.

The CZE electropherograms from a rectangular capillary are shown in Figs. 13 and 14. The rectangular capillary column used in Figs. 13 and 14 has 25 times

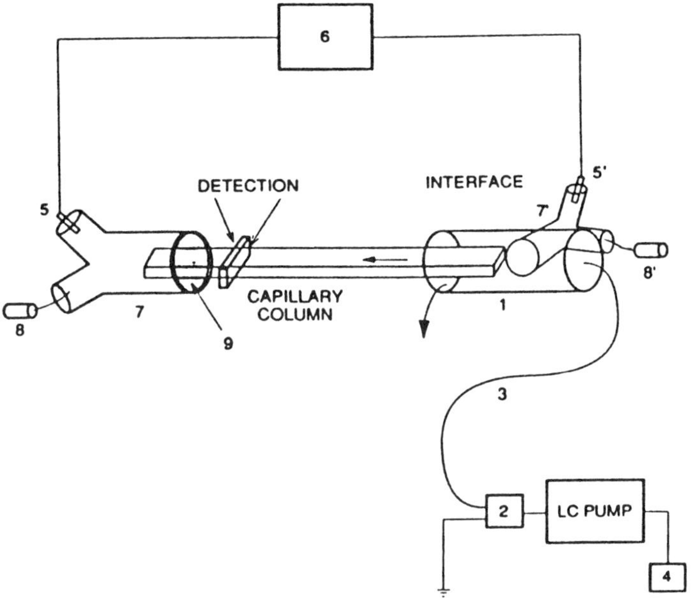

Figure 10 Schematic diagram of capillary electrophoretic system with split injector. 1, interface; 2, injection valve; 3, delivery tube of fused silica capillary, 50- or 100-μm id and 1.5–2 m in length; 4, reservoir for LC pump; 5 and 5′, electrodes; 6, high-voltage supply; 7 and 7′, reservoirs using three-way PTFE connectors; 8 and 8′, syringes for filling reservoirs; 9, epoxy seal. (From Ref. 16)

larger cross-sectional area compared with that of a 50-μm id round fused silica capillary. As the sample-loading capacity will be increased with the cross-sectional area, 0.25 μL or more will be possible. This injection amount corresponds to that used in large-bore capillary gas chromatography.

The number of theoretical plates varies from 20,000 to 60,000 for solutes in Fig. 13. Although not high by CZE standards, much of the width of these peaks may be attributed to the width of the injection during the moving process in the interface (see Fig. 11).

The other advantage of rectangular capillaries is the gain in sensitivity. To determine the sensitivity gain obtained with the increased path length of the rectangular capillary, the sensitivity of the system is determined by using two different detection geometries for the 50 × 1000-μm capillary, shown in Fig. 14. Figures 14a and b show the absorbance electropherogram obtained across the 50- and 1000-μm path lengths, respectively. In the latter, the peak is off-scale, and a lower sensitivity trace is shown in Fig. 14c. The applied voltage was 7.9 kV in part a

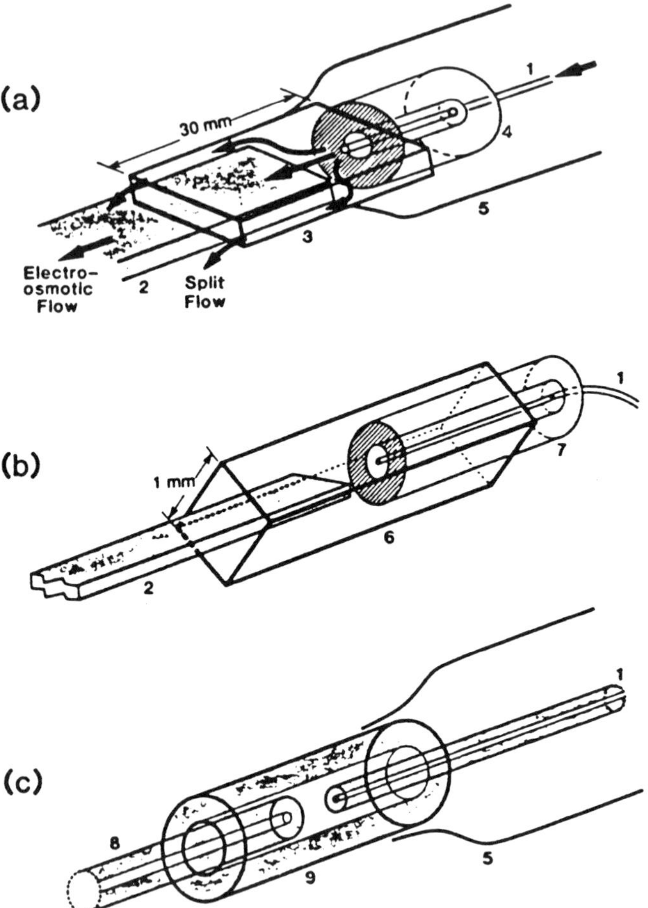

Figure 11 Interface for split injector having (a) "rectangular" interface for rectangular capillary (50 × 1000 μm); (b) "square" interface for rectangular capillary (50 × 1000 μm); (c) "circular" interface for circular capillary (50–100-μm id). Arrows show directions of sample and solvent flow. 1, fused silica capillary from injection valve (2 in Fig. 10); 2, rectangular capillary; 3, rectangular 200 × 2000-μm capillary; 4, PTFE tubing, 0.5-mm id, 2.0 mm od, and 10 mm in length; 5, cover; 6, square tubing (side 1 mm long); 7, PTFE tubing, 0.5-mm id and 1-mm od; 8, circular capillary; 9, circular capillary tubing, 0.5-mm id × 0.8-mm od. (From Ref. 16)

Table 2 Efficiency and Separation Conditions of the Pyridoxine Peak for a Variety of Capillary Cross-Sectional Geometries

Capillary size (μm)	L_{total} (cm)	$L_{(inj.-det.)}$ (cm)	Applied voltage (kV)	Current (μA)	N
		Round			
51	53	34	15	27	2.1×10^5
		Square			
50 × 50	60	43	15	39	2.8×10^5
103 × 108	65	48	12	91	1.3×10^5
		Rectangular			
16 × 195	53	34	12	27	1.6×10^5
27 × 340	83	63	15	62	2.1×10^5

Source: Ref. 15.

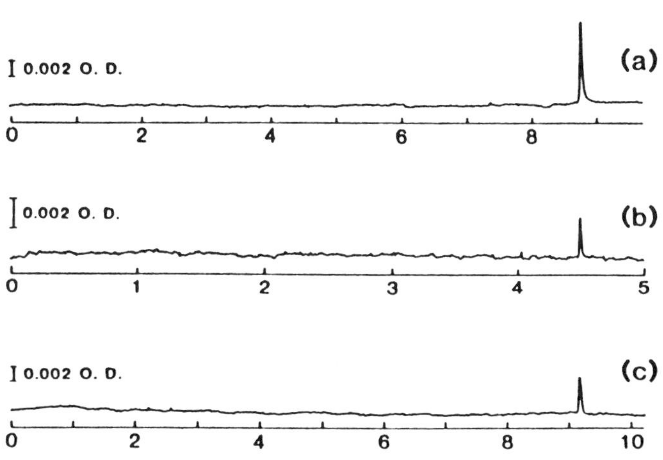

Figure 12 CZE electropherograms of pyridoxine for rectangular, square, and round cross-sectional geometry capillaries: (a) 27 by 340-μm rectangular; (b) 52-μm square; and (c) 51-μm round. The separation conditions and peak efficiencies are listed in Table 2. (From Ref. 15)

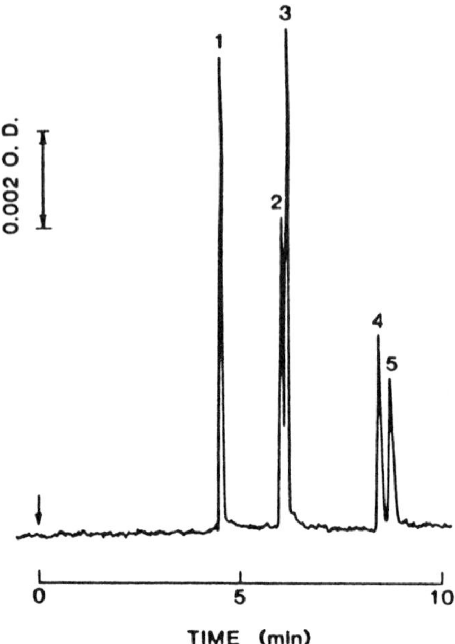

Figure 13 CZE electropherogram of (1) pyridoxine, (2) dansyl-L-valine, (3) dansyl-L-serine, (4) dansyl-L-aspartic acid, and (5) dansyl-L-cysteic acid, each injected at 1.3×10^{-5} M. The separation conditions are a 57-cm long and 50 by 1000-μm id capillary, 10-kV applied voltage, 132-μA current, and 310-nm detection. (From Ref. 15)

and 9 kV in parts b and c, with a capillary length of about 60 cm and a current of 110 μA. The current density of 110 μA is similar to a current of 4 μA in a 50-μm circular capillary. As can be seen in this figure, an approximate 15-fold increase in sensitivity is gained by the longer path length.

The improvement in detection sensitivity from the cell path length increase is most pronounced when the concentration of sample is low. For instance, when the concentration of sample is just sufficient to be detectable in a 50-μm path length rectangular capillary, then a gain of nearly 20 is observed by employing a rectangular capillary with 1000 μm. Pyridoxine and dansyl-L-serine (8×10^{-7} M) are just above the limit of detection with 310-nm light. The 310-nm light is used instead of 254 or 280-nm because of the high absorbance of the borosilicate glass capillary walls for shorter wavelengths [15]. If we have a quartz glass capillary or quartz window for detection, the foregoing detection limits will be improved more than ten times.

The proposal to use a rectangular capillary as a column in CZE would be the first study of direct semipreparative CZE. If larger-scale (preparative) separations

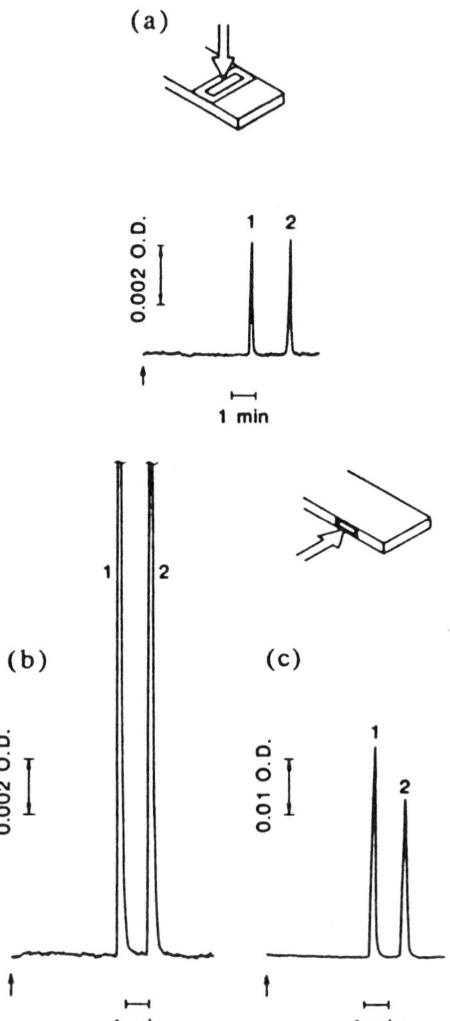

Figure 14 CZE electropherogram of (1) pyridoxine and (2) dansyl-L-serine, each at 4.2×10^{-5} M, using two different detector arrangements of the rectangular capillary in the UV-vis absorbance detector. In part a, the capillary was positioned so that detection was across the 50-μm axis of the capillary, and in parts b and c, across the 1000-μm axis of the capillary. In part b, the electropherogram is recorded by using the same detector sensitivity as in part a, and the peaks are off-scale whereas in part c, the sensitivity has been reduced by a factor of 5. (From Ref. 15)

are needed, one can envision separation channels formed between two flat plates separated by a spacer, with the ultimate limits of volume limited only by geometric constraints [15].

REFERENCES

1. F. M. Everaerts, J. L. Beckers, and P. E. M. Verheggen, *Isotachophoresis*, *J. Chromatogr. Libr.*, Vol. 6, Eleservier, Amsterdam (1976).
2. H. Miyazaki and K. Kato *Isotachophoresis*, Koudansha, Tokyo (1980).
3. V. Dolnik, K. A. Cobb, and M. Novotny, *J. Microcolumn Separation*, *2*:127 (1990).
4. S. Hjerten, K. Elenbring, F. Kilar, J.-L. Liao, A. J. C. Chen, C. J. Siebert, and M.-D. Zhu, *J. Chromatogr.*, *403*:47 (1987).
5. R. Aebersold and H. D. Morrison, *J. Chromatogr.*, *516*:79 (1990).
6. M. Merion, R. H. Aebersold, and M. Fuchs, *Life Science Application*—3, Reports of Waters Japan (1991).
7. N. A. Guzman, M. A. Trebilcock, and J. A. Advis, *J. Liq. Chromatogr.*, *14*:997 (1991).
8. N. A. Guzman and F. E. Regnier, First symposium of the Protein Society, San Diego, August 9–13, Abstr. 823 (1987).
9. N. A. Guzman, M. A. Trebilcock, and J. P. Advis, *Anal. Chim. Acta.*, *249*:247 (1991).
10. C. Fujimoto, Y. Muramatsu, M. Suzuki, and K. Jinno *J. High Resolut. Chromatogr.*, *14*:178 (1991).
11. T. Anbe, H. Tsuruta, and T. Hanai, Tenth Symposium on Capillary Electrophoresis, Dec. 13–14, Osaka, Japan, Abstr. p. 29 (1990).
12. T. Akiyama and A. Omori, Eleventh Symposium on Capillary Electrophoresis, Dec. 12–13, Tokyo, Abstr. p. 73 (1991).
13. J. Gagnon and P. Goeltz, Application brief No. DS-801, Beckman Instruments, Inc. (1991).
14. A. Tisellius, *Trans. Faraday Soc.*, *33*:524 (1937).
15. T. Tsuda, J. V. Sweedler, and R. N. Zare, *Anal. Chem.*, *62*:2149 (1990).
16. T. Tsuda and R. N. Zare, *J. Chromatogr.*, *599*:103 (1991).
17. M. Martin, J.-L. Jurado-Baizaval, and G. Guiochon, *C. R. Acad. Sci. Ser. 2*, *295*:579 (1982).
18. M. Martin and G. Guiochon, *Anal. Chem.*, *56*:614 (1984).
19. M. Martin, G. Guiochon, Y. Walbroel, and J. W. Jorgenson, *Anal. Chem.*, *57*:561 (1985).
20. T. Tsuda, K. Nomura, and G. Nakagawa, *J. Chromatogr.*, *264*:385 (1983).
21. T. Tsuda, M. Ikedo, G. Jone, R. Dadoo, and R. N. Zare, *J. Chromatogr.*, *632*:201 (1993).

15

Micropreparative Capillary Electrophoresis

Chuzo Fujimoto and Kiyokatsu Jinno

Toyohashi University of Technology
Toyohashi, Japan

High-performance liquid chromatography (HPLC) and electrophoresis in flat gel media have long assumed a central role in biochemical research. These high-efficiency separation techniques are essential for purification of target biomolecules, although traditional separation methods (e.g., precipitation) are still widely used for the initial isolation of biomolecules from various biological materials. These modern techniques have permitted separations of biomolecules at microgram levels on a routine basis. With the increasing sophistication of biochemical research, however, the separation requirements are more demanding. Probably, the materials to be examined in the future will be those that are present at even smaller levels in extraordinarily complex mixtures or those that are too labile to be amenable to conventional methods. Therefore, the preparation and isolation of small amounts of materials in a form compatible with subsequent analysis represent a major technical challenge.

Recently, handling sample volumes of nanoliters in HPLC and electrophoresis has become feasible by use of narrow-bore columns having inner diameters (id) of say 10–100 μm. For example, columns of 50-μm id can accommodate sample sizes of 4 nL. Another important advantage of this method is the potential for high-separation efficiencies. With narrow id capillaries, hundreds of thousands of theoretical plates are easily attainable in a short time. The two microscale separation techniques [i.e., capillary electrophoresis (CE) and capillary LC] have become valuable separation techniques for a large variety of substances. Both techniques have generally been considered to be analytical tools. However, as they become routine

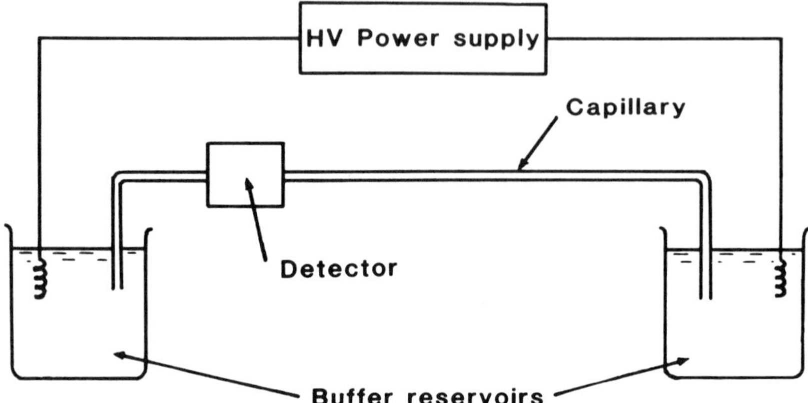

Figure 1 A schematic representation of basic CE instrumentation. Each end of the electrophoretic capillary is dipped into a buffer reservoir to make a closed electrical circuit.

techniques for analytical determination of many biological substances, questions are increasingly being asked concerning the ability to collect the separated species for subsequent characterization. Of the two techniques, the development of micropreparative CE systems seems more pressing, since molecules of biological importance mostly possess charges in buffer solution, and CE appears more suitable for the separation of such species because the separation principle of CE is based on the effective charge densities of species.

Unfortunately, fraction collection in CE is not as straightforward as it is with HPLC. In CE, both ends of the capillary have to be immersed in buffer reservoirs containing electrodes during the fraction collection to complete the electrical circuit (Fig. 1). All species in the injected sample migrate out of the capillary into the outlet reservoir. This results in the remixing of the separated components. Furthermore, the reservoirs are filled with a large volume of electrolyte solution relative to the column volume to minimize the effects of electrolysis: the volumes of the reservoirs are typically several milliliters, whereas those of the sample zones are nanoliters. Consequently, each sample component is diluted by a factor of 10^6.

In recent years, several approaches have been taken to overcome the problems associated with micropreparative CE. These can be categorized into two types: interrupted (or discrete) and continuous fraction collection methods. Both approaches can provide at least nanogram quantities of isolated components.

SAMPLE-LOADING CAPACITY

Polyimide-clad fused silica capillaries, which can be available from several sources, are the most common separation columns for CE. The internal diameter (id) and length of the capillary column can be variable, depending on the application. In open-tubular CE, columns between 25- and 100-μm id and between 25 and 100 cm in length are now in common use. Obviously, the maximum sample load decreases with reduction of the column diameter, as shown in Table 1. When it is assumed that the sample enters the capillary as a square plug and that diffusion is the only source of peak broadening, the acceptable length of the injected sample plug can be approximated [1] as

$$l_{inj} = (24DEt)^{1/2}$$

where D is the diffusion coefficient of solute, E is the acceptable increase in plate height (H) relative to the theoretical minimum H of the system, and t is the elution time. This equation shows that the injection length is a function of the solute's diffusivity (and, therefore, molecular size) and the elution time. Calculations show that for a 10-minute analysis time and $E = 0.1$, the injection plug would be less than 0.4 mm for macromolecules, whereas it would be approximately 1.2 mm for small molecules. In reality, the length of the capillary that is filled with sample solution can be a multiple of the solute diffusion length [2,3]. To minimize peak distortion, the solute concentration should be as low as possible: a region of high-solute concentration changes the local conductivity of the buffer and, therefore, yields a change in the local potential gradient. These effects lead to uneven electrophoretic migration of the solute zone [4,5]. Roughly, the concentration of solute should be about 100 times less than that of the background electrolyte. Since the electrolyte concentration is normally in the range of 10^{-1}–10^{-3} M, the solute concentration should be not more than 10^{-3} M. Consequently, the maximum quantity of solute that can be loaded on a 100-μm id column will be 16 pmol (i.e. 16 ng for a relative molecular mass of 1000). For micropreparative runs, the capillary may be overloaded and the highest-concentration region of the component of interest may be collected. It should be noted that automated high-sensitivity methods allow the sequence determination of peptide samples at the 10- to 20-pmol levels [6].

Table 1 Column and Sample Sizes in the Standard Operating Mode[a]

Column id (μm)	Column volume (μL)	Sample volume (nL)	Sample load (pmol)
25	0.4	1	1
50	1.5	4	4
75	3.3	9	9
100	6.0	16	16
200	23.6	63	63

[a]Calculations are based on the following assumptions: the column length is 75 cm; injection length, 2 mm; sample concentration, 1 mg mL^{-1}; molecular mass, 1000.

It appears likely that employing a larger id capillary will increase the sample quantity to be isolated by CE. Unfortunately, the largest column diameter that may be used in electrophoresis is limited by joule heating of the buffer. For a given field and buffer system, wider columns produce higher current. According to Knox [7,8] and Grushka et al. [9], if self-heating is not to produce serious distortion of solute bands and loss of resolution, the product of the field strength (in kV m^{-1}), the column id (in μm), and the cube root of the solute concentration (in mol L^{-1}) should not exceed 1500. For example, for 10^{-2} M electrolyte and an electrical field strength of 500 V cm^{-1}, the maximum id should be 140 μm. This value is somewhat larger than those that have been used in open-tubular CE. Wider-diameter columns are usable for CE by using an anticonvective medium such as polyacrylamide. Capillaries with internal diameters of 150–200 μm have been used for recovery of nearly 1 μg of material [10].

Another avenue is to use multicapillaries arranged in bundles. By using four to eight capillaries (75- to 100-μm id) in the open-tube form, Guzman and colleagues [11] collected purified proteins and peptides the quantities of which were sufficient for microsequencing and other testing. One feature of the multicolumn operation is that the maximum number of capillaries depends on the capacity of the power supply used. For the arrangement shown in Fig. 2a, the capacity can be expected from the equivalent circuit shown in Fig. 2b. Under a constant field V, the current I is given by

$$I = V\sum_{i=1}^{n} \frac{1}{R_i} \qquad (i = 1, 2, 3, \cdots, n)$$

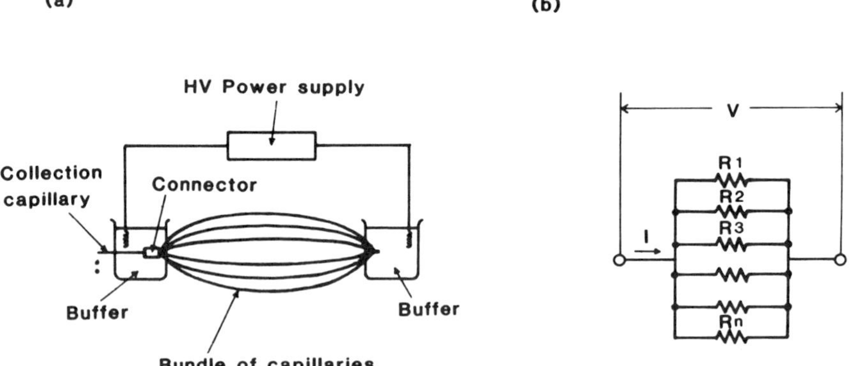

Figure 2 (a) A schematic representation of a multicapillary column electrophoretic system and (b) its equivalent electrical circuit. In this example, the connector also serves as an electrical joint.

where the resistance $R = L/\kappa S$, κ = specific conductance of buffer, L = capillary length, and S = cross-sectional area of a capillary. It is clear that if each capillary has the same resistance, the total current will be n times the current for single column operation.

When designing the multicapillary bundle system, considerable care must be taken to retain electrophoretic resolution and peak symmetry [12]. These include:

1. Mixed volume in the coupling device (see Figs. 2 and 7) should be minimized.
2. The capillaries being merged should be aligned in a uniform and rigid manner.
3. The capillaries should be identical in internal diameter and length.
4. The state of the capillary surface should not differ greatly between the capillaries.
5. All the capillaries should be maintained at the same temperature.

Considering the high-speed separation ability of CE, a multiple injection method may be an effective way to maximize the sample throughput: several consecutive injections may be possible before the arrival of the first-eluting peak at the end of the column. To put it into practice, prior knowledge of elution times and peak widths for the solutes would be needed. Probably, the utilization of this approach will largely depend on the complexity of the sample to be separated.

INTERRUPTED FRACTION COLLECTION SYSTEMS

In open-tubular CE, the presence of electroosmotic bulk flow renders the fraction collection at the capillary exit easy. Several reports have appeared involving fraction collection methods based on the interruption of applied voltage. In these methods, as the outlet end of the capillary is transferred from one collection reservoir to another, the electric field is broken and the electroosmotic flow is interrupted. After the capillary is positioned in a new reservoir, the electrophoretic process is continued again.

The collection system developed by Rose and Jorgenson [13], shown in Fig. 3, consists of a collection tray (CT) which contains ten collection cones (CC), with 10–15 μL of buffer, and a 4-mL rectangular slot reservoir (BR). The cones and reservoir are grounded by platinum electrodes (P). The system is equipped with three digital linear actuators (DLAs) to allow precise movement of the capillary (D1 and D2) and of the collection tray (D3, not shown in Fig. 3). The inlet end of the electrophoretic capillary (EC) (75-μm id, 60- to 80-cm long) is mounted in an autosampler with the end dipped into the anode buffer. The anode electrode is held at 20 kV. The outlet end of the capillary is placed in the buffer slot reservoir. For collection, a series of events occur: (a) D3 moves the collection tray so that the capillary is positioned opposite the cone into which the fraction will be collected; (b) D1 lifts the capillary installed in the holder (CH) up into the tapered glass capillary (TC), which contains a grounded electrode and buffer; (c) D2 moves the capillary guide (G) so that the end of the capillary is positioned over the cone; and

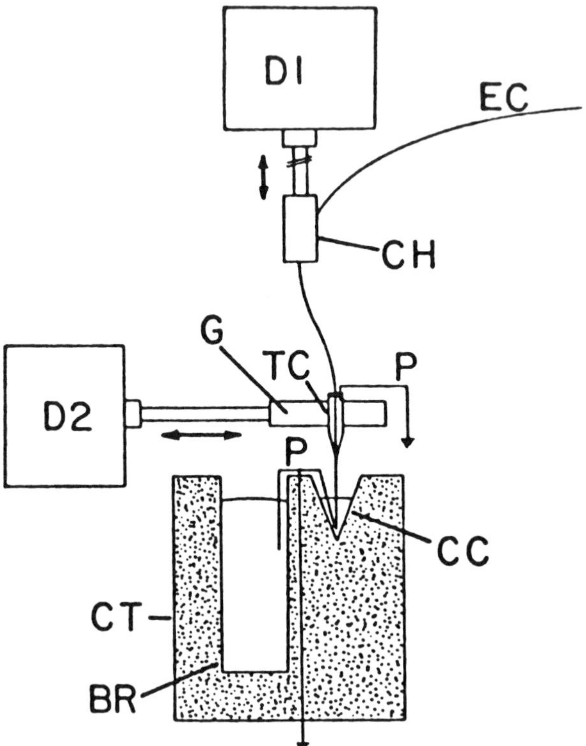

Figure 3 A cross-sectional schematic representation of the fraction collector, being capable of collecting fractions from a CE apparatus in a discrete manner. See text for abbreviations. (From Ref. 13)

(d) D1 lowers the end of the capillary into the cone. During the collection, each component migrates out of the capillary into the buffer of the cone. After the peak to be collected has eluted, steps b–d are retraced and the end of the capillary is moved back to the slot buffer reservoir. The movements of the DLAs and the autosampler, and on/off of the power supply were computer-controlled. The use of the tapered glass capillary reduced the field interruption time to a great extent. This system allowed α-chymotrypsin to be collected without losing the enzymatic activity. The concentration of adenosine in the collected fraction was approximately 3.5 orders of magnitude lower than that of the original sample.

Proteins have been successfully collected with such a discontinuous technique and microsequenced at the picomole level. Hecht et al. [14] reported the manual collection of recombinant human interleukin-3 (rhIL-3), which contains 152 residues [15]. The interleukin samples were introduced into a 75-μm id capillary column at 2 mg mL^{-1}. Twenty-four runs yielded a recovery of about 1 μg of the material. From 48 pmol of purified rhIL-3, the first ten residues were determined as expected. Akiyama and Omori [16] used a commercial CE instrument to isolate β-lactoglobulin, a standard for protein sequencing. A 100-nL injection of 1 μg μL^{-1} β-lactoglobulin was made, representing 5 pmol injected; the protein was collected into 20 μL of distilled water. With use of an improved microsequencing method, it was possible to go through up to 18 cycles or more.

Collection of substances migrating from gel-filled capillaries has been demonstrated by the group of Karger. Electrophoresis of a crude 20-mer oligodeoxynucleotide was performed with a 75-μm id column containing polyacrylamide gel (8% T and 3.3% T) in 7 M urea/0.1 mM Tris/0.25 mM borate, pH 8.3 [17]. The high-voltage end (anode) was placed in a microcentrifuge tube containing 2–5 μL of water. It should be noted that no electroosmotic flow generates in capillary gel electrophoresis. Therefore, under the influence of an electric field, the migrating buffer and sample ions are collected into the tube, whereas the uncharged species is not electrophoresed into the collection tube. With this method, about 800 ng of the oligonucleotide was isolated in a single run and identified by a standard dot-blot analysis. There was no significant broadening of sample zones for at least 1 hour, with no applied field, owing to the high viscosity of the gel matrix and, as a consequence, the sample was collected with little or no loss in resolution, even with such a discontinuous method.

In addition, small peptides (angiotensin I and an ACTH fragment) were collected from sodium dodecyl sulfate (SDS) polyacrylamide gel capillary columns [18]. The buffer system used, which consisted of 0.1% SDS, 7 M urea and 5 mM glycine (pH 4.1), affected the subsequent characterization at a minimum. For both peptides, low picomole sequencing was achieved.

Later, Guttman et al. [19] applied field programming methods to micropreparative capillary gel electrophoresis. In general, CE peak widths are on the order of seconds, which makes collection procedure unmanageable, particularly in manual operation. The principle of the field programming involved reducing the applied potential during fraction collection and operating at a high potential at all other times. Because the zone velocity is directly proportional to the applied field strength, the zone will migrate slowly into the collecting reservoir. By using this method, they were able to isolate an individual peak from a polydeoxyadenylic acid test mixture, $p(\mathrm{dA})_{40-60}$.

CONTINUOUS FRACTION COLLECTION SYSTEMS

Introduction

Despite the great promise of the interrupted approaches, they appear to impose some limitations on micropreparative collection. For example, samples are always significantly diluted. Even if the volume of liquid held in the collecting reservoir is reduced to nanoliters, the dilution factor remains as high as 10^3. Worse yet, it the collection volume is too small, then the interferences from electrochemical reactions could occur at the outlet electrode in the reservoir. Obviously, any interrupted fraction collection method needs sophisticated hardware: the real value will be proved with an entirely automated system.

A good solution to these problems is to complete the electrical circuit at a point a short distance from the end of the capillary. In CE, using an open tube, the electroosmotic flow generated in the separation column will transport buffer and solute ions through the electrical connection point to the capillary outlet, where small droplets will be formed during the CE run. Several approaches to continuous fraction collection have been described. To achieve such electrical connection, several devices have been implemented.

Porous Glass Joint

The first method for isolating the end of capillary from the applied electrical field was through a porous glass joint assembly as shown in Fig. 4. The assembly has been developed by Wallingford and Ewing [20,21] for off-column electrochemical detection. The conductive joint was made near the outlet end of the column by fracturing the separation column and encasing the fracture in porous Vycor glass. By placing the joint assembly in a buffer reservoir along with the ground electrode, the potential can be applied across only the first section of capillary. The porous glass joint allows ions to move into the reservoir, thereby permitting electrical connection to ground; the substances to be analyzed are forced to move past the electrical joint and the second section of the capillary. A carbon fiber electrode for electrochemical detection can be inserted into the open end of the second section of capillary. Guzman et al. [22–24] incorporated the porous glass joint method

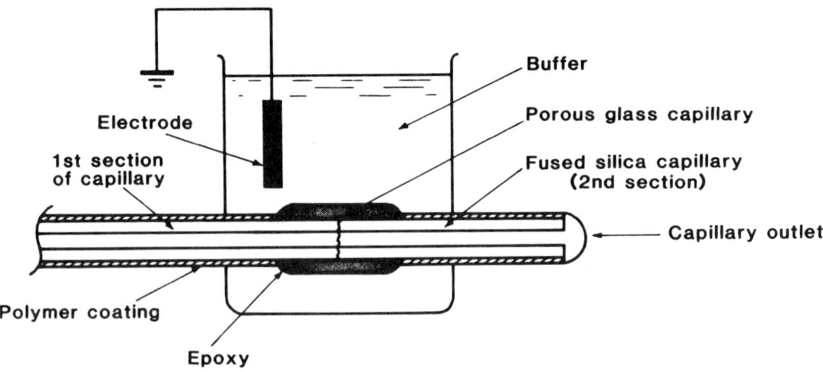

Figure 4 A schematic representation of the porous glass joint assembly.

into an automated CE apparatus (Fig. 5). The other features of the apparatus included an autosampler, a capillary column holder, an adjustable column length box, an on-column detector, a fraction collector, an on-line degasing system, and protective covers. The apparatus can be either modular (as shown in Fig. 5) or a completely integrated system [24]. The usefulness of the porous glass joint assembly was also demonstrated for off-line coupling of CE and desorption mass spectrometry [25]. A recovery study using a radioactive-labeled compound showed that the leakage of analyte through the joint was negligible [25].

On-Column Glass Frit

Huang and Zare [26] preferred a frit structure made in the wall of the separation column (Fig. 6). This approach is different from the others in that the capillary has one-piece construction. A hole of about 40 μm in diameter was drilled in the capillary (75-μm id) by use of a high-power CO_2 laser equipped with a computer-controlled $x - y$ translation stage, and then filled with a mixture of Coning solder glass 7723 and fused silica powder (particle size, 1–10 μm). The powder mixture forms a frit structure after sintering at an appropriate temperature. The frit structure is fixed in a protective jacket to reduce its fragility. A hole drilled in the protective jacket allows the frit structure to be in contact with the electrically grounded buffer. They were able to make this frit structure in 50% yield. The dilution factors ranged

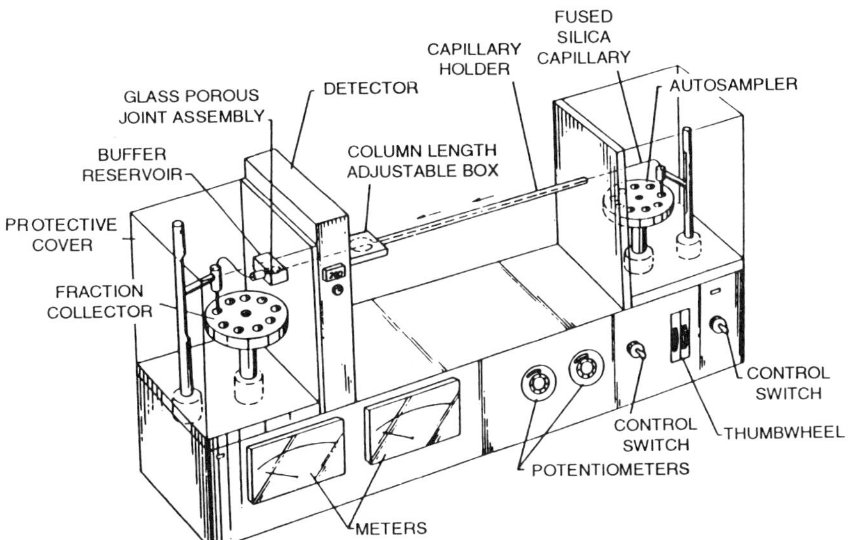

Figure 5 A schematic diagram of an automated CE apparatus. (From Ref. 11)

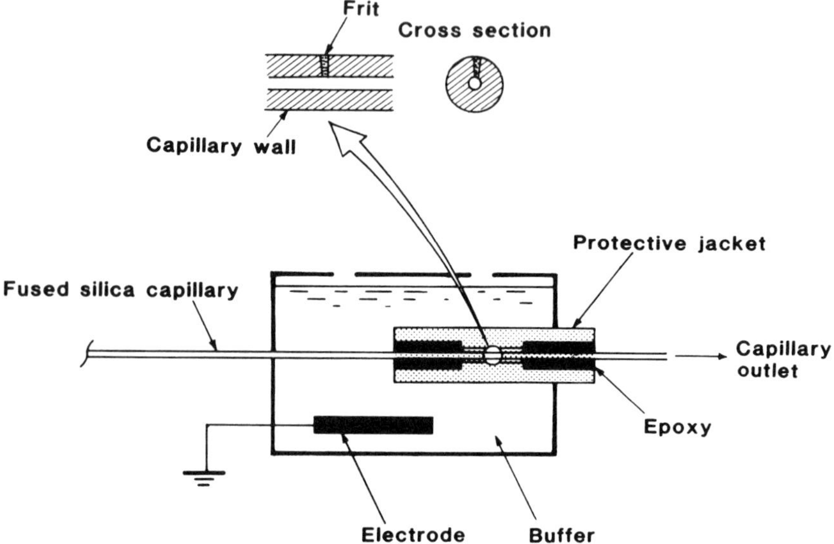

Figure 6 A schematic representation of the on-column frit structure covered with a protective jacket.

from 1:10 to 1:130, depending on the volume of the sample injected. At least 90% of an injected sample could be collected with the on-column glass frit method.

The effluent emerging from a capillary with an on-column frit was deposited in a continuous manner on a moving surface [27]. Since the separated components are retained on the surface as a record of the separation, it is possible to examine the "electropherogram" by some detection scheme. A mixture of three dansylated amino acids, a few picomoles each, was separated and deposited on a filter paper that was placed on the periphery of a rotating plastic disk. The fluorescent spots corresponding to the individual components could be observed under UV light.

Micro-Tee Device

Fujimoto et al. [28] described a method in which connection to ground is provided by the buffer-filled channel of a small id capillary that forms one branch of the tee. Details of the device are shown in Fig. 7. A short piece (about 12 mm) of PTFE tubing was chemically treated and used for supporting the three capillary tubes (i.e. the separation, collection, and grounding capillaries). The separation capillary used was either a single capillary (100-μm id) or a bundle of four capillaries (as shown in Fig. 7). Each of the three components of the tee assembly was glued into a short

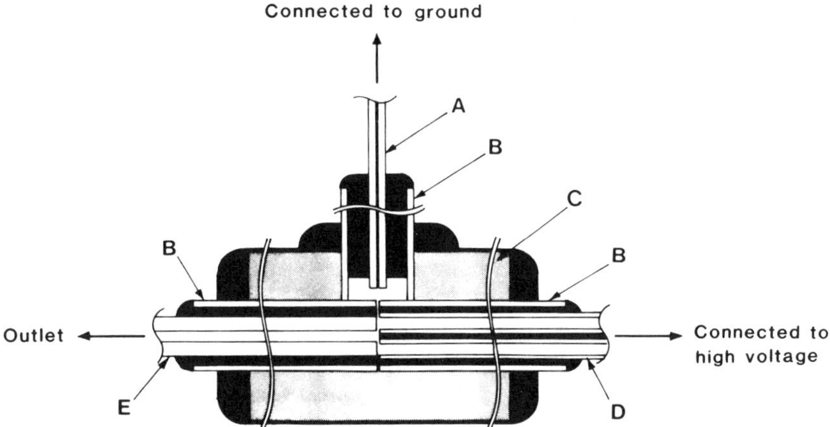

Figure 7 A schematic representation of the micro-tee device. (A) 10-μm id, 150-μm od fused silica capillary; (B) 530-μm id, 660-μm od fused silica capillary; (C) PTFE tubing; (D) 100-μm id, 200-μm od fused silica capillaries; (E) 100-μm id, 375-μm od fused silica capillary.

piece of fused silica tubing having an id larger than the od of each capillary. A vertical hole was drilled at the midpoint of the PTFE tubing, into which the smaller-id capillary (10 μm) was inserted and secured with epoxy. The separation and collection capillaries were then forced into the PTFE tube so that their ends touched each other in the center of the PTFE tubing. To make the assembly more rigid, glue was put on their exposed interfaces. The 10-μm id capillary was filled with buffer and dipped into an electrically grounded buffer-containing reservoir. Upon the application of an electrical field, the grounding process was initiated through the narrow-id capillary, the end of which was positioned at the junction of the other two capillaries. The bulk flow (and most of the analyte) moved to the collection capillary because of the extremely high resistance to flow in the direction of the grounded buffer-containing reservoir. The quantity of the solute that may be lost through the narrow capillary should depend on the ratio of the flow resistance in the two branches.

Guzman et al. [29] used a sodium silicate link-tube to merge multiple capillaries into a single capillary (Fig. 8). A tee assembly was constructed by the method of Fujimoto et al. [28], with minor modifications and located 24 cm after the coupling device (Fig. 9). In this case, two 75-μm id capillaries, rather than several capillaries and one capillary, meet a third capillary (10-μm id) in the tee assembly. Ultraviolet or fluorescence monitoring systems were placed halfway between the coupling

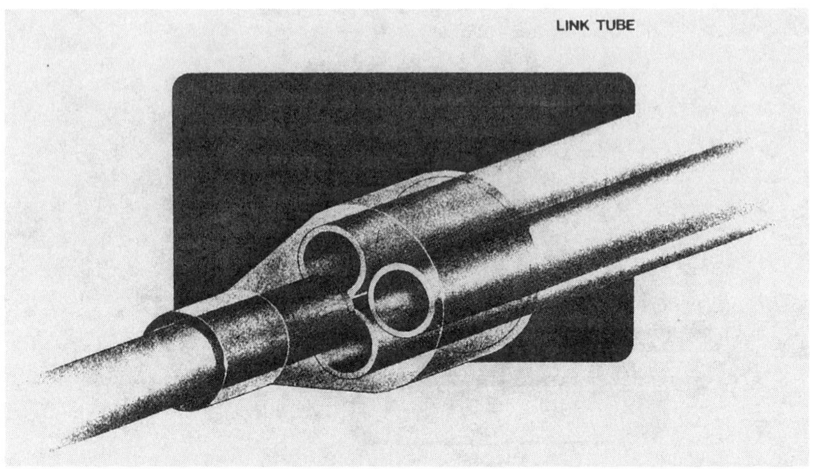

Figure 8 A schematic representation of the link tube. The diagram depicts three inlet capillaries fused into one outlet capillary.

device and the tee assembly. The results, obtained using one to five capillaries with the tee assembly for CE separations of monoclonal antibody samples, showed that the detector response increased with the number of capillaries.

The micro-tee assembly was also used to collect urinary components [29]. A bundle of five capillaries or an analyte concentrator was used to increase the sample load to the CE system. The latter method involved the use of an antibody covalently bound to glass beads that were held inside a portion of a fused silica capillary column. The recovery yield was at least 70% for the capillary bundle method, whereas it varied from 20 to 65% for the analyte concentrator method, depending on many factors associated with its preparation process. Both methods allowed production of more than 1 μg of pure urinary constituents in 1 day. The purified compounds were further analyzed by mass spectrometry.

Polyacrylamide Gel Joint

More recently, Fujimoto [unpublished results] examined another continuous fraction collection method. A short section of polyimide coating was removed and glued to a Plexiglas mount (Fig. 10). The area on which a small scratch had been made was then pushed slowly until a fracture was produced. Subsequently, the region of the fracture structure was surrounded by a certain thickness of polyacrylamide gel. The polyacrylamide gel joint was placed in a buffer reservoir that was connected to a high-voltage power supply. It was found that the construction of the gel joint was

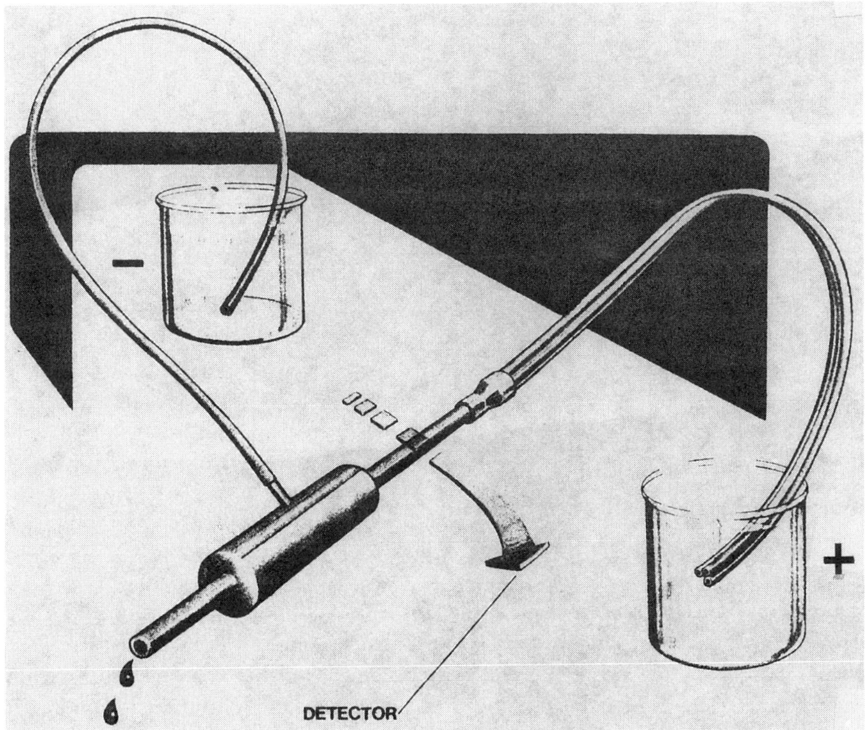

Figure 9 A schematic representation of the continuous fraction collection system equipped with the micro-tee device (see Fig. 7) and the link tube (see Fig. 8).

very easy. Preliminary experiments showed that stable electroosmosis and high recovery (>96%) were accomplished with this method.

SUMMARY

Several attempts have been made to collect nanogram to microgram amounts of substances separated by CE. It appears that completing an electrical circuit before the capillary outlet for fraction collection is more promising than with the collection methods involving interruption of applied voltage. To achieve this, four types of electrical connectors have been developed: some of the connectors are at a preliminary stage of development. In addition to the electrical connection ability, the recovery of analyte, the reproducibility of elution time, the ease of construction, the mechanical durability, and the connector contribution to extracolumn zone broadening, all are important practical considerations for a "good" electrical connector. At the

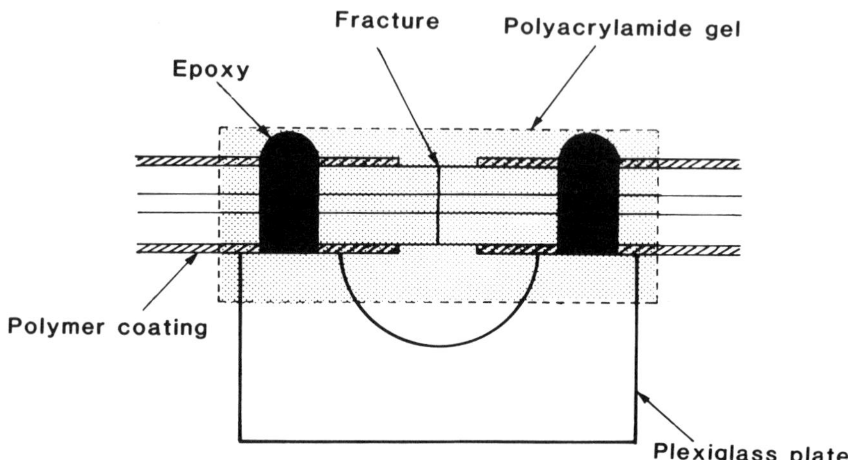

Figure 10 A schematic representation of the polyacrylamide gel joint.

moment, it is difficult to recommend the best method for electrical connection. We forecast that a micropreparative CE system based on one of the methods described in this chapter or any other method will become available commercially within several years. When it has been realized, some analyte concentration technique or multiple parallel capillaries will be a powerful option.

REFERENCES

1. E. Grushka and R. M. McCormic, *J. Chromatogr.*, *471*:421 (1989).
2. X. Huang, W. F. Coleman, and R. N. Zare, *J. Chromatogr.*, *480*:95 (1989).
3. K. Otsuka and S. Terabe, *J. Chromatogr.*, *480*:91 (1989).
4. F. E. P. Mikker, F. M. Everaerts, and P. E. M. Verheggen, *J. Chromatogr.*, *169*:1 (1979).
5. J. Vacik, *Electrophoresis: A Survey of Techniques and Applications, Part A: Techniques* (Z. Dyle, ed.), Elsevier, Amsterdam, p. 23 (1979).
6. B. L. Karger, A. S. Cohen, and A. Guttman, *J. Chromatogr.*, *492*:585 (1989).
7. J. H. Knox, *Chromatographia*, *26*:329 (1988).
8. J. H. Knox and K. A. McCormac, *J. Liq. Chromatogr.*, *12*:2435 (1989).
9. E. Grushka, R. M. McCormic, and J. J. Kirkland, *Anal. Chem.*, *61*:241 (1989).
10. A. S. Cohen, A. Paulus, and B. L. Karger, *Chromatographia*, *24*:15 (1987).
11. N. A. Guzman and F. E. Regnier, First Symposium of the Protein Society. San Diego, California, Aug. 9-13, Abstr. 823 (1987).
12. N. A. Guzman, M. A. Trebilcock, and J. P. Advis, *J. Liq. Chromatogr.*, *14*:997 (1991).
13. D. J. Rose and J. W. Jorgenson, *J. Chromatogr.*, *438*:23 (1988).

14. R. I. Hecht, J. F. Coleman, Jr., J. C. Morris, F. S. Stover, and C. Demarest, *Prep. Biochem.*, *19*:363 (1989).
15. Y. C. Yang, A. B. Ciarietta, M. P. Chung, S. Kovaic, J. S. Witek-Giannotti, A. C. Leary, R. Kris, R. E. Donahue, G. G. Wong, and S. C. Clark, *Cell*, *47*:3 (1986).
16. T. Akiyama and A. Omori, Eleventh Symposium on Capillary Electrophoresis. Tokyo, Japan, Dec. 12-13, Abstr. 39 (1991).
17. A. S. Cohen, D. R. Najarian, A. Paulus, A. Guttman, J. A. Smith, and B. L. Karger, *Proc. Natl. Acad. Sci. USA*, *85*:9660 (1988).
18. A. Guttman, A. Paulus, A. S. Cohen, and B. L. Karger, "Electrophoresis '88," Proceedings of the Sixth Meeting of the International Electrophoresis Society (C. Schaefer-Nielsen, ed.), VCH Publishers, New York, p. 151 (1988).
19. A. Guttman, A. S. Cohen, D. N. Heiger, and B. L. Karger, *Anal. Chem.*, *62*:137 (1990).
20. R. A. Wallingford and A. G. Ewing, *Anal. Chem.*, *59*:1762 (1987).
21. R. A. Wallingford and A. G. Ewing, *Anal. Chem.*, *60*:258 (1988).
22. N. A. Guzman and L. Hernandez, *Techniques in Protein Chemistry* (T. E. Hugli, ed.), Academic Press, San Diego, p. 456 (1989).
23. N. A. Guzman, L. Hernandez, and B. G. Hoebel, *Biopharmacology*, *2*:22 (1989).
24. N. A. Guzman, L. Hernandez, and S. Terabe, *Analytical Biochemistry: Capillary Electrophoresis and Chromatography* (*ACS Symp. Ser. 434*) (C. Horváth and J. G. Nikelly, eds.), American Chemical Society, Washington, D.C., p. 1 (1989).
25. R. Takigiku, T. Keough, M. P. Lacey, and R. E. Schneider, *Rapid Commun. Mass Spectrom.*, *4*:24 (1990).
26. X. Huang and R. N. Zare, *Anal. Chem.*, *62*:443 (1990).
27. X. Huang and R. N. Zare, *J. Chromatogr.*, *516*:185 (1990).
28. C. Fujimoto, Y. Muramatsu, M. Suzuki, and K. Jinno, *J. High Resolut. Chromatogr.*, *14*:178 (1991).
29. N. A. Guzman, M. A. Trebilcock, and J. P. Advis, *Anal. Chim. Acta.*, *249*:247 (1991).

16

Mass Spectrometric Detection for Capillary Electrophoresis

Richard D. Smith and Harold R. Udseth

Pacific Northwest Laboratory
Richland, Washington

The increasing application of capillary electrophoresis (CE) has led to a growing interest in detection methods that are more sensitive, selective, or structurally informative for separated species. Mass spectrometry (MS) has long been recognized as perhaps the most selective and broadly applicable detector for analytical separations. Recent and on-going developments in MS are resulting in continuing improvements in the sensitivity, speed, level of structural detail obtainable, and flexibility of this powerful "detector." Concurrent developments involving tandem MS methods and higher-order MS methods (i.e., MS^n where $n \geq 3$) promise that even greater selectivity and structural information will be routinely achievable in the future, particularly when combined with rapidly improving ion-trapping MS methods [1]. Although MS is still one of the most complex and expensive CE detectors, costs continue to decrease and user "friendliness" to increase, and the density of information that can be gained from this combination increasingly mitigates these disadvantages.

The mass spectrometer provides the equivalent of up to several thousand "detectors" functioning in parallel and capable of providing both molecular weight (mass) information from the intact "molecular ion," or structurally related information from the ions dissociation in the mass spectrometer. The CE–MS combination is, in one sense, nearly ideal: CE is based on the differential migration of ions, whereas MS analyzes ions by their mass/charge (m/z) ratio. On the other hand, these two highly orthogonal analytical methods exploit ion motion in two quite different environments; conducting liquids and high vacuum, respectively. Obviously, a primary role of the CE–MS interface is to bridge these different environments. Fortunately, for historical reasons, the combination of liquid chromatography (LC) with MS has long been of interest, and a variety of clever LC–MS interfaces have been developed over the last 25 years [2]. However, the CE–MS combination places significantly different demands on an interface than those of LC–MS. The CE flow rates are quite low or negligible, the buffer is moderately conductive, electrical contact must be maintained with both ends of the capillary and, if the high-separation efficiencies possible with CE are to be realized, any "dead volumes" associated with detection and band broadening owing to pressure differences between the capillary termini need to be avoided or minimized. Interfaces have been developed that, to one degree or another, meet all these requirements.

In this chapter, we will review CE–MS interfacing methods, experimental considerations, and applications. We emphasize CE separations by free-solution electrophoresis [i.e., capillary zone electrophoresis (CZE)] as well as the more limited work based on capillary isotachophoresis (CITP). Although we discuss both major interfacing methods, based upon continuous-flow, fast-atom bombardment (FAB) and electrospray ionization (ESI), the greater emphasis of this review is on the latter. The FAB interfaces and their application are discussed in more detail in Chapter 17. This chapter's bias also reflects the explosive growth in ESI–MS application, the increasing availability of such instrumentation, and our belief that the ESI interface is the easier to implement, offers greater flexibility, and is capable of providing more structural information. (This emphasis is not meant to diminish the impressive results obtained with FAB interfaces, particularly for small-molecule applications for which results comparable with ESI interfaces in terms of sensitivity have been achieved.) Finally, we wish to communicate the current status of CE–MS, a highly promising combination, but still somewhat limited by imperfect control of the ESI interface, sensitivity, and the scan speed constraints of current MS technology. Since these MS limitations are almost sure to be overcome in the next few years (in our optimistic assessment), the future for CE–MS appears to hold exceptional promise.

CAPILLARY ELECTROPHORESIS–MASS SPECTROMETRY INTERFACING METHODS

Two on-line interfacing methods have been used for CE–MS, based upon continuous-flow, fast-atom bombardment and electrospray ionization introduction methods for mass spectrometry. The two methods differ in the process used for ionization (or for transfer of ions from the liquid to the gas phase) to allow the subsequent *m/z* analysis in the MS. ESI–MS interfaces require an atmospheric pressure ionization inlet and the availability of MS instruments with such inlets is rapidly increasing. However, they are generally incompatible with FAB–MS methods, which operate through conventional probe inlets and produce ions under vacuum. There have been two variations on both the FAB and ESI interfaces reported, based upon "liquid junction" and "sheath flow" (or "coaxial") approaches. These variations are differentiated by the methods of establishing electrical contact at the capillary terminus and for providing the simultaneous introduction of "makeup" liquid generally needed to support the ionization method, and are discussed in later sections. In addition to these approaches, off-line methods, involving the collection of CE effluents and subsequent MS analysis, unconstrained by interfacing methods, have been reported and are briefly discussed in the following section.

Off-Line Methods

The advantages of off-line CE–MS include the flexibility of using any available ionization method or MS. The primary disadvantages of off-line methods include limited resolution (owing to practical constraints upon sample collection), reduced sensitivity, and the need for extra sample handling. The latter is particularly problematic because of the small sample sizes and injection volumes for CZE ($\lesssim 10$ nL). Consequently, the earliest reported off-line CE–MS studies involved isotachophoresis for which sample sizes are generally much larger than those encountered with CZE. One of the earliest reports is from Kenndler and Kaniansky [3] and demonstrated the identification of pesticides using conventional electron ionization (EI)–MS. Later work by Kenndler and Haidl [4] demonstrated the potential for quantitative application for amenable compounds, such as hydrogenation products of quaternary ammonium salts. A major limitation of EI methods is that ionization occurs in the gas phase, and the general problem of getting thermally labile or nonvolatile analytes into the gas phase is not solved.

A much more powerful approach for off-line CZE–MS has recently been described by Takigiku et al. [5] that used plasma desorption (PD)–MS. This method allows analysis of nonvolatile compounds and has been successfully applied to proteins with average relative molecular masses (M_r) as large as 45 kDa [6]. These workers used a porous glass joint, similar to that first described by Ewing and coworkers [7], and functionally similar to the liquid junction-interfacing methods described later. Electrical contact with the capillary terminus was established through the porous glass capillary, and the eluents were collected (after flowing through another 0.5-cm–long capillary segment) on nitrocellulose-coated aluminum foil. Samples were then lyophilized, redissolved in compatible solution, and again deposited on nitrocellulose-coated foils and washed with deionized water before analysis [5]. Picomole level sensitivities were obtained for several peptides and proteins, and improved sensitivity and M_r range can be anticipated using newer matrix-assisted laser desorption methods [8]. However, some buffer systems are incompatible with PD–MS, and off-line methods reported thus far remain constrained owing to the limited number of fractions generally collected, small sample sizes, and extensive sample manipulation.

Continuous Flow, Fast-Atom Bombardment Interfaces

Fast-atom bombardment ionization is the most widely used MS method for analysis of labile and nonvolatile compounds and, as such, provides a basis for CE–MS. The liquid junction and coaxial sheath flow variations upon CE–FAB–MS interfacing have been reported, each imposing its own constraints.

The liquid junction approach couples the analytical capillary with an additional length of capillary used for transfering analytes in a makeup flow to the FAB "tip," at which ionization occurs. A variety of liquid–junction FAB interfaces have been reported [9–18]. A variation on this interface for the higher-flow rates of combined simultaneous capillary chromatography and electrophoresis (termed *psuedoelectrochromatography*) has been described [19]. The FAB detection method is most appropriate for detection of smaller-molecular weight species than the ESI interfaces discussed in the next section. With the exception of the report by Wolfe et al. [15], all CE–FAB–MS has employed multisector (magnetic–electric) MS instruments, rather than the quadrupole instruments that have dominated ESI work.

The FAB method generally requires a viscous liquid component, and 5–25% glycerol in a variety of solvent systems is commonly used with the liquid–junction interface. Generally, 50- to 100-μm capillaries are used for CZE, and the small CZE flow rate (generally <0.1 μL min^{-1}) is much smaller than the liquid flow introduced through the junction (~5–10 μL min^{-1}). The long transfer capillary (~1 m) results in some degradation of CZE separation efficiency, and about an order of magnitude decrease in the number of theoretical plates for FAB–MS detection has been noted, compared with detection before the liquid junction [16,17]. Suter and Caprioli have also noted loss of separation efficiency as well as reduced sensitivity compared with sheath flow FAB interface [13]. They note, however, that the liquid–junction interface is easier to handle and to set up.

A sheath flow or *coaxial* continuous flow FAB interface for CE–MS was developed by Deterding, Moseley, Tomer, and co-workers [20–26]. The sheath flow arrangement here is restricted to relatively small-diameter CE capillaries (10- to 15-μm id) for conventional low-viscosity buffers, because of the low pressure at the capillary terminus and the desire to avoid a large component of pressure-driven flow in the capillary. Thus, sample injection volumes are quite small (often <1 nL). The CZE flow rates are extremely small, and smaller sheath flows (~0.5–2 μL min^{-1}) containing up to 25% glycerol have been used. This results in improved detection limits compared with the liquid junction approach. Although absolute detection limits are lower, and perhaps better suited for extremely limited sample volumes, the smaller-injection volumes result in higher-detection limits when expressed in terms of sample concentration. Thus, dynamic range is often more limited. The sheath flow arrangement appears to preserve CZE separation efficiency better than the liquid junction arrangement, not surprising owing to the long transfer capillary used in the latter. Indeed, Moseley et al. have obtained up to 4×10^5 theoretical plates using 10–100 fmol of sample injection (and somewhat larger samples for tandem MS) [19]. Suter and Caprioli have compared the sheath flow and liquid junction interfaces [13], finding the former more difficult to operate, and the latter less sensitive. They note that a significant loss in resolution is experienced with both FAB interfaces, but also note that this restriction is now of limited practical significance owing to the trade-offs in MS detection. The reader is referred to Chapter 17 for a further discussion of CE–FAB–MS methods and applications.

Electrospray Ionization Interfaces

Although the choice of interface will ultimately be governed by the MS instrumentation available, the clear trend now favors ESI methods. The ESI method is well-suited for CE–MS interfacing, since it produces ions directly from liquid solutions at atmospheric pressure. Considerations for interfacing generally derive from some limitations upon buffer composition for direct ESI and the desire to position the ESI source (i.e., the point of charged droplet formation) as close as possible to the analytical capillary terminus, avoiding lengthy transfer lines.

The ESI source is at elevated voltage of 2–5 kV relative to an orifice, into which ions are entrained by a flow of gas entering the mass spectrometer [27]. Many of the practical constraints and considerations for CE–MS interfacing derive from the voltage bias between these components. Some ESI sources allow the liquid effluent to be at ground potential, whereas others require the capillary terminus to be operated at 3–5 kV relative to ground potential. The ESI source requires production of a high electric field, which causes charge to accumulate on the liquid surface at the capillary terminus, thereby disrupting the liquid surface. The ESI liquid nebulization process can be pneumatically assisted, using a high-velocity annular flow of gas at the capillary terminus, and it sometimes is referred to as *ion spray* [28], an approach originally described for ESI by Dole and co-workers [29].

As with the CE–FAB–MS combination, both liquid junction and sheath flow variations of the ESI interface have been reported. The first CZE–MS was based on an electrospray ionization interface developed at our laboratory [30,31]. This initial approach used a metalized capillary terminus to make contact with the CZE eluent under conditions of high electroosmotic flow. We later reported an interface using a flowing sheath liquid [32] and also demonstrated the combination of capillary isotachophoresis (CITP) with MS [33,34] and CZE–MS/MS [34,35]. More recently, other researchers have reported similar interfaces [36–48]. The liquid junction variation was developed by Henion and co-workers [49,50], who have reported extensively upon its applications [51–57], and more recently by Pleasance et al. [37], who have compared its performance with the sheath flow approach.

Figure 1 shows a recent version of the liquid junction interface [55]. Electrical contact with the capillary terminus is established through a liquid reservoir that surrounds the junction of the analytical capillary and a transfer capillary. The gap between the two capillaries is typically adjusted to 10–20 μm, a compromise resulting from the need for sufficient makeup liquid being drawn into the transfer capillary, while avoiding analyte loss by diffusion into the reservoir. The flow of makeup liquid arises from a combination of gravity-driven flow, owing to the height of the makeup reservoir, and flow induced in the transfer capillary by a mild vacuum generated by the Venturi effect of the nebulizing gas used at the ESI source [37].

In the sheath flow (or coaxial) electrospray interface an organic liquid (typically pure methanol, methoxyethanol, or acetonitrile, but frequently augmented by as much as 10–20% formic acid, acetic acid, water, or other reagents), flows through the annular space between the ~200-μm od CE capillary and a fused silica or stainless steel capillary (≳250-μm id). A detailed view of a recent version of this interface developed at our laboratory is shown in Fig. 2. In this design, a 300-μm id fused silica capillary surrounds the smaller-diameter electrophoresis capillary. For

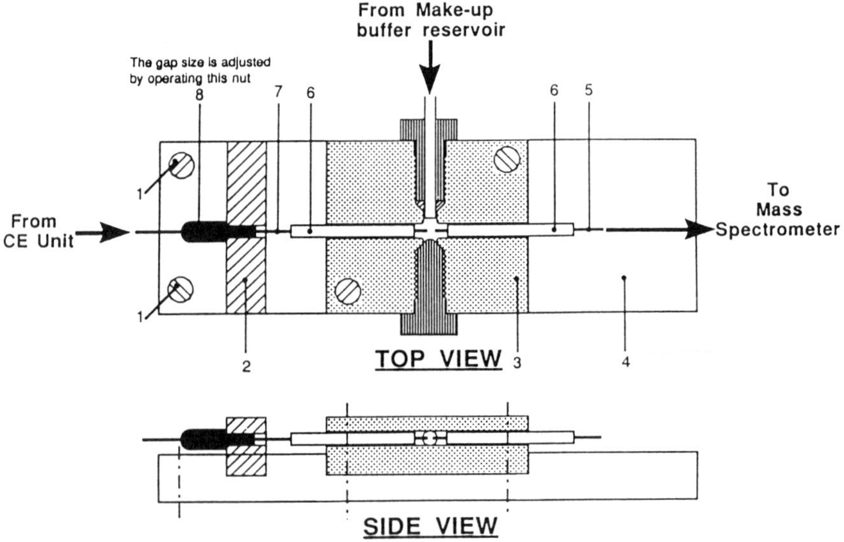

Figure 1 Pneumatically assisted electrospray ionization liquid junction interface for CE–MS developed by Henion and co-workers. (1, mounting screws (nylon); 2, capillary holder (plexiglass); 3, liquid-junction block (plexiglass); 4, support (plexiglass) mounted on a X-Y-Z positioner; 5, transfer line: 75 μm × 375 μm; 6, capillary guides (PEEK tubing cut out of Upchurch capillary sleeves); 7, CE column: 75 μm × 375 μm; 8, Micro-Union.) (From Ref. 55)

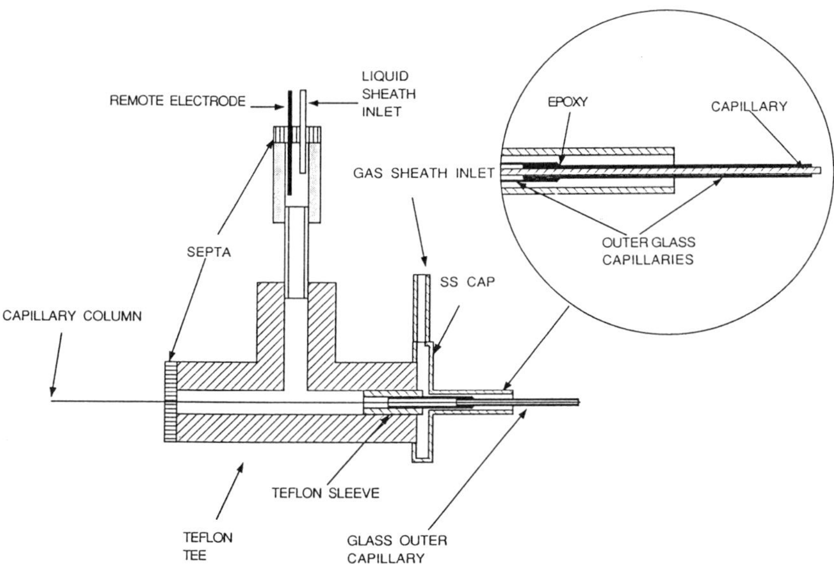

Figure 2 One version of a sheath flow (or coaxial) electrospray ionization interface for CE–MS developed at the authors' laboratory. (From Ref. 58)

positive-ion ESI, a voltage of +3 to +5 kV is applied indirectly to the sheath liquid from which the fused silica tube protrudes approximately 0.2 mm. A syringe pump delivers flow of sheath liquids at 2–5 μL min^{-1}. Potential problems from the formation of bubbles in the connecting lines are minimized by the inclusion of a trapping volume. Enhanced stability can sometimes be obtained by degasing the organic solvents used in the sheath and by minimizing heating (e.g., from the ion source or a countercurrent gas flow). Cooling of the ESI source or the use of less volatile sheath solvents, such as methoxyethanol [46], have also been useful in some situations. A schematic representation of a tandem (triple) quadrupole mass spectrometer coupled with a modified commercial CE instrument is shown in Fig. 3 [58]. In addition to those noted earlier, several reports from our laboratory on the use of the sheath flow approach are now in the literature [58–64].

It appears that most of the distinctions between the sheath flow and liquid junction ESI interfaces are purely mechanical; the performance in terms of spectral quality is similar, and most other MS-related considerations are quite general. However, some differences have been noted. Thibault et al. have compared the "liquid junction" variation of Henion and co-workers with the sheath flow interface (using a pneumatically assisted ESI interface), and have noted that the latter provided a "more robust and reproducible interface with the added advantage of offering an

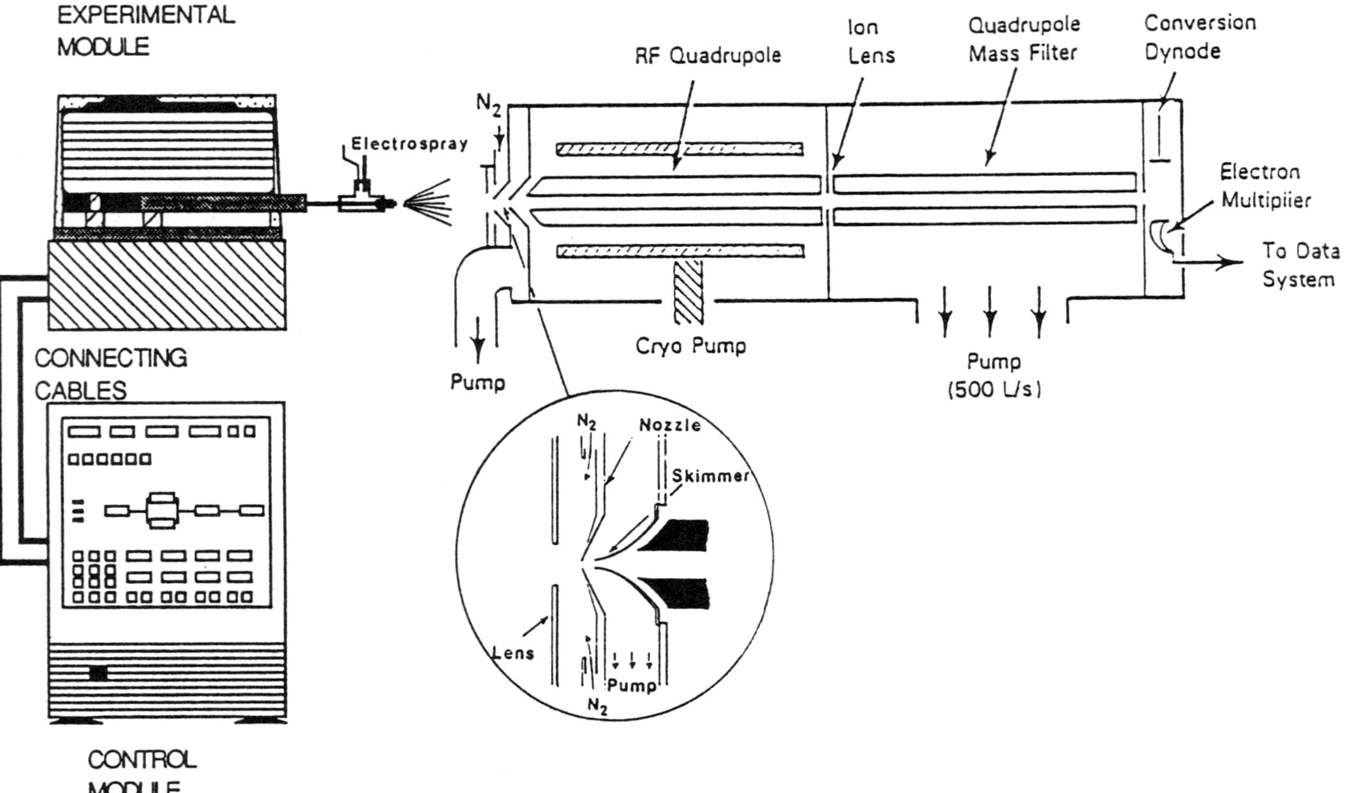

Figure 3 Overall schematic representation of instrumentation for combined CE–MS showing the separated experimental and control modules assembled from commercially available CE instrumentation (Beckman Instruments P/ACE 2000) at the authors' laboratory. Major physical concerns derive from the desire for positioning close to the ESI source to minimize capillary length, as well as to eliminate laminar flow caused by vertical displacement of the buffer reservoir.

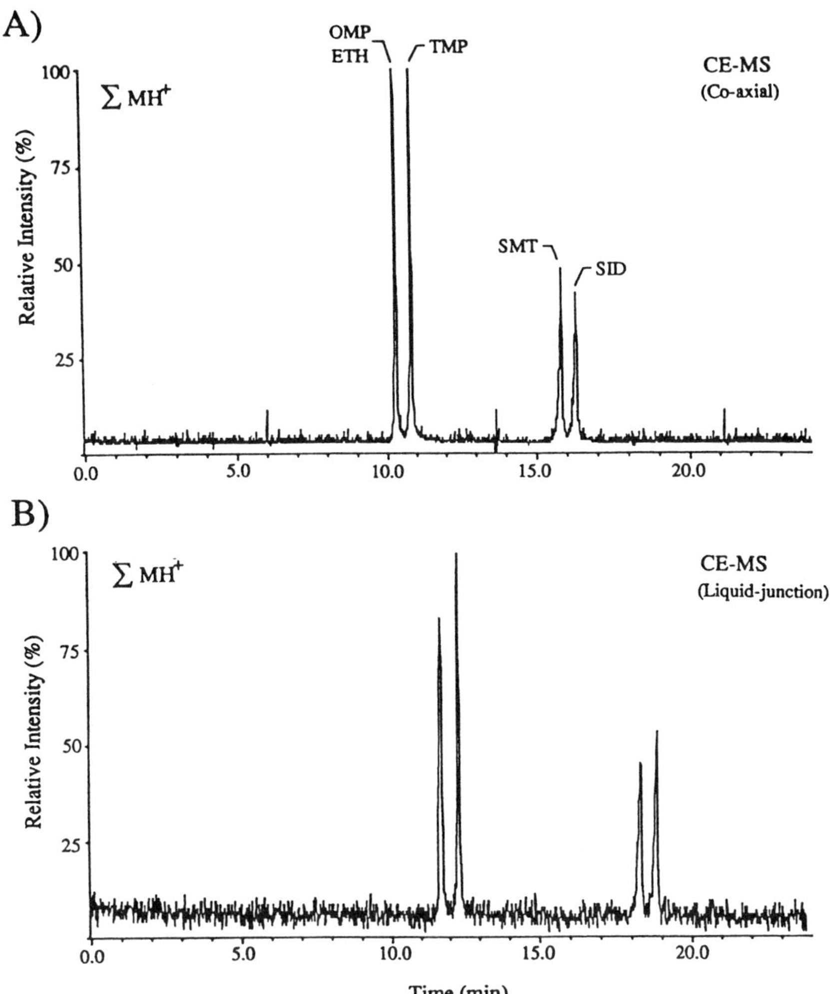

Figure 4 Comparison of CE–MS separations of antibacterial drugs obtained with the (A) sheath flow (coaxial) and (B) liquid junction ESI–MS interfaces, reported by Pleasance et al. [37]. Improved sensitivity was obtained with the sheath interface. The difference in separation times noted for the two interfaces was due to modification of the junction liquid during capillary "flushing" between separations.

independent means of calibration and quantitation . . . through the sheath" [41]. Figure 4 shows a comparison for a separation of marine toxins, during which care was taken to obtain comparable conditions [37]. Improved signal/noise ratios and slightly improved separations were obtained with the sheath flow interface. The difference in elution times evident in Fig. 4 was attributed to modification of the liquid junction buffer caused by flushing of the CE capillary between analyses.

The dependence of ESI ion current upon solution conductivity is relatively weak, generally 0.1–0.3 μA at atmospheric pressure, and about 10–200 pA integrated ion current (i.e., the sum of all ions) is actually focused into the MS and detected. The ESI currents for a typical water/methanol/5% acetic acid solution are in the range of 0.1–0.3 μA, only a small portion of which ($<10^{-3}$) is generally transmitted for MS analysis. An electron scavenger is often used to inhibit electrical discharge at the capillary terminus, particularly for ESI of aqueous solutions. Sulfur hexafluoride has also proved particularly useful for suppressing corona discharges and improving the stability of negative-ion ESI. Introduction of this gas is most effectively accomplished using a gas flow (~100–250 mL min^{-1}) through an annular volume surrounding the sheath liquid (see Fig. 2).

The formation of molecular ions by ESI requires conditions affecting solvent evaporation from the initial droplet population. Droplets must shrink to the point at which repulsive coulombic forces approach the level of droplet cohesive forces (e.g., surface tension). This evaporation can be accomplished at atmospheric pressure by a countercurrent flow of dry gas at moderate temperatures (~80°C), by heating during ion transport through the sampling region, or by (particularly in the case of ion-trapping MS methods) energetic collisions at relatively low pressure.

An important attribute of a CE–MS interface based upon ESI is the efficiency of sampling and transport of ions into the mass spectrometer. The simplest commercially available instrument uses a 100- to 130-μm pinhole sampling orifice to a vacuum region maintained by a single stage of high-speed cryopumping (Sciex, Thornhill, Ontario, Canada). Charged droplets formed by ESI drift against a countercurrent flow of dry N_2, which serves to speed desolvation and exclude high *m/z* residual particles and solvent vapor. As the ions pass through the orifice into the vacuum region, further desolvation is accomplished as the gas density decreases from collisions as the ions are accelerated by the ion optics of the mass spectrometer.

Alternatively, many instruments have been developed based on differentially pumped interfaces. The ESI capillary inlet–skimmer interface developed by Fenn and co-workers [65] is widely used and is commercially available in versions compatible with many mass spectrometers (Analytica of Branford, Connecticut). A countercurrent flow of N_2 bath gas assists solvent evaporation, similar to the approach described above. Ions migrate toward the sampling orifice where some small fraction is entrained in the gas flow entering a glass capillary (metalized at both ends to establish well-defined electric fields). Although some loss occurs, ions are transmitted through capillaries with high efficiency. Ions emerge from the capillary, in the first differentially pumped stage of the MS (~1 torr), as a component of a free-jet expansion. A fraction of the ions is then transmitted through a "skimmer" and additional ion optics into the *m/z* analyzer (i.e., the MS). The electrically insulating nature of the glass capillary provides considerable flexibility because ion transmission does not depend strongly on the voltage gradient between the conducting ends of the capillary. This approach has the advantage that the CE capillary terminus can be at ground potential, simplifying current measurements and voltage manipulation during injection.

An alternative approach to droplet desolvation for ESI relies solely on heating during droplet transport through a heated metal [66,67] or glass capillary [65,68]. A countercurrent gas flow is not essential and, in fact, may decrease obtainable ion currents. The electrospray source can be closely positioned to the sampling capillary orifice (typically 0.3–1.0 cm), since the larger spacing is not used for desolvation, resulting in more efficient charge transport. The charged droplets from the ESI source are swept into the heated capillary, which heats the gas sufficiently to provide effective ion desolvation, particularly when augmented by a voltage gradient in the capillary–skimmer region. The disadvantage of this approach, and that of all sources not using a gas curtain or countercurrent flow, is that much more solvent and residual material (i.e., solute particles) enter the mass spectrometer, resulting in the need for a much more frequent disassembly and cleaning. Advantages of the heated capillary, however, are the ease with which it can be adapted to a variety of MS configurations, and the ability to "heat" ions for either desolvation or dissociation [67] in a manner largely independent of *m/z*.

Electrospray ionization is an extremely "soft" ionization technique and will yield, under appropriate conditions, intact molecular ions. Molecular mass measurements for large biopolymers that exceed the MS *m/z* range can be obtained because ESI mass spectra generally consist of a distribution of molecular ion charge states, without contributions from dissociation (unless induced during transport into the MS vacuum). The envelope of charge states for proteins, arising generally from protonation for positive-ion ESI, yields a distinctive pattern of peaks owing to the discrete nature of the electronic charge; that is, adjacent peaks vary by addition or subtraction of one charge. For noncovalently bonded species, such as multimeric proteins, mass spectra typically show only the individual subunits.

As applications for CE have developed, the advantages of CE combined with MS have been increasingly recognized and the merits of various interfaces have been identified. When compared with FAB, the ESI methods appear to offer advantages, in most instances, because of better sensitivity, reduced background, and interface designs that do not require long transfer lines or incur a pressure drop across the capillary. Perhaps the most significant advantage of the ESI interface to CE is its applicability to higher-molecular-mass compounds, impractical by CF–FAB.

EXPERIMENTAL METHODS AND CONSIDERATIONS

Separation Conditions

In general, all the concerns in CE relevant to sample injection, buffer composition, capillary surface interactions, and separation efficiency apply to CE–MS, and the reader is referred to other chapters in this volume. Most CE–MS has been with 50- to 100-μm id diameter capillaries, representing a compromise related to a number of factors. Both the liquid junction and sheath flow ESI interfaces allow essentially any CE buffer to be successfully electrosprayed [27]. Since both the liquid junction and sheath flow ESI use an effective dilution of the low CE elution flow by a much larger volume of liquid, considerable flexibility exists, despite some constraints upon the composition of samples, which can be successfully addressed (high salt and surfactant concentrations are problematic). Buffer concentrations of at least 0.1 M can be used for CE–MS owing to the dilution step inherent in present interfaces. However, because of practical sensitivity constraints, CE buffer concentrations are generally minimized, and it has also been found that sensitivity can vary significantly with buffer composition. In general, the best sensitivity is obtained with the use of volatile buffer components, such as acetic acid at the lowest practical concentration, and by minimizing nonvolatile components. In addition, buffer components that interact strongly with the sample (e.g., denaturants) substantially degrade sensitivity. In particular, the use of surfactants, sodium dodecyl sulfate (SDS), for example, gives rise to intense signals in both positive- and negative-ion ESI, presenting a major barrier for capillary electrokinetic chromatography–MS applications.

The MS sensitivity, which is ultimately limited by the ability to efficiently analyze ions produced at atmospheric pressure by ESI, is probably the most important factor related to MS detection and will be discussed at more length in the two following sections. Sensitivity does not appear to depend substantially on the mechanical details of the electrospray interface, although such factors can influence ESI stability and ease of operation.

Increasingly, CE–MS instrumentation benefits from the features of the automated CE instruments, including precise electrokinetic or hydrostatic injection, on-column spectrophotometric detection, and capillary temperature control [58]. The advantages of computer-controlled injection are significant for both accuracy and precision, as well as freedom from artifacts arising from less reliable manual injection methods [64]. Temperature control has also been recognized as important in obtaining good reproducibility. Gas generated in either the CE capillary or the ESI sheath electrode (which often consists of more volatile organic solvent) can create a region of high resistance and effectively terminate a separation. The use of a less volatile sheath liquid, such as methoxyethanol, has been found attractive for reducing such difficulties and may even have sensitivity advantages, depending on interface design (e.g., desolvation conditions).

Proper CE capillary surface coatings and conditioning are essential for many applications. As an example, our initial CZE–MS of proteins in acidic buffers (pH 3–5) with uncoated capillaries produced poor separations for proteins such as myoglobin and cytochrome *c* owing to interactions with capillary surfaces [61]. However, excellent separations of myoglobin mixtures have been obtained at higher pH (>8) with uncoated capillaries owing to the net negative charge of the protein, and reasonable detection limits (<100 fmol) obtained using multiple ion monitoring in 10 mM Tris [58]. In this example, methanol–water sheath liquid, containing 5–10% acetic acid, was used to obtain the acidic conditions for ESI–MS. However, the Tris buffer component, and perhaps limitations upon mixing of the CE and sheath flows, resulted in reduced sensitivity, compared with that obtained for the same separations conducted in volatile acidic buffers [58]. Recently, Thibault et al. have demonstrated that protein separations conducted in acidic solution (e.g., 10 mM acetic acid, pH 3.4) using amino-based coated capillaries results in substantially improved sensitivity, and allows useful full-scan mass spectra to be obtained on similar amounts of analyte [41]. Similar CE–MS results have been demonstrated by others [45,46].

Mass Spectrometric Considerations

Since ESI takes place at atmospheric pressure, the pressure drop across the CE capillary can be made precisely zero, and ideally no degradation of separation efficiency should arise because of laminar flow. The ESI method is expected to be amenable to essentially any charged species in solution, but this in itself does not guarantee success. Gas-phase ions formed must be substantially desolvated, free of adducted buffer-related species, and must have an m/z within the capabilities of the MS. The flow rate necessary to sustain a stable ESI source from a capillary of conventional dimensions is generally 1–10 $\mu L\ min^{-1}$, but can be higher or lower, depending on source tip design, the use of nebulizing gas, and so on. A crucial point is that CE flow rates are generally much less than desired for ESI, and the buffers used are often substantially more conductive. Thus, both the sheath flow and liquid junction interfaces introduce a makeup flow of a less conductive (and generally more volatile) solution, which serves to dilute the CE buffer and assist ESI. The result of this additional flow is that a wide range of CE buffer compositions and solvents can be effectively addressed, but with some constraints owing to the volatility and any inherent ESI response (i.e., background) caused by buffer components or their adduction with other solution species.

The major considerations relevant to MS detection are generally caused by the nature and complexity of the particular sample, and pragmatic constraints because of MS detection sensitivity, resolution, and related scan-speed compromises. For quadrupole mass spectrometers, single or multiple ion monitoring, sometimes referred to as selected ion monitoring (SIM), leads to significantly enhanced detection limits compared with scanning operation relying on the greater dwell times at specific *m/z* values. With samples for which analyte molecular masses are known and *m/z* values can be predicted, SIM detection is an obvious choice. If sufficient sample is available, a direct infusion experiment can be used to produce a mass spectrum of the unseparated mixture and the results used to select a suite of *m/z* values for SIM detection. This approach can be problematic, however, since relative ionization efficiencies can be different for direct infusion of the mixture compared with those of the separated species, owing to the competition for charge that occurs in ESI. Thus, important components may be overlooked during the direct introduction of mixtures. One of the advantages of CE–MS is that mixture components strongly discriminated against in such direct infusion experiments often show much more uniform response after CE separation.

The realization of high-efficiency CE–MS separations, leading to peak widths of only a few seconds or less, presents challenges for most mass spectrometers. The use of full-scan detection with quadrupole mass spectrometers often results in either limited signal intensities (because of the extremely short dwell time at each *m/z*) or too few scans obtained during peak elution. One option is to reduce MS resolution, which increases signal intensities and allows faster scan speeds. This is not an ideal solution, since mass measurement accuracy and other valuable information is often lost and chemical noise from background signals may mitigate any significant gain. Consequently, obtaining the maximum number of theoretical plates possible with combined CE–MS has not been a major concern. Thus, longer separations and larger injection volumes than are optimum, and yielding broader peaks, are often used. Similar but magnified constraints arise for the use of tandem MS methods for obtaining structural information, as in polypeptide sequencing. These limitations are a major-driving force for the implementation of improved MS instrumentation using array detection or ion-trapping methods (see later section on future developments).

CAPILLARY ZONE ELECTROPHORESIS–MASS SPECTROMETRY APPLICATIONS

To this point, much of the reported CE–MS has been concerned with investigating and demonstrating the methodology and comparing it with other techniques (particularly microcolumn LC–MS). Since CE–MS is a less familiar technique to the analytical chemist, the first wave of reports are generally those arising from laboratories more interested in new methods and instrumental approaches, as opposed to routine applications or specific analytical problems. Gradually, however, more real problems are being examined (the second wave) and, as CE becomes more widespread, such reports should increase greatly.

In this section we discuss the range of CE–MS examples and applications reported to date. We arbitrarily divide this discussion into two sections based upon molecular size.

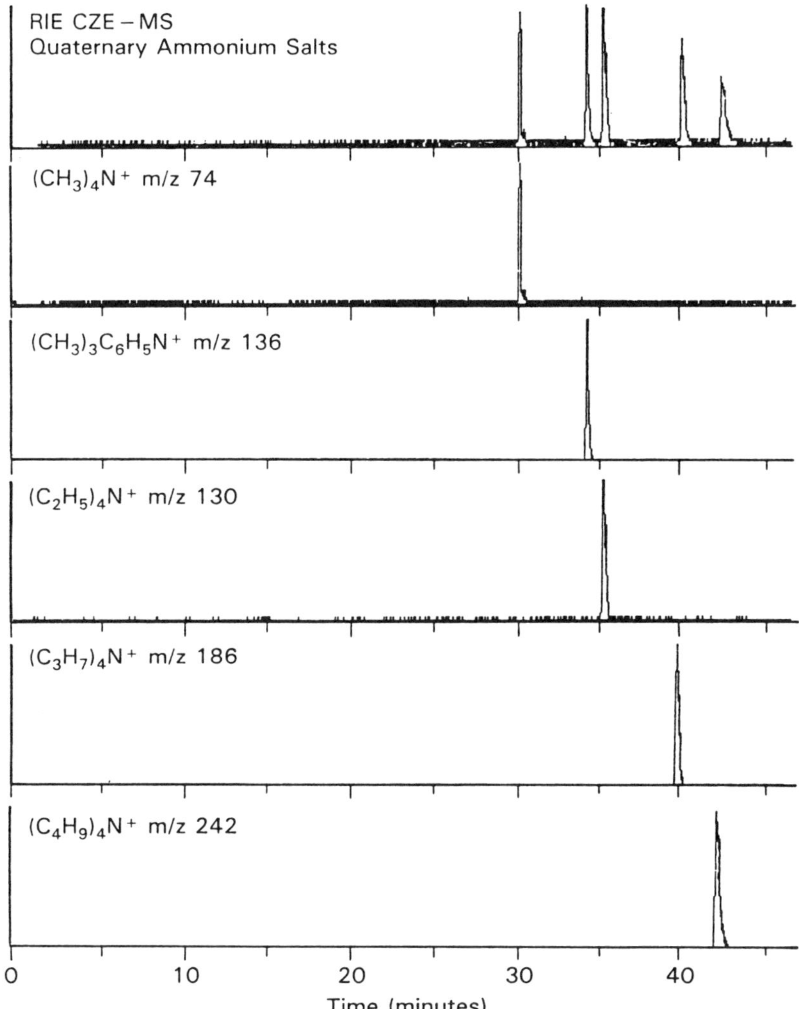

Figure 5 Total and reconstructed single ion electropherograms for CZE–MS of a mixture of five quaternary ammonium salts. The separation was conducted on a 2-m × 100-μm id uncoated capillary in a 10-mM phosphate buffer at pH 4, with an electric field of 200 V cm^{-1}.

Small-Molecule Applications

The on-line combination of capillary zone electrophoresis with mass spectrometry was first demonstrated in this laboratory for synthetic mixtures of quaternary ammonium salts [30,31]. These initial separations used 100-μm–id × 100-cm–long uncoated fused silica capillaries and a "buffer" of 10^{-4} M KCl dissolved in water/methanol 1:1 (v/v), with a field gradient of 370 V cm^{-1} [30]. As shown in Fig. 5 for a subsequent study using a phosphate buffer [31], a mixture of five quaternary ammonium cations (*m/z* from 74 to 242) were detected at subfemtomolar levels, with separation efficiencies ranging from 35,000 to 140,000 theoretical plates. The first work with organic molecules of biological significance followed shortly thereafter [31,32]. An eight-component mixture containing vitamins, amino acids, and dipeptides was separated in a 10^{-2}-M phosphate buffer at a field strength of 360 V cm^{-1} using pHs from 3 to 9 [32]. Low picomolar sensitivities were reported at signal/noise ratios greater than 30, suggesting that subpicomolar amounts could have been detected, even for these buffers, which are far from ideal.

Figure 6 shows a separation of four phosphonium ions obtained in a short 60-cm × 100-μm id capillary, using SIM detection. Under these conditions, the structurally similar ethyl- and vinyltriphenylphosphonium ions (*m/z* 289 and 291) have very similar electrophoretic mobilities and are not separated. Similarly, the tetrabutyl- and tetraphenylphosphonium ions coelute. A small amount of tailing is evident, which likely arises from interaction with the untreated capillary walls. A separation in a longer 2-m capillary at similar electric field strength gave the separation shown in Fig. 7. Here, the ethyl and vinyl triphenylphosphonium ions still coelute (although a very slight shift in the peak apex can be discerned), but the tetraphenyl- and tetrabutylphosphonium ions are now partially resolved. These results highlight a key advantage of MS detection; all mixture components need not be resolved. Indeed, for simple mixtures, such as this, separation may not be required unless the *m/z* values are similar. These results also highlight the stability, signal/noise ratios, and sensitivity that can be obtained for optimized ESI–MS conditions.

Henion and co-workers, using the pneumatically assisted liquid junction electrospray interface, have demonstrated several applications of CZE–MS. Using a 100-μm id × 100-cm column, a field gradient of 270 V cm^{-1}, and an acetonitrile/30-mM acetate (pH 4.8) buffer, they demonstrated a detection of 24 fmol for leucine (leu)-enkephalin [56]. With larger injections (5 pmol per component) a mixture of four dynorphins was separated and detected with high signal/noise ratios while scanning a 450 *m/z* range. The ability to conduct CID with MS/MS using a triple quadrupole instrument during the separations was also demonstrated. Here, they continuously passed a selected parent ion for one mixture component through the first quadrupole (Q_1) and dissociated it in Q_2 while scanning Q_3 to obtain a daughter spectrum of leu-enkephalin from a 5 pmol injection. Further applications to mono-

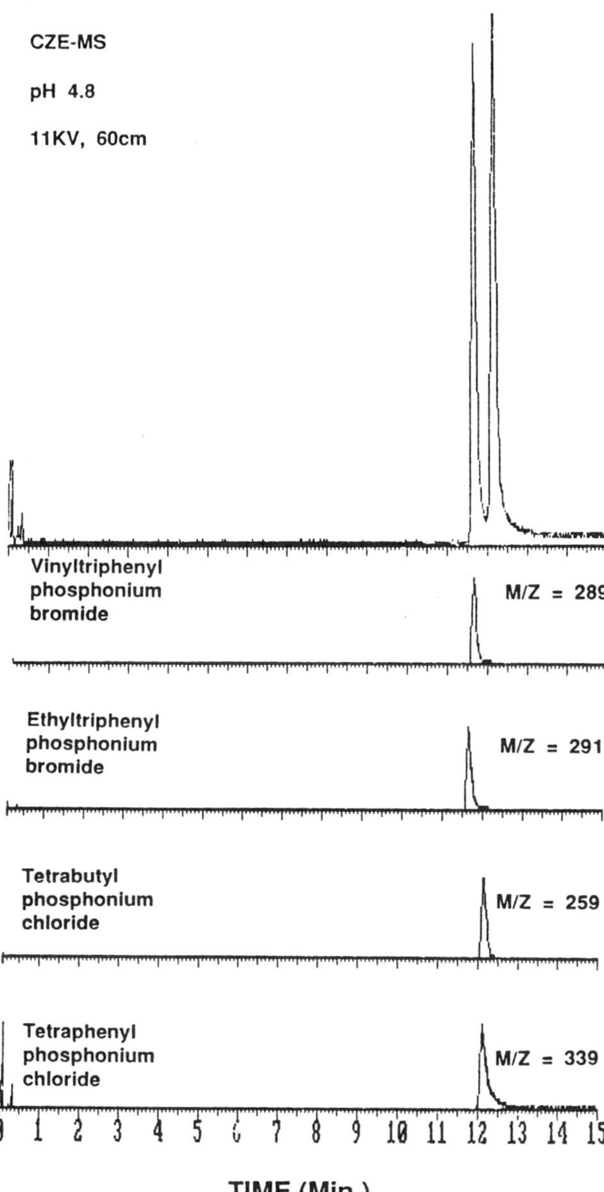

Figure 6 Total ion and reconstructed single ion electropherograms for CZE–MS separation of a mixture of quaternary phosphonium ions separated in a 0.6-m × 100-μm id uncoated capillary at pH 4.8 with an electric field of 183 V cm^{-1}.

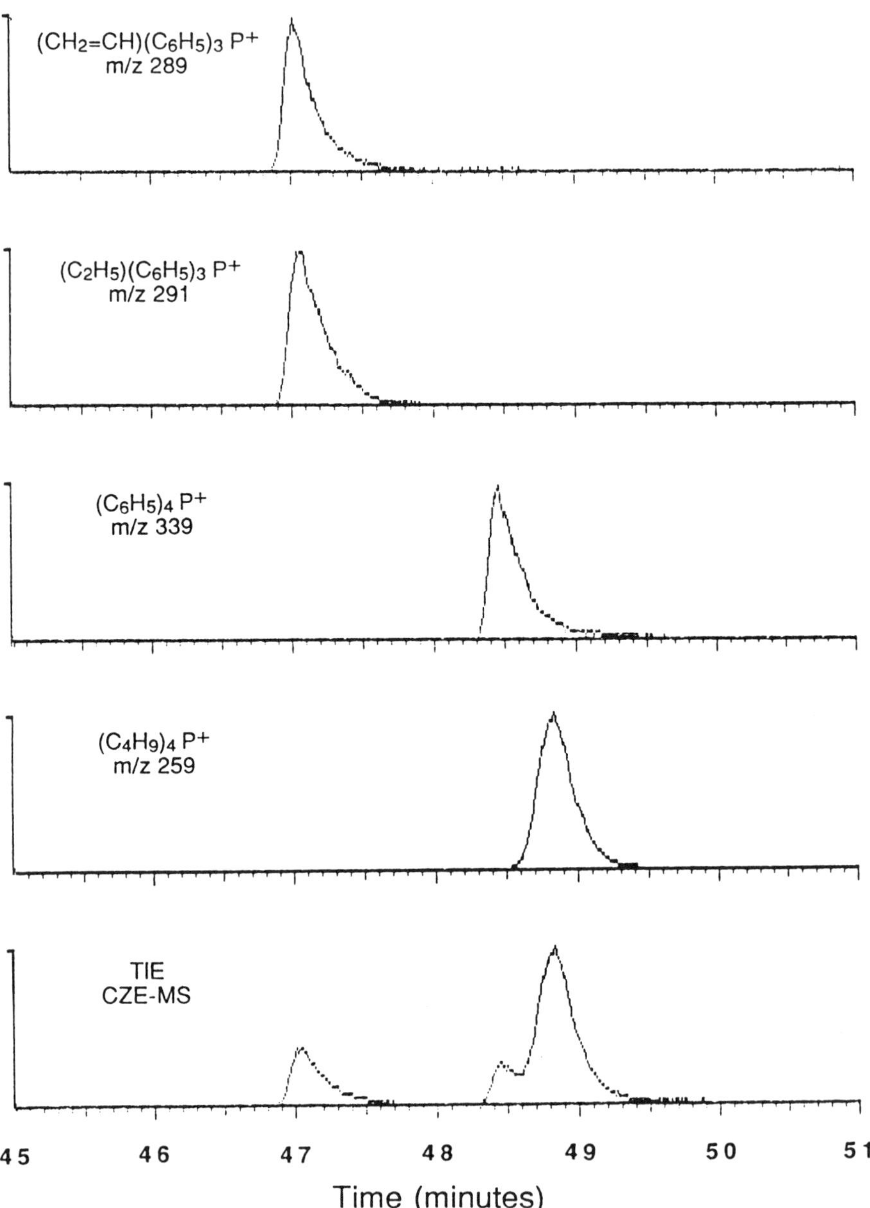

Figure 7 CZE–MS separation of the quaternary phosphonium ion mixture shown in Fig. 6, but using a 2-m × 100-μm id capillary at an electric field of 175 V cm^{-1}. A better separation is obtained than in Fig. 6, primarily because of the longer separation time and the reduced effective contribution of injection volume.

and polysulfonated azo dyes [49,50], acid pesticides [50], sulfonamides and benzodiazepines [52], and endogenous quantities of enkephalins followed in subsequent work [53]. The azo dyes were separated using a 60:40 acetonitrile/20-mM ammonium acetate (pH 6.8) buffer, a 100-μm id capillary, at a field gradient of 330 V cm^{-1}. These compounds produce intense signals using ESI in the negative-ionization mode (i.e., operating the ESI source, or often capillary terminus, at an inverted polarity relative to the sampling orifice, compared with that used for positive-ion production). From 10 to 30 pmol of each dye component was sufficient to allow scanning detection and CID. In later work [54], anionic aromatic sulfonates were separated and detected as negative-ions using a 75-μm id capillary, 330 V cm^{-1}, and 10-mM ammonium acetate, adjusted to pH 9.3 with TEA, in 30% acetonitrile buffer. A mixture of five acid pesticides was separated at the same field strength using a 90:10 acetonitrile/0.1-M ammonium acetate buffer [54]. Negative-ion ESI was again used for these compounds, and 1.2 pmol per component was sufficient to detect all of them using SIM detection. The usefulness of CE–MS for the analysis of benzodiazepines and sulfonamide drug metabolites in urine samples was examined [52]. With a 75-μm id capillary, a field gradient of 260 V cm^{-1}, and a buffer of 20 mM ammonium acetate containing 20% methanol (pH 6.8), a separation of a synthetic mixture of six sulfamides was reported using SIM detection. Two picomoles of each drug were injected and adequate signal/noise was obtained for the detection of all six drugs. For the sulfamethazine 2 pmol were sufficient to obtain a daughter ion spectrum using CID. Using the same conditions (but with 15 mM ammonium acetate adjusted to pH 2.5 by the addition of TFA and containing 15% methanol as the buffer), a separation of a synthetic mixture of four benzodiazepines was shown. In a SIM experiment 1.4–2 pmol of each component (injected) was separated and detected. From extracts of urine samples, metabolites of one of these drugs that had been administered orally (flurazepam dichloride) was detected from a separation using similar conditions (a 0.2-mM ammonium acetate buffer adjusted to pH 1.3 with TFA containing 15% methanol). The metabolites *N*-1-hydroxyethylflurazepam, didesethylflurazepam, and monodesethylflurazepam were detected, and a detection limit on a concentration of 0.5 μg mL^{-1} was reported for the first of these, although the authors indicated that this could be lowered by concentrating the sample. The usefulness of CZE–MS for the detection of trace amounts of leu-enkephalin and met-enkephalin found in horse cerebrospinal fluid was also investigated [53]. By using a 100-μm id capillary, a field strength of 300 V cm^{-1}, and a buffer of 50:50 acetonitrile/30 mM ammonium acetate (pH 6.8) they determined a detection limit of 2 ppm (60 fmol) for these enkephalins. Although the absolute amount detected was quite low, the concentration detection limit was too high for detection of these enkephalins in horse cerebrospinal fluid. This detection problem is due to the small volume injected (15 nL) and the absence of prior sample enrichment or concentration during injection. In later work [54], with 75-μm id capillary, 15 mM ammonium acetate adjusted to pH 2.5 with TFA in 15% methanol at 260 V cm^{-1}, low picomole amounts of quaternary ammonium salts were separated and detected while scanning. Recently, Henion and co-workers used a modified commercial CE instrument (Beckman P/ACE 2000) interfaced to a triple quadrupole mass spectrometer to analyze a mixture of nine peptides; bradykinin, angiotensin II, α-melanocyte-stimulating hormone, thyrotropin-releasing hormone, luteinizing hormone-releasing hormone, oxytocin, leu-enkephalin, met-enkephalin, and bombesin. A variety of separations were conducted using both 75- and 100-μm id capillaries at a field strength of 260 V cm^{-1} in two buffer systems. A 15-mM ammonium acetate buffer adjusted to pH 2.5 by the addition of trifluoroacetic acid and modified by the addition of 15% of either methanol or acetonitrile, or adjusted to pH 5.0 by the addition of acetic acid and 50% acetonitrile. They report detection limits in the low picomolar range for both SIM experiments and scanning experiments, and both faster analysis times and better efficiencies using acetonitrile, but reduced sensitivity, compared with methanol. With slightly larger sample sizes (10–50 pmol) CZE–MS/MS experiments were demonstrated.

Tomer and co-workers have recently applied CZE–MS to the analysis of macrolide antibiotics [38]. These workers used 75-μm id capillaries that were both uncoated and treated with 3-aminopropyl trimethoxysilane (APS). The APS coating reverses the charge at the column–buffer interface and, thereby, the direction of the electroosmotic flow. This has been useful for acidic buffers, when ESI is often most sensitive for polypeptides, and when many ions of interest are positively charged and may interact strongly with the negatively charged silanols on untreated fused silica. A field strength of 300 V cm^{-1} was used and the buffers employed were 0.01 M acetic acid or ammonium acetate, for which the pH was adjusted by the addition of ammonium hydroxide or acetic acid down to pH 3.4, and by the addition of formic or trifluoroacetic acid for pH below 3.4, using a Vestec ESI–MS instrument with a sheath flow (coaxial) interface. The CZE–MS experiments were done on two mixtures. The first contained the six macrolides, erythromycin A, spiramycin, tylosin, midecamycin, miokamycin, and rokitamycin, and the second contained the four macrolides oleandomycin, josamycin, troleandomycin, and leucomycin-A5. These workers noted that at the lowest pHs used, the addition of trifluoroacetic acid decreased the sensitivity. They also experimented with the addition of organic modifiers in the buffer (up to 10% of 2-propanol), a strategy that Henion and co-workers had found effective in improving some of their separations. They found no improvement, but observed the expected decrease in electroosmotic flow and correspondingly longer analysis times. The best separations were obtained at either pH 8.5 or pH 3.4, with the lower-pH buffer yielding the better separation efficiency (~190,000 theoretical plates). There is little difference in the electrophoretic mobilities of these compounds, since they differ only slightly in structure and functionality, and only partial separation was achieved. An injection of 44 fmol per component was detected using SIM detection and 300 fmol per component for scanning detection. However, separation efficiency for the larger injections was noticeably reduced, primarily because of the larger injection volume. These workers found that dynamic range was limited, with peak broadening at only slightly higher loadings than their detection limit.

Thiabault and co-workers have recently shown the application of CZE–MS to compounds of interest to the aquaculture industry such as antibiotics and marine toxins [37]. One purpose was to compare the performance of the liquid junction and sheath flow interfaces (see Fig. 4 and section on ESI interfaces). They demonstrated separations of saxitoxin, neosaxitoxin, C_{11}-sulfated gonyautoxins (GTX-2 and GTX-3), erythromycin, ometoprim, trimethoprim, sulfisomidine, sulfamethazine, and okadaic acid in 50-μm id capillaries at a field strength of 270 V cm^{-1} using a Trizma pH 7.2 buffer. These samples are difficult compounds to analyze by more conventional means. In both interfaces, a 0.2% formic acid solution was used as the sheath or junction solution. Both CZE–MS and CZE–MS/MS experiments were performed with detection limits down to the 1 ppm range. They report the expected discrimination effects during electrokinetic injection, finding hydrostatic injections to be much more useful than electrokinetic injections for quantitation.

Recent work by Moseley et al. [45] demonstrated encouragingly low detection limits for analysis of peptide mixtures, as shown in Figure 8. With 75-μm id APS-treated columns in a sheath flow interface and a buffer of 10 mM acetic acid at pH 3.4, 160 fmol of neurotensin and b-casmorphin were detected, with efficiencies of as much as 130,000 theoretical plates and with high signal/noise ratios. Other synthetic mixtures of opioid peptides, angiotensin-related peptides, and neuropeptides were similarly analyzed. Injections of as little as a few tenths of a picomole provided high signal/noise ratios for scanning as much as 1100 Da and separation efficiencies of up to 200,000 theoretical plates. From slightly larger injections (780 fmol) MS/MS spectra were obtained for the mixture of neuropeptides.

Application to more complex polypeptide mixtures (i.e., tryptic digests of proteins) has also been investigated by several researchers, potentially an extremely important area of application because of the significance of peptide mapping and sequencing using ever smaller sample sizes. With these mixtures, high-efficiency separations have been demonstrated, generating over 250,000 theoretical plates. Complex mixtures of peptides generated from tryptic digestion of large proteins present a difficult analytical challenge, since the fragments cover a large range of both pI and hydrophobicity. Since trypsin specifically cleaves peptide bonds COOH-terminally at lysine and arginine, the resulting peptides generally form doubly charged as well as singly charged molecular ions by positive-ion ESI. Such doubly charged tryptic peptides generally fall within the m/z range of quadrupole mass spectrometers. The resolving power of these methods is illustrated in Fig. 9, which shows a UV electropherogram (top), obtained in conjunction with a CZE–MS analysis [58], for a separation of a tryptic digest of tuna cytochrome *c* in a 50-mM acetate buffer (at pH 6.1), mixed with an equal volume of acetonitrile. Shown in Figure 10 are single-ion electropherograms for two charge states of two tryptic

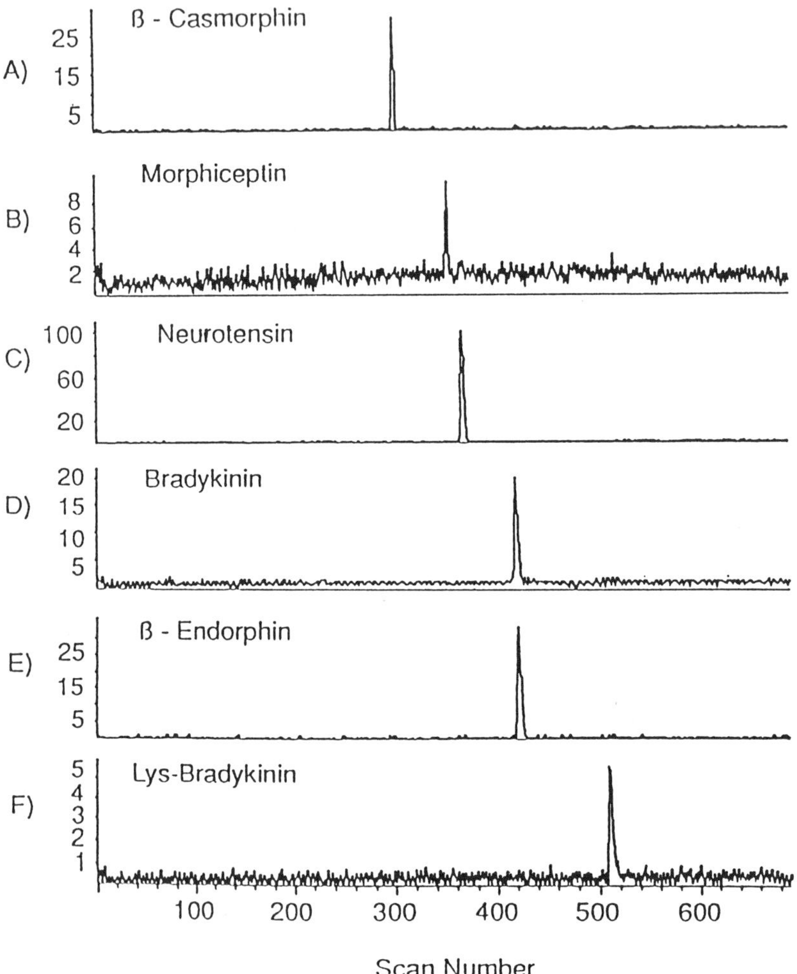

Figure 8 Selected ion electropherograms from the CZE–ESI–MS analysis of a mixture of neuropeptides [45]. The electropherograms were constructed for the molecular ions [i.e., $(M + H)^+$] from *m/z* 400–1500 scanned at a rate of two scans per second. (Data courtesy of M. A. Moseley et al.)

fragments. The mass spectrum for one polypeptide component of this mixture is shown in Figure 11. In this work, a commercial CE instrument (Beckman P/ACE 2000) was modified for CE–MS to allow UV-detection of the separation at one-half the MS detection time, providing a preview of ESI–MS detection at twice the separation time for UV-detection. The individual peaks indicate a separation efficiency corresponding

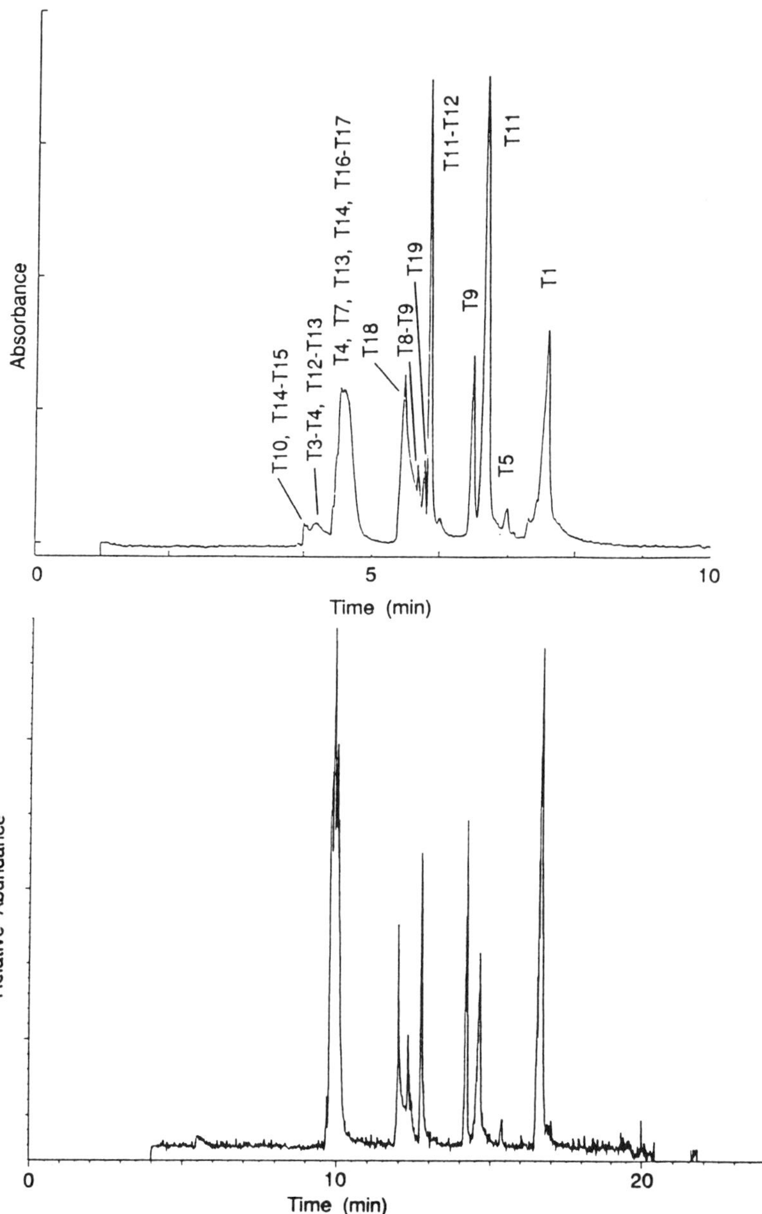

Figure 9 Comparison of the UV (214 nm) (top) and reconstructed ion electropherograms (bottom), with time axes adjusted for the corresponding peaks, for a CZE–MS separation of a tryptic digest of tuna cytochrome *c*. (From Ref 58)

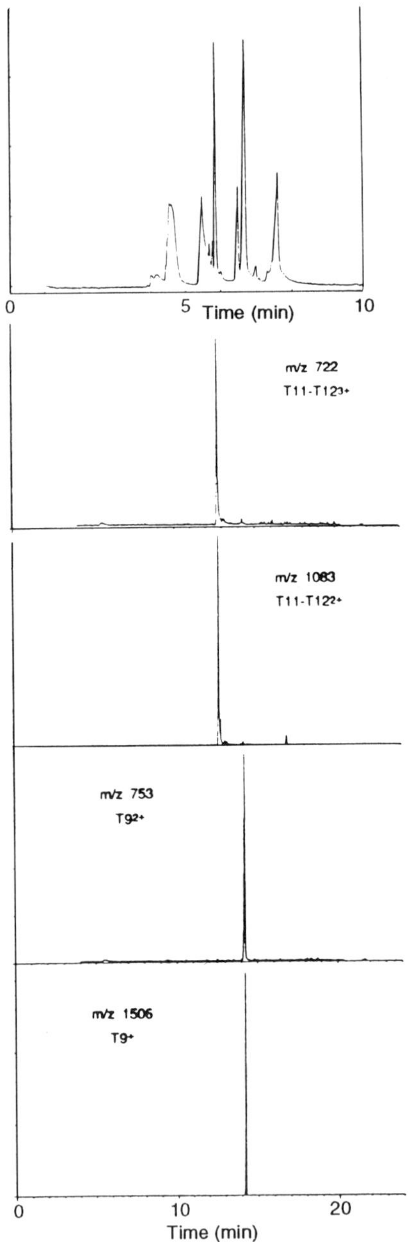

Figure 10 Comparison of the UV and four single-ion electropherograms from the tryptic digest of tuna cytochrome *c* (see Fig. 9) showing the doubly and singly charged ions of two different tryptic fragments (with the UV and MS time-axis owing to location of the UV detector at the capillary midpoint).

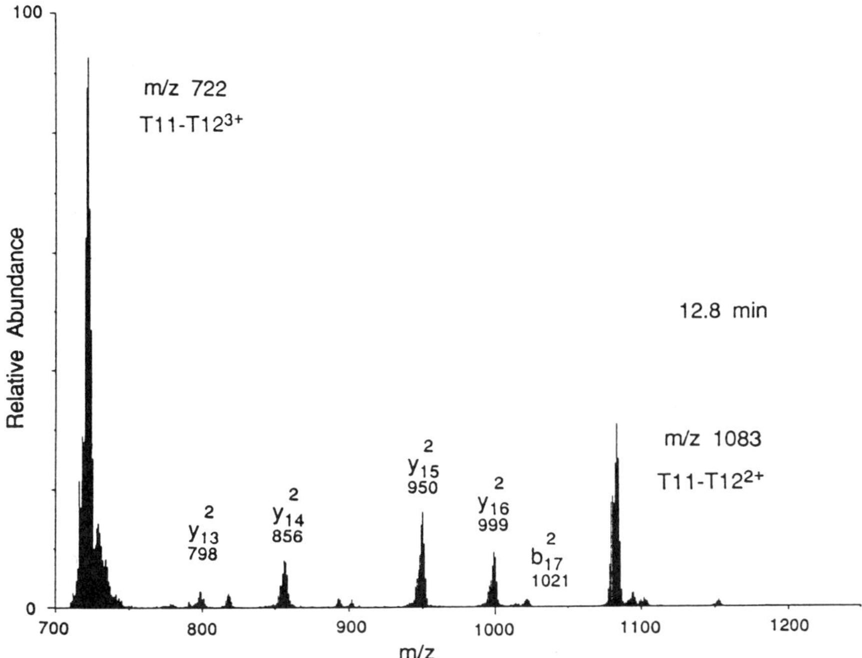

Figure 11 Mass spectrum acquired for the $T_{11}-T_{12}$ peptide component from the tryptic digest of tuna cytochrome *c* at ~12.8 minutes using MS conditions in which partial parent ion dissociation is induced in the interface. The spectrum shows the doubly and triply charged ions corresponding to the molecular ion, and a series of dissociation products consistent with the amino acid sequence.

to approximately 4×10^5 theoretical plates. It was noted that the apparent efficiency for MS detection is significantly greater than for UV detection (1.2×10^5 theoretical plates for the same peak), a fact that can be largely attributed to the longer separation time and the very small effective dead volume, or residence time, in the ESI interface [58].

Lee et al. [50] showed the detection and identification of 11 components of a tryptic digest of recombinant bovine somatotropin using SIM detection. Their separations were conducted in a 100-μm id capillary using a 50:50 acetonitrile/5-mM ammonium acetate buffer, and efficiencies as high as 300,000 theoretical plates were obtained. In separate experiments, CZE–MS/MS was demonstrated for one of the components, providing information on the amino acid sequence. Subsequently, Johansson et al. [51] reported the analysis of a hemoglobin trytic digest. They use a 100-μm id capillary, a field strength of 260 V cm^{-1}, and a 50:50 acetonitrile/

15-mM ammonium acetate buffer, adjusted to pH 5.0 with acetic acid. Approximately 10 pmol was injected electrokinetically (25 s at 10 kV from a 154 pmol/ μL^{-1} digest). The digest was expected to have 28 components and a complex total ion electropherogram with several coeluting peaks is seen from a scanning experiment covering a 600 *m/z* range. The authors report that all of the expected enzymatic products were detected. In separate experiments a 10-pmol injection of the digest was used to obtain a MS/MS spectrum of the T_3 fragment. The product daughter spectrum showed fragment ions that provided information on the peptide. In another experiment these workers used a 75-μm id capillary with a 15-mM ammonium acetate buffer, adjusted to pH 2.5 by the addition of TFA and 15% methanol. They monitored eight ions (SIM) characteristic of the expected fragments identified from their previous experiments and demonstrated efficiencies from 85,000 to 135,000 theoretical plates for an 8-pmol injection.

Thibault et al. [41] examined a separation of the tryptic fragments of a glucagon digest. The separation was done in a 50-μm id capillary at 390 V cm^{-1}, using either a 10-mM, pH 3.4, or 2.5-M acetic acid buffer (pH 2.1). They used a commercial (Microcoat) polymeric-coating reagent. Two picomoles of glucagon (digested) were injected. The four expected fragments (yielding both singly and doubly charged ions) were observed using MS scanning (950 *m/z* range). In addition, evidence for the partial oxidation of the sample was obtained, along with a small amount of CID fragmentation that occurred in the ESI–MS interface. Higher efficiency (~34,000 theoretical plates), but longer analysis times, were obtained with the lower pH buffer.

Experience now suggests that nearly all tryptic fragments can be effectively detected by ESI–MS methods. Fragments detected by the UV detector are almost always detected by ESI–MS. However, detection has sometimes been problematic for very small fragments (i.e., amino acids and dipeptides) owing to difficulties in obtaining optimum ESI–MS conditions over a wide *m/z* range. Excessive internal excitation or large solvent-related background peaks at low *m/z* appear to be the origin of some difficulties. Available evidence indicates that if tryptic fragments are not detected, they were generally "lost" before the separation, a problem particularly evident with nanoscale sample handling. As observed in the past, full realization of more sensitive analytical methods will depend on concomitant improvements in the earlier stages of sample handling and preparation.

Large-Molecule Applications

The ESI–MS interface provides the basis for detection of molecules having an M_r that exceeds 100,000. The usefulness of CE–ESI–MS for such applications is being studied, and the combination is of growing interest. As noted earlier, a major consideration in such applications is detection sensitivity. Methods to alleviate these concerns appear to be on the horizon (see final section). The analysis of proteins by

CZE–MS is also challenging because of the well-established difficulties associated with protein interactions with capillary surfaces. Often, capillaries must be coated to minimize, or prevent, wall interactions [41,45]. It is unlikely that one capillary or buffer system will be ideal for all proteins, but rather, as for LC separations, procedures optimized for specific classes of proteins will be developed. Most CZE separations require the ionic strength of the buffer to be about 10^2 greater than that of the sample, to prevent degradation of the separation by perturbation of the local electric field in the capillary. An exception to this is the instance in which much greater concentrations of volatile buffer components (such as acetic acid) are acceptable. Detection sensitivity for proteins is generally lower than for small peptides owing to the greater number of charges per molecule and the greater number of charge states.

An important attribute of the ESI–MS interface is the capability to obtain structurally related information on large molecules by dissociation in the interface by a *collisional heating* process (induced by a simple electric field gradient). This capability was first demonstrated for large molecules by our laboratory in 1988 [69] and has the advantage, compared with CID MS/MS methods, of much greater selectivity, but the disadvantage of greater ambiguity, since only one MS step is used, and a good separation is generally required. The complexity of large-molecule dissociation spectra precludes obtaining complete sequence information with current MS quadrupole technology, but its potential for fingerprinting has been established [27,70]. As shown in Fig. 12, extensive sequence-related fragmentation can be obtained for molecules as large as serum albumins (~66 kDa), yielding spectra suitable for fingerprinting purposes. We have also shown that direct infusion experiments using low picomolar sample quantities provide spectra for which a limited amount of sequence information can be obtained for materials of substantially known sequence. This suggests the potential utility of direct CE–MS analysis of proteins for evaluation of suspected primary structural or posttranslational modifications.

Figure 13 shows a separation of three myoglobins obtained on an uncoated 50-μm capillary of 80-cm length [58]. The separation was obtained in 10-mM Tris buffer at pH 8.3 with an electric field strength of 120 V cm^{-1}. Approximately 100 fmol per component was injected onto the capillary column. The top of the figure shows the UV trace for the separation and the bottom shows two single-ion electropherograms from MS detection for the same separation. Whale myoglobin (M_r 17,199) was detected at *m/z* 861 (owing to loss of the heme) and carried 20+ charges. The resolution of the mass spectrometer was too low to separate the 19+ molecular ion of horse myoglobin (M_r 16,950) from sheep myoglobin (M_r 16,923), and both contribute to the *m/z* 893 ion in the single-ion electropherogram. The ion at *m/z* 617 is due to the heme group of the myoglobins and clearly remains associated with the protein during CZE separation [58]. The heme is not covalently bound to these proteins, and it is labile under the ESI conditions used. Of interest is that an additional myoglobin-related compound is observed that also retains the heme func-

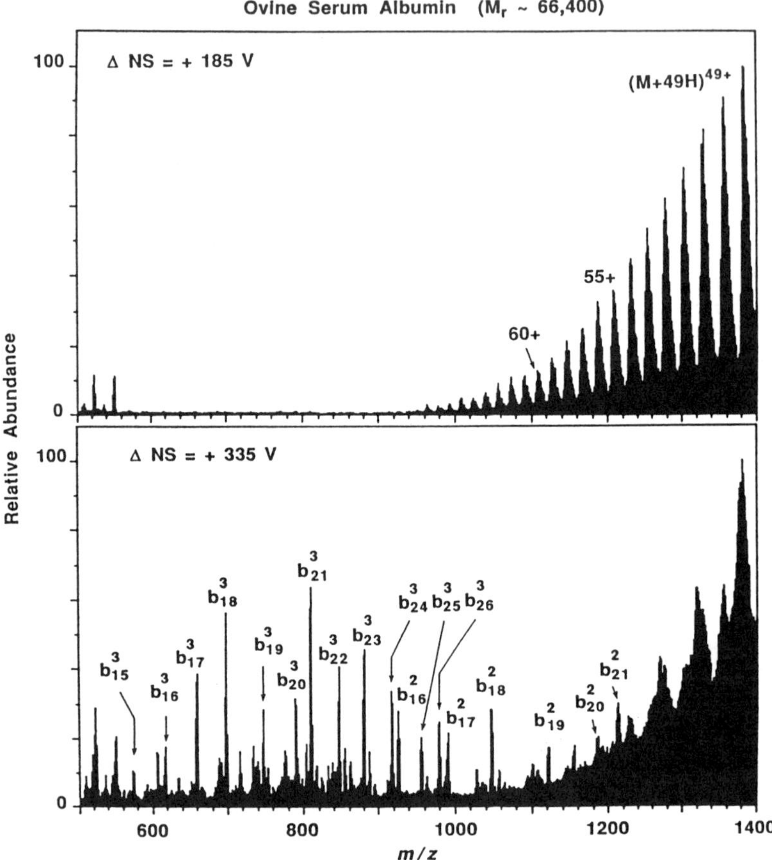

Figure 12 Electrospray ionization mass spectra of ovine serum albumin (M_r 66,400) obtained with a nozzle–skimmer voltage (ΔNS) of 185 V (top) and +335 V (bottom) obtained with a modified Sciex TAGA 6000E triple quadrupole mass spectrometer. The high ΔNS spectra provide structurally related information, with much greater sensitivity than possible with triple quadrupole (MS/MS) methods, that is useful for fingerprinting.

tionality during CE and that has an M_r similar to whale myoglobin. Measurements of the peak widths in this experiment show that peaks in the UV trace are 3.5 seconds wide (half-height) at half the MS detection time, compared with 4.5 second for MS detection. This suggests that, under optimum conditions, more efficient injection methods designed to concentrate the analyte at the head of the column, in conjunction with smaller sample loading, could potentially reduce the peak width in these separations to approximately 1 second.

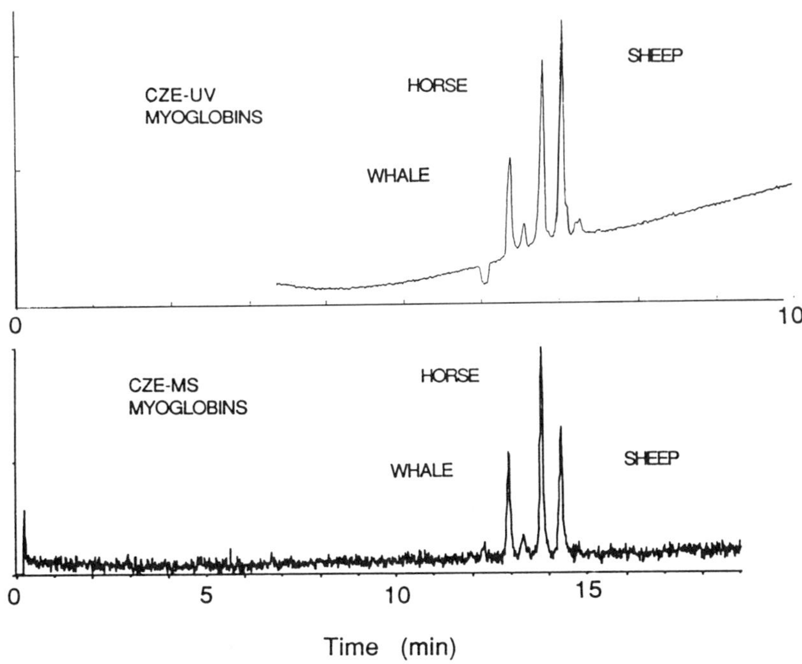

Figure 13 CE–MS separation of a mixture of whale (M_r 17,199), horse (M_r 17,951), and sheep (M_r 16,923) myoglobins. The on-line UV response and later MS response (at approximately twice the separation time) are shown with adjusted time scales. The separation was conducted in 10 mM Tris (pH 8.3) at 120 V m^{-1} with a 1-m × 50-μm id uncoated capillary. A sheath liquid incorporating acetic acid was used to assist positive ion ESI. Approximately 100 fmol per component was injected.

The peak widths obtained in these studies are already sufficiently narrow to exceed current capabilities for obtaining high-quality full-scan spectra when signal intensities are low and the mass range is large. For targeted compound analysis, where a single ion could be monitored continuously, detecting peaks of less than 1 second in width is quite practical. Obtaining full-scan spectra for fast or high-resolution separations will obviously benefit from ion trap–MS, time-of-flight, or array-based MS instrumentation.

In addition to our laboratory, others have also demonstrated CZE–MS protein separations. Particularly successful have been separations in formic or acetic acid solutions using capillary coatings designed to reverse (make positive) surface charge. Thibault and associates [41] have obtained mass spectra while scanning (m/z 1400–2350 range) using 0.2- to 0.8-pmol injections of the proteins lysozyme, α-lactalbumin, chymotrypsinogen A, and bovine apotransferrin, as shown in Fig. 14. In a

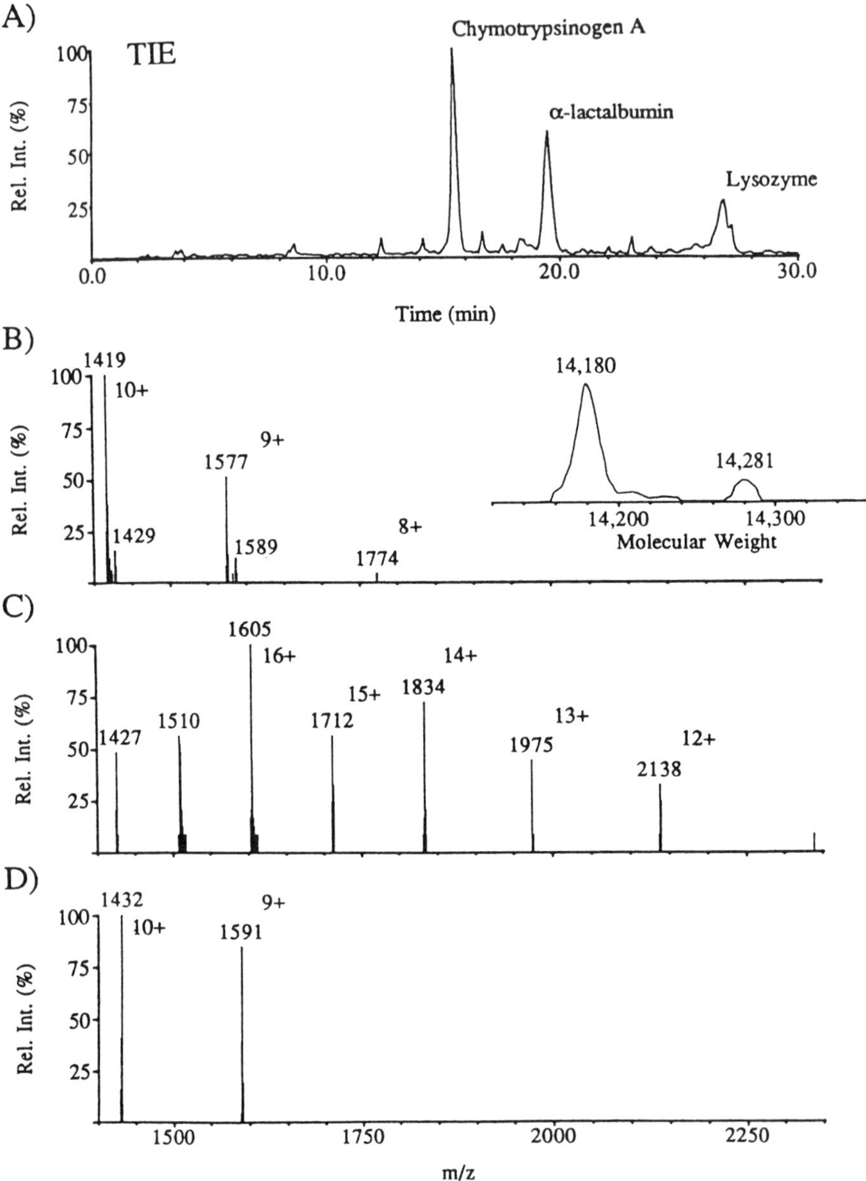

Figure 14 Total ion electropherogram (*m/z* 1400–2350 Da) for CE–MS analysis of a protein mixture for an 18-nL injection of 0.7-mg/mL protein solution, with a 90-cm × 50-μm id capillary using a 10-mM acetic acid buffer (pH 3.4) at −35 kV (effective), using a sheath flow of 0.2% formic acid at 3–5 μL min^{-1}. Mass spectra: (B) α-lactalbumin (pI 4.8), (C) chymotrypsinogen A (pI 9.1), and (D) chicken-egg lysozyme (pI 11.1).

more recent study, Moseley et al. [45] used an APS-treated capillary (75-μm id) with a 10-mM, (pH 3.4) acetic acid buffer, and a field strength of 300 V cm^{-1} to evaluate a separation of the proteins myoglobin and cytochrome *c*. Scanning over an *m*/*z* 1200 range allowed proteins to be detected and resolved in the total ion electropherogram for a 130-fmol injection.

The CZE–MS method can also provide information on impurities, or possibly multimeric species, specifically noncovalently associated dimers or larger complexes stable in solution [70]. For example, the CE–MS analysis of a sample of B-chain insulin in 10-mM Tris (at pH 8.3), at an electric field strength of 150 V cm^{-1}, using 0.7-pmol injection of a sample showed a second smaller, slightly lower peak eluting by both MS and UV detection. The second peak was attributed to either an impurity of very similar M_r or a dimer (and perhaps larger multimers) which would dissociate to multiply charged monomers by collisional heating in the ESI interface [58]. Such species can generally be separated by differences in electrophoretic mobility and, depending on interface conditions and multimer stability in the CE buffer, can be detected as either the monomer or intact multimer.

MASS SPECTROMETRIC DETECTION WITH OTHER CAPILLARY ELECTROPHORESIS FORMATS

Capillary Isotachophoresis–Mass Spectrometry

Isotachophoresis in a capillary is an alternative electrophoretic method that provides an attractive complement to CZE. The instrumentation used for capillary isotachophoresis (CITP) can be nearly identical with that of CZE. In CZE, analyte bands are separated on the basis of the differences in their electrophoretic mobilities in an electric field gradient, which ideally is unperturbed by their presence (owing to the much lower effective concentrations). In contrast, in isotachophoresis, the analyte mobilities define the electric field strength. Two different electrolyte solutions are chosen as the leading and terminating electrolyte solutions, which have sufficiently high and low electrophoretic mobilities, respectively, to bracket the electrophoretic mobilities of the sample components of interest. The sample is inserted in the capillary between the two electrolyte solutions, and the sample components are separated into distinct bands on the basis of their electrophoretic mobilities. The electric current in the capillary is determined by the leading electrolyte, which defines the ion concentration in each band. The electric field strength varies in each band, the highest field strength being found in the bands with the lowest mobility (i.e., greatest resistance). The length of a band is then proportional to the concentration of its ions in the sample. As with CZE, resolution is ultimately limited by band broadening caused by molecular diffusion, joule heating, and (in contrast with CZE) electroosmosis.

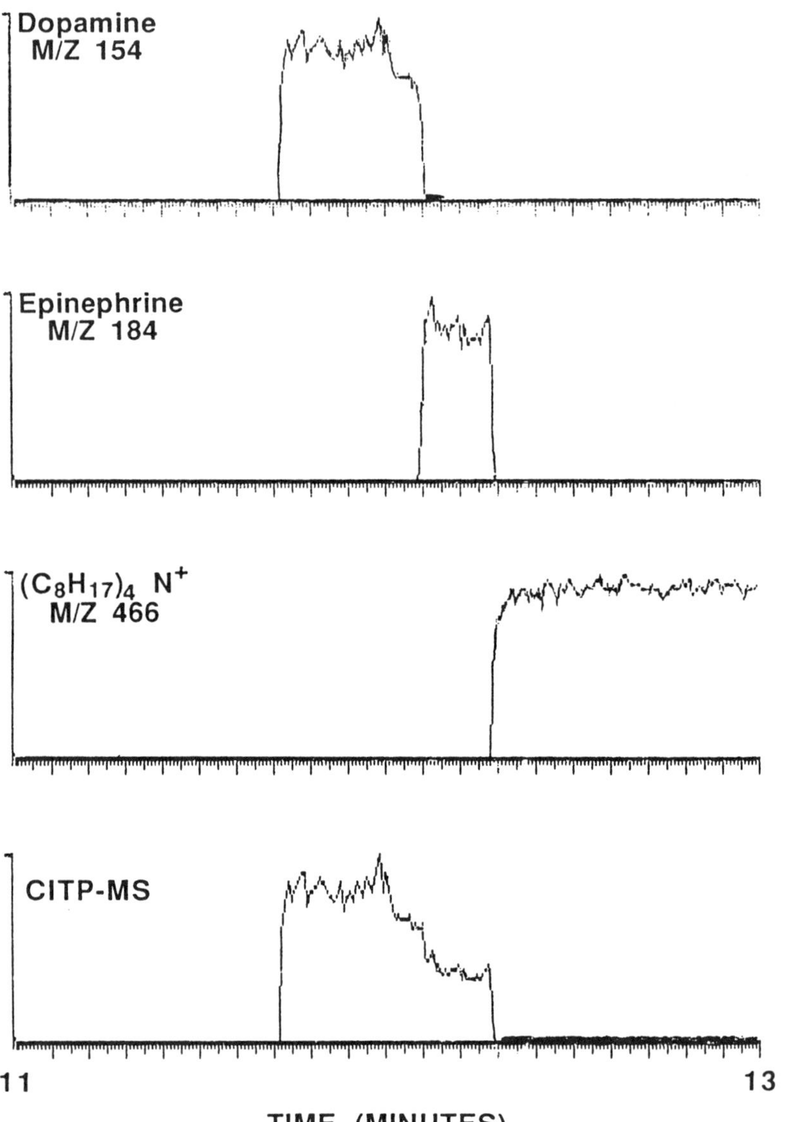

Figure 15 A CITP–MS separation of a dopamine and epinephrine in a 0.8-m capillary at 30 kV. Tetraoctylammonium bromide was used for the terminating electrolyte.

A direct CITP–MS combination is an attractive complement to CZE for several reasons. First, the sample size which can be introduced is greater by several orders of magnitude than can be tolerated in CZE, and is limited primarily by the volume of the capillary and the separation time allowed (which is inversely related to the voltage). Second, CITP generally results in concentration of analyte bands (depending on their concentration relative to the leading electrolyte), that is in contrast with the inherent dilution obtained with CZE. These two features can provide a substantial improvement in concentration detection limits compared with CZE. It must be remembered that although 10-amol detection limits have been reported for CZE, the 10-nL injection volume typical of CZE corresponds to a concentration detection limit (i.e., sensitivity) of about 10^{-7}–10^{-9} M with SIM detection (depending on ESI and instrumental characteristics). If the ionization efficiency is lower, or full-scan mass spectra are required, the CITP–MS detection limits will be substantially better than those obtainable by CZE–MS. A third benefit of CITP is derived from the nature of the separations, compared with CZE or chromatographic methods. The peak capacity needed to separate all the components of a sample generally greatly exceeds the number of components, and for much of the time, the detector is observing only the baseline. In contrast, with isotachophoresis, one sample band immediately follows another and detector time is not wasted (once a separation is obtained). In addition, all the bands have similar ion concentrations, so there should ideally be no large differences in signal intensities between bands and a broad-dynamic-range detector is generally not required. In addition, information on the amount of analyte is conveyed by the length of the analyte band.

Figure 15 demonstrates an early CITP–MS separation of dopamine and epinephrine, showing the characteristic handlike structure of the two components. The separation was conducted at 175 V cm^{-1}, with a 2-m × 100-μm id untreated fused silica capillary. The leading electrolyte was 10^{-3} M ammonium acetate and the trailing electrolyte was 10^{-3} M tetraoctylammonium bromide in a 1:1 v/v water/methanol mixture as solvent. The application of CITP–MS to the phosphonium salt mixture used for earlier CZE–MS experiments (see Figs. 6 and 7) provides an interesting comparison of these electrophoretic methods. The concentrations of the four quaternary phosphonium salts were all 10^{-5} M, in a 1:1 v/v water/methanol mixture. The sample was loaded by electromigration for 60 seconds at 30 kV. Figure 16 shows the CITP–MS single- and total-ion isotachopherograms. The separation is well developed, with minimum band overlap and sharp edges. The widths of the four bands are nearly equivalent, consistent with the similar sample concentrations and the minor discrimination expected from use of electromigration for introduction of the mixture.

The most striking feature of the separation shown in Fig. 16 is the excellent resolution of the four phosphonium salts, which were not resolved in the shorter CZE separations. The fact that the vinyltriphenyl- and ethyltriphenylphosphonium

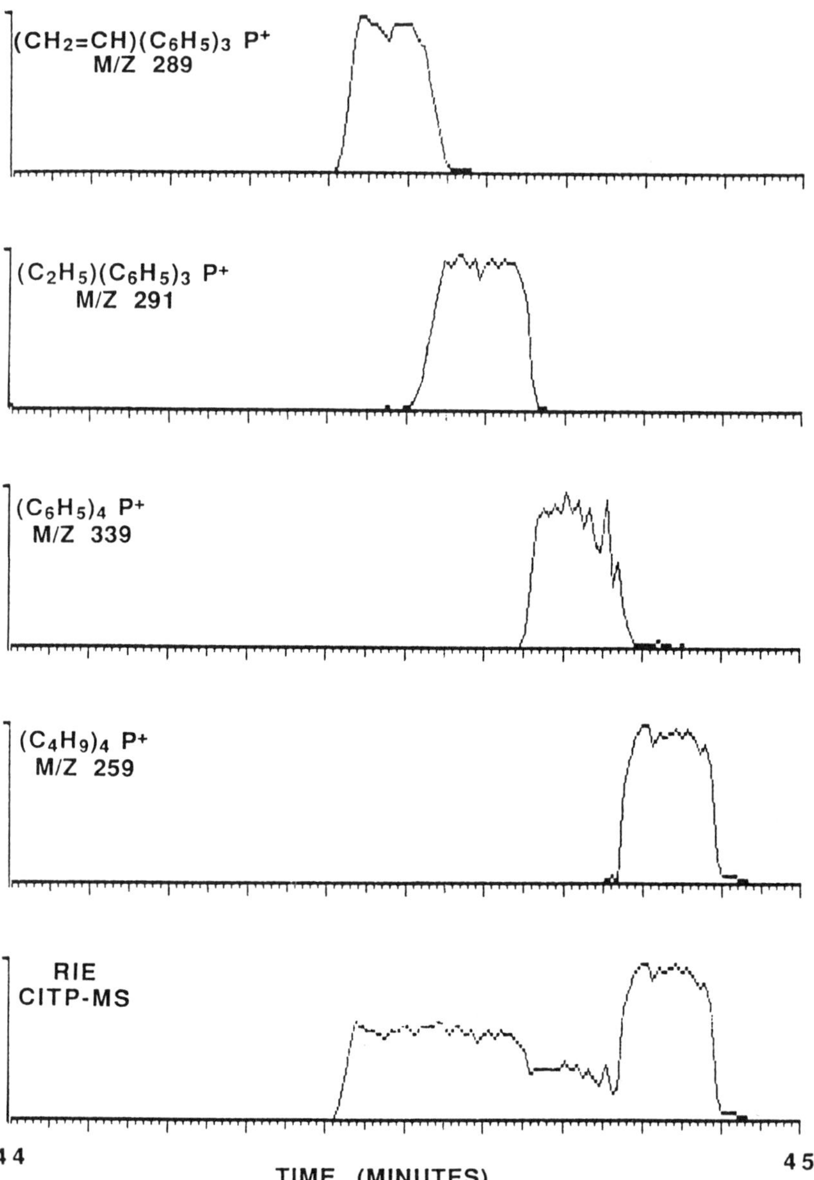

Figure 16 A CITP–MS separation of a mixture of four quaternary phosphonium salts showing the single-ion plots for the four cations and reconstructed ion electropherogram (RIE). A 2-m capillary was used at 35 kV, and the sample was loaded by electromigration at 35 kV for 60 seconds. Short and long CZE–MS separations for this same mixture are shown in Figs. 7 and 8.

ions are clearly resolved illustrates the potential of CITP–MS for high-resolution separations. Another advantage is the greater signal/noise ratio.

Figure 17 shows the total reconstructed selected ion isotachopherograms for the CITP–ESI–MS analysis of a mixture of peptides, derived from the enzymatic digestion of glucagon with trypsin. The time scale shown on the axis refers to the total time from the beginning of the separation. Separation development in CITP may be accomplished before MS detection starts and, as shown here, the actual time for CITP–MS analysis is determined from the time of application of hydrostatic flow, owing to elevation of the trailing electrolyte reservoir, causing elution of the focused bands (here at 27 minutes). The isotachopherograms show typical bands and some

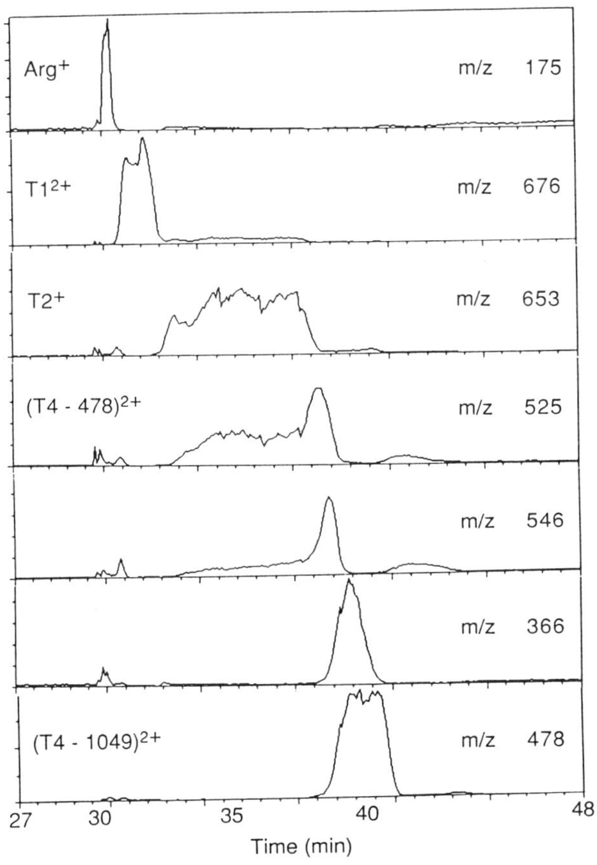

Figure 17 Selected ion current isotachopherograms for the CITP–MS analysis of an enzymatic digest of glucagon with trypsin (and contaminating chymotrypsin).

step changes at boundaries between bands, but is far from ideal, with the centers of sample bands showing no obvious structure. Strong absorbance at 280 nm with the UV detector indicated the presence of tryptophan in the final band eluting. The selected ion isotachopherograms shown in Fig. 17 include the tryptic fragments arginine (Arg, *m/z* 175), T_1^{2+} (*m/z* 676), and T_2^+ (*m/z* 653), as well as other products. Additional peptide fragments apparently arise from the action of contaminating chymotrypsin on the predicted tryptic fragments $(T_4\text{-}478)^{2+}$ and $(T_4\text{-}1049)^{2+}$, and on intact glucagon (giving *m/z* 546 and 366). The ESI mass spectrum of the T_1 peptide contains intense 3^+ and 2^+ peaks, whereas the singly charged molecular ion is observed to have much lower intensity. The CID tandem mass spectra for the T_1 tryptic fragment of glucagon in various charge states is given in Fig. 18.

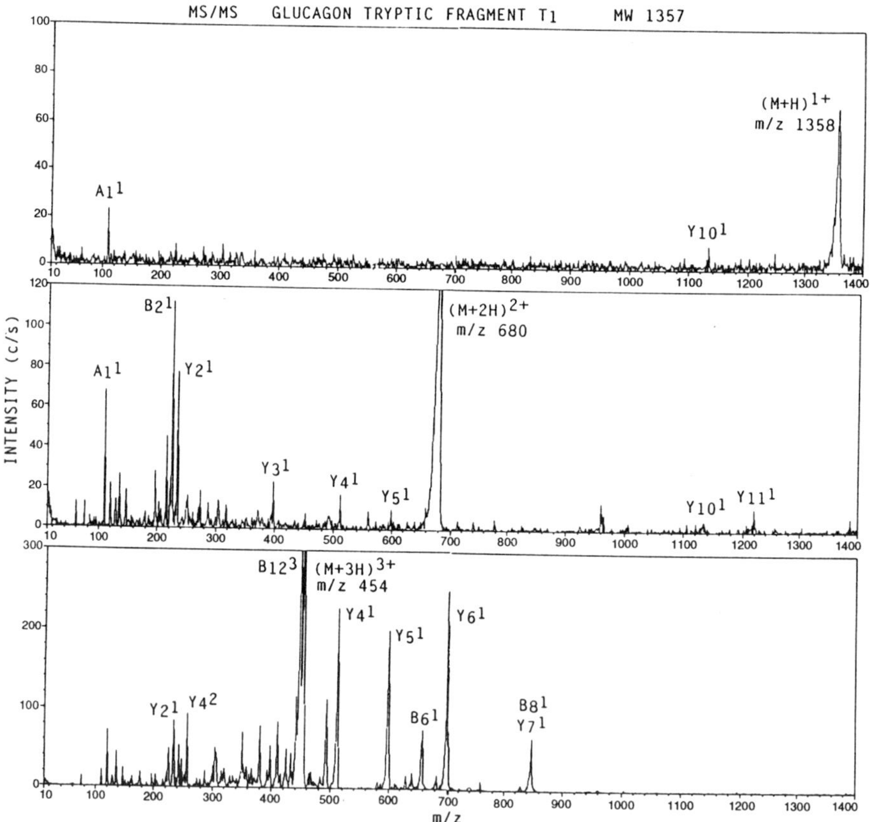

Figure 18 Daughter ion tandem MS/MS mass spectra for the 1^+ to 3^+ charge states of the molecular ion for the glucagon T_1 tryptic fragment.

A potentially important capability of CE–MS (and CE–MS/MS) is a method that will yield primary structural information (i.e., sequence) for polypeptides and small proteins. The ESI–MS method affords unique opportunities here, since ionization efficiencies are high and good results can be obtained, even for large proteins. The fact that ESI mass spectra generally consist of only intact multiply charged molecular ions is sometimes cited as a disadvantage of this method, since it is claimed that structural information cannot be obtained. However, effective dissociation of molecular ions can be induced in the nozzle–skimmer regions of the ESI interface [27,69,70]. A more powerful approach is tandem MS of molecular ions from several of the major charge states. The relatively long and stable period of elution of separated bands in CITP would facilitate MS/MS experiments requiring longer integration, signal averaging, or more concentrated samples than provided by CZE.

The CID tandem mass spectra for the various tryptic fragments (currently obtained in separate experiments owing to limitations of our data acquisition) can be assembled as demonstrated previously for the smaller polypeptide, melittin. The CID fragments observed from MS/MS spectra, such as in Fig. 18, provide a large amount of redundant information concerning the peptide sequence. For smaller peptides, such as glucagon, major fragment ions are observed that provide information on the entire glucagon sequence [34]. CITP provides a relatively pure analyte band to the ESI source, without the large concentration of a supporting electrolyte demanded by CZE. This latter characteristic circumvents the principal disadvantage of CZE–ESI–MS: the reduced sensitivity from ionization of supporting electrolyte. Thus, CITP–MS has the potential of allowing much greater sensitivities (and analyte ion currents) than feasible with CZE–MS, facilitating MS/MS experiments. Although CITP–MS has as yet attracted relatively little attention, these characteristics make CITP–MS/MS potentially well-suited for characterization of enzymatic digests of proteins.

Capillary Gel Electrophoresis—Mass Spectrometry

The use of gel-filled capillaries has attracted extensive interest, since it provides a basis for separation of compounds that otherwise have similar electrophoretic mobilities in solution. In particular, impressive results have been obtained for large biopolymers, particularly DNA. Our early experience highlighted several difficulties associated with CGE–MS interfacing related to gel stability and to sensitivity constraints arising owing to typical buffer components (e.g., urea, Tris). Over the last few years, the stability and reproducibility of CGE has advanced significantly. Recently, Henion and co-workers [54] have successfully demonstrated CGE–MS with 50- to 75-μm id capillaries using a nebulization-assisted ESI interface for synthetic mixtures of aromatic sulfonates, dansylated amino acids, and low-molecular-mass polyacrylic acids (7 mer to 14 mer). The liquid junction interface used employed a 10–20%

methanol/water solution containing 1.0% triethylamine, and the inlet buffer consisted of 0.1 M Tris-borate and 7 M urea. A significant limitation of this coupling is the reduced sensitivity compared with CZE–MS, which required the use of single-ion monitoring methods. The reduced sensitivity was attributed to the low sample capacity of gel-filled capillaries. In earlier studies in our laboratory, we noted another possible contribution to reduced sensitivity with gel-filled capillaries that is due to the ionization of gel buffer components. In particular, urea and Tris-borate buffer components produced an extensive background, accounting for a substantial portion of the total ESI current. Capillary gel electrophoresis (CGE) based upon more volatile buffer components not resulting in reduced sensitivity, or gels that more strongly retain these components, may yet provide a basis for more sensitive CGE–MS detection. As MS instrumental methods become more sensitive, the CGE–MS combination should become more broadly useful.

FUTURE DEVELOPMENTS

Most ESI–MS instruments, to date, have been based on quadrupole mass spectrometers. The limitations on obtaining high-quality fast full-scan mass spectra for CE–ESI–MS experiments with quadrupole mass spectrometers, and even more demanding tandem MS/MS (perhaps for peptide sequencing), require further discussion, since the constraints arise directly from the present levels of ESI–MS performance. Consideration and understanding of the nature of these constraints is important for understanding the trade-offs between scan speed, resolution, *m/z* range, and sensitivity, as well as for the development of improved methods and instrumentation that lead to their resolution or circumvention.

The greatest limitation of present ESI–MS instrumentation arises because of the small fraction of the total ESI ion current actually introduced and analyzed by the mass spectrometer. At present, a typical ESI–MS quadrupole instruments will provide an ion current of approximately 10^{-11} A to the detector, corresponding to $\sim 6 \times 10^7$ singly charged ions per second. Although not always true, the buffer ionic strength generally must exceed the sample by about 10^2 to obtain the high efficiency possible with CE. Thus, if the ESI process does not discriminate greatly between buffer ions and sample, only 6×10^5 sample ions per second will be detected (at most). (In practice, ESI generally discriminates in favor of higher-molecular-mass species, and somewhat higher sample ion currents might be obtainable using "volatile buffers" such as acetic acid. Examination of full-scan spectra, say from *m/z* 10 to 2000, under conditions that minimize discrimination, generally shows that even the most intense peaks correspond to $<<10\%$ of the *total* ion current, consistent with these expectations.) The best-resolved and fastest eluting peaks for capillaries of conventional dimensions have widths at half-height of 1–3 seconds. If ten scans are required during peak elution, and the region *m/z* 100–2000 is scanned in 0.2 *m/z* steps, then the most intense peak that can be detected would correspond to only ~20 ions. Even with pulse-counting detection more than three to five ions are required to provide minimal confidence of detection, suggesting a dynamic range of ~5. Clearly, this quality of MS performance is impractical with present quadrupole instruments. For scans at a reduced width of 1000-*m*/*z* units and 1-*m*/*z* steps, ~200 ions can result for major peaks, giving a more reasonable dynamic range of ~50. Slower scans are also useful in practice. For a 1.4-second scan over a 500-*m*/*z* unit window, with 1-*m*/*z* step size, a dynamic range of ~400 would be estimated, qualitatively consistent with the present results. Clearly, although results obtained at slower scan speeds, lower resolution, and a smaller *m*/*z* range are far less than ideal, we believe that simultaneous optimization of both CE and MS components is impractical with existing instrument performance. Thus, ESI–MS detection for CE requires a careful consideration of the experiment and the necessary compromises. The situation is more severe for tandem MS/MS using triple quadrupole instrumentation, owing to the much lower signal intensities after collisional dissociation (generally at least one order of magnitude smaller, often a factor of 100).

There are several possible approaches to resolving these limitations. One is to increase the efficiency of ion transport from the ESI source into the mass spectrometer, now only ~0.01% in overall efficiency [27]. A second approach is to analyze all the ions that enter the mass spectrometer, using either ion-trapping methods, orthogonal time-of-flight, MS, or array detection. Indeed, several workers are already beginning to examine the quadrupole ion trap mass spectrometer (ITMS) for this purpose [1]. Since the ITMS can accumulate ions over a wide *m*/*z* range, and then rapidly conduct a swept ejection–detection step, the fraction of total ions detected (compared with that entering the ITMS) is much greater, allowing the utilization of a much larger fraction of the total ESI current (assuming efficient ion injection and trapping). Much faster scan speeds over extended *m*/*z* ranges are thus possible. The best MS performance, in principle, would be obtained with full-range array detection, but practical instrumentation for this purpose is not yet available, and would likely be much more expensive than either ITMS or existing instrumentation. Projections concerning ITMS performance must be made with caution, since the trapping efficiency upon injection from external ESI sources is not yet established, and the *m*/*z* measurement accuracy currently obtainable has been reported to be significantly lower than for quadrupole instruments, a particular concern for multiply charged species and sequencing applications. However, preliminary results indicate the sensitivity obtainable with ESI is indeed better than for quadrupole instrumentation [1], although concerns about mass measurement accuracy have not been completely addressed.

Other developments are promising, but less certain. McLafferty and co-workers have pioneered the combination of ESI with Fourier transform-ion cyclotron resonance mass spectrometry (FT-ICR) [71–73]. Laude and co-workers have demonstrated that substantial gains in sensitivity are obtainable by having the ESI interface in the high magnetic field of an ICR superconducting magnet [74]. Recent results from our laboratory have demonstrated that both high resolution ($>10^5$) and high sensitivity (<4 fmoles) are obtainable for proteins using FT-ICR [75]. The potential for higher-order mass spectrometry, MS^n ($n \geq 3$), using ICR or ITMS instruments appears promising. High-resolving power at moderate *m/z*, a capability of ion-trapping devices, has been demonstrated for proteins and appears to have considerable potential, although technical difficulties remain to be solved, particularly in conjunction with CE methods. Very recent results have shown that greatly improved sensitivity can be obtained using smaller diameter capillaries, even with ESI-quadruple instruments [76]. At least for the next few years, it appears that quadrupole mass spectrometers will be the most widely used analyzers with CE because of factors that include lower cost and ease of interfacing.

ACKNOWLEDGMENTS

We thank the US Department of Energy, Office of Health and Environmental Research (Contract DE-AC06-76RLO 1830), and Beckman Instruments for support of this research. We also thank Drs. M. A. Moseley (Glaxo), K. B. Tomer (NIEHS), P. Vouros (Northeastern), J. D. Henion (Cornell), P. Thibault (National Research Council of Canada), and R. M. Caprioli (University of Texas), for providing manuscripts and figures before publication.

REFERENCES

1. G. J. Van Berkel, G. L. Glish, and S. A. McLuckey, *Anal. Chem.*, *62*:1284 (1990).
2. P. Arpino, *Fresenius J. Anal. Chem.*, *337*:667 (1990).
3. E. Kenndler and D. Kaniansky, *J. Chromatogr.*, *209*:306 (1981).
4. E. Kenndler and E. Haidl, *Fresenius J. Anal. Chem.*, *322*:391 (1985).
5. R. Takigiku, T. Keough, M. P. Lacey, and R. E. Schneider, *Rapid Commun. Mass Spectrom.*, *4*:24 (1990).
6. J. A. Loo, C. G. Edmonds, R. D. Smith, M. P. Lacey, and T. Keough, *Biomed. Environ. Mass Spectrom.*, *19*:286 (1990).
7. R. A. Wallingford and A. G. Ewing, *Anal. Chem.*, *59*:1762 (1987).
8. F. Hillenkamp, M. Karas, R. C. Beavis, and B. T. Chait, *Anal. Chem.*, *63*:1193 (1991).
9. R. M. Caprioli, W. T. Moore, M. Martin, and B. B. DaGue, *J. Chromatogr.*, *480*: 247 (1989).
10. R. M. Caprioli, In *Proc. Jpn. Soc. Biomed. Mass Spectrom.* (1991).
11. M. J.-F. Suter, B. B. DaGue, W. T. Moore, S.-N. Lin, and R. M. Caprioli, *J. Chromatogr.*, *553*:101 (1991).
12. W. T. Moore and R. M. Caprioli, In *Techniques in Protein Chemistry II*, Academic Press, New York, p. 511 (1991).
13. M. J.-F. Suter and R. M. Caprioli, *J. Soc. Mass Spectrom.*, *3*:198 (1992).

14. R. D. Minard, D. Luckenbill, R. Curry, Jr., and A. G. Ewing, *Proceedings of the 37th ASMS Conference on Mass Spectrometry and Allied Topics*, p. 950 (1989).
15. S. M. Wolf, P. Vorous, E. Jackim, and C. Norwood, Presented at Int. Conf. in Amsterdam; (1991). *J. Am. Soc. Mass Spectrom.*, *3*:757 (1992).
16. N. J. Reinhoud, W. M. A. Niessen, and U. R. Tjaden, *Rapid Commun. Mass Spectrom.*, *3*:348 (1989).
17. N. J. Reinhoud, E. Schroder, U. R. Tjaden, W. M. A. Niessen, M. C. Ten Noever de Brauw, and J. van der Greef, *J. Chromatogr.*, *516*:147 (1990).
18. M. Tehrani, R. Macomber, and L. Day, *J. High Resolut. Chromatogr.*, *14*:10 (1991).
19. E. R. Verheij, U. R. Tjaden, W. M. A. Niessen, and J. Van Der Greef, *J. Chromatogr.*, *554*:339 (1991).
20. J. S. M. DeWit, L. J. Deterding, M. A. Moseley, K. B. Tomer, and J. W. Jorgenson, *Rapid Comun. Mass Spectrom.*, *2*:100 (1988).
21. M. A. Moseley, L. J. Deterding, K. B. Tomer, and J. W. Jorgenson, *J. Chromatogr.*, *480*:197 (1989).
22. M. A. Moseley, L. J. Deterding, K. B. Tomer, and J. W. Jorgenson, *Rapid Commun. Mass Spectrom.*, *3*:87 (1989).
23. M. A. Moseley, L. J. Deterding, K. B. Tomer, and J. W. Jorgenson, *Anal. Chem.*, *63*:109 (1991).
24. L. J. Deterding, M. A. Moseley, K. B. Tomer, and J. W. Jorgenson, *J. Chromatogr.*, *554*:73 (1991).
25. L. J. Deterding, C. E. Parker, and J. R. Perkins, *J. Chromatogr.*, *554*:329 (1991).
26. L. J. Deterding, C. E. Parker, J. R. Perkins, M. A. Moseley, J. W. Jorgenson, and K. B. Tomer, *J. Chromatogr.*, *554*:329 (1991).
27. R. D. Smith, J. A. Loo, C. G. Edmonds, C. J. Barinaga, and H. R. Udseth, *Anal. Chem.*, *62*:882 (1990).
28. A. P. Bruins, T. R. Covey, and J. D. Henion, *Anal. Chem.*, *59*:2642 (1987).
29. L. L. Mack, P. Kralik, A. Rheude and M. Dole, *J. Chem. Phys.*, *52*:4977 (1970).
30. J. A. Olivares, N. T. Nguyen, C. R. Yonker, and R. D. Smith, *Anal. Chem.*, *59*:1230 (1987).
31. R. D. Smith, J. A. Olivares, N. T. Nguyen, and H. R. Udseth, *Anal. Chem.*, *60*:436 (1988).
32. R. D. Smith, C. J. Barinaga, and H. R. Udseth, *Anal. Chem.*, *60*:1988 (1988).
33. H. R. Udseth, J. A. Loo, and R. D. Smith, *Anal. Chem.*, *61*:228 (1989).
34. R. D. Smith, S. M. Fields, J. A. Loo, C. J. Barinaga, H. R. Udseth, and C. G. Edmonds, *Electrophoresis*, *11*:709 (1990).
35. C. G. Edmonds, J. A. Loo, S. M. Fields, C. J. Barinaga, H. R. Udseth, and R. D. Smith, In *Biological Mass Spectrometry* (A. L. Burlingame and J. A. McCloskey, eds.), Elsevier Science Publishers, Amsterdam, p. 77 (1990).
36. M. Hail, J. Schwartz, I. Mylchreest, K. Seta, S. Lewis, J. Zhou, I. Jardine, J. Liu, and M. Novotny, In *Proceedings of the 38th ASMS Conference on Mass Spectrometry and Allied Topics*, Tucson, Ariz., p. 353 (1990).
37. S. Pleasance, P. Thibault, and J. Kelly, *J. Chromatogr.*, *591*:325 (1992).
38. C. E. Parker, J. R. Perkins, and K. B. Tomer, *J. Am. Soc. Mass Spectrom.*, *3*:139 (1992).

39. D. F. Hunt, J. E. Alexander, A. L. McCormack, P. A. Martino, H. Michel, J. Shabanowitz, M. A. Moseley, J. W. Jorgenson, and K. B. Tomer, In *Techniques in Protein Chemistry II* (J. J. Villafranca, ed.), Academic Press, San Diego, p. 441 (1991).
40. D. F. Hunt, J. E. Alexander, A. L. McCormack, P. A. Martino, H. Michel, J. Shabanowitz, N. Sherman, M. A. Moseley, J. W. Jorgenson, and K. B. Tomer, In *Techniques in Protein Chemistry II* (J. J. Villafranca, ed.), Academic Press, San Diego, p. 441 (1991).
41. P. Thibault, C. Paris, and S. Pleasance, *Rapid Commun. Mass Spectrom.*, *5*:484 (1991).
42. P. Thibault, S. Pleasance, and M. V. Laycock, *J. Chromatogr.*, *542*:483 (1991).
43. P. Thibault, S. Pleasance, and M. V. Laycock, (1991). In *Proceedings of the 39th ASMS Conference on Mass Spectrometry and Allied Topics*, p. 593 (1991).
44. S. Pleasance, S. W. Ager, M. V. Laycock, and P. Thibault, *Rapid Commun. Mass Spectrom.*, *6*:14 (1992).
45. M. A. Moseley, J. W. Jorgenson, J. Shabanowitz, D. F. Hunt, and K. B. Tomer, *J. Am. Soc. Mass Spectrom.*, *3*:289 (1992).
46. I. Mylchreest and M. Hail, *Proceedings of the 39th ASMS Conference on Mass Spectrometry and Allied Topics*, Nashville, Tenn., p. 316 (1991).
47. Y. Shida, C. E. Parker, J. R. Perkins, K. O. O'Hara, and K. B. Tomer, *Proceedings of the 39th ASMS Conference on Mass Spectrometry and Allied Topics*, Nashville, Tenn., p. 587 (1991).
48. K. A. Halm, S. E. Unger, R. L. St. Claire, R. F. Arendale, and M. A. Moseley, *Proceedings of the 39th ASMS Conference on Mass Spectrometry and Allied Topics*, Nashville, Tenn., p. 591 (1991).
49. E. D. Lee, W. Mück, J. D. Henion, and T. R. Covey, *Biomed. Environ. Mass Spectrom.*, *18*:253 (1989).
50. E. D. Lee, W. Mück, J. D. Henion, and T. R. Covey, *Biomed. Environ. Mass Spectrom.*, *18*:844 (1989).
51. I. M. Johansson, E. C. Huang, J. D. Henion, and J. Zweigenbaum, *J. Chromatogr.*, *554*:311 (1991).
52. I. M. Johansson, R. Pavelka, and J. D. Henion, *J. Chromatogr.*, *559*:515 (1991).
53. W. M. Mück and J. D. Henion, *J. Chromatogr.*, *495*:41 (1989).
54. F. Garcia and J. D. Henion, *Anal. Chem.*, *64*:985 (1992).
55. F. Garcia and J. D. Henion, *J. Chromatogr.*, *606*:237 (1992).
56. E. D. Lee, W. Mück, J. D. Henion, and T. R. Covey, *J. Chromatogr.*, *458*:313 (1988).
57. T. Wachs, J. C. Conboy, F. Garcia, and J. D. Henion, *J. Chromatogr. Sci.*, *59*:357 (1991).
58. R. D. Smith, H. R. Udseth, C. J. Barinaga, and C. G. Edmonds, *J. Chromatogr.*, *559*:197 (1991).
59. C. G. Edmonds, J. A. Loo, C. J. Barinaga, H. R. Udseth, and R. D. Smith, *J. Chromatogr.*, *474*:21 (1989).
60. J. A. Loo, C. G. Edmonds, and R. D. Smith, *Anal. Chem.*, *63*:2488 (1991).
61. J. A. Loo, H. K. Jones, H. R. Udseth, and R. D. Smith, *J. Microcolumn Separations*, *1*:223 (1989).

62. R. D. Smith, J. A. Loo, C. J. Barinaga, C. G. Edmonds, and H. R. Udseth, *J. Chromatogr.*, *480*:211 (1989).
63. R. D. Smith, J. A. Loo, C. G. Edmonds, C. J. Barinaga, and H. R. Udseth, *J. Chromatogr.*, *516*:157 (1990).
64. R. D. Smith, H. R. Udseth, J. A. Loo, B. W. Wright, and G. A. Ross, *Talanta*, *36*:161 (1989).
65. J. B. Fenn, M. Mann, C. K. Meng, S. F. Wong, and C. M. Whitehouse, *Mass Spectrom. Rev.*, *9*:37 (1990).
66. S. K. Chowdhury, V. Katta, and B. T. Chait, *Rapid Commun. Mass Spectrom.*, *4*:81 (1990).
67. A. L. Rockwood, M. Busman, and R. D. Smith, *Rapid Commun. Mass Spectrom.*, *5*:582 (1991).
68. R. R. Ogorzalek Loo, H. R. Udseth, and R. D. Smith, *J. Phys. Chem.*, *95*:6412 (1991).
69. J. A. Loo, H. R. Udseth, and R. D. Smith, *Rapid Commun. Mass Spectrom.*, *2*:207 (1988).
70. R. D. Smith, J. A. Loo, R. R. Ogorzalek Loo, M. Busman, and H. R. Udseth, *Mass Spectrom. Rev.*, *10*:359 (1991).
71. K. D. Henry, E. R. Williams, B. H. Wang, F. W. McLafferty, J. Shabanowitz, and D. F. Hunt, *Proc. Natl. Acad. Sci. USA*, *86*:9075 (1989).
72. K. D. Henry and F. W. McLafferty, *Org. Mass Spectrom.*, *25*:490 (1990).
73. K. T. Henry, J. P. Quinn, and F. W. McLafferty, *J. Am. Chem. Soc.*, *113*:5447 (1991).
74. S. A. Hofstadler and D. A. Laude, *Anal. Chem.*, *64*:569 (1992).
75. B. E. Winger, S. A. Hofstadler, J. E. Bruce, H. R. Udseth, and R. D. Smith, *J. Am. Soc. Mass Spectrom.* (in press).
76. J. H. Wahl, D. R. Goodlett, H. R. Udseth, and R. D. Smith, *Anal. Chem.*, *64*:3194 (1992).

17

Capillary Zone Electrophoresis–Mass Spectrometry: *Continuous Flow Fast-Atom Bombardment and Electrospray Ionization*

Kenneth B. Tomer

National Institute of Environmental Health Sciences
Research Triangle Park, North Carolina

The high separation efficiencies, short analysis time, and high sensitivities associated with capillary electrophoretic techniques have generated a great deal of interest in the field [1–5]. As with any exciting area of separation research and application, several detection systems, including laser-induced fluorescence, electrochemical detection, as well as ultraviolet (UV) detection, have been investigated for their suitability for incorporation into capillary electrophoresis (CE) separations.

Each one of the methods that has been investigated has a weak point: the detection technique may be applicable to only certain types of compounds; the method may necessitate chemical derivatization; detection limits may be relatively high; reproducible migration times, especially of components in complex mixtures, may be difficult to attain; and little structural information about the analyte, other than its presence or the presence of a specific functional group, is provided. For these reasons, one of the most exciting detection methods to be incorporated into CE separations is mass spectrometry (MS).

The mass spectrometer as a detector offers several advantages over other detection methods: (a) It approaches being a universal detector. With modern instrumentation, analytes varying from small inorganic species to large proteins or DNA strands can be detected. (b) Mass spectrometry can provide high sensitivity. If the nature of the analyte is known, attomole sensitivity can sometimes be achieved. (c) Mass spectrometry provides a high degree of structural information about the analyte. This includes relative molecular mass (M_r) information obtained from the molecular ion species and structural information obtained from fragment ions. The wealth of structural information achievable also provides the ultimate level of confidence in analyte identification. This latter aspect also alleviates many of the problems associated with irreproducible migration times. Rapid, accurate identification of components based on molecular mass or on characteristic fragment ions, or both, can readily pinpoint which of several closely eluting components is the analyte of interest. (d) Mass spectrometry provides a second stage of analyte separation because mass spectrometry itself is a separation technique, with separation based on mass. Thus, overlapping components that differ in molecular ion or in a characteristic fragment ion can be differentiated. This is illustrated in Fig. 1 for the partial separation of two peptides, α-melanocyte-stimulating hormone (α-MSH) and substance P free acid, in a mixture of six peptides. Looking at the total ion current (see Fig. 1a), there is no indication of the presence of two components in the peak eluting at 7 minutes 50 seconds (scan 223), whereas the selected ion traces of the protonated molecules (see Fig. 1b and 1c) show that there are two components eluting at that time.

Although the major concern often expressed about using mass spectrometry for detection in capillary electrophoresis is the cost, coupling of mass spectrometry and

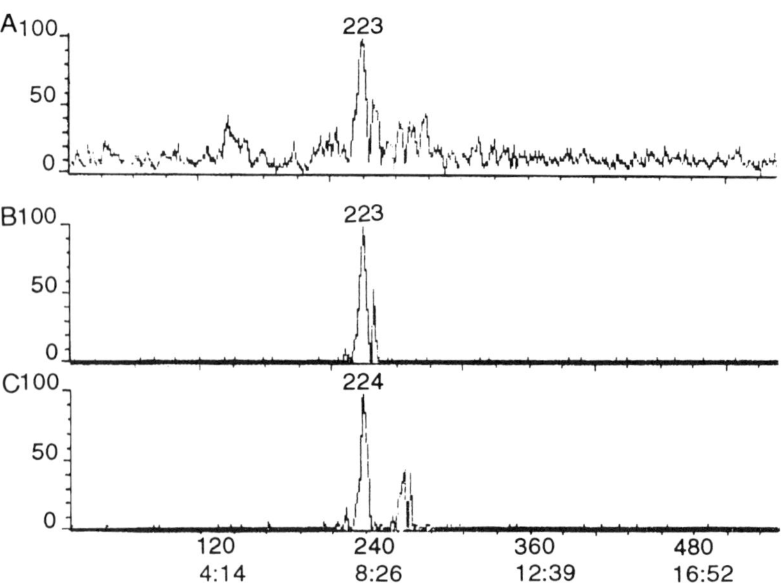

Figure 1 (A) Reconstructed total ion current of a separation of six neuropeptides. (B) Reconstructed electropherogram of the $(M + 3H)^{+3}$ ion of α-MSH. (C) Reconstructed electropherogram of the $(M + 2H)^{2+}$ ion of substance P free acid.

capillary electrophoresis engenders other difficulties as well. These problems may be viewed as arising from pushing the mass spectrometer to its limits.

One problem arises from the high separation efficiencies associated with capillary electrophoresis. Most current mass spectrometers are scanning instruments; that is, a range of masses are scanned over a set time period, which means that the mass spectrometer passes ions of a given mass to the detector for only a very small percentage of the scan time. If the CE peak is sufficiently narrow, the mass spectrometer may not be passing ions of appropriate mass to the detector when an analyte is eluting, which may lead to missed components. If the mass spectrometer scans too fast, problems with ion statistics may be encountered; that is, the instrument may be scanning fast enough to pass the appropriate mass when the compound elutes, but too fast to pass sufficient ions to the detector to obtain a reasonable signal/noise ratio (insufficient swell time). An alternative mass spectrometric technique, selected ion monitoring, (SIM) is available. In this technique, the mass spectrometer is cycled continuously and rapidly between a few predetermined ions. This eliminates the time spent on scanning masses of no interest. This technique can increase the detection level by over an order of magnitude. A prerequisite for its use is that the analyst must know the mass of the analytes of interest, thus negating one of the advantages of the mass spectrometer. An example of the efficacy of the SIM technique is discussed later.

A second problem is the incompatibility of nonvolatile buffers with the mass spectrometers. The use of nonvolatile buffers can lead to clogging of vacuum orifices and probe tips and can form adduct ions that effectively decreases the sensitivity of the instrument by spreading the ion abundances among several masses; for example, through formation of $(M + Na)^+$, $(M - H + 2Na)^+$, and so on, as well as $(M + H)^+$ ions. Thus, we are typically restricted to volatile buffers such as ammonium acetate.

Although the use of MS for CE detection does engender difficulties, the advantages are more than sufficient to encourage its use. The first published account of MS detection for CE was that of Smith and co-workers in 1987 [6] using electrospray ionization (ESI)–MS [7–11]. A discussion by Smith and Udseth of this approach can be found in Chapter 16. Henion and co-workers were the second group to report on the coupling of CE with MS in 1988 [12]. Henion's group reported the use of ion spray MS and a liquid junction interface. Also in 1988, Minard [13] presented the use of a liquid junction interface with continuous flow fast atom bombardment (CF–FAB)–MS. In 1989, this author published a report of the use of a coaxial CF–FAB–MS interface [14]. Later that year Caprioli and co-workers [15] and van der Greef and co-workers [16] published papers on the use of liquid junction interfaces with CF–FAB–MS. We have recently upgraded an older quadrupole mass spectrometer for electrospray ionization and have also interfaced CE with this ionization technique.

CAPILLARY ELECTROPHORESIS–MASS SPECTROMETRY INTERFACES

In the remainder of this chapter, the techniques we have used for interfacing CE with both CF–FAB–MS and electrospray ionization (ESI) MS will be described and examples that highlight both the benefits and the problems of CE–MS will be given. Before the descriptions of the interfaces, a very brief description of the ionization technique will be given.

Capillary Zone Electrophoresis Coaxial Continuous-Flow Fast-Atom Bombardment Mass Spectrometry

In fast-atom bombardment, first developed by Barber et al. [17], the analyte is dissolved in a volatile matrix, typically glycerol, a mixture of dithiothreitol and dithioerythritol, or *m*-nitrobenzyl alcohol, on the end of a probe tip inside the source of the mass spectrometer. The probe tip is bombarded with fast atoms, typically 8–10 keV xenon atoms formed in an ion gun. Bombardment of the surface by the fast atoms desorbs the upper layer of analyte and matrix into the vacuum system. Ionization can occur either in the matrix or in the selvage above the matrix. Once the ions are in the gas phase, they are extracted into the analyzer. The purpose of the matrix is to allow replenishment of the surface by analyte from within the sample drop. In its absence, radiation damage rapidly builds up and signal lifetime is quite short. Fast-atom bombardment is generally suitable for ionization of thermally labile or polar molecules.

There are several disadvantages associated with FAB, namely, the high chemical background owing to matrix ions, the dilution of the analyte within the matrix drop, and the decreasing sensitivity and ion suppression effects that are observed because of varying solubility of analytes at the surface or within the body of the droplet. Continuous-flow–FAB [18] was developed to try to alleviate these problems. In CF–FAB, the analyte dissolved in the matrix is delivered continuously to the probe tip through a fused silica capillary line that terminates at the probe tip. Flow rates are about 5–10 μL min^{-1}, with the mobile phase typically being 2–5% glycerol in water.

As part of a collaborative project with Professor James Jorgenson at the University of North Carolina–Chapel Hill, we were involved in interfacing open-tubular LC (OTLC) columns (10-μm id) with mass spectrometry [19]. Our initial approach used a heated interface, which was suitable for thermally stable compounds, but was not as well-suited for polar or thermally labile compounds. To extend the range of compounds that we could analyze, we turned to CF–FAB.

In standard CF–FAB, the glycerol required by the FAB process is added to the mobile phase or added postcolumn. Neither approach was optimum for the OTLC experiments because either the flow rate of the OTLC column (low nL min^{-1}) was too low to deliver sufficient matrix to the probe tip, or the percentage of glycerol in the mobile phase would be too high for optimum chromatography, or because the dead volumes associated with postcolumn addition could seriously compromise the OTLC separation efficiency.

To alleviate these problems, we adapted the coaxial system developed by Rose and Jorgenson [20] for postcapillary fluorescent detection to interfacing OTLC columns with CF–FAB. The same interface design, with only minor modifications, was found to be suitable for use with CE–MS (Fig. 2) [14,21–26]. The significant features of this interface are as follows: (a) The separation column ends at the probe tip, which acts as the "ground" for the system. This means that separation is occurring all the way to the probe tip. For the CE coaxial CF–FAB experiments the probe tip is at the MS source voltage, 8 kV, resulting in an 8-kV "ground". (b) There are no dead volumes in the system. (c) The column has a 13-μm id, which is necessary to reduce vacuum-induced flow. (d) The CE column is pulled back slightly into the matrix column, which also helps reduce the vacuum-induced flow. (e) The flow of matrix around the analyte flow may help constrain the analyte to the center of the probe tip and, thus, to the center of the FAB beam, resulting in improved sensitivity. (f) The CE conditions (buffer and such) and the FAB matrix composition can be independently optimized. For example, separation can be carried out on anionic species at basic pHs, followed by rapid protonation to positively charged species at the probe tip (if the FAB matrix is acidic).

A typical set of operating variables for the CZE coaxial CF–FAB experiments is shown in Table 1. Details of the design of the coaxial interface have been published [14,23,26]. All of the coaxial continuous-flow work has been carried out on a VG ZAB 4F tandem mass spectrometer.

Capillary Zone Electrophoresis–Electrospray Ionization–Mass Spectrometry

The second ionization mode that we have interfaced to MS is electrospray ionization (ESI). In electrospray ionization, a liquid stream flows through a needle electrode held at a high voltage above ground (typically 3 kV). This produces charged droplets at atmospheric pressure. Evaporation of these charged droplets leads to increased charge density, resulting in a coulombic explosion, forming additional, smaller charged droplets. This procedure is repeated until there is sufficient charge density to desorb ions directly into the gas phase. These ions are drawn into the mass spectrometer's analyzer. This technique has two significant features. It is a soft ionization technique, leading to primarily protonated or deprotonated intact molecules, and it results in multiply charged species. The latter feature enables compounds with a high relative molecular mass (M_r), such as proteins, to be mass analyzed because mass spectrometers analyze compounds based on their mass/charge (m/z) ratio. Increasing the charge on an analyte means that the mass can be increased proportionally and still be within the mass range of the instrument. Thus, a quad-

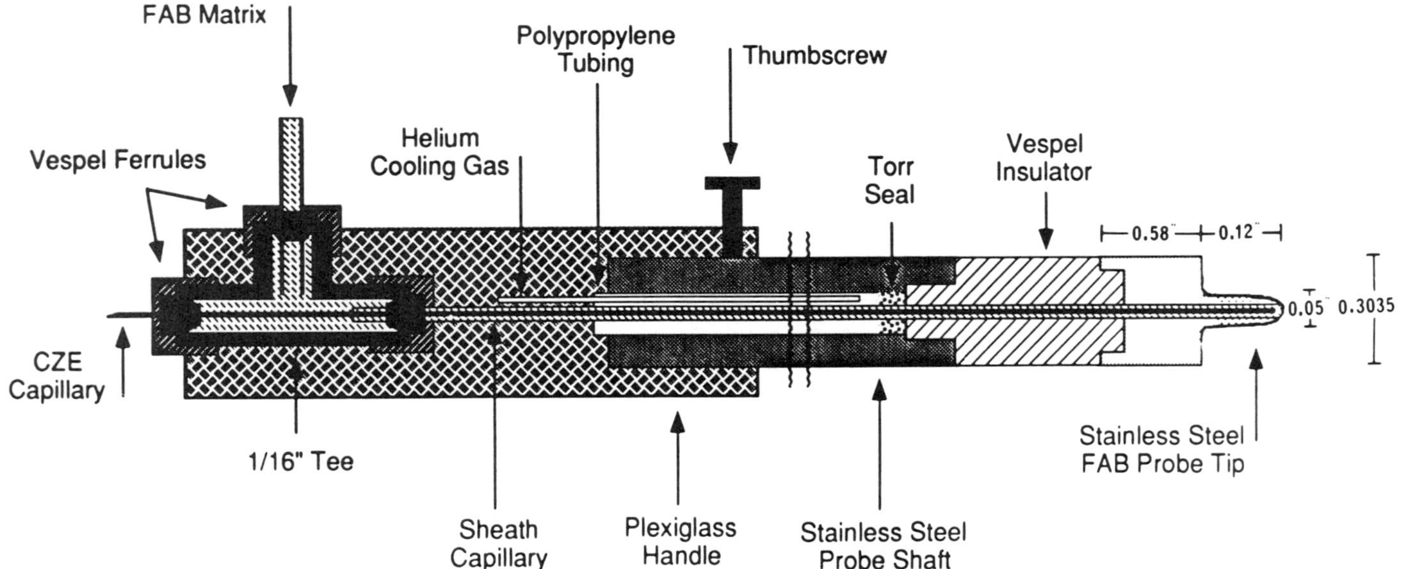

Figure 2 CZE–coaxial continuous-flow FAB interface. (From Ref. 23)

Table 1 CZE–CF–FAB–MS Parameters

Column	1 m × 13-μm id/150-μm od, untreated or coated with aminopropyltrimethoxysilane
Columns flow rates	30 nL min^{-1}
Sample injection volumes	0.25–5 nL
Sheath column	160 μm id/365 μm od
Sheath composition	25:75 glycerol/0.5 mM aqueous heptafluorobutyric acid
Sheath flow rate	0.5–0.3 μL min^{-1}
Voltage drop	±30 kV
Buffer composition	5 mM ammonium acetate adjusted to pH 8.5 with ammonium hydroxide, or 0.01 M acetic acid at pH 3.4–3.5

rupole mass spectrometer with a mass range of 1200 can analyze a protein with an M_r of 12,000, if the molecule carries ten or more protons.

The instrument we are using for the CE–ESI–MS experiments is a VG 12-250 quadrupole that has been updated with a Vestec electrospray source [27]. As indicated in the foregoing, ESI is an atmospheric pressure ionization source and works best with flow rates of 5–10 μL min^{-1}, significantly greater than the electroosmotic flow found in CE. Therefore, a makeup flow needs to be incorporated into the ESI interface design. From our experience with the coaxial CF–FAB interface for use with CE and the success of Smith's use of a coaxial sheath flow in his CE–ESI interface [28–31], we modified the commercial Vestec ESI probe to incorporate a sheath flow for use with CE (Fig. 3) [32]. As in the coaxial CF–FAB probe design, the separation column ends at the probe tip, eliminating any dead volumes. Typical experimental variables are listed in Table 2.

The major advantage of the interfaces just described, other than their conceptual simplicity, is that they can provide optimum separation efficiency. They can do this because (a) separation occurs all the way to the probe tip; (b) no dead volumes are involved; (c) no compromises are necessary between the CE buffer and the matrix–sheath composition; and (d) separation and detection can be optimized independently.

Sample Injection Techniques

There are three injection techniques that have been used for both the coaxial CF–FAB and the ESI–CE–MS systems. These are electromigration, hydrostatic injection, and pressure-induced injection. Although electromigration may be the simplest technique to employ, it does suffer from discrimination effects (i.e., compounds with differing charges will migrate onto the column at different rates) and, thus, is not quantitative.

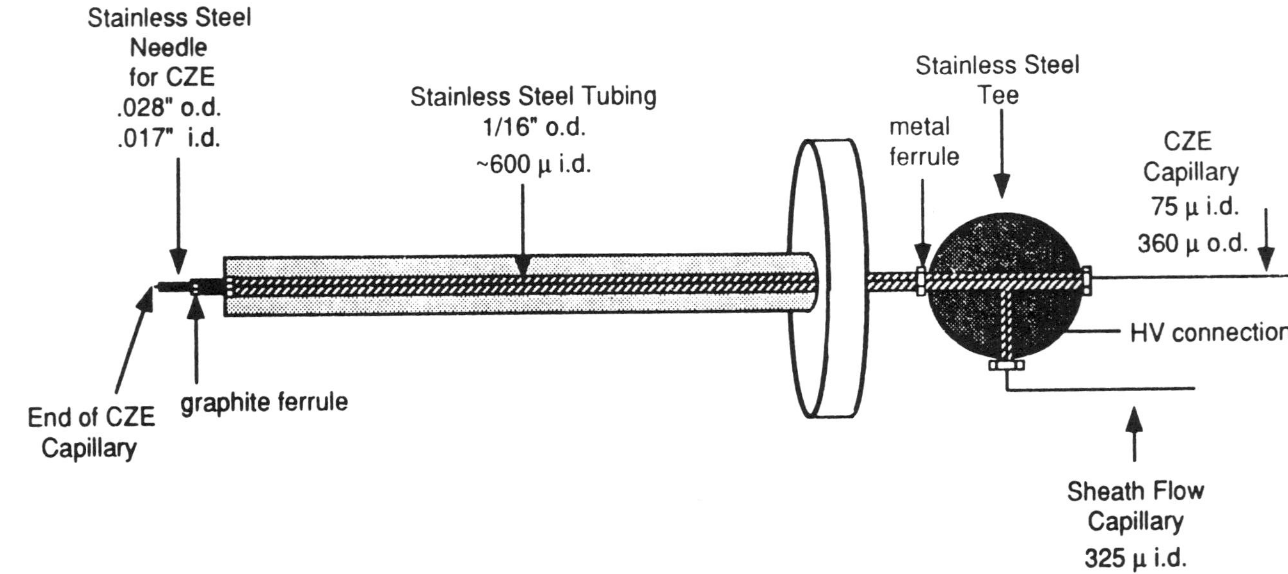

Figure 3 CZE–electrospray ionization probe. (Adapted from Ref. 32)

Table 2 CZE–ESI–MS Parameters

Column	1 m × 75-μm id/375-μm od, untreated or coated with aminopropyltrimethoxysilane
Columns flow rates	300–400 nL min^{-1}
Sample injection volumes	0.25–5 nL
Sheath column	600-μm id stainless steel
Sheath composition	50:50 methanol/3% acetic acid
Sheath flow rate	5–10 μL min^{-1}
Voltage drop	±30 kV
Buffer composition	10 mM ammonium acetate adjusted to pH 8.5 with ammonium hydroxide, or 0.01 M acetic acid at pH 3.4

For the wider-bore columns used with the ESI interface, hydrostatic injection works very well. A vial containing the sample solution is raised a set distance above the ground end of the column and the column end is placed into the vial for a set time. The column is then removed and placed into the buffer reservoir and the separation is initiated. The amount placed onto the column is determined from the flow rate of the sample solution through the column and the amount of time the column is in the solution. The flow rate through the column can be determined by placing and keeping the column in either an analyte solution or a standard solution containing a known analyte and measuring the time it takes for an analyte to appear in the mass spectrometer. Given the calculated column volume and the migration time, a flow rate can be determined.

Pressure-induced injections are especially useful when using the narrower-bore columns (10-μm id) in which the hydrostatic flow rate is very low. We have developed a pressurized injection vessel that can be used for the pressurized injections [25,33]. Reproducible nanoliter volumes can be injected using this device. Flow rates can be determined in the same manner as for hydrostatic injections. Large amounts of sample can also be injected for survey experiments, and large volumes of buffer for column conditioning can be passed through the column in a short period.

EXAMPLES: CAPILLARY ELECTROPHORESIS COAXIAL CONTINUOUS-FLOW FAST-ATOM BOMBARDMENT–MASS SPECTROMETRY

The application of CE interfaced with coaxial CF–FAB to the determination of a series of chemotactic peptides (Fig. 4) illustrates the power of the technique. This separation was performed in a basic buffer (pH 8.5) in which the analytes are negatively charged. At the probe tip, the analytes are doubly protonated by the

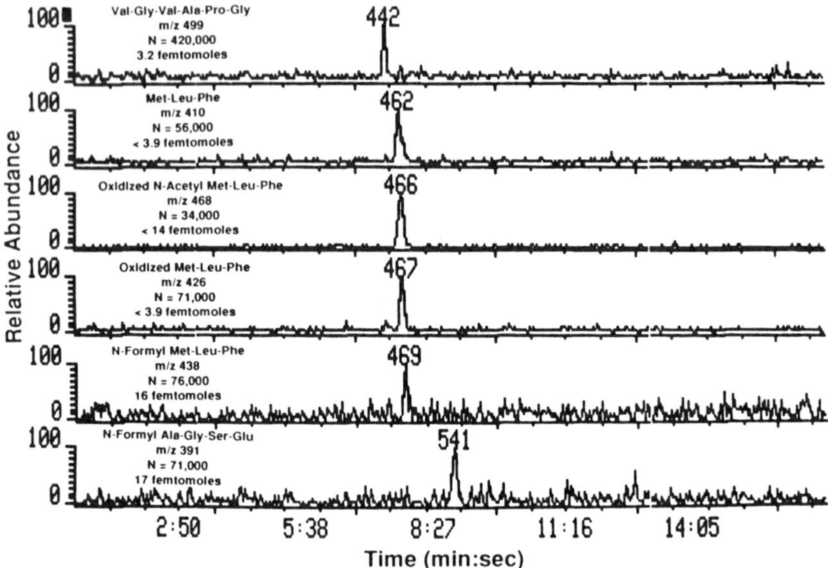

Figure 4 Reconstructed ion electropherograms of $(M + H)^+$ ions of chemotactic peptides separated at pH 8.5 in 0.005 M ammonium acetate containing 1% 2-propanol. (From Ref. 23)

acidic matrix to form protonated molecules. These protonated molecules are then focused and detected by the mass spectrometer. Detection levels are in the low femtomole range, and signal/noise ratios in the reconstructed ion chromatograms of the protonated molecules are typically better than 10:1. Separation efficiency is also very high. The number of theoretical plates for the peaks range from 34,000 to 420,000. The concentrations of the analytes range from 10^{-4} to 10^{-5} M.

This same mixture was also separated and detected as negative ions (Fig. 5). In this experiment, the FAB matrix was at pH 9. Under these conditions, the sensitivity of the FAB–MS was approximately an order of magnitude lower than under positive-ion conditions.

A major disadvantage in working at these low analyte levels is that the total ion current is dominated by the FAB matrix and is typically saturated, especially at masses below about 600. This can definitely create problems in the analysis of unknowns. The use of modern computer technology can circumvent this problem, however. Thus, a three-dimensional (3-D) plot of the data, even over a wide mass range, can be generated on modern workstations in less time than it takes to draw one spectrum on earlier-generation computers. Rapid background subtraction provides the

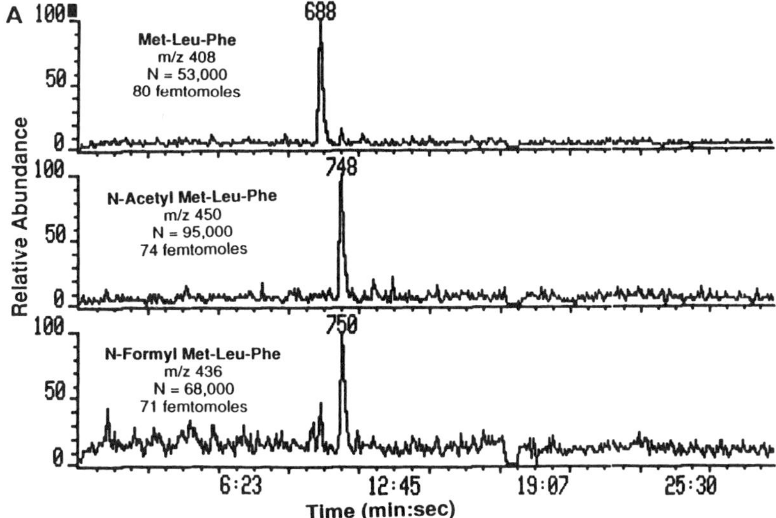

Figure 5 Reconstructed ion electropherograms of $(M - H)^-$ ions of chemotactic peptides separated at pH 8.5. (from Ref. 23)

information that is missing from the total ion current. This is illustrated for the separation of five neuropeptides shown in Fig. 6 [34]. The analytes are readily observable at the 70–90 fmol level using the 3-D capabilities of a Sun workstation. The number of theoretical plates calculated for the components of this separation ranges from 27,000 up to 200,000, while sensitivity is very high, in the low femtomole range.

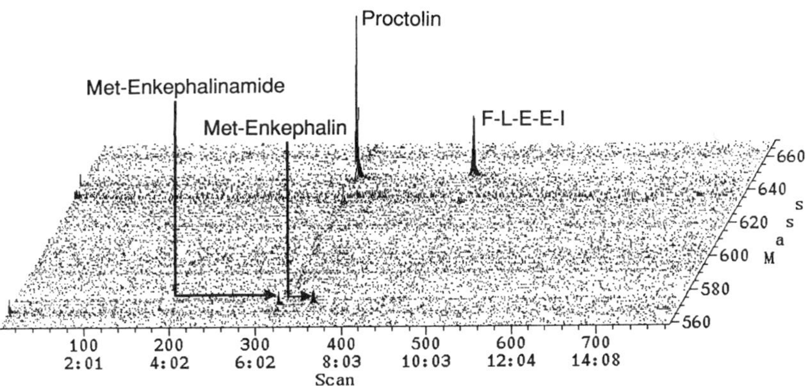

Figure 6 Three-dimensional representation of the spectra obtained during a separation of neuropeptides (From Ref. 34)

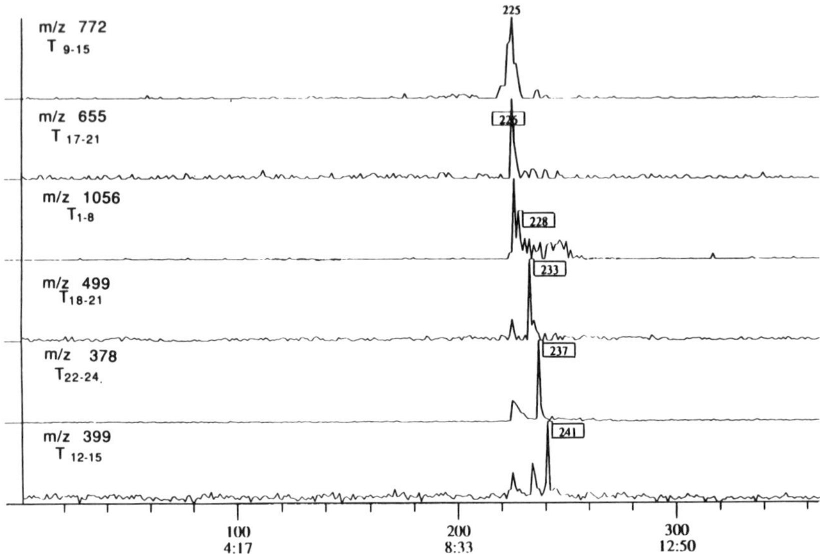

Figure 7 Reconstructed ion electropherograms of the $(M + H)^+$ ions of tryptic fragments of ACTH. The amount injected corresponds to less than 2 pmol of undigested protein.

An example of the separation of a tryptic digest is shown in Fig. 7. All the major tryptic fragments of the peptide, ACTH (1–24), are detected. Under these conditions, the peaks corresponding to three of the fragments, T_{9-15}, T_{17-21}, and T_{1-8}, show significant overlap. They are identifiable and distinguishable on the basis of their mass spectra. All three have distinctly different masses. Better separation can be achieved with lower amounts of analyte being injected, but at the expense of dropping below the detection limits of some of the components.

EXAMPLES: CAPILLARY ELECTROPHORESIS–ELECTROSPRAY IONIZATION–MASS SPECTROMETRY

Earlier in this chapter, it was mentioned that the dynamic range on CE–MS can be limited, with MS detection limits and CE column overloading being similar. This limitation was encountered during a series of experiments on the separation of several macrolide antibiotics. These compounds consist of a 14- or 16-membered lactone, an amino sugar, and one or two sugars. The aminosugar is bound to the lactone ring, whereas the sugars can be bound to the lactone ring or to the aminosugar. To obtain full scan mass spectral data from a separation of these compounds 2-ng per component was needed. This led to column overloading and broad peaks (Fig. 8). The experiments were repeated using selected–ion-monitoring conditions

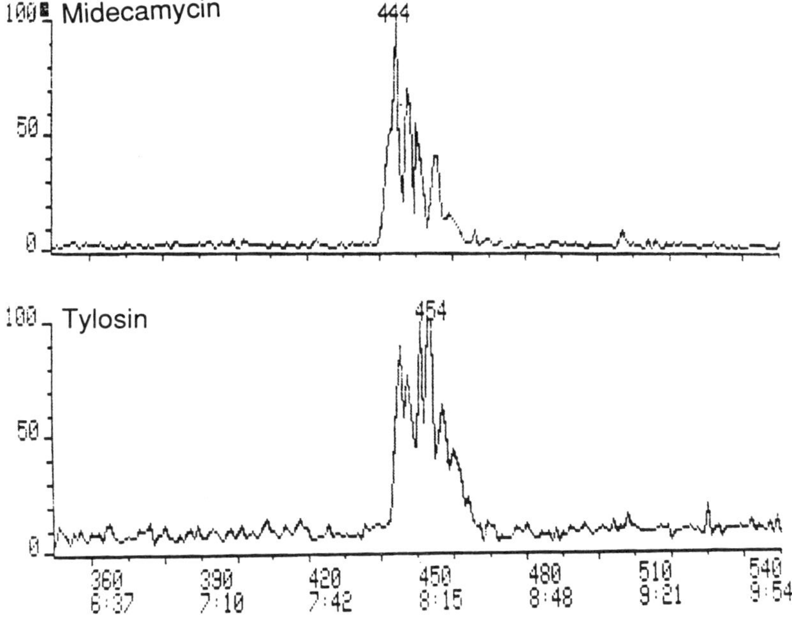

Figure 8 Reconstructed ion electropherograms of the $(M + H)^+$ ions of 2 ng of midecamycin and tylosin obtained in the full-scan mode.

rather than full-scan conditions, and the separation shown in Fig. 9 was obtained from 250 pg per component. In this way, column overloading can be avoided.

Additional improvements in sensitivity can be achieved by using surface-modified columns. In unmodified fused silica columns, positively charged analytes can interact with the negatively charged interior surface of the column. This can lead to adsorption and peak tailing, with concomitant loss of sensitivity and resolution. If the surface is treated so that the interior wall is positively charged, rather than negatively charged, adsorption and peak tailing of positively charged analytes can be reduced. We have treated fused silica columns with aminopropyltrimethoxysilane (APS) to form a positively charged wall [23]. An illustration of the usefulness of APS columns is seen in Fig. 10. This separation is, again, of two macrolide antibiotics. At pH 8.5, the dimethylamino group is still expected to be protonated. Significant adsorption and peak tailing can be seen for the erythromycin and tylosin peaks at this pH. The same separation was repeated on an APS-treated column at pH 3.4. The macrolide peaks are now sharp and show little or no adsorption. Both sensitivity and resolution are enhanced. Sensitivity is improved by a factor of 50, whereas the number of theoretical plates for the peaks is improved by a factor of 2.

In the brief description of electrospray ionization given earlier, it was mentioned that one feature of electrospray ionization is that multiply charged ions can be

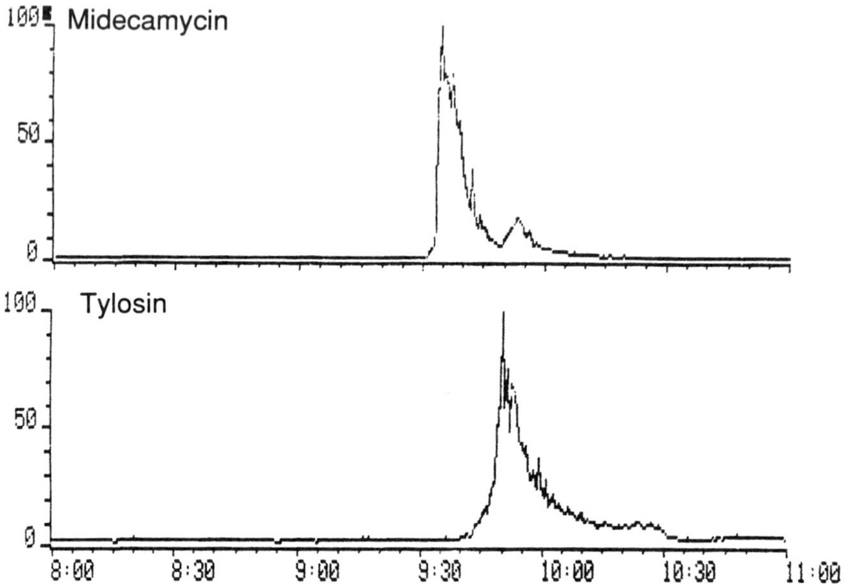

Figure 9 Selected ion electropherograms of the $(M + H)^+$ ions of 250 pg of midecamycin and tylosin obtained by selected ion monitoring.

formed. This is illustrated for simple peptides in the separation of six neuropeptides in Fig. 11. The number of protons on the molecules in the dominant molecular species ranges from one for proctolin to three for α-MSH. This separation was obtained from 800 fmol of each component (50-μM solution) and was obtained in the full-scan mode.

The major advantage of the formation of multiply protonated molecules in ESI is that high-molecular-mass compounds, such as proteins, are determinable on low-mass (*m/z*) range instruments. This permits the use of the mass spectrometer as the detector for the CE separation of complex biological mixtures, such as protein mixtures, oligonucleotides, and glycoproteins. An example of the use of MS as a detector in protein separations is shown in Fig. 12 for the separation of myoglobin, horse heart cytochrome *c*, and pigeon heart cytochrome *c* [35]. All three proteins are well resolved and readily detected under full-scan conditions at the 400-fmol level. The spectra of the horse heart cytochrome *c* obtained from this separation is shown in Fig. 13. Note that the major ion in the spectrum is due to the protein molecule plus 16 protons.

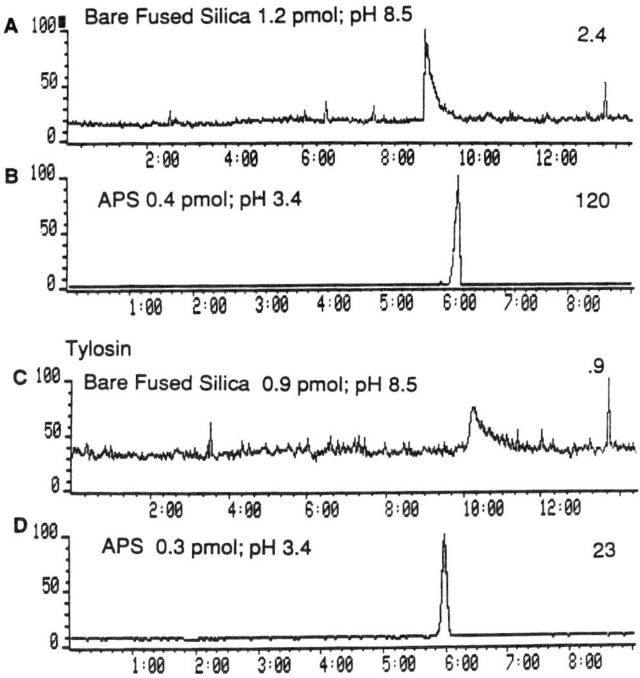

Figure 10 Separation of two macrolide antibiotics: (A) selected ion electropherogram of the $(M + H)^+$ ion of erythromycin at pH 8.5 on fused silica; (B) selected ion electropherogram of the $(M + H)^+$ ion of erythromycin at pH 3.4 on an APS-modified column; (C) selected ion electropherogram of the $(M + H)^+$ ion of tylosin at pH 8.5 on fused silica; and (D) selected ion electropherogram of the $(M + H)^+$ ion of tylosin at pH 3.4 on an APS-modified column.

SUMMARY

Capillary electrophoresis can be readily interfaced with mass spectrometry in combination with either fast-atom bombardment ionization or electrospray ionization. Use of a coaxial approach to the interface, in which the separation column terminates at the probe tip, while any makeup flow is delivered independently, permits independent optimization of separation conditions and ionization variables. The combination of CE with MS provides a sensitive and powerful combination of separation and detection techniques for a wide range of compounds, including high-molecular-mass biomolecules, such as proteins.

ACKNOWLEDGMENT

The author gratefully acknowledges the work performed by Leesa J. Deterding, M. Arthur Moseley, Carol Parker, and John Perkins, and the collaborative efforts of

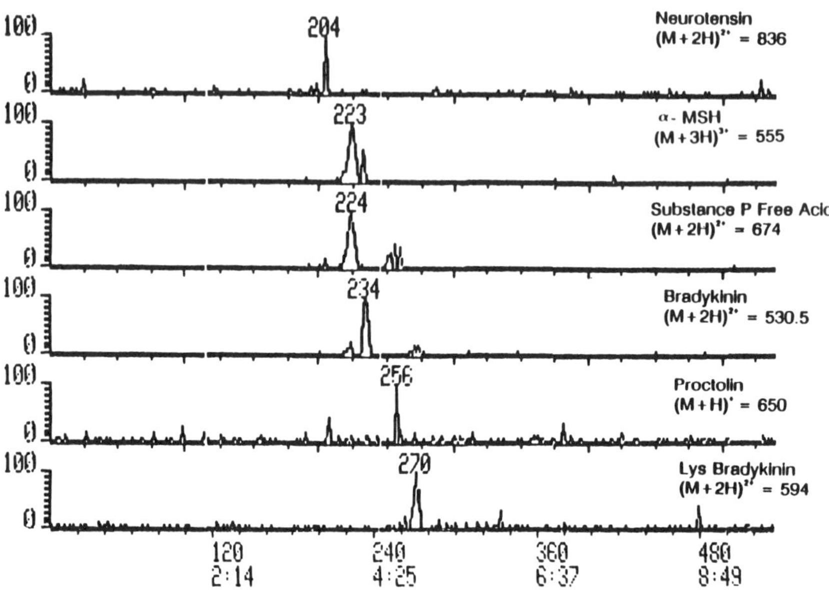

Figure 11 Reconstructed ion electropherograms of the major protonated molecules of six neuropeptides obtained by CZE–ESI–MS. This is the same separation as shown in Fig. 1.

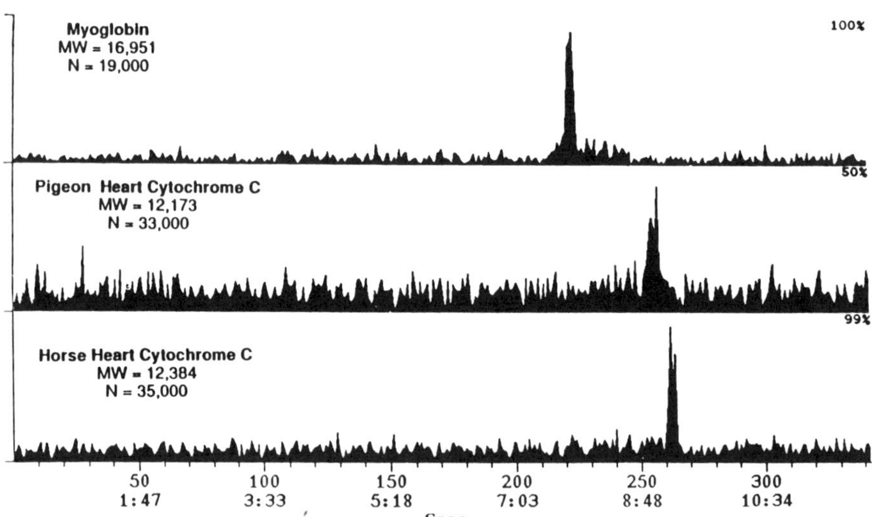

Figure 12 Reconstructed ion electropherograms of the major ion obtained in the separation of 400 fmol each of myoglobin, horse heart cytochrome *c*, and pigeon heart cytochrome *c*, with detection by ESI–MS (From Ref. 35)

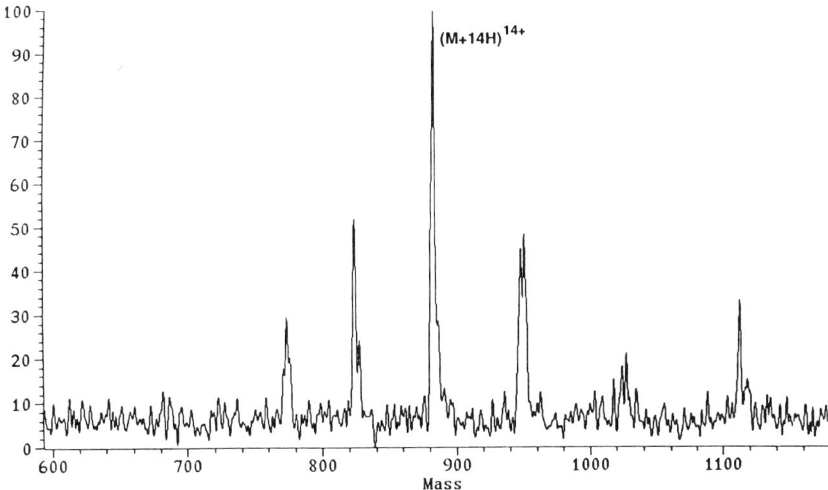

Figure 13 Mass spectrum of 400 fmol of horse heart cytochrome *c* obtained from the separation in Fig. 11. Note that the major ion $(M + 14H)^{14+}$, is that shown in Fig. 11. (From Ref. 35)

Professor James Jorgenson of the University of North Carolina and Professor Yasuo Shida of the Tokyo College of Pharmacy. Without these persons' efforts, this work could not have been done.

REFERENCES

1. J. W. Jorgenson and K. D. Lukacs, *Anal. Chem.*, *53*:1298 (1981).
2. J. W. Jorgenson and K. D. Lukacs, *J. Chromatogr.*, *218*:209 (1981).
3. J. W. Jorgenson and K. D. Lukacs, *Science*, *222*:266 (1984).
4. J. W. Jorgenson, *Anal. Chem.*, *58*:754A (1986).
5. B. Nickerson and J. W. Jorgenson, *J. High Resolut. Chromatogr. Chromatogr. Commun.*, *11*:553 (1988).
6. J. A. Olivares, N. T. Nguyen, C. R. Yonker, and R. D. Smith, *Anal. Chem.*, *59*:1230 (1987).
7. M. Dole, L. L. Mack, R. L. Hines, R. C. Mobley, L. D. Ferguson, and M. B. Alice, *J. Chem. Phys.*, *49*:2240 (1968).
8. M. Dole, H. L. Cox, Jr., and J. Gleniec, *Adv. Chem. Ser.*, *125*:73 (1973).
9. J. B. Fenn, M. Mann, C. K. Meng, S. F. Wong, and C. Whitehouse, *Science*, *246*:64 (1989).
10. C. M. Whitehouse, R. N. Dreyer, M. Yamashita, and J. B. Fenn, *Anal. Chem.*, *57*:675 (1985).
11. S. F. Wong, C. K. Meng, and J. B. Fenn, *J. Phys. Chem.*, *92*:546 (1988).

12. E. D. Lee, W. Mueck, J. D. Henion, and T. R. Covey, *J. Chromatogr.*, *458*:313 (1988).
13. R. D. Minard, D. Chin-Fatt, P. Curry, and A. G. Ewing, *Proceedings of the 36th Annual Conference on Mass Spectrometry and Allied Topics*, San Francisco, June 5-10, p. 950 (1988).
14. M. A. Moseley, L. J. Deterding, K. B. Tomer, and J. W. Jorgenson, *Rapid Commun. Mass Spectrom.*, *3*:87 (1989).
15. R. M. Caprioli, W. T. Moore, M. Martin, B. B. DaGue, K. Wilson, and S. Moring, *J. Chromatogr.*, *480*:247 (1989).
16. N. J. Reinhoud, W. M. A. Niessen, V. R. Tjaden, L. G. Gramberg, C. R. Verheij, and J. van der Greef, *Rapid Commun. Mass Spectrom.*, *3*:348 (1989).
17. M. Barber, R. S. Bordoli, R. D. Sedgwick, and A. N. Tyler, *J. Chem. Soc. Chem. Commun.*, p. 325 (1981).
18. R. M. Caprioli, *Anal. Chem.*, *62*:477A (1990).
19. J. S. M. de Wit, C. E. Parker, K. B. Tomer, and J. W. Jorgenson, *Anal. Chem.*, *59*:2400 (1987).
20. D. J. Rose, Jr. and J. W. Jorgenson, *Anal. Chem.*, *60*:1840 (1988).
21. M. A. Moseley, L. J. Deterding, K. B. Tomer, R. T. Kennedy, N. L. Bragg, and J. W. Jorgenson, *Anal. Chem.*, *61*:1577 (1989).
22. M. A. Moseley, L. J. Deterding, K. B. Tomer, and J. W. Jorgenson, *J. Chromatogr.*, *480*:197 (1989).
23. M. A. Moseley, L. J. Deterding, K. B. Tomer, and J. W. Jorgenson, *Anal. Chem.*, *63*:109 (1991).
24. J. S. M. de Wit, L. J. Deterding, M. A. Moseley, K. B. Tomer, and J. W. Jorgenson, *Rapid Commun. Mass Spectrom.*, *2*:100 (1988).
25. L. J. Deterding, M. A. Moseley, K. B. Tomer, and J. W. Jorgenson, *Anal. Chem.*, *61*:2504 (1989).
26. K. B. Tomer and M. A. Moseley, In *Continuous-Flow Fast Atom Bombardment Mass Spectrometry* (R. M. Caprioli, ed.), John Wiley & Sons, New York p. 121 (1990).
27. M. H. Allen and M. L. Vestal, *J. Am. Soc. Mass Spectrom.*, *3*:18 (1992).
28. R. D. Smith, J. A. Olivares, N. T. Nguyen, and H. R. Udseth, *Anal. Chem.*, *60*:436 (1988).
29. R. D. Smith, C. J. Barinaga, and H. R. Udseth, *Anal. Chem.*, *60*:1948 (1988).
30. R. D. Smith, J. A. Loo, C. J. Barinaga, C. G. Edmonds, and H. R. Udseth, *J. Chromatogr.*, *480*:211 (1989).
31. J. A. Loo, H. R. Udseth, and R. D. Smith, *Anal. Biochem.*, *179*:404 (1989).
32. C. E. Parker, J. R. Perkins, K. B. Tomer, Y. Shida, K. O'Hara, and M. Kono, *J. Am. Soc. Mass Spectrom.*, *3*:563 (1992).
33. R. M. Caprioli and K. B. Tomer, In *Continuous-Flow Fast Atom Bombardment Mass Spectrometry*, (R. M. Caprioli, ed.), John Wiley & Sons, New York, p. 93 (1990).
34. K. B. Tomer, K. B. Moseley, L. J. Deterding, C. Parker, J. Perkins, and J. W. Jorgenson, *Nippon Iyo Masu Supekutoru Gakkai Koenshu*, *15*:87 (1990).
35. L. J. Deterding, C. E. Parker, J. R. Perkins, M. A. Moseley, J. W. Jorgenson, and K. B. Tomer, *J. Chromatogr.*, *554*:329 (1991).

18

Optical Detection Schemes for Capillary Electrophoresis

Edward S. Yeung

Iowa State University
Ames, Iowa

Detection of the separated components in capillary electrophoresis (CE) poses some unique challenges to analytical chemists. Three important factors must be considered. These are the small volume, the high data rate, and the special solvent environment that are characteristic of CE separations.

Although practitioners of liquid chromatography (LC) are accustomed to volumes of the eluted bands in the 100–1 μL range for analytical scale and microcolumn instruments, bands of 10 nL–10 pL are common in CE. This is a direct result of the high-separation performance, up to millions of theoretical plates [1], and the small internal diameters of the columns used, from 100 μm down to 2 μm. The small diameters are required to limit joule heating, which can lead to nonuniform temperature profiles and turbulent mixing of the well-defined zones. The most popular capillary dimension in commercial instruments is 50-μm id. This has allowed adaption of absorbance detectors in LC for CE monitoring. In fact, all that is required is additional optics based on ball (spherical) lenses to collimate light across the capillary (i.e., along a diameter). Because the path lengths through the quartz and through the solvent are quite short, wavelengths down to 200 nm can be effectively used. For organic compounds with any type of functionality, this detection mode is then almost universal. The simplicity and close analogy to LC instrumentation led to the success of the first wave of commercial instruments, as well as the first series of publications on diverse applications of CE as an analytical tool.

Unfortunately, as the field advances, more demands are being placed on the detection power of CE instruments. One quickly faces the limits set by Beer's law. It is easy to see that for a 50-μm absorption path length, the best concentration limit of detection (LOD) even for compounds with large molar absorptivities (ε = 10^4 L mol^{-1} cm^{-1}) is about 2 μM. Actually the LOD is not even as good as this because not all the light passes through the middle of the capillary and vibrations in the presence of the electric field bring additional alignment noise. In some instruments, one also has contributions from shot-noise owing to inefficient coupling of light for the system as a whole. The amount of light available is not easily increased by lengthening the imaged area along the capillary, since the zones are only a few millimeters long. It is unlikely that the standard absorbance detector can be redesigned to provide more sensitive detection.

High separation speed [2] is another attractive feature of CE. This is a benefit of electrophoretic and electroosmotic transport, which can be substantial at high fields, compared with pressure-driven flow in LC. Data acquisition, therefore, must keep up with these high rates. Given the standard guideline of requiring at least ten data points for each peak [3], a reasonable time constant for the detector is 0.1 second. This means that measurement times are roughly ten times shorter than in LC. With the advance of computer technology, data storage, transfer, and analysis are not bottlenecks. However, the amount of light (or electrons, in the case of electrochemical detection) reaching the sensor is correspondingly less for each integration period, so the signal/noise ratio (S/N) will be about three times (= $\sqrt{10}$) lower for shot-noise-limited detection. The situation actually is more favorable for flicker noise resulting from misalignment or source intensity fluctuations. These fluctuations vary as $1/f$, where f is the frequency relevant to the measurement and, therefore, are suppressed by about ten times.

The inherent ionic nature of the separations restricts the choice of solvents for CE. Although organic systems can be used, in principle [4], most of these call for water-based salt buffer systems. A totally organic system would allow simple vaporization of the solvent and eliminate the major component to avoid potentially high background signals and enhance detection in general. For water-based systems, even if the solvent is vaporized, the buffer salts can still overwhelm the analytes. This has led to specialized buffer systems in mass spectrometric (MS) applications [5]. Micelles in CE further complicates the situation because the monomers are at high concentrations. The need for a conducting electrolyte in CE also presents a problem for conductivity detection, in which a difference in conductance is sensed, and even a small voltage drop between the detection electrodes is undesirable [6]. Finally, extra attention must be given to the compatibility of the buffer type and ionic strength for electrochemical detection. Also, one must always consider the relative mobilities of the analyte ions and the buffer ions to maintain sharp peaks [7].

ABSORBANCE DETECTION

Absorbance detection in CE is limited primarily by the available path length and, to a smaller extent, by the efficiency of light coupling. A possible approach is to use a light source with higher intensity and with better focusing characteristics. Lasers are likely candidates. However, the intensity stability of typical lasers are in the ± 1 to $\pm 0.1\%$ range, which at best translates to an absorbance noise of 0.001. Diode lasers can be significantly more stable, but spectral output is presently unavailable in the most useful UV wavelengths. The mode structure of diode lasers is also not optimal for focusing to 50 μm and below.

The obvious way to enhance performance in absorption detection is to increase the absorption path length. The length of bands in CE is a few millimeters. Even though this is short compared with bands in LC, it is substantially longer than the 50-μm path lengths along the diameter of the capillary tubes. This aspect ratio suggests orienting absorption detection along the capillaries, rather than across them. This scheme has been demonstrated for capillary LC by locking the flexible capillary in a Z-configuration [8]. An increase in absorbance was, in fact, demonstrated. The key is that the irradiated path length must be short compared with the band length, typically one-tenth, to avoid band broadening. This is much more difficult to implement in CE, for which the bands are only a few millimeters long. A submillimeter Z band is beyond the flexibility limit of capillaries. An additional problem is introduced in this simple arrangement because there is a substantial fraction of light guided down the quartz capillary walls, decreasing the effective absorbance and increasing nonlinearity in response. A fluorescence-based absorption scheme was reported, based on the decrease in fluorescence intensity (source depletion) as a laser beam travels from the exit end of a capillary into the column [9]. Photobleaching of the fluorophore and stray light severely limited the performance of this system. Another related scheme is based on multiple reflection of the light from a laser across the capillary column [10]. The silvered surface of the capillary traps the light beam and effectively increases the absorption path length. Unfortunately, even though the laser beam starts out as a well-focused source, it rapidly diverges on reflection, and only a few passes are possible. The particular geometry also means that a section of the capillary a few millimeters in length contributes to the signal. It is difficult to achieve the submillimeter total interaction length that is required for the highest separation performance.

To sidestep the limitation of having an interaction length much shorter than the band lengths, we devised a scheme for shining light down the axis of the entire capillary [11]. This axial-beam geometry then takes advantage of the entire length of each band for absorption measurement. Under normal operation in CE, light transmission down the center of the capillary column is extremely inefficient. However, by using a laser source and by pulling the capillary straight, adequate intensity can be detected at the opposite end for 75-μm capillaries. This is in effect a very poor but finite *f* number. At the same time, light is also transmitted from one end to another through the quartz walls. This can be discriminated against by suitable imaging of the exit end of the capillary. As the internal diameter of the capillary decreases to 25 μm, we found that the decreased center intensity and increased wall coupling make it impractical to perform absorption measurements in CE.

Efficient transmission of light can be accomplished in the total internal reflection mode. Here, the refractive index (RI) of the core must be larger than the RI of the outer jacket. The growing importance of fiber optics in communications is based on this phenomenon. The parallel is to provide a solvent medium (liquid core) that has a higher RI than the column walls (jacket). In LC this is fairly easy because organic solvents generally have a high RI. For example, a pyridine eluent for LC satisfies the requirement for total internal reflection in a quartz capillary column [11]. Water, unfortunately, has a low RI among common liquids. Still, the use of a 50:50 ethylene glycol/water mixture produces total internal reflection in Teflon capillary tubes [12]. There are materials for capillaries that possess even lower RI [13], but there are as yet no commercial sources for these. The optical quality (internal smoothness) of these capillaries also leaves much to be desired. To work with standard quartz capillaries, we have used a 98% dimethyl sulfoxide (DMSO) modifier for the water–salt buffer mixture for CE [14]. There will be a change in fractional ionization (p*K*a) and solubility, but the basic CE mechanism is still operative. In principle, it may be possible to coat the interior of the quartz capillaries with a material with low RI, such as saturated hydrocarbons (e.g., wax), to force total internal reflection. The additional benefit is the reduction of electroosmotic flow by shielding the surface silanol groups. It must be noted, however, that the coating must be thick compared with the wavelength of light for implementation. This can be difficult for very small capillaries.

At first sight, it seems strange to monitor absorption along the entire column for any separation scheme. All components contribute to absorption at all times and are not separately detected. A typical electropherograms obtained in the axial-beam absorption mode is shown in Fig. 1. Before sample introduction, there is a finite amount of light, I_0, passing to the detector, dependent on the source intensity and the coupling efficiency. On injection of a sample, this intensity is reduced because of absorption of any or all of the components. This level of transmission is maintained as the components migrate down the column, providing no information about the separation. When the components elute out of the column one at a time, the transmitted intensity increases in steps until it returns to the original level when all components are eluted. It can be shown that such an electropherogram gives information identical with that derived in standard absorption detectors. The step height is equivalent to the integrated peak area. The inflexion point of the step is equivalent to the peak maximum. The slope of the step is related to the peak width and peak shape. In fact, this is no more than the integral electropherogram [15],

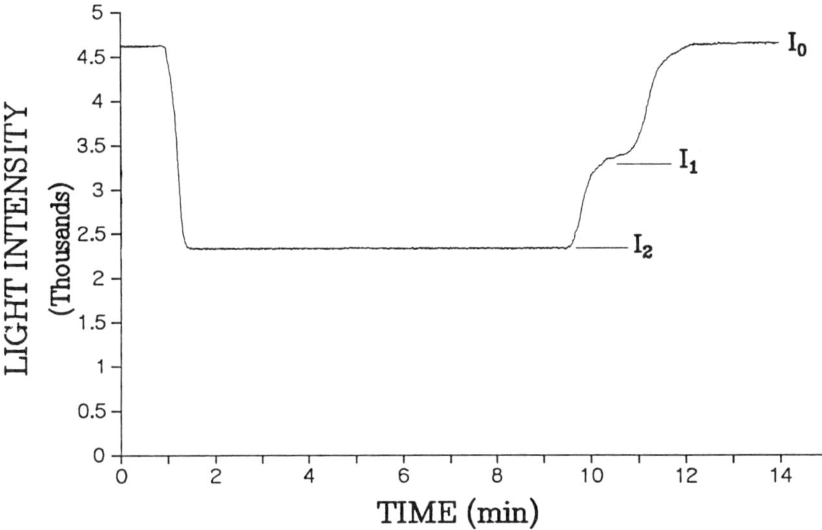

Figure 1 Axial-beam on-column absorption detection of azulene (AZ) and bromocresol green (BG). I_0, baseline intensity; I_1, intensity with BG; I_2, intensity with AZ and BG. (From Ref. 11)

and can be confirmed by transforming this to the traditional plots by numerical derivatization. It may initially appear that such an electropherogram degrades the resolution because the components are simultaneously monitored. However, since detection occurs only when the components are on the column, this scheme is truly on-column and no extracolumn broadening is involved.

The initial arrangement relies on a laser source for measurement. This is unnecessary for absorption detection because only relative intensities are involved. There needs to be only enough light to overcome the shot-noise limit.

Lasers can actually degrade performance, since the available intensity stability is far worse than from conventional light sources. We have, therefore, modified a commercial CE absorption detector for axial-beam monitoring [14]. The ball-lens optics in these instruments already provides a reasonably collimated beam for 50- to 75-μm capillaries. One can simply reorient the capillary column so that one end (entrance or exit) is directly irradiated through the ball lens. This is not efficient coupling, but it is quite adequate for axial-beam monitoring when total internal reflection is in effect. The multiple wavelength capability of commercial absorption detectors can then be retained.

The gain in sensitivity in the axial-beam mode is simply the ratio of the zone length to the column diameter. It is irrelevant whether the zone is broadened during separation. The lower concentration will be exactly compensated by the longer interaction length. One gains some additional advantage because in standard absorption detectors the effective path length is actually smaller than the column diameter owing to imperfect collimation. However, one loses some detection power owing to a reduced and a less stable light intensity. The latter is a result of the physical stability of the capillary column, altering the path of the light beam inside the capillary.

FLUORESCENCE DETECTION

The high sensitivity of fluorescence detection in LC, especially schemes involving lasers, has been well documented in the literature [16]. Fluorometry is basically an absorption technique, except that reemitted light, rather than the transmitted light, is monitored. The gain in LOD is a direct result of the low-background levels at the fluorescence wavelength and in the direction of observation. Low-background levels generally also mean low-noise levels, enhancing the ability to detect small changes in intensity. Concentration LOD in cuvettes is in the 10^{-12} M range upon laser excitation. For detection in capillaries, although the amount of light absorbed scales with path length (to a first approximation) just like absorption measurements, the decrease in fluorescence intensity can be compensated for by increasing the excitation intensity. This is why laser-excited fluorescence has been so successful as a detector in microcolumn LC [17] and in CE [18]. In fact, single-molecule detection has recently been demonstrated [19].

Several factors need to be considered in fluorescence detection in CE. Stray light from the optical components and the capillary walls leads to increased noise levels. This is especially important for UV excitation, since there is a higher probability for materials to fluoresce. Stray light rejection has been the subject of extensive research, even for LC applications. These include a windowless cell [20], a flowing jet [21], imaging by a microscope objective [21], and sheath flow isolation [22]. Of these, the last two are applicable to the small dimensions and low-flow rates of CE. It is not surprising that similar principles were used early on for CE detection. The process of scaling downward in size means that mechanical mounting must be more stable and finer adjustments have to be made. Here, the high degree of collimation of the laser beam, which readily allows focusing down to micron sizes, is a major advantage. Stray light from the solvent (CE buffer) is often the limiting factor in detection performance. This originates from the same physical region and cannot be discriminated against by apertures and good collimation. At the picomolar detection level, trace fluorescent contaminants that are normally negligible can become major contributors to the background. Special treatment of water, buffer salts, glassware, and even high-grade reagents is necessary for ultimate performance. Contaminants can include submicron particles, which act as scattering centers and produce spikes in the electropherogram. These cannot be removed completely by normal filtration. The solvent also contributes by Raman scattering. This can be at the exact wavelength of fluorescence detection, ruling out the use of spectral discrimination. Fortunately, water is a particularly poor Raman scatterer and most CE applications benefit from this fact. The only remedy is to force spectral separation between the Raman bands and fluorescence, such as by using fluorophores with a large Stokes shift or by using D_2O. The latter is easy to implement because only very small volumes are used in CE. Finally, although increased excitation intensity is in general desirable, it can cause some problems. Highly focused laser beams can damage the capillary walls, especially if they are from pulsed sources. Bleaching of the fluorophores can occur at high-power densities, putting an upper limit on the available fluorescence intensity. Minor impurities in the solvent or dust particles in the air can become fused to the capillary walls, destroying the optical quality of the detector and increasing stray light. Experience shows that output from continuous lasers in the order of 1–5 mW, focused down to a 5-μm beam, is optimal for fluorescence detection in CE.

In comparing the various optical arrangements for fluorescence detection in CE, each has some advantages and disadvantages. Sheath flow offers the lowest reported LOD because rejection of stray light from the optical components is essentially perfect. Detection down to a few hundred molecules has been achieved [23]. However, there is one more parameter that requires adjustment in the system. This arrangement also is not easily compatible with sample collection after CE, or for operation at nonstandard configurations, such as a high potential at the detector end, or when manipulation of the buffer at the exit end is needed. On-column imaging is relatively easy to do, but difficult to do well because of the presence of the quartz capillary wall that is cylindrical. A window has to be introduced by removing the polyimide coating (which fluoresces). This makes the capillaries brittle and is a complication in gel-filled capillaries, since the window has to be created before filling. Light coupling by an optical fiber and a ball lens, as in one commercial fluorescence detector, is the most rugged and the easiest to align. However, the quartz capillary wall is illuminated nondiscriminately, and the background level is higher. This will be particularly critical if UV excitation is used. In general, detection power decreases from sheath flow, to on–column-imaging, to on–column-nonimaging schemes.

Unlike conventional light sources, laser radiation is not readily available in a broad range of wavelengths. Tunable lasers are as yet simply too expensive and too complicated for use with CE. In general, continuous lasers are more stable in intensity, easier to focus to micron dimensions, and less likely to cause damage to the optics. The exception is the waveguide excimer laser at 248-nm or 308-nm operation. Rare gas ion lasers are capable of producing several discrete laser lines that can be selected either internally or externally. A dual-wavelength system has been successfully used for exciting different tags for DNA sequencing. Table 1 summarizes the different types of lasers commonly used for CE detection. The

Table 1 Lasers Used for Fluorometric Detection in CE

λ (nm)	Type	Cost ($)	Stability	Comments
830	Diode	300	+ +	Poor focusing
760	Diode	300	+ +	Poor focusing
670	Diode	300	+ +	Poor focusing
633	HeNe	300	0	
543	HeNe	1,500	0	
514	Ar	5,000	+	
488	Ar	5,000	+	
459	Ar	8,000	+	Cheaper if less stable
442	HeCd	8,000	0	Short lifetime
364	Ar	20,000	+	
350	Ar	20,000	+	
337	N_2	15,000	−	Very poor focusing
334	Ar	30,000	0	
325	HeCd	13,000	0	Short lifetime
308	XeCl	15,000	−	Pulsed
305	Ar	50,000	0	
275	Ar	50,000	0	
257	Ar	40,000	− −	Frequency doubled
248	KrF	15,000	−	Pulsed

comparison is based on commercial products that will produce 1–5 mW of light. Selection of the proper source naturally depends on the particular application, cost, and convenience. Still, the performance of the detector can vary substantially, as indicated by the last two columns in Table 1. One additional note is that diode lasers are being actively developed for UV output, owing to the needs in data storage and in communications. This may become the laser source for CE applications in the future.

Although it is easy to show good LOD at most of these laser wavelengths in fluorometric detection in CE, developing a procedure for a particular application may be fairly involved. Good fluorophores are more abundant in the UV region, at which lasers are the most expensive and least reliable. Except for certain drugs, porphyrins, and flavins, very few compounds of interest are naturally fluorescent under visible or near-infrared (IR) irradiation. Much effort has been directed toward the design of fluorescent tags for much broader classes of compounds. Of these, derivatives that can be excited at 325 nm, 442 nm, and 488 nm are most common. For amino acids, these wavelengths nicely fit the absorption bands of dansyl, napthylphthaldehyde, and fluorescein derivatives, respectively. Applications to a broad range of biological materials are thus possible. In considering LOD for derivatized material, one must be cognizant of the exact experimental conditions used. In most reports in the literature, derivatization was performed at high concentrations, and the reaction products were then diluted to very low levels for CE studies. This underscores the sensitivity of laser fluorometric detection, but may have limited usefulness for actual applications. When the amount of analyte available is small, chemical manipulation may be very difficult. When the concentration of the analyte is low, strong adsorption and slow diffusion may prevent complete derivitization. What one can actually measure in CE, therefore, may be substantially less than what is required of the starting material.

An interesting application of laser fluorometric detection is in the study of modified nucleotides in DNA [24], which usually precedes carcinogenesis. The level of modification can be as low as $1{:}10^8$ bases. If one starts with 1 μg of DNA, this implies that only femtogram levels of modified material will be found. Fluorometric detection is an obvious solution. To convert the DNA bases into fluorescent compounds, one can use the dansyl derivative. This is conveniently excited at 325 nm and detected at 400 nm. For the separation of standard nucleotides, the necessary detection power at the $1{:}10^8$ level of modification is confirmed (Fig. 2A). The crucial test comes in the study of modified nucleotides in the presence of a large excess of normal nucleotides, (see Fig. 2B). Enrichment by high-performance LC (HPLC) in calf thymus DNA digest can provide a factor of 10^4 enhancement. That, combined with the CE scheme, should make it possible to study actual DNA damage. The unique feature in Fig. 2B is that CE zones are very sharp compared with LC peaks, enabling the detection of minor components eluted adjacent to the major components.

The tremendous potential of laser-excited fluorescence detection in CE is best illustrated by the recent study of proteins by their native fluorescence [25]. Fluorescence derivatization of proteins is well known in LC and in CE. Unfortunately, proteins have multiple sites for derivatization. Often a series of derivatives are produced, rather than a single derivative. This complicates the chromatogram or electropherogram and leads to irreproducible quantitation. Improved manufacturing techniques in the last few years have produced commercial argon ion lasers that operate at 275 nm and other UV wavelengths. This is a good match for exciting tyrosine and tryptophan groups in peptides and proteins. These are not extremely efficient fluorophores, but reasonable sensitivity might be expected using laser excitation. Before our work, native protein fluorescence has been excited by the frequency-doubled output of an argon ion laser [26]. Because of mismatch in the excitation wavelength and the unstable laser output, detection required 15-nM concentrations of proteins.

Figure 3 shows a CE peak corresponding to injection of a 0.5-nM sample of conalbumin. The LOD is thus around 0.1 nM. The improvement is the combined result of more efficient excitation, high-intensity stability, careful imaging for rejecting fluorescence from the capillary walls, and the use of a high pH (10) to increase the quantum yield. Since typical protein samples are at pH 7, there is a preconcentration effect at injection, similar to that of stacking in electrophoresis. In that case, the LOD becomes 3 pM injected. This is as low as any precolumn derivatization scheme, but without the complications. Figure 3 also addresses one of the important questions about

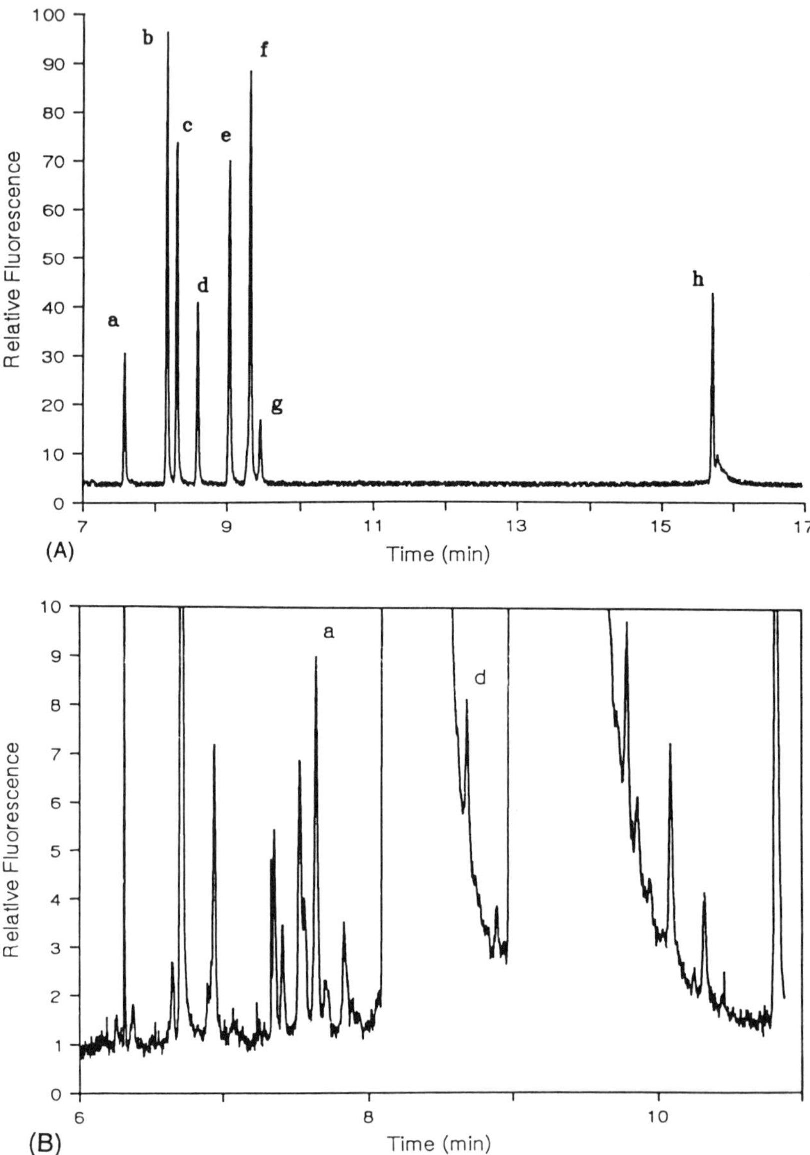

Figure 2 MECC–LIF analysis of deprotonated dansyl hydroxide (a), dansyl dTMP (b), dansyl dCMP (c), dansyl 5-MedCmp (d), dansyl dAMP (e), dansyl dGMP (f), dansyl 8-OHdGMP (g) and dansyl 8-AAFdGMP (h). (A), standards; (B), 10^4 excess of dNMP.

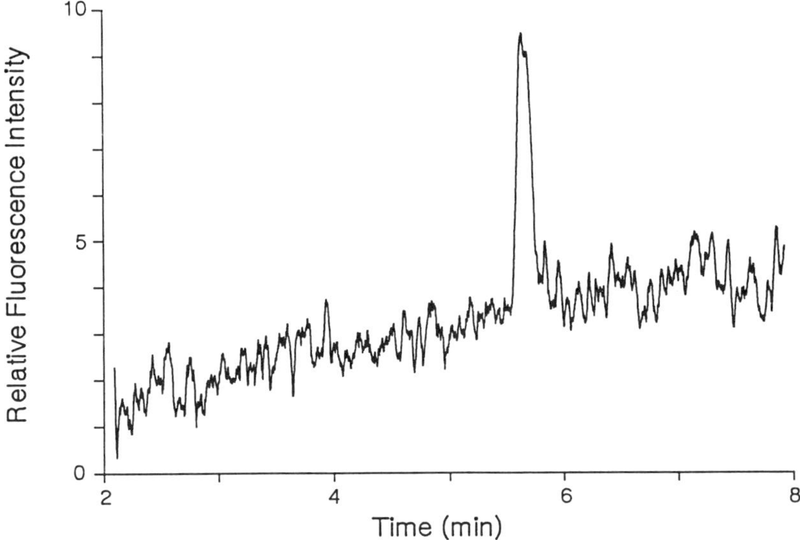

Figure 3 Electropherogram of 5×10^{-10} M conalbumin injected based on native fluorescence detection.

recovery. It shows that, under these conditions, wall adsorption of these proteins is not a problem. The observed LOD is, therefore, of practical value. In fact, the calibration plot is linear from the detection limit up to the micromolar range, indicating >90% recovery of the injected material at attomolar quantities.

This development opens up a broad range of biochemical and clinical applications, that is, for proteins and peptides containing tyrosine or tryptophan groups. For biotechnological products, one can detect trace impurities much more readily. The clinical trials of these products can be facilitated by the ability to follow metabolites over long periods, when their concentrations are low. Direct monitoring of peptides and proteins in single cells should be feasible. For protein characterization, tryptic digest maps can be obtained on much smaller quantities of proteins, provided that digestion in microchambers can be developed. Other information about proteins can become accessible, such as their on-column conformation during CE, through spectral shifts in fluorescence and even fluorescence-detected circular dichroism [27].

INDIRECT FLUORESCENCE DETECTION

Since most compounds of interest do not fluoresce appreciably, it is important to develop techniques that are suitable for detecting these at the small volumes and the low concentrations typically found in CE. The advantages of the laser for coupling to small columns have already been mentioned. It should be useful if laser-excited fluorescence can be adapted for detecting nonfluorescent molecules. This can be accomplished in a scheme known as indirect fluorometry [28].

Indirect detection methods have been known for a long time. Most of the reports have been in LC, particularly ion chromatography [29], and the adaptation for CE is fairly recent [30]. The phenomenon is fairly easy to understand. In CE, we always have an electrolyte in the column to provide electrical conductivity. Otherwise the applied potential will not create a uniform field inside the column for electrophoresis. This electrolyte typically also acts as a buffer for controlling the pH and, thereby, the fractional charge on the analytes. For indirect fluorometric detection, we deliberately choose the electrolyte to include a fluorescing ion; hence, a large fluorescence background is always present when the excitation laser interacts with the liquid passing through the observation region. Of itself this is not interesting; however, when the analytes of interest appear in the observation region, there will be mutual interaction with the fluorescing ion in the electrolyte, leading to a change in the background signal (i.e., a peak will be observed; Fig. 4).

The simplest type of mutual interaction is charge displacement. A rough picture is that the charged analyte ions displace the same-charge fluorescing ions on a one-to-one basis because of local charge neutrality requirements. A more exact picture is provided by the Kohlrausch regulating functions [31]. The decrease in fluorescence intensity translates to a negative peak, but that is just as useful for identification and for quantitation. Other displacement schemes also exist, but in general, these are not as efficient as charge interaction and, consequently, would not provide comparable performance. It is fortuitous that most CE separations involve charged

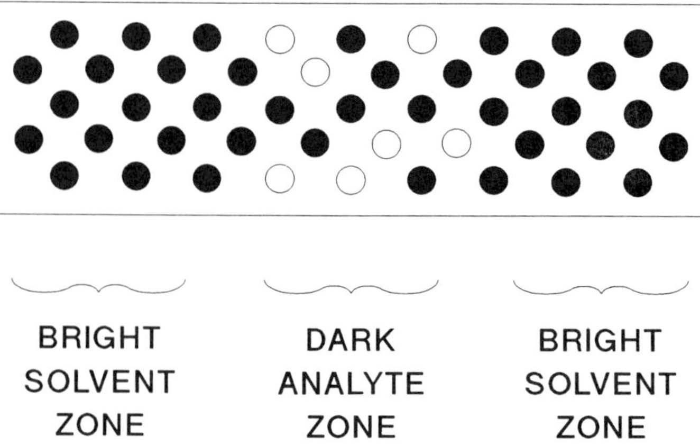

Figure 4 General scheme for indirect fluorescence detection. The mobile-phase additive provides a large background signal at the phototube. In the analyte zone, displacement occurs, and a lower fluorescence intensity is observed.

species, so that charge displacement is an inherent feature of the analytes. This then becomes a universal detection scheme for CZE.

Optimization of indirect fluorescence detection involves both instrumental and chemical considerations. Normal fluorescence detection depends on having very low stray light levels for good performance. For indirect detection, the fluorescence background is inherently high. So, stability in the background signal is the major concern. This can be achieved by using stabilized laser sources, which generally perform at the $\pm 0.1\%$ level for continuous lasers [32]. Optical misalignment can lead to background fluctuations; hence, the optical components, especially the capillary columns, must be held rigidly. Fluorophores also tend to bleach out when irradiated by an intense laser beam; accordingly, a high flow rate of the fluorophore across the optical region is desirable. Other stray light is less important, as long as it is kept substantially below the background fluorescence signal. This is not hard to do, since the fluorophores are generally present above the micromolar level. The concentration of the fluorophore and its mobility are also important factors. Charge displacement is nonselective (as modified by the Kohlrausch function) [31]. To create a large fractional intensity change for a given concentration of analyte, the fluorophore should be fully charged and be present at as low a concentration as possible. This means that low-ionic strength systems are favored. This may require reoptimization of the separation conditions. A significantly large concentration of the fluorophore over the analyte concentration is also not possible because one is looking for a fractional intensity change. Any mismatch in ionic mobilities will result in distorted peak shapes [7]. Also, fluorophores are generally large organic molecules that have a high tendency to adsorb onto the walls of the capillary. Any adsorbed material will not participate in the charge displacement scheme. Finally, the low fluorophore concentration limits the detection mode to those that are highly sensitive to begin with. Hence, indirect absorption detection [33] is possible, but the performance will never approach those of indirect fluorescence detection in small capillaries because one cannot create a large enough background signal to observe the small fractional changes in the signal.

Indirect fluorescence detection in CE has been applied to native amino acids [30], proteins [32], simple inorganic anions [34], cations [35], nucleotides [32], and even sugars [36]. In the last example, the electrolyte was at a pH of 11.5, imposing fractional charges on the sugars for separation and for detection. For fully charged analytes, the LOD is about 10^{-7} M, or 5×10^{-17}-mol injected material. This is substantially better than absorbance detection, but not as good as direct laser-excited fluorometry, as expected. The observed LOD is also consistent with the prediction of an intensity stability of 0.1% combined with a 100-μM concentration of the fluorescing ion. Linearity of peak areas as a function of analyte concentrations was achieved over a factor of 500 [34].

The unique capability of indirect fluorescence detection in CE is illustrated in the study of the intracellular fluid of single cells [37]. The ability to perform chemical measurements on smaller and smaller samples has been a major challenge to analytical chemists. We recently reached a milestone in ultramicrotechniques—the development of a procedure for studying the contents of a single human red blood cell. These are entities of about 0.1 pL in volume, containing subfemtomolar quantities of materials of interest. Before this development, the "world record" was for samples 50–1000 times larger, such as snail neurons [38]. Even there, only a few compounds that happen to be electroactive or can be derivatized have been detected.

Capillary electrophoresis can be performed in capillaries that are 10-μm id, which is an ideal fit for the 7-μm erythrocytes. The high-separation efficiency of CE also provides identification of the species being monitored. We developed a procedure that allows individual red blood cells to be transferred into the capillary column under a microscope. These lyse on contact with the buffer solution, providing a discrete injection plug for electrophoresis. For nonfluorescing materials such as Na^+ and K^+, physiological concentrations are sufficient for detection by indirect fluorescence.

One of the reasons to study single cells, as opposed to using standard blood tests, is that there is a wide variation in the chemical composition among cells. Aging, disease, mutation, and cell cycle all affect chemical composition. The average content, therefore, is much less informative than individual contents. If there are indeed chemical markers, owing to disease or cancer, that show up before physical changes occur in cells, single-cell measurements can lead to early diagnosis. It should also be possible to study pharmacokinetics and drug efficacy on a single-cell basis, avoiding many legal and moral issues. Results obtained for single cells so far include the determination of Na^+/K^+ ratios and a study of oxidation–reduction of glutathione in vivo. In the future, we hope to study other important chemical systems, such as lactate–pyruvate, ATP–ADP, and the transport of small peptides across cell membranes.

REFRACTIVE INDEX DETECTION

Normally, one would not classify refractive index (RI) detection as a sensitive technique. This is mainly from our experience with RI detectors in LC, where many other sensitive detectors are available, including the absorbance detector. However, in many of the RI-monitoring schemes (e.g., refraction), the signal is independent of the optical path length and, therefore, of the probe volume. At some point, the decreasing absorption sensitivity in going to smaller and smaller capillaries will be exceeded by the relatively constant concentration sensitivity of RI detection. Refractive index also is a universal detector and would give good complementary information in CE.

Refraction of laser beams in a cylindrical flow cell has been known for some time. Different beam paths ultimately produce a set of interference fringes at far field. The shift in the locations of these fringes is consequently a measure of the changing RI of the liquids inside. Just like RI detectors for LC, temperature control is an important design parameter. The good performance of this detection scheme is depicted in Fig. 5. A linear dynamic range of 100 and a noise level of 3×10^{-8} RIU was achieved. The simplicity of the device and the ability to monitor non-absorbing compounds should make this a powerful scheme for CE applications.

The change in refractive index can be used to monitor other solute properties as well. In photothermal deflection spectrometry a second laser beam is used to excite the absorbing analyte. Heat is generated eventually, leading to a change in the local refractive index. Even though this is an absorbance detector, the reduced path length available in small capillaries can be compensated by increasing the incident power of the second (excitation) laser beam. Critical alignment of the two laser beams and the capillary is required, but excellent detection limits have been reported—down to the 5×10^{-8} M or 37 amol range [40].

Another interesting application of beam deflection is based on the density or RI gradient generated between zones in the capillary tubes. Because of the very sharp bands in CE, the gradients are particularly large compared with LC. The laser

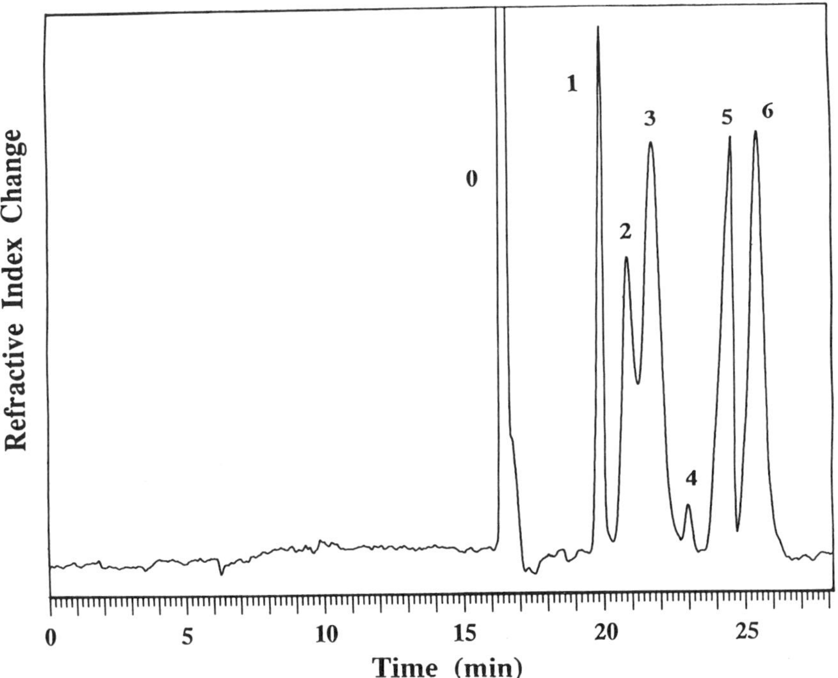

Figure 5 Electropherogram of a mixture of five underivatized saccharides. The peaks are identified as buffer (0), saccharose (1), *N*-acetylglucosamine (2), cellobiose (3), impurity (4), *N*-acetylgalactosamine (5), and lactose (6). (From Ref. 39)

beam is thus deflected when a zone boundary crosses it. This works very well in capillary isotachophoresis, in which detection of a few picomoles of analyte has been demonstrated [41]. For isoelectric focusing in 20-μm capillary tubes, the LOD is 10^{-6} M of analyte [42].

SUMMARY

The small volumes and low concentrations typical of CE separation are major limitations to detector performance. Fortunately, many new schemes for dealing with these situations have recently been demonstrated. The next generation of applications in CE will require much more sensitive detectors. It is clear that detector development is keeping pace with these new demands. In fact, at least two commercial laser–fluorometric detectors are now being marketed. It is the unified systems approach to CE separation and detection that is needed to advance the field.

ACKNOWLEDGMENTS

The author thanks the many coworkers in his laboratory for contributions to parts of this work. The Ames Laboratory is operated by Iowa State University for the US Department of Energy under contract W-7405-Eng-82. This work is supported by the Director of Energy Research, Office of Basic Energy Sciences and Office of Health and Environmental Research.

REFERENCES

1. H. E. Schwartz, K. Ulfelder, F. J. Sunzeri, M. P. Busch, and R. G. Brownlee, *J. Chromatogr.*, *559*:267 (1991).
2. C. A. Monnig and J. W. Jorgenson, *Anal. Chem.*, *63*:802 (1991).
3. K. A. Duell, J. P. Avery, K. L. Rowlen, and J. W. Birks, *Anal. Chem.*, *63*:73 (1991).
4. Y. W. Walbroehl and J. W. Jorgenson, *Anal. Chem.*, *58*:479 (1986).
5. R. D. Smith, H. R. Udseth, C. J. Barinaga, and C. G. Edmonds, *J. Chromatogr.*, *559*:197 (1991).
6. X. Huang, T. K. J. Pang, M. J. Gordon, and R. N. Zare, *Anal. Chem.*, *59*:2747 (1987).
7. F. E. P. Mikkers, F. M. Everaerts, and T. P. E. M. Verheggen, *J. Chromatogr.*, *169*:1 (1979).
8. J. P. Chervet, R. E. J. van Soest, and C. J. Meijvogel, *J. High Resolut. Chromatogr.*, *12*:278 (1989).
9. I. H. Grant and W Steuer, *J. Microcolumn Separation*, *2*:74 (1990).
10. T. Wang, J. H. Aiken, C. W. Huie, and R. A. Hartwick, *Anal. Chem.*, *63*:1372 (1991).
11. X. Xi and E. S. Yeung, *Anal. Chem.*, *62*:1580 (1990).
12. J. A. Taylor and E. S. Yeung, *J. Chromatogr.*, *550*:831 (1991).

13. K. I. Tsunoda, A. Nomura, J. Yamada, and S. Nishi, *Appl. Spectrosc.*, *44*:163 (1990).
14. X. Xi and E. S. Yeung, *Appl. Spectrosc.*, *45*:1199 (1991).
15. R. E. Synovec and E. S. Yeung, *Anal. Chem.*, *57*:2162 (1985).
16. E. S. Yeung and M. J. Sepaniak, *Anal. Chem.*, *52*:1465A (1980).
17. E. S. Yeung, *Microbore Column Chromatography* (F. J. Yang, ed.), Marcel Dekker, New York, p. 117 (1989).
18. P. Gozel, E. Gassmann, H. Michelsen, and R. N. Zare, *Anal. Chem.*, *59*:44 (1987).
19. S. A. Soper, E. B. Shera, J. C. Martin, J. H. Jett, J. H. Hahn, H. L. Nutter, and R. A. Keller, *Anal. Chem.*, *63*:432 (1991).
20. G. J. Diebold and R. N. Zare, *Science*, *196*:1439 (1977).
21. S. Folestad, L. Johnson, B. Josefsson, and B. Galle, *Anal. Chem.*, *54*:925 (1982).
22. L. W. Hershberger, J. B. Callis, and G. D. Christian, *Anal. Chem.*, *51*:1444 (1979).
23. D. Y. Chen, H. P. Swerdlow, H. R. Harke, J. Z. Zhang, and N. J. Dovichi, *J. Chromatogr.*, *559*:237 (1991).
24. T. Lee, E. S. Yeung, and M. Sharma, *J. Chromatogr.*, *565*:197 (1991).
25. T. Lee and E. S. Yeung, *J. Chromatogr.*, *595*:319 (1992).
26. D. F. Swaile and M. J. Sepaniak, *J. Liq. Chromatogr.*, *14*:869 (1991).
27. P. L. Christensen and E. S. Yeung, *Anal. Chem.*, *61*:1344 (1989).
28. E. S. Yeung and W. G. Kuhr, *Anal. Chem.*, *63*:275A (1991).
29. H. Smith and T. E. Miller, *Anal. Chem.*, *54*:462 (1982).
30. W. G. Kuhr and E. S. Yeung, *Anal. Chem.*, *60*:1832 (1988).
31. H. Poppe, *J. Chromatogr.*, *506*:45 (1990).
32. W. G. Kuhr and E. S. Yeung, *Anal. Chem.*, *60*:2642 (1988).
33. M. T. Ackermans, F. M. Everaerts, and J. L. Beckers, *J. Chromatogr.*, *549*:345 (1991).
34. L. Gross and E. S. Yeung, *J. Chromatogr.*, *480*:169 (1989).
35. L. Gross and E. S. Yeung, *Anal. Chem.*, *62*:427 (1990).
36. T. W. Garner and E. S. Yeung, *J. Chromatogr.*, *515*:639 (1990).
37. B. L. Hogan and E. S. Yeung, *Anal. Chem.*, *64*:2841 (1992).
38. R. T. Kennedy, M. D. Oates, B. R. Cooper, B. V. Nickerson, and J. W. Jorgenson, *Science*, *246*:57 (1989).
39. A. E. Bruno, B. Krattiger, F. Maystre, and H. M. Widmer, *Anal. Chem.*, *63*:2689 (1991).
40. M. Yu and N. J. Dovichi, *Anal. Chem.*, *61*:37 (1989).
41. T. McDonnell and J. Pawliszyn, *Anal. Chem.*, *63*:1884 (1991).
42. J. Wu and J. Pawliszyn, *Anal. Chem.*, *64*:219 (1992).

19

Laser-Induced Fluorescence Detection for Capillary Electrophoresis: *A Powerful Analytical Tool for the Separation and Detection of Trace Amounts of Analytes*

Luis Hernandez and Narahari Joshi

Los Andes University
Mérida, Venezuela

Phillipe Verdeguer

Europhor Instruments
Toulouse, France

Norberto A. Guzman*

Princeton Biochemicals, Inc.
Princeton, New Jersey

Capillary electrophoresis (CE) is probably one of the fastest-growing analytical techniques in the separation sciences. During the last decade the number of scientific papers published in this field has grown exponentially (Fig. 1). In almost all of these articles, the method used for detection of analytes has been based primarily on use of ultraviolet light. Currently, there are approximately ten companies commercializing capillary electrophoresis instruments equipped with ultraviolet (UV) detectors and only two selling a system equipped with laser-induced fluorescence detection (LIFD). As shown in Fig. 1, the LIFD technology has also grown exponentially throughout the years.

**Current affiliation*: The R. W. Johnson Pharmaceutical Research Institute, a Johnson & Johnson Company, Raritan, New Jersey.

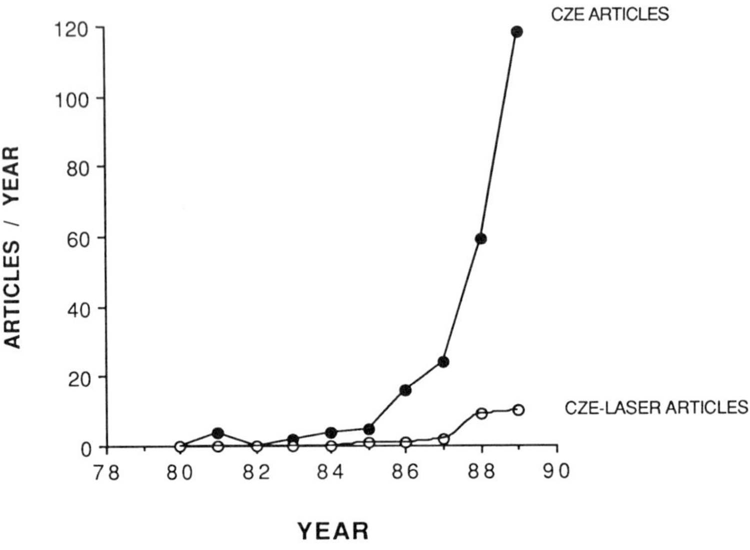

Figure 1 Articles published on the subject of capillary electrophoresis relative to the general field (closed circles), versus articles published on the subject of capillary electrophoresis coupled with laser-induced fluorescence detection system (open circles).

Although use of LIFD as a detection method for capillary electrophoresis began as early as 1985, a substantial interest has recently developed based on the limitation of concentration sensitivity of other detection methods, such as UV detection. For example, biologically active peptides, which are found in body fluids and tissues at extremely low concentrations, are able to exert their physiological–pharmacological actions at these low concentrations because of their high biological potency. The range of concentration of biologically active peptides usually varies between 10^{-11} and 10^{-13} M. It would be almost impossible for such peptides to be detected at these concentrations by capillary electrophoresis when using UV detection, especially if only a few nanoliters of the substance to be analyzed is injected into the capillary. A typical calibration curve for neurotensin, a biologically active neuropeptide, shows a maximum concentration sensitivity at 10^{-6} M when using capillary electrophoresis and UV detection (Fig. 2).

The only way to determine biologically active neuropeptides, when analyzed by capillary electrophoresis coupled with UV detection, is to improve detection sensitivity at least five to seven orders of magnitude. Since such an improvement is unlikely in the near future, LIFD has gradually become the method of choice for detection of substances found in extremely low concentrations in tissues and biological fluids. In the present chapter, we examine the state of development of LIFD

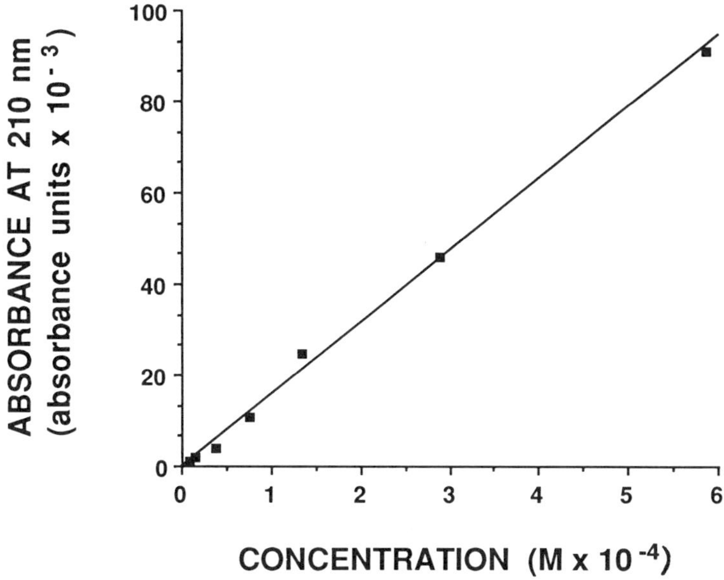

Figure 2 Calibration curve generated for neurotensin using capillary electrophoresis and UV detection. The square dots located at the origin of the curve represent a detection limit of 9×10^{-6} M for neurotensin.

and the main difficulties to be overcome to accommodate CE–LIFD as a popular tool among the several analytical techniques available in separation sciences that are necessary in research–clinical laboratories.

The first successful use of LIFD for capillary electrophoresis was accomplished by Gassmann and associates [1] for the detection of dansylated amino acids by means of the 325-nm line of a He–Cd laser. This accomplishment was carried out by using fiber optics to excite a segment of a fused silica capillary, and the emitted radiation was collected with an appropriate set of lenses (Fig. 3). With the introduction of CE–LIFD, it was possible to measure fluorescent compounds, such as the drug methotrexate [1] and certain vitamins [2]. At this early stage of the dual technology of CE–LIFD, having high-power resolution and high-power detection, it was already possible to improve the minimum detected concentration by three orders of magnitude (when compared with UV detection) reaching detection sensitivities of 10^{-9} M.

During the next few years, several attempts to expand the use of CE–LIFD were made. One example of new applications for CE–LIFD was the measurement of nonfluorescent compounds that could be determined by indirect fluorescence. The rationale for applying this technique was based on the use of a fluorescent buffer

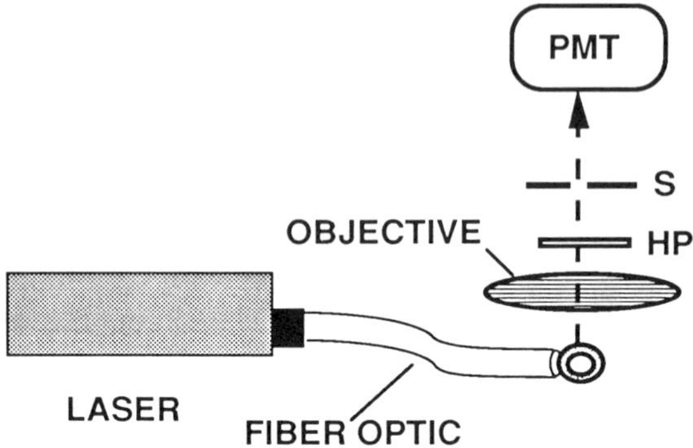

Figure 3 Schematic representation of the first laser-induced fluorescence detection system designed for capillary electrophoresis. PMT, photomultiplier; S, spatial; HP, high pass.

(as separation buffer) and measuring the separated zones by a decrease of fluorescence intensity [3–5]. Another interesting application was the improvement in the speed of separation of fluorescent amino acids; by using a 10-μm-id capillary column, it was possible to separate several amino acids in less than 2 minutes [6].

New approaches to even further improve detection sensitivity were attempted by introducing new laser lines and new fluorescent chromophores. One of the first successful experiments was the introduction of the 442-nm line of He–Cd laser for the analysis of naphthalene dialdehyde (NDA) derivatized amino acids by capillary electrophoresis [7]. Under these experimental conditions, it was possible to reach detection sensitivity levels of 10^{-10} M. Later, Dovichi and co-workers [8,9] introduced a longer-wavelength laser, the 488-nm line of argon ion laser. Also, they used fluorescein isothiocyanate (FITC) as a chromophore, a fluorescent tag that has a higher quantum efficiency than NDA. In addition, the fluorescence emission was detected not by using the conventional on-column detection system for capillary electrophoresis, but rather, outside the capillary, in a flow-sheath cuvette, thereby achieving a remarkable noise reduction (Fig. 4).

The new improvements in detection sensitivity for capillary electrophoresis allowed the possibility of reaching a new dimension in detection. It was now possible to reach 10^{-12} M detectable concentration for fluorescein thiocarbamyl (FTC)–amino acids [9], a number that suggested the possibility of building a molecular-counting machine, the dream of many analytical chemists. At this point, the improvement of the minimum detected amount was six orders of magnitude better than UV detection, indicating that a new era had started for detection of trace amounts of substances present in biological fluids, increasing the probability for expansion of capillary electrophoresis applications to the field of biomedicine.

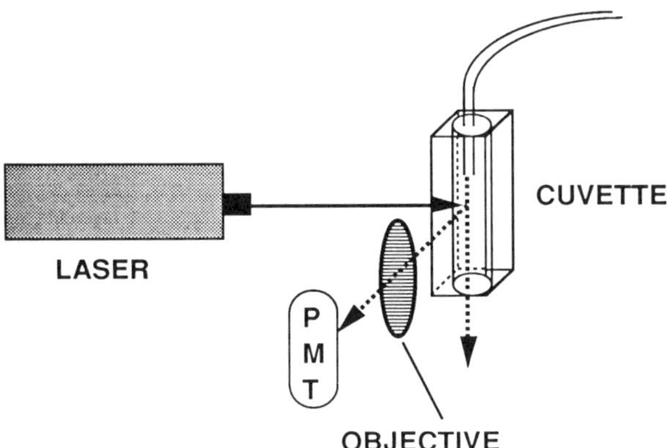

Figure 4 Schematic representation of a laser-induced fluorescence detection system for capillary electrophoresis using a flow sheath cuvette. The analytes are excited in the postcolumn detection mode in a quartz cuvette.

Our laboratory in Venezuela also contributed to further improvement of the minimum detected amount of an analyte. It was possible to change the optical arrangement of the detection system for CE–LIFD. Rather than using two optic lines, one for excitation and another for collection, as was traditionally up to this point, we used one optic line for both excitation and emission [10,11]. This new arrangement, named *colinear* or *confocal*, was a detection system based on the principle of the epillumination fluorescence microscope [10]. Figure 5 shows the main features of a colinear system. In this system, the laser is reflected by a dichroic mirror, and the objective of a microscope focuses the laser beam on the capillary column. The emitted fluorescence is collected by the same objective. Since the wavelength of the fluorescence is longer than the wavelength of the laser, the fluorescence is refracted by the dichroic mirror. After proper filtering, the fluorescence is detected by a photomultiplier and recorded in either a digital or an analog mode.

We will discuss the main features of a colinear device that can be used as a detection system for capillary electrophoresis. The main characteristics of a laser-induced fluorescence detection system are described in more detail elsewhere by Cheng and Dovichi [8]. In general, the main difficulty in optical analysis is the alignment of the optical components. The process requires that different optical components be placed in a precise sequence, such that excitation and emission are optimal. In a colinear device, the same optic line is used for focusing the laser beam and for collecting the emitted fluorescence. The spatial geometry of the colinear alignment of the laser beam and the capillary column seems to be much simpler than other geometric arrangements. The main advantage of the colinear geometry

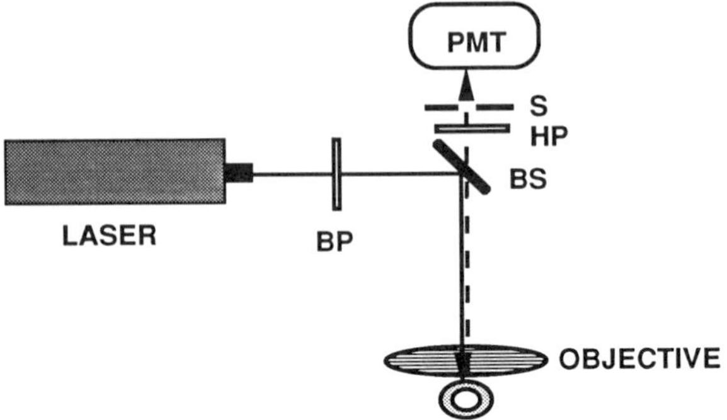

Figure 5 Schematic representation of capillary electrophoresis coupled with a laser-induced fluorescence detection system using the colinear or confocal optic configuration arrangement. Both the excitation and the emission radiation travel in the same plane but in opposite directions.

is that the collection of the emitted fluorescence is located on the side of the capillary at which the fluorescence is maximal. Maximal emission of fluorescence occurs in the superficial layers of the fluorescence material. This layer is located on the side that the laser penetrates into the fluorescence solution, creating an area on the surface of a cylinder. It is on this side that the collection of the emitted fluorescence must be maximal. Therefore, the position of the detector is crucial to obtain the maximum collected fluorescence, since less fluorescence is obtained when the detector is placed at an angle. For example, if the detector is placed opposite the laser spot, it will see a minimal amount of fluorescence. By contrast, a detector placed on the same side as the laser spot should collect the maximum amount of fluorescence.

The colinear arrangement permits the use of lenses with very high numerical apertures. Since only one optic line is used, then both the focusing of the laser beam and the collection of the emitted fluorescence are performed through the same objective, positioning the lens as close as possible to the capillary column. Typically, large numerical aperture lenses are best for fluorescence collection, partly because these lenses collect light emitted at angles quite well. This particular light cannot reach the lenses if they are far away from the object to be analyzed. In other words, large numerical aperture lenses have a very low working distance, making it necessary to locate them very close to the object, thereby allowing more collection of light. With this in mind, it has been possible to develop a mathematical model for the collection of light that is based on the working distance of the lenses, rather than on the numerical aperture. The calculations of the collection efficiency of the lenses, based on the numerical aperture, do not take into account the emission surface, whereas the calculations based on working distance do.

The calculations based on numerical aperture are given by the following equation:

$$\text{Collection efficiency} = \sin^2\left[\frac{\arcsin \times \left(\dfrac{\text{NA}}{\eta}\right)}{2}\right] \tag{1}$$

where NA is the numerical aperture and η is the refractive index of the material of the lens. As mentioned before, this equation does not take into account the irradiance of the emitting surface.

The equation that does take into consideration the irradiance of the emitting surface is the following:

$$\Delta\Phi = \frac{L \times \Delta A_1 \times \Delta A_2}{D^2} \tag{2}$$

$\Delta\Phi$ = power of the light incident on the objective
ΔA_1 = illuminating area
ΔA_2 = area of the objective
L = irradiance of the photoilluminating area
D = distance between illuminating area and objective

For practical purposes, the product $L \times \Delta A_1$ is considered constant when the numerical aperture of the lenses varies. This assumption is reasonable, because the irradiance and the illuminating area change in the opposite direction as a function of the numerical aperture. For instance, if the numerical aperture of the lens increases, the solid angle for light collection becomes larger, and the irradiance increases. At the same time, the illuminating area decreases, because the spot of the laser becomes smaller. Therefore, the product $L \times \Delta A_1$ remains constant. Finally, we have assumed that $\Delta\Phi$ is directly proportional to the current generated by the incident light on the photomultiplier (PMT) tube. Therefore, it is possible to obtain the following mathematical formula for PMT current caused by fluorescence:

$$\text{PMT current} = 15.7 + \frac{14}{(D + 0.2)^2} \tag{3}$$

According to Eq. 3 the constant 15.7 is interpreted as the background current for lenses with extremely large working distances.

Another phenomenon to be taken into account, when considering the design of a laser-induced fluorescence instrument, is the background noise. In our experience, the main source of noise in LIFD is laser scattering. This parameter becomes of particular relevance when we want to detect very low-intensity light, which is true for fluorescence intensity emitted in very small capillaries and for almost all analytes present in biological fluids. The laser is scattered by the fused silica, which contains impurities that fluoresce with visible and UV lasers. This source of noise is bleached by the laser itself. A 30-minute waiting period is enough to decrease the fluorescence of the capillary wall significantly. The other source of noise is the laser itself. In cylindrical capillaries, the laser is scattered as a circle centered in the capillary and in a plane perpendicular to the main axis of the capillary. This geometry of scattering depends on the shape of the capillary. The laser finds a curved surface and is reflected around it. As the size of the spot of the laser beam decreases, the laser finds a surface closer to flat. Therefore, as the numerical aperture increases, the magnification of the lens increases, and the laser spot approaches that of a flat surface, diminishing the laser scattering and the background noise. With this in mind, it is possible to generate the following mathematical formula that relates PMT current from laser scattering to magnification size:

$$\text{PMT current} = 14.7 + \frac{3744.5}{M} \tag{4}$$

An evaluation of Eqs. 2 and 4 shows that the two independent mathematical terms are very close. In a series of experiments conducted to investigate the influence of numerical aperture of the lenses, it was found that an increase in the numerical aperture of the lenses produces an increase in the collection of fluorescence and a decrease in the background noise owing to a lowering in laser scattering. The results of this experiment are presented in Fig. 6.

An important issue is that a colinear arrangement of the optical system facilitates the use of multiple capillaries in the on-column detection mode. The use of several capillaries in a single run for the analysis of multiple analytes in very short periods offers an attractive feature of capillary electrophoresis technology. If we consider each capillary as a column, then it is possible to compare it with another separation system. For example, if we use ten capillaries for one capillary electrophoresis system, it can be compared with an equivalent of using ten high-performance capillary chromatography (HPLC) instruments (i.e., 10 pumps, 10 columns, 10 detectors), or the equivalent of using one HPLC instrument containing ten columns, and controlled by a complex system composed of a computer, switch valves, and buffer systems. From an economical point of view, the latter can be a very costly system. Capillary columns are significantly more inexpensive than HPLC columns, and basically the only requirement to make the ten capillaries operational as a separation system is to have an *xy* displacer to enable all columns to be moved very fast in a synchronized motion in front of the laser beam. The rest of the components of the capillary electrophoresis instrument will be the same. A schematic diagram showing the arrangement of the capillaries in front of the laser detection system is presented in Fig. 7. In fact, this concept of using multiple capillaries to analyze substances by capillary electrophoresis coupled with LIFD has recently been demonstrated by Huang et al. [12]. These investigators were able to fabricate a DNA sequenator based on the use of multiple gel-filled capillary columns. The orthogonal geometric

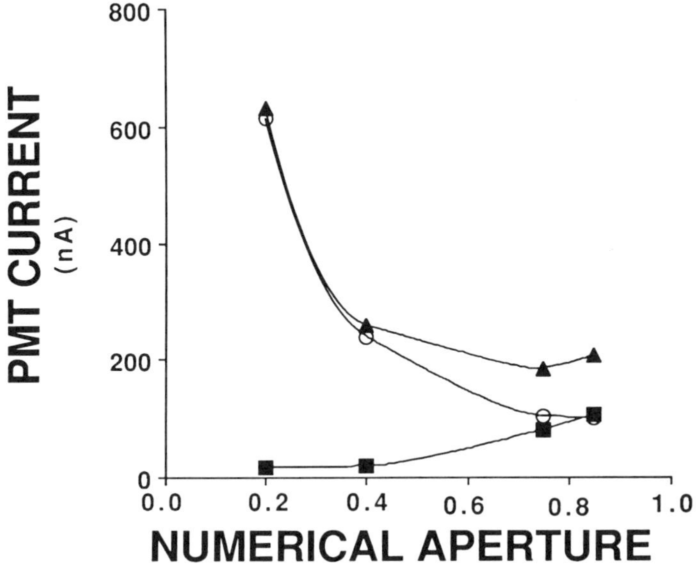

Figure 6 A curve representing a relation between PMT current and numerical aperture. An increase in numerical aperture appears to increase the collected fluorescence (black squares) and to decrease the background noise (open circles). The black triangle curve represents the PMT current generated by the signal plus the noise.

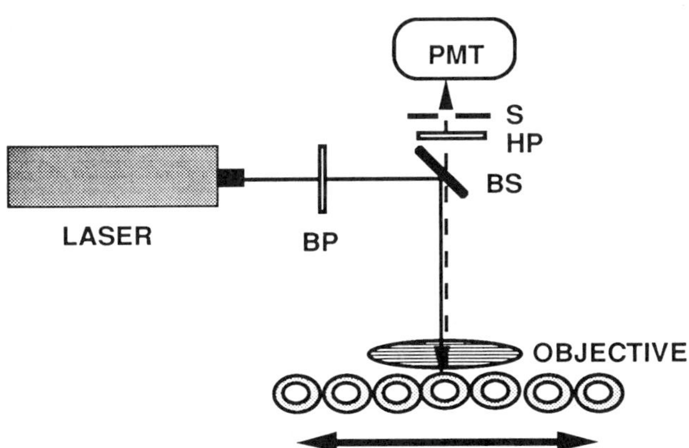

Figure 7 Schematic representation of capillary electrophoresis coupled with a laser-induced fluorescence detection using a colinear optical arrangement and multiple capillaries. PMT, photomultiplier; S, spatial; HP, high pass; BS, beam splitter.

arrangement will be almost impossible to use for multiple capillaries because the capillaries block the laser.

In conclusion, the use of capillary electrophoresis technology coupled with a laser-induced fluorescence detection system can generate a very competitive instrumentation system for the analytical techniques used in the separation sciences. Many application examples can be mentioned, but some of the most attractive are the sequencing of macromolecules, such as proteins and DNA, and the separation and identification of trace amounts of analytes present in tissues, biological fluids, and individual cells.

REFERENCES

1. E. Gassmann, J. E. Kuo, and R. N. Zare, *Science*, *230*:813 (1985).
2. D. E. Burton, M. J. Sepaniak, and M. P. Maskarinec, *J. Chromatogr. Sci.*, *24*:347 (1986).
3. W. G. Kuhr, and E. S. Yeung, *Anal. Chem.*, *60*:2642 (1988).
4. W. G. Kuhr, and E. S. Yeung, *Anal. Chem.*, *60*:1832 (1988).
5. L. Gross, and E. S. Yeung, *J. Chromatogr.*, *480*:169 (1989).
6. B. Nickerson, and J. W. Jorgenson, *J. High Resolution Chromatogr. Commun.*, *10*:533 (1988).
7. B. Nickerson and J. W. Jorgenson, *J. High Resolution Chromatogr. Commun.*, *11*:878 (1988).
8. Y.-F. Cheng, and N. J. Dovichi, *Science*, *242*:562 (1988).
9. S. Wu and N. J. Dovichi, *J. Chromatogr.*, *480*:141 (1989).
10. L. Hernandez, J. Escalona, N. Joshi, and N. A. Guzman, *J. Chromatogr.*, *559*:183 (1991).
11. L. Hernandez, N. Joshi, E. Murzi, P. Verdeguer, R. Kerr, and N. A. Guzman, *J. Chromatogr.*, (in press; 1993).
12. X. C. Huang, M. A. Quesada, and R. A. Mathies, *Anal. Chem.*, *64*:967 (1992).

V

APPLICATIONS

20

Chiral Separation by Capillary Electrophoresis and Electrokinetic Chromatography

Koji Otsuka

Osaka Prefectural College of Technology
Neyagawa, Osaka, Japan

Shigeru Terabe

Himeji Institute of Technology
Kamigori, Hyogo, Japan

High performance capillary electrophoresis (HPCE) [1–3], which is an instrumental version of electrophoresis, has become popular in various analytical fields owing to its high-resolving power. At the present stage, various modes of HPCE have been developed: capillary zone electrophoresis (CZE), capillary gel electrophoresis (CGE), capillary isotachophoresis, capillary isoelectric focusing, and electrokinetic chromatography (EKC). In any of these modes, a capillary smaller than 100-μm id is usually used for separation. The foregoing four modes, except EKC, can separate ionic or charged substances only. This limitation was a serious problem of HPCE, compared with conventional high performance liquid chromatography (HPLC). In such circumstances, the technique of EKC [4] has been developed as a method that can be applied to the analysis of neutral solutes. Electrokinetic chromatography is based on chromatographic principles using homogeneous solutions and has the unique characteristic that both neutral and charged analytes can be separated electrophoretically. Among some modes of EKC, micellar EKC (MEKC) [5–7], which uses micellar solutions of ionic surfactants, has become the most popular method for separating small neutral molecules and the fundamental characteristics of MEKC have been published.

Chiral separation is an important objective, especially in the pharmaceutical, medical, and biological fields, and has become one of major applications of chromatography. Recently, many studies on optical resolution by chromatographic techniques have appeared. Similarly to the application of HPCE, some papers on this area have also been published.

In this chapter, we will briefly review the application of HPCE to chiral separation. First, chiral separation by capillary electrophoresis (CE), which includes CZE and CGE, will be described. It should be noted that we use the terms of CE and EKC separately in this chapter. By using the CE technique, optical resolution is performed mainly by the following three procedures: addition of chelating chiral reagents to electrolyte solutions in CZE, addition of cyclodextrins in CZE, and using a cyclodextrin-immobilized gel in CGE. Then, we will describe chiral separations by EKC. In EKC, three methods are carried out to achieve the chiral recognition: MEKC with chiral micelles, cyclodextrin-modified MEKC (CD–MEKC), and cyclodextrin EKC (CD-EKC).

CHIRAL SEPARATION BY CAPILLARY ELECTROPHORESIS

Addition of Chelating Reagents

The first report on optical resolution by CE was published in 1985 by Gassmann et al. [8]. They have achieved chiral separations of some dansylated DL-amino acids (Dns-DL-AAs) by adding copper(II)–L-histidine (Cu/L-His) complexes to electrolyte (buffer) solutions. Under experimental conditions used (pH 7–8), the electroosmotic flow was observed toward the negative electrode. Since the Cu/L-His complex had the positive charge, Dns-AA bound to the Cu/L-His complex moved toward the cathode faster than the free Dns-AA having no charge. Dns-DL-AA formed two diastereomeric ternary complexes with the Cu/L-His (i.e., Cu/L-His/Dns-D-AA and Cu/L-His/Dns-L-AA). The formation constants of these two complexes were different; hence, the separation of D- and L-forms of Dns-AA could be achieved.

By using L-aspartyl-L-phenylalanine methyl ester (aspartame) instead of aforementioned L-His, chiral separation could be achieved more successfully. Gozel et al. [9] have reported on enantiomeric resolution of 14 Dns-DL-AAs using copper(II)–aspartame (Cu/aspartame) complexes. In Fig. 1, chiral separation of four Dns-DL-AAs is shown. The Dns-DL-AA formed two disastereomeric ternary complexes having positive charges with the Cu/aspartame. In this system, the D-enantiomer of each Dns-AA migrated faster than the L-form, as the stability constant of the Cu/aspartame/Dns-D-AA ternary complex is greater than that of the Cu/aspartame/Dns-L-AA ternary complex: Specific hydrophobic interactions between the aspartame phenylalanine residue (R) and the AA side chain (R′) can occur only in the LD configuration, as shown in Fig. 2. Since the separations of different Dns-AAs were not successful with this approach, the sodium tetradecyl sulfate (STS) micelle was added to the electrolyte solution to improve the resolution by using the technique of MEKC.

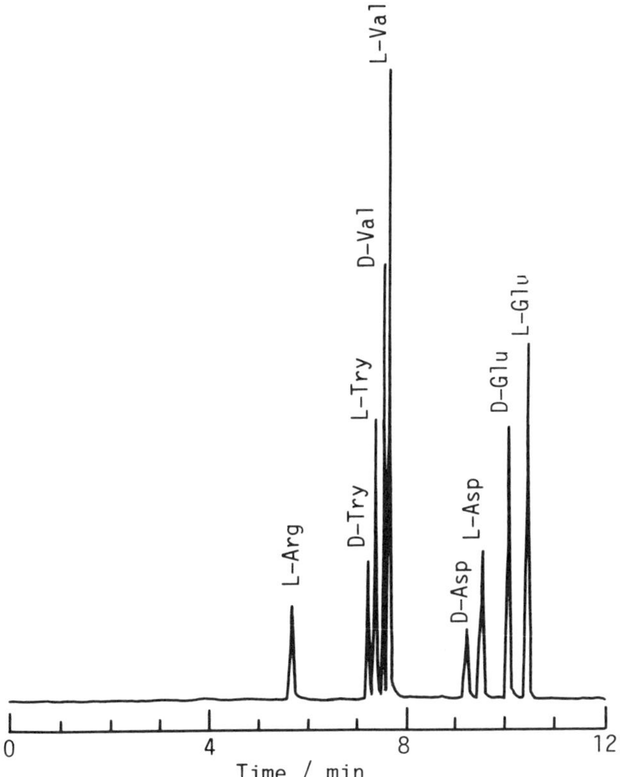

Figure 1 Chiral separation of four Dns-DL-AAs by CE. Electrolyte solution, 2.5 mM copper sulfate, 5.0 mM aspartame, and 10 mM ammonium acetate (pH 7.4); separation capillary, 100 cm × 75-μm id; effective length of the capillary, 75 cm; detection, fluorescence; applied voltage, 30 kV; current, ~33 μA. (From Ref. 9)

There is no additional report on this method, but this can be easily applied to chiral separations of compounds having amino groups. Use of other chiral chelating reagents will also be possible.

As a similar method, Fanali et al. [10] have reported on resolution of enantiomeric and diastereomeric cobalt(III) complexes with ethylenediamine (en) and amino acid ligands, (e.g., $[Co(en)_3]^{3+}$) by using L-(+)-tartaric acid as a complexing counterion. Here, Co complexes were injected as optical isomeric samples. For example, in the electrolyte solution, the sample $[Co(en)_3]^{3+}$, which has two possible structures, coordination octahedron or Δ and Λ, formed two complexes with tartrate. The difference in formation constants brought about the difference in the electrophoretic mobility.

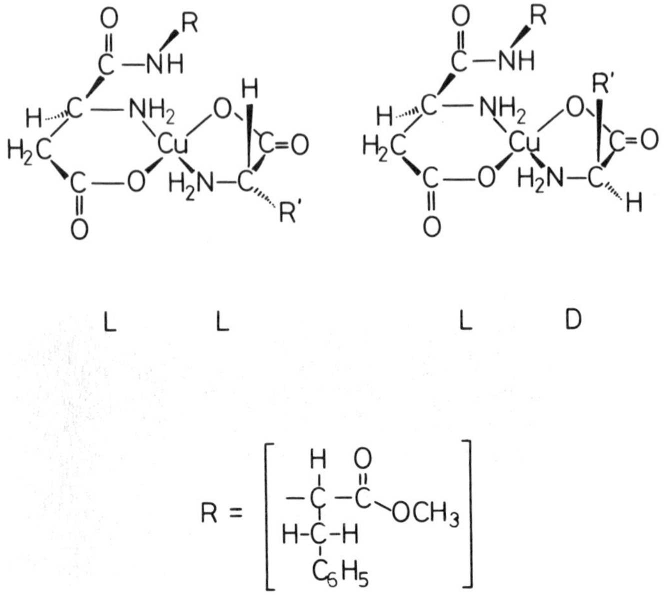

Figure 2 Configuration of the Cu/aspartame/AA ternary complex. (From Ref. 9)

Addition of Cyclodextrins

Cyclodextrins (CDs) are capable of incorporating various compounds into their cavities forming inclusion complexes, and CDs also have an ability for chiral recognition. Fanali [11] has reported on optical resolution of catecholamines, such as ephedrine, norephedrine, and epinephrine, by CE with electrolyte solutions containing β-CD or some β-CD derivatives. Since neither CD nor the CD derivatives have any charge, they are not affected by electrophoresis and migrate with an identical velocity as bulk solutions or at the electroosmotic velocity. An ionic solute migrates by the effect of its electrophoretic mobility only when it is not included by CD, whereas the effect is that of the electrophoretic mobility of an inclusion complex when included by CD. Since the ratio of mass or molecular weight to charge for the CD inclusion complex is larger than that of the solute itself, the electrophoretic mobility of the CD inclusion complex becomes small. Hence, when there is a difference in formation constants between two inclusion complexes, these two solutes can be separated by the difference in migration velocities, even though the original electrophoretic velocities of the two solutes were identical. Commonly, electrophoretic mobilities do not differ between enantiomers, but formation constants of inclusion complexes with CD may be different. Therefore, it is possible that the enantiomers are resolved by the CE systems modified by CDs.

Recently, these CD-modified CE systems have been used by some investigators for chiral separations, which include the works by Fanali and Bocek [12], Sydor and Mularz [13], and Swartz [14] on enantiomeric separations of some pharmaceutical compounds.

Capillary Gel Electrophoresis with Immobilized Cyclodextrins

In capillary gel electrophoresis (CGE), the electroosmosis is completely suppressed and only an ionic solute migrates with the electrophoresis. When using a capillary filled with CD-immobilized polyacrylamide gel, an ionic solute does not migrate when it is incorporated by CD in the gel. This is the same situation as in HPLC using a CD stationary phase: In the former case, the sample migrates by electrophoresis, whereas in the latter, it migrates by the flow of the mobile phase.

With this system, a difference in formation constants of inclusion complexes between enantiomers allows optical resolution. Guttman et al. [15] have successfully achieved chiral separations of 12 Dns-DL-AAs. They used a 30-cm × 75-μm capillary filled with a 5% T, 3% C gel, containing 7 M urea and β- or γ-CD. Three Dns-DL-AAs were successfully separated and optically resolved with the gel capillary containing 75 mM β-CD (Fig. 3).

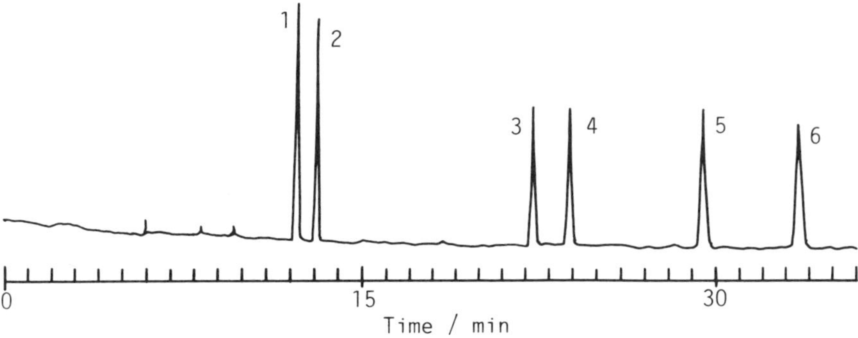

Figure 3 Chiral separation of three Dns-DL-AAs by CGE with immobilized CD: Corresponding AAs: (1) L-Glu, (2) D-Glu, (3) L-Ser, (4) D-Ser, (5) L-Leu, (6) D-Leu; buffer solution, 0.1 M Tris–0.25 M boric acid (pH 8.3), 7 M urea, 75 mM β-CD; effective length of the capillary, 15 cm; applied voltage, 12 kV; detection wavelength, 254 nm. (From Ref. 15)

CHIRAL SEPARATION BY ELECTROKINETIC CHROMATOGRAPHY

Micellar Electrokinetic Chromatography With Chiral Micelles

Micellar Electrokinetic Chromatography Using Metal Complexes

In 1987, Cohen et al. [16] reported on the chiral separation of three Dns-DL-AAs by MEKC using Cu(II) attached to a chiral ligand of a chelate detergent, *N*,*N*-didecyl-L-alanine. The surfactant was soluble in aqueous solutions only in the presence of sodium dodecyl sulfate (SDS); that is, mixed micelles were formed in the solutions. The principles of optical resolution in this system are the same as those in the paper by Gassmann et al. [8]: enantiomers form the metal chelate complexes on the surface of micelles. A difference in formation constants causes the difference in the migration velocities.

Micellar Electrokinetic Chromatography With Sodium N*-Dodecanoyl-L-valinate*

Dobashi et al. [17,18] have used an anionic chiral surfactant, sodium *N*-dodecanoyl-L-valinate (SDVal), for enantiomeric resolution of amino acid derivatives such as *N*-(3,5-dinitrobenzoyl)-*O*-isopropyl esters and their analogues. By using SDVal solutions without any additives (e.g., SDS or organic modifiers) chiral separation was possible, although resolution was not good. They also used SDVal/SDS mixed micellar solutions, with or without a methanol addition. In either circumstance, chiral separations were achieved, but resolution was still low because of low efficiencies.

We have also investigated enantiomeric resolution of phenylthiohydantoin-DL-amino acids (PTH-DL-AAs) by MEKC with SDVal. By using SDVal alone, poor peak shapes were observed [19]. The peak shapes were improved by adding methanol or urea [20]. Addition of SDS to the micellar solutions could manipulate separations and change the selectivity. Six PTH-DL-AAs were successfully separated from each other, and each enantiomeric pair was optically resolved with SDVal/SDS/urea/methanol solutions (Fig. 4) [21]. Effects of urea addition to micellar solutions on electrokinetic migrations and separation characteristics have been reported elsewhere [22].

Micellar Electrokinetic Chromatography With Bile Salts

Bile salts are well known as natural-occurring anionic surfactants. We have reported on some MEKC applications, which include separations of the ingredients of a cold medicine [23] and lipophilic compounds [24], using bile salts. Bile salts are also popular as chiral compounds, and we have tried optical resolution by MEKC using bile salts in a manner similar to that of using SDVal.

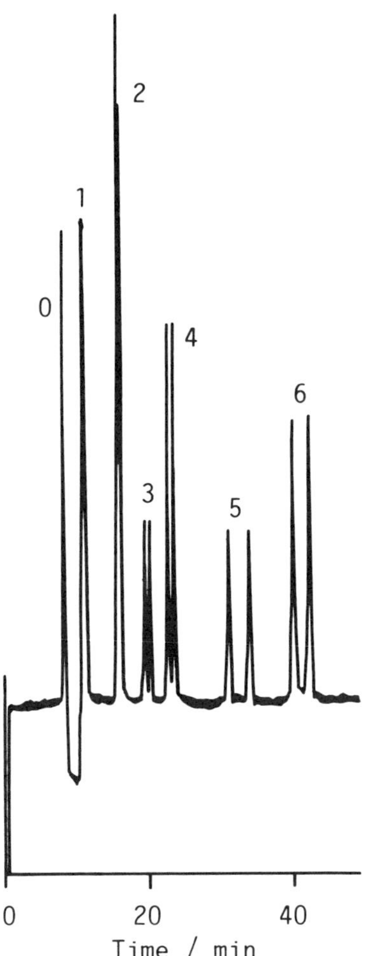

Figure 4 MEKC separation of six PTH-DL-AAs with SDVal. Corresponding AAs: (1) Ser, (2) Aba, (3) Nva, (4) Val, (5) Trp, (6) Nle, (0) acetonitrile. Micellar solution, 50 mM SDVal/30 mM SDS/0.5 M urea (pH 9.0) containing 10%(v/v) methanol; separation capillary, 65 cm × 50-μm id; effective length of the capillary, 50 cm; applied voltage, 20 kV; current, 17 μA; detection wavelength, 260 nm; temperature, ambient. (From Ref. 21)

By using sodium taurocholate (STC) or sodium taurodeoxycholate (STDC), some Dns-DL-AAs were separated from each other, and each pair was optically resolved (Fig. 5) [25]. Although a long separation time was required, the time could be shortened by adding SDS. Since the pH of buffers used here was 3.0, the electroosmotic flow was substantially suppressed compared with that in the neutral solutions: the absolute value of the electroosmotic velocity was lower than that of the electrophoretic velocity of the mixed micelle. Therefore, the migration direction of the micelle was the same as that of the electrophoresis of the micelle; in other words, the opposite of the electroosmosis, or from the negative to the positive electrodes. The effects of pH on electrokinetic velocities, such as the electroosmotic velocity,

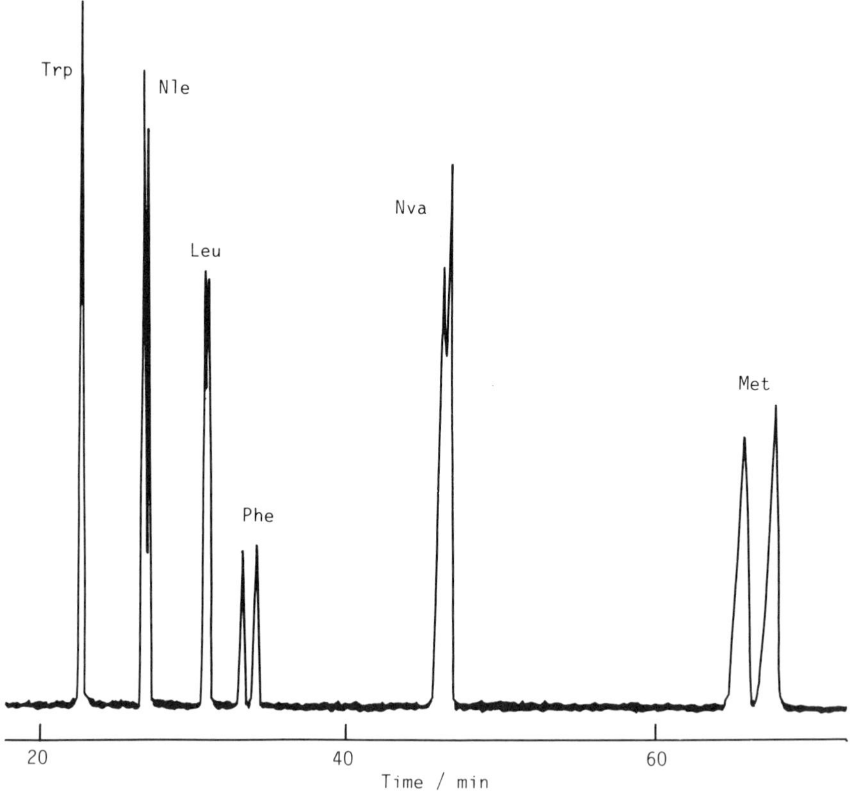

Figure 5 MEKC separation of six Dns-DL-AAs with STDC. Micellar solution, 50 mM STDC in 50-mM phosphate buffer (pH 3.0); separation capillary, 70 cm × 50-μm id; effective length of the capillary, 50 cm; current, 50 μA; detection wavelength, 210 nm; temperature, 40°C. (From Ref. 25)

electrophoretic velocity of the micelle, and migrating velocity of the micelle, and on the chromatographic parameters in MEKC in the presence of SDS solutions have been reported previously [26].

Bile salt systems have also been applicable to chiral separation of optical isomeric drugs [27,28], such as diltiazem hydrochloride, trimethoquinol (tretoquinol) hydrochloride, tetrahydropapaveroline, and so on, and determination of their optical purity [29]. For these compounds, STC and STDC gave good resolution under the neutral or basic conditions.

Cole et al. [30] have also reported on enantiomeric separation of binaphthyl analogues, by MEKC with bile salts. In their work, an SDC solution containing methanol gave good results.

It is noteworthy that the bile salt micelles possess unique structures. Campanelli et al. [31] have reported that micellar aggregates of SDC, STDC, and also sodium glycodeoxycholate, in aqueous solutions, have helical structures.

Micellar Electrokinetic Chromatography With Digitonin

Digitonin, which is a glycoside of digitogenin, is popular as a chiral surfactant. We tried to use digitonin as a chiral separation carrier in MEKC [19]. Since digitonin is electrically neutral, SDS was added to digitonin solutions to form chiral mixed micelles having negative charges.

First, we used a digitonin–SDS solution of pH 7.0, and no enantiomeric resolution was achieved for any PTH-DL-AAs. In this case, the electroosmotic velocity was high compared with the electrophoretic velocity of the digitonin–SDS mixed micelle; hence, the migration time range was narrow, and sufficient resolution could not be attained.

To suppress the electroosmotic flow and to extend the migration time range, an acidic micellar solution (pH 3.0) was used, as mentioned earlier. Under the acidic condition, six PTH-DL-AAs were separated from each other, and each enantiomeric pair was optically resolved (Fig. 6), although a long separation time was required. Here, the direction of migration of the mixed micelle was toward the positive electrode and was opposite that of the electroosmotic flow.

Many chiral surfactants are available from natural sources, and it might be possible to use those compounds as separation carriers in MEKC for optical resolution, even though they are nonionic.

Cyclodextrin-Modified Micellar Electrokinetic Chromatography

We have reported that the addition of CD to micellar solutions is effective to separate lipophilic compounds [32,33], and we called this technique *cyclodextrin-modified MEKC* (CD–MEKC). Since CD itself is electrically neutral and it will not interact with the micelle, CD in the micellar solution should behave as another phase against the micelle, and it migrates with a velocity identical with that of the bulk solution.

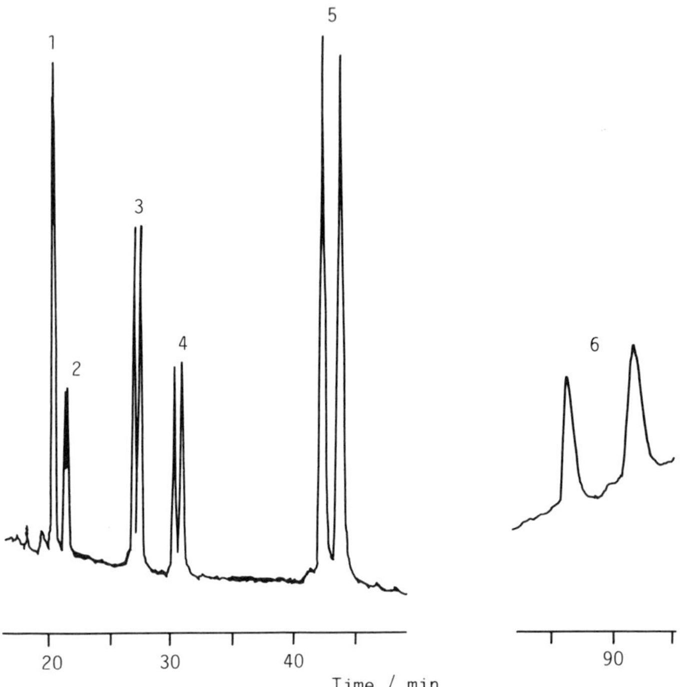

Figure 6 MEKC separation of six PTH-DL-AAs with digitonin. Corresponding AAs: (1) Trp, (2) Nle, (3) Nva, (4) Val, (5) Aba, (6) Ala. Micellar solution, 25 mM digitonin/50 mM SDS (pH 3.0); separation capillary, 63 cm × 50-μm id; effective length of the capillary, 49 cm; applied voltage, 20 kV; detection wavelength, 260 nm; temperature, ambient. (From Ref. 19)

When a highly hydrophobic solute is injected into the CD–MEKC system, it will distribute itself between the micelle and CD. The highly lipophilic solute is incorporated by either the micelle or by CD, but it does not exist in the aqueous phase. Less hydrophobic substances, on the other hand, may be partitioned among three, the micelle, the CD, and the aqueous phase. In either event, because the micelle migrates differentially with CD or with the aqueous phase, chromatographic separation can be made in the CD–MEKC system. Any CD derivatives can also be used in CD–MEKC, although theoretical treatments may become difficult if they have charges.

Cyclodextrin can achieve chiral recognition; hence, the CD–MEKC system can also be applied to optical resolution. Enantiomeric resolution of some Dns-DL-AAs by CD–MEKC using SDS solutions containing β- or γ-CD has been accomplished [34].

As shown in Fig. 7, five Dns-DL-AAs were successfully separated and optically resolved with a 60-mM γ-CD/100-mM SDS solution. By using not only β- or γ-CD, but 2,6-di-*O*-methyl-β-CD or 2,3,6-tri-*O*-methyl-β-CD as modifiers of SDS micellar solutions, optical isomers of some pharmaceutical compounds, such as thiopental, barbital, pentobarbital, phenobarbital, and so on, have been resolved [35]. With these agents, addition of some chiral compounds (e.g., *d*-camphor-10-sulfonate or *l*-menthoxyacetic acid) to micellar solutions enhanced the enantioselectivity.

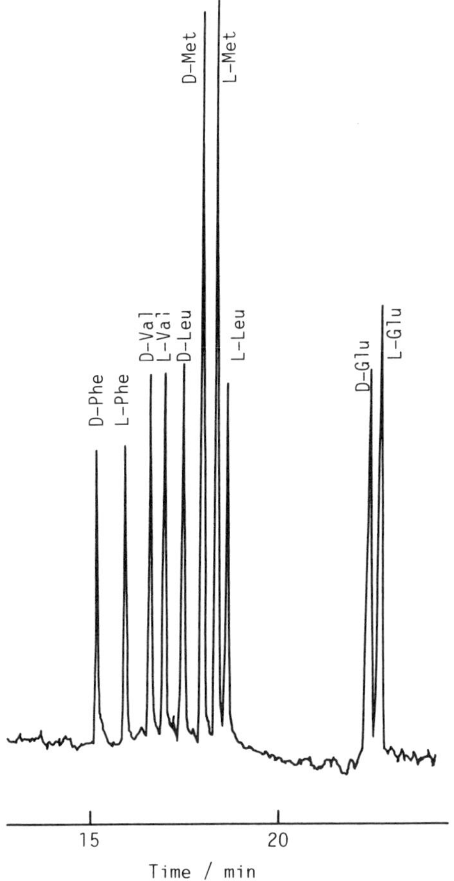

Figure 7 CD–MEKC separation of five Dns-DL-AAs. Electrolyte solution, 60 mM γ-CD/100 mM SDS (pH 8.3); separation capillary, 57 cm × 75-μm id; effective length of the capillary, 50 cm; applied voltage, 12 kV. (From Ref. 34)

Cyclodextrin Electrokinetic Chromatography

Cyclodextrin derivatives having ionic functions can behave as the separation carrier in EKC, and the system is called CD-EKC. Although the term CD-EKC might be sometimes confused with CD-MEKC, they are strictly different. In the CD-EKC system, ionic micelles never exist, but ionic CD does, whereas in the CD–MEKC system, ionic micelles must always exist, and ionic or nonionic CD is required. We have reported CD-EKC separation of some aromatic compounds [36].

As with CD–MEKC, CD–EKC can be applied to optical resolution. We have investigated chiral separation by CD–EKC using an ionic β-CD derivative, mono-(6-β-aminoethylamino-6-deoxy)-β-CD. In this system, enantiomeric resolution of Dns-DL-AAs, PTH-DL-AAs, or mandelic acid were successfully achieved [4; Y. Miyashita and S. Terabe, unpublished data].

CONCLUSION

Although there have not yet been many applications of CE and EKC on chiral separation, separation characteristics of CE or EKC are usually superior to HPLC and, therefore, an increase in the number of reports in this area is expected in the near future. Among the various methods available to chiral separation, as outlined in the foregoing, both the addition of chiral chelating reagents and the use of CD-immobilized gel capillaries may be especially effective in the CE mode. Since even neutral samples can be analyzed by EKC, the application range of EKC is wider than that of CE. In EKC, MEKC using chiral surfactants, and CD–MEKC are easily carried out for chiral separation.

REFERENCES

1. F. E. P. Mikkers, F. M. Everaerts, and T. P. E. M. Verheggen, *J. Chromatogr.*, *169*:11 (1979).
2. J. W. Jorgenson and K. D. Lukacs, *Anal. Chem.*, *53*:1298 (1981).
3. S. Hjertén, *J. Chromatogr.*, *270*:1 (1983).
4. S. Terabe, *Trends Anal. Chem.*, *8*:129 (1989).
5. S. Terabe, K. Otsuka, K. Ichikawa, A. Tsuchiya, and T. Ando, *Anal. Chem.*, *56*:111 (1984).
6. S. Terabe, K. Otsuka, and T. Ando, *Anal. Chem.*, *57*:834 (1985).
7. S. Terabe, K. Otsuka, and T. Ando, *Anal. Chem.*, *61*:251 (1989).
8. E. Gassmann, J. E. Kuo, and R. N. Zare, *Science*, *230*:813 (1985).
9. P. Gozel, E. Gassmann, H. Michelsen, and R. N. Zare, *Anal. Chem.*, *59*:44 (1987).
10. S. Fanali, L. Ossicini, F. Foret, and P. Bocek, *J. Microcolumn Separations*, *1*:190 (1989).
11. S. Fanali, *J. Chromatogr.*, *474*:441 (1989).
12. S. Fanali and P. Bocek, *Electrophoresis*, *11*:757 (1990).
13. W. Sydor and E. Mularz, *Abstracts of Papers, Third International Symposium on High Performance Capillary Electrophoresis*, San Diego, Calif., LW-9 (1991).

14. M. E. Swartz, *Abstracts of Papers, Third International Symposium on High Performance Capillary Electrophoresis*, San Diego, Calif., PT-54 (1991).
15. A. Guttman, A. Paulus, A. S. Cohen, N. Grinberg, and B. L. Karger, *J. Chromatogr.*, *448*:41 (1988).
16. A. S. Cohen, A. Paulus, and B. L. Karger, *Chromatographia*, *24*:15 (1987).
17. A. Dobashi, T. Ono, S. Hara, and J. Yamaguchi, *Anal. Chem.*, *61*:1984 (1989).
18. A. Dobashi, T. Ono, S. Hara, and J. Yamaguchi, *J. Chromatogr.*, *480*:413 (1989).
19. K. Otsuka and S. Terabe, *J. Chromatogr.*, *515*:221 (1990).
20. K. Otsuka and S. Terabe, *Electrophoresis*, *11*:982 (1990).
21. K. Otsuka, J. Kawahara, K. Tatekawa, and S. Terabe, *J. Chromatogr.*, *559*:209 (1991).
22. S. Terabe, Y. Ishihama, H. Nishi, T. Fukuyama, and K. Otsuka, *J. Chromatogr.*, *545*:359 (1991).
23. H. Nishi, T. Fukuyama, M. Matsuo, and S. Terabe, *J. Chromatogr.*, *498*:313 (1990).
24. H. Nishi, T. Fukuyama, M. Matsuo, and S. Terabe, *J. Chromatogr.*, *513*:279 (1990).
25. S. Terabe, M. Shibata, and Y. Miyashita, *J. Chromatogr.*, *480*:403 (1989).
26. K. Otsuka and S. Terabe, *J. Microcolumn Separations*, *1*:150 (1989).
27. H. Nishi, T. Fukuyama, M. Matsuo, and S. Terabe, *J. Microcolumn Separations*, *1*:234 (1989).
28. H. Nishi, T. Fukuyama, M. Matsuo, and S. Terabe, *J. Chromatogr.*, *515*:233 (1990).
29. H. Nishi, T. Fukuyama, M. Matsuo, and S. Terabe, *Anal. Chim. Acta*, *236*:281 (1990).
30. R. O. Cole, M. J. Sepaniak, and W. L. Hinze, *Abstracts of Papers, 11th International Symposium on Capillary Chromatography*, Monterey, Calif., p. 890 (1990).
31. A. R. Campanelli, S. Canderloro De Sanctis, E. Chiessi, M. D'Alagni, E. Giglio, and L. Scaramuzza, *J. Phys. Chem.*, *93*:1536 (1989).
32. S. Terabe, Y. Miyashita, O. Shibata, E. R. Barnhart, L. R. Alexander, D. G. Patterson, B. L. Karger, K. Hosoya, and N. Tanaka, *J. Chromatogr.*, *516*:23 (1990).
33. H. Nishi and M. Matsuo, *J. Liq. Chromatogr.*, *14*:973 (1991).
34. Y. Miyashita and S. Terabe, *Application Data, High Performance Capillary Electrophoresis*, DS-767, Beckman (1990).
35. H. Nishi, T. Fukuyama, and S. Terabe, *J. Chromatogr.*, *553*:503 (1991).
36. S. Terabe, H. Ozaki, K. Otsuka, and T. Ando, *J. Chromatogr.*, *332*:211 (1985).

21

The Sheath-Flow Cuvette in DNA Sequencing by Capillary Gel Electrophoresis and Two-Spectral–Channel, Laser-Induced Fluorescence Detection

Jianzhong Zhang, Da Yong Chen,
Heather R. Harke, and Norman Dovichi

University of Alberta
Edmonton, Alberta, Canada

Sanger reported a powerful DNA sequencing technique based on DNA polymerase and dideoxynucleotides [1]. In this technique, a primer is attached to a single-strand DNA template. The template–primer mixture is divided into four containers, and DNA polymerase and deoxynucleotide triphosphates are added to each container. Finally, dideoxyadenosine triphosphate is added to the first container, dideoxyguanosine triphosphate is added to the second container, dideoxycytidine triphosphate is added to the third container, and dideoxythymidine triphosphate is added to the last container. In the absence of the dideoxynucleotide triphosphates, the polymerase enzyme catalyzes the extension of the primer, producing a DNA strand that is complementary to the template. In the presence of a small amount of a dideoxynucleotide, the chain extension will occasionally be halted upon incorporation of that nucleotide. The product of the reaction in the first tube is a set of DNA fragments that have an identical origin (the primer) and extend to various lengths, always ending in dideoxyadenosine. Similarly, the products of the other reactions are sets of DNA fragments that end in the other nucleotides.

Sanger incorporated a small amount of radioactive nucleotide in the sequencing reaction mixture. The radioactive products of the sequencing mixture were separated in adjacent lanes of a polyacrylamide gel. The gel separated the fragments based on size: mobility decreases monotonically with fragment length. After the electrophoresis was complete, x-ray film was placed in contact with the gel, and the location of the DNA was detected by autoradiography. Finally, the sequence was interpreted from the pattern of alternating bands in the lanes corresponding to the terminal base of the fragment.

This technology represents a spectacular improvement over earlier technology used to sequence DNA. Classic organic chemical techniques produced miniscule sequence information with heroic effort. After Sanger's technology became widely available, DNA sequencing became a routine, albeit tedious, technique in the biochemical laboratory. Sanger's method requires roughly 8–16 hours for the electrophoretic separation, 8–16 hours for autoradiography, and perhaps another 4 hours for visual inspection of the autoradiogram. In addition, Sanger's method requires use of radionucleotides; the expense of disposal of low-level waste is rapidly becoming a significant item in the sequencing budget.

Advances in sequencing technology occurred in 1986–1988 when Smith et al. [2] in Hood's laboratory, Ansorge [3], Prober et al. [4] at DuPont, and Kambara et al. [5] at Hitachi reported DNA sequencers that replaced the radioactive labels and autoradiography in Sanger's method with fluorescent labels and laser-based detection. By use of fluorescence detection, the difficulties associated with autoradiography are eliminated. Furthermore, by use of computerized data collection, the interpretation of the electropherogram to obtain the sequence information may be automated. As a result, these fluorescence instruments determine DNA sequence within an 8- to 12-hour period after introduction of analyte into the slab gel. Gels of 0.25- to 0.5-mm thickness are employed in these slab-gel sequencers. Joule heating limits the applied electric field to about 75 V cm^{-1}; sequencing rates are about 50 bases per lane at this electric field strength. However, 12–24 samples may be separated simultaneously, producing sequencing rates of about 1000 bases per hour for each slab gel. Sequence may be determined by use of computer algorithms to about 450 bases, which is limited by the signal/noise ratio of the data and the resolution of the gel.

Smith's, Ansorge's, Prober's, and Kambara's fluorescence sequencers have been commercialized by Applied Biosystems, Pharmacia, DuPont, and Hitachi, respectively; as an example of typical performance, the Applied Biosystems model 373A instrument runs 24 lanes simultaneously on a slab gel to produce sequencing rates of 50 bases per hour per lane or 1200 bases per hour per slab. Similar sequencing rates are produced by other instruments.

Increased sequencing rates are produced by operating the gels at high electric field strength. Unfortunately, the finite resistance of the separation buffer leads to unacceptable heating of conventional 200- to 400-μm–thick gels at electric field strengths much greater than 100 V cm^{-1}. Gel-filled capillaries have attracted interest because their high surface/volume ratio provides excellent heat transport properties, allowing use of very high electric field strengths. Typical capillaries are 50- to 75-μm inner diameter (id) and 25- to 100-cm long, although early work on RNA separations used cellulose fibers of 10 μm in diameter and 25 mm in length [6,7]. More recently, Hjertén's and Karger's groups have reported the use of gel-filled capillaries for protein separations [8,9]. Several groups have reported the use of gel-filled capillaries for separation of oligonucleotide standards with detection by ultraviolet (UV) absorbance [10–13]. An important issue is the performance of the gel-filled capillaries under high electric field strength; patents have been issued on the use of a bifunctional silane reagent to bind the acrylamide to the wall of the capillary to improve the gel stability [14,15].

Capillary gel electrophoresis has been applied to the separation of fluorescently labeled DNA sequencing fragments. Swerdlow and Gesteland reported the separation of a single dideoxynucleotide reaction mixture with a single spectral channel laser-induced fluorescence detector [16]. Drossman and co-workers in Smith's laboratory reported the separation of a single reaction mixture in gel-filled capillaries [17]. When operated at 400 V cm^{-1}, the system produced sequencing rates of 1000 bases per hour after elution of the primer, a roughly 25 times higher-sequencing rate than produced by conventional slab gel electrophoresis. A subsequent report from Smith's laboratory described the use of capillary gel electrophoresis (CGE) for the separation of the reaction products of the four-spectral–channel, single–lane-sequencing system [18]. Resolution was between a factor of 1.5 and 2 times superior to slab-gel data. Karger's laboratory reported separation of DNA fragments, labeled with a single fluorophore, at 350 V cm^{-1}, yielding sequencing rates of 450 bases per hour; baseline resolution was obtained for fragments up to 330 nucleotides in length [19]. Gesteland's laboratory has reported the use of a charge coupled device for fluorescence detection in DNA sequencing by capillary gel electrophoresis [20]. Our laboratory has reported several designs for fluorescent detection in capillary gel electrophoresis for DNA sequencing [21–23].

High-sensitivity detection appears to be required for DNA sequencing by capillary gel electrophoresis. To ensure that the local electric field is not perturbed by the ionic strength of analyte, it is necessary that the ionic strength of the sample be less than 1% of the ionic strength of the buffer. Although the buffers used in gel electrophoresis have concentration in the high-millimolar range, they are poorly ionized and the ionic strength of the buffer is about 10 mM. Therefore, the total ionic strength of the sample in the electrophoresis gel must be less than 100 μm, to minimize electric field artifacts. At best, half the analyte molecules will be fluorescently labeled; at least half the analyte molecules are unlabeled template. Because the template, which is ~7500 bases long and has a negative charge per base, carries a high charge, most of the ionic strength of the sample is associated with the template. If one assumes that the average analyte fragment is 300 bases in length, only 4% of the ionic strength of the sample is associated with analyte; the analyte ionic strength is 4 μM. However, since each analyte molecule has, on average, 300 bases, the concentration of fluorescently labeled analyte is about 1×10^{-8} M. Clearly, very sensitive detection is required for DNA sequencing.

The sample must occupy a volume less than 10^{-9} L in the gel to avoid excessive band broadening; an average of 10 amol of DNA will be present in each band.

Although additional material can be loaded onto the column through use of stacking injection conditions, only a minute amount of material is available for sequencing. To obtain a decent signal/noise ratio and to account for those fragments that are inefficiently labeled it is necessary to use a system with zeptomole (1 zeptomole = 1 zmol = 10^{-21} mol = 600 molecules) detection limits in DNA sequencing by capillary gel electrophoresis. On-column fluorescence detectors, for which the capillary serves as the detection chamber, produce detection limits of ~200 zmol of fluorescently labeled product [16,17,19,21]. These on-column detectors are dominated by a large background signal generated by light scatter from the capillary walls. This light scatter is generated by refraction and reflection from the capillary walls, along with scatter from scratches and pits in the wall. Scatter takes the form of a fan of light in the plane perpendicular to the capillary axis. Although detection of the scattered light is minimized by tilting the capillary relative to the collection optic, use of a high numerical aperture objective to increase collection efficiency of fluorescence also enhances sensitivity to scattered light. Refraction and reflection may be minimized by use of a tightly focused laser beam so that only a small fraction of the capillary cross section is illuminated; although background is reduced, only a small fraction of analyte molecules eluted from the electrophoresis column is illuminated, decreasing sensitivity. Furthermore, the high irradiance produced by the focused beam leads to enhanced photobleaching of the analyte, an important problem with photolabile dyes used in the sequencing reactions. To eliminate light scatter associated with the capillary windows, a postcolumn fluorescence detector can be used. The detector, based on a sheath-flow cuvette, can produce excellent detection performance at modest cost and effort.

EXPERIMENTAL

Our laboratory has developed a postcolumn fluorescence detector for capillary electrophoresis [24–28]. A sheath-flow cuvette is used as the detection chamber (Fig. 1). This chamber is typically 200-μm square in cross section, 20-mm long, and has 2-mm thick windows. The quartz windows have good optical quality and generate very little light scatter. The 190-μm outer diameter (od) capillary is placed within the square flow chamber. As analytes elute from the bottom of the capillary, they are swept downstream by a sheath stream, which is usually composed of the same buffer as is used for the separation. Sheath flow rates are low, ~0.15 mL h^{-1}. The sheath flow can be generated either with a high-precision (and high-cost) syringe pump or with a simple siphon, with about 7-cm head above the bottom of the cuvette. Disruption of the flow pattern within the cuvette occurs when drops dislodge from the bottom of the cuvette. A piece of plastic tubing can be used to direct the waste flow from the cuvette to a receiving vessel, eliminating the formation of droplets. In zone electrophoresis, it is important that the sample and waste vials are at the same height; otherwise, a siphon can form between the two reservoirs.

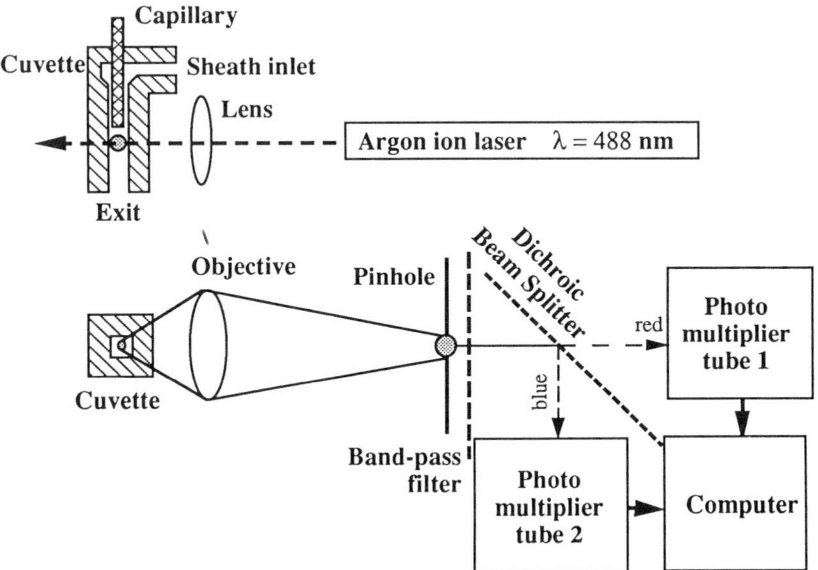

Figure 1 Laser-induced fluorescence detector for two-spectral–channel sequencing. A 5-mW argon ion laser beam (λ = 488 nm) was focused with a 4 × microscope objective about 200-μm below the exit of the capillary in a sheath-flow cuvette with 200-μm square flow chamber and 2-mm thick windows [21]. Fluorescence was collected at right angles with a 32 ×, 0.6 numerical aperture microscope objective and imaged onto a 0.75-mm–diameter pinhole. A single interference filter (525-nm center wavelength, 40-nm band pass) transmitted fluorescence. A dichroic filter, centered at 525 nm, was used to direct fluorescence in two spectral channels to two R1477 photomultiplier tubes operated at 1200 V; the transmitted light was centered at 535 nm (red) and the reflected light was centered at 515 nm. The photomultiplier tubes' signals were passed through a 1-Hz low-pass filter, digitized at 10 Hz with a computer, and passed through a quadratic-cubic polynomial filter before display.

The siphon will lead to bulk flow that both biases the injection volume and introduces a parabolic flow profile in the separation capillary, distorting the separation. The high resistance to flow found in gel electrophoresis eliminates hydrodynamic flow in the separation capillary when the receiving and injection reservoirs are not at the same height.

At the very low-flow rates employed in the sheath-flow cuvette, flow is laminar. Analyte forms a narrow stream in the center of the flow chamber, far from the cuvette walls. By restricting the field of view of the photodetector to the sample stream, light scatter from the interface of the cuvette window and the sample stream is not detected, greatly reducing the background signal. In our zone electrophoresis detector, the best detection limits (3 σ) are 60 yoctomole (1 yoctomole = 1 ymol =

10^{-24} mol = 0.5 analyte molecule) of tetramethylrhodamine isothiocyanate injected onto the capillary. In gel electrophoresis, for which hydrodynamic effects are not as favorable as in zone electrophoresis, detection limits are between 2 and 20 zmol (1200–12,000 analyte molecules).

The DuPont-sequencing technique uses succinylfluorescein dyes to label the four dideoxynucleotides. These dyes have relatively closely spaced absorbance and emission bands. As a result, a single argon ion laser wavelength, 488 nm, is used to excite fluorescence from all dyes. Emission is distributed between two spectral channels centered at 510 and 540 nm. The ratio of the fluorescence intensity in the two spectral channels is used to identify the terminating dideoxynucleotide. Because of photobleaching, modest laser power is used in these experiments. The longer-wavelength dyes used in the DuPont-sequencing protocol are particularly sensitive to bleaching; we typically use 5-mW laser power to excite fluorescence from these dyes.

Fluorescence is collected with a high-efficiency microscope objective and imaged onto a pinhole to restrict the field of view of the photodetector to the illuminated sample stream. A single band-pass filter is used to block scattered laser light, whereas a dichroic filter is used to split the fluorescence into two spectral channels, followed by detection with two photomultiplier tubes. The detection limit for this instrument is 20 zmol for a labeled dideoxynucleotide triphosphate and 2 zmol (3000 molecules) for a 100-mer fluorescein-labeled oligonucleotide. The difference in detection limit between the monomer and a 100-mer fragment is related to the behavior of the sheath-flow cuvette when applied to gel electrophoresis. A very low sheath flow rate is employed to transport analyte from the capillary to the illumination region. During the transit time, analyte can diffuse an appreciable amount, decreasing the effective concentration of analyte in the detector. However, higher-molecular-mass fragments undergo less diffusion and are more concentrated in the illumination volume. As a result, the detection limit for the system improves for larger DNA fragments, which are typically produced in sequencing reactions at lower concentration compared with early-eluting fragments.

RESULTS

A typical separation on a 4%T gel, at 465 V cm^{-1}, of the fragments from an M13mp18 template is shown in Fig. 2a. The first large peak, eluting at 12.8 minutes is associated with unincorporated dideoxynucleotides. The peaks from 12.8 to 13.5 minutes are difficult to interpret and are associated with the primer. The large peak at 14.2 minutes is a severe compression associated with a GC-rich portion of the sequence. The remainder of the electropherogram is consistent with the known sequence of M13mp18 to about 275 bases. The signal amplitude for longer fragments is quite low, and the data are difficult to interpret. The decrease in signal for longer fragments is associated with exhaustion of the dideoxynucleotides from the reaction mixture. The fluorescently labeled dideoxynucleotides are expensive; hence, the reaction conditions are designed to minimize their use. These conditions lead to termination of the reaction at relatively short fragments, so that there is very poor signal amplitude for the longer fragments. On the other hand, alternative sequencing protocols can be employed that allow much longer sequencing runs to be generated. The data presented here are simply illustrative of the two-spectral–channel detector employed with this particular sequencing protocol.

The peak spacing in Fig. 2a is quite uniform; a plot of retention time versus fragment length yields a straight line ($r > 0.999$). *Sequencing rate* is defined as the slope of this line. For this data, the sequencing rate is 1000 bases per hour. More precisely, the sequencing rate is 17 bases per minute for fragments from 25 to 300 bases in length. Very long fragments tend to comigrate, particularly at high electric field strength. There is much effort in the design of sequencing gels that produce efficient separation of long fragments at high-sequencing rates.

In the region from 60 to 100 nucleotides, the peaks are nearly baseline resolved (see Fig. 2b). Resolution ranges from 1 to 2 and typically is 1.5. In this sequencing protocol, the relative peak amplitude in the two spectral channels is used to identify the terminal nucleotide. The solid curve corresponds to a fluorescent signal detected in the red spectral channel centered at 540 nm, whereas the dashed curve corresponds to a fluorescent signal detected in the blue spectral channel centered at 510 nm. In this system, T generates peaks that are much stronger in the red spectral channel compared with the blue channel, C generates peaks at which the red channel is slightly larger than the blue channel, A generates peaks with roughly equal amplitude in the two spectral channels, and G generates peaks at which the blue channel amplitude is larger than the red channel. Careful inspection of the data reveals that the accuracy of the sequence data is not high; several errors can be noted in the data, particularly bases 63 and 83, which produce similar peak ratios, but are associated with C and A, respectively. The DuPont-sequencing protocol suffers from limited accuracy because a relatively small difference in peak amplitude is used to identify each peak. Noise corrupts sequence identification. Improved accuracy is produced by alternative sequencing chemistry.

We tested the behavior of the gel at higher electric fields. First, the separation was allowed to run to 60 bases at an electric field of 150 V cm^{-1}. The separation was stopped, the background signal was measured, and the electric field was ramped to 500 V cm^{-1} over a 30-second period. Figure 3 presents the separation of fragments from 61 to 100 bases in length. The sequencing rate increases to 30 bases per minute (1800 bases per hour) as the electric field was increased. An increased sequencing rate arises from both the increased electric field and the decreased capillary length. The resolution at this high electric field strength is poorer than that observed at the lower electric field, with resolution between 0.5 and 1. This poorer resolution is due to two factors. First, this separation was the third performed on this capillary; after several injections at high electric fields, the gel tends to degrade, and the

50
100
Fluorescence Intensity (Arbitrary Units)
Time (Min)
150
200
Fluorescence Intensity (Arbitrary Units)
Time (Min)
250
300
Fluorescence Intensity (Arbitrary Units)
Time (Min)

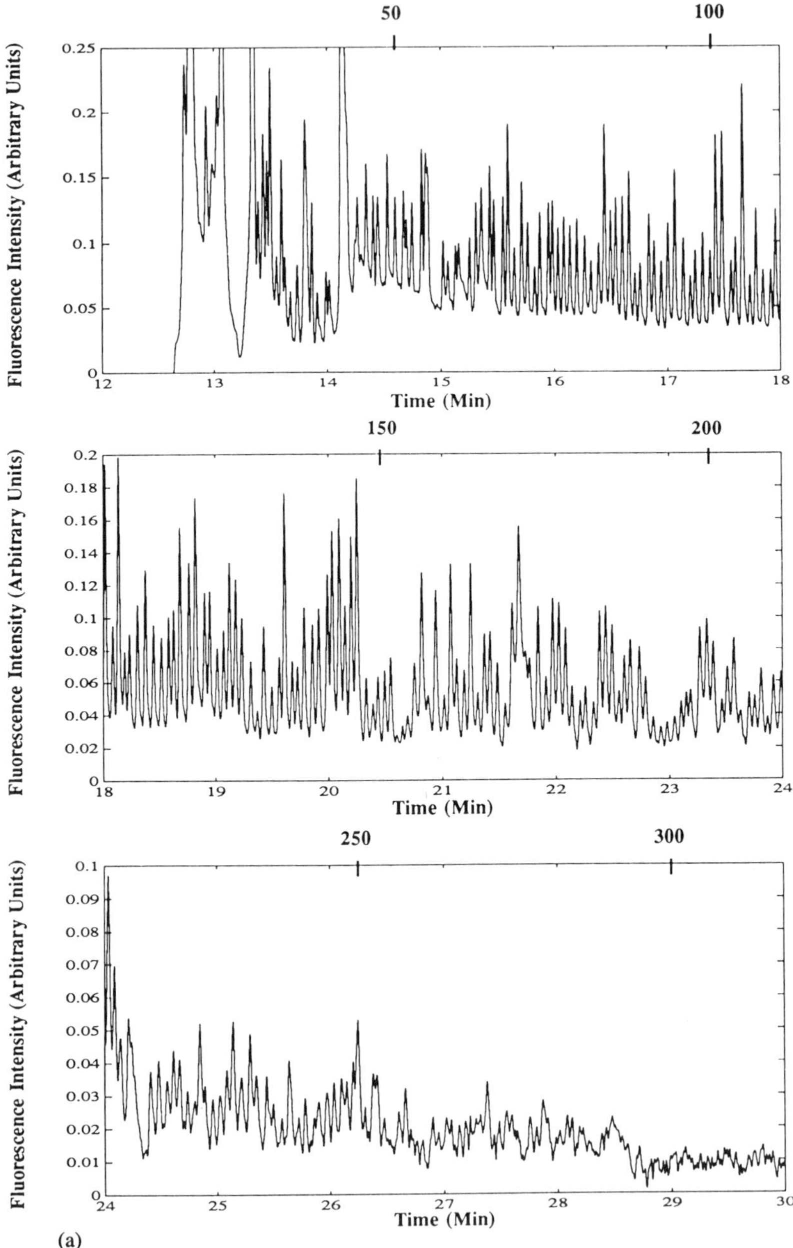

(a)

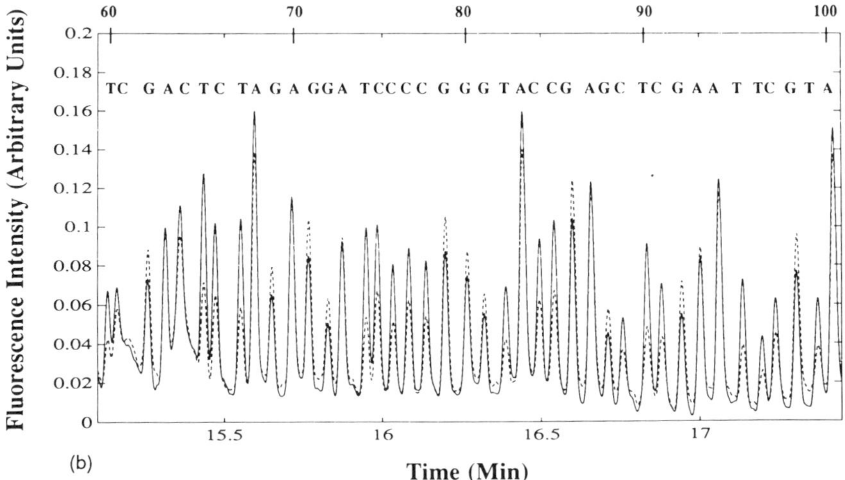

Figure 2 Two-spectral–channel sequencing of M13mp18. The sample was prepared from a DuPont Genesis 2000 protocol using 3 μM M13mp18 single-stranded DNA, 15 ng -40 17-mer M13 primer, and 1 μL Sequenase (US Biochemicals) in a standard reaction mix. The reaction products were ethanol precipitated, washed, and resuspended in 3 μL of a 49:1 mixture of formamide/0.5 M EDTA at pH 8.0. The 34-cm long, 50-μm id, 190-μm od capillary was filled with 4%T, 5%C acrylamide gel that was covalently bound to the first and last 2 cm of the capillary wall through use of γ-methacryloxypropyltrimethoxysilane. The sample was injected at 100 V cm^{-1} for 30 seconds; after injection, the sample was replaced with a fresh vial of 1 × TBE. The electrophoresis continued for 1 minute at 100 V cm^{-1}. The capillary was then trimmed at the injection end by 1 mm, and the voltage was increased in 1-kV increments over a 14-minute period to a total electric field strength of 465 V cm^{-1}. The sheath stream was 1 × TBE at a flow rate of 0.16 mL hr^{-1}. Time is measured from the injection; the numbers at the top of the figure correspond to the nucleotide length, including the -40 17-mer primer. (a) Extended run, only the long wavelength data are shown. (b) Expanded region corresponding to nucleotides 60–100; the known sequence is printed above. The solid trace corresponds to emission at wavelengths longer than 525 nm and the dashed trace correspond to emission at wavelengths shorter than 525 nm.

separation efficiency suffers. Second, at the high electric field, thermal gradients across the capillary axis lead to decreased efficiency. The decreased resolution degraded the sequencing accuracy, as peak overlap distorted the relative amplitudes observed in the two spectral channels. Although this sequencing accuracy is quite poor, the data do show that high-speed sequencing is produced by polyacrylamide gels operating at high electric field strengths. Improved sequencing techniques, for example, based on improved gels and alternative fluorophores, will lead to higher-sequencing accuracy at high-sequencing rates.

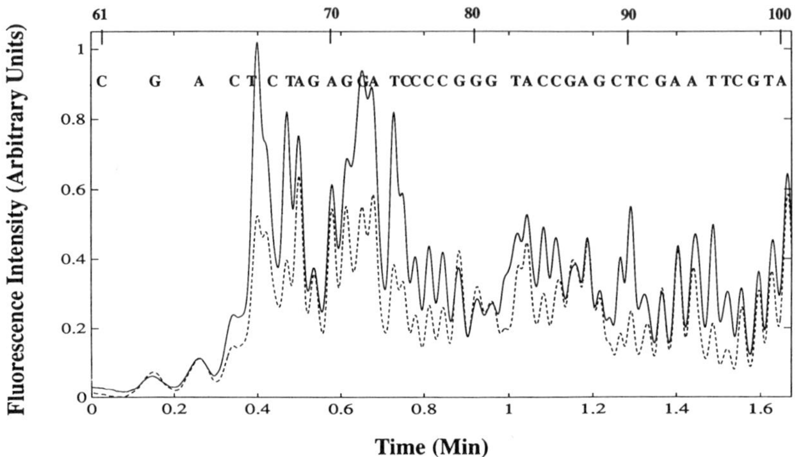

Figure 3 High-Speed DNA Sequencing. The 25-cm, 50-μm id capillary was filled with a 4%T, 5%C gel that was covalently bound to the first and last 2 cm of the capillary through use of γ-methacryloxypropyltrimethoxysilane. The sample was injected at 150 V cm^{-1} for 1 minute. After injection, the sample was replaced with a fresh vial of 1 × TBE. Electrophoresis proceeded at 150 V cm^{-1} for 24 minutes, until base number 60 eluted. The electric field was removed, the injection end of the capillary was trimmed by ~1 mm, and the electric field was ramped to 500 V cm^{-1} over ~30 seconds. The sheath flow rate was 0.16 mL h^{-1}.

ACKNOWLEDGMENTS

This work was funded by the Natural Sciences and Engineering Research Council of Canada, the Department of Chemistry of the University of Alberta, Pharmacia Inc., and Waters Division of Millipore Inc. HRH acknowledges a predoctoral fellowship from the Alberta Heritage Foundation for Medical Research. NJD acknowledges a Steacie fellowship from the Natural Sciences and Engineering Research Council. G. Tyce of DuPont kindly supplied the DNA sample.

REFERENCES

1. F. Sanger, S. Nicklen, and A. R. Coulson, *Proc. Natl. Acad. Sci. USA*, *74*:5463 (1977).
2. L. M. Smith, J. Z. Sanders, R. J. Kaiser, P. Hughes, C. Dodd, C. R. Connell, C. Heiner, S. B. H. Kent, and L. E. Hood, *Nature*, *321*:674 (1986).
3. W. Ansorge, B. S. Sproat, J. Stegemann, and C. Schwager, *J. Biochem. Biophys. Methods*, *13*:315 (1986).
4. J. M. Prober, G. L. Trainor, R. J. Dam, F. W. Hobbs, C. W. Robertson, R. J. Zagursky, A. J. Cocuzza, M. A. Jensen, and K. Baumeister, *Science*, *238*:336 (1987).

5. H. Kambara, T. Nishikawa, T. Katayama, and T. Yamaguchi, *BioTechnology*, *6*:816 (1988).
6. J. E. Edström, *Nature*, *172*:809 (1953).
7. J. E. Edström, *Biochim. Biophys. Acta*, *22*:378 (1956).
8. S. Hjertén, *J. Chromatogr.*, *270*:1 (1983).
9. A. S. Cohen and B. L. Karger, *J. Chromatogr.*, *397*:409 (1987).
10. A. S. Cohen, D. R. Najarian, A. Paulus, A. Guttman, J. A. Smith, and B. L. Karger, *Proc. Natl. Acad. Sci. USA*, *85*:9660 (1988).
11. A. Guttman, A. S. Cohen, D. N. Heiger, and B. L. Karger, *Anal. Chem.*, *62*:137 (1990).
12. A. Paulus, E. Gassmann, and M. J. Field, *Electrophoresis*, *11*:702 (1990).
13. H. F. Yin, J. A. Lux, and G. Schomburg, *J. High Resolut. Chromatogr.*, *13*:624 (1990).
14. P. F. Bente and J. Myerson, *U.S. Pat.* 4,810,456 (1989).
15. B. L. Karger and A. S. Cohen, *U.S. Pat.* 4,865,706 (1989).
16. H. Swerdlow and R. Gesteland, *Nucleic Acids Res.*, *18*:1415 (1990).
17. H. Drossman, J. A. Luckey, A. J. Kostichka, J. D'Cunha, and L. M. Smith, *Anal. Chem.*, *62*:900 (1990).
18. J. A. Luckey, H. Drossman, A. J. Kostichka, D. A. Mead, J. D'Cunha, T. B. Norris, and L. M. Smith, *Nucleic Acids Res.*, *18*:4417 (1990).
19. A. S. Cohen, D. R. Najarian, and B. L. Karger, *J. Chromatogr.*, *516*:49 (1990).
20. A. E. Karger, J. M. Harris, and R. F. Gesteland, *Nucleic Acids Res.*, *19*:4955 (1991).
21. H. Swerdlow, S. Wu, H. R. Harke, and N. J. Dovichi, *J. Chromatogr.*, *516*:61 (1990).
22. H. Swerdlow, J. Z. Zhang, D. Y. Chen, H. R. Harke, R. Grey, S. Wu, N. J. Dovichi, and C. Fuller, *Anal. Chem.*, *63*:2835 (1991).
23. D. Y. Chen, H. P. Swerdlow, H. R. Harke, J. Z. Zhang, and N. J. Dovichi, *J. Chromatogr.*, *559*:237 (1991).
24. N. J. Dovichi, J. C. Martin, J. H. Jett, and R. A. Keller, *Science*, *219*:845 (1983).
25. N. J. Dovichi, J. C. Martin, J. H. Jett, M. Trkula, and R. A. Keller, *Anal. Chem.*, *56*:348 (1984).
26. Y. F. Cheng, and N. J. Dovichi, *Science*, *242*:562 (1988).
27. Y. F. Cheng, S. Wu, D. Y. Chen, and N. J. Dovichi, *Anal. Chem.*, *62*:496 (1990).
28. S. Wu and N. J. Dovichi, *J. Chromatogr.*, *480*:141 (1989).

22

The Use of Capillary Electrophoresis in Clinical Diagnosis

Norberto A. Guzman*, Carmen L. Gonzalez, and Maria A. Trebilcock

Princeton Biochemicals, Inc.
Princeton, New Jersey

Luis Hernandez

Los Andes University
Mérida, Venezuela

Clifford M. Berck

SUNY-Health Science Center at Brooklyn
Brooklyn, New York

Juan P. Advis

Rutgers University
New Brunswick, New Jersey

INTRODUCTION

Basic knowledge in immunology, biochemistry, physiology, and pharmacology related to normal physiological activities and human diseases has expanded greatly in the past two decades. Consequently, advancements in basic scientific developments are a necessity for the analysis of biochemical substances present in biological fluids, blood cells, and tissue specimens. In turn, the presence or absence of one or more biochemical substances can help us determine the "state of health" of an individual. Rational prognosis and therapy of a disease can follow only after a precise diagnosis. Diagnosis, in turn, depends primarily on skillful history-taking and physical examination of a patient, with subsequent help from a variety of laboratory, radiological, and clinical tests.

**Current affiliation*: The R. W. Johnson Pharmaceutical Research Institute, a Johnson & Johnson Company, Raritan, New Jersey.

Much information about a disease state has been gained by determining a certain number of biochemical parameters that are used as indicators of disease activity, as well as providing information about the intensity of the whole pathological process. Although the molecular basis of many diseases is not clearly understood, it has been recognized that the analysis of biochemical constituents of body fluids, blood cells, and tissues is important in making a diagnosis. Therefore, the clinical chemistry laboratory is an important contributor to the medical team involved in the diagnosis and treatment of disease, and physicians rely heavily on laboratory test results before making decisions.

Clinical chemistry, however, involves more than analysis of biochemical constituents of body fluids, blood cells, and tissues. *Clinical chemistry* has been defined as the study of the chemical aspects of human life in health and illness and the application of chemical laboratory methods to diagnosis, control of treatment, and prevention of disease. Thus, clinical chemistry is a fundamental science when it seeks to understand the physiological and biochemical processes operating both in the normal state and in disease; it is an applied science when analyses are performed on body fluids, blood cells, or tissue specimens to provide useful information for the diagnosis or treatment of disorders.

INSTRUMENTATION IN A CLINICAL LABORATORY

Today, the clinical chemistry laboratory could not operate without the widespread use of instrumentation. The dream of all physicians is to have the maximum amount of useful clinical and laboratory information in the shortest time. A variety of techniques have been used throughout the years, including most electrophoretic and chromatographic techniques [1–3]. Simple spectrophotometric apparatus as well as more sophisticated instruments, such as atomic absorption spectrophotometers and gas chromatography–mass spectrometers, are now part of a modern laboratory. In addition, a variety of immunochemical techniques using radioactive or fluorescent probes are used routinely in the laboratory. However, since hundreds of thousands of tests are performed in the chemistry laboratory of a large hospital, automated instruments are required so that multiple specimens can be analyzed, and more than one substance can be determined simultaneously with high precision and accuracy. The use of properly controlled, automated instruments can minimize errors of technique and measurement.

Nevertheless, many problems arise when dealing with body fluids, blood cells, or tissue specimens. One of the most common problems is the interference of other substances in determining one particular compound, potentially generating false-positive or false-negative results. That assay specificity cannot be fully characterized creates the need for confirmatory tests. The method used for confirmation should employ a different analytical principle from that used in the primary screen. Therefore, a battery of different analytical and immunochemical tests are required in modern hospital laboratories.

It is difficult to estimate the number of substances that are present in body fluids, blood cells, or tissue specimens and many substances cannot be detected because of their low concentration in the sample. Therefore, the need for more efficient methods of separation and of highly sensitive techniques is also an important issue in the laboratory.

The purpose of this chapter is to review the use of capillary electrophoretic analysis of biochemical substances found in biological fluids of physiological and pathophysiological states. Analysis of substances administered from outside the body, by any delivery system, such as drugs, toxins, or other chemicals, will not be discussed here. For the capillary electrophoresis analysis of drugs in body fluids, see Chapter 25.

CAPILLARY ELECTROPHORESIS

Capillary electrophoresis (CE) is a relatively young branch of separation science. Many observations of importance to capillary electrophoresis were made by analytical chemists during the last decade, usually in the course of active research in the separation and characterization of simple standard samples [4–8]. Recently, however, more complex substances present in solution mixtures have been separated by capillary electrophoresis, opening possibilities for a variety of new applications [9–15].

The explosive increase in fundamental information has made capillary electrophoresis a unique branch of separation science. It is now possible to determine nanoliter to picoliter quantity of material and to reach detection sensitivity levels of less than 3 zeptomoles (1 zeptomole = 10^{-21} mol = 600 molecules) in relatively short times [16,17; see Chap. 19]. Because of the separation power of capillary electrophoresis, structurally closely related substances, such as enantiomers and isomers, can be baseline separated [18,19; see Chap. 21]. The basic schematic of a capillary electrophoresis instrument is depicted in Fig. 1.

Since the first electropherogram of urinary components [23], several papers have reported the use of capillary electrophoresis to analyze various constituents present in biological fluids, blood cells, and tissue biopsies (Table 1) [20–35]. Thus, it is possible to assume that this technique can be used in a clinical laboratory, both as a complementary technique to the many existing diagnostic methods and as a confirmatory diagnostic technique. Although, in many instances, capillary electrophoresis might not be practical for routine sample analysis (i.e., mass screening of urine or serum specimens), its unprecedented resolution, specificity, and accuracy recommends it as a reference method for validating other methods.

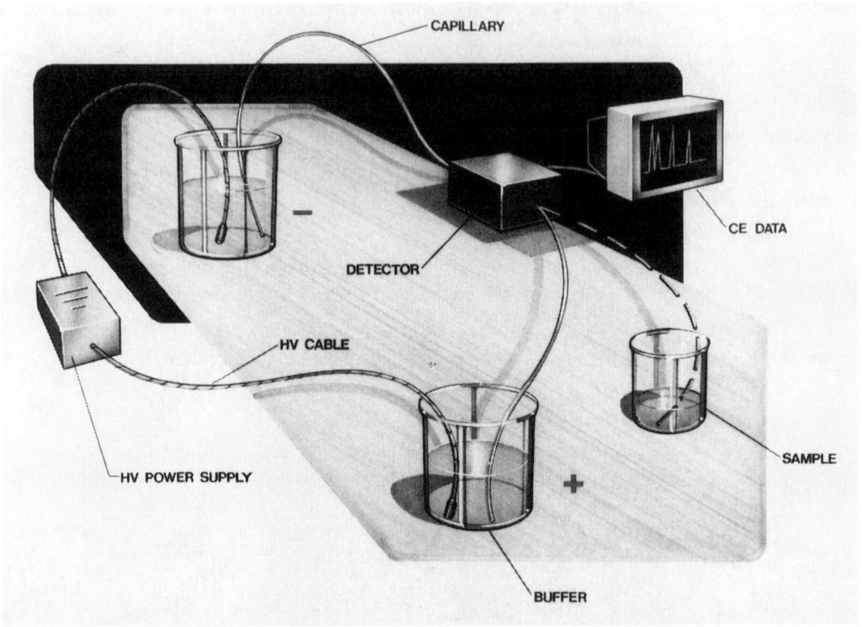

Figure 1 A schematic diagram of a capillary electrophoresis instrument with on-column detection.

Capillary Electrophoresis of Urine Specimens

In general, examination of urine can be considered from two standpoints: (a) diagnosis and management of renal or urinary tract disease, and (b) detection of metabolic or systemic diseases not directly related to the kidney. The usefulness of urinary data for clinical judgment depends on many factors, one of the most important being the accuracy and precision of the method.

Chemical analyses of urine specimens are routinely performed in 24-hour collection samples, since many urine constituents exhibit a diurnal variation [3]. Although urine specimens are usually collected in the presence of an appropriate preservative, refrigerated samples, with no preservatives between voidings, are common. Furthermore, refrigeration is critical to avoid urine contamination, since the analysis is usually performed at the end of the day, or even a few days after the collection of the specimen. However, refrigeration or freezing of these samples causes precipitation of some urine components. In addition, thawing after freezing may reveal some turbidity (possibly colloidal protein) that does not redissolve [3]. Therefore, appropriate experimental conditions should be considered for the determination of certain urine analytes by capillary electrophoresis.

Table 1 Determination of Biological Fluid, Blood Cells, and Tissue Biopsy Constituents by Capillary Electrophoresis

Specimen	Year	Ref.
Urine	1981	20
Serum	1983	4
Serum	1987	21
Urine	1988	22
Serum	1989	23
Urine	1989	9
Urine	1989	24
Urine	1990	25
Urine, serum	1990	26
Plasma	1990	27
Urine, serum, cerebrospinal fluid	1991	28
Urine	1991	29
Urine	1991	30
Urine	1991	31
Urine, blood cells, tissue biopsy	1991	32
Urine, blood cells, tissue biopsy	1992	33

Figure 2a depicts a typical electropherogram of normal human urine using an ultraviolet (UV) detection method. The urine specimens, freshly collected as clean-catch (of fasting morning urine), were obtained from six normal adult men of approximately 70- to 90-kg body weight. The time of collection of biological fluids is critical, since dietary constituents may significantly influence the profile of the electropherogram. Depending on the wavelength used to monitor the urinary constituents and the attenuation used in the instrument, there will be variability in the profile of the electropherogram obtained. But in general, approximately ten major and several minor components are identified in normal urine specimens of fasting morning urine. Figure 2B depicts the effect of low-temperature storage on the profile of the electropherogram of a normal urine specimen. When the urine sample was stored at $-70°C$ and then thawed at 4°C, a strong precipitate was formed. This precipitate was separated from the sample by centrifugation and the clear supernatant was analyzed by capillary electrophoresis. As shown in Fig. 2B, peak 8 was significantly decreased in the supernatant of the thawed urine, when compared with peak 8 (see Fig. 2A) of a normal control urine obtained and immediately analyzed at room temperature.

Our results indicate that by using capillary electrophoresis (as a high-resolution biochemical separation technique), urinary constituents can be identified in a simple manner. This identification is achieved in a short time, without any processing of

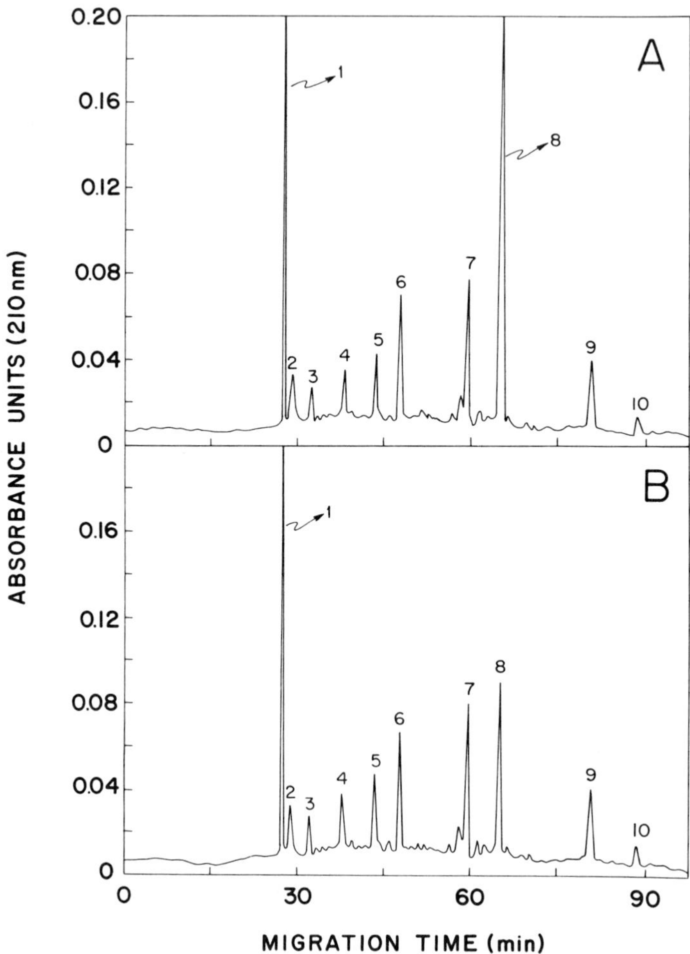

Figure 2 Typical electropherogram of human urine using an ultraviolet detection method. Approximately ten major components were observed in the capillary electrophoretic profile of a clean-catch urine specimen obtained from a pool of six normal persons. (A) Electropherogram of a sample maintained at ambient temperature; (B) electropherogram of the supernatant (after thawing and centrifugation) of a sample maintained at −70°C. (From Ref. 24)

collected urine other than a filtration or centrifugation step, to remove cells and particulated matter. Only a few papers have as yet dealt with the use of capillary electrophoresis to analyze urinary components in physiological or pathophysiological processes (see Table 1). The potential of using capillary electrophoresis in the differentiation and analysis of urinary constituents will contribute substantially to our knowledge of physiological and pathophysiological processes associated with glomerular filtration and tubular catabolism of plasma constituents.

So far, three approaches have been used to identify urinary components by capillary electrophoresis and additional approaches will probably be developed in the future. The first approach consists of direct UV-absorbing analysis of urinary constituents, without any chemical processing of urine (see Fig. 1). The second approach is the determination of derivatives formed between a urine constituent and specific derivatizing agents. The precolumn-formed derivatives are then separated and analyzed by a UV or fluorescence detection system. Both of these methods use standard substances to identify the separated urinary components by comparing migration times or spiked peak areas. The third approach is to use a collection device for capillary electrophoresis eluates in order to collect separated urinary constituents for further identification by other methods, such as mass spectrometry. Figures 2, 3, and 6 show identification of several peaks present in the typical electropherogram, using these three approaches. Peak 1 (see Fig. 2A) has been identified as creatinine, under the experimental conditions described here. However, urea also migrated in the same position (see Fig. 3). Although the molecular structure of urea and creatinine are quite different (Fig. 4), both substances migrated at the speed of electroosmotic flow markers. The only major difference between the two substances is that the absorbance of creatinine at 210 or 214 nm is significantly larger than the absorbance of urea.

Ideally, the presence of a detection system with fast-scanning capabilities or a diode-array detector should be a main feature for a capillary electrophoresis instrument and, in turn, useful in clinical diagnosis for identification purposes. Such a detection paradigm is important, since substances have different spectral properties at various wavelengths. Figure 5 shows the capillary electrophoresis profile of a normal urine at three different wavelengths, 280, 254, and 214 nm. As shown in Fig. 5C, an electropherogram with a profile containing more peaks was obtained at 214 nm. Analyses of peak area ratios of a particular urine constituent, at various wavelengths, might be useful in monitoring a disease state. Since at 214 nm, or lower wavelengths, it was possible to detect more urinary components, it was convenient to use 210 or 214 nm as the monitoring wavelengths for subsequent experiments.

Since capillary electrophoresis separates a variety of components in a single sample (see foregoing), this technique might be useful to determine fingerprint patterns in samples of different origins. Such fingerprint profiles might have important applications in clinical diagnosis, as well as in the characterization of biom-

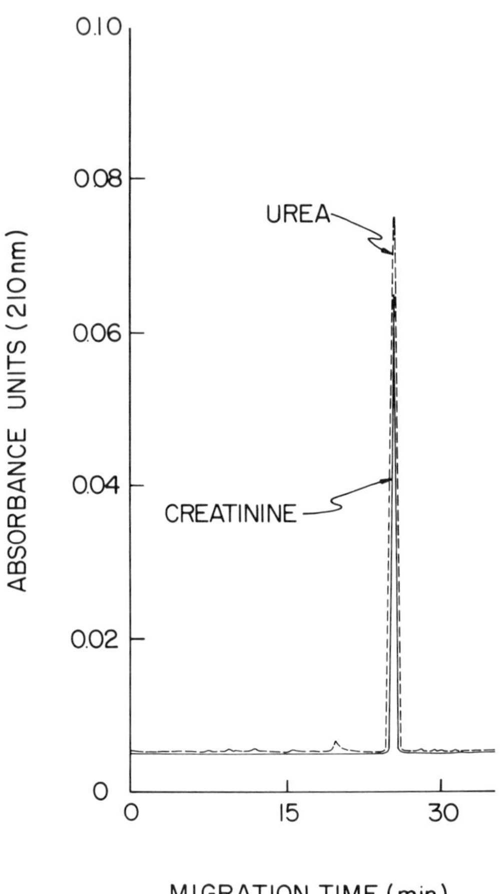

Figure 3 Comigration of urea and creatinine peaks. Both metabolites were injected into the capillary column under identical experimental conditions, except that the concentration of metabolites were different (1 M urea, 1×10^{-4} M creatinine). (From Ref. 25)

olecules, such as proteins. With this idea in mind, we are in the process of assessing human urine samples obtained under different physiological and pathophysiological conditions, as well as urine samples derived from different animal species. These studies have been performed using a capillary electrophoresis-based analytical assay and 210 or 214 nm as the monitoring wavelengths. Figures 6, 7, and 8 are examples of these studies. Figure 6 depicts a comparative capillary electrophoretic profile of urines derived from two physiological conditions. In this particular example, creatinine was the urine constituent of interest. As shown in Fig. 6C and 6D, higher

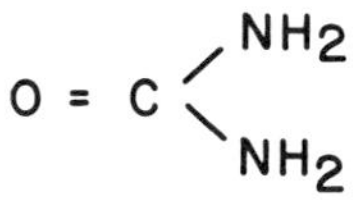

$O = C(NH_2)_2$

UREA

HN
C = NH
H_3C — N — CH_2 — C = O

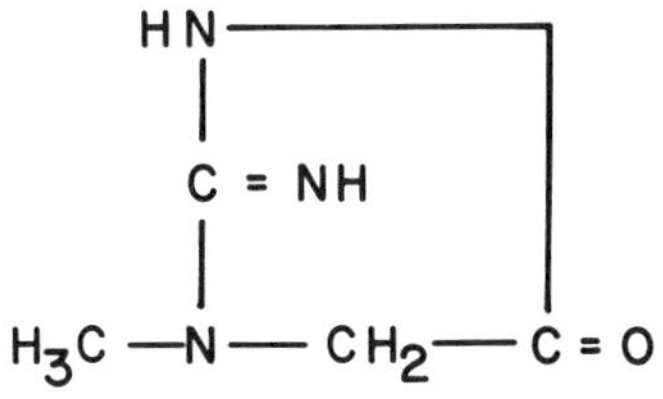

CREATININE

Figure 4 Molecular structures of urea and creatinine. Creatinine has more double bonds, which is probably one of the main reasons for its strong absorbance at 210 nm.

levels of creatinine (peak 1) were observed in the urines derived from a dehydrated patient (concentrated urine) and from a patient under strenuous exercise, when compared with normal controls (Fig. 6A and 6B, peak 1). Similarly, Fig. 7 depicts a comparative capillary electrophoretic profile of urines derived from one normal and two pathophysiological conditions. Figure 7A depicts the capillary electrophoretic profile of a normal urine; Fig. 7B depicts the capillary electrophoretic profile of a patient with Kaposi's sarcoma; and Fig. 7C depicts the capillary electrophoretic profile of a patient with gout. The three electropherograms show significant differences in the profile of the separated peaks. Although the two diseases present a unique pattern, it is impossible to draw a final conclusion about the fingerprint of the disease at this time. For example, some peaks might be related to therapeutic drugs, rather than to a metabolic change in the patient. More patients are now being investigated. Finally, since veterinary medicine has also been a growing field of investigation, we examined a capillary electrophoretic profile of urine specimens derived from domestic animals. Figure 8A shows a capillary electrophoretic profile of a normal human urine specimen; Fig. 8B shows one obtained from a ewe; and Fig. 8C shows one obtained from a mare. It seems that every animal species might have a unique separation pattern of urinary constituents when analyzed by capillary electrophoresis.

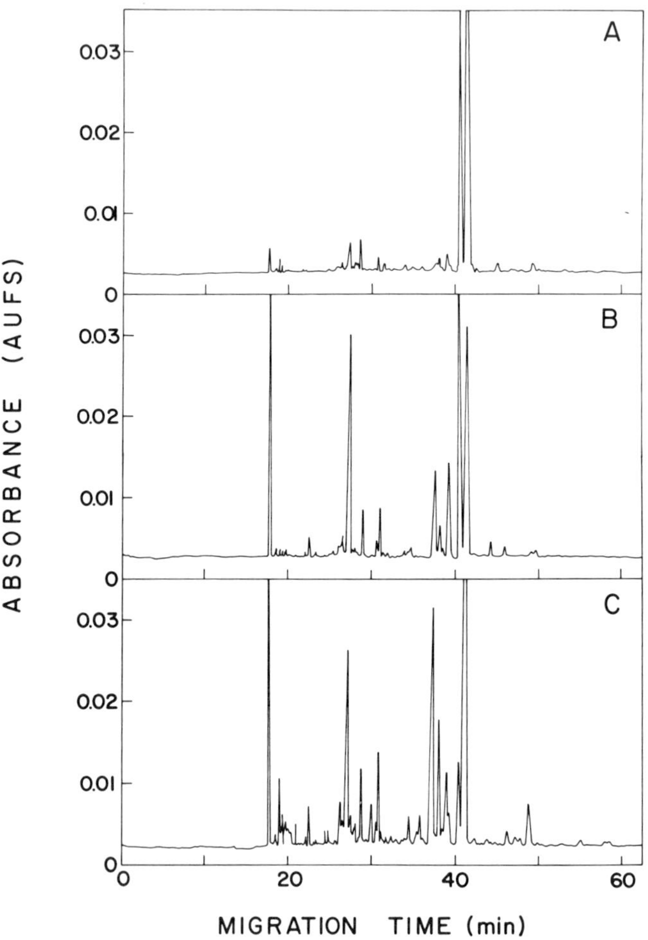

Figure 5 Capillary electrophoretic profile of a clean-catch urine specimen obtained from a normal person. Electropherograms generated at (A) 280 nm; (B) 254 nm; and (C) 214 nm. (From Ref. 52)

The typical capillary electrophoretic profile of urinary specimen, using UV detection at 210 or 214 nm as the monitoring wavelengths, clearly shows the presence of biological components found in large quantities. Nevertheless, constituents found at low concentrations may not be detected by UV. To overcome this problem and to be able to detect substances that are present in small quantities, a different approach was employed. This approach is based on the derivatization of urinary constituents with the reagent fluorescamine to form stable derivatives. The purpose

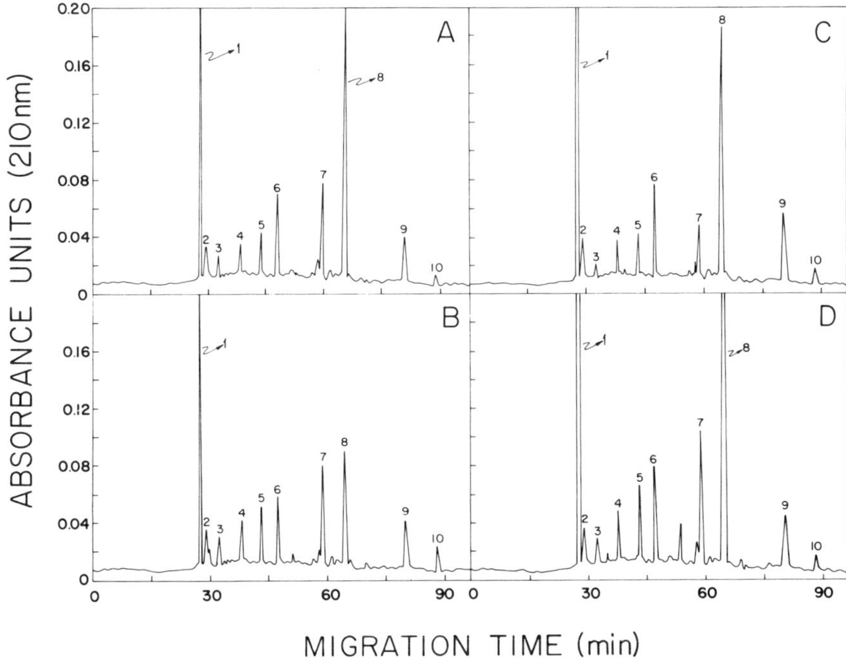

Figure 6 Electropherograms of urine obtained from persons under certain physiological conditions. (A) Urine profile from normal person-1; (B) urine profile from normal person-2; (C) urine profile from dehydrated patient; and (D) urine profile from patient undergoing strenuous exercise. (From Ref. 25)

of this derivatization was twofold: first, to increase the sensitivity levels of constituents found at low concentrations and, second, to improve their separation. Figure 9 shows the proposed chemical reaction scheme of fluorescamine with primary and secondary amines, and Figs. 10 and 11 show electropherograms of fluorescamine-derivatized urinary constituents of pathological urine specimens (see following).

Fluorescamine (4-phenylspiro[furan-2(3*H*), 1′-phthalan]-3,3′-dione) has been commonly used as a fluorogenic reagent. This reagent reacts readily and rapidly at alkaline pH with primary amines to form intensely fluorescent substances, providing the basis for a rapid and sensitive assay of amino acids, peptides, proteins, and other primary amines [36–38]. In addition to fluorescence detection, the fluorescamine-derivative has a much higher absorbance in the low-UV spectra than the nonderivatized counterpart substance. This property was taken into consideration to detect urinary constituents at 214 nm when derivatized with fluorescamine. At 214 nm, derivatives formed with substances containing either primary or secondary amines

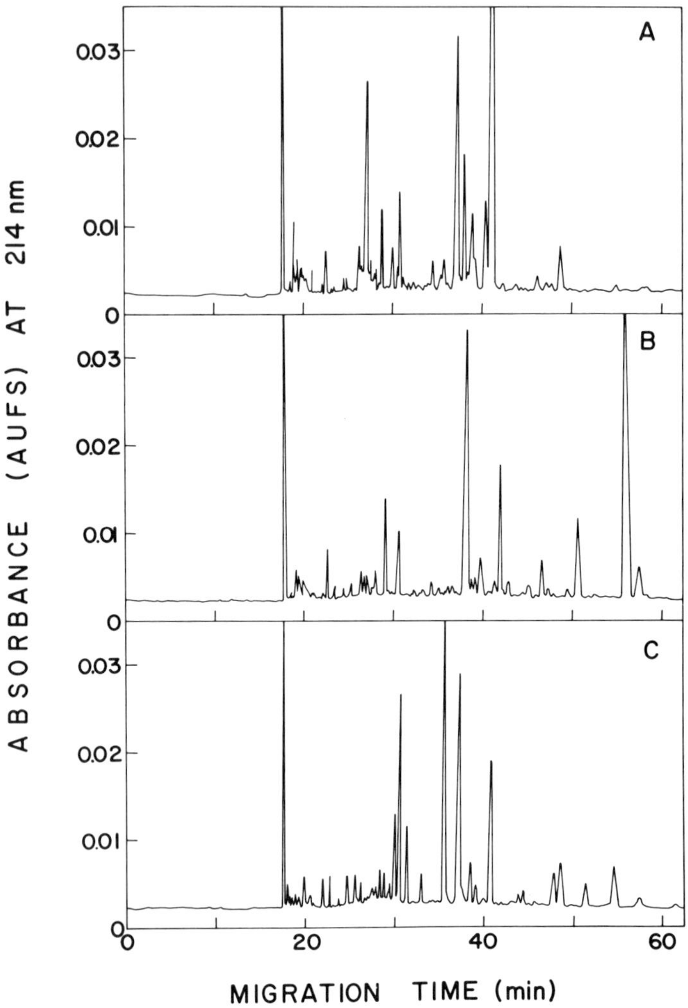

Figure 7 Comparative capillary electrophoresis profile of three different human urine specimens. Electropherograms generated from (A) a normal person; (B) from a patient with Kaposi's sarcoma; and (C) from a patient with gout. (From Ref. 52)

can be detected. Imino acids containing secondary amines can also form nonfluorescent-type aminoenone fluorescamine derivatives with high absorbance in the low-UV spectra. The use of fluorescamine-derivatized substances, analyzed by UV detection, has been previously demonstrated [36–38], and the proposed chemical reaction scheme of fluorescamine with primary and secondary amines is shown in Fig. 9.

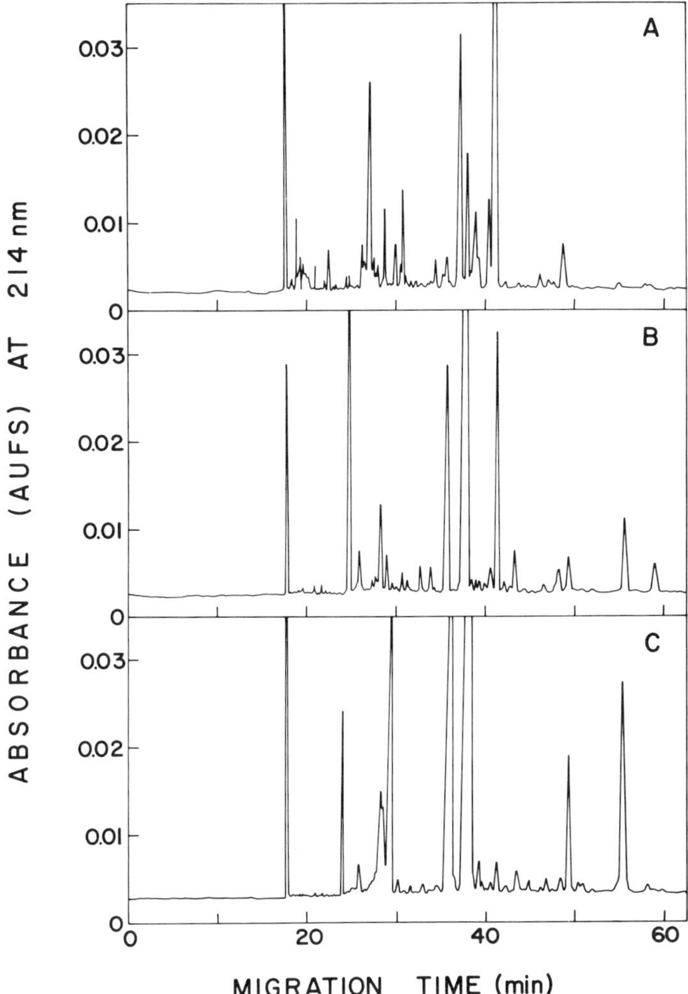

Figure 8 Comparative capillary electrophoresis profile of human and domestic animal urine specimens. Electropherograms generated from (A) a normal person; (B) a ewe; and (C) a mare. (From Ref. 57)

Electropherograms of fluorescamine-derivatized urinary constituents of pathological urine specimens are shown in Figs. 10, and 11. Figure 10 compares the electropherogram obtained from an untreated urine specimen obtained from a patient with Kaposi's sarcoma and that of a fluorescamine-derivatized urine specimen from the same patient. Analytes were detected at 214 nm. As shown in Fig. 10B, the

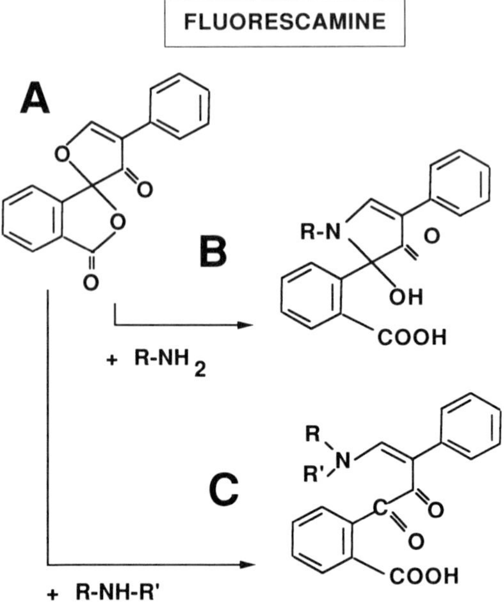

Figure 9 Schematic representation of the molecular structure of (A) fluorescamine; (B) the derivatized reaction product involving a reacting primary amine functional group-containing analyte; and (C) the derivatized reaction product involving a reacting secondary amine functional group-containing analyte.

profile of the electropherogram obtained with fluorescamine-derivatized urinary constituents is different from that of the profile of untreated control urine. Similarly, fluorescamine-derivatized urinary constituents were obtained from a patient with gout (Fig. 11). As mentioned before, the derivatization of functional amine groups in urinary constituents with fluorescamine has double importance. It improves sensitivity and has a marked effect on analyte mobility, producing an enhanced separation as well. Therefore, from these results, it is possible to create a unique fingerprint for a particular disease by derivatizing urinary constituents with fluorescamine. This might potentially lead to detection of a marker for some disease states. This marker might be either a single substance or a combination of urinary constituents. The final goal will be to identify such markers.

Regardless of whether potential markers are in abundance or at low concentrations in urinary samples, the electropherographic profile of substances is as yet only qualitative. Because capillary electrophoresis deals with very small amounts of analytes, usually a few nanograms of material, it is still a challenge to collect separated

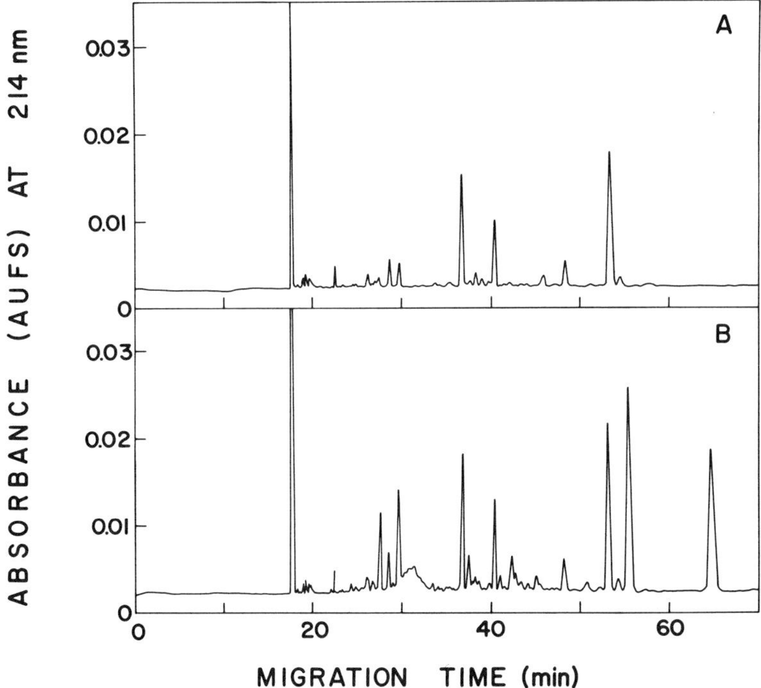

Figure 10 Capillary electrophoretic profile of underivatized and fluorescamine-derivatized urine specimen obtained from a patient with Kaposi's sarcoma. (A) Control underivatized urine specimen; and (B) fluorescamine-derivatized urine. (From Ref. 52)

and purified samples for further analyses, such as mass spectrometry. A few approaches have already been considered for the collection and identification of constituents that might be potential markers: (a) the availability of an instrument capable of delivering the same amount of material in the collecting reservoir, at the same migration time, all the time. Unfortunately, this approach does not eliminate, for example, the nature of the analyte, capillary column, buffer, and so on. Therefore, it is unrealistic to expect the maximum reproducibility, precision, and accuracy, constantly, and to persistently collect pure material without contamination. (b) The use of multiple or a bundle of capillaries to load and collect more sample could provide 1 μg of collected material in a working day. (c) The use of an analyte concentrator and (d) the use of other preconcentration procedures [39] are other possible approaches.

Figure 12 shows a microcollection system for capillary electrophoresis. This microcollection system has been previously described [40], and it is also reviewed

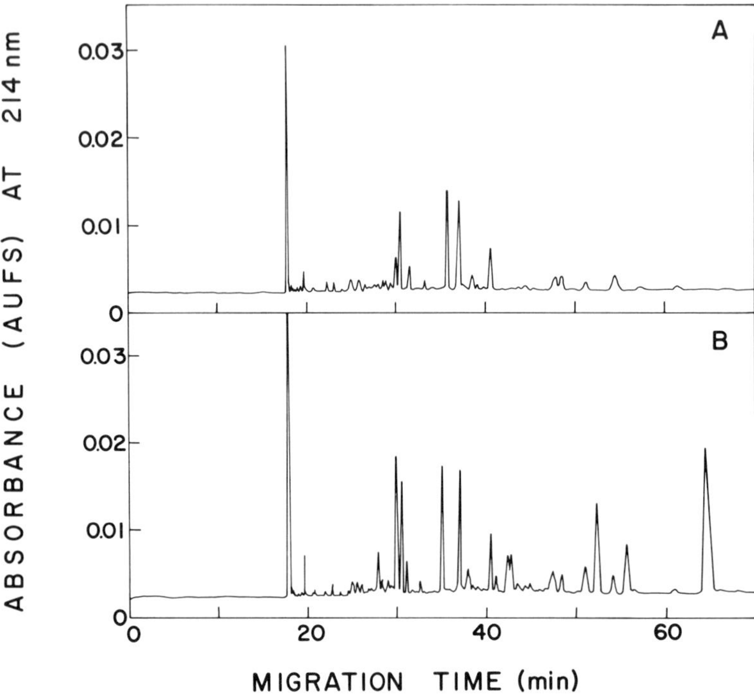

Figure 11 Capillary electrophoresis profile of underivatized and fluorescamine-derivatized urine specimen obtained from a patient with gout. (A) Control underivatized urine specimen; and (B) fluorescamine-derivatized urine. (From Ref. 52)

in this book in Chapters 14 and 15. In general, two or more capillaries are used to increase both the detector response and sample load in direct proportion to the number of capillaries. Multiple capillaries are joined together through a link tube. (Fig. 13) and merged to a single outlet capillary. In addition, a "tee assembly" unit (Fig. 14) is needed as a grounding device to enable the system to maintain a normal electroosmotic flow through the outlet capillary containing a free terminal end. This way of collecting samples is much simpler than previously described systems (see Chaps. 14 and 15).

By using a bundle of capillary columns as a collection method for capillary electrophoresis, it has been possible to collect 1 μg of a purified urinary constituent. This amount of material was enough to obtain the identification of uric acid by mass spectrometry [31] (Fig. 15).

Another useful method to concentrate and purify analytes is use of the so-called analyte concentrator (Fig. 16). This device consists of a small piece of fused silica capillary, localized close to the injection terminal, containing a solid support com-

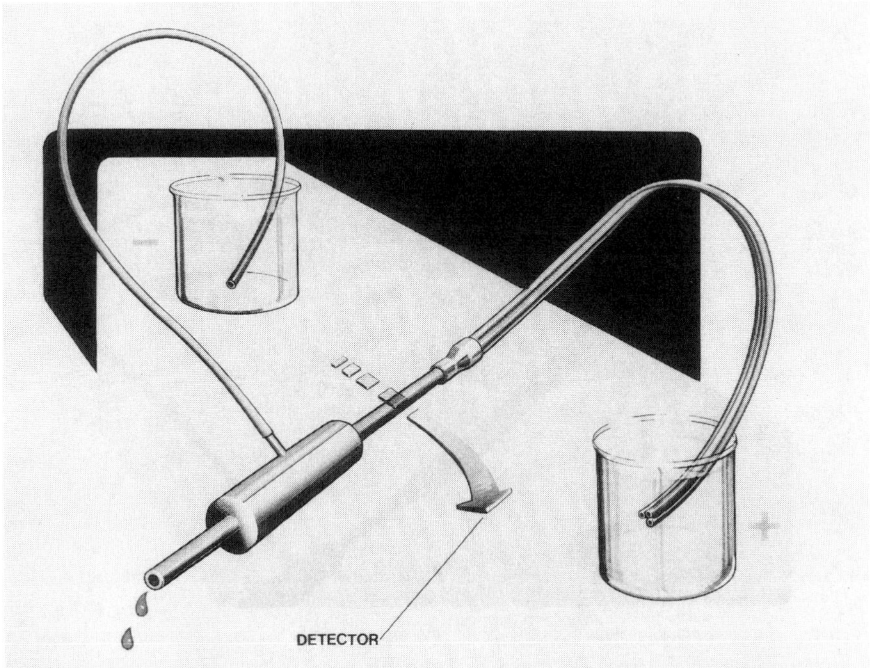

Figure 12 Schematic diagram of a continuous fraction collection system for semipreparative capillary electrophoresis. The system includes both the link tube (see Fig. 13) and the microgrounding device (see Fig. 14). (From Ref. 40)

posed of either a chromatographic material or controlled-porous glass beads, with the entire package held by two frit structures. The chromatographic material now used is C-18–coated silica beads [41], and it allows one to concentrate samples that can be detected at nanogram per milliliter concentrations. The other approach is the use of controlled-porous glass beads, which can be coated chemically with various substances that, in turn, will capture other substances present in the biological fluids in a very specific manner (see Fig. 16). Both approaches are now being used for various purposes. If we can combine the approach of concentrating samples with a high-powered detection system, such as laser-induced fluorescence detection, we probably will be in a position to analyze samples that previously have been difficult to separate or detect.

Capillary Electrophoresis of Serum Specimens

Determination of the presence and concentration of various compounds in the blood is of increasing importance in the diagnosis and treatment of disease. The blood

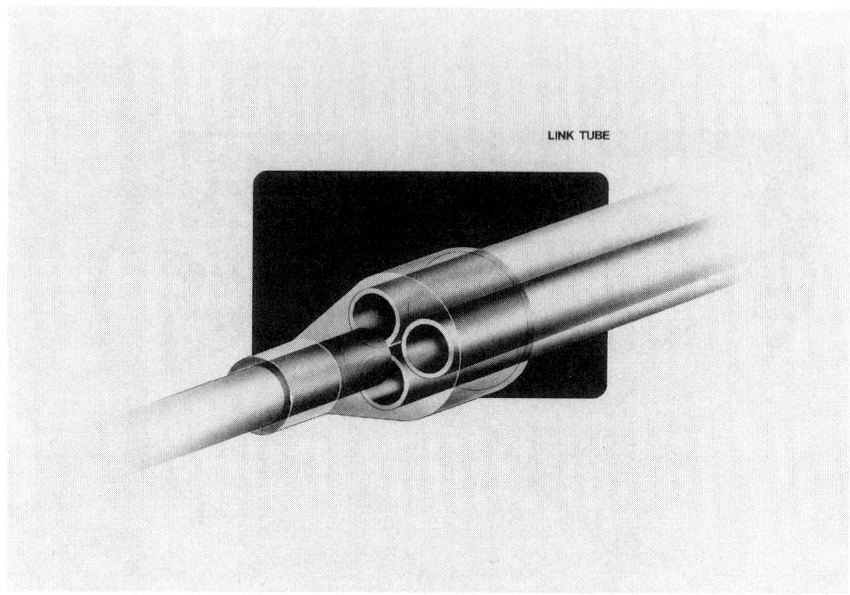

Figure 13 Schematic representation of the link tube. The diagram depicts three inlet capillaries fused into one outlet capillary. A minimum distance between inlet and outlet capillaries was maintained to avoid distortion of the electropherographic peaks (see Ref. 40 for further details). (From Ref. 40)

not only reflects the overall metabolism of the tissues, but affords the most accessible method for the sampling of body fluids.

For serum specimens, the separation of constituents by capillary electrophoresis is more complex. Since albumin and immunoglobulins are serum components present in relatively large quantities, it becomes more difficult to separate and detect other serum components of low molecular weight present in small quantities. Usually, low molecular weight serum components are masked, comigrating with the major serum components, such as albumin, immunoglobulins, hemoglobin, lipoproteins, fibrinogen, transferrin, and others. Similarly important, detection by conventional UV-visible detection systems of low molecular weight serum components, using capillary electrophoresis, has limited usefulness, since the concentration in which these substances are found in serum is often quite low.

To overcome these limitations, several approaches have already been developed for capillary electrophoretic analysis of low molecular weight serum constituents. Many of the methods of blood chemistry require the preparation of a protein-free filtrate, which is then analyzed for those constituents that remain. Among them

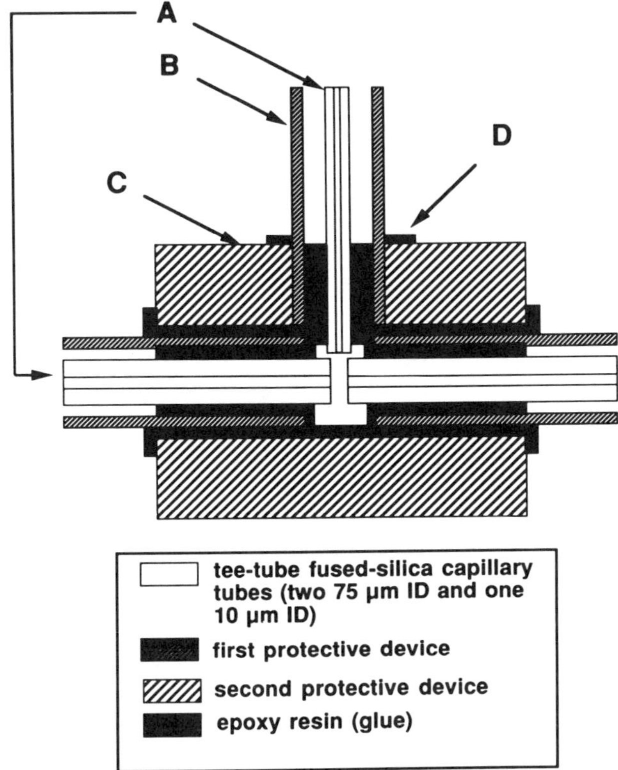

Figure 14 Schematic diagram of the tee-assembly unit used as a grounding device. (A) Three fused silica capillary tubes were used to form the tee-assembly unit. Two of these tubes were generated by breakage of a single capillary column (75-μm–id, 375-μm–od, fused silica capillary) in which the separation process occurred. The third capillary tube (10-μm–id, 150-μm–od, fused silica capillary) was placed at the narrow interconnection point of the other two capillaries. At this point the electrical grounding process was initiated. (B) The first protective device was composed of a 530-μm–id, 660-μm–od, fused silica capillary, covering and protecting both the 75- and 10-μm columns. (C) The second protective device was composed of Teflon tubing acting as a supportive structure to give strength to the connecting components of the tee-assembly unit. (D) Epoxy resin, to glue all spaces and produce a leakage-free system (see Ref. 40 for further details).

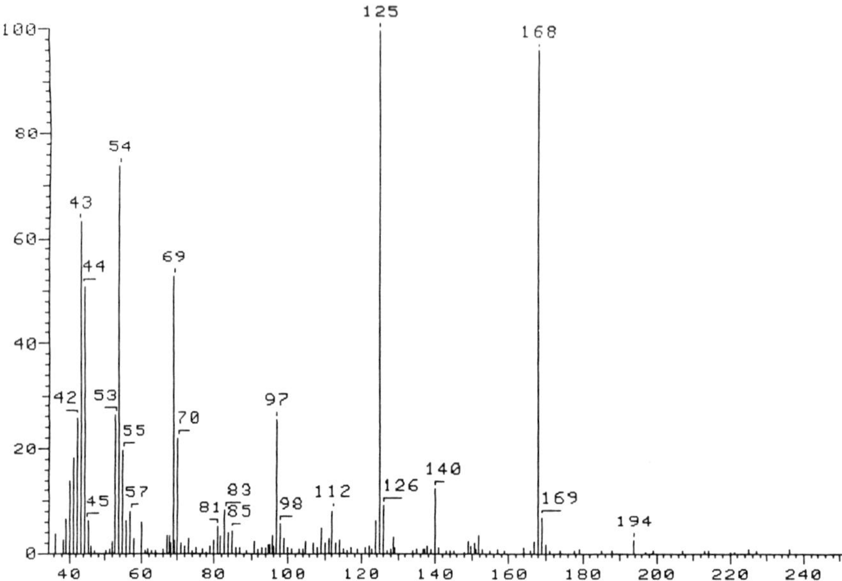

Figure 15 Mass spectrometry data using the electron ionization method. Purified urine component was separated, collected, and processed as described in Ref. 31. This electron ionization spectra of the substance collected by the bundle of multiple capillaries method corresponds to that of a uric acid standard. (From Ref. 31)

are the deproteinization of blood by (a) hemodialysis, (b) precipitation by strong acids, (c) use of chromatographic media to separate the analytes, and (d) use of extraction methods with organic solvents. The serum substances are then analyzed by capillary electrophoresis in the protein-free dialysates, supernatants, column eluates, or extracted solutions. Similarly, these procedures can also be applied to other biological fluids.

One simple approach is the analysis of samples obtained from artificial blood purification methods, such as hemodialysis and continuous ambulatory peritoneal dialysis. The use of a selective semipermeable membrane allows the discrimination of chemical substances (i.e., proteins), based on the porosity size of the membrane. Hemodialysis with the artificial kidney or self-dialysis at home are well-established methods for treatment of chronic renal failure. The development of permanent inlying plastic arteriovenous "shunts" has made the technique relatively painless. One of the advantages of self-dialysis at home is that patients may arrange their schedule to suit themselves. The convenience of these methods permits the study of various substances present in the dialysate without the interference of albumin and immunoglobulins. Schoots et al. have reported the quantification of hippuric acid from ultrafiltered uremic serum samples

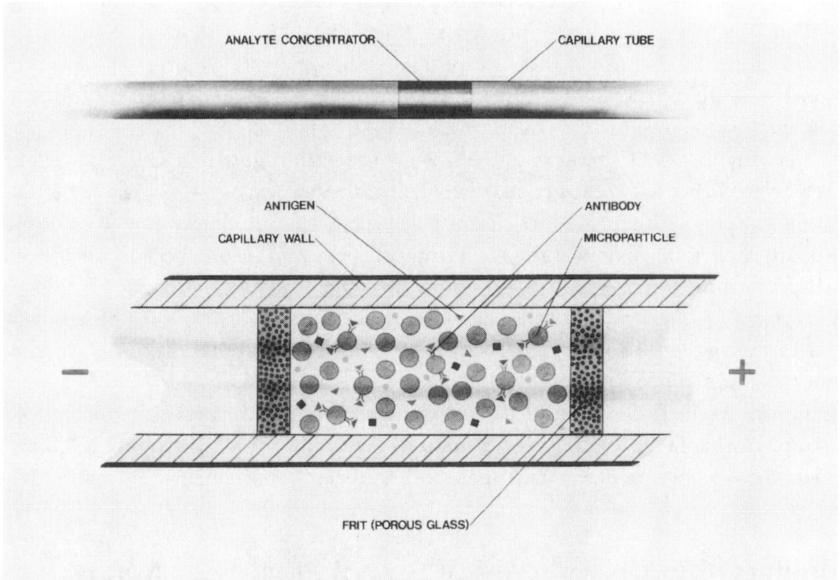

Figure 16 A schematic diagram of an analyte concentrator. Controlled-pore glass beads were activated, conjugated with a purified antibody, and installed inside a portion of a fused silica capillary as described in Ref. 31.

by capillary electrophoresis [26]. Accuracy and precision of the capillary electrophoretic method were similar to those reported for high-performance liquid chromatography. Nevertheless, the analysis time of hippuric acid in hemodialysates by capillary electrophoresis required only 8 minutes, in comparison with 90 minutes by high-performance liquid chromatography.

The use of dialysis membranes, as it is used in microdialysis probes, seems to be an important method now being applied in the medical field (see later).

Another approach for deproteination of blood specimens is the precipitation of proteins from blood by a variety of experimental conditions, which include acid or alkaline precipitation, heat, or a combination of these. The most common technique for the preparation of a protein-free filtrate is the method of Folin and Wu, which uses sodium tungstate and sulfuric acid to make tungstic acid for the precipitation of the plasma proteins. Other protein precipitants include trichloroacetic acid or picric acid and a mixture of sodium hydroxide and zinc sulfate (the Somogyi precipitants). In blood substance analysis by capillary electrophoresis, for example, Roach et al. have reported the precipitation of blood proteins by addition of a strong acid to serum [42]. To 200 μL of serum, 200 μL of 20% trichloroacetic acid was added. The sample was heated at 100°C for 5 minutes, cooled and then centrifuged for 10 minutes to remove precipitated proteins. The supernatant can then be neutralized, injected directly into the capillary electrophoresis column, or before this stage, the sample present in the supernatant can be further processed by additional steps, such as oxidation or derivatization.

In many instances, because of the low concentration in which analytes are found in serum (or other biological fluids) it is convenient to extract the analyte by using organic solvents. This procedure permits the separation of the substance of interest from the high molecular weight macromolecules. In addition, because of the possibility of using larger volumes of serum (or other biological fluids), samples are concentrated by evaporation of the organic solvent. The residue can then be suspended in an appropriate solution or buffer and injected into the capillary column at a much higher starting concentration level.

Figure 17 depicts a typical electropherogram of a serum specimen. The main electropherographic peak represents albumin and the other peaks are globulins. Under these experimental conditions, other serum components are difficult to analyze.

Capillary Electrophoresis of Spinal Fluid Specimens

The cerebrospinal fluid is formed as an ultrafiltrate of the plasma by the choroid plexuses of the brain. The process is not one of simple filtration, since active secretory processes are involved. The normal fluid is water-clear with low-protein content (20–45 mg/dL) and with an albumin/globulin ratio of 3:1. However, in disease,

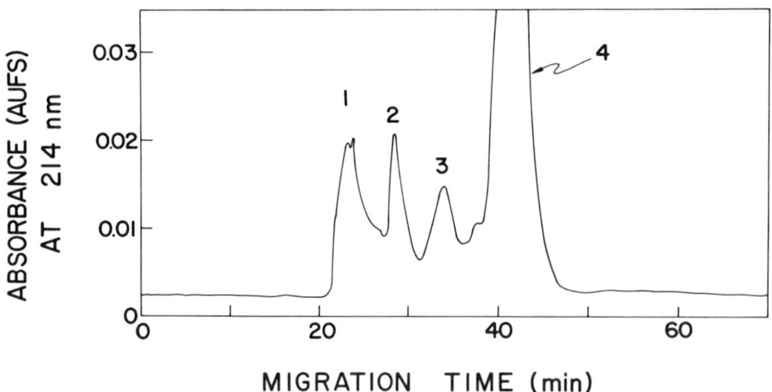

Figure 17 Typical electropherogram of human serum using an ultraviolet detection method. Approximately four major components were observed in the capillary electrophoresis profile of a human serum specimen. Peaks 1, 2, and 3 represent serum immunoglobulin components, and peak 4 represents human serum albumin.

an increase in protein, particularly in globulin, is characteristic. In meningitis, for example, the protein may rise as high as 125 mg/dL to 1 g/dL or more.

Figure 18 depicts a typical electropherogram of spinal fluid derived from a normal person. Figure 18A shows an electropherogram of a nonderivatized sample. Figure 18B shows an electropherogram of a fluorescamine-derivatized sample.

Capillary Electrophoresis of Blood Cell Specimens

Chemical analysis of blood cells can be considered from two standpoints: the determination of analytes from a specific population of cells or from a defined single cell. Examples have been performed with red blood cells. Also individual neuronal cells have been used as a model system to analyzed their chemical contents by capillary electrophoresis because of the large size of certain neuronal cells.

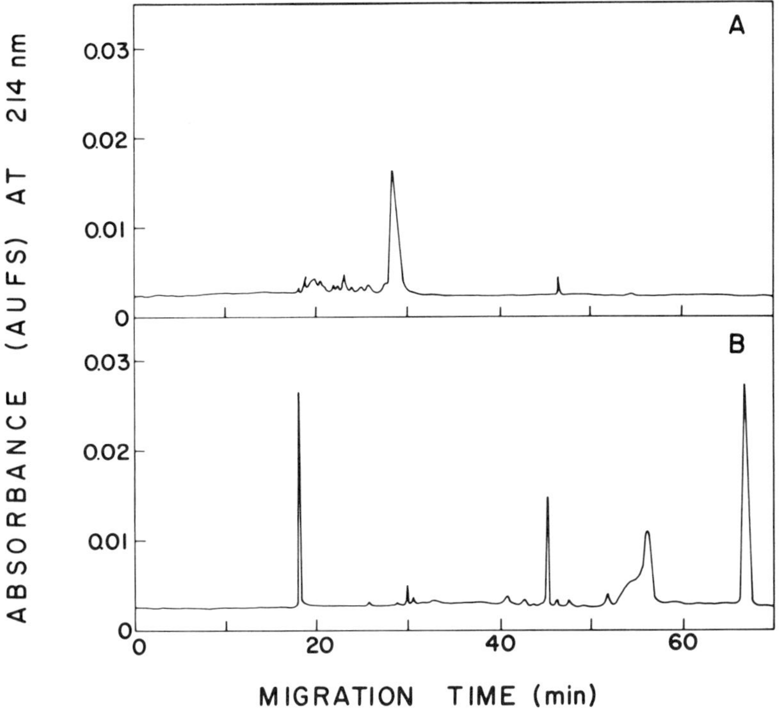

Figure 18 Capillary electrophoretic profile of underivatized and fluorescamine-derivatized cerebrospinal fluid specimen obtained from a normal patient. (A) Control underivatized cerebrospinal fluid specimen; and (B) fluorescamine-derivatized cerebrospinal fluid. (From Ref. 52)

For a population of cells, prior preparation steps are required. Red blood cells, isolated usually in isotonic buffers, can easily be lysed by diluting the cells in distilled–deionized water and further freezing and thawing steps. Furthermore, proteins can be precipitated by acid precipitation, and the protein-free lysate can then be obtained by centrifugation. The analytes present in the supernatant can then be analyzed directly by capillary electrophoresis and detected by various detection methods, or they can be further derivatized and detected by fluorimetric or other detection methods.

For a single cell, chemical measurements have been carried out using capillary electrophoresis [43,44]. The content of an individual red blood cell is directly transferred into a capillary column (under a microscope) after the cell is lysed on contact with a buffer solution. After separation, laser excitation is used to detect fluorescent species (down to attomole or zeptomole range), and nonfluorescent species are detected indirectly (down to femtomole or attomole range) by a scheme based on charge displacement. Figure 19 shows the separation and identification of monobromobimane-derivatized glutathione from a single erythrocyte.

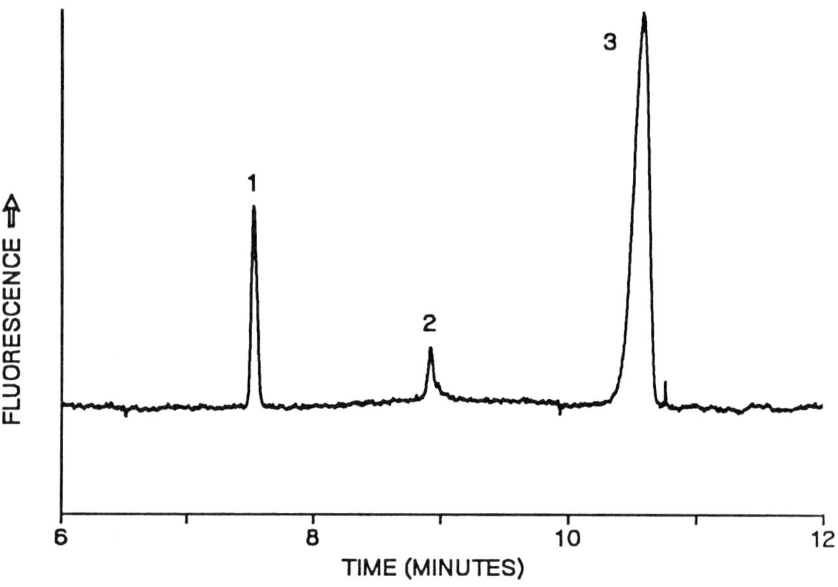

Figure 19 Capillary electrophoretic profile of a single erythrocyte. The intracellular content of a single red blood cell was derivatized with monobromobimane and then separated by capillary electrophoresis as described in Ref. 43. Peak 1 represents excess of reagent; peak 2 represents an unidentified peak; and peak 3 represents derivatized glutathione. (From Ref. 44)

The purpose of analyzing cells is at least twofold: (a) to look for markers that appear before physical changes occur, leading to early diagnosis; and (b) to study pharmacokinetics and drug efficacy on a single-cell basis, avoiding many legal and moral issues.

Capillary Electrophoresis of Tissue Specimen Biopsies

Biopsy has traditionally been an important test to obtain a definitive diagnosis when other methods have proved inconclusive or if histological appearance is the sole means of diagnosis. For example, cutaneous or visceral carcinomas or sarcoma, liver cirrhosis, and hematological and bone malignancies meet these criteria.

The use of tissue specimens for capillary electrophoresis requires prior preparative steps, such as homogenization of the tissue in a specially designed glass microhomogenizer, with a ground-glass and manually operated pestle. An internal standard for the tissue is normally added before homogenization, as a control of all the mechanical manipulations of the process. Proteins are precipitated by addition of acid (e.g., perchloric acid) to the homogenate, which can then be removed by centrifugation. The supernatant is neutralized and the analytes in the sample can be directly analyzed by capillary electrophoresis, using various detectors, or they can be derivatized with a chromophore and analyzed by a fluorimetric detector or other detection systems.

It is not uncommon, despite proper pathological and immunopathological examination, for the diagnosis to be still in doubt. In those cases, the only recourse is observation and further biopsy at a later day until the disease "presents" itself. An example of this problem is the slow progression of parapsoriasis en plaque into cutaneous T-cell lymphoma. The use of capillary electrophoresis with its unprecedented resolution may yet find markers of disease not previously discovered.

Capillary Electrophoresis of In Vivo Perfusate Specimens

Until a few years ago, perfusion techniques, such as those using microdialysis membranes for the analysis of tissue constituents, were limited only to brain tissue. Now, the use of microdialysis probes to determine substances in tissues other than the brain is having more acceptance. For example, it is possible to continuously monitor substances (for prolonged periods) from blood in freely moving animals by using a flexible intravenous microdialysis probe [45,46]. These microdialysis probes are constructed from Silastic tubing (0.5-mm id and 1.0-mm od), with a cellulose hollow fiber tip 0.2 mm in diameter and 25 mm long, having a 6000 molecular weight cutoff limit [46]. In general, these new methods of analysis of blood constituents, free of proteins, are much simpler than the traditional catheterization procedures. Because the catheter was opened at both ends, there was a high probability of infection [47]. Although additional sampling probes to obtain in vivo perfusates are currently being used only under research conditions, as for example

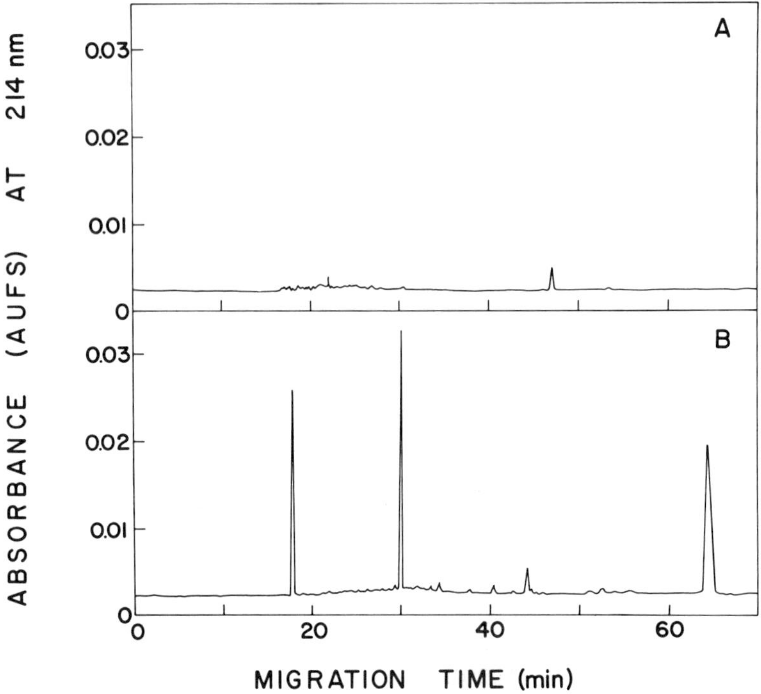

Figure 20 Capillary electrophoretic profile of underivatized and fluorescamine-derivatized saliva specimen obtained from a normal person. (A) Control saliva specimen; and (B) fluorescamine-derivatized saliva. (From Ref. 52)

brain microdyalysis and push-pull cannula sampling in specific brain areas, they might identify potential diagnostic markers for brain diseases. For example, the link between brain and reproductive activity, in all animals including humans, is the decapeptide gonadotropin-releasing hormone (GnRH). In vivo release of this neuropeptide from the hypothalamic median eminence, as well as of its local putative neuronal inputs, can be determined by push-pull cannula sampling and capillary electrophoresis [48–51].

Capillary Electrophoresis of Other Tissue and Fluid Specimens

In addition to the commonly used biological fluids for analysis of substances, there are also other tissues and fluids that are of importance in determining changes in physiological or pathophysiological states. For example, hair, tears, saliva, sweat, and fluids derived from organs or tissues, such as lungs, peritoneal cavity, or others.

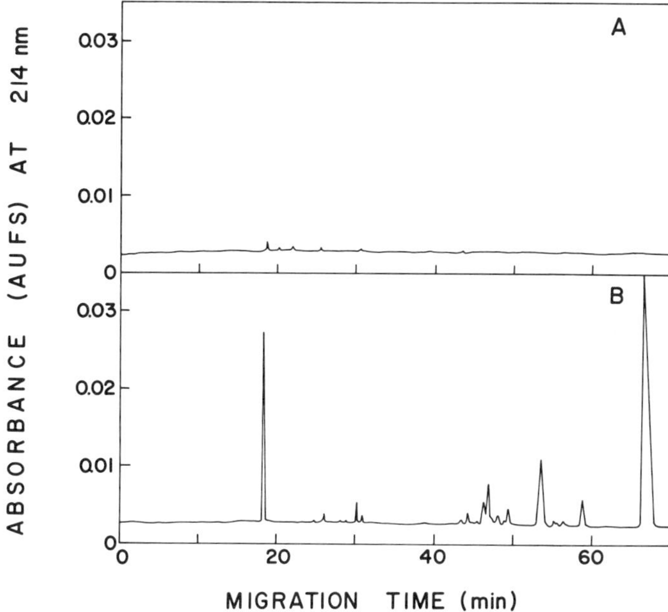

Figure 21 Capillary electrophoretic profile of underivatized and fluorescamine-derivatized tears specimen obtained from a normal person. (A) Control tears specimen; and (B) fluorescamine-derivatized tears. (From Ref. 52)

Accurate profiles of readily accessible samples, such as tears, saliva, and sweat, might be important for the widespread use of screening tests for drugs of abuse or disease states. Figures 20 and 21 show the capillary electrophoresis profile of saliva and tears.

CONCLUSIONS

Since its development, capillary electrophoresis has become one of the most important and powerful techniques of the separation sciences. However, as shown in Table 1, to date very few studies have been reported relative to the medical sciences. Even at these early stages of the application of capillary electrophoresis to biomedical problems, the method shows promise as a potential tool to be used in medicine.

Clinical laboratories will make a crucial contribution to the attainment of national health objectives for the next century. Emphasis is slowly, but definitely, shifting from the traditional diagnosis and treatment approach to health care toward preventive medicine. Laboratory technologies must advance to make possible the detection of presymptomatic disease. Early measurement of familial risk factors will suggest life-style changes soon enough to delay or even eliminate the onset of disease. There will also be increasing emphasis on a safe and healthy environment, with laboratories being called upon to detect and monitor an increasing number of environmental risk variables. Measurement capabilities will need to be expanded and analytical limits extended to provide risk assessment from conception to the grave. We will need to reach beyond our current level of analytical excellence to improve communications so that laboratory data will facilitate the targeted health improvements.

Capillary electrophoresis can contribute significantly to the achievement of these goals as it gains acceptance by a wider research community and is employed in other disciplines, including clinical laboratories and biomedicine in general.

REFERENCES

1. P. M. S. Clark and L. J. Kricka, *Adv. Clin. Chem.*, *22*:247 (1981).
2. S. S. Raphael, *Lynch's Medical Laboratory Technology*. W. B. Saunders, Philadelphia (1983).
3. J. B. Henry and G. A. Threattle, *Clinical Diagnosis and Management by Laboratory Methods*, 17th ed. (J. B. Henry, ed.), W. B. Saunders, Philadelphia, pp. 380–458 (1984).
4. J. W. Jorgenson and K. D. Lukacs, *Anal. Chem.*, *27*:1298 (1981).
5. J. W. Jorgenson and K. D. Lukacs, *Science*, *222*:266 (1981).
6. H. H. Lauer and D. McManigill, *Trends Anal. Chem.*, *5*:11 (1986).
7. D. Burton, M. J. Sepaniak, and M. P. Maskarinec, *J. Chromatogr. Sci.*, *24*:347 (1986).
8. M. J. Gordon, X. Huang, S. L. Pentoney, Jr., and R. N. Zare, *Science*, *242*:224 (1988).
9. N. A. Guzman, B. G. Hoebel, and L. Hernandez, *BioPharm*, *2*:22 (1989).
10. N. A. Guzman, L. Hernandez, and S. Terabe, *Analytical Biotechnology. Capillary Electrophoresis and Chromatography* (C. Horváth, and J. G. Nikelly, eds.), ACS Symp. Ser. *434*:1 (1990).
11. B. L. Karger, A. S. Cohen, and A. Guttman, *J. Chromatogr.*, *492*:585 (1989).
12. R. A. Wallingford and A. G. Ewing, *Adv. Chromatogr.*, *29*:1 (1990).
13. M. Novotny, K. A. Cobb, and J. Liu, *Electrophoresis*, *11*:735 (1990).
14. Z. Deyl and R. Struzinsky, *J. Chromatogr.*, *569*:63 (1991).
15. W. Kuhr and C. A. Monnig, *Anal. Chem.*, *64*:389R (1992).
16. Y. F. Cheng, and N. J. Dovichi, *Science*, *242*:562 (1988).
17. L. Hernandez, J. Escalona, N. Joshi, and N. Guzman, *J. Chromatogr.*, *559*:183 (1991).
18. M. J. Sepaniak, A. C. Powell, D. F. Swaile, and R. O. Cole, *Capillary Electrophoresis. Theory and Practice* (P. D. Grossman, and Joel C. Coburn, eds). Academic Press, San Diego, p. 159 (1992).
19. J. Vindevogel and P. Sandra, *Introduction to Micellar Electrokinetic Chromatography* (J. Vindevogel, and P. Sandra), Hüthig Buch Verlag, Heidelberg (1992).
20. J. W. Jorgenson and K. D. Lukacz, *Clin. Chem.*, *27*:1551 (1981).
21. X. Huang, T.-K. J. Pang, M. J. Gordon, and R. Zare, *Anal. Chem.*, *59*:2747 (1987).
22. X. Huang, M. J. Gordon, and R. Zare, *J. Chromatogr.*, *425*:385 (1988).

23. N. A. Guzman, L. Hernandez, C. M. Berck, and J. P. Advis, "CIS '89 Tokyo," International Symposium on Chromatography. Tokyo, Japan, October 17-20, Abstract I-P-32 (1989).
24. N. A. Guzman, L. Hernandez, and J. P. Advis, *J. Liquid Chromatogr.*, *12*:2563 (1989).
25. N. A. Guzman, C. M. Berck, L. Hernandez, and J. P. Advis, *J. Liquid Chromatogr.*, *13*:3833 (1990).
26. A. C. Schoots, T. P. E. M. Verheggen, P. M. J. M. De Vries, and F. M. Everaerts, *Clin. Chem.*, *36*:435 (1990).
27. M. Miyake, A. Shibukawa, and T. Nakagawa, Twelfth International Symposium on Capillary Chromatography. (K. Jinno, and P. Sandra, eds.), Industrial Publishing & Consulting, Tokyo, (1990).
28. F.-T. A. Chen, C.-M. Liu, Y.-Z. Hsieh, and J. C. Sternberg, *Clin. Chem.*, *37*:14 (1991).
29. B. J. Wildman, P. E. Jackson, W. R. Jones, and P. G. Alden, *J. Chromatogr.*, *546*:459 (1991).
30. N. A. Guzman and C. M. Berck, International Meeting of the Electrophoresis Society, Washington D.C., March 19–21, Abstract 105 (1991).
31. N. A. Guzman, M. A. Trebilcock, and J. P. Advis, *J. Liquid Chromatogr.*, *14*:997 (1991).
32. E. Jellum, A. K. Thorsrud, and E. Time, *J. Chromatogr.*, *559*:455 (1991).
33. E. Jellum, Fourth International Symposium on High Performance Capillary Chromatography, Amsterdam, The Netherlands, February 9–13, p. 28 (1992).
34. D. Perret and G. Ross, *Trends Anal. Chem.*, *11*:156 (1992).
35. Z. K. Shihabi and M. S. Constantinescu, *Clin. Chem.*, *38*:2117 (1992).
36. N. A. Guzman, J. Moschera, K. Iqbal, and A. W. Malick, *J. Liquid Chromatogr.*, *15*:1163 (1992).
37. N. A. Guzman, J. Moschera, C. A. Bailey, K. Iqbal, and A. W. Malick, *J. Chromatogr.*, *598*:123 (1992).
38. N. A. Guzman, J. Moschera, K. Iqbal, and A. W. Malick, *J. Chromatogr.*, *608*:197 (1992).
39. F. Foret, E. Szoko, and B. L. Karger, *J. Chromatogr.*, *608*:3 (1992).
40. N. A. Guzman, M. A. Trebilcock, and J. P. Advis, *Anal. Chim. Acta*, *249*:247 (1991).
41. M. Merion, R. H. Aebersold, and M. Fuchs, First Conference on Capillary Electrophoresis, National Cancer Institute–Frederick Cancer Research and Development Center, Frederick, Md., October 15–16, Abstract 21 (1990).
42. M. C. Roach, P. Gozel, and R. Zare, *J. Chromatogr.*, *426*:129 (1988).
43. E. S. Yeung and B. L. Hogan, *Chem. Eng. News*, *70*:29 (1992).
44. B. L. Hogan and E. S. Yeung, *Anal. Chem.*, *64*:2841 (1992).
45. M. Telting-Diaz, D. O. Scott, and C. E. Lunte, *Anal. Chem.*, *64*:806 (1992).
46. P. Rada, M. Parada, and L. Hernandez, *J. Applied Physiol.*, *74*:466 (1993).
47. B. H. Migdalof, *Drug Metab. Rev.*, *5*:295 (1976).
48. J. P. Advis, L. Hernandez, and N. A. Guzman, *Peptide Res.*, *2*:389 (1989).
49. N. A. Guzman, L. Hernandez, and J. P. Advis, *Current Research in Protein Chemistry: Techniques, Structure, and Function* (J. J. Villafranca, ed.), Academic Press, San Diego, p. 203 (1990).

50. T. J. O'Shea, P. L. Weber, B. P. Bammel, C. E. Lunte, and S. M. Lunte, *J. Chromatogr.*, *608*:189 (1992).
51. J. P. Advis, K. Iqbal, A. W. Malick, and N. A. Guzman, *Handbook of Hormonal Assay Techniques* (F. de Pablo, C. G. Scanes, and B. D. Weintraub, eds.), Academic Press, San Diego, 1993 (in press).
52. N. A. Guzman, and C. L. Gonzalez, *J. Chromatogr.*, (1993) (submitted).

23

The Utility of Capillary Electrophoresis in Forensic Science

David M. Northrop

Washington State Patrol Crime Laboratory
Kennewick, Washington

The analytical chemistry needs of the forensic laboratory differ from those of the typical research-oriented laboratory in several ways. First, the primary function of the forensic laboratory is to provide expert analysis of samples collected in criminal investigations. Forensic samples are often a heterogeneous mixture of organic and inorganic materials, ranging in size from small molecules (e.g., drugs) to macromolecules (e.g., DNA). Resolution of these complex samples into identifiable constituents free from the interference of other substances is essential to any forensic method. It is also common to have a limited amount of sample for analysis (e.g., microgram quantities), necessitating methods capable of sampling and analyzing small quantities. Second, analysis time is an important factor in a forensic laboratory. Rapid analysis is often necessary to provide vital information for use either by investigators or in court proceedings. Third, the analytical procedures must be reliable and provide information that is unequivocal. Finally, because of limited instrumentation budgets and case load demands, forensic laboratories are dependent on economical, efficient, reliable, and versatile analytical methods.

The development of capillary electrophoresis (CE) over the last 15 plus years, as discussed in previous chapters, shows great promise as an analytical tool and is well suited to forensic needs. The excellent mass detection limits, potentially economical cost, rapid analysis time, exceptional separation and resolution of complex mixtures, and extremely small sample requirements make CE an attractive method for a wide range of forensic problems. The purpose of this chapter is to demonstrate some of the ways in which CE can be used in the forensic laboratory.

GUNSHOT RESIDUE ANALYSIS BY CAPILLARY ELECTROPHORESIS

Forensic investigators can use the analysis of gunshot residues (GSRs) to identify materials and individuals involved in crimes during which firearms have been used. Routine application has been limited because of interferences, high blanks, prohibitive instrumentation costs, and lengthy analysis time. The following discussion will demonstrate how we have used CE [1,2] as an analytical method for GSR analysis, and how this method addresses the shortcomings of previously used techniques.

Gunshot Residues: Composition, Collection, and Analysis

Gunshot residues are traces of unburned and burned gunpowder, primer, and bullet material (as well as their combustion products) that escape from the muzzle, breach, or other leakage areas of a firearm when it is discharged. The two major sources of GSRs are the primer and the propellant used in the ammunition. Some traces may also come from the bullet or jacketing material, as a result of frictional contact with the barrel of the weapon. Primers may consist of several different substances, with the most common, currently used primers consisting of various combinations of lead styphnate, barium nitrate, antimony sulfide, and other compounds [3]. Current propellant formulations consist of various combinations of nitrocellulose (NC), nitroglycerin (NG), and nitroguanidine (NGU), with stabilizers and plasticizers to form single-, double-, and triple-base smokeless powders [4]. A study of 33 smokeless powders [5] identified four of the most commonly encountered additives as dibutylphthalate (DBP), 2,4-dinitrotoluene (2,4-DNT), diphenylamine (DPA), and ethylcentralite (EC). According to Hatcher et al. [4], bullets, for the most part, are composed of lead; however, other elements such as antimony and tin are frequently alloyed into the material, with copper being a common jacketing material. When a weapon is fired, the combustion inside the chamber results in the bullet being expelled from the barrel. At the same time combustion products as well as partially burned and unburned primer and propellant are forced out of the weapon and may be deposited on the hands, face, and clothing of the individual who discharged the weapon. Thus the GSRs deposited on the body or clothing of an individual may be used to connect that person to a weapon or, at least, to demonstrate that he or she had fired a weapon.

Collection methods for GSR have been developed based on the location of the sample and the method of analysis. An extensive study by Harrison and Gilroy [6] has shown that the greatest concentrations of GSRs are likely to be found on the back of the thumb, along the web between the thumb and forefinger, and on the back of the forefinger on the shooting hand. The GSRs are also commonly found in a suspect's pockets if samples are not collected before the suspect has time for hand contact with his or her pockets [26]. The two most commonly used sample collection methods are solvent swabbing and adhesive-film lifts. Brihaye et al. [7] used cotton swabs moistened with dilute nitric acid to collect GSRs for analysis of the inorganic constituents by atomic absorption spectroscopy (AAS). Adhesive tape has been used to collect GSRs for scanning electron microscopy (SEM) [8] and for neutron activation analysis (NAA) [9] of the inorganic constituents of the primer residues. Work has been done to analyze the organic constituents of gunshot residues by high-performance liquid chromatography (HPLC) [10,11]. The organic residues have also been analyzed by gas chromatography–mass spectrometry (GC–MS) [5], and by gas chromatography–thermal energy analysis (GC/TEA) [12]. In each of these procedures, sample collection was done using cotton swabs moistened with an organic solvent, such as methanol or acetone.

Given the many analytical methods available, forensic gunshot residue analysis should be well established. However, research for better techniques continues for several reasons. A study by Havekost et al. [13] of the environmental background levels of the inorganic GSR constituents found levels to be significant enough to interfere with unequivocal GSR analysis. Also, for environmental reasons, manufacturers have been moving away from the use of lead in the primers, opting instead for organic-based materials [14]. Both the time and monetary costs of AAS, SEM, and NAA tend to prohibit routine use by many forensic laboratories. Although analysis time and cost is less prohibitive for HPLC and GC techniques, a survey of forensic laboratories [15] indicates that neither method has gained widespread use. Some of the reasons may be due to the thermal lability of some of the analytes and the lack of sufficient separation (resolving power) for court-defensible results by these techniques. As will be discussed, CE is an analytical method that is capable of overcoming many of these difficulties.

Capillary Electrophoresis of Gunshot Residues

A method has been developed for the analysis of gunshot residue constituents by micellar electrokinetic capillary electrophoresis (MECE) [1]. Since most of the organic constituents of GSRs are uncharged species (Table 1), a micellar phase is added to the buffer (as discussed in Chap. 2) to provide an interactive medium that allows these noncharged species to be separated using CE.

The development of MECE for GSR analysis involved the optimization of buffer conditions, detection, analyte resolution, and sample collection. Optimization of these conditions must be balanced by the need to keep analysis time to a minimum.

The importance of the buffer in CE has been discussed in earlier chapters, but must be reiterated. The ionic concentration of the buffer must be kept such that the current through the capillary does not create more joule heat than can be readily dissipated by the system. For GSR analysis, a dilute sodium tetraborate/boric acid buffer (pH 8.5 ± 0.5) maintained the current at acceptable levels. The effect of sodium dodecylsulfate (SDS; the micellar agent) concentration on the separation of

Table 1 Gunpowder Constituents

Constituent	Abbreviation
Dibutylphthalate	DBP
N,*N*′-Diethyl-*N*,*N*′-diphenylurea (ethylcentralite)	EC
2,3-Dinitrotoluene	2,3-DNT
2,4-Dinitrotoluene	2,4-DNT
2,6-Dinitrotoluene	2,6-DNT
3,4-Dinitrotoluene	3,4-DNT
Diphenylamine	DPA
Glycerol trinitrate (nitroglycerin)	NG
Nitroguanidine	NGU
2-Nitrodiphenylamine	2-nDPA
N-Nitrosodiphenylamine	N-nDPA

Source: Ref. 1.

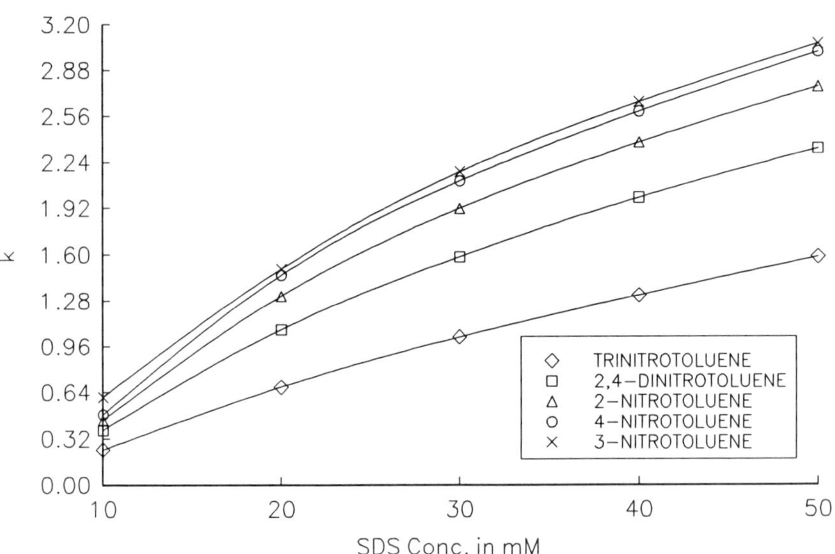

Figure 1 Effect of SDS concentration on k'. Conditions: 2.5 mmol L^{-1} borate buffer at pH 8.9; detection at 220 nm; injection at 5 kV for 10 seconds; run at 20 kV at ambient temperature in a 50-μm id × 67-cm capillary. (From Ref. 1)

several standards is seen in Fig. 1. The improvement in analyte resolution with SDS concentrations beyond about 20–25 mmol L^{-1} is not significant, particularly for the effect on the time of analysis and the contribution to the ionic strength. This last fact was particularly important in the optimization of the detection limits. By keeping the buffer ionic strength low (2.5 mmol L^{-1} borate and 25 mmol L^{-1} SDS), the capillary diameter could be increased from 50 to 100 μm while maintaining resistive heating at acceptable levels. Doubling the capillary diameter increased the volume of sample introduced by a factor of 4 and the path length for the absorbance detector by a factor of 2, thus improving concentration detection limits by almost one order of magnitude. A further increase in capillary diameter resulted in currents that were too large, making the analysis impossible.

Ultraviolet (UV) absorbance detection can be used for GSR analysis by MECE, since most of the GSR constituents have characteristic UV absorbance profiles. Table 2 lists the absorbance maxima of the GSR constituents of interest. Nitroglycerin is the only component that is a poor UV absorber, thus necessitating its detection at 200 nm. Complete absorbance profiles of each component have been compiled [16].

Resolution of analytes that eluted close to each other was improved by increasing the separation efficiency. This was accomplished by decreasing the injection time, which resulted in the introduction of a more narrow sample plug into the capillary, thereby decreasing peak widths and improving efficiency. This effect can be seen from the data in Table 3.

Table 2 GSR Component UV Absorbance Maxima[a]

Sample[b]	Estimated wavelength maxima (nm)
DBP	>200, 240
2,3-DNT	210, 270
2,4-DNT	260
2,6-DNT	>200, 240
3,4-DNT	220, 270
DPA	205, 285
EC	>200, 250
2-nDPA	>200, 280, 260, <450
N-nDPA	>200, 230, 300
NG	>200
NGU	270, 220

[a]First value is largest maxima, with proceeding values in descending order of intensity.
[b]See Table 1 for abbreviations.
Source: Ref. 1.

Table 3 Injection Time vs. Efficiency (*N*)

Time (s)	Efficiency (*N*) × 10^{-3} TNT	2,4-DNT	2-NT
10	36.4	43.6	66.2
5	60.9	97.0	124
2	171	165	283
1	213	256	368

Source: Ref. 1.

The method developed from these studies was used to separate a mixture of GSR constituent standards (Fig. 2). Approximate detection limits were about 5×10^{-6} mol L^{-1} for the aromatic compounds, and about 1×10^{-5} mol L^{-1} for the aliphatic compounds. On the basis of an injection volume of approximately 10 nL, these would correspond to mass detection limits of about 10–30 pg. Identification of constituents can be made based on UV absorbance characteristics and migration

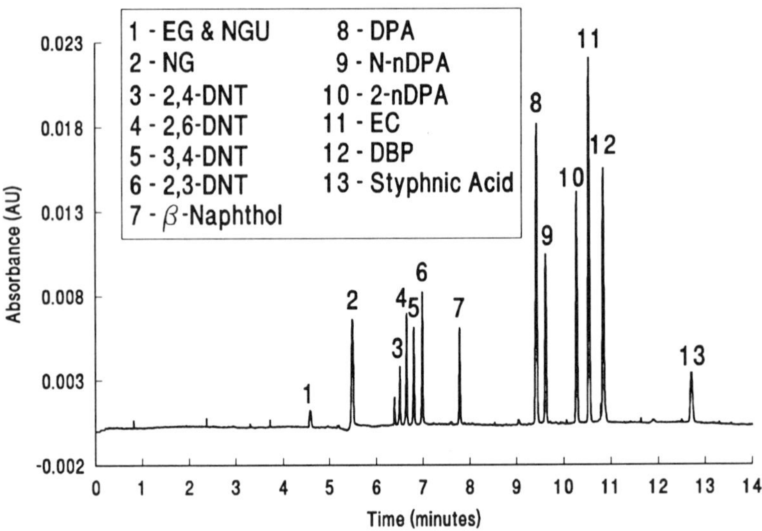

Figure 2 MECE separation of gunshot residue constituent standards at 10^{-4} mol L^{-1}. Conditions: 2.5 mmol L^{-1} borate, 25 mmol L^{-1} SDS, pH 8.5; gravity injection, 50 mm, 10 seconds; voltage, 25 kV; capillary, 82.5-cm × 100-μm id; detection, 200 nm.

behavior. Absolute migration times tend to be variable; however, this variability can be reduced through calculation of the capacity factor, k':

$$k' = \frac{t_r - t_o}{t_o(1 - t_r/t_m)} \tag{1}$$

where t_o is the migration time of any compound that migrates with the electroosmotic flow (e.g., ethanol), t_m is the migration time of any compound that migrates with the micelle front (e.g., DBP), and t_r is the migration time for the analyte of interest.

Appropriate sample collection and handling techniques were developed by Northrop and MacCrehan [2] for MECE analysis to be applied to actual GSR samples. Several requisites must be met. Northrop et al. [1] and Jorgenson [17] have demonstrated that samples must (a) have an analyte concentration range of $1-100$ μmol L^{-1} (owing to the relatively poor concentration detection limits when using absorbance detection); (b) have an ionic strength similar to the CE electrolyte buffer (to avoid poor peak shape); and (c) have been dissolved in a solvent similar in both polarity and viscosity to the MECE buffer (to avoid peak distortion, migration time shift, and injection volume variability). In addition, there are several difficulties in applying MECE to the determination of GSR components collected from human skin. The two major problems are (a) analyte losses by adsorption during collection and handling, and (b) effects of the dissolution solvent and matrix on the sample viscosity. A careful choice of the sample collection and extraction protocol circumvented these difficulties.

Northrop and MacCrehan [2] found solvent swabbing to be a poor approach to collecting the organic GSRs. Although different swabbing materials were tested (cotton, polyester, and polytetrafluoroethylene) using ethanol for the recovery of standards from the skin, all demonstrated large analyte losses. The losses were primarily adsorptive, since analyte losses from volatilization or decomposition were minimal under the experimental conditions. In addition, swabbing recovers excessive quantities of skin fats and oils, causing a matrix interference when evaporative concentration is used to remove the excess organic solvent (as required for MECE analysis). The remaining gelatinous concentrate prevents the hydrophobic analytes from redissolving in the MECE buffer. In addition, a significant concentration of gelatinous material creates a high sample viscosity, resulting in analyte peak shape distortion for any redissolved residue, thereby interfering with MECE quantitation.

Collection of the GSR particulate material using adhesive tape lifts provides a much more appropriate sampling approach for organic residue analysis and takes advantage of the very small sample size requirements of CE. It is not necessary to collect all of the GSR from a hand, since a single particle provides sufficient material for MECE analysis. In addition, only minute quantities of the interfering fats and oils are collected using tape lifts.

The sample collection film needs to meet several criteria. It must (a) possess adequate adhesive character for collection of GSRs from sweaty hands, (b) provide

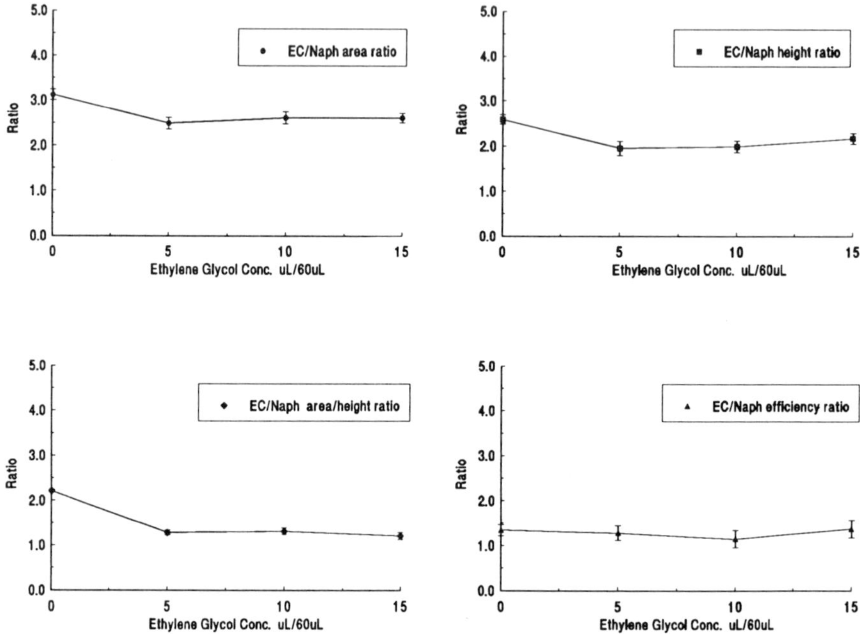

Figure 3 Effect of ethylene glycol (EG) concentration on the ratio of analyte peak areas, heights, area/height ratios, and efficiencies to internal standard peak areas, heights, area/height ratios, and efficiencies. Conditions: 5 μL β-naphthol + 2.5 μL ethylcentralite + EG in buffer. Capillary, 100-μm id × 82.5 cm; buffer, 25 mmol L^{-1} SDS, 2.5 mmol L^{-1} borate, pH 8.5; gravity injection, 50 mm, 10 seconds; voltage, 25 kV; detection, 200 nm. (From Ref. 2)

resistance to extraction solvents, and (c) exhibit a MECE blank that is free from coeluting interferences. The adhesive used on cellophane-type adhesive tape is readily extracted with ethanol. Although masking-tape extraction blanks showed the presence of plasticizers, such as dibutylphthalate (DBP), which migrates with the micelle front ($k' = \infty$), no interference was observed with the GSR components of interest. Given the negligible interferences from the tape blanks, it was possible to extract residues from sections of the tape that may not contain visible residue particles.

Analyte losses may also occur during evaporative concentration and reconstitution of GSR samples. These losses can be avoided by using glass sample vials (instead of the adsorptive polyethylene type), and by the addition of a small percentage of a nonvolatile "keeper" to prevent the sample from going to dryness. Ethylene glycol (EG) can be used as a keeper because of its high boiling point (BP = 198°C) and high polarity. The use of EG as a keeper has the added advantage

Table 4 GSR Samples Analyzed

Sample no.	Weapon	Hand	Firing mode	Shots	Sample type[a]
1	None	L	Before firing		T blank
2	None	R	Before firing		T blank
3	None	L	Before firing		T blank
4	None	R	Before firing		T blank
5	None	L	Before firing		T blank
6	None	R	Before firing		T blank
7	45	L	Right Hand	3	T
8	45	R	Right Hand	3	P
9	45	R	Right Hand	3	T
10	44	L	Two-Handed	3	T
11	44	R	Two-Handed	3	P
12	44	R	Two-Handed	3	T
13	9	L	Two-Handed	3	T
14	9	R	Two-Handed	3	T
15	9	R	Two-Handed	3	P
16	45		Unfired powder		P
17	44		Unfired powder		P
18	9		Unfired powder		P
19	None	L	After washing		T blank
20	None	R	After washing		T blank
21	None	L	After washing		T blank
22	None	R	After washing		T blank
23	None	L	After washing		T blank
24	None	R	After washing		T blank

[a]T, 2-cm^2 section of tape; P, One particle taken from tape.
Source: Ref. 2.

that in the MECE procedure it is not retained in the micelles, thus it migrates with the electroosmotic flow, providing a convenient marker for determining k'. Although EG and nitroguanidine (NGU) do coelute, NGU can be distinguished by its strong UV absorbance at 270 nm, whereas EG is transparent at that wavelength.

Another issue in CE analysis is both qualitative and quantitative reproducibility. In GSR analysis, the use of k' to normalize migration times eliminates any qualitative variance [1]; however, variance in the quantitative measurement of peak areas and heights can be in the range of 5–25%, compared with better than 1% for normal free-zone CE under similar conditions. The source of the quantitative variance is largely indeterminate, being independent of the CE instrument used or mode of column cooling. By normalizing peak areas and heights, using β-naphthol as an internal standard, the quantitative precision can be improved [2]. The addition of EG to the sample as a keeper modifies the sample viscosity, which can affect both the qualitative migration behavior of the analytes and the quantitative results. With ethylcentralite (EC) as a representative GSR component, the effect of EG concentration on quantitation can be largely eliminated by normalizing to the internal standard (Fig. 3). The relative standard deviations (RSDs) are improved to better than 5%. Ethylene glycol concentrations of up to 25% (v/v) in the sample can be tolerated; however, the best results are obtained when the EG concentration is kept below 8% (v/v). An added benefit of using EG as a keeper is that, even though it changes the viscosity of the sample, at constant concentrations it acts as a viscosity buffer. As a result, small amounts of fats and oils extracted from human skin do not have as dramatic an effect on the viscosity of the sample.

Application of the MECE method to actual GSR analysis has been demonstrated [2]. Firing range experiments were conducted by volunteers using a Beretta 9-mm semiautomatic pistol, a Smith and Wesson, Bulldog 44 special revolver, and a Colt Series 80, 45-caliber semiautomatic pistol. Sample blanks were collected from the cleaned hands of each volunteer before weapon firing. Each individual fired only one weapon to prevent cross-contamination. Each weapon was fired three times, after which samples were collected from the firing hand, as well as from the hand that did not hold the weapon. The sample area of interest was the back of the hands along the thumb and forefinger and the webbing between those two digits, as suggested earlier. The samples were collected using 1-in.^{2} (6.5-cm^2) sections of masking tape that were placed in separate sealed glass vials and refrigerated until analysis. Table 4 provides a list of samples collected. Qualitative MECE analysis conducted on each of the samples provided the results shown in Table 5. Characteristic GSR constituents were found on each adhesive film lift on which GSRs would have been expected, and no evidence of GSR components was found on prefiring blanks and postfiring blanks. Figure 4 shows a comparison of electropherograms from MECE runs of GSR standards, an extract of a particle from sample 11, and an extract of a section of the tape from sample 12. Identification of GSR constituents was made by comparison with the migration of standards, using EG and DBP as markers for the determination of k', and by spectral characterization, using sequential MECE analysis at several wavelengths.

One possible use of MECE GSR analysis is to provide information concerning the commercial source of gunpowder and perhaps specific batch or lot information. Compositional differences between powders from different manufacturers can be seen in Fig. 5. Stine has shown that propellants decompose with age [18]. Some of the compositional differences can be due to aging or as a result of the mixing of various lots and batches with old feedstock to produce the appropriate propellant behavior. MECE examinations of the qualitative characteristics of individual unburned propellant grains from a single 9-mm cartridge showed as much as 10–20% variation

Table 5 GSR Constituents in Analyzed Samples

	GSR constituents[a]					
Sample no.	NG	DPA	N-nDPA	2-nDPA	EC	DBP[b]
1						x
2						x
3						x
4						x
5						x
6						x
7						x
8	xxx[c]				x	x
9	xxx				x	x
10	x					x
11	xxx	x	x			x
12	xxx	x	x			x
13	x					x
14	x					x
15	xxx	x	x	x	x	x
16	xxx				x	x
17	xxx	x	x	x	x	x
18	xxx	x	x	x	x	x
19						x
20						x
21						x
22						x
23						x
24						x

[a]See Table 1 for abbreviations.
[b]DBP or any phthalate plasticizer.
[c]xxx, excess quantity.
Source: Ref. 2.

in the quantities of the different GSR constituents from grain-to-grain [19]. In fact, one grain was completely lacking in one GSR constituent that all of the other grains possessed, and there were no decomposition products from that constituent. Thus, the use of quantitative data to provide investigators with identification of the ammunition source must be used with extreme caution. These findings do not affect the qualitative value of MECE analysis of GSRs as a method to establish that a firearm has been discharged by a suspect. MECE analysis for the presence of various organic GSR constituents should have greater probative value than the previously used inorganic GSR analyses.

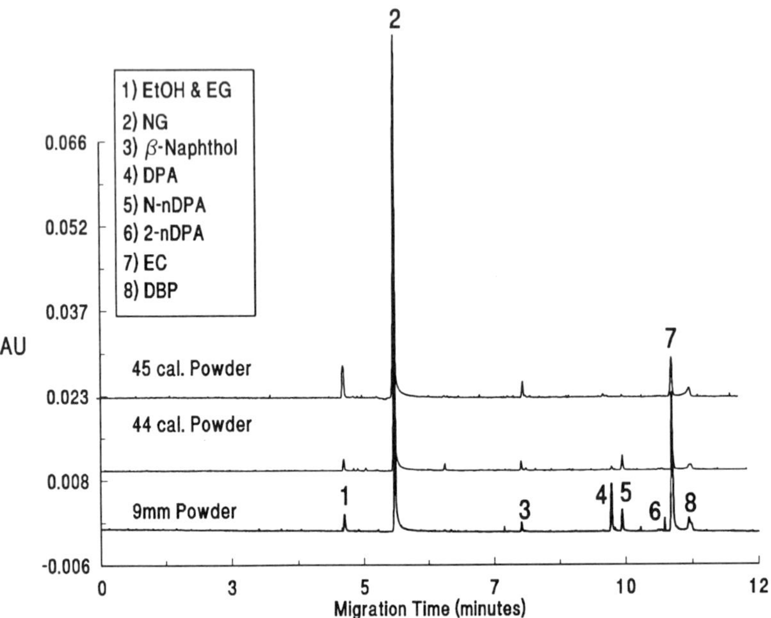

Figure 4 MECE analysis of a solution of GSR standards, an extract of a single GSR particle on a film-lift from a hand used to fire a 44-caliber revolver, and an extract from a film-lift blank. Capillary, 100-μm id × 82.5 cm; buffer, 25 mmol L^{-1} SDS, 2.5 mmol L^{-1} borate, pH 8.5; gravity injection, 50 mm, 5 seconds; voltage, 25 kV. (From Ref. 2)

OTHER FORENSIC ANALYSES BY CAPILLARY ELECTROPHORESIS

Explosive Analysis by Micellar Electrokinetic Column Electrophoresis

The analysis of explosive materials may take several forms. The most common situation involves the analysis of debris from the scene of an explosion to determine if the explosion was accidental or the result of the intentional detonation of an explosive device. The analyst may need to examine various types of material (soil, clothing, building materials, etc.) for the presence of residues from explosive materials or their decomposition products. Cases sometimes occur in which an undetonated device is discovered. In this instance, information concerning the constituents used in the device is needed so that investigators can connect possible suspects to materials used in the construction of the device. The thermally stable organic constituents of explosive materials are commonly analyzed by GC–MS [20], whereas liquid chromatography (LC) with electron-capture detection [21], or thermal energy

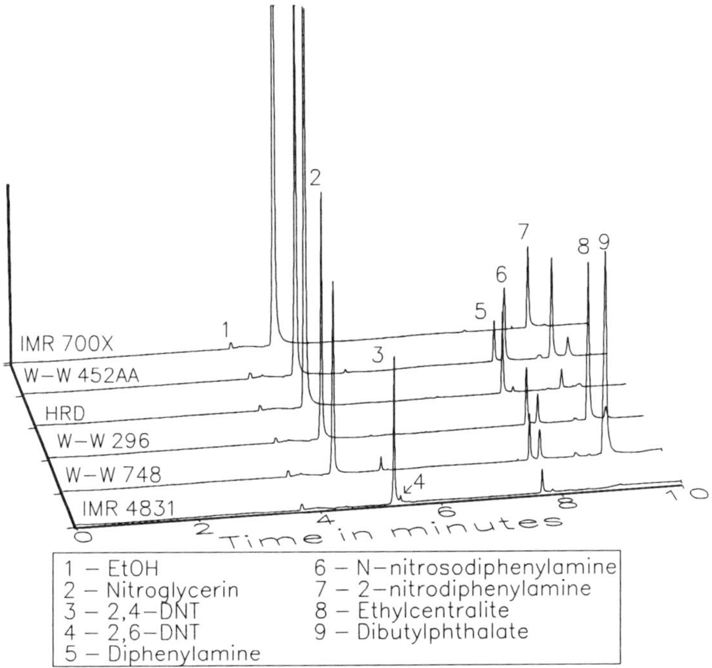

Figure 5 Ethanol extracts of reloading powder. Conditions: capillary, 100-μm id × 67 cm; buffer, 2.5 mmol L^{-1} borate, 25 mmol L^{-1} SDS, pH 8.5; injection, 5 kV, 2 seconds; voltage, 20 kV; detection, 200 nm. (From Ref. 1)

analysis [22], and supercritical fluid chromatography with UV detection [23] are used for the more thermally labile constituents. Since many of the constituents of explosive formulations are similar to those found in gunpowder, the use of MECE for the analysis of explosive constituents has been readily demonstrated [1]. Table 6 lists some of the organic compounds that can be encountered in explosive formulations, and Fig. 6 shows the MECE separation of a sample containing standards of those compounds. All of the components are well resolved, with the exception of the 1,5- and 1,8-isomers of dinitronaphthalene. Application of this method is seen in Fig. 7 in which compositional differences in the extracts from four different plastic explosives are shown. Sample collection and handling procedures developed for GSR analysis should also prove useful in explosive analysis.

Drug Analysis by Micellar Electrokinetic Capillary Electrophoresis

There is probably no area of forensic science that currently accounts for more casework than drug analysis. The number of samples that are confiscated as suspected con-

Table 6 Explosive Constituents

Constituent	Abbreviation
Dibutylphthalate	DBP
Diethylene glycol dinitrate	DEGDN
1,3-Dinitronaphthalene	1,3-DNN
1,5-Dinitronaphthalene	1,5-DNN
1,8-Dinitronaphthalene	1,8-DNN
Ethylene glycol dinitrate	EGDN
Glycerol trinitrate (nitroglycerin)	NG
Nitroguanidine	NGU
2-Nitronaphthalene	2-MNN
Pentaerythritol tetranitrate	PETN
Picric acid	PA
2,4,6,*N*-Tetranitro-*N*-methylaniline	Tetryl
1,3,5,7-Tetranitro-1,3,5,7-tetrazacyclooctane	HMX
2,4,6-Trinitrotoluene	TNT
1,3,5-Trinitro-1,3,5-triazacyclohexane	RDX

Source: Ref. 1.

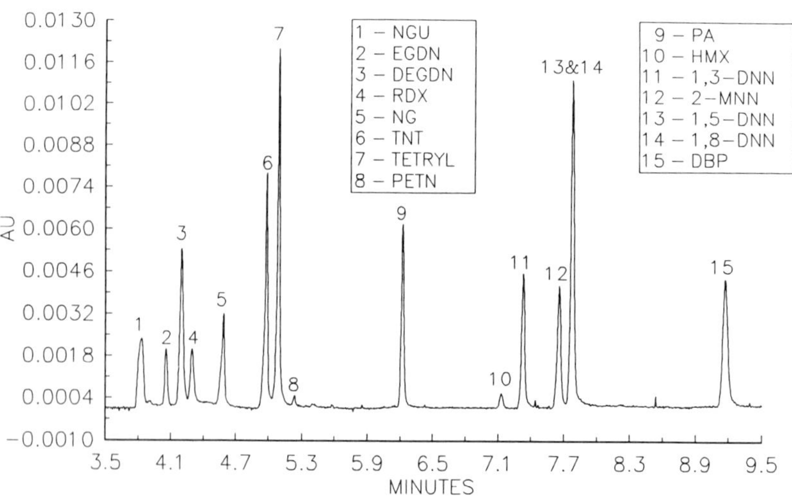

Figure 6 High-explosive constituent standards mixture at 10^{-4} mol L^{-1}. Conditions: capillary, 100-μm id × 67 cm; buffer, 2.5 mmol L^{-1} borate, 25 mmol L^{-1} SDS, pH 8.5; injection, 5 kV, 2 seconds; voltage, 20 kV; detection, 220 nm. (From Ref. 1)

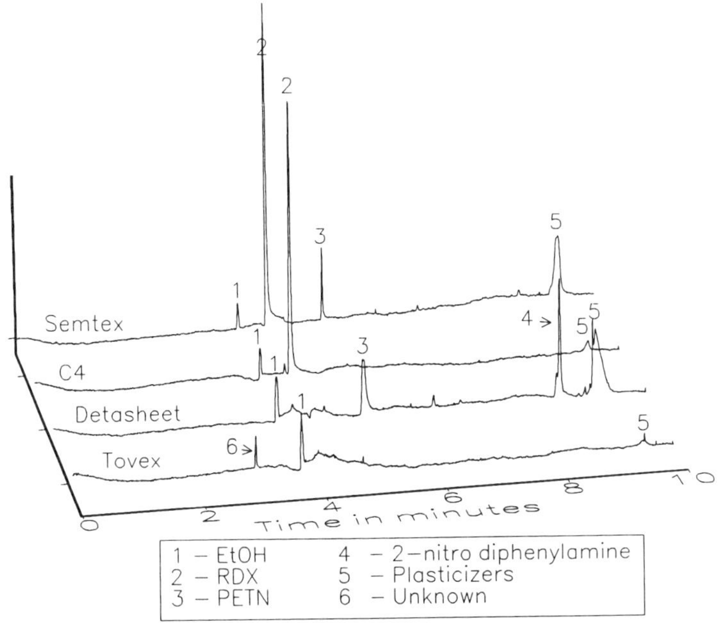

Figure 7 Ethanol extracts from plastic explosives. Conditions: capillary, 100-μm id × 67 cm; buffer, 2.5 mmol L^{-1} borate, 25 mmol L^{-1} SDS, pH 8.5; injection, 5 kV, 2 seconds; voltage, 20 kV; detection, 200 nm. (From Ref. 1)

trolled substances are often submitted faster than the forensic laboratory can analyze them. Toxicological analysis of body fluids also represents many cases handled by forensic laboratories or the medical examiner's office (this subject is discussed is more detail in chapter 24). Rapid and absolute identification of controlled substances in both instances is important.

Two ways in which MECE can be used in the forensic examination of controlled substances have been demonstrated [24]. First, MECE analysis can be used as a general drug-screening tool. By using a phosphate/borate buffer, with SDS as a micellar agent and acetonitrile (15%) as an organic modifier, a separation of 18 common drugs of abuse was accomplished in less than 40 minutes (Fig. 8). A list of the drugs plus some drug impurities that were separated using this method are shown in Table 7. The advantage of this method over other general-screening techniques is that acidic, basic, and neutral drugs can be screened for, using a single analytical method, without the need for class separation. The second way in which this method can be used is for establishing the source of bulk drug samples. This can be done by concentrating extracts of a drug sample and analyzing for the presence

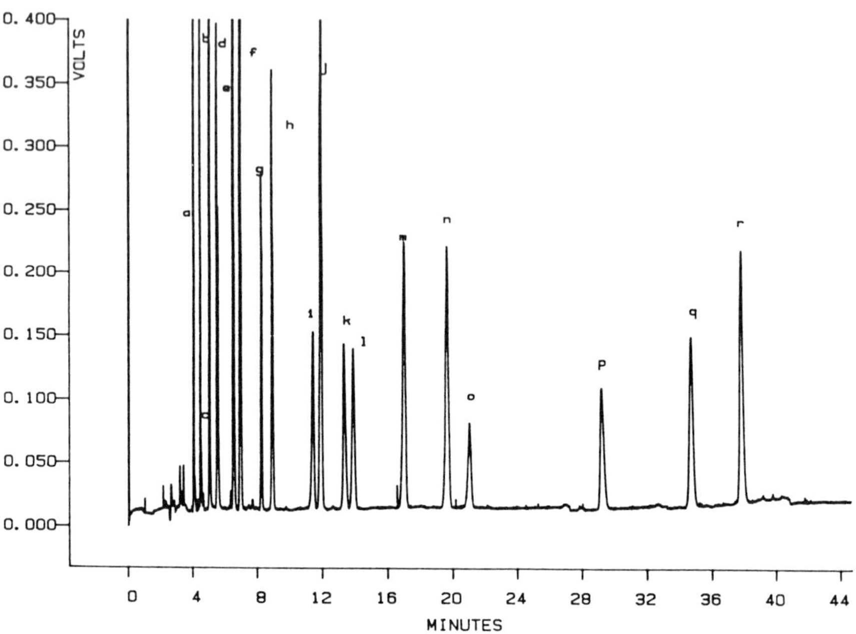

Figure 8 MECE forensic drug screen. Conditions: capillary 50-μm id × 25 cm; voltage, 20 kV; temperature, 40°C; buffer, 85 mmol L^{-1} SDS, 8.5 mmol L^{-1} phosphate, 8.5 mmol L^{-1} borate, 15% acetonitrile, pH 8.5; detection, 210 nm; sample concentration 250 μg mL^{-1}. Key: (a) psilocybin, (b) morphine, (c) phenobarbitol, (d) psilocin, (e) codeine, (f) methaqualone, (g) LSD, (h) heroin, (i) amphetamine, (j) librium, (k) cocaine, (l) methamphetamine, (m) lorazepam, (n) diazepam, (o) fentanyl, (p) PCP, (q) cannabidiol, (r) Δ^9-THC. (From Ref. 24)

of various adulterants, decomposition products, and synthesis by-products. The work by Weinberger and Lurie [24] has demonstrated the usefulness of the MECE method for the analysis of illicit heroin samples, as part of a program to identify the source of the drug, based on the composition and quantities of substances other than heroin in the bulk sample. Application of CE to other forensic drug analyses is likely to be accomplished soon.

Analysis of Pen Inks by Capillary Electrophoresis

Ink analysis is another area in the field of forensic science in which CE has been useful. There are several ways that ink analysis is used by the forensic scientist. The composition of an ink can be used to determine the source of counterfeit money, to validate the age of and thus authenticity of an important document, and to match a writing instrument with a particular document. An article by Fanali and Schudel

Table 7 Illicit Drug and Drug Impurities Separated by MECE[a]

Benzoylecgonine	Librium
Psilocybin	Cocaine
Morphine	MDMA
Hydromorphone (Dilaudid)	Methamphetamine
Phenobarbital	Phentermine
Psilocin	Noscapine
0^6-Monoacetylmorphine	*cis*-Cinnamoylcocaine
0^3-Monoacetylmorphine	Lorazepam
Codeine	Diazepam
Methaqualone	*trans*-Cinnamoylcocaine
Mescaline	Fentanyl
LSD	Flurazepam
LAMPA	PCP
Heroin	Cannabidiol
Acetylcodeine	Cannabinol
Papaverine	Δ^9-THC
MDA	
Amphetamine	

[a]Listed in order of elution.
Abbreviations:
LSD, lysergic acid diethylamide; LAMPA, lysergic acid-N-(methylpropyl)amide; MDA, methylenedioxymethamphetamine; MDMA, methylenedioxymethamphetamine; PCP, phencyclidine; Δ^9 THC, Δ^9-tetrahydrocannabinol.
Source: Ref. 24.

[25] describes the use of CE to obtain a separation of the different dye components in the inks used in four red and three black fiber-tipped pens. The separation procedure used a 3:1 ammonium acetate (0.1 mol L^{-1}, pH 4.5):methanol buffer, and a 25-μm−id × 20-cm capillary. The applied potential was 8 kV and detection made at 206 nm. The results showed the separation of several different components in each of the inks studied. From these results, qualitative differentiation of inks from different pens was readily demonstrated. Further application to the examination of inks from documents and other writing instruments should be forthcoming.

THE FUTURE OF CAPILLARY ELECTROPHORESIS IN FORENSIC SCIENCE

What is the future for CE in forensic science? Capillary electrophoresis is an analytical technique that should be a valuable tool in the forensic laboratory. It has several features that make it a more versatile technique than many other separation methods. First, the forensic applications that have already been discussed demonstrate the ability of CE to provide rapid separations (usually under 10 minutes) and high resolution (efficiencies as high as 10^6 plates per meter) of multiple components in complex mixtures. Second, the quantity of sample needed for analysis is extremely small. Injection volumes of about 10 nL allowed for CE analysis of GSR constituents obtained from samples weighing less than 0.1 μg. Third, a wide variety of analytes (including organic, inorganic, charged, and neutral) can be separated by CE, and analytes need not be UV active, since various methods of detection from mass spectrometry to electrochemical have been developed (as discussed in previous chapters). As an example of CEs versatility, it can easily handle thermally labile compounds and salts that cannot be analyzed by GC. In LC, the use of large quantities of solvents, including potentially hazardous solvents, such as methylene chloride, provides the laboratory with an expensive waste disposal problem. In contrast, in CE the small volume of buffers that are used (typically milliliters per day) can be safely disposed of through the sewer system. Finally, CE is relatively economical, compared with many other analytical methods. Autosampling, commercial instruments are currently selling for between 20,000 and 40,000 dollars, and it is also possible to construct simple units from commercially available components for well under 10,000 dollars. Buffer materials, solvents, and other consumable items are used at a much slower rate than with most other separation methods, and analytical columns for most applications are very inexpensive (about 5 dollars per meter for fused silica capillaries). Even the 80–150 dollar per column cost of commercial gel-filled or derivitized capillaries is as good or better than the cost of columns for most chromatographic methods.

DNA sequencing and toxicological analysis by CE are both rapidly progressing (as discussed in other chapters) and show great promise. Beyond these areas of interest, it would seem that methods for the analysis of fiber dyes, paints, arson accelerants, soil samples, and many other samples could be readily developed. The potential uses of CE in forensic science are only limited by forensic researchers' imaginations.

ACKNOWLEDGMENTS

The author would like to thank Dr. William A. MacCrehan for his inciteful review and discussion of this work.

REFERENCES

1. D. M. Northrop, D. E. Martire, and W. A. MacCrehan, *Anal. Chem.*, *63*:1038 (1991).
2. D. M. Northrop and W. A. MacCrehan, *J. Liq. Chromatogr.*, *15* (6):1041 (1992).
3. M. Tassa, N. Adan, N. Zeldes, and Y. Leist, *J. Forensic Sci.*, *27*:671 (1982).
4. J. Hatcher, F. Jury, and J. Weller, *Firearms Investigation, Identification, and Evidence* (T. Samsworth, ed.), The Stackpole Co., Harrisonburg, Penn.

5. M. Mach, A. Pallos, and P. Jones, *J. Forensic Sci.*, *23*:433 (1978).
6. H. C. Harrison and R. Gilroy, *J. Forensic Sci.*, *4*:184 (1959).
7. C. Brihaye, R. Machiroux, and G. Gillian, *Forensic Sci. Int.*, *20*:279 (1982).
8. M. Tassa, Y. Leist, and M. Steinberg, *J. Forensic Sci.*, *27*:677 (1982).
9. S. S. Krishnan, *J. Forensic Sci.*, *22*:304 (1977).
10. D. B. Dahl, J. C. Cayton, and P. F. Lott, *Microchem. J.*, *35*:360 (1987).
11. E. O. Espinoza, *Diss. Abstr. Int. B.*, *49*:4779 (1989).
12. J. M. F. Douse, *J. Chromatogr.*, *464*:178 (1989).
13. D. G. Havekost, C. A. Peters, and R. D. Koons, *J. Forensic Sci.*, *35*:1096 (1990).
14. J. S. Wallace, *AFTE J.*, *22*:364 (1990).
15. D. DeGaetano, J. A. Siegel, *J. Forensic Sci.*, *35*:1087 (1990).
16. D. M. Northrop, Dissertation, Georgetown University, Washington, D.C. (1991).
17. J. W. Jorgenson, *New Directions in Electrophoretic Methods*, *ACS Symp. Ser.*, Washington, D.C., p. 182 (1987).
18. G. Y. Stine, *Anal. Chem.*, *63*:475A (1991).
19. D. M. Northrop, Unpublished data (1991).
20. W. Gielsdorf, *Z. Anal. Chem.*, *308*:123 (1981).
21. J. B. F. Lloyd, *J. Energy Mater.*, *4*:239 (1986).
22. E. C. Bender, *Proc. Int. Symp. Anal. Detect. Explos.*, p. 309 (1983).
23. W. H. Griest, C. Guzman, and M. Dekker, *J. Chromatogr.*, *467*:423 (1989).
24. R. Weinberger and I. S. Lurie, *Anal. Chem.*, *63*:823 (1991).
25. S. Fanali and M. Schudel, *J. Forensic Sci.*, *36*:1192 (1991).
26. R. S. Nesbitt, J. E. Wessel, G. M. Wolten, and P. F. Jones, *J. Forensic Sci.*, *22*:304 (1977).

24

Capillary Electrophoresis and Electrokinetic Capillary Chromatography of Drugs in Body Fluids

Wolfgang Thormann

University of Bern
Bern, Switzerland

With the more efficient therapeutic application of various drugs, there has evolved a need for reliable analytical procedures. Studies on patient compliance, bioavailability, pharmacokinetics and genetics, organ function, and the influences of co-medication and underlying disease can be based only on accurately measured drug levels in body fluids, including blood (serum, plasma), urine, and saliva. Furthermore, screening and confirmation of drugs in body fluids is important for the investigation of intoxications, the detection of potential abusers of drugs, and the control of drug addicts following withdrawal therapy. Currently used methods are based on the principles of spectrophotometry, immunoassays [radioimmunoassay, enzyme multiplied immunoassay (EMIT), fluoroimmunoassay, fluorescence polarization immunoassay (FPIA), nephelometric immunoassay] and chromatography [thin-layer chromatography, gas chromatography (GC), and high-performance liquid chromatography (HPLC)] [for reviews see Refs. 1–3]. All of these techniques have advantages and disadvantages. The reagents for many of the immunological assays are available in kit form, together with highly automated instrumentation. This permits such analyses to be performed easily and efficiently, as well as with high sensitivity and precision. They provide the most rapid (high sample throughput) analytical procedures yet available. However, immunological techniques are prone to disturbances by molecules of similar structure (cross-reactivity). Many antibodies involved do not recognize only the drug of interest, but also some of its metabolites. Moreover these techniques are, by nature, unsuited to simultaneous monitoring of several drugs and metabolites. The chromatographic assays, on the other hand,

Table 1 Selected Applications of CZE/MECC Drug Determinations in Body Fluids

Drug/metabolite	Matrix	Method	Sample preparation	Detection method	Ref.
Methotrexate/7-OH-met.	Serum	CZE	p-p or s-l before ox	LIF	9
Lithium	Serum	CZE	Ultrafiltration	conductivity	10
Anthracyclines	Plasma	CZE	l-l	LIF	11
Cefixime	Urine	CZE	None	ABS	12
Ara-C	Plasma	CZE	s-l	ABS	13
Benzodiazepines	Urine	CZE	Hydrolysis, l-l	ABS, MS	14
Ferulic acid	Dog plasma	CZE	l-l	ABS	15
B_6 vitamers	Urine	MECC	l-l	LIF	16
Cefpiramide	Plasma	MECC	None	ABS	17, 18
Aspoxicillin	Plasma	MECC	None	ABS	19
Cicletanine enantiomers	Plasma	MECC	l-l	ABS	20
Antiepileptics	Plasma	MECC	l-l	ABS	21
Barbiturates	Serum/urine	MECC	None, l-l, s-l	ABS	22
Thiopental	Serum	MECC	l-l	ABS	23
Drugs of abuse	Urine	MECC	s-l (hydrolysis)	ABS	24
THC	Urine	MECC	Hydrolysis, s-l	ABS	25
Substituted purines	Serum/saliva/urine	MECC	None, l-l, s-l	ABS	26
AFMU/1-methylxanthine	Urine	MECC	None	ABS	27
Cimetidine	Serum	MECC	Electrochromatography	ABS	28

Abbreviations: THC, 11-nor-Δ^9-tetrahydrocannabinol-9-carboxylic acid; ara-C, cytosine-β-D-arabinoside; AFMU, 5-acetylamino-6-formylamino-3-methyluracil; p-p, protein precipitation; l-l, liquid–liquid extraction; s-l, solid-phase extraction; ABS, absorbance; LIF, laser-induced fluorescence; MS, mass spectrometry; ox, oxidation. None refers to direct sample injection, but may include simple fluid handling, such as dilution, centrifugation, or filtering.

provide specific results of multiple compounds, but typically require extensive sample preparation and derivatization. The sample throughput is low because of sequential injection of the samples. Furthermore, complete automation is difficult.

Recently, instrumentation for electrokinetic separations in capillaries of very small internal diameter (id) (25–75 μm) have become available [4–8], and first papers reporting its use for drug monitoring in body fluids appeared in the literature (Table 1). Electrokinetic separations or analyses in capillaries should be regarded as complementary or as attractive alternatives to other capillary separation techniques, such as GC, HPLC, supercritical fluid chromatography, and field flow fractionation. The advantages of electrokinetic capillary analyses are high resolution, efficiency and speed, full automation, small sample size, rapid method development, small amounts of inexpensive and nonpolluting chemicals, as well as simple adaptation for micropreparative work. Electrokinetic capillary techniques can exploit numerous separation principles, making them flexible and easily applicable to a variety of separation problems. The specific techniques comprise capillary zone electrophoresis (CZE), capillary isotachophoresis, capillary isoelectric focusing, and a range of electrokinetic capillary chromatography methods [e.g., micellar electrokinetic capillary chromatography (MECC)]. The differentiating features of these methods lie in the initial and boundary conditions applied, which determine the character of the migrating sample zones [8,29].

The purpose of this chapter is to provide a brief, yet concise, review on capillary electrophoresis and electrokinetic capillary chromatography of drugs in body fluids. The review is restricted to CZE and MECC separations of drugs and metabolites in body fluids, particularly serum (plasma), saliva, and urine (see Table 1). It does not deal with the determination of endogenous compounds, such as creatinine and uric acid [30], or the diagnosis of human diseases from metabolic disorders [31], nor does it review the innumerable papers reporting the use of this emerging technology for the analysis of drugs in pharmaceutical preparations [see, e.g., 32,33]. Applications employing capillary isotachophoresis are also not covered, and interested readers are referred to the monograph of Boček et al. [34].

PRINCIPLES OF SEPARATION AND INSTRUMENTATION

Capillary zone electrophoresis is a one-phase process with separation based on differences in electrophoretic mobilities. In MECC, an interface between electrophoresis and chromatography, two distinct phases are used, an aqueous and a micellar phase or pseudostationary phase. These two phases are established by employing buffers containing surfactants (e.g., sodium dodecyl sulfate; SDS), which are added above their critical micellar concentration. Micelles are dynamic structures that are in equilibrium with the monomer. An MECC analysis is performed in equipment designed for capillary electrophoresis (i.e., in an open, tubular capillary of very small id). A high-voltage DC electric field is applied along the column, thereby

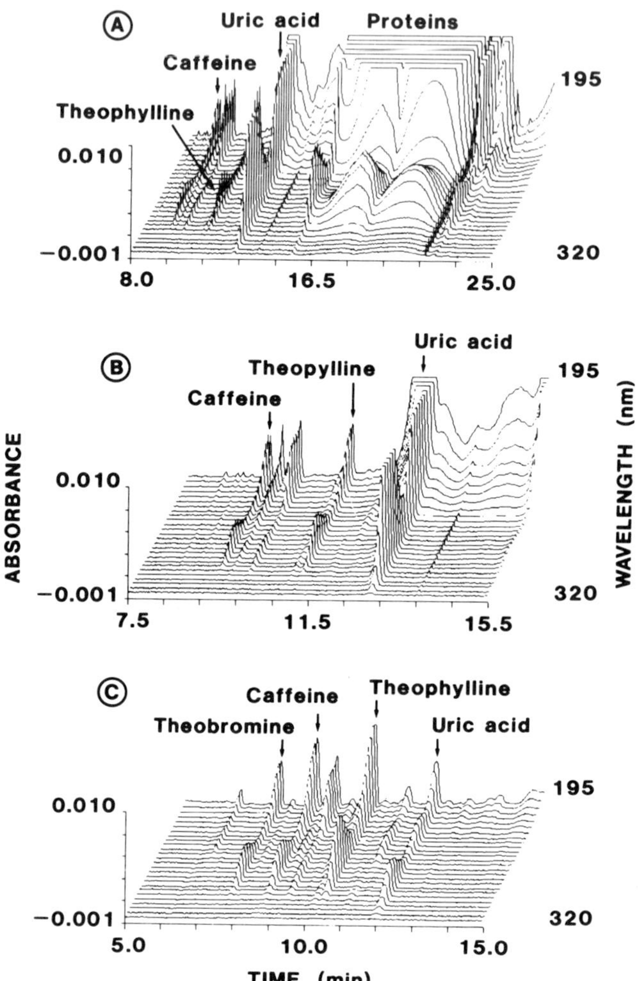

Figure 1 Three-dimensional MECC data of (A,B), a directly injected, and (C) Sep-Pak C_{18}-extracted serum sample of a patient undergoing theophylline pharmacotherapy. For extraction, theobromine (40 μM) was added as internal standard. Note that panel B represents a section of the data presented in panel A. The instrument employed is the same as described previously by Thormann et al. [22,26]. Briefly it featured a 75-μm–id fused silica capillary of about 90-cm length (Product TSP/075/375, Polymicro Technologies, Phoenix, Arizona), together with a fast-scanning multiwavelength detector model UVIS 206 PHD with on-column capillary detector cell No. 9550-0155 (both from Linear Instruments, Reno, Nevada) toward the capillary end. The effective separation distance was 70 cm. A constant voltage of 20 kV was applied. The cathode was on the detector side. Sample application occurred manually by gravity through lifting the anodic capillary end, dipped into the sample vial, about 34 cm for 5 seconds. Multiwavelength data were read, evaluated, and stored employing

causing both a movement of the entire liquid (the so-called electroosmotic flow) and migration of the charged micelles. As a result, the two phases migrate at different velocities, thereby permitting chromatographic separations. Nonionic solutes partition between the two phases and elute with zone velocities between those of the two phases. For that case, separation is solely chromatographic, and elution order is based on the degree of partitioning. For ionic solutes, separation is based on chromatography and electrophoresis. The MECC separations are dependent on micellar formation, buffer composition, and pH (typically between 7 and 10), as well as the presence of modifiers, such as small amounts of organic solvents (typically acetonitrile or methanol). Variation of the surfactant concentration has effects similar to changes of surface structure in HPLC (e.g., change from C_8 to C_{18}). The elution sequence in MECC is typically different from that in reversed-phase liquid chromatography. MECC is unsuitable for the separation of large molecules, such as proteins, but it allows the resolution of chiral compounds [32,35].

Instrumentation for CZE and MECC is identical. It features fused silica capillaries of very small, 25- to 75-μm id that are connected to buffer reservoirs containing the driving electrodes. Depending on surface structure and coatings, capillary lengths of about 20–120 cm are employed. Small amounts of samples (nanoliter quantities) are introduced by electrokinetic or hydrodynamic techniques. Upon application of power (10–30 kV, 30–120 μA) samples are transported through the capillary by the combined action of electrophoresis and electroosmosis. Detection principles applied include, on-column absorbance, fluorescence and, occasionally, conductivity or radiometry, as well as off-column monitoring employing mass spectrometry (MS) or amperometry. Although most of the currently employed instruments have been assembled in the researchers' laboratories, there are several commercial instruments available. Most of these apparatuses are fully automated, comprising an autosampler as well as data gathering and evaluation with a computerized data station [7]. These instruments have the potential of becoming important tools for the drug assay laboratories, but to my knowledge, have not yet been introduced for routine use. For the analysis of drugs in body fluids, ultraviolet (UV) absorbance is currently the most popular method, particularly in the approach of scanning multiwavelength monitoring, with which eluting zones are characterized by both retention and absorption behavior [22] (Figs. 1 and 2). Other detection techniques

---

a Mandax AT 286 computer system and running the 206 detector software package version 2.0 (Linear Instruments, Reno, Nevada) with windows 286 version 2.1 (Microsoft, Redmont, Washington). Conditioning for each experiment occurred by rinsing the capillary with 0.1 M NaOH for 3 minutes and with buffer for 5 minutes. The 206 detector was employed in the high-speed polychrome mode by scanning from 195 to 320 nm at 5-nm intervals (26 wavelengths). A buffer composed of 75 mM SDS, 6 mM $Na_2B_4O_7$, and 10 mM Na_2HPO_4 (pH about 9.1) was used. The current was about 75 μA.

reported in the literature include laser-induced fluorescence (LIF), and MS, and conductivity (see Table 1).

SAMPLE PREPARATION

Typically, on receipt of a sample, it must be prepared for analysis. This stage is intended to improve specificity of the assay by removing interfering matrix compounds, while concentrating the analyte; to stabilize the analyte; and to remove matrix particles that would block instrumental parts (e.g., syringe, tubing, column). Sample preparation methods include simple liquid-handling procedures (e.g., centrifugation, dilution, filtering), release of the analyte from the biological matrix (e.g., hydrolysis, sonication), removal of endogenous compounds (e.g., precipitation, ultrafiltration, and extraction), and enhancement of selectivity and sensitivity by analyte derivatization [for a comprehensive review refer to Ref. 36]. For CZE or MECC of drugs in body fluids most of these techniques have been applied. Selected examples are listed in Table 1.

Both liquid–liquid and solid-phase (solid–liquid) extraction schemes have been applied to a number of drug determinations (see Table 1). The two approaches are attractive for two reasons. First, they are typically quite selective, thereby greatly simplifying the sample matrix, and second, the analyte can simultaneously be concentrated by one to two orders of magnitude. Because of the relatively high sample concentrations required for electrokinetic capillary analysis, the latter effect is very important. On the other hand, liquid–liquid and, somewhat also, solid–phase extractions are quite time-consuming. Typical protocols are listed in Fig. 3. In most of the applications reported thus far, standard approaches were employed. Liquid–liquid extractions were executed in glass tubes, and solid-phase extraction was performed with disposable cartridges, processes that are standard practice in chromatography [36]. Recent investigations mention the in-column use of solid-phase material for on-line preconcentration of the analyte [37] and the application of solid-phase cartridges for electrochromatographic extraction [28]. Alternatively, the sample matrix can be simplified by protein precipitation (e.g., with trichloroacetic acid or methanol) as well as ultrafiltration [determination of the free (unbound) fraction of a drug].

Most importantly, MECC with SDS allows the direct injection of proteinaceous fluids, such as serum, an approach that bears great similarity to HPLC with micellar mobile phases [38]. The proteins are solubilized by SDS and elute (as a very broad zone) after uric acid (see Fig. 1A). With human serum or plasma, several drugs, including cefpiramide [17,18], aspoxicillin [19], phenobarbital [22], caffeine, and theophylline [26], were shown to elute in an interference-free or analytical window in front of uric acid, allowing these substances to be analyzed by direct sample injection. Similarly, drugs can be analyzed in saliva [26] and, in special cases, also in urine [27,39]. For detection by on-column absorbance, direct injection of a body

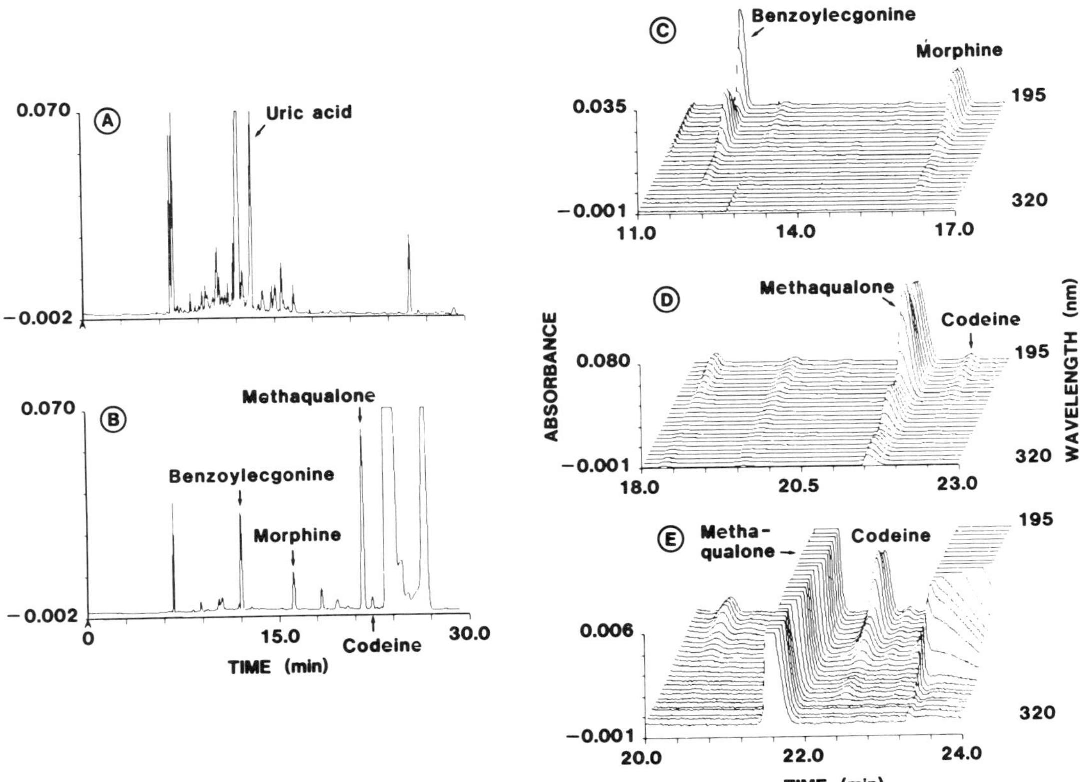

Figure 2 An MECC analysis of an opiate- and cocaine-positive urine specimen (A) by direct injection, and (B) after Bond Elut Certify pretreatment. The electropherograms depicted in panels C–E represent sections of the three-dimensional data obtained with the cleaned up sample. Data of panels B–E all stem from the same experiment. The instrumental configuration was the same as that of Fig. 1. (Adapted and reprinted with permission from Ref. 24. Copyright 1991 American Chemical Society).

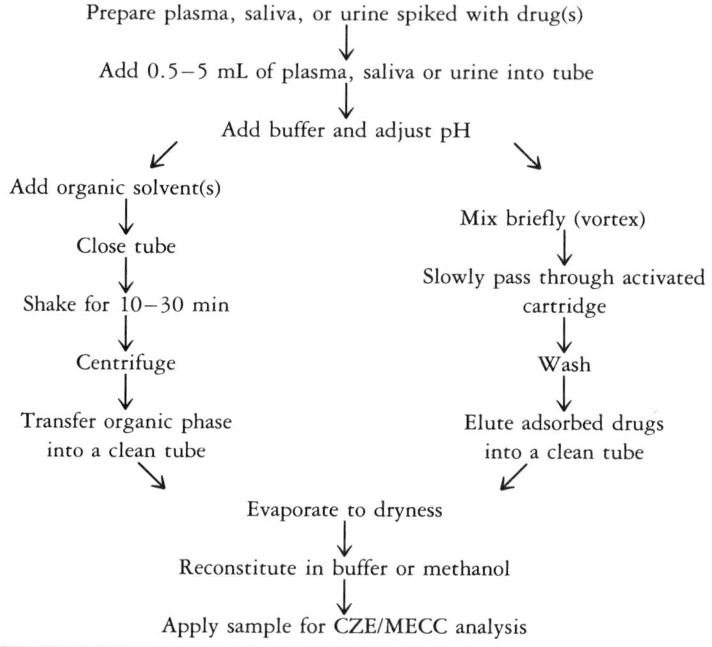

Figure 3 Typical flow diagrams for drug extraction.

fluid requires that drug concentrations be at or higher than the microgram per milliliter (μM) level.

QUANTITATION, REPRODUCIBILITY, AND DETECTION LIMIT

In most cases of drug monitoring, the response has to be quantified after analysis. In CZE and MECC, peak heights and peak areas can be determined with a standalone integrator or a PC integration pack, both developed for the evaluation of chromatograms. Calibration graphs using bovine plasma, or human serum or urine, spiked with the drugs, showed good linear correlations when assessed over a concentration range of not more than about two orders of magnitude. Quantitation was performed by the internal standard method or by direct calibration with the peak area of the drug of interest (i.e., without inclusion of the internal standard). Although some deviations were observed, the two approaches produced clinically meaningful data. The reproducibility of retention times and peak areas were about 1–2% and 3–7%, respectively [23,26].

Detection limits are primarily dependent on the type of detector used and the analyte concentration in the applied sample. With UV absorption detection and samples without preconcentration, the limit of detection is in the low microgram per milliliter (μM) range. With preconcentration by sample extraction, concentrations as low as a few tens of nanograms per milliliter could be unambiguously monitored [25]. With LIF detection, a detection limit of about 200 pg mL^{-1} was attained [11].

TWO EXAMPLES

The three-dimensional data depicted in Fig. 1 were obtained with serum from a patient undergoing theophylline pharmacotherapy. The suitability of direct sample introduction for drug monitoring is demonstrated with the example shown in panel (A). A clear theophylline zone within the analytical window is produced. An expanded section of the same data is depicted in panel (B), providing an improved insight into the zones in front of uric acid as well as revealing the formation of a caffeine zone. Panel (C) shows the three-dimensional pherogram of the same serum, but after extraction with Sep-Pak C_{18}, according to Thormann et al. [26]. These data clearly show that well-resolved peaks for both theophylline and caffeine are obtained with either method. The serum concentration of theophylline was determined to be 79 μM using Sep-Pak cleanup and 90 μM employing EMIT or FPIA. Thus, MECC has the potential to analyze for serum theophylline on a pharmacologically interesting concentration level (therapeutic range of this compound is 55–110 μM) and without elaborate sample pretreatment. Similar findings were found to apply to caffeine [26].

The effects of sample cleanup by solid-phase extraction are further illustrated with the data presented in Fig. 2. The analyzed patient urine was markedly positive for cocaine and opiates using EMIT assays. Direct injection (panel A) provided a complex electropherogram in which uric acid (marker of body fluids) could easily be identified. No peak, however, could be assigned to one of the illicit compounds of interest. In MECC, with on-column UV absorption detection, sample concentrations have to be at least on the microgram per milligram (μM) concentration level. For direct urine injection, this sensitivity limit is not as good as that of the commonly used immunological-screening methods. Therefore, extraction of the drugs or their metabolites remains essential for their confirmation by MECC. Panel (B) of Fig. 2 depicts single-wavelength data obtained after Bond Elut Certify extraction of the same patient's urine using the procedure for cocaine and metabolites [24]. A much smaller number of peaks are detected, despite that with this approach and 100% recovery a 50-fold concentration of the extracted compounds can be achieved. By having data between 195 and 320 nm, as shown in panels (C)–(E), as well as reference spectra, a quick and reliable confirmation was permitted of the presence of benzoylecgonine (cocaine metabolite), morphine, codeine, and methaqualone in that sample. The former three compounds extracted with efficiencies of about 50, 60, and 90%, respectively, and spiking of urine blank on the 200 ng mL^{-1} level still provided small peaks that could be unambiguously assigned. From calibration graphs, based on peak heights (linear, four-point calibration with standards between 0.25 and 10 $\mu g\ mL^{-1}$), the concentrations of benzoylecgonine, morphine, and codeine in that patient's urine were estimated to be 6.5, 4, and 1 $\mu g\ mL^{-1}$, respectively. It is important to add that these data were obtained without hydrolysis and derivatization of the sample.

ACHIEVEMENTS AND FUTURE OUTLOOK

The monitoring of drugs in body fluids using CZE or MECC has several areas of interest, the most important being (a) the determination of specific drugs or metabolites for therapeutic or research reasons, (b) the confirmation of drugs of abuse and metabolites in a specimen that tested positively employing a routine screening procedure, and (c) the determination of metabolic ratios for pharmacogenetic purposes. Thiopental HPLC and MECC data for some 66 patients' samples [23], as well as CZE or MECC and immunoassay data of methotrexate [9] and theophylline–caffeine [26] compared well on linear regression analysis. The graphs, however, deviated somewhat from the line of equality. The elucidation of the reasons for this behavior are subject of ongoing investigations. For therapeutic drug monitoring, however, such differences are relatively insignificant. The value of combining solid-phase extraction and MECC analysis with fast-scanning multiwavelength detection for the confirmation of opioids (heroin metabolites), the cocaine metabolite benzoylecgonine, amphetamines, barbiturates, and hypnotics [24], as well as 11-nor-Δ^9-tetrahydrocannabinol-9-carboxylic acid (THC-COOH; a metabolite of the psychotropic drug Δ^9-tetrahydrocannabinol present in marijuana) [25], in human urine has been reported. Most importantly, this new method did not require sample hydrolysis (exception THC-COOH) and derivatization for monitoring these compounds at concentrations equal to or lower than the cutoff levels used in routinely applied immunoassays. Last, but not least, Lloyd et al. [27] were able to demonstrate the value of MECC with direct injection of body fluid for the determination of the acetylator phenotype by measuring peak height ratios of two urinary metabolites of caffeine, 5-acetylamino-6-formylamino-3-methyluracil (AFMU) and 1-methylxanthine.

The use of CZE or MECC instead of GC or HPLC for therapeutic and diagnostic drug monitoring has several important advantages. These are high efficiency and degree of automation, small sample size, potential for direct sample injection, no requirement for large amounts of organic solvents, and rapidity of analysis. Furthermore, an electrokinetic capillary setup permits a quick change from one buffer configuration to another, thereby making method development simple and cost-effective. The direct injection of proteinaceous fluids, such as plasma and serum, in MECC provides a special feature of electrokinetic capillary analyses. It allows rapid investigations and can be performed on very small sample volumes (a few microliters), such as serum samples of prematurely born infants [26], droplets of tears, sweat, or other samples that are typically too small to be pretreated. All data gathered thus far are most encouraging and demonstrate the high potential of electrokinetic capillary methods of analysis. However, further investigations are needed for adoption of CZE–MECC as routine methods in a drug assay laboratory. Important areas of research include (a) the application of reliable automated operation over long periods (overnight), (b) the establishment of automated data evaluation and recognition of unexpected interferences and capillary fouling, as well as (c) the elucidation of optimized protocols for the analysis of specific drugs.

ACKNOWLEDGMENTS

I would like to acknowledge the valuable contributions made by all research colleagues and students who are coauthors of our publications in the area of electrokinetic capillary analysis of drugs, as well as by Mrs. Yolanda Aebi and her co-workers in the departmental drug assay laboratory. The generous loan of the 206 UVIS detector by its manufacturer, Linear Instruments, Reno, Nevada, is gratefully acknowledged. This work was sponsored partly by the Research Foundation of the University of Bern, the Liver Foundation in Bern, and the Swiss National Science Foundation.

REFERENCES

1. K. Aziz, *Am. Lab.*, *Apr.*:78 (1988).
2. R. DeCresce, A. Mazura, M. Lifshitz, and J. Tilson, *Drug Testing in the Workplace*, ASCP Press, Chicago, (1989).
3. W. J. Taylor and M. H. Diers Caviness, *A Textbook for the Clinical Application of TDM*, Abbott Laboratories, Irving, Ill., (1986).
4. J. W. Jorgenson and K. D. Lukacs, *Science*, *222*:266 (1983).
5. B. L. Karger, A. S. Cohen, and A. Guttman, *J. Chromatogr.*, *492*:585 (1989).
6. N. A. Guzman, L. Hernandez, and B. G. Hoebel, *BioPharm*, *2*:22 (1989).
7. Z. Deyl and R. Struzinsky, *J. Chromatogr.*, *569*:63 (1991).
8. W. Thormann and M. A. Firestone, *Capillary Electrophoretic Separations, in Protein Purification* (J. C. Janson and L. Rydén, eds.), VCH Verlag Weinheim, pp. 469–492 (and references cited therein) (1989).
9. M. C. Roach, P. Gozel, and R. N. Zare, *J. Chromatogr.*, *426*:129 (1988).
10. X. Huang, M. J. Gordon, and R. N. Zare, *J. Chromatogr.*, *425*:385 (1988).
11. N. J. Reinhoud, U. R. Tjaden, H. Irth, and J. van der Greef, *J. Chromatogr.*, *574*:327 (1992).
12. S. Honda, A. Taga, K. Kakehi, S. Koda, and Y. Okamoto, *J. Chromatogr.*, *590*:364 (1992).

13. D. K. Lloyd, A. M. Cypess, and I. W. Wainer, *J. Chromatogr.*, *568*:117 (1991).
14. I. M. Johansson, R. Pavelka, and J. D. Henjon, *J. Chromatogr.*, *559*:515 (1991).
15. S. Fujiwara and S. Honda, *Anal. Chem.*, *58*:1811 (1986).
16. D. E. Burton, M. J. Sepaniak, and M. P. Maskarinec, *J. Chromatogr. Sci.*, *24*:347 (1986).
17. T. Nakagawa, Y. Oda, A. Shibukawa, and H. Tanaka, *Chem. Pharm. Bull.*, *36*:1622 (1988).
18. T. Nakagawa, Y. Oda, A. Shibukawa, H. Fukuda, and H. Tanaka, *Chem. Pharm. Bull.*, *37*:707 (1989).
19. H. Nishi, T. Fukuyama, and M. Matsuo, *J. Chromatogr.*, *515*:245 (1990).
20. J. Pruñonosa, R. Obach, A. Diez-Cascón, and L. Gouesclou, *J. Chromatogr.*, *574*:127 (1992).
21. K. Lee, G. S. Heo, N. J. Kim, and D. C. Moon, *J. Chromatogr.*, *608*:243 (1992).
22. W. Thormann, P. Meier, C. Marcolli, and F. Binder, *J. Chromatogr.*, *545*:445 (1991).
23. P. Meier and W. Thormann, *J. Chromatogr.*, *559*:505 (1991).
24. P. Wernly and W. Thormann, *Anal. Chem.*, *63*:2878 (1991).
25. P. Wernly and W. Thormann, *J. Chromatogr.*, *608*:251 (1992).
26. W. Thormann, A. Minger, S. Molteni, J. Caslavska, and P. Gebauer, *J. Chromatogr.*, *593*:275 (1992).
27. D. K. Lloyd, K. Fried, and I. W. Wainer, *J. Chromatogr.*, *578*:283 (1992).
28. H. Soini, T. Tsuda, and M. V. Novotny, *J. Chromatogr.*, *559*:547 (1991).
29. R. A. Mosher, D. A. Saville, and W. Thormann, *The Dynamics of Electrophoresis*, VCH Publishers, Weinheim, (1992).
30. M. Miyake, A. Shibukawa, and T. Nakagawa, *J. High Resolut. Chromatogr.*, *14*:181 (1991).
31. E. Jellum, K. Thorsrud, and E. Time, *Chromatographia*, *559*:455 (1991).
32. H. Nishi and S. Terabe, *Electrophoresis*, *11*:691 (1990).
33. A. Pluym, W. Van Ael, and M. De Smet, *Trends Anal. Chem.*, *11*:27 (1992).
34. P. Boček, M. Deml, P. Gebauer, and V. Dolník, *Analytical Isotachophoresis*, VCH Publishers, Weinheim, (1988).
35. S. Terabe, *Trends Anal. Chem.*, *8*:129 (1989).
36. R. D. McDowall, *J. Chromatogr.*, *492*:3 (1989).
37. M. E. Swartz, M. Merion, and R. Pfeifer, Fourth Int. Symp. High Performance Capillary Electrophoresis, Feb. 10–13, Amsterdam, Abstr. M-43 (1992).
38. M. J. Koenigbauer, *J. Chromatogr.*, *531*:79 (1990).
39. Y. Tanaka and W. Thormann, *Electrophoresis*, *11*:760 (1990).

25

Quantitative Analysis with Capillary Zone Electrophoresis

Arthur M. Hoyt, Jr.

University of Central Arkansas
Conway, Arkansas

Quantitative methods employing capillary zone electrophoresis (CZE) represent an applications area still in the early stages of development. The separating power and extremely sensitive mass detection limits of the technique are well documented, but the literature is dominated by qualitative uses of the technique. Early investigators were, as expected, concerned with optimizing parameters and investigating the strengths and weaknesses of the procedure. Only recently have the possibilities for quantitative applications been explored. This report, although not exhaustive, gives an overview of the successes and frustrations workers have experienced while attempting to employ CZE as the basis of assays in a variety of matrices.

COLUMNS AND BUFFERS

In virtually all of the quantitative references cited, the columns were 50- to 75-μm id. One procedure for methotrexate, a drug used in cancer treatment, measures fluorescent radiation from a 200-μm id column [1]; this column size increases the signal level, but the diameter is atypical.

Columns used in quantitative analysis are almost uniformly silica and vary in length from 20 to 100 cm, with 25–75 cm being the most common. As Jorgenson and Lukacs [2] point out, column length is not a factor in separation efficiency, but, as a practical matter, lengths of less than 20 cm are not generally employed.

Polyacrylamide-packed columns were used to separate and quantify recombinant proteins [3]; this paper also points out how separations could be based on molecular mass (similar to slab gels), as well as on electrophoretic mobility. The packed columns in these experiments had to be water-jacketed to maintain gel consistency. In the fluorescent procedure cited earlier, the laser source heated the column excessively, and cooling was needed for this as well. Heat is not usually a problem (except in such special instances), because of the high surface area afforded by the capillary, an often-noted advantage of CZE.

Buffers used by various experimenters in assay procedures vary to a greater degree than column type and length. Buffers of all types have been successfully employed; phosphate and phosphate–borate buffers are used often, but there appears to be a great deal of flexibility allowed. Borate buffers have been used in the assay of carbohydrates [4–6], since the borate adduct gives electrophoretic mobility to the compounds separated. In the determination of neutral compounds, micellar electrokinetic capillary chromatography (MECC) can be employed. Emulsifier-type compounds must be added to generate micelles, and this technique has been used successfully in a method for thiopental in serum and plasma [7]. Other than the borate/carbohydrate and MECC/emulsifier examples buffer composition seems to be a trial-and-error discovery. Buffer pH, which varies from 2 to 11 over the examples cited, also seems to be an empirical choice, and generalizations of predictive value are difficult to make.

INJECTION MODES

Hydrostatic injection, which is called suction, pressure, gravity, and siphon injection in various reports, is employed in about 80% of the techniques I have noted. Electrokinetic injection, which allows more uniform sample size introduction, was found to suffer from the disadvantage of selecting for the components of the sample with the highest electrophoretic mobility [8]. Given this drawback, and the easy compensation for the nonuniform sample size in hydrostatic sample introduction, most workers have chosen hydrostatic methods.

The obvious compensation for erratic sample size in hydrostatic injection is the use of an internal standard, and this is very common in the techniques cited. An interesting variation is standard addition [9], used to quantitate purines in fish tissue. Lookabaugh, and co-workers [10] point out that to use external standards, the matrices for standard and sample must be the same; internal standards compensate for this problem, as well as for sample size differences.

ANALYTE DETECTION

Direct Ultraviolet

Ultraviolet (UV) detectors, operated from 200 to 300 nm, are used in almost all assays, regardless of sample type. With the very short path length commonly available, UV detection generally allows work in the 10^{-4} to 10^{-8} M concentration range, depending on the chromophore employed and the conditions in the analysis. The linearity range has been as small as one order of magnitude [4], or as large as three to four orders of magnitude [11]; in the former, calibration plots were not always linear, but the analyses could still be carried out.

Sample Concentration Techniques

Preconcentration has been employed in both qualitative and quantitative procedures [12–15] to extend the lower limits of detection, but it is cumbersome for routine use. Stacking was supplementary to preconcentration in one of the foregoing references, and application of this phenomenon is simple to carry out. Applied Biosystems, in their applications literature [16], states that stacking occurs on-column when the analyte solution has low conductivity, compared with the separating buffer; further investigation of this ability to concentrate may be warranted in future method development. Use of tandem techniques, in which liquid chromatography (LC) or isotachophoresis serves to concentrate or purify the sample for CZE, have also appeared [11,17] but are still a curiosity in methods applications.

Ultraviolet Variations

Indirect detection, in which the sample reduces UV absorption, has been used to determine organic acids in food and wine samples [18]. Compounds that do not absorb UV strongly have been determined by coupling them with reagents possessed of higher absorptivities; in one example [19], the reagent also added electrophoretic mobility.

Conductometric Methods

Krivankova and associates, in their isotachophoresis–CZE procedure, used conductometric detection on the isotachophoresis portion of the apparatus [11]. They did, however, use UV absorbance at 254 nm for detection of the analyte in CZE. Huang et al. used conductometric detection for the carboxylic acids in their determination [20], and pointed out that with this detector, it was unnecessary to calibrate the response for each component of a mixture. Also of interest in this paper was the use of tetradecyltrimethylammonium bromide to control electroosmotic flow. Ackerman and associates considered quantitative conductometric and indirect UV detection in a paper that is largely theoretical [21], but point out some interesting relations between mobility and peak area.

Fluorescence

Fluorescence has the distinct advantage that it is capable of allowing work at 10^{-12}-M levels [22], but has been used primarily in qualitative separations. Guzman

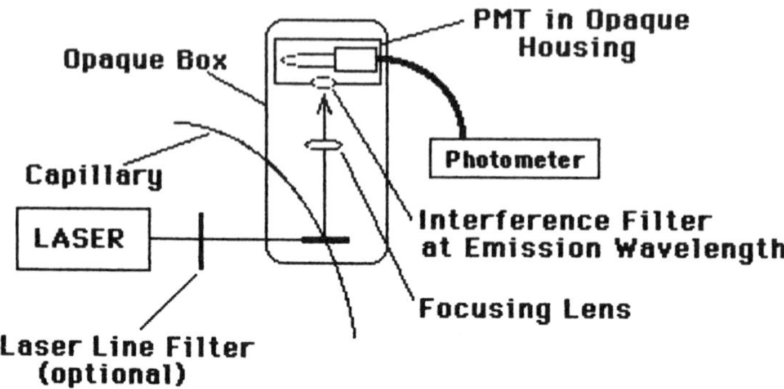

Figure 1 Block diagram of apparatus for CZE fluorescence analysis.

and co-workers [12] used a mercury lamp for excitation in their qualitative monoclonal antibody separations; reaction of the antibody proteins with fluorescein isothiocyanate (FITC) was used to provide fluorescence capability. The methotrexate assay of Roach et al. [1] used laser-induced fluorescence in a quantitative method, and emission was enhanced by oxidation of the drug. Some authors mentioned this mode of detection as a possibility, and others even tried unsuccessfully to employ it, but the possibilities are largely unexplored and underutilized in quantitative methods. As these examples point out, compounds are capable of being adapted to fluorescence by derivatization or chemical modification. Figure 1 shows a simple system able to perform adequate CZE fluorescence analysis. This setup is not only simple, but realistic; rarely is it necessary to use a more sophisticated apparatus. Although the sensitivity is a desirable (and in sometimes necessary) advantage of fluorescence, alignment of optics, expense of a laser source (if used), and general complexity of the complete laboratory apparatus required make it somewhat unattractive if alternatives exist.

SIGNAL PRESENTATION AND EVALUATION

Data Collection Devices and Methods

Commercial ultraviolet CZE detectors generally have two outputs: a low-voltage (generally 10 mV) output for use with a stripchart recorder, and a second one (generally 1 V) for use with an integrator. The integrator is sometimes replaced with an interface that allows data to be fed directly to a computer. Both micro and mainframe varieties have been used, but whether an integrator or computer is present, one output device is almost always a stripchart recorder.

The recorder alone has been used to provide peak height data in quantitative methods. In the fluorescence methotrexate procedure mentioned earlier, peak height was compared directly with concentration as a measure of drug level. In a penicillin method developed by this writer [23], drug level was measured by comparing the ratio of the peak height of the antibiotic with that of an internal standard.

Integration of peak area has been used as a direct measure of analyte level by several authors. Aguilar and co-workers determined ferrous and ferric iron by computer integration of the peak areas for the cyanide complexes [24]. The procedures cited for halofuginone [11], recombinant proteins [3], cytarabine (ara-C) [15], nucleotides [9], and disaccharides [6] also measured peak area directly, with the packed-column procedure using VAX integration. To be noted is the problem of comparing areas in possibly nonidentical matrices of sample and standard [10].

Most of the quantitative procedures cited in this report use internal standards and peak area ratios; some workers have tried both peak area and peak height ratios [18,19]. It is interesting that there is direct conflict between measurement techniques that were purported to give the best precision.

Precision of Measurements

In our laboratory, as well as those of Kenney [18] and Honda et al. [25], peak area ratio values showed less variation than peak height ratios. While developing a method for cimetidine in pharmaceuticals [26], we found (Fig. 2) that 20 repetitions of an injection on the same sample gave much more consistent results with peak area ratios. Peak heights were measured from the recorder trace, relative peak areas from an integrator.

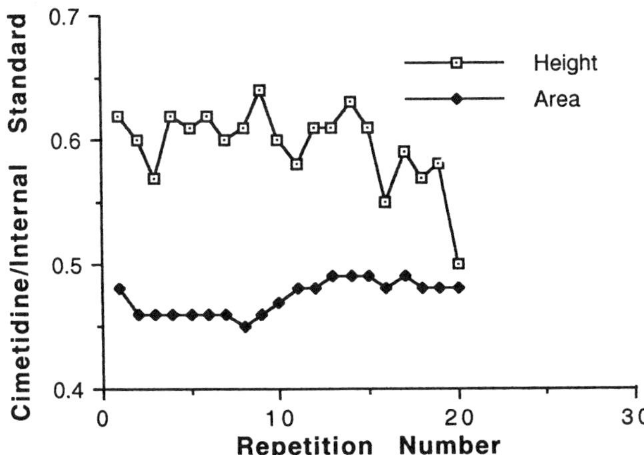

Figure 2 Comparison of precision: peak height ratio versus peak area ratio measurements on a single cimetidine sample.

Inconsistencies appeared in experiments that reported results with and without an internal standard. An internal standard divided the relative standard deviation in half for Honda et al. [4], and by a factor of five for Lookabaugh et al. [10]. Tsuji anticipates improvement is relative standard deviation with an internal standard [3]. Although not directly comparable with CZE, the MECC procedure for thiopental, mentioned earlier, is troubling in that an internal standard caused poorer precision. Ling and associates found that the presence or absence of an internal standard was not a major factor in their results [19]; their prime interest, however, was in a comparison of micro-LC with CZE, rather than a comparison of internal standard effect.

The heterogeneity of micelles may be a factor in the MECC results, or the partitioning process itself could lead to the scatter. In any event, ascertaining the source of the poor results would be helpful in future method development. In principle, an internal standard should compensate for the inconsistent volumes in manual injection and should negate media effects by placing the standard and sample in the same matrix. It should be expected, then, that the use of internal standards will appear regularly in procedures involving analyses in biological or other complex matrices, and in those that use hydrostatic injection.

Table 1 shows relative standard deviations in some of the methods discussed. These data show that CZE is typically as precise as high-performance liquid chromatography (HPLC), a technique with which it is most often compared. CZE also compared favorably with micro-LC [19]. One study compared HPLC and HPCE, and HPCE quantitated palytoxin at lower levels than HPLC [27]. In the cimetidine method developed in our laboratory [26], a comparison showed precision is similar in the two techniques. A report by Dose and Guichon [28] states that relative standard deviations of 1% are possible with use of two internal standards. If dual references are feasible in a given application, the precision possible should satisfy the most stringent requirement.

Table 1 Precision of Some Representative CZE Methods

Ref.	Internal standard	Rel. std. dev. (%)
11	Butyric acid	1.3–14.1
10	None	0.44–13.1
10	Dansyl amino acid	0.37–2.51
3	None	3–7
15	None	9–15
19	Mercaptoethanol	3.5–4.5
4	Cinnamic acid	~5
4	None	~10
26	Phenol	1.9–6.4

APPLICATIONS

Two interesting quantitative applications of CZE were in inorganic analysis. One procedure [24] determined the levels of iron(II), iron(III), and zinc in plating solutions. The second follows the periodate oxidation of carbohydrates by measurement of iodate and periodate in the reaction mixture [25]. In both of these procedures, the analytes have electrophoretic mobilities that exceed the electroosmotic flow; therefore, samples are injected at the end of the capillary which is the electroosmotic exit. The paper by Ackermans and co-workers, which was primarily concerned with conductometric and indirect UV detection, reported separation of chloride, chlorate, fluoride, and organic acids from C_1 to C_6 [21].

The use of CZE to follow progress of reactions has been the focus of several quantitative studies. Acrylamido acids and bases (immobilines) are used in isoelectric focusing; the degradation of these compounds has been monitored in three studies [29–31]. Nucleotide degradation [9] and periodate oxidation [25] are reactions that have appeared in methods already reported. Potentially useful in analysis of organic reaction mixtures is the use of cyclodextrins in the resolution of enantiomers. In a study by Fanali and Bocek, tryptophan and epinephrine enantiomers were separated [32]; the epinephrine compounds were in a commercial pharmaceutical preparation.

Drugs and compounds of pharmaceutical importance have been assayed in several CZE studies. Methods for methotrexate, benzylpenicillin, halofuginone, cytarabine (ara-C), thiopental, and cimetidine have already been cited. Two methods of interest that have not been reported are the determination of arbutin, a diuretic glucoside, in the leaves of bearberry [33], and the determination of quinine and proflavine in pharmaceuticals [34]. A paper qualitatively surveying a number of compounds was published by Wainright [35]. This series of experiments established variables for future quantitative analysis of pharmaceuticals, both by CZE and MECC. Capillaries (both size and presence of coating), pH and ionic strength of buffer, and the presence or absence of micelles were evaluated for separation of anti-inflammatory drugs, antibacterials, peptides, and barbiturates.

Biochemical compounds were determined in several methods, including those for carbohydrates [4,5], aminoglycans [6], insulin [10], and recombinant proteins [3]. Monoclonal antibodies have been measured [36] and preparatively collected [12]. Hernandez and associates were able to separate and quantitate the components of a mixture of cyclic nucleotides [37]. Paralytic shellfish poisons were separated and determined in a method by Thibault and co-workers [38]. This work also confirmed structures for the toxins by using mass spectrometric detection through an ion-spray interface.

Several authors have developed analyses for compounds in urine; Guzman et al. devised a procedure for determining the levels of urea and creatinine in human urine [39]. Capillary electrophoresis was used to concentrate uric acid and methamphetamine from urine before mass spectrometry [13]. Recent experiments in our laboratory resulted in a method for determination of an antiulcer medication, cimetidine, in the urine of rats [40]. Research efforts in other body fluids, such as serum and plasma, has allowed development of the cited procedures for methotrexate, thiopental, cytarabine and the micro-LC–CZE methods for thiols.

SPECIAL CONSIDERATIONS

The purifications necessary for CZE analyses are difficult to generalize, except for two expected observations: more complex matrices require more elegant separations, and compounds that interact with the capillary must be removed (or the capillary regenerated after electrophoresis). Simple backgrounds, such as pharmaceutical samples, or reaction mixtures, may require only dilution and assay [14], or at most, an extraction [26] or similar cleanup.

Samples derived from biological sources are very common, but pose the greatest challenge, because they contain such high backgrounds and a wide variety of interfering materials (Fig. 3). It is not uncommon for capillaries exposed to such contamination to become useless after one injection. Proteins are ubiquitous in these situations, but adsorb strongly to silica and generally need to be removed. Most commonly, the protein is precipitated with trichloracetic acid, followed by centrifugation or syringe filtering to remove the precipitate [9,19]; however, other techniques, such as isoelectric precipitation, are possibilities for protein removal.

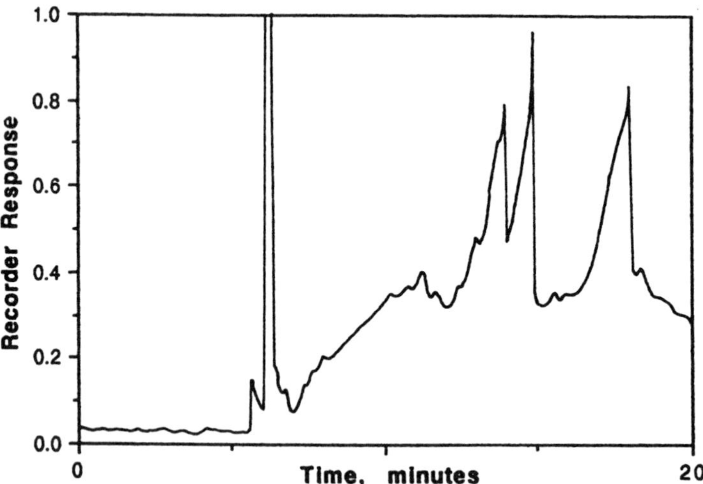

Figure 3 Electropherogram of raw human urine at 25 KV in 20 mM phosphate buffer pH 7.

Towns and Regnier have used nonionic surfactant–alkylsilane column coatings to prevent protein adsorption to the column wall [41]. Guzman and Hernandez quantified antibody proteins [36], and Tsuji described a method for recombinant proteins [3]; these reports indicate removal is not universally necessary.

Extractions have been useful for biological sample cleanup before CZE [26,40], but schemes devised thus far have been empirical; steps are tried and added until sufficient for the purpose. The ideal result of steps before assay would be removal of everything except analyte, but this is rarely practical. It is sufficient to remove those compounds that cause capillary performance to deteriorate and those that interfere with measurements of the analyte or internal standard. Even this level of preparation has been avoided by some authors [4,9,15] by using methanol or NaOH, or other reagents; these compounds were used to purge the capillary after analyses to restore performance.

In summary, analysts have been quick to recognize the resolving power, versatility and tiny sample requirement of this technique. This report has shown a variety of methods used to solve quantitation problems unique to CZE, while exploiting the foregoing advantages. Analytical chemists have shown imagination, insight, and creativity in trying to increase optical path length or sample concentration, while coping with the sample characteristics required and the interferences encountered. This early work is exciting in that it shows such power and potential, pointing the way toward a promising area in separations technology. Capillary zone electrophoresis is in the process of taking its place alongside classic analytical procedures as a viable option for quantitative analysis.

REFERENCES

1. M. C. Roach, P. Gozel, and R. N. Zare, *J. Chromatogr.*, *426*:129 (1988).
2. J. W. Jorgenson and K. D. Lukacs, *Science*, *222*:266 (1983).
3. K. Tsuji, *J. Chromatogr.*, *550*:823 (1991).
4. S. Honda, S. Iwase, A. Makino, and S. Fujiwara, *Anal. Biochem.*, *176*:72 (1989).
5. S. Honda, K. Suzuki, M. Kataoka, A. Makino, and K. Kakehi, *J. Chromatogr.*, *515*:653 (1990).
6. A. Al-Hakim and R. J. Linhardt, *Anal. Biochem.*, *195*:68 (1991).
7. P. Meier and W. Thormann, *J. Chromatogr.*, *559*:505 (1991).
8. X. Huang, M. J. Gordon, and R. N. Zare, *Anal. Chem.*, *60*:375 (1988).
9. A. L. Nguyen, J. H. T. Luong, and C. Masson, *Anal. Chem.*, *62*:2490 (1990).
10. M. Lookabaugh, M. Biwas, and I. S. Krull, *J. Chromatogr.*, *549*:357 (1991).
11. L. Krivankova, F. Foret, and P. Bocek, *J. Chromatogr.*, *545*:307 (1991).
12. N. A. Guzman, M. A. Trebilcock, and J. P. Advis, *Anal. Chim. Acta*, *249*:247 (1991).
13. N. A. Guzman, M. A. Trebilcock, and J. P. Advis, *J. Liq. Chromatogr.*, *14*:997 (1991).

14. A. M. Hoyt, Jr., J. W. Jorgenson, and J. P. Larmann, Jr., "Conference on Capillary Electrophoresis," NCI-FCRDC, Frederick, Md, Oct. 15–16, Abstr. #8 (1990), submitted to *J. Microcolumn Sep.*
15. D. K. Lloyd, A. M. Cypess, and I. W. Wainer, *J. Chromatogr.*, *568*:117 (1991).
16. Model 270A Application Update No. 6, Applied Biosystems, Foster City, Calif. (1989).
17. M. M. Bushey and J. W. Jorgenson, *Anal. Chem.*, *62*:978 (1990).
18. B. F. Kenney, *J. Chromatogr.*, *546*:423 (1991).
19. B. L. Ling, W. R. G. Baeyens, and C. Dewaele, *J. High Resolut. Chromatogr.*, *14*:169 (1991).
20. X. Huang, J. A. Luckey, M. J. Gordon, and R. N. Zare, *Anal. Chem.*, *61*:766 (1989).
21. M. T. Ackermans, F. M. Everaerts, and J. L. Beckers, *J. Chromatogr.*, *549*:345 (1991).
22. N. J. Dovichi, *Chem. 13 News*, *190*:7 (1989).
23. A. M. Hoyt, Jr. and M. J. Sepaniak, *Anal. Lett.*, *22*:861 (1989).
24. M. Aguilar, X. Huang, and R. N. Zare, *J. Chromatogr.*, *480*:427 (1989).
25. S. Honda, K. Suzuki, and K. Kakehi, *Anal. Biochem.*, *177*:62 (1989).
26. S. Arrowood and A. M. Hoyt, Jr., *J. Chromatogr.*, *586*:177 (1991).
27. K. A. Mereish, S. Morris, G. McCullers, T. J. Taylor, and D. L. Bunner, *J. Liq. Chromatogr.*, *14*:1025 (1991).
28. E. V. Dose and G. A. Guichon, *Anal. Chem.*, *63*:1154 (1991).
29. M. Chiari, C. Ettori, and P. G. Righetti, *J. Chromatogr.*, *559*:119 (1991).
30. M. Chiari, M. Giacomini, C. Micheletti, and P. G. Righetti, *J. Chromatogr.*, *558*:285 (1991).
31. P. G. Righetti, C. Ettori, and M. Chiari, *Electrophoresis*, *12*:55 (1991).
32. S. Fanali and P. Bocek, *Electrophoresis*, *11*:757 (1990).
33. E. Kenndler, C. Schwer, B. Fritsche, and M. Poehm, *J. Chromatogr.*, *514*:383 (1990).
34. K. D. Altria and C. F. Simpson, *J. Pharm. Biomed. Anal.*, *6*:801 (1987).
35. A. Wainright, *J. Microcolumn Separation*, *2*:166 (1990).
36. N. A. Guzman and L. Hernandez, *Techniques in Protein Chemistry* (T. E. Hugli, ed.), Academic Press, San Diego, p. 456 (1989).
37. L. Hernandez, B. G. Hoebel, and N. A. Guzman, "Analytical Biotechnology: Capillary Electrophoresis and Chromatography," *ACS Symp. Ser. 434*:50 (1990).
38. P. Thibault, S. Pleasance, and M. V. Laycock, *J. Chromatogr.*, *542*:483 (1991).
39. N. A. Guzman, C. M. Berck, L. Hernandez, and J. P. Advis, *J. Liq. Chromatogr.*, *13*:3833 (1990).
40. S. Arrowood and A. M. Hoyt, Jr., *Microchem. J.*, (in press).
41. J. K. Towns and F. E. Regnier, *Anal. Chem.*, *63*:1126 (1991).

26

Capillary Polyacrylamide Gel Electrophoresis

András Guttman

Beckman Instruments, Inc.
Fullerton, California

In bioanalytical chemistry, separation methods fall into three major categories: those based on size differences (gel filtration, ultrafiltration, centrifugation), electrical charge differences (ion-exchange chromatography, electrophoresis), and differences in specific chemical or biological properties (affinity chromatography, affinity electrophoresis, immunoadsorption). Since most biopolymers are electrically charged, they will move under an applied electric field. The transport that takes place in a solution or a solid matrix is called *electrophoresis* [1].

Recently, separation and purification methods handling small sample amounts are necessary for bioanalysis [2]. Capillary electrophoresis (CE) is a rapid and powerful separation technique that offers all of the main advantages of high-performance liquid chromatography (HPLC) and is fully complementary with that method [3–7]. This instrumental approach to electrophoresis can be characterized as a rapid, high-resolution analytical technique and an effective tool for the characterization and isolation of biomolecules. The technique offers different modes of selectivity, compared with HPLC, as well as generation of very high efficiency (i.e., large plate numbers) [8]. Since the method employs high electric fields (up to the range of 10^3 V cm^{-1}), a significant joule heating results that must be dissipated [9].

Historically, moving boundary electrophoresis was used first in free solution, but more recent methods frequently employ supporting media, such as polyacrylamide gels (PAGs) [10]. This matrix acts as an anticonvective medium, reducing convective transport and diffusion so that the separated components remain in sharp zones in the gel [11]. The filling of narrow-bore fused silica capillary tubes with polyacrylamide gels was introduced first by Karger [12] and Hjerten [13]. The molecular sieving rate can be controlled by the concentration of the gel [14].

THEORY

Migration Properties

The electrophoretic migration velocity of an ion in solution depends on the applied electric field, the charge and the radius of the particle, the composition and the viscosity of the liquid phase, the porosity of the gel matrix used, and the temperature [2].

In an ideal case, when there is no porous phase involved, the electrophoretic migration velocity of an M ion (v_{eM}) is:

$$v_{eM} = l/t_m = \frac{\mu_M V}{L} = \mu_M E \tag{1}$$

where l is the effective migration distance from the injection point to the detector, t_m is the migration time of the ion, μ_M is the mobility of the M ion, V is the applied voltage to the L total column length between the buffer reservoirs, and E is the applied electric field.

According to Stokes law the mobility of the solute depends on the charge (q) and the radius (r) of the hydrated particle, as well as on the viscosity (η) of the system:

$$\mu = \frac{q}{6\pi\eta r} \tag{2}$$

When the electrolyte is in a gel matrix, then the migration distance of the ion is increased from l to l' because of the tortuous passages of the pores [1]:

$$v_{eM} = \mu_M E\left(\frac{l}{l'}\right) = \mu_M E f \tag{3}$$

where $f = l/l'$ is the pore size factor of the gel used. The value of l' can be determined from the resistance (R) of the gel-filled capillary column containing V_p volume of buffer solution with k conductivity:

$$l' = RV_p k \tag{4}$$

If polyacrylamide gel is used as an anticonvective medium in the tubing, there is practically no charge on the inside surface of the capillary [3]. Therefore, the electroosmotic mobility of the M ion (μ_{oM}) is assumed to be zero, thus:

$$\mu_M = \mu_{eM} + \mu_{oM} = \mu_{eM} \qquad (\text{since } \mu_{oM} = 0) \tag{5}$$

where μ_{eM} is the electrophoretic mobility of M ion and mobilities are added vectorially.

If there is any special additive in the gel, such as a complexing agent, the solute will have a distribution between the complex and the electrolyte:

$$v_{eM} = \mu_{eM}Ef\left(\frac{1}{K_M + 1}\right) \tag{6}$$

where K_M is the complex formation constant of the M ion. Depending on the charge and size of the complexing agent, the complex may migrate faster or slower than the free solute. When the complexing agent has a charge opposite that of the solute, then an uncharged complex may be formed so that the higher the concentration of the complexing agent, the slower will be the migration velocity of the solute:

$$v_{eM} = \mu_{eM}EfR_p\left(\frac{1}{K_M + 1}\right) \tag{7}$$

where R_p is the molar ratio of the free solute [15].

Efficiency and Resolution

In capillary electrophoresis, the major contributor to band broadening, besides the injection and detection extracolumn broadening effects, is the longitudinal diffusion of the solute in the capillary tube [16]. The theoretical plate number (N) achieved is characteristic of column efficiency:

$$N = \mu_e\left(\frac{El}{2D}\right) \tag{8}$$

where D is the diffusion coefficient of the solute in the gel–buffer system.

Resolution (R_s) between two peaks can be calculated from the differences of their electrophoretic mobility ($\Delta\mu_e$) [2]:

$$R_s = 0.18\Delta\mu_e\left(\frac{El}{D\overline{\mu}_e}\right)^{1/2} \tag{9}$$

where $\overline{\mu}_e$ is the mean mobility of those species.

Equations 8 and 9 demonstrate that a higher applied electric field and a lower diffusion coefficient of the solute produce higher efficiency (N) and higher resolution (R_s). The limiting factor is mainly the heat generation rate (Q_j, joule heating) from the power ($P = VI$) applied to the narrow-bore capillary tubing with a radius of r [9]:

$$Q_j = \frac{P}{r^2IL} \tag{10}$$

Because of the sensitivity of the electrophoretic mobility to temperature, as well as the temperature dependence of the complex formation constant, good temperature control is extremely important for achieving good migration reproducibility [17].

INSTRUMENTATION

The basic instrumental configuration for capillary electrophoresis is shown in Fig. 1. It has many similarities to HPLC. A power supply is used in the range of ±30 kV. The polyimide coating of the fused silica capillary column is removed from a small section to provide on-column detection [4]. Since the polyacrylamide gel has a light transparency cutoff point of about 230 nm, UV detection is used at 260 nm for nucleotides and 280 nm for proteins. Commercially available instruments provide various on-column injection methods such as pressure, vacuum, gravity, or electrokinetic. Some instruments are capable of controlling the temperature of the capillary column, which is very important for reproducibility [9]. Another important feature is the possibility of fraction collection and field programing for collecting nanomolar amounts of biopolymer samples for subsequent microsequencing [18]. Similarly to HPLC, on-line system control and computerized data processing are available.

POLYACRYLAMIDE CAPILLARY GEL COLUMNS

Polyacrylamide gels were originally introduced as anticonvective media for zone electrophoresis [10]. The adsorption of biomolecules to the gel matrix is negligible [17]. Polyacrylamide is made up entirely of synthetic monomers and cross-linkers, such as acrylamide and *N*,*N*′-methylene-bis-acrylamide. Depending on the concentration and the ratio of the monomer and cross-linker, the pore size of the gel can be well controlled [19]. The polyacrylamide gel matrix has a molecular sieving effect in which small molecules migrate faster than larger-sized particles.

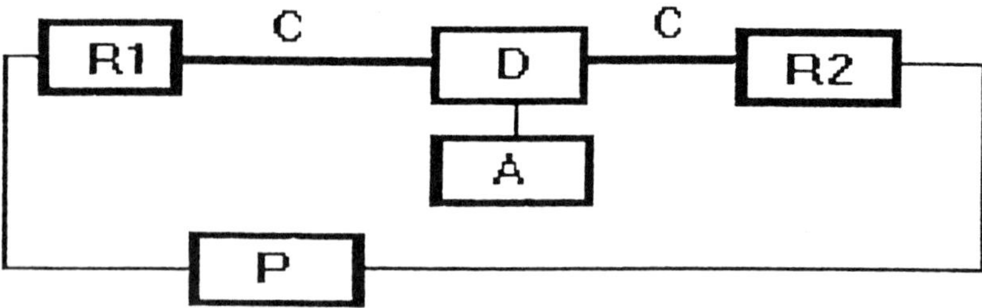

Figure 1 Diagram of the capillary electrophoresis instrument: C, fused silica capillary column; P, power supply; R1, buffer reservoir/autoinjector; R2, buffer reservoir/fraction collector; D, on-column UV detector; A, computerized data acquisition system.

Nondenaturing Polyacrylamide Gel Capillary Columns

In nondenaturing polyacrylamide gel electrophoresis, the separation is based on size and charge. Since proteins carry a net charge at any pH other than their isoelectric point, they too will migrate, and their rate of migration will depend on the charge density (the ratio of charge to mass) of the specific proteins. The application of an electric field to a protein mixture in solution, therefore, will result in different proteins migrating at different rates toward one of the electrodes. Figure 2 shows high-performance capillary gel electropherograms of two native proteins: lysozyme and β-lactoglobulin A, using the *isoelectrostatic* (constant applied electric field) separation mode. Since the direction of migration is based on the net charge of the molecule, lysozyme (pI 11.0) must be injected from the positive end of a pH 9.0 polyacrylamide gel column because it is positively charged at the pH of the column. On the other hand, the β-lactoglobulin (pI 5.1) must be injected from the negative side, because it is negatively charged at the pH of the gel-filled capillary. Otherwise their migration rate will be controlled by their charge/mass ratio using a given pore size and pH gel [20].

Molecules having the same charge/mass ratio can also be separated on nondenaturing polyacrylamide gel columns. Here, the principle of separation is purely the sieving effect of the gel matrix. For polynucleotides, the phosphate group of each nucleotide carries a strong negative charge that is much greater than any of the charges on the bases above pH 7.0 [21]. Since the charge/mass ratio of all polynucleotides is independent of the base composition, it is nearly the same for all polynucleotides. The molecular sieving effect will cause small molecules to move faster than larger ones. Figure 3 shows an electropherogram of a DNA restriction fragment mixture over a broad relative molecular mass (M_r) range (78 kDa–1.2 MDa) using the *isorheic* (constant applied current) separation mode. All of the five sample components have the same charge/mass ratio. A typical semilogarithmic plot of molecular mass versus migration time (see insert Fig. 3) is helpful for the determination of the size and M_r of an unknown DNA within the range of the plot [22]. The peaks were identified on the basis of their area, which correlates with the lengths of the fragments.

Denaturing Polyacrylamide Gel Capillary Columns

For the separation of single-stranded DNAs, chemical denaturants, such as urea, formamide, or other, must be used in the polyacrylamide gel to prevent the formation of DNA secondary structures [23]. In capillary gel electrophoresis of polynucleotides, 7–9 M urea is used as denaturing agent [24]. Figure 4 shows an isorheic separation of a polydeoxyadenylic acid mixture, $p(dA)_{5-60}$, base by base on an eCAP U100P (Beckman Instruments Inc., Palo Alto, California) prepacked gel column. This figure demonstrates the extremely high-resolving power of the method, especially in the base number range of DNA primers (20–30 mers) and DNA probes (30–80 mers).

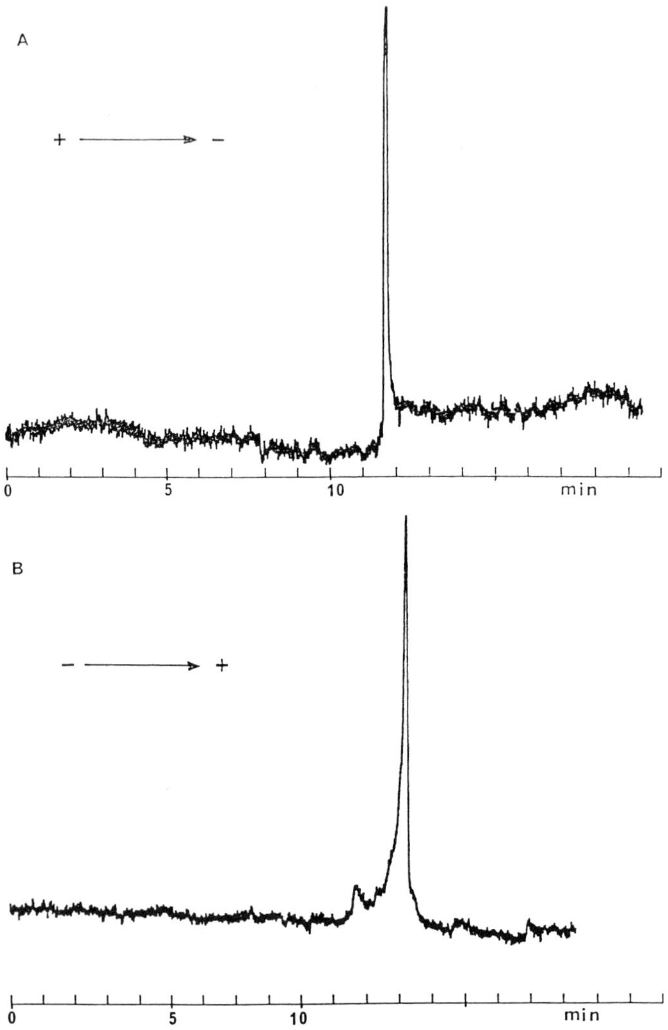

Figure 2 Nondenaturing capillary polyacrylamide gel electrophoresis of (A) lysozyme, M_r 14,300; and (B) β-lactoglobulin A, M_r 35,000. Conditions: isoelectrostatic, 600 V cm^{-1}, 25°C, effective column length, 15 cm; fused silica capillary filled with 10% polyacrylamide gel; buffer: 0.1 M Tris, 0.25 M boric acid, 5 mM diethylamine, pH 9.0. Lysozyme (pI 11.0) was injected from the + high-voltage end (5 seconds, 3 kV); β-lactoglobulin A (pI 5.1) was injected from the ground side (5 seconds, 6 kV). Monitoring at 280 nm.

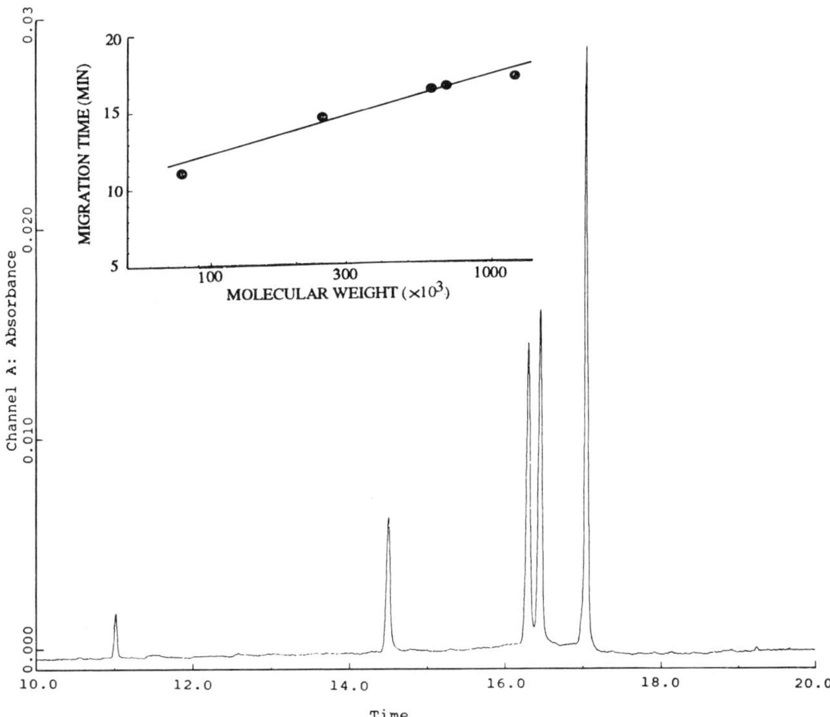

Figure 3 Capillary gel electrophoresis separation of *Bst* NI digest of pBR322 DNA on a replaceable polyacrylamide gel column. Peaks were identified by their increasing area, which correlates to the chain length: (1) 121, (2) 383, (3) 929, (4) 1060, (5) 1857 base pairs. Conditions: isorheic, 38 μA, 20°C, effective column length, 40 cm; buffer, 0.1 M Tris-borate, pH 8.3; electrokinetic injection, 3 seconds, 5 kV. Insert: size calibration curve for identification of DNA fragments. Monitoring at 254 nm by P/ACE 2000 system.

The average resolution in the 40- to 60-mer range is $R_s > 2.0$. This kind of gel column is now being used for DNA sequencing of up to several hundreds of bases using laser-induced fluorescence detection methods [25].

Another important application of denaturing capillary gel electrophoresis is the separation of proteins according to their relative molecular mass using sodium dodecyl sulfate (SDS) as the denaturing agent [26]. At neutral pH in the presence of SDS (1.4 g of SDS per gram of protein) and mercaptoethanol, most multichain proteins bind SDS, disulfide linkages are broken by the reducing agent, secondary structure is lost, and the SDS–protein complex is assumed to be a random coil configuration. Proteins treated in this way have uniform shape and identical charge/mass ratios, so separation is based on molecular sieving. Because of the association–

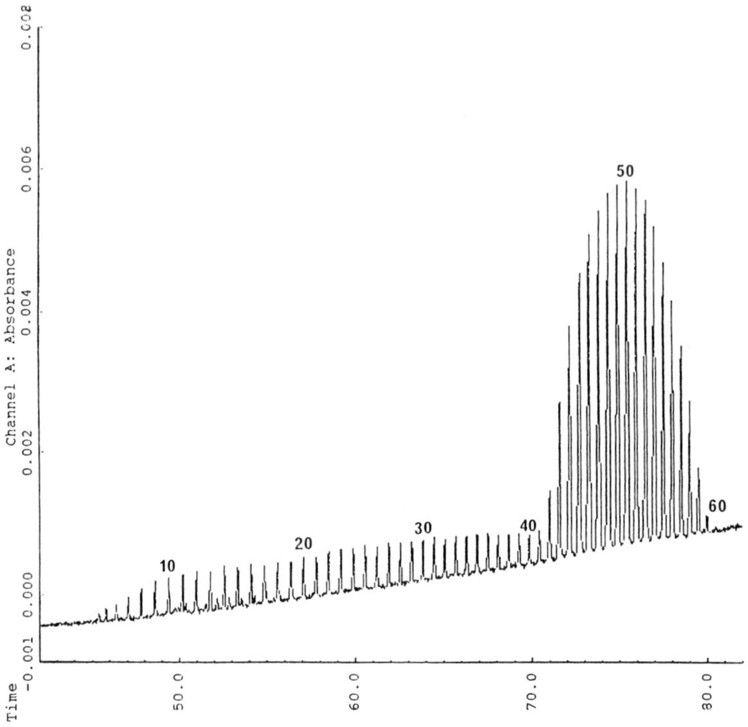

Figure 4 High-resolution capillary gel electrophoresis of polydeoxyadenylic acid mixture, $p(dA)_{5-60}$, on an eCAP U100P prepacked capillary column. Conditions: isorheic, 4 μA, 20°C, effective column length, 100 cm; buffer, 0.1 M Tris-borate, 7 M urea, pH 8.8; monitoring at 254 nm by P/ACE 2000 system.

dissociation phenomenon of the SDS–protein complex; the gel–buffer system must contain a sufficient amount of SDS (usually 0.1%) to stabilize the SDS–protein complex. Figure 5 shows an isorheic separation of a prestained protein mixture by SDS–polyacrylamide capillary gel electrophoresis over the M_r range of 16–110 kDa, using a multiphasic buffer system [27,28] to achieve sharper peaks. The insert of Fig. 5 shows that the plot of the migration time of the sample components versus log M_r gives a straight line. Hence, if a protein of unknown M_r is analyzed by being electrophoresed together with the standards, then the M_r of the unknown can be calculated to an accuracy ranging between 5 and 10% [29]. A similar linear relation was found by Ferguson [30] for proteins, by varying the monomer concentration of the polyacrylamide gels. By means of Ferguson plots, it can be demonstrated that a pure size separation is achieved [12,31].

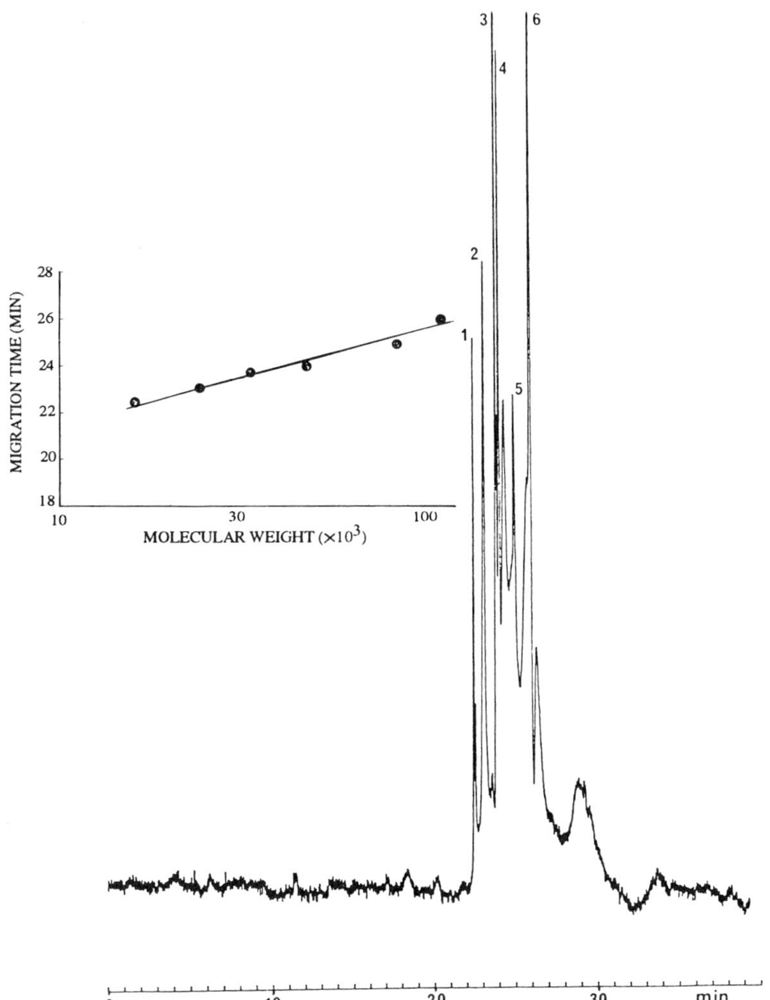

Figure 5 SDS–PAGE separation of prestained M_r markers on a 12% polyacrylamide capillary gel column. (1) lysozyme, (2) soybean trypsin inhibitor, (3) carbonic anhydrase, (4) ovalbumin, (5) bovine serum albumin, (6) phosphorylase B. Conditions: isorheic, 30 μA, 20°C, effective column length, 15 cm; gel-buffer, 25 mM Tris, 192 mM glycine, 0.1% SDS, pH 8.3; running-buffer: 0.375 M Tris-HCl, 0.1% SDS, pH 8.8; monitoring at 280 nm.

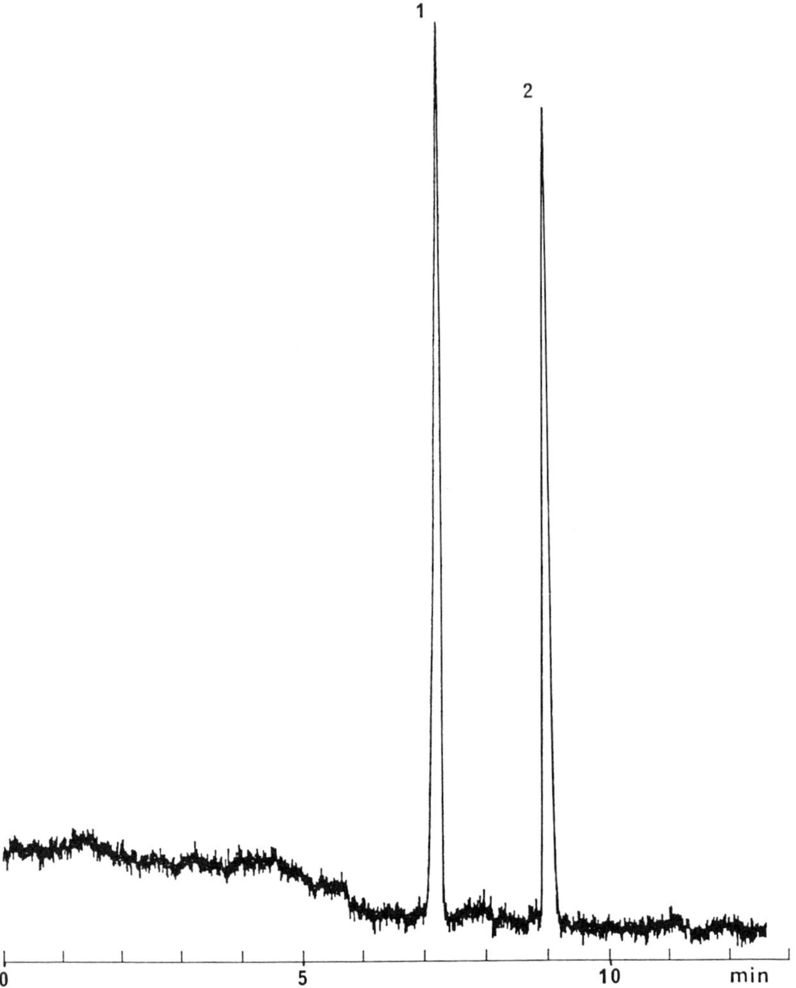

Figure 6 Separation of barbiturates by β-cyclodextrin/10% polyacrylamide capillary gel column. (1) Barbital, (2) buthetal. Conditions: isoelectrostatic, 500 V cm^{-1}, 25°C, effective column length, 15 cm; buffer: 0.1 Tris, 0.25 boric acid, 7 urea, 75 mM β-cyclodextrin, 10% MeOH; monitoring at 254 nm.

Polyacrylamide Capillary Gel Electrophoresis in the Presence of Complexing Agents

Capillary polyacrylamide gel columns can also be used with complexing agents to achieve special selectivities. As in affinity electrophoresis [32], the additive can be either covalently bound to the gel matrix or just incorporated into the polymeric fiber gel matrix [15]. The latter method is always preferable, since it requires no special chemistry. Under an applied electric field, the complex will migrate according to its overall charge in the gel when the complexing agent is not bound to, or entrapped in, the matrix. Thus, it can slow down or speed up the separation procedure, depending on the complex formation constant and the electrokinetic behavior of the complex (see Eq. 7).

As an example of selectivity manipulation, cyclodextrins (CD) can be used as complexing agents in polyacrylamide capillary gel electrophoresis [33]. The CDs are nonionic cyclic polysaccharides of glucose, with the shape of a toroid or hollow truncated cone. The cavity is relatively hydrophobic, whereas the external faces are hydrophilic. The torus of the larger circumference contains chiral secondary hydroxyl groups [34]. By incorporation of β-cyclodextrin into the gel matrix, the complexing agent is practically immobilized (since the CD has no charge), particularly when the pore size of the gel is smaller than the size of the β-CD (<14 nm). By using the special selectivity of CDs, closely related species can be separated. Figure 6 shows an isoelectrostatic separation of two barbiturates using a CD additive in a capillary polyacrylamide gel column. Here, the hydrophobic interaction between the sample components and the cavity of the CD additive was used to enhance the separation achieved purely on the basis of electrophoretic mobility differences. Cyclodextrins can be used for chiral separations, as well, since chiral selectivity can arise from the entrance of the cavity with the chiral glucose moiety. Figure 7 shows an isoelectrostatic chiral separation of a mixture of dansylated amino acids, based on the chiral recognition of the cyclodextrin incorporated into the polyacrylamide gel matrix.

As an another example to the approach of using complexing agents, high resolution of DNA restriction fragments by capillary gel affinity electrophoresis has been achieved by adding a soluble intercalating agent, ethidium bromide, to the gel–buffer system [15]. Ethidium bromide has been used previously in affinity chromatography [35], in affinity electrophoresis [36], and in capillary electrophoresis [37] as a selective intercalating agent for double-stranded DNA molecules to increase selectivity. Since the ethidium bromide is positively charged, it causes reduction in electrophoretic mobility of DNA when intercalating into the two strands (see Eq. 7). Figure 8 compares the separations of the ΦX174 DNA *Hae* III digest restriction fragments mixture on replaceable polyacrylamide gel-filled capillaries with and without ethidium bromide. The peaks were identified similarly to those in Fig. 3. We used a high M_r range replaceable polyacrylamide gel–buffer (optimized above 1-kb

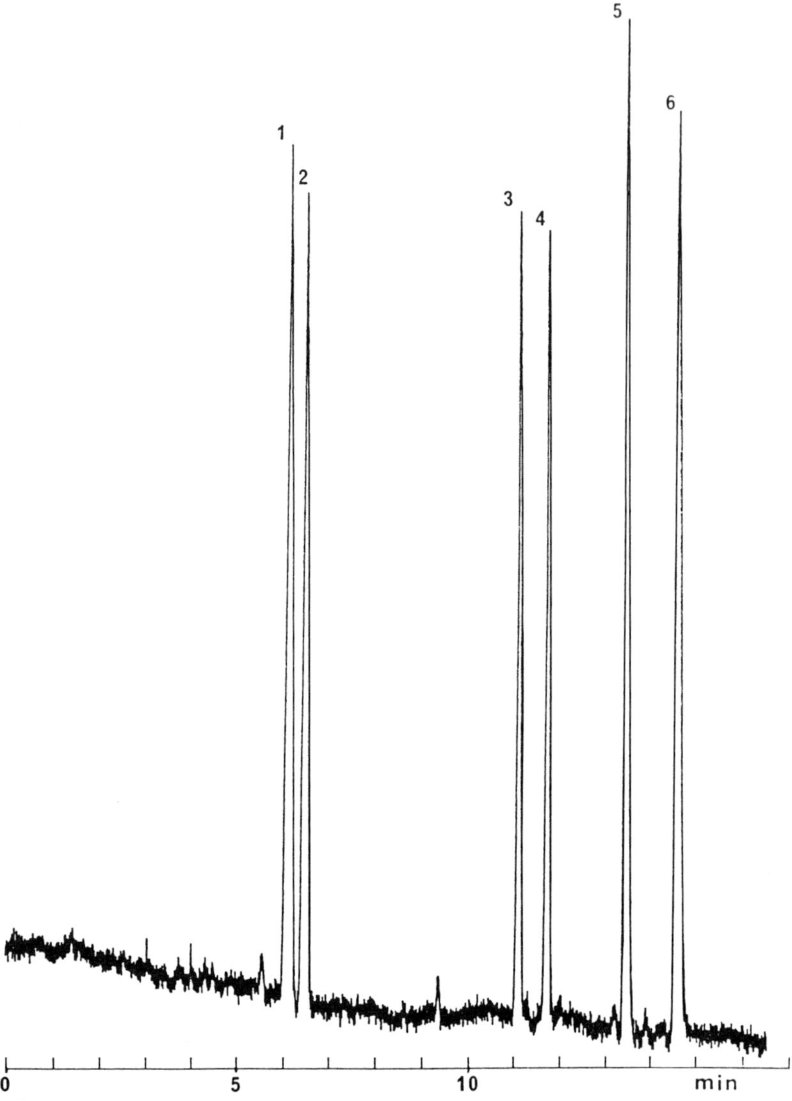

Figure 7 Chiral separation of dansylated (Dns) DL-amino acids by β-cyclodextrin polyacrylamide gel electrophoresis. (1) Dns-L-Glu, (2) Dns-D-Glu, (3) Dns-L-Ser, (4) Dns-D-Ser, (5) Dns-L-Leu, (6) Dns-D-Leu. All the conditions are the same as in Fig. 6.

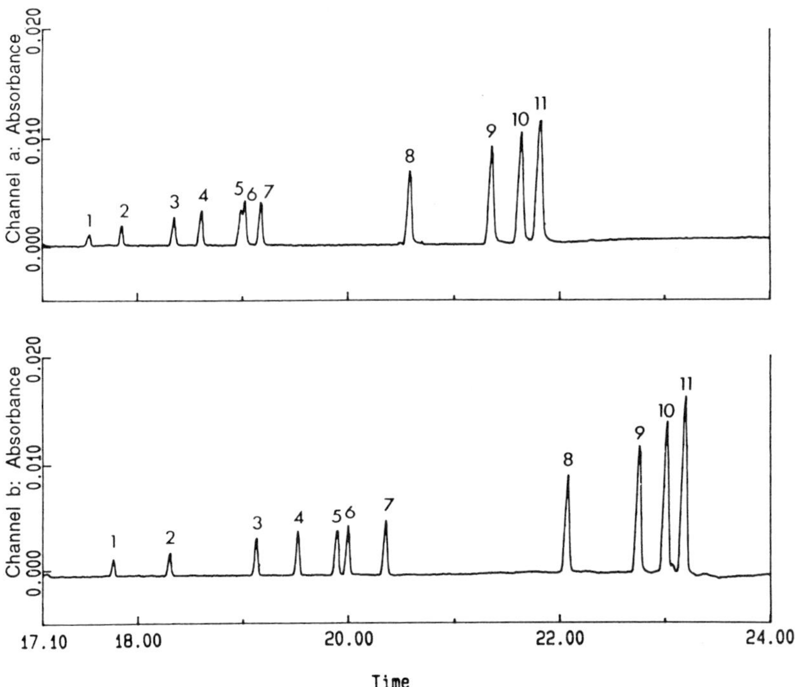

Figure 8 Capillary gel electrophoretic separation of ΦX174 DNA *Hae* III digest restriction fragment mixture in the absence (a) and in the presence (b) of ethidium bromide (1 μg mL^{-1}) in the gel–buffer system. Peaks were identified by their increasing area, which correlates with the chain length: (1) 72; (2) 118; (3) 194; (4) 234; (5) 271; (6) 281; (7) 310; (8) 603; (9) 872; (10) 1078; (11) 1353 base pairs (bp). Conditions: linear gradient voltage separation mode, 0–400 V cm^{-1} in 30 minutes. Replaceable polyacrylamide gel column: effective length, 30 cm, total length 37 cm; buffer, 0.1 M tris-borate, 2 mM EDTA, pH 8.5; injection, 0.25 mW.

pair) in which there is no separation of 10-bp difference below the 1-kb pair (see peaks 5 and 6, Fig. 8a). Adding only 1 μg mL^{-1} ethidium bromide (complexing agent), the same gel–buffer system can be used in a wider M_r range, as shown in Fig. 8b, and here the 271/281-bp peaks are separated. To achieve better separation the P/ACE System 2000 was used in gradient voltage separation mode [38].

CONCLUSION

As this chapter demonstrates, capillary gel electrophoresis is rapidly becoming an important separation tool in analytical biochemistry and molecular biology. By using narrow-bore fused silica capillary columns, filled with polyacrylamide gels, and high voltages, separations of DNA and protein molecules up to the megadalton range, with extremely good resolutions, are possible in minutes. As an instrumental approach to electrophoresis, the method offers the capability of on-column detection, micropreparative operation [39], and automation with on-line data processing. Finally, capillary electrophoresis is not a tool to replace other separation methods such as HPLC, slab gel electrophoresis, or ultracentrifugation; rather, it is a complementary technique.

ACKNOWLEDGMENT

The author expresses his gratitude to Professor Barry L. Karger for stimulating discussions. The help of Phyllis Browning in the preparation of the manuscript is also highly appreciated.

REFERENCES

1. H. Kunkel and H. Tiselius, *J. Gen. Physiol.*, *35*:89 (1951).
2. B. L. Karger, A. S. Cohen, and A. Guttman, *J. Chromatogr.*, *492*:585 (1989).
3. F. E. P. Mikkers, F. M. Everaerts, and T. P. E. M. Verheggen, *J. Chromatogr.*, *19*:1 (1979).
4. J. W. Jorgenson and K. D. Lukacs, *Science*, *222*:266 (1983).
5. S. Hjerten, *J. Chromatogr.*, *270*:6 (1983).
6. R. A. Wallingford and A. G. Ewing, *Anal. Chem.*, *58*:258 (1988).
7. M. J. Gordon, X. Hung, S. L. Pentoney, Jr., and R. N. Zare, *Science*, *242*:224 (1988).
8. A. Guttman, A. S. Cohen, D. N. Heiger, and B. L. Krager, *Anal. Chem.*, *62*:137 (1990).
9. R. J. Nelson, A. Paulus, A. S. Cohen, A. Guttman, and B. L. Karger, *J. Chromatogr.*, *480*:111 (1989).
10. S. Raymond and L. Weintraub, *Science*, *130*:711 (1959).
11. B. D. Hames and D. Rickwood, *Gel Electrophoresis of Proteins*, IRL Press, Oxford (1981).
12. A. S. Cohen and B. L. Karger, *J. Chromatogr.*, *397*:409 (1987).
13. S. Hjerten, K. Elenbring, F. Kilar, J. Liao, A. J. Chen, C. J. Sibert, and M. D. Zhu, *J. Chromatogr.*, *403*:47 (1987).
14. S. Hjerten, *Chromatogr. Rev.*, 9:122 (1962).
15. A. Guttman and N. Cooke, *Anal. Chem.* *63*:2038 (1991).
16. S. Terabe, K. Otsuka, and T. Ando, *Anal. Chem.*, *61*:251 (1989).
17. R. J. Wieme, *Chromatography: A Laboratory Handbook of Chromatography and Electrophoretic Methods*, Van Norland, New York (1975).
18. A. Guttman, A. Paulus, A. S. Cohen, B. L. Karger, H. Rodriguez, and W. S. Hancock, *Electrophoresis '88* (C. Shafer-Nielsen, ed.), New York, p. 151 (1988).
19. A. Chrambach, *The Practice of Quantitative Gel Electrophoresis*, VCH Publishers, Deerfield Beach, Fl. (1985).
20. P. G. Righetti, *Isoelectric Focusing Theory, Methodology and Applications*, Elsevier, New York (1983).

21. A. T. Andrews, *Electrophoresis*, 2nd ed., Claredon Press, Oxford, (1986).
22. D. N. Heiger, A. S. Cohen, and B. L. Karger, *J. Chromatogr.*, *516*:33 (1990).
23. J. Sambrook, E. F. Fritsch, and T. Maniatis, *Molecular Cloning*, Cold Spring Harbor Press, Cold Spring Harbor, N.Y. (1989).
24. A. S. Cohen, N. T. Najarian, A. Paulus, A. Guttman, J. A. Smith and B. L. Karger, *Proc. Natl. Acad. Sci. USA*, *85*:9660 (1989).
25. S. L. Pentoney and K. D. Konrad, "Genome Sequencing Conference II," Hilton Head Island, S.C., Sept. 30–Oct. 3, p. 3 (1990).
26. K. Weber and M. Osborn, *J. Biol. Chem.*, *244*:4406 (1969).
27. L. Orstein, *Ann. N. Y. Acad. Sci.*, *121*:321 (1964).
28. B. J. Davis, *Ann. N. Y. Acad. Sci.*, *121*:404 (1964).
29. D. Freifelder, *Physical Biochemistry*, W. H. Freeman & Co., New York (1982).
30. K. A. Ferguson, *Metabolism*, *13*:21 (1964).
31. K. Ganzler, A. S. Cohen, A. Guttman, and B. L. Karger, *Anal. Chem.*, *64*:2665 (1992).
32. V. Horejsi and M. Ticha, *J. Chromatogr.*, *216*:43 (1981).
33. A. Guttman, A. Paulus, A. S. Cohen, N. Grinberg, and B. L. Karger, *J. Chromatogr.*, *448*:41 (1988).
34. J. Szejtli, *Cyclodextrins and Their Inclusion Complexes*, Akadémiai Kiadó, Budapest (1982).
35. A. Vacek and D. P. Bourque, *Anal. Biochem.*, *124*:414 (1982).
36. D. H. Flint and R. E. Harrington, *Biochemistry*, *11*:4858 (1972).
37. H. E. Schwartz, K. Ulfelder, F. J. Sunzeri, M. P. Busch, and R. G. Brownlee, *J. Chromatogr.*, *559*:267 (1991).
38. A. Guttman, B. Wanders and N. Cooke, *Anal. Chem.*, *64*:2348 (1992).
39. A. Guttman and I. Mazsaroff, In *New Approaches in Liquid Chromatography '91*, (H. Kalasz and L. Ettre, eds.), Akadémiai Kiadó, Budapest, (in press).

27

Use of Cyclodextrins in Capillary Electrophoresis

Salvatore Fanali

Consiglio Nazionale delle Ricerche
Monterotondo Scalo, Italy

Cyclodextrins (CD) are cyclic oligomers of several D-(+)-glucopyranose units. Although cyclodextrins containing 6–12 D-(+)-glucopyranose units are known, 6, 7, and 8 unit-formed molecules, named α-, β-, and γ-CDs, are in frequent use [1]. Their shape is similar to a truncated cone, with a relatively hydrophobic cavity, able to include several types of compounds (e.g., organic, inorganic, other). The mechanism involved in this process is called host–guest complexation or inclusion complexation. Chemical compounds, with higher hydrophobicity than the solvent molecule in the cavity, easily replace the latter.

Cyclodextrins have been applied successfully in analytical chemistry, for example, in the separation of several isomers; positional, structural and enantiomer isomers have been separated by gas chromatography (GC), thin layer chromatography (TLC), high-performance liquid chromatography (HPLC), isotachophoresis (ITP), micellar electrokinetic chromatography (MECC), and capillary zone electrophoresis (CZE) [1–4].

The aim of the present chapter is to discuss the use of cyclodextrins in capillary electrophoresis as selective agents for analytical separation of isomers. A brief consideration of the structure and main properties is also given to explain the mechanism involved in the analysis.

STRUCTURE AND PROPERTIES OF CYCLODEXTRINS

Cyclodextrins are natural, neutral oligosaccharides produced by enzymatic reaction on starch. The source of the enzyme used for the preparation of these compounds is a bacillus that Schardinger called *Bacillus macerans* [2]. Other bacteria containing the enzyme used for producing CDs have been studied. Even though other names have been suggested to indicate the different CDs, the most common designation is that using the Greek alphabet. Thus, α-, β-, γ-, and δ-CDs, with 6, 7, 8, and 9 glucose units in the ring, respectively, are those most frequently used [2,4].

Their shape is similar to a truncated cone, with a relatively hydrophobic cavity, and two openings of different sizes [4] that are relatively hydrophilic owing to the presence of hydroxyl groups (primary and secondary). As an example Fig. 1 illustrates the shape of β-CD and one of its derivatives.

Table 1 gives the main properties of the cyclodextrins most commonly used in analytical chemistry.

From the data shown in Table 1, we can note that β-CD possesses the lowest solubility in water (0.02 M), which can be a problem, especially when the CD is used as an additive in solution [e.g., in the mobile phase (HPLC), or in the background electrolyte (BGE) (CZE)]. However, the solubility of β-CD can be increased by dissolving it in an appropriate solvent mixture; for example water/ethanol show the maximum solubility for β-CD at a ratio of 70:30 (v/v) [1,5]. Furthermore, Pharr et al. [6] have demonstrated that when β-CD is dissolved in water/urea solutions of 4 M and 8 M, it was possible to prepare 0.089 M and 0.226

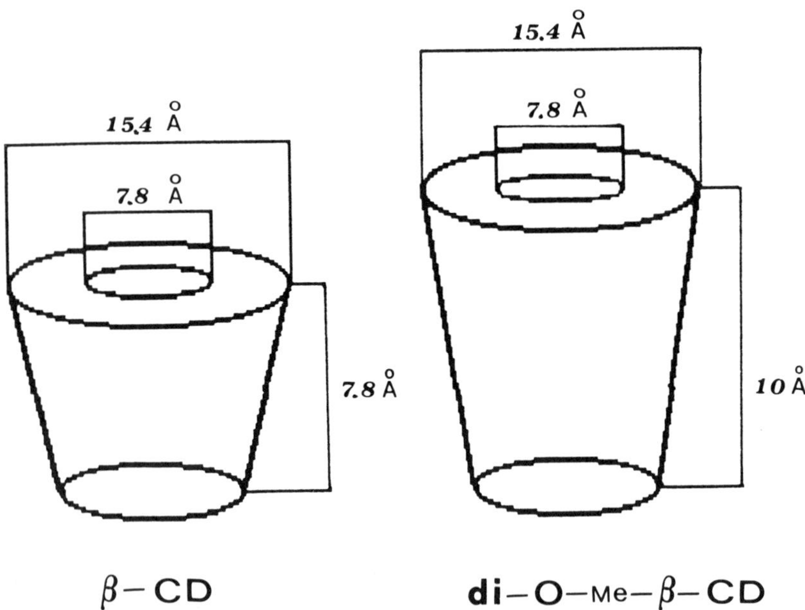

Figure 1 The shape of β-cyclodextrin compared with that of its dimethylated derivative.

Table 1 Properties of Cyclodextrins

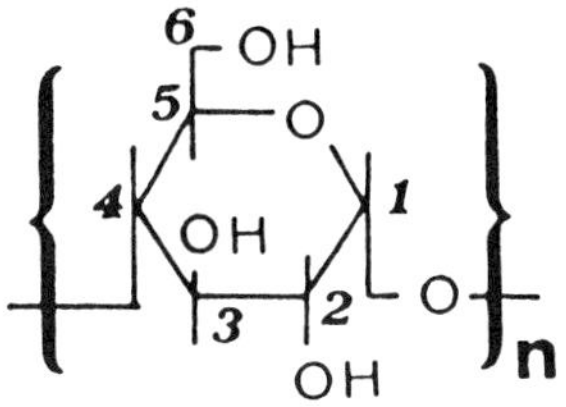

Cyclodextrins	α-CD	β-CD	di-OMe-β[a]	tri-OMe-β[b]	γ-CD
Number of D(+)glucopyranose units (n)	6	7	7	7	8
Molecular weight	973	1135	1331	1430	1297
Internal diameter (A)	5.7	7.8	7.8	7.8	9.5
Depth (A)	7.8	7.8	10–11[c]	—	7.8
Solubility water g/100 mL, 25°C	14.5	1.85	high	17	23.2

[a]Heptakis-2,6-di-*O*-methyl-β-CD.
[b]Heptakis-2,3,6-tri-*O*-methyl-β-CD.
Source: Ref. 4; [c]Ref. 7.

M of β-CD, respectively. Interestingly, the use of urea solutions also increases the solubility of γ-CD, but not that of α-CD.

MODIFIED CYCLODEXTRINS

The hydroxyl groups at positions 2, 3, and 6 of each glucose can react with electrophilic compounds (e.g., dimethylsulfate, acylchlorides, tosylchlorides, α-halogen carboxylic acids) to give stable modified cyclodextrins that can, in turn, undergo subsequent modifications. Cyclodextrans with such modifications have been used successfully in analytical chemistry (e.g., dodecakis-2,6-*O*-dimethyl-α-CD, heptakis-2,3,6-tri-*O*-methyl-β-CD) [for a description of CD derivatives see Refs. 1,4,7].

When CD derivatives are synthesized, the main problem arises from the amount of substitution. In fact, several compounds are prepared with modification of one or more hydroxyl groups. Thus, in this synthesis it is necessary to carefully select the preparation method and to purify the obtained compound. The modified CD can exhibit properties different from those of the parent CD. Accordingly, 2,6-di-*O*-methyl-β-CD possesses a deeper cavity and higher solubility in water than the parent β-CD [1,8,9]. When CDs are modified with charged groups (e.g., sulfate, phosphate, amino groups), the solubility is improved and the p*K* values are different.

When the cyclodextrins are modified, several advantages can be expected both for higher solubility and for stabilizing the inclusion complexes formed [4]. Furthermore, the use of charged CDs is important in the separation of neutral compounds by capillary electrophoresis. Here, the uncharged compounds are moved toward the detector by electroosmotic flow and the selectivity is improved by the introduction of a modified CD [10].

ANALYTICAL CONTROL OF CYCLODEXTRINS

Cyclodextrins are prepared by enzymatic reaction on starch. During the reaction a mixture of CDs is obtained that needs ultimate separation and purification.

Since CDs are widely used in different application fields (e.g., enzymic reactions [8], analytical chemistry [5,11,12], and pharmaceutical science [13]), their purity has to be carefully controlled to optimize the results.

Considering that several pharmaceutical formulations contain CDs as solubilizing and carrier agents [14], toxicological problems can arise if the appropriate amount and type of complexing agent is not used. Therefore, analytical methods for checking the synthetic steps and the final products are needed.

Several analytical methods have been used for the analysis of CDs, namely, paper or thin-layer chromatography, gas chromatography, photometry [1]; high-performance liquid chromatography [15]; and capillary zone electrophoresis [16].

As an example, Fig. 2 shows a separation of α-, β-, and γ-cyclodextrins with a UV indirect detection method, by using capillary zone electrophoresis. Cyclodextrins, as neutral compounds, are moved toward the detector by electroosmotic flow, and the selectivity of the separation is improved by using an anion, moving in the opposite direction of the CD, that forms inclusion complexes with the compounds to be analyzed. In this way, the CDs will have different charges and will be retarded selectively. The anion in the BGE, as well as having the characteristic to be complexed, must absorb at a certain wavelength to allow the indirect detection of the cyclodextrins (compounds with poor absorbing power in the commonly used UV visible wavelength).

Figure 3 shows the scheme for the separation mechanism of cyclodextrins in CZE.

MECHANISM OF INCLUSION COMPLEXATION

Because of their structure, cyclodextrins interact with chemical compounds mainly by an inclusion complexation mechanism, even if adsorption with the substituent groups at the cavity openings has been demonstrated in HPLC when organic solvents are used [17].

A wide variety of both organic and inorganic compounds, charged and uncharged, the size and type of which can range from small to relatively large molecules,

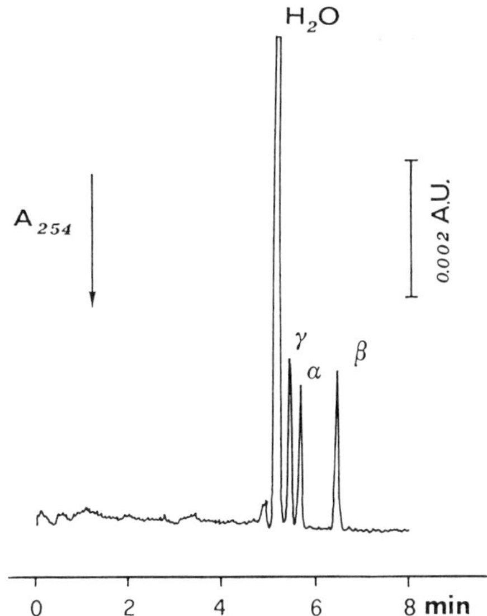

Figure 2 Electrophoretic separation of α-, β-, and γ-cyclodextrins. Background electrolyte, 30 mM benzoate/Tris-(hydroxymethyl)aminomethane pH 6.2 (indirect UV 254-nm detection); capillary, fused silica 50 cm × 75-μm id (36 cm to the detector); applied voltage, 20 kV, 7 μA; sampling siphoning 5 seconds 10 cm of a mixture 10^{-3} M for each CD. (From Ref. 16)

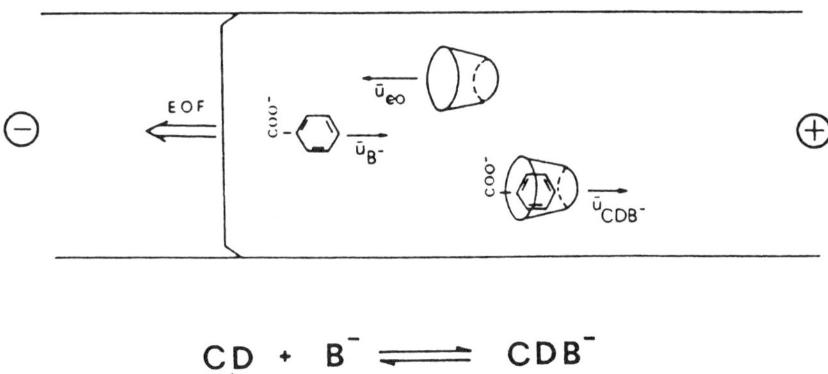

Figure 3 The separation mechanism of CDs by using indirect UV detection in capillary zone electrophoresis. (From Ref. 16)

such as methanol and fatty acyl-coenzyme A compounds [1,4], can be selectively included into the cavity of the CDs.

To explain the mechanism of inclusion complexation let us consider that, after dissolution of a CD in water, its cavity will "host" the relatively polar molecule of water called "guest."

As the CD cavity is relatively hydrophobic and the surface is hydrophilic, the water is in an energetically unfavorable position and, thus, can easily be replaced by other compounds of a different nature (e.g., aromatic) that will fit better in the cavity by forming nonpolar–nonpolar interactions with the cavity [4].

If the guest compound possesses not only a hydrophobic, but also a hydrophilic, group, the latter will remain outside the cavity and will thereby stabilize the inclusion complex by bonding the hydroxyls or their modified groups on the openings of the CD.

The degree of complexation is influenced by several variables, such as size and nature of the substituent groups of both the guest and the CD used; for example, naphthalene (a compound with two condensed aromatic rings) can form inclusion complexes with β- and α-CDs, but anthracene (three aromatic rings) only with γ-CD; biphenyl (two aromatic rings with a different shape than the previously cited compounds) can be included by α-, β-, and γ-CD [4,18].

Generally, the ratio CD/guest is 1:1, although the ratios 2:1 and 1:2 also have been reported [1]. The 2:1 complex of CD and a guest compound can be obtained when the latter is too large to fit the cavity; consequently, only a part of its molecule will be included, and the remaining one possesses the properties to form inclusion complexes [4].

Inclusion complex formation is a dynamic equilibrium that can be used selectively for analytical applications. Each equilibrium is characterized by a constant formation equilibrium $K_f(M^{-1})$ that can be different for each analyzed compound; for example, benzoic acid shows a $K_f = 700$ and 125 when it interacts with α-CD and β-CD, respectively, and the constant of *m*-hydroxybenzoic acid with α-CD is 172. Another important example is the data for enantiomeric compounds, such as (+)- and (−)-sarin, L- and D-phenylalanine /K_f of sarin (+) = 166, (−) = 25 at pH 9, and for phenylalanine L = 15.8, D = 20.6 at pH 11 when α-CD was used [4]. The latter example represents a very important field of application in analytical chemistry, because enantiomer resolution is very difficult owing to similar properties of this class of compounds.

From the foregoing data as an example, we can forecast the type of CD to be used for our analytical separation. The chosen CD can selectively influence those parameters involved in the separation mechanism of our analytical method (e.g., in capillary electrophoresis the effective mobility of an analyte can be modified by selective complex-forming equilibria [19]).

THE ROLE OF CYCLODEXTRINS IN CAPILLARY ELECTROPHORESIS

The use of cyclodextrins in analytical chemistry is well known. In fact the inclusion complexation mechanism has been used successfully for the separation of several classes of compounds, such as positional, geometric, and optical isomers. The main effort was made in HPLC, and CDs have been used either bonded to the stationary phases or dissolved in the mobile phase [20,21]. A wide number of applications are available also in TLC and GC [22,23].

Capillary electrophoresis is a modern electrophoretic technique with a great potential for the separation of both charged and uncharged compounds. Its power comes from the high electric field that can be applied to the background electrolyte (BGE) used for the electrophoretic runs. The separation is performed in a capillary (usually fused silica) with a relatively small internal diameter (10–100 μm) that allows efficient dispersal of the joule heat generated during the electrophoretic process, thus permitting the use of relatively high voltages (10–30 kV). Such voltages are necessary to obtain fast separations with high efficiency and high resolution, which are the most important characteristics of this technique for analytical use.

Analytes can be separated in capillary electrophoresis if they possess sufficient differences in effective electrophoretic mobility. Consequently, compounds with similar physicochemical properties (e.g., enantiomers) show the same mobility and cannot be separated by this technique unless an additional mechanism that is able to change the mobility is introduced.

Complex-forming equilibrium is well known in capillary electrophoresis; for example, additions to the BGE complexing agents, such as the cupric ion, with a chiral selector for the resolution of dansylated amino acids [24]; metal ions [Cu(II) and Zn(II)] for the resolution of peptides [25]; or hydroxyisobutyric acid for the separation of inorganic cations [26–28]. Considering the wide complexing ability of CDs toward several classes of compounds, it is evident that they can be used successfully in capillary electrophoretic separations. They can be easily employed with these techniques because they can be added directly to the electrolyte and, being neutral compounds, they will not change the conductivity of the background; the viscosity is the only parameter that will be modified.

ELECTROPHORETIC SEPARATIONS

Enantiomers

The separation of racemic enantiomeric mixtures is an interesting field of application in analytical chemistry, especially in pharmaceutical analysis, for which very often the same compound can show a different physiological effect if administered as (+)- or (−)-isomer. Consequently, rapid, sensitive, and selective analytical methods are required for pharmacokinetic studies, purity control of drugs, or other assays.

The separation mechanism of enantiomers is based on the formation of the correspondent diastereoisomers, which can be easily separated because they possess different physicochemical properties.

Two different procedures have been used for enantiomer resolution: the indirect and the direct resolution methods [29]. In the indirect type, the racemic mixture interacts with a chiral compound, producing a mixture of two diastereoisomers that will be separated by capillary electrophoresis [30] or other techniques [29]. The interactions involved in this separation method are relatively strong; in fact, covalent bonds are formed between enantiomers and the chiral reagent. In the direct resolution method, widely used in electrophoretic techniques, the racemic mixture is injected directly into the capillary containing a chiral environment that will interact reversibly by forming diastereoisomers during the electrophoretic run. Here, relatively weak bonds are involved in the separation procedure (e.g., dipole–dipole, hydrogen bonding, or hydrophobic interactions).

The resolution process can be described by the following equation:

$$\underbrace{[(R)\text{–E} + (S)\text{–E}]}_{\text{racemic mixture}} + \underbrace{\begin{matrix}(R)\text{–C}\\ \text{or}\\ (S)\text{–C}\end{matrix}}_{\text{chiral environment}} \rightleftharpoons \underbrace{\begin{matrix}[(R)\text{–E—}(R)\text{–C}] + [(S)\text{–E—}(R)\text{–C}]\\ \text{or}\\ [(R)\text{–E—}(S)\text{–C}] + [(S)\text{–E—}(S)\text{–C}]\end{matrix}}_{\text{diastereoisomers}}$$

Cyclodextrins are dextrorotatory, owing to the chiral D-(+)-glucopyranose units present in their structure and, thus, being chiral compounds, they can be used as a chiral environment for enantiomers resolution. The two enantiomers can form inclusion complexes with CD and, if the association constants are sufficiently different, they will move toward the detector with different velocities and thus their separation is possible by using electrophoretic techniques. The CDs usually are added to the BGE or to the LE even if they have been incorporated into the gel [31].

A very interesting study [32] of ephedrine alkaloid enantiomers' resolution by using capillary isotachophoresis was carried out in which β-CD and two of its derivatives, 2,6-di-*O*-methyl-β-CD and 2,3,6-tri-*O*-methyl-β-CD, were added separately to the leading electrolyte at pH 5.48 to resolve the enantiomers of the investigated compounds. Increasing the amount of either β-CD or its derivatives caused a general increase of the relative step heights $(h_i)_{rel}$ for all the analyzed compounds, showing an increasing complexing effect. The use of β-CD allowed the enantiomeric resolution of racemic norpseudoephedrine (a), *p*-hydroxynorpseudoephedrine (b), pseudoephedrine (c), and *o*-acetylpseudoephedrine (d), but not ephedrine (e), norephedrine (f), or *p*-hydroxynorephedrine (g). When 2,6-di-*O*-methyl-β-CD was used, a higher complexing effect than that with β-CD was observed, but also this chiral compound was a good enantioselective agent for compounds (a), (b), (c), and (d), and it also gave a slight indication for chiral resolution of compounds (e), (f), and (g). Comparative studies at higher pH did not give satisfactory results. The authors confirmed the outstanding role of a weak hydrogen bond between analytes and CDs. Furthermore, the selectivity is influenced by the size and hydrophobicity of the cavity collar of CD.

The possibility for resolving racemic sympathomimetic compounds, namely ephedrine (E), norephedrine (n-E), epinephrine (Ep), norepinephrine (n-Ep), and isoproterenol (i-P) was studied [33] by using CZE with a phosphate buffer, at pH 2.4, containing β-CD, 2,6-di-*O*-methyl-β-CD, or 2,3,6-tri-*O*-methyl-β-CD. The use of the three CDs caused a general decrease in the effective mobility for all the analyzed compounds. The type of CD, the hydrophobicity of the collar cavity, the presence of the hydroxyl groups on the opening, and the shape of the analyzed compounds played an important role in the enantiomeric resolution of the studied alkaloids. In fact, β-CD poorly resolved only ephedrine and isoproterenol at a relatively high concentration (20 mM), and 2,3,6-tri-*O*-methyl-β-CD did not resolve any enantiomer. The complete resolution of the five analyzed compounds was obtained with dimethylated CD when the buffer was supported with the chiral additive (18 mM).

α-Cyclodextrin was selected as a chiral additive to the BGE for the separation of DL-tryptophan enantiomers by using capillary zone electrophoresis [34]. The electrophoretic runs were carried out in a coated fused silica tube 20 cm × 25-μm id, containing 0.1 M phosphate buffer, at pH 2.5, and the appropriate amount of α-CD. At the operating conditions, tryptophan was moving as a cation in a relatively short time (less than 8 minutes), and the resolution of the two enantiomers was obtained when 10 mM of α-CD was added to the BGE. Increasing the amount of the chiral additive increased both migration times and resolution of the two isomers. Figure 4 shows the baseline resolution of racemic tryptophan $R > 2$ when the BGE contained 0.040 M of α-CD. The use of a different counterion showed that it is possible to optimize the resolution of a racemic mixture. In fact, the best resolution of racemic tryptophan was obtained when ClO_4^- was selected. As a continuation of this study, several tryptophan derivatives were studied by capillary zone electrophoresis [35] with the aim of verifying those variables affecting the resolution of enantiomers when cyclodextrins were added to the BGE. To illustrate the shape of the analyzed compounds, Fig. 5 shows the chemical structure of tryptophan.

The studied racemic mixtures (1-, 4-, 5-, and 7-methyltryptophan; 5-hydroxytryptophan; tryptophan-methyl-, ethyl-, butyl-, and octyl-ester) were run in a coated capillary, 20 cm × 25-μm id, containing a phosphate buffer solution, at pH 2.5, and the appropriate amount of CD. The use of α-CD in the BGE caused a general increase in the migration time for all the analyzed compounds owing to the ability of α-CD to include them selectively into its cavity. Complete resolution was obtained for compounds with methyl-substitution on the indole ring at positions 6, 1, and

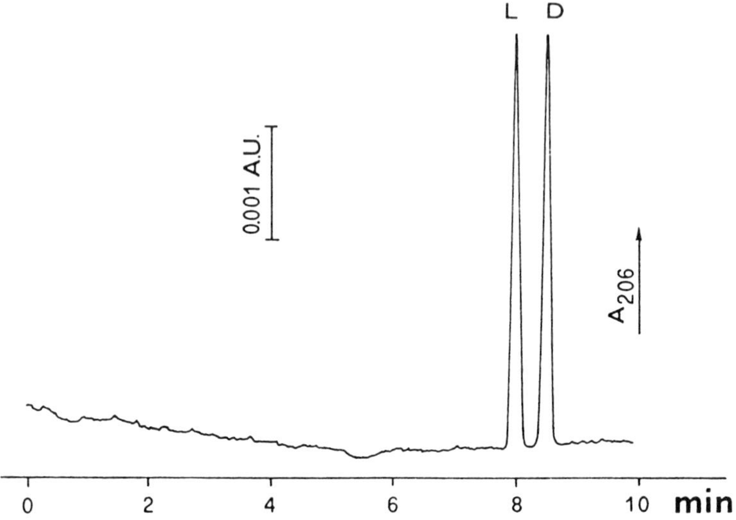

Figure 4 Resolution of a racemic mixture of tryptophan enantiomers. Apparatus Bio-Rad HPE 100, coated fused silica capillary 20 cm × 25-μm id; background electrolyte, 0.1 M phosphate/perchloric acid pH 2.5 and 40 mM of α-CD; applied voltage, 9 kV, 25 μA; sampling: electromigration 7 kV, 10 seconds of 10^{-4} M DL-tryptophan. (From Ref. 34)

5. When the methyl group at the 5-position was replaced by a hydroxyl, with higher hydrophilic properties, no resolution was recorded. When the electrophoretic analysis of tryptophan ester derivatives, namely, methyl-, ethyl-, butyl- and octyl-, was carried out, the complexation increased by increasing the amount of α-CD; chiral resolution was obtained only for DL-tryptophan octyl ester. No enan-

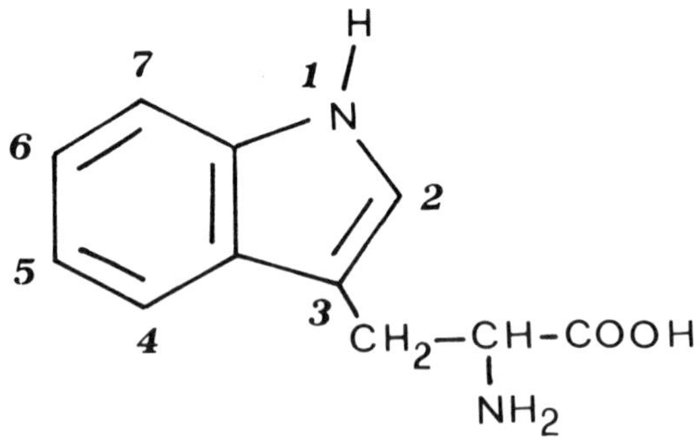

Figure 5 Chemical structure of tryptophan.

tiomeric resolution was obtained when β-CD and γ-CD were used. The effective mobility of all analyzed compounds was reduced by increasing the amount of dimethylated β-CD, showing a higher complexing effect than when the parent CD was added to the BGE. The use of the modified CD allowed the resolution of all racemic mixtures studied. Furthermore, in this paper the effect of the composition of the BGE, by adding methanol, was also reported.

The use of charged modified CDs has been shown by Terabe [36]. A phosphate buffer at pH 3, containing 0.025 M of mono-(6-β-aminoethylamino-6-deoxy)-β-cyclodextrin (CDen) and 0.1% hydroxypropylcellulose (HPC) was used for the resolution of several dansyl amino acids. Under the operating conditions, the electroosmotic flow was negligible and, thus, none of the neutral solutes were detected. Strong amino acids, D-glutamic and D-threonine, had short migration times. At pH 3 the dansyl group was positively charged. (For other enantiomeric separations also see the sections on drugs and on amino acids.)

Inorganic Ions

Tazaki et al. [37] successfully used α-CD for the separation of inorganic anions by capillary isotachophoresis. In this paper, the authors discussed, but did not show, isotachopherograms, for which they used 2% α-CD solution for the separation of iodide ion from a mixture of chloride and bromide. Furthermore, 1% α-CD allowed the analysis of chlorate in the presence of ferricyanide and ferrocyanide ions.

Positional Isomers

Because of the physicochemical properties of positional isomers, these compounds are difficult to separate from each other under conventional electrophoretic operating conditions. The use of modified electrolytes (e.g., buffer water–organic modifier mixtures) has been shown to be effective for the separation of positional isomers of aminobenzoic, hydroxybenzoic, and methylbenzoic acids [38]. The use of cyclodextrins as complex-forming equilibria added to the electrolyte seems to be advantageous for the resolution of these kinds of compounds.

The first application for the separation of structural isomers by using cyclodextrins in electrophoretic techniques was demonstrated by Terabe et al. [10], who separated several positional isomers by using micellar electrokinetic chromatography. The *o*-, *m*-, and *p*-isomers of cresol, nitroaniline, chloroaniline, nitro- and dinitrophenol were separated in a phosphate buffer at pH 7; the electroosmotic flow allowed the neutral compounds to reach the detector, and the selectivity of the separation was improved by adding a charged cyclodextrin (2-*O*-carboxymethyl-β-CD) to the electrolyte.

Tazaki et al. [39] added α-CD to a leading electrolyte, 10 mM acetic acid-potassium acetate (1:1), in capillary isotachophoresis. The terminator was a solution of 5 mM of 1-naphthylamine hydrochloride. The study was carried out in PTFE capillary tube of 0.5-mm id and 15-cm long by applying a current of 50 μA. The effect of the amount of α-CD on the effective mobility of several structural isomers of quaternary ammonium ions containing an aromatic nucleus—namely, *o*-, *m*-, and *p*- (methylbenzyl)trimethylammonium ions; methyl- and nitro-substituted (benzyl)trymethylammonium ions; and quaternary pyridinium ions—was studied. The *para*-isomers showed the highest mobility decrease with the increasing amount of α-CD, which means that this compound can better fit the CD cavity than the *o*- and *m*-isomers owing to steric hindrance from the CD wall. K^+, Na^+, Li^+ and TEA^+ were used as calibration ions; when separated by ITP by adding increasing amount of α-CD in the leading electrolyte they showed a slight decrease in effective electrophoretic mobility owing to differences in viscosity of the electrolytes. Furthermore, the authors evaluated the binding equilibrium constant between the analyzed compounds and α-CD, correlating the effective mobilities and the concentrations of analytes, CD, and the leading and complexed CD analytes. The binding constants expressed as log *K* between α-CD and *o*-, *m*-, and *p*-(methylbenzyl)trimethylammonium were calculated to be 0.8, 1.1, and 1.6, respectively.

Another application to study the effect of cyclodextrins on the separation of positional isomers by using capillary isotachophoresis was carried out by Snopek [40]. The relative step height (h_i) was correlated to the amount of α-CD added to the leading electrolyte. h_i is a parameter measured in ITP and is influenced by the electrophoretic mobility calculated by using Eq. 1.

$$(h_i)_{rel} = \frac{h_i - h_L}{h_L - h_T} \tag{1}$$

where *h* is the step height of *i* (analyte), *L* (leading electrolyte) and *T* (terminating electrolyte).

Several halogen benzoate ions and benzoic acid were studied by adding separately to the LE different amounts of α- and β-CD. The LE used consisted of 5 mM hydrochloric acid, titrated with 6-aminocaproic acid (EACA) to pH 4.7, and 0.2% hydroxypropyl methylcellulose (HPMC); as terminator, 5 mM of 4-morpholinoethanesulfonic acid, at pH 5.5. The study was done in a PTFE capillary, 400 mm × 0.5-mm id, at 18°C; the detection current was 50 μA. By using increasing amounts of both α-CD and β-CD in LE a general increase of $(h_i)_{rel}$ (lowered mobility) was recorded for all analyzed compounds. When *o*-, *m*-, and *p*-fluorobenzoate were analyzed the *o*-isomer was not complexed by α-CD; weak retardation with this isomer was obtained when β-CD was used. The complexing power of both α- and β-CD increased by increasing the dimensions of halogen-substituents iodo- > bromo- > chlorobenzoate. In all cases the *o*-isomer showed a slight decrease in effective mobility. The use of γ-CD as a complexing agent was effective only for *m*- and *p*-iodobenzoate, but no resolution of these isomers was recorded. Only weak complexes were observed when *m*- and *p*-bromobenzoate were analyzed. Thus several parameters influence the complexation between CD and halogen benzoates, namely, dimension of the CD cavity, position of the halogen substituent on the aromatic ring, and diameter of the substituent. This correlation is shown in Fig. 6.

Fanali and Sinibaldi [41] studied the effect of CDs added to the leading electrolyte on the mobility of several positional isomers of benzoic acid derivatives [39]. Isotachophoretic experiments were carried out by using an anionic leading electrolyte at pH 5.1 and morpholinoethane sulfonic acid as terminator: α-, β-, and γ-CD were added separately in the LE for this study.

Resolution of positional isomers of methoxyphenyl acetates and naphthyl phosphates was also recently reported [42]. The separation was performed by capillary

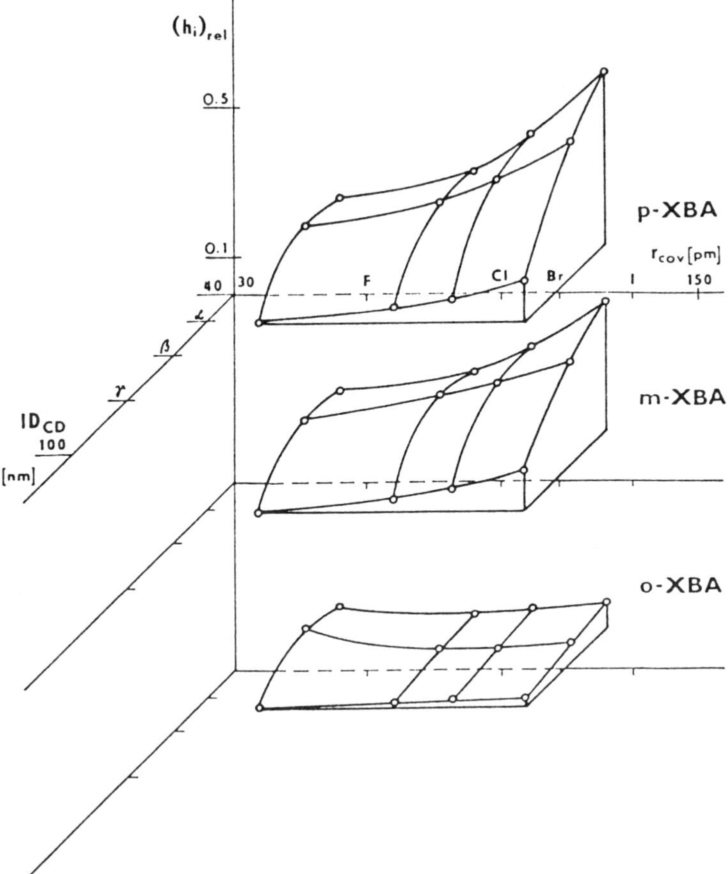

Figure 6 Correlation between dimension of CD cavity, covalent radius of the halogen substituent and relative step height $(h_i)_{rel}$. (From Ref. 40).

zone electrophoresis at pH 8.06 and 8.50 with β-CD and α-CD, respectively. By increasing the amount of α-CD, the resolution of naphthyl phosphate isomers is improved.

Fanali studied the separation of *o*-, *m*-, and *p*-aminophenol and 1,2-, 1,3-, and 1,4-diaminobenzene by using capillary isotachophoresis [43]. To obtain complete resolution of the six analyzed cations, β-cyclodextrin was added to the leading electrolyte. The optimized method was used for the determination of aminophenols and diaminobenzene isomers in permanent hair colorant creams.

Drugs

Mixtures of sulbactam synthesis intermediates and sulbactam and its alkaline degradation products have been separated by using capillary isotachophoresis [44]. The separation of related penicillins was performed at pH 4.8, at which the analyzed compounds are stable. The stability of these drugs was studied in solutions saturated with β-CD, and the authors found good stability at pH 3.8–4.8. The complete separation of (a) penicillanic acid sodium salt, (b) 6,6-dibromopenicillanic potassium salt, (c) 6,6-dibromopenicillanic acid sulfone potassium salt, and (d) penicillanic acid sulfone sodium salt was obtained by supporting the LE with 10 mM of β-CD. To show the role of the cyclic structure of CD, comparative experiments were carried out by adding separately to the LE D-glucose, starch, and β-CD. The resolution of the four analyzed compounds was improved only by using β-CD. The use of α- and γ-CD did not provide satisfactory results for the separation of the mixture, showing that the cavity dimension of the CD is of paramount importance. Furthermore, the pH of the LE can play a very important role in this separation; in fact, electrophoretic runs performed at pH 3.50, 4.00, 4.30, 4.55, and 4.65, and 4.80 showed that the best results were obtained at pH 3.50 and 4.80.

The purity control of nafronyl oxalate (naftidrofuryl) was shown by capillary isotachophoresis [45] when an LE containing 5 mM sodium acetate at a pH 5.5, 0.2% HPMC, and 0.5 mM of β-CD; as terminator 10 mM of 6-aminocaproic acid was used.

Nafronyl oxalate is a drug with antisclerotic effect and, because of the impurity the reagent used in the synthesis—1-(chloromethyl)naphthalene, which can contain 7% of 2-(chloromethyl)naphthalene—can produce a mixture of two positional isomers 1-NHF (the active drug) and 2-NHF. The chemical structures are shown in Fig. 7.

The two isomers interact with β-CD, with the apolar part (naphthyl group) lowering selectively their effective mobilities, thereby allow the separation. The axial 2-NHF–β-CD complex was more stable than the equatorial 1-NHF–β-CD one. The method was optimized, with good results, and then applied to the analysis in real samples. It was possible to analyze as little as 0.1% of 2-NHF in the presence of 99.9% 1-NHF. In the same paper, the authors reported the relative step height

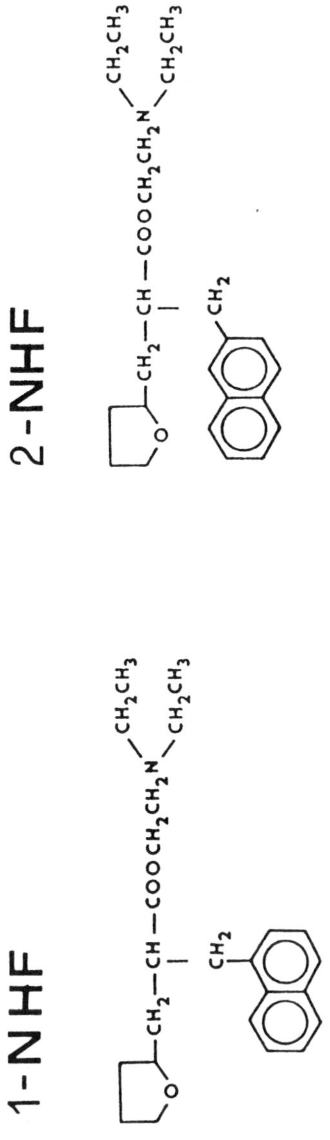

1-NHF

2-NHF

Figure 7 Chemical structure of 1-naftidrofuryl (1-NHF) and 2-naftidrofuryl (2-NHF).

for four nafronyl oxalate intermediates, in an anionic system electrolyte, by using isotachophoresis.

The chiral resolution of 4-(1-methylpiperidylene)-4*H*-benzo[4,5] cyclohepta-[1,2*b*]thiophen-10(9*H*)-one (ketotifen) was obtained by Jelinek [46] by capillary isotachophoresis. The studied drug is used as a nonspecific mast cell-stabilizing agent, and the existence of its enantiomers had not been previously reported in the literature. The study was carried out by adding, separately to the LE at pH 5.50, β-CD and 2,6-di-*O*-methyl-β-CD.

Secobarbital, pentobarbital, and hexobarbital have been resolved in their enantiomeric forms by capillary zone electrophoresis [47]. The results have been expressed as $R' = 100 \times (H - H')/H$ where H and H' are the height of the first peak and that of the valley between the two peaks, respectively; $R' = 100$ represents a baseline separation of the two enantiomers. The BGE used consisted of a phosphate/borate buffer at pH 9 containing 10 mM of different CDs; the capillary was 60 cm × 50-μm id, the applied voltage and the detection were 15 kV and 220 nm, respectively. Secobarbital and pentobarbital enantiomers were resolved ($R' = 100$) by using 2,3,6-tri-*O*-methyl-α-CD. 2,3,6-Tri-*O*-methyl-β-CD was able to resolve racemic pentobarbital ($R' = 94.3$), whereas hexobarbital enantiomers were resolved by using 2,6-di-*O*-methyl-β-CD. The use of unmodified CDs α- and β-CD gave no satisfactory results.

The separation of the four barbiturates, thiopental, pentobarbital, phenobarbital, and barbital, has been recently achieved by using capillary zone electrophoresis [48]. A 20-mM phosphate-borate buffer, at pH 9, was used for the separation of the four barbiturates in less than 10 minutes. The addition of 50 mM of SDS and 30 mM of γ-CD to the BGE allowed separation of the four compounds and partial resolution of racemic thiopental and pentobarbital enantiomers. To obtain the complete resolution of the two barbiturates, it was necessary to use 40 mM of γ-CD and 40 mM of sodium *d*-camphor-10-sulfonate.

Epinephrine is a sympathomimetic drug, commercially available in vials, used for cardiac stimulation and treatment of bronchial asthma. This compound shows optical activity and (−)-epinephrine is ten times more potent than its enantiomer [49]. The possibility of using CZE for purity control of this drug was demonstrated by analyzing two different commercial samples in a 0.1 M phosphate buffer, at pH 2.5, and supported with 20 mM of 2,6-di-*O*-methyl-β-CD [34].

Verapamil and norverapamil isomers have been resolved by CZE after the addition to the BGE of 2,3,6-tri-*O*-methyl-β-CD. 2,6-Di-*O*-methyl-β-CD, used as a chiral additive, was able to resolve racemic disopyramide. Instead, the enantiomeric resolution of SC-48274 (3-[3-(dimethylamino)propyl]-3-(3-methoxyphenyl-4,4-dimethyl-2,6-piperidinedione) was obtained by supporting the BGE with isopropanol and 2,3,6-tri-*O*-methyl-β-CD. The presence of a phenyl or methoxyphenyl group in the chemical structure of the analyzed compounds was important for the interaction with the modified CDs to obtain chiral resolution [50].

The enantiomeric resolution of 2-*tert*-butylamino-1-(3,5-dihydroxyphenyl)ethanol (terbutaline) and 1-(isopropylamino)-3-(1-naphthyloxy)-2-propanol (propranolol) have been studied [11]. Terbutaline is a sympathomimetic drug used in the treatment of asthmic disease, and propranolol is a β-blocker used in the treatment of angina pectoris. The two compounds have several differences in their shape (e.g., the chiral center of terbutaline is closer to the aromatic ring than that of propranolol, and there are different substituents on the nitrogen). The separations were carried out with a BGE at pH 2.5 in which both the analyzed compounds moved as cations. The addition of different CDs (α-, γ-, β-, 2,6-di-*O*-methyl-, and 2,3,6-tri-*O*-methyl-β-CD) caused a general decrease in the migration time. Good resolution of racemic terbutaline enantiomers was obtained with β- and 2,6-di-*O*-methyl-β-CD. Figure 8 shows the separation of the two enantiomers of terbutaline. In this study, the effect of the amount of CD on the resolution was also studied. The resolution of racemic propranolol enantiomers was obtained only by increasing the amount of β-CD (40 mM) and supporting the BGE with 30% (v/v) of methanol.

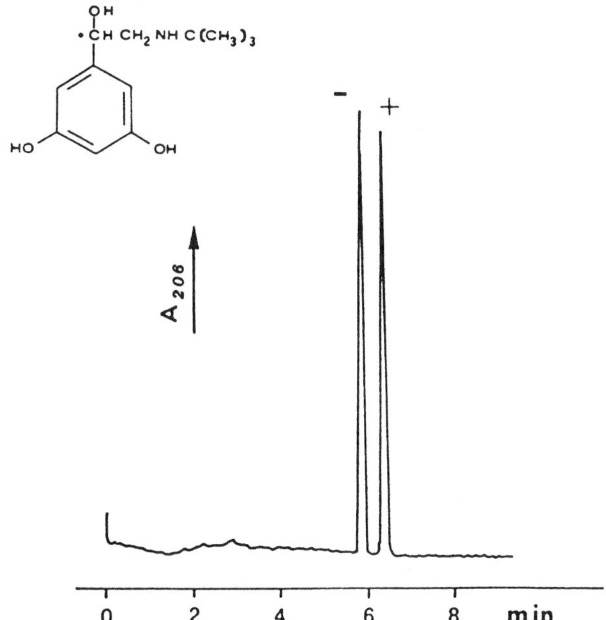

Figure 8 Electrophoretic resolution of (−)- and (+)-terbutaline. Apparatus, BIO-RAD HPE 100; coated capillary tube 20 cm × 25-μm id; background electrolyte: 0.1 M phosphate buffer, pH 2.5, containing 5 mM 2,6-di-*O*-methyl-β-CD; sampling, electromigration 8 kV, 10 second of 10^{-5} M racemic terbutaline; electrophoresis, 19 μA (const.), 9 kV. (Adapted from Ref. 11)

Enantiomeric resolution of several racemic ergot alkaloid derivatives has been shown [51]. It is well-known that this type of drug exhibits strong biological and pharmaceutical activity: inhibition of prolactin release [52], antagonism to adrenoreceptors [53], and use in treatment of both migraine and orthostatic hypotension. The activity depends on the configuration of the drug. As an example, Fig. 9 shows the chemical structure of nicergoline.

Racemic terguride, lisuride, nicergoline, isolysergic acid, and 1-methyl-10α-methoxydihydrolysergol (meluol) have been resolved from their enantiomers by CZE in a coated capillary tube 20 cm × 25-μm id, containing a 0.1 M phosphate buffer, at pH 2.5, and the appropriate amount of CD. Even if 2,6-di-*O*-methyl-β-CD was able to resolve only some of the analyzed compounds (e.g., nicergoline and meluol) the use of γ-CD as a chiral additive to the BGE allowed the resolution of the five racemic ergot alkaloid derivatives. The optimized method was applied for the purity control of commercial formulations containing L-nicergoline. The same authors showed the possibility of improving the selectivity of the separation of epimeric mixtures (ergotamine–ergotaminine and ergometrine–ergometrinine) by CZE in a BGE containing 30 mM of γ-CD.

Chiral resolution of chloramphenicol by CZE, with a BGE at pH 3.5, containing di-*O*-methyl-β-CD has been shown [42]. The addition of 0.1% of methylhydroxy propylcellulose (MHEC) to the BGE improved the resolution. A 5 mM solution of γ-CD allowed the resolution of thioridazine enantiomers, either by ITP or CZE at pH 5.47 and 2.5, respectively. The same authors resolved optical isomers of ketotifen and enantiomers of structurally related compounds; here, CZE obtained better results than those of ITP.

Amino Acids

β-Cyclodextrins have been incorporated in a polyacrilamide gel, and the electrophoretic system was used for the resolution of several dansyl-amino acid enantiomers

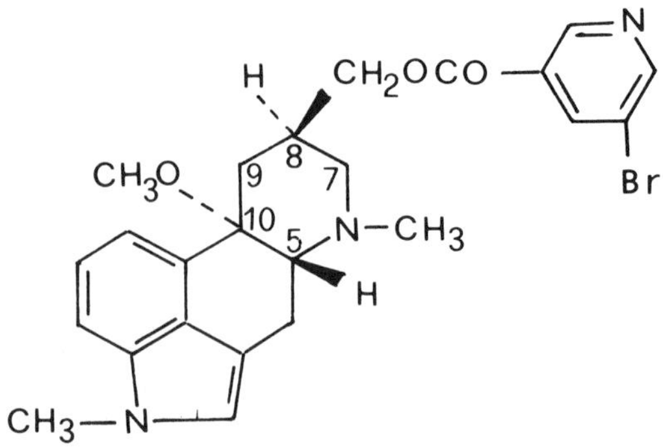

Figure 9 Chemical structure of nicergoline.

[31]. The use of α- or γ-CD gave poor chiral resolution. To study factors influencing chiral resolution of derivatized amino acids, several variables were changed; for example, the amount of β-CD, the composition of the BGE (by adding methanol), and the temperature of the capillary. Furthermore, the association constants between β-CD and several dansyl-amino acid enantiomers have been calculated.

Racemic mixtures of amino acids, such as aspartic and glutamic acids, leucine, norleucine, norvaline, phenylalanine, and valine, as their dansyl derivatives, have been separated in free-zone capillary electrophoresis by using different CDs in a BGE at pH 9 [47]. α-Cyclodextrin was able to only partially resolve DL-aspartic acid, instead β-CD and γ-CD exhibited good enantioselectivity for all compounds. With unmodified CDs the D-isomer moved toward the detector with higher velocity than the L-isomer. The migration order was reversed when modified CDs were used. Baseline resolution was obtained only for leucine and phenylalanine when 2,3,6-tri-*O*-methyl-α-CD and 2,3,6-tri-*O*-methyl-β-CD were used.

Other Compounds

The separation of bile acids in aqueous electrolyte, modified with cyclodextrins, has been studied [54]. The α-, β-, γ-CD, 2,6-di-*O*-methyl-β-CD, and 2,3,6-tri-*O*-methyl-β-CD were added, separately, to an LE at pH 6.4. The use of β- and γ-CD showed a good solubilization effect, but did not help the selectivity of the separation. 2,3,6-Tri-*O*-methyl-β-CD was the best LE additive for the separation of the studied compounds, and an interesting result, from the analytical point of view, was the resolution of the two diastereomers (chenodeoxycholic and ursodeoxycholic acids).

The separation of several corticosteroids and aromatic hydrocarbons, neutral compounds, was obtained by using micellar electrokinetic chromatography (MEKC) with a buffer at pH 9 containing sodium dodecyl sulfate (SDS). To improve the selectivity of the separations Nishi recently added cyclodextrins to the SDS solution for electrophoretic experiments [55]. In such a system, the solute is distributed among three phases: the aqueous buffer, the micelle, and the CD.

Because CDs are relatively hydrophilic (outside the cavity), they will not be solubilized by SDS, and their velocity is in accord with the electroosmotic flow, relatively high at this pH. Thus, by adding 30 mM of β-CD or 15 mM of γ-CD to the SDS buffer solution a shorter migration time was recorded for the analyzed compounds, showing a different inclusion mechanism among the tested solutes.

Recently, Terabe et al. studied the separation of neutral highly hydrophobic compounds by electrophoretic technique that consisted of a CD-modified electrokinetic micellar chromatography [56]. Neutral solutes are incorporated into the micelle or form inclusion complexes with CD. With such a system, containing 60 mM of γ-CD, it was possible to separate 11 of the 24 possible isomers of trichlorobiphenyls. The same BGE was able to separate three pairs of closely related tetrachloro-*p*-dioxin isomers. Furthermore, the method was used for the separation of chlorinated benzene congeners (12) and several polycyclic aromatic hydrocarbons, namely, naphthalene, acenaphthene, anthracene, fluorene, phenanthrene, chrysene, pyrene, and fluoranthene. Nevertheless, the separation has not yet been used for environmental or biological analysis. Thus, the method may be used successfully for microanalytical applications.

Tanaka et al. studied the effect of CDs, added to the LE at pH 5.0, on the mobility of several catecholamines [47]—3,4-dihydroxybenzylamine, dopamine, normetanephrine, metanephrine, norepinephrine, epinephrine, and isoproterenol—by capillary isotachophoresis. Among the CDs used, β-CD was the most effective complexing agent for the analyzed compounds. The substituent groups on the aromatic ring of the analytes and the neutral surfactant in the LE played an important role on the complexation mechanism.

CONCLUSIONS

The use of cyclodextrins as an additive to the background electrolyte in capillary electrophoretic techniques extends the potential of these relatively new methods and enables one to perform fast separations of either positional or enantiomeric isomers. The method proves to be easy to perform and is inexpensive. In fact, for the analyses, only a few milliliters of electrolyte are needed.

For optimization of the separation, several variables have to be kept under control [e.g., shape and amount of the chiral additive, the shape of the analyzed compounds, the composition of the electrolyte (pH, ionic strength, organic solvent) used for the analysis].

REFERENCES

1. W. L. Hinze, *Sep. Purif. Methods*, *10*:159 (1981).
2. T. J. Ward and D. W. Armstrong, In *Chromatographic Chiral Separations* (M. Zief and L. J. Crane, eds.), Marcel Dekker, New York, p. 131 (1988).
3. J. Snopek, I. Jelinek, and E. Smolkova-Keulemansova, *J. Chromatogr.*, *452*:571 (1988).
4. J. Szejtli, *Cyclodextrins and Their Inclusion Complexes*, Akademiai Kiado, Budapest (1982).
5. J. Debowski and D. Sibilska, *J. Chromatogr.*, *353*:409 (1986).
6. D. Y. Pharr, Z. S. Fu, T. K. Smith, and W. L. Hinze, *Anal. Chem.*, *61*:275 (1989).
7. A. P. Croft and R. A. Bartsch, *Tetrahedron*, *39*:1417 (1983).
8. F. Cramer and G. MacKensen, *Angew. Chem.*, *78*:641 (1966).
9. R. K. McMullan, W. Saenger, J. Fayos, and D. Mootz, *Carbohydr. Res.*, *31*:37 (1973).
10. S. Terabe, H. Ozaki, K. Otsuka, and T. Ando, *J. Chromatogr.*, *332*:211 (1985).
11. S. Fanali, *J. Chromatogr.*, *545*:437 (1991).
12. M. Gazdag, G. Szepesi, and L. Huszar, *J. Chromatrogr.*, *351*:128 (1986).
13. D. Duchene, and D. Wouessidjeve, *Drug Dev. Ind. Pharm.*, *16*:2487 (1990).
14. D. Duchene, and D. Wouessidjeve, *Acta Pharm. Technol.*, *36*:1 (1990).
15. T. Takeuchi, M. Murayama, and D. Ishij, *J. Hgh. Resolut. Chromatogr.*, *13*:69 (1990).

16. A. Nardi, S. Fanali, and F. Foret, *Electrophoresis*, *11*:774 (1990).
17. C. A. Chang, H. Ji, and G. Lin, *J. Chromatogr.*, *522*:143 (1990).
18. H. J. Schneider, T. Blatter, and S. Simova, *J. Am. Chem. Soc.*, *113*:1996 (1991).
19. P. Bocek, M. Deml, P. Gebauer, and P. Dolnik, *Analytical Isotachophoresis*, VCH Verlag, Weinheim (1988).
20. J. Debowski, J. Jurczak, and D. Sybilska, *J. Chromatogr.*, *282*:83 (1983).
21. K. Fujimura, S. Suzuki, H. Hayashi, and S. Masuda, *Anal. Chem.*, *62*:2198 (1990).
22. W. L. Hinze, D. Y. Pharr, Z. S. Fu, and W. G. Burkert, *Anal. Chem.*, *61*:422 (1989).
23. V. Schuring and H. P. Nowotny, *Angew. Chem. Int. Ed. Engl.*, *29*:939 (1990).
24. P. Gozel, E. Gassmann, H. Michelsen, and R. N. Zare, *Anal. Chem.*, *59*:45 (1987).
25. R. A. Mosher, *Electrophoresis*, *11*:765 (1990).
26. F. Foret, S. Fanali, A. Nardi, and P. Bocek, *Electrophoresis*, *11*:780 (1990).
27. T. Hirokawa, N. Aoki, and Y. Kiso, *J. Chromatogr.*, *312*:11 (1984).
28. I. Nukatsuka, M. Taga, and H. Yoshida, *J. Chromatogr.*, *205*:95 (1981).
29. W. Lindner, In *Chromatographic Chiral Separations* (M. Zief and L. J. Crane, eds.), Marcel Dekker, New York, p. 91 (1988).
30. A. N. Tran, T. Blanc, and E. J. Leopold, *J. Chromatogr.*, *516*:241 (1990).
31. A. Guttman, A. Paulus, A. S. Cohen, N. Grinberg, and B. L. Karger, *J. Chromatogr.*, *448*:41 (1988).
32. J. Snopek, I. Jelinek, and E. Smolkova-Keulemansova, *J. Chromatogr.*, *438*:211 (1988).
33. S. Fanali, *J. Chromatogr.*, *474*:441 (1989).
34. S. Fanali and P. Bocek, *Electrophoresis*, *11*:757 (1990).
35. A. Nardi, L. Ossicini, and S. Fanali, *Chirality*, (in press; 1992).
36. S. Terabe, *Trends Anal. Chem.*, *8*:129 (1988).
37. M. Tazaki, M. Takagi, and K. Ueno, *Chem. Lett.*, p. 639 (1982).
38. S. Fujiwara and S. Honda, *Anal. Chem.*, *59*:487 (1987).
39. M. Tazaki, T. Hayashita, Y. Fujino, and M. Takagi, *Bull. Chem. Soc. Jpn.*, *59*:3459 (1986).
40. J. Snopek, J. Jelinek, and E. Smolkova-Keulemansova, *J. Chromatogr.*, *411*:153 (1987).
41. S. Fanali and M. Sinibaldi, *J. Chromatogr.*, *442*:371 (1988).
42. J. Snopek, H. Soini, M. Novotny, E. Smolkova-Keulemansova, and I. Jelinek, *J. Chromatogr.*, *559*:215 (1991).
43. S. Fanali, *J. Chromatogr.*, *470*:123 (1989).
44. I. Jelinek, J. Snopek, and E. Smolkova-Keulemansova, *J. Chromatogr.*, *405*:379 (1987).
45. I. Jelinek, J. Dohnal, J. Snopek, and E. Smolkova-Keulemansova, *J. Chromatogr.*, *435*:496 (1988).
46. I. Jelinek, J. Snopek, and E. Smolkova-Keulemansova, *J. Chromatogr.*, *439*:386 (1988).
47. M. Tanaka, S. Asano, M. Yoshinago, Y. Kawaguchi, S. Tanaka, T. Kaneta, M. Taga, H. Yoshida, and H. Ohtaka, *J. Chromatogr.*, *587*:364 (1991).
48. H. Nishi, T. Fukuyama, and S. Terabe, *J. Chromatogr.*, *553*:503 (1991).
49. I. R. Innes and M. Nickerson, In *The Pharmacological Basis of Therapeutics*, 5th ed. (L. S. Goodman and A. G. Gilman, eds.), Macmillan, New York, p. 477 (1975).
50. P. J. Gilbert, J. M. Dethy, and M. L. Lesne, Presented at the Second Int. Symp. Chiral Discrimination, Rome, Italy, May 27–31, Abstr. P40 (1991).

51. S. Fanali, M. Flieger, N. Stereinova, and A. Nardi, *Electrophoresis*, (in press; 1992).
52. E. Brambilla, E. DiSalle, G. Briatico, A. Mantegani, and A. Temperelli, *Eur. J. Med. Chem.*, *24*:421 (1989).
53. G. Arcari, L. Bernardi, G. Bosisio, S. Coda, G. Fregan, and A. H. Glaesser, *Experimentia*, *28*:819 (1972).
54. J. Snopek, E. Smolkova-Keulemansova, I. Jelinek, J. Dohnaln, J. Klinot, and E. Klinotova, *J. Chromatogr.*, *450*:373 (1988).
55. H. Nishi and M. Matsuo, *J. Liq. Chromatogr.*, *14*:973 (1991).
56. S. Terabe, Y. Miyashita, O. Shibata, E. R. Barnhart, L. R. Alexander, D. G. Patterson, B. L. Karger, K. Hosoya, and N. Tanaka, *J. Chromatogr.*, *516*:23 (1991).
57. T. Tetsumi and T. Shono, *Fresenius J. Anal. Chem.*, *339*:63 (1991).

28

Prospects for the Use of Capillary Electrophoresis in Neuroscience

Mark A. Hayes, S. Douglass Gilman, and Andrew G. Ewing

The Pennsylvania State University
University Park, Pennsylvania

The functional chemistry of the brain is exceedingly difficult to study. Not only are neurotransmitters, metabolites, and such present in minute quantities, but also brain tissue is so complex that isolation of a given neurotransmitter is difficult [1]. An understanding of neural physiology at the single-cell level requires chemical probes for specific cellular regions. Toward this requirement, many chemical methods have been developed. Early methods to separate and determine neurochemicals involved column chromatography, paper chromatography, and thin-layer chromatography. The inherent inefficiency and poor sensitivity of these techniques was improved with the development of high-performance liquid chromatography (HPLC) in the 1960s. Since this development, many advances have occurred in the application of HPLC to neurochemical analysis. For instance, tissue extracts have been chromatographed using ion-exchange [2,3], adsorption [4,5], and soap [5] chromatography. Detection for these separations are accomplished by UV absorption, radioisotope monitoring, fluorescence, or electrochemistry. Of these, electrochemical detectors are commonly used in the neuroscience field because many biologically important molecules can easily undergo electron transfer.

A relatively new technique, electrophoresis in capillary tubes, offers several exciting methods for fast, efficient separations of ionic species, separations of macromolecules important in the area of analytical biotechnology, and development of small-volume separations-based sensors. In 1974, Virtanen reported on potentiometric detection of electrophoretically separated solutes in 200- to 500-μm id glass tubes [6]. This work dealt with zone electrophoresis and pointed out many of the unique advantages when small-diameter tubes are used. In 1979 Mikkers et al. refined this technique to 200-μm–id Teflon tubes and obtained separations with plate heights of less than 10 μm [7]. Jorgenson and Lukacs [8–11] advanced the area to smaller capillaries (75 μm) and predicted plate heights of less than 1 μm for separations of proteins. Actual experimental values of only a few micrometers for dansylated amino acids were obtained. This work, coming at the beginning of a major thrust in biotechnology, has become the landmark for work in capillary electrophoresis (CE). The small capillaries used in this work offer a unique opportunity for the separation of solutes in ultralow volumes. To ensure accuracy with these small injection volumes, careful control of the flow in the capillary system becomes important.

Electroosmotic flow is the movement of bulk solvent in a capillary when a potential field tangential to the capillary wall is applied [12]. This flow, if not deliberately altered, is often strong enough to cause all solutes to elute at one end of the capillary. Thus, capillary electrophoresis is more easily automated than large-scale electrophoresis, which can be very labor-intensive and irreproducible. In addition, the ultrasmall volume flow rates obtainable in capillary electrophoresis permit sampling from microenvironments (i.e., single cells), which is of great importance to the area of neurobiology and neurochemistry.

This chapter describes the development and application of capillary electrophoresis for neurochemical analysis. These experiments include separations of neurochemicals obtained from microdialysis and push–pull cannulae, from sampling single whole nerve cells and, finally, from sampling single-cell cytoplasm. The key features of all these experiments are the selectivity and low-volume capabilities of capillary electrophoresis. The chapter is divided into two major sections. The first part covers methodology for sampling neurochemical environments, CE separation schemes, and detection methods employed for neurochemically important compounds. The second section covers the application to neurochemical systems including peptides, catecholamines, and cyclic nucleotides.

DESCRIPTION OF METHODS

The description of methods will be divided into three sections: injection techniques, separations, and detection schemes. Each section will have a short description of the apparatus or technique and deal, in detail, with published work relating to neuroscience in each specific area.

Injection Techniques

Injection for CE, as applied to neuroscience, can be divided between traditional solution-sampling techniques and those techniques that are designed for direct sampling of biological environments. Traditional solution-sampling techniques, including electromigration and pressure, have been applied to the neurosciences by sampling solutions that have been physically removed from the biological system and prepared before injection. The preparation may include purification or derivatization. On the other hand, direct-sampling techniques for CE include electromigration (at ultrasmall injectors), microdialysis, and perfusion with push–pull cannulae. These direct-sampling techniques offer varying degrees of spatial resolution and biocompatibility, and the latter two provide real-time, dynamic experimental information.

Overview of *In Vitro* Solution Injection Techniques

Much work has been done to demonstrate unique separations of neurochemicals. These separation schemes use traditional injection methods involving the sampling of derivatized or synthetic neurochemicals from in vitro solutions [13–24]. In addition, early work in the HPLC of neuroactive chemicals [25–27] was an important stepping stone in the more recent developments involving the use of CE for small-scale neurochemical analysis.

PRESSURE INJECTION The pressure injection technique involves the placement of the injection end of the capillary into the sample and application of negative pressure to force the sample solution into the tip. This pressure may be created either hydrostatically (raising and then lowering the injection buffer), by pressure control of the chamber holding the sample, or by the application of vacuum at the anode reservoir. In one study, amino acids from a sampled neuron of the land snail *Helix aspersa* were injected into the CE capillary by pressure [22]. This was after dissection, isolation, and homogenization of the neuron and a derivatization procedure with the fluorophore naphthalene-2,3-dicarboxyaldehyde (NDA). Minimum injection volume for this pressure technique (Fig. 1), whether pneumatically or hydraulically operated, is reported as approximately 1 nL [22]. This volume may be considered large from the standpoint that volumes as small as 270 fL have been reproducibly injected with the electromigration technique [28]. Also, from the number of literature reports, the electromigration injection technique is a more popular style [29]. This is apparently due the extra equipment needed for pressure injection, whereas electromigration is accomplished with the existing separation apparatus. However, the pressure injection technique delivers a more representative sample, without bias, according to the charge on the injected species; a phenomenon associated with the electromigration technique [30].

ELECTROMIGRATION INJECTION Electromigration injection is the most commonly used technique for samples that have been derivatized or purified from biological sources [20,31]. This technique is also used for the injection of standards for methods development [19,21]. The technique typically consists of the placement of the cathodic end of the capillary into the prepared sample, application of a voltage for a short period (usually 5–30 seconds) and then placement of the same end of

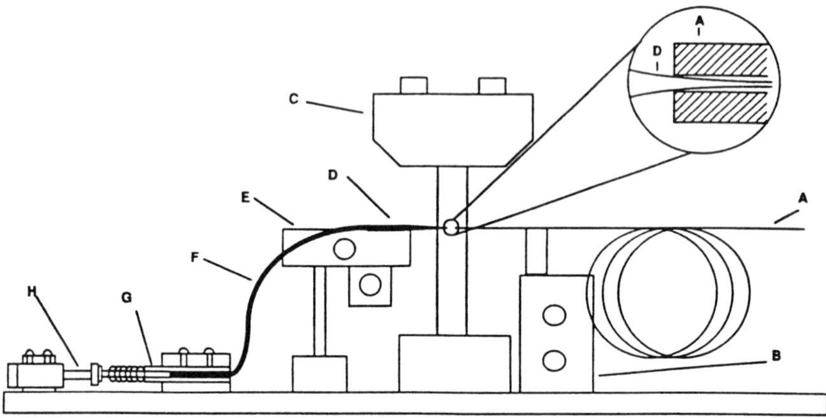

Figure 1 Schematic diagram of microinjector. The labels correspond to the following parts: A, chromatographic column; B, column positioning stage; C, stereomicroscope; D, micropipette; E, micromanipulator; F, connector tubing; G, spring-loaded syringe; and H, micrometer. The darkened areas are those that are filled with mercury. The inset shows the pipette inserted into the column ready for injection. Note that the pipette fits tightly against the column. (From Ref. 49)

the capillary in the operating buffer reservoir before application of the separation voltage.

One example of the use of electromigration injection is the analysis of leucine enkephalin and methionine enkephalin from equine cerebrospinal fluid [20]. In this method, cerebrospinal fluid (CSF) was collected postmortem and stored at $-60°C$. Isotopic internal standards were then added, the sample diluted with acetonitrile, and centrifuged (precipitating the larger molecular mass proteins). The evaporated supernatant was then loaded onto a hydrophobic solid-phase extraction column for further purification and, finally, evaporated to dryness. The final step was to reconstitute the purified sample with the operating buffer before injection. Detection was accomplished with mass spectrometry, allowing easy detection of the underivatized analytes. Table 1 shows the recovery percentages for this preparation procedure. These data indicate a consistent recovery for all the analytes. The important aspects of this sampling method are that (a) the analytical apparatus is far removed from the biological system, (b) the sample has been manipulated for the benefit of the separation and detection methods, and (c) the sampling of the biological system is in no way dependent on the injection scheme for the analytical system.

Sample derivatization and preparation has also played a key role in neurochemical analysis of tissues by CE. As an example of this, ovine brain tissue was removed from the medial preoptic area and the median eminence and examined for luteinizing

Table 1 Recovery by LC–MS–MS (SRM) Determinations of Enkephalins

Spiked sample	Compound[a]	Recovery (%) 10 ng mL^{-1}	Recovery (%) 50 ng mL^{-1}
Water (n = 5)[b]	LE	64.4 ± 14.2	63.7 ± 11.4
	ME	62.8 ± 17.5	61.4 ± 20.8
	IS	65.2 ± 10.9	71.3 ± 8.8
CSF (n = 5)[c]	ME	67.2 ± 29.7	79.4 ± 12.4
	ME[d]	67.1 ± 12.4	79.0 ± 15.7
	IS	74.9 ± 13.0	76.8 ± 7.0
	IS[d]	67.4 ± 9.0	73.4 ± 16.3

All calculations are based on ion counts of peak area measurements. LE was excluded from the CSF experiments owing to its endogenous presence in equine CSF.
[a]LE, leu-enkephalin; ME, met-enkephalin; IS, internal standard.
[b]All values are evaluated against external standards of 200 ng mL^{-1} and 1 μg mL^{-1}. The LC–MS–MS reproducibilities of the external standard injections at the 200 ng mL^{-1} level were 10.3% (n = 5) for LE, 6.0% (n = 5) for ME, and 12.4% (n = 16) for the IS.
[c]As in a, except that ME data are based on ratio calculation with the IS.
[d]The second data sets were obtained 8 weeks later.
Source: Ref. 20.

hormone-releasing hormone (LHRH) [23]. These brain regions are discrete hypothalamic areas, described as being rich in peptidergic-containing neuronal cell bodies and terminals. Tissues removed were pooled, concentrated, and derivatized with fluorescein isocyanate (FITC) isomer 1 and, subsequently, injected into the CE capillary by electromigration for fluorescence detection (Fig. 2). This sample preparation scheme yielded two to three orders of magnitude better sensitivity than UV detection without the sample preparation.

Sampling by Microdialysis and Push–Pull Cannulae

Microdialysis is a technique whereby a small probe tip with a semipermeable membrane is placed inside the tissue of interest [32]. On the inside of the probe a solution is pumped past the inner surface of the membrane. Analytes are osmotically transported across this membrane from high concentration to low. This is shown schematically in Fig. 3. This technique offers the advantage that no volume of fluid is being removed from the system. The probe may also be used to deliver drugs or bioactive species to the local tissue, with subsequent sampling by the same probe. The push–pull cannulae (PPC) technique (Fig. 4) is similar to dialysis; however, the probe tip is not covered by a dialysis membrane [33,34]. Hence, artificial cerebral spinal fluid is pumped into the sampled tissue and an equal amount of fluid is pulled out for analysis. Both probes can be constructed with total tip dimensions in the 100- to 500-μm range and can be used to sample discrete regions

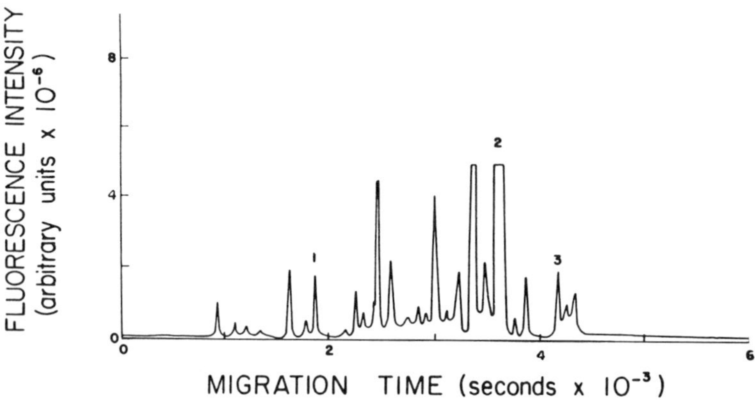

Figure 2 Electropherogram of a median eminence tissue sample conjugated to FITC. Peak 1 represents LHRH. Peak 2 represents FITC–isomer 1. Peak 3 represents either a second isomer of FITC or a contaminant. (From Ref. 23)

of brain tissue. However, the push–pull cannulae technique may cause more tissue damage during sampling.

Both microdialysis and the push–pull cannulae techniques have been used extensively in combination with HPLC. Recently, microdialysis has been used to directly collect samples for detection by mass spectrometry (MS) and radioimmunoassay (RIA) [35,36]. Of course, these techniques are ultimately limited by the sensitivity and resolution of the detection scheme. For example, RIA used with microdialysis typically requires the entire dialysate [20–200 μL] to detect a single component. The use of CE does not improve detector sensitivity, but it does improve component resolution and reduce sample volume requirements. There is a real need for more sensitive detectors before the microdialysis and push–pull cannulae sampling techniques will be able to take full advantage of the resolution and speed of CE.

Capillary electrophoresis, used in conjunction with microdialysis and push–pull cannulae uses the ability of CE to separate many components in a small volume and short period of time. Presently, two studies serve as examples of the usefulness and limitations of this injection technique. First, CE has been used to sample dialysates from ovine brain for LHRH [37]. In this study, CE detection schemes were not sensitive enough to directly sample the dialysate. Rather, six dialysate sample volumes of 200 μL were combined and concentrated, and finally injected to a CE system for separation (Fig. 5). This injection process, will improve dramatically when detector sensitivity is improved to the level at which the dialysate can be directly sampled and detected. This will allow real-time, dynamic in vivo studies with a relatively noninvasive biological probe.

A second application of this technique involved effusion of several drugs from the microdialysis probe, with subsequent monitoring of dopamine levels in the CSF

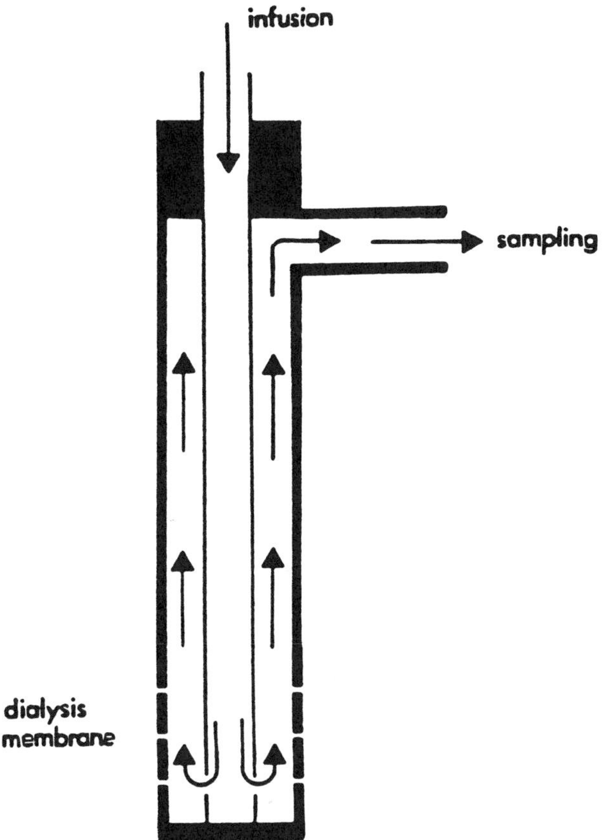

Figure 3 Radial section of a microdialysis probe. (Ref. 36)

[38]. Known amounts of either cocaine, procaine, or lidocaine were added to the entering dialysis solution, which was delivered to the nucleus accumbens in the rat brain. Here, the concentration gradient is high to low from the inside of the dialysis probe to the tissue and, hence, drug is effectively delivered to the tissue around the probe. The dialysate is then monitored by CE for the concentration of drug to determine the amount delivered. In addition this study monitored the CSF dopamine concentration in the dialysate by HPLC with electrochemical detection. In summary, microdialysis and the push–pull cannulae techniques can be used for both sampling and for delivering drugs to specific neuronal tissue environments.

Sampling by Direct Electromigration

To take full advantage of the ultrasmall volume separation capabilities of CE, it is important to minimize sample handling and dilution. This can be accomplished by

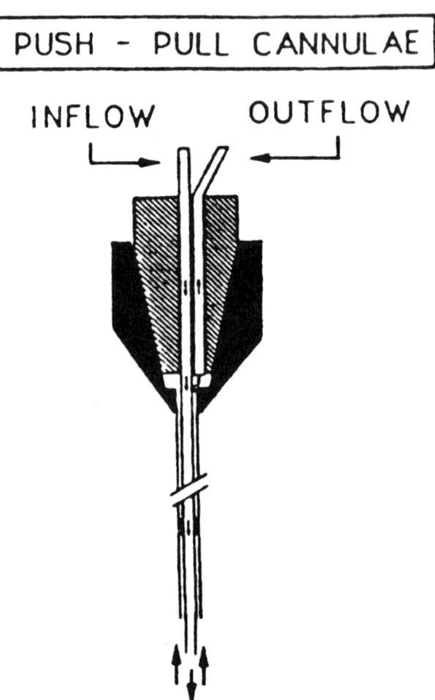

Figure 4 Radial section of a PPC. Artificial CSF is pushed through the inner tube (inlet) and pulled through the outer tube (outlet). Both the push and the pull side of this system are controlled by separate peristaltic pumps, equilibrated to push and pull 600 μL of CSF per hour. (From Ref. 37)

directly sampling the biological matrix in vivo with a microinjector as an integral part of the CE capillary. Electromigration injection techniques involve both electrophoresis and electroosmosis, according to the following equation:

$$Q = \frac{(\mu_e + \mu_{eo})VACt_1}{L}$$

where Q is the quantity injected, μ_e is the electrophoretic mobility of the analyte, μ_{eo} is the electroosmotic flow coefficient, V is the potential field, A is the cross-sectional area of the capillary, C is the concentration, t_i is the injection time and L is the capillary length. This results a bias in the sampling of charged species. Under normal electrophoretic conditions, cations will move ahead of the volume flow and anions behind. Thus, the apparent injection volume for cations and anions differs from the volume removed from the system. This is only a minor problem, since electrophoretic mobilities for each component may easily be calculated from the resulting electropherogram, and the apparent injection volume, V_{app}, for each com-

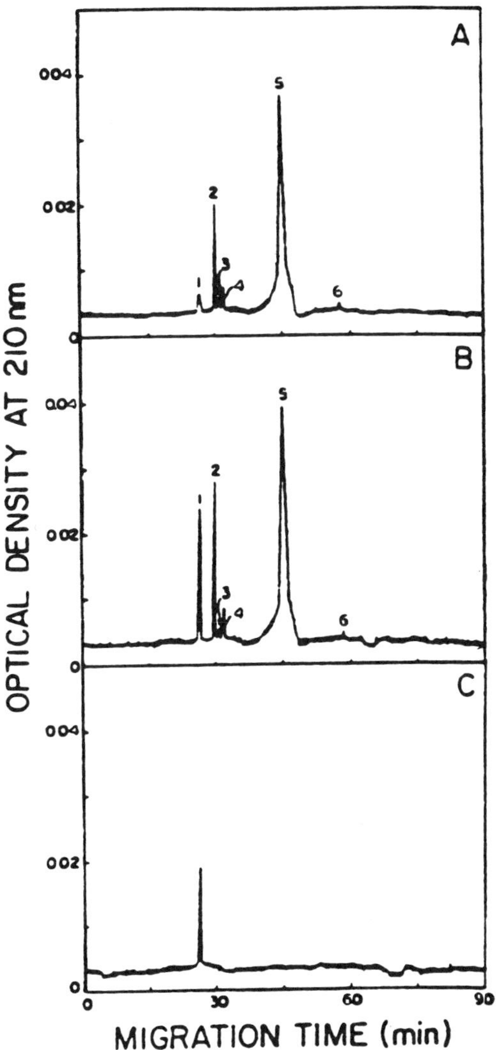

Figure 5 Capillary electrophoresis results from PPC release samples. Electropherograms of ME in vivo release samples (pooled and concentrated) collected using push–pull perfusion. (A) Electropherogram of components eluted with acetonitrile from the Sep-Pak column. (B) Electropherogram of components eluted with acetonitrile from the Sep-Pak column and spiked with LHRH before injecting the sample to the CE apparatus. (C) Electropherogram of homogeneous synthetic LHRH (peak 1).

ponent may be calculated by the following equation:

$$V_{app} = \frac{(\mu_e + \mu_{eo})VAt_i}{L}$$

Early techniques for direct sampling into CE used principles developed for iontophoresis. In this technique, materials are ejected from a small micropipette tip by electromigration [39]. Interestingly, this concept can be used to inject samples into a CE capillary. Dual-bore pipettes have been pulled down to approximately 10-μm tips for injection into a CE capillary [40]. An electrode is placed in one barrel and the injection tip of a CE capillary into the other barrel (Fig. 6). In this scheme, the biological sample is electroosmotically pumped through the micropipette into the CE capillary. This technique suffers from band broadening, attributed to eddy currents and dead volume at the injection tip. In addition, these

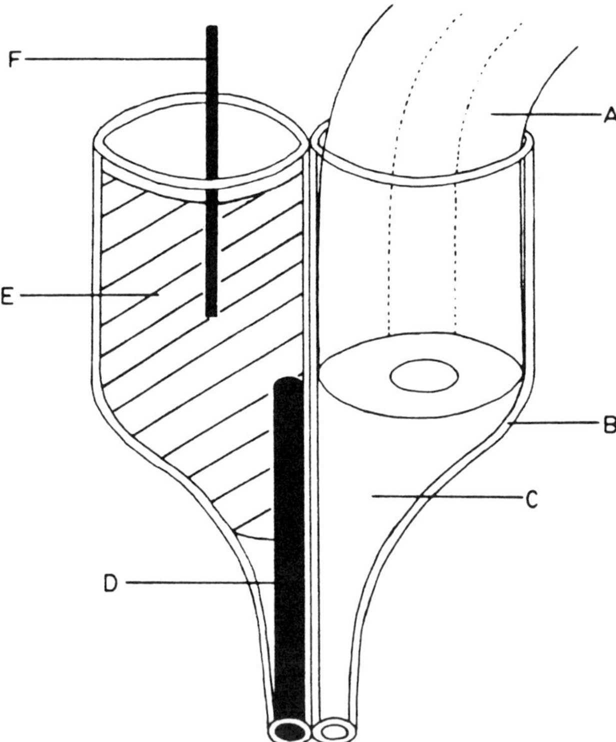

Figure 6 Schematic of dual-barreled microinjector: A, fused silica capillary column; B, glass dual-barreled micropipet; C, buffered solution; D, carbon fiber; E, mercury; F, platinum wire. (From Ref. 40)

microinjectors have been reported to be difficult to fabricate and suffer from irreproducible injections owing to current leaks across the glass wall between the barrels in the tip. Despite these minor problems, this technique has been used to remove and directly inject a small portion of cytoplasm (Fig. 7) from the large dopamine cell of the pond snail *Planorbis corneus* [41].

A significant advance in injection of cytoplasmic samples involves construction of a microinjector from the CE capillary itself [28]. This has been accomplished by removing a small portion of the polyimide coating near the tip of a 5-μm–id capillary and etching to an od of 8–10 μm (Fig. 8A). An integrated electrode is

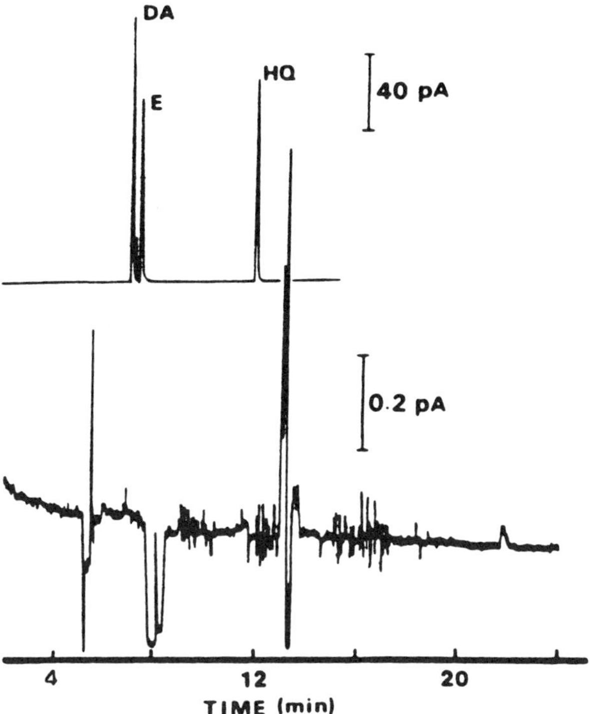

Figure 7 Electropherogram of a sample taken from the giant dopamine neuron of *Planorbis corneus*. The following experimental conditions were used: capillary, 12.7-μm id × 79 cm; detection, electrochemical (carbon fiber) at 0.7 V vs a sodium-saturated calomel reference electrode; buffer, 0.025 M 2-(*N*-morpholino)ethanesulfonic acid (pH 5.55); injection, 25 seconds at 15 kV through a 6-μm–od microinjector; separation potential, 25 kV. Top: electropherogram of 1×10^{-4} M dopamine (DA), epinephrine (E), and hydroquinone (HQ) assayed under similar conditions, except that the injection was from a standard solution for 1 second at 30 kV. (From Ref. 41)

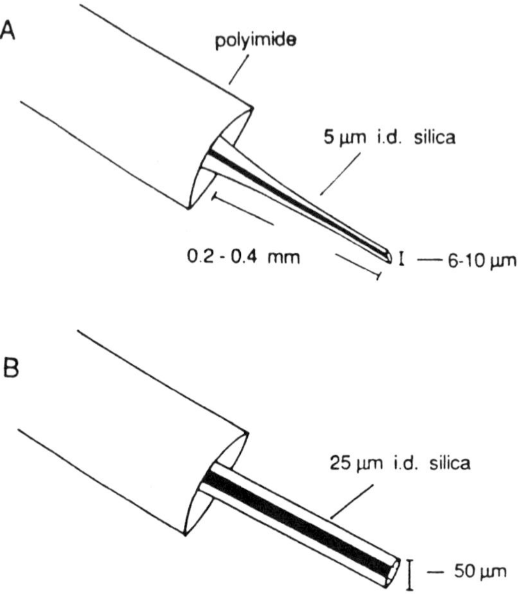

Figure 8 Schematic depiction of two etched microinjectors. (A) Injector for acquiring and injecting cytoplasmic samples, constructed at the high-voltage end of a 5-μm–id electrophoresis capillary. (B) Injector for acquiring and injecting whole cells, constructed at the high-voltage end of a 25-μm–id electrophoresis capillary. (From Ref. 42)

not required because of the high resistance of the buffer-filled capillary (10^{12} Ω). This prevents any appreciable current flow in the sampled region (total currents are typically 5–15 nA). Injection volumes as small as 270 fL have been obtained from in vitro solutions. In addition, the injector has been used to sample cytoplasm from the large dopamine and serotonin cells in *P. corneus*. This will be discussed in greater detail in the section on applications.

Capillary electrophoresis has also been used to sample whole invertebrate neurons [42] and whole-cell homogenates [22]. Whole cells can be injected into larger capillaries by electromigration. This has been accomplished by etching 25-μm–id capillaries to 50-μm od with hydrofluoric acid (see Fig. 8B) [42]. The smaller tip permits viewing under a microscope as the capillary tip is manipulated to a specific cell. Injection of the cell is accomplished by electroosmotic suction that is followed by cell lysing and subsequent separation of the released components. A recent improvement of this technique involves etching both the outside and a small portion of the inside of the capillary tip to provide a conical external structure and a funnel-like internal structure to capture the cell. Applications of these techniques to the analysis of whole dopamine cells will be discussed the section on applications.

Separation Schemes

Capillary electrophoresis separations that are directly related to neuroscience have appeared, analyzing peptides and proteins and easily oxidized catechols and indoles. Zone electrophoresis in capillaries is often sufficient to resolve mixtures of catechols and indoles, although some work has been carried out to improve selectivity for these separations by use of borate complexation, micellar systems, mixed micellar systems, and nonpolar additives [14,15,18,43]. Separation schemes for neuropeptides are often more problematic, and several examples are discussed in the following.

Methods for separation of peptides and proteins have been widely developed for capillary electrophoresis, although only a few of these have been directly applied to the neurosciences [16,19,22,36]. The separation of proteins and peptides is often problematic owing to solute adsorption on the walls of the capillary. This can result in severe band broadening and the loss of resolution and sensitivity. General approaches to solving these problems include adjusting the pH of the buffer [44,45], use of high-ionic–strength buffers [46], and coating the surface of the capillary with nonionizable materials [11,47,48]. Adsorption is not always a problem, however, and publications of CE separations of neurochemicals contain only one example of adsorption problems.

This example studies the effects of column derivatization for the separation of basic neuropeptides [21]. This is shown by the two separations of bradykinin and lys-bradykinin shown in Fig. 9. The first separation was conducted with a pH 7.5 buffer in an untreated fused silica capillary. The adsorption of the positively charged neuropeptides to the negatively charged silica surface results in severe band broadening, as shown in the resulting electropherogram. The second separation shows the results of column derivatization. Here, the column was silylated with (3-aminopropyl)trimethoxysilane, resulting in the replacement of surface silanol groups on the column wall with aminopropyl groups. At the buffer pH used in this separation (pH 3.4), the aminopropyl groups are positively charged. This change in surface charge reduces the adsorption of the neuropeptides, giving a much more efficient separation. Additionally, electroosmotic flow is reversed by this treatment, and the polarity of the CE power supply must be reversed.

In another study, the separation of neurotensin, angiotensin II, met-enkephalin, substance P, and cholecystokinin-8 in a 0.05 M sodium tetraborate buffer at pH 8.3 has been reported [16,36]. Capillary electrophoresis with UV detection was used to quantitate these neuropeptides in ovine brain dialysates. A series of unidentified peaks, separated by CE, are assumed to be peptides and proteins extracted from the dialysate. A subtlety of this data is that many peptides of neurological importance can easily be separated without elaborate buffers or column-coating schemes.

Capillary electrophoresis has been used to investigate the binding of Ca^{2+}, denaturation, and the effects of sample additives on the protein calmodulin [19].

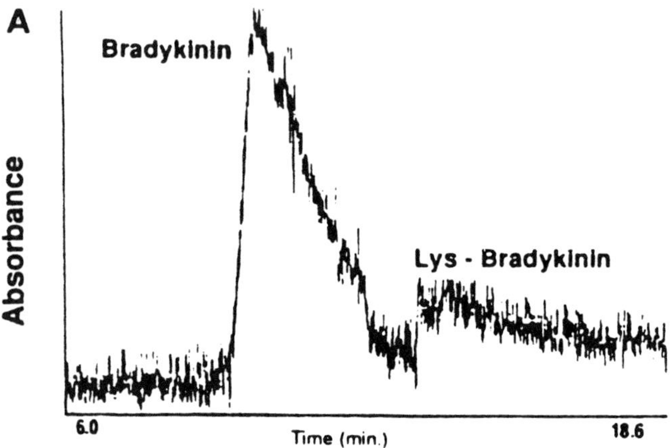

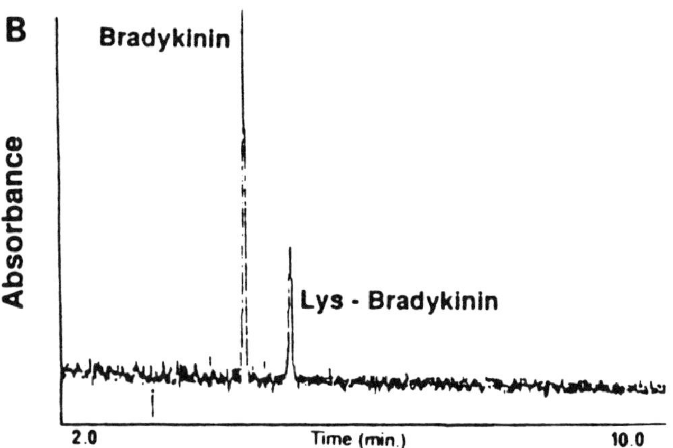

Figure 9 CZE separation of two basic neuropeptides, detected by UV absorption at 200 nm: (A) separation with bare fused silica CZE column, (B) separation with APS-treated CZE column. (From Ref. 21)

Interestingly, the addition of Ca^{2+} to the system alters the electrophoretic mobility of calmodulin and, hence, ion binding can be a useful tool to manipulate the selectivity of protein separations in CE. Also, the addition of [ethylene-bis(oxyethylene nitrile)]-*N*,*N*,*N*′,*N*′-tetraacetic acid to the protein sample causes the disappearance of all the peaks in the resulting electropherogram. This has been explained as resulting from irreversible adsorption to the wall of the capillary. Finally, the alteration of buffer pH to 8.0 caused the calmodulin to split into three subunits, which were separated by CE. Gel electrophoresis under denaturing and nondenaturing condition does not indicate the presence of more than one species of calmodulin. In summary, CE can be used to perform high-resolution separations of neurochemically important peptides and proteins in free solution.

Detection Methods

There are many schemes that have been used for detection of neurochemically important compounds separated by CE. These schemes include absorption, fluorescence, laser-induced fluorescence (LIF), mass spectrometric, and electrochemical detection. These detection schemes have been developed for use with other microcolumn separation techniques [49–52]. Discussion of these detection modes is presented, with reference to their use with neurochemically important compounds.

Absorption Detection

Detection by UV or visible absorption is the most common, commercially available technique used with CE. Typically, a small portion of the polyimide coating on the capillary is removed by heat or acid leaching to form an on-column detection window. Detection most often is carried out by absorbance measurements at 200 or 254 nm, but many array and scanning detection systems are available. Multiwavelength systems provide added qualitative information, but at the cost of reduced sensitivity.

Absorbance detection is relatively insensitive when compared with most other techniques [53] and, therefore, is usually not useful for the low levels of neurochemicals extracted from typical neuroscience experiments. One area for which absorbance detection has some usefulness is in the detection of drugs used at relatively high levels. Capillary electrophoresis with absorbance detection at 200 nm has been used to determine cocaine, procaine, and lidocaine concentrations following microdialysis in the nucleus accumbens of the rat [38]. A polypeptide, luteinizing hormone-releasing hormone (LHRH), has also been determined in samples obtained by push–pull cannulae and microdialysis from ovine brain, using CE with absorbance detection at 210 nm. Ultraviolet detection is a well-developed technique that is simple and inexpensive to use for CE and neuroscience applications. However, because of the short path length resulting from the narrow bore of the on-column detection cell, it is a relatively insensitive technique. Several orders of magnitude lower detection limits have been achieved by other detection methods.

Fluorescence Detection

The fluorescence detection technique has proved to be the most sensitive detection scheme for CE [54,55]. Despite the exceptional detection limits available, this technique is only beginning to be applied to neurochemically interesting materials [23,24,56]. In one study, a microscope was altered to fit a detection window of a capillary near the objective (Fig. 10). The capillary was illuminated by a mercury

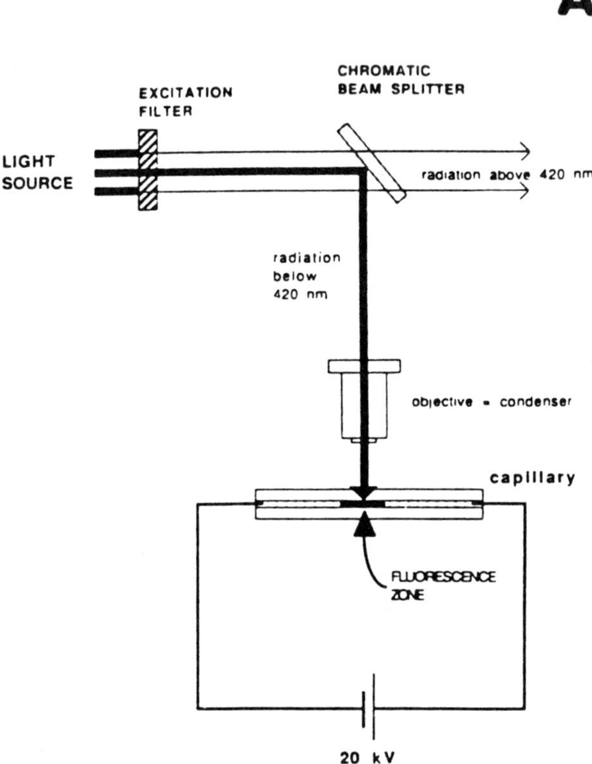

Figure 10 The principles of adaptation of fluorescence microscopy. (A) Excitation of the sample with an epillumination fluorescence microscope. The UV beam is shown as the thick black line reflected at right-angles by the chromatic beam-splitter. The fluorescence zone is indicated by the dark dotted area in the center of the capillary. (B) Emission and measurement of fluorescence. The emitted light is indicated by the thick dashed line that crosses the chromatic beam-splitter and is detected by the PMT. (From Ref. 23)

lamp and signal/noise levels were enhanced by a computer-driven single-photon detection system. The samples were drawn from the ovine brain with a push–pull cannulae and, after derivatization with FITC, analyzed for LHRH. In a second study, neuropeptides from the bag cells of *Aplysia californica* were derivatized with FITC-isomer 1 and detected with an axial illumination–charge-coupled device scheme [24]. This study reported detection limits in the 10^{-19} mole range. One other example of this detection scheme recently appeared in which several neuropeptides were derivatized with fluorescamine and detected with the fluorescence microscope [23].

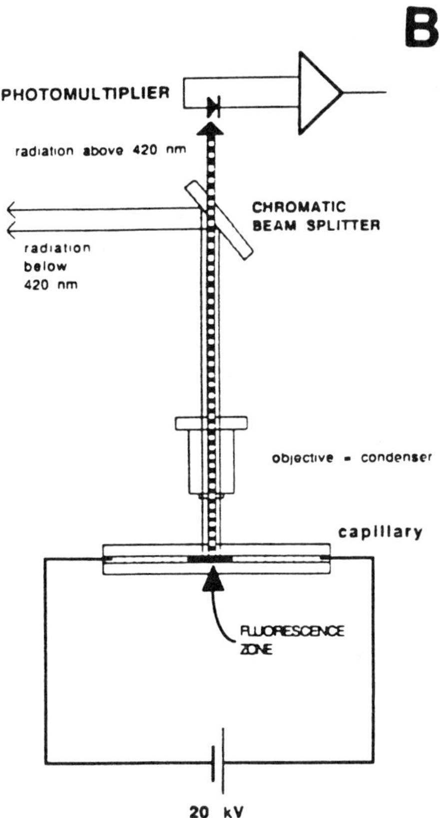

Mass Spectrometric Detection

Capillary electrophoresis with mass spectrometry has been used in several different configurations to separate neurochemically important constituents. Work directly relating to the neurosciences has been limited to the development of appropriate on-line interfaces to handle the effluent of the CE capillary and the application of these interfaces to a limited number of peptides and proteins. These interfaces include atmospheric ionization (both electrospray [57,58] and ion-spray [13,20]) and continuous-flow fast-atom bombardment [21]. Each of these techniques are in common use and are commercially available [59,60]. All of these interfaces provide for "soft ionization" only; that is, there is very little fragmentation of the solute, and generally, only molecular ions are produced. More information may be obtained by using a tandem MS–MS system in which further fragmentation takes place to provide more structural information. This is at the cost of sensitivity.

Electrochemical Detection

Capillary electrophoresis with electrochemical detection has been widely used to examine neurologically interesting questions. Fortunately, many neurochemicals of interest are easily oxidized in their natural state. This gives electrochemical detection two distinct advantages; first no derivatization is needed and, second, as a consequence of the first advantage, ultrasensitive detection (near that offered by fluorescence) is available for direct measurements in vivo. Amperometric detection has been accomplished both on-column [61,62] and off-column [14,28,30,41–43]. The experimental setup for off-column detection is shown in Fig. 11. This apparatus employs a piece of porous glass tubing over a small crack in the capillary to allow electrical isolation of the separation potential field from the potential applied to the amperometric electrode. This detection scheme has allowed the use of extremely small capillaries (to 2-μm id). Detection of neurotransmitters from samples of cytoplasm of single cells and whole single cells will be discussed in the applications section.

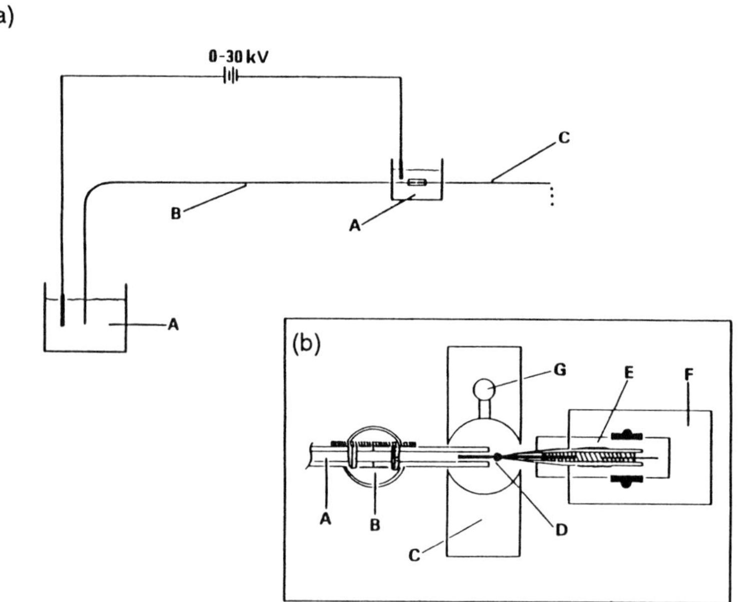

Figure 11 (a) Schematic of CZE system with electrically conductive joint: A, buffer reservoirs; B, separation capillary; C, detection capillary. (b) Top view of amperometric detection system: A, column; B, porous glass joint assembly; C, Plexiglas block; D, carbon fiber working electrode; E, microscope slide; F, micromanipulator; G, reference electrode port. (From Ref. 14)

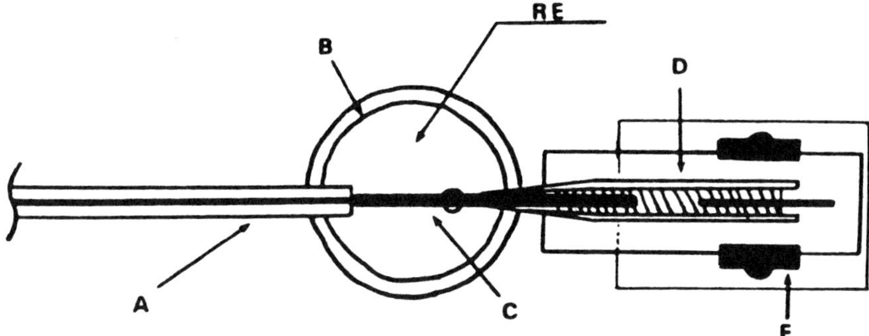

Figure 12 Schematic drawing of CZE with end-column amperometric detection: A, capillary; B, cathodic buffer reservoir and electrochemical cell; C, carbon fiber electrode; D, electrode assembly; E, micromanipulator; RE, reference electrode. (From Ref. 61)

The off-column detection scheme has recently been improved and optimized [61,62]. This new detection scheme has been called end-column detection and is more reproducible and easier to construct. A schematic of this detection scheme is shown in Fig. 12. Amperometric detection is possible because a large resistance exists across the narrow-bore buffer-filled capillary. The potential field near the detection electrode is only in the millivolt range and this allows the placement of the electrode directly at the capillary end to facilitate good positioning and mass transfer of solutes to the surface of the electrode.

APPLICATIONS TO NEUROSCIENCE

Separation and Detection of Neuropeptides and Neuroactive Proteins

The application of CE to the analysis of neuropeptides and neuroactive proteins has been the object of a sizable amount of research in recent years. Two critical problems to solve for this type of analysis are the development of sensitive detection methods for peptides, and sampling of these molecules from neurochemical environments. This area of research has seen the use of a variety of detection methods, including mass spectrometric [13,20,21,63] UV absorbance [17,37,64], and laser-induced fluorescence detection [23,24,56]. Only a few examples of analysis of neuropeptides and neuroactive proteins from real neurobiological environments have been reported [23,24,36,37,56,64].

Laser-induced fluorescence (LIF) has proved to be the most sensitive detection method for CE analysis of neuropeptides [23,24,56]. Sweedler et al. have used CE

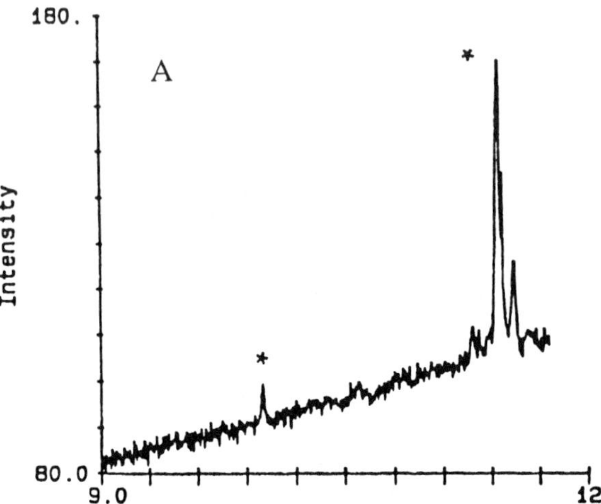

Figure 13 Electropherogram of approximately 2 millionths of a single bag cell neuron: (A) extract, (B) spiked with γ, and (C) spiked with both γ and ELH. Separation conditions: 63-μm–id capillary; 60 mM borate buffer, pH 9.0. (From Ref. 24)

with CCD-based axial multichannel LIF detection to detect low levels of neuropeptides from the bag cells of the marine mollusk *Aplysia californica* [24]. Mass detection limits of 2×10^{-20} to 8×10^{-20} mol were reported for amino acids derivatized with FITC; however, detection limits were not reported for the neuropeptides analyzed in this work.

The analysis of neuropeptides from a small sample from homogenized *A. califonica* bag cells was used to demonstrate the potential of CE with LIF detection to eventually analyze neuropeptides from a small portion of a single cell [24]. Several hundred cells from a single *A. califonica* bag cell cluster were homogenized and diluted to 2 mL. The purified cell extract was then derivatized with FITC and diluted to 1 L. An electropherogram of a 10-nL sample of the diluted extract is shown in Fig. 13, with two peaks tentatively identified as egg-laying hormone and γ bag cell peptide. The 10-nL injection corresponds to the amount of material contained in approximately 2 millionths of a single *A. californica* bag cell. However, the difficulty in derivatization of small-volume samples is a key limitation of these experiments.

Guzman and co-workers have used CE coupled to fluorescence microscopy to detect LHRH and neuropeptides from discrete hypothalamic areas in the ewe brain [23,56]. Luteinizing hormone-releasing hormone is a neuropeptide that transmits reproductive information from the central nervous system to the endocrine system in humans and farm animals [37]. It is released from neuronal terminals in the median eminence into primary portal vessels, and from there it reaches the anterior

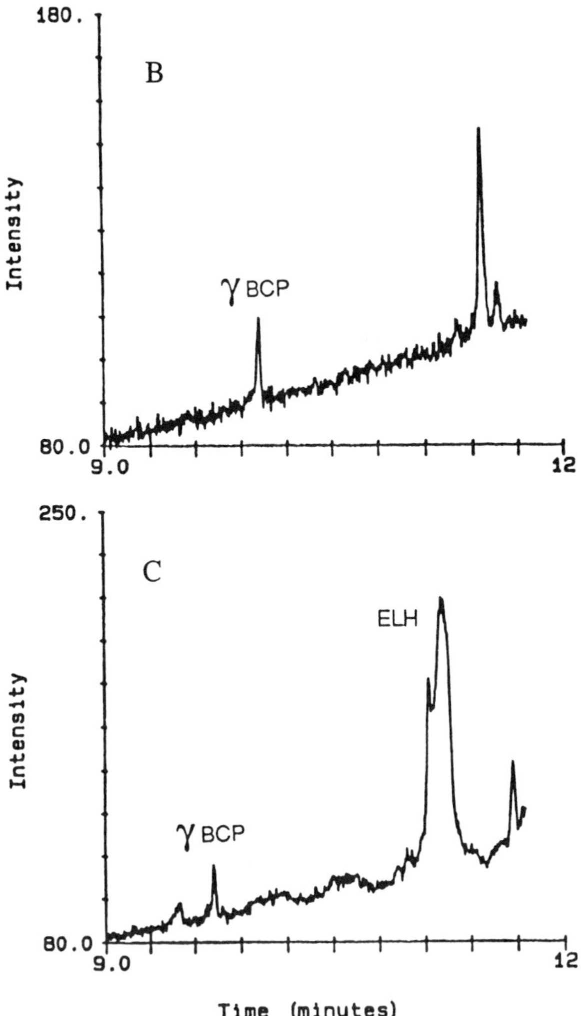

pituitary gland [37]. Release of LHRH and the release of other neuropeptides in the median eminence are measured primarily by RIA analysis of tissue samples from brain areas near the median eminence and in vivo, using push–pull cannulae-sampling techniques with RIA analysis. Although RIA is very sensitive for LHRH and other neuropeptides, RIA procedures are not generally able to determine more than one analyte at a time [23]. Consequently, nearly all of a single push–pull cannulae sample volume is consumed to analyze only one compound in the sample [37]. Capillary electrophoresis was examined as an alternative to RIA analysis of

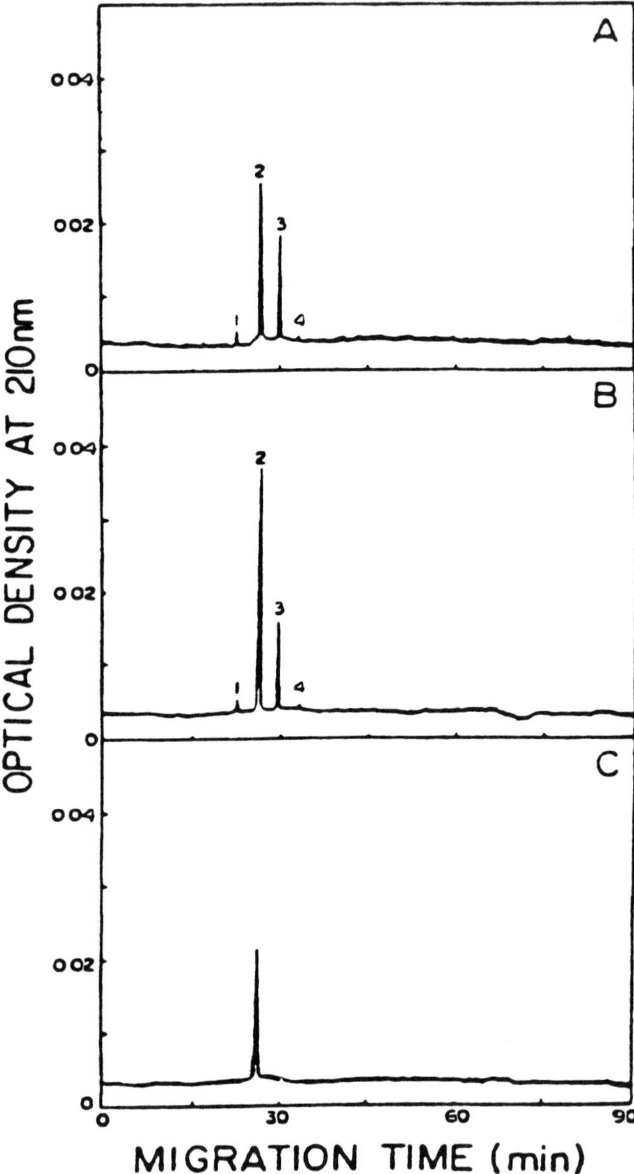

Figure 14 CE profile of tissue content homogenate samples from the ME: (A) electropherogram of a representative ME sample, (B) electropherogram of the same ME sample spiked with LHRH standard, (C) electropherogram of LHRH standard (peak 2). Experimental conditions: 75-μm–id, capillary; 50 mM sodium tetraborate buffer, pH 8.3; field strength 100 V cm^{-1}; 10 nL injection. (From Ref. 37)

LHRH release because of the low sample volume requirements and because the technique is capable of determining several components from one sample simultaneously [23,36,37,56]. Successful use of CE for this type of analysis could allow the determination of several neuropeptides at once, possibly shedding light on the interdependent nature of the actions of these compounds.

An electropherogram of a tissue homogenate from a ewe median eminence is shown in Fig. 14. Tissue samples were removed by microdissection, homogenized in 0.1 N HCl, and centrifuged. Lipids were removed from the sample before injection into the CE system. The electropherogram in Fig. 14 shows several major peaks. Peak 2 was identified as LHRH by the observation of an increase in the peak size upon spiking the sample with an LHRH standard. The remaining peaks in the electropherograms were not identified. Luteinizing hormone-releasing hormone was quantitated by CE for four different tissue samples; median eminence, accurate nucleus area, para accurate nucleus area, and lateral hypothalamus. The values obtained with CE were compared with those obtained with an LHRH RIA. Capillary electrophoresis with UV detection was found to be an adequate alternative to RIA for these tissue sample experiments. However, the sensitivity of CE with UV detection was inadequate for in vivo monitoring of LHRH release with push–pull cannulae sampling [37]. Guzman and co-workers have recently overcome some of the sensitivity limitations of UV detection in these push–pull cannulae experiments using LIF detection [23,56]. Figure 15 shows electropherograms using LIF detection of a tissue sample (see Fig. 15**A**) and a push–pull cannulae sample (see Fig. 15**B**) from the median eminence. Peaks 1–3 were identified as LHRH, neuropeptide Y, and β-endorphin, respectively.

Although mass spectrometric detection with CE is not as sensitive as CE with LIF detection, the structural information provided by mass spectrometry has led to its use for the analysis of several neuropeptides [13,20,21,63]. Muck and Henion have used CE, coupled to a triple-quadrapole mass spectrometer with an atmospheric pressure ionization interface, to examine several enkephalins and dynorphins with femtomole sensitivity [20]. Figure 16 shows the CE separation of three dynorphins, leucine enkephalin, and an internal standard, (D-Ala^2)-leucine enkephalin, as detected by mass spectrometry in the selected ion mode. A detection limit of 2 μg mL^{-1} (60 fmol) was obtained with this instrument for methionine enkephalin. This concentration detection limit was about 100 times larger than that for microbore-HPLC equipped with a similar MS detector and did not permit the use of CE to determine leucine enkephalin and methionine enkephalin in equine cerebrospinal fluid. However, microbore-HPLC with MS detection, which requires larger sample volumes, was sufficiently sensitive for those experiments. The CE–MS instrument was also used to obtain an MS–MS spectrum for (D-Ala^2)-leucine enkephalin.

Smith and co-workers have separated a mixture of four enkephalins and two proteins using CE with MS detection and an atmospheric pressure ionization (API) interface

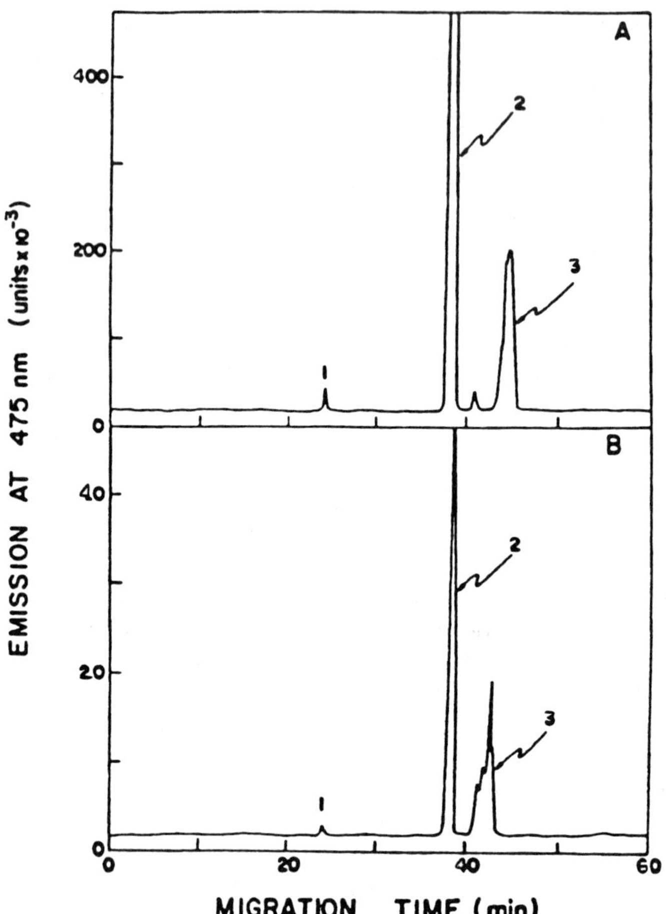

Figure 15 Capillary electrophoresis profile of fluorescamine-derivatized PPC and tissue sample analyzed by fluorescence detection. (A) Electropherogram of a representative PPC sample. Peak 1 represents LHRH; peak 2, NPY; and peak 3 βEND. (B) Electropherogram of tissue sample. Peak 1 represents LHRH; peak 2 NPY; and peak 3, βEND. (From Ref. 56)

(Fig. 17) [63]. This experiment demonstrates the increased selectivity that can be achieved by coupling CE to a mass spectrometer. Although the enkephalins appear as one peak in the total ion electropherogram, single ion electropherograms allow the mass spectral-based resolution of all four enkephalins, whereas two myoglobins were resolved by CE. Detection limits with this instrument were approximately 100 fmol.

Moseley et al. [21] have used CE with fast-atom bombardment mass spectrometry to separate and detect several neuropeptides. This instrument has demonstrated

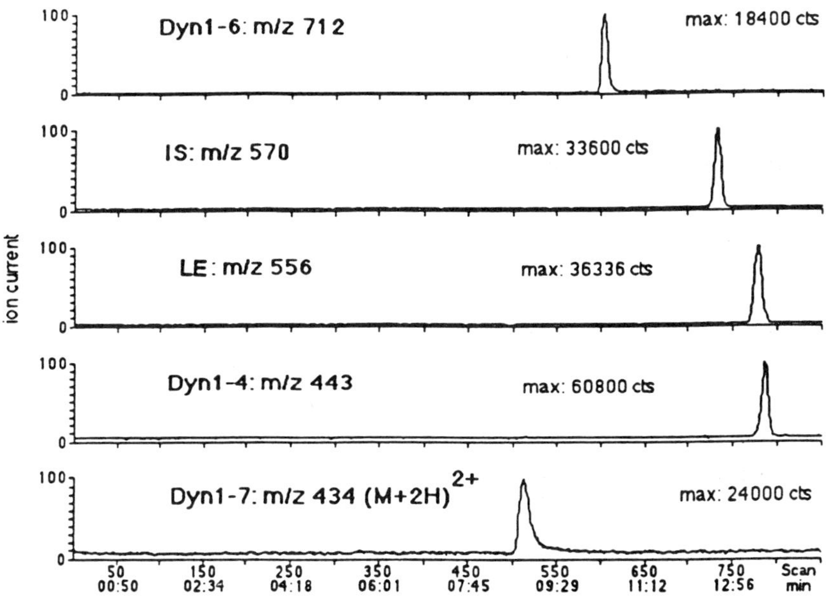

Figure 16 CZE–MS (SIM) of dynorphin standards. Experimental conditions: 100-cm × 100-μm–id fused silica capillary; eluent, acetonitrile–25 mM ammonium acetate, pH 6.8 (50:50, v/v); applied voltage, 30 kV; about 15 nL were loaded by hydrostatic injection (5 cm for 10 seconds). The amount injected was calculated as 2.3 ng each, corresponding to ca. 5 pmol. (From Ref. 20)

somewhat better detection limits for neuropeptides than the API instruments [20,63], but it is not capable of analyzing large neuroactive proteins. Figure 18 shows a series of single-ion electropherograms from a separated mixture of four neuropeptides: leucine enkephalinamide, morphiceptin, proctolin, and FLEEI (Phe-Leu-Glu-Glu-Ile). The low amounts of neuropeptides injected in this separation (60–75 fmol) indicate that the system has very good mass sensitivity, although the authors did not give specific detection limits for these analytes.

The CE–MS instrumentation with a fast-atom bombardment ion source has also been used to conduct CE–MS–MS analysis for three neuropeptides (Fig. 19). The mixture of methionine enkephalin, methionine enkephalinamide, and FLEEI was initially analyzed at low femtomolar levels (21–42 fmol) by CE–MS to determine migration times and m/z of the molecular ions of the compounds. Because of the poor sensitivity of CE–MS–MS, analysis required injections of 420–840 fmol, which severely overloaded the column. Three separate runs had to be conducted to obtain MS–MS data for all three neuropeptides because of limitations of the data acquisition software. These results, as well as the MS–MS data acquired by Muck

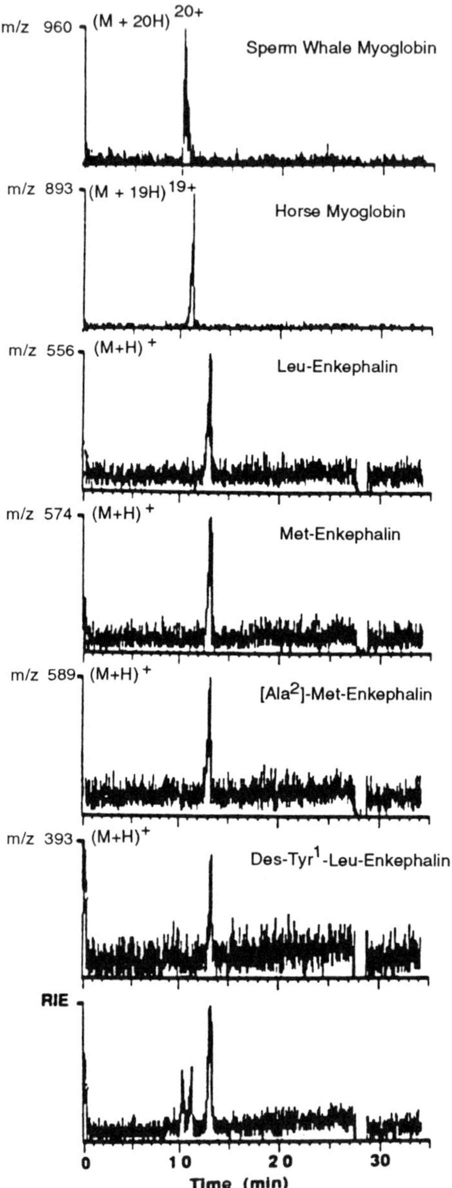

Figure 17 CZE–MS separation of a mixture of (0.1 mM each) whale myoglobin (M_r 17,199), horse myoglobin (M_r 16,950), des-tyr^1-leu-enkephalin (M_r 391), leu-enkephalin (M_r 555), met-enkephalin (M_r 573), and [ala^2]-met-enkephalin (M_r 588) obtained in a 20 mM Tris buffer (at pH 8.25) in a 100-cm × 50-μm fused silica capillary at 30 kV (17 μA). (From Ref. 63)

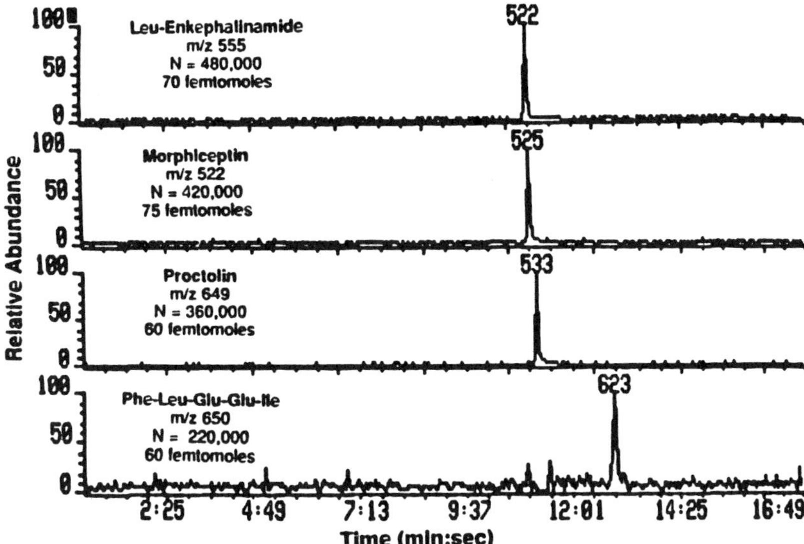

Figure 18 Single-ion electropherograms of the protonated molecular ions of a mixture of neuropeptides. (Numbers associated with peaks are scan numbers.) The peptides were separated as negative ions (CZE buffer pH of 8.5) and were detected as positive ions (FAB matrix pH of 3.5). (From Ref. 21)

and Henion [20], demonstrate the potential of CE with MS–MS detection to identify the structures of analytes as they elute from a CE column, taking advantage of the structural determination capabilities of tandem mass spectrometry.

Several neuropeptides and neuroactive proteins have been separated by CE and detected by UV absorbance, primarily at wavelengths near 210 nm. Neurotensin, angiotensin II, met-enkephalin, unsulfated cholecystokinin, and sulfated cholecystokinin have been separated in an untreated 75-μm–id fused silica capillary with a 50 mM sodium tetraborate (pH 8.3) buffer [16]. Five endorphins (β-, Des-Tyr-β-, *N*-acetyl-β-, γ-, and Des-Tyr-γ-) have been separated in an untreated silica capillary with a 20 mM citric acid, pH 2.5, buffer [17]. An electropherogram of a human cerebrospinal fluid sample is shown in Fig. 20 before and after dialysis through a 14-kDa cutoff membrane [64]. Two peaks were identified as albumin and prealbumin.

Separation of Catecholamines, Cyclic Nucleotides, and Other Compounds

Catecholamine neurotransmitters and other neurochemically important molecules can be separated with CE. Catecholamines have been detected primarily by electro-

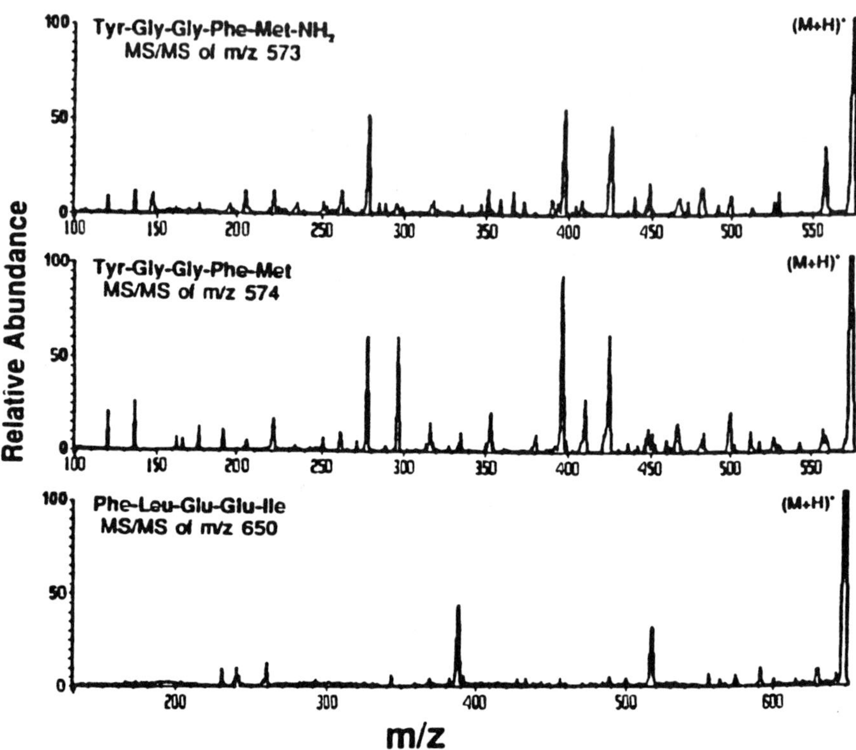

Figure 19 CZE–MS–MS analysis of the three neuropeptides. The peptides were separated as negative ions (CZE buffer pH of 8.5) and desorbed, fragmented, and detected as positive ions (FAB matrix pH of 3.5). (From Ref. 21)

chemical detection [14,15,18]. Figure 21 shows a separation of dopamine, norepinephrine, epinephrine, and catechol, with amperometric detection. Amperometric detection has produced detection limits as low as 0.5 amol for serotonin and 0.6 amol for catechol in a 5-μm–id column [28]. Catechols and serotonin have also been separated in micellar buffers [18]. Separation of norepinephrine, epinephrine, dopamine, catechol, and serotonin as borate complexes in a micellar SDS buffer is shown in Fig. 22. Detection limits are increased in the micellar buffers to about 20 fmol for catechol [14]. Catechols have also been separated and detected by CE–MS [58]. Dopamine, serotonin, norepinephrine, and epinephrine were detected with this system, but no detection limits were reported for these compounds.

Hernandez et al. [31] have separated and detected cyclic nucleotides at picomolar levels using CE with UV detection. A separation of cyclic-AMP, cyclic-GMP, and cyclic-IMP with detection at 210 nm is shown in Fig. 23. The separation was carried

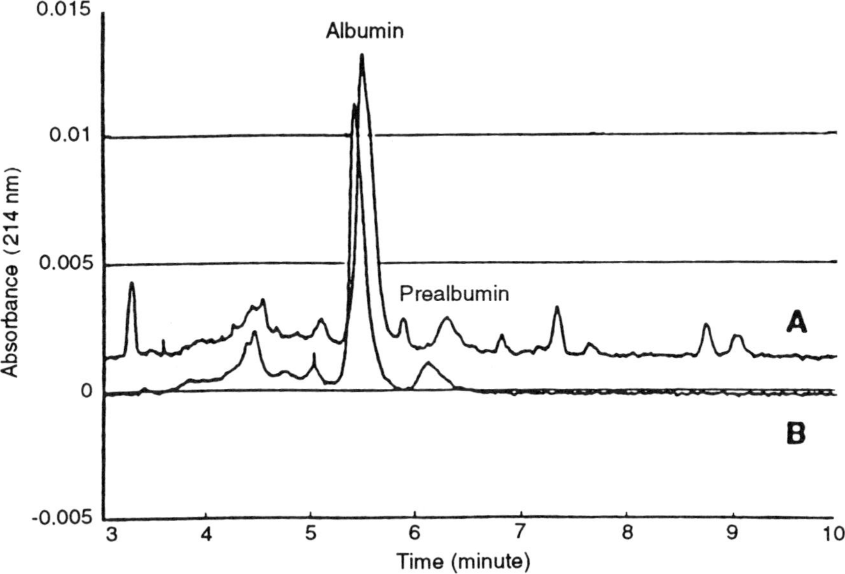

Figure 20 Capillary electropherogram of a CSF sample (A) before and (B) after dialysis through a membrane having a molecular mass cutoff of 14 kDa. Separation buffer, pH 10.0. Beckman proprietary buffer. (From Ref. 64)

out in an untreated fused silica capillary using a 50-mM sodium tetraborate buffer at pH 8.3.

Analysis of Single Whole Nerve Cells

Capillary electrophoresis has recently begun to fulfill some of the expectations for neurochemical analysis resulting from the technique's high selectivity and low-volume–sampling ability. This has led to the separation of components in single whole neurons from snails [22,42,65]. Neurons from mollusks are well suited to initial whole-cell studies with CE owing to their large size relative to mammalian neurons and because mollusk neurons are part of simpler brains, making interpretation of data from single-cell analysis slightly more manageable. Two distinctly different approaches to CE analysis of whole neurons have been taken. Jorgenson and co-workers have used CE with LIF detection to obtain electropherograms of derivatized components from single whole neurons from the land snail *H. aspersa* [22]. Individual snail neurons were removed by microdissection, homogenized, centrifuged, and injected onto the column with a micropipette, as described earlier. The components of the single-cell sample were derivatized with NDA before injection. Several peaks from the resulting electropherogram were identified as labeled

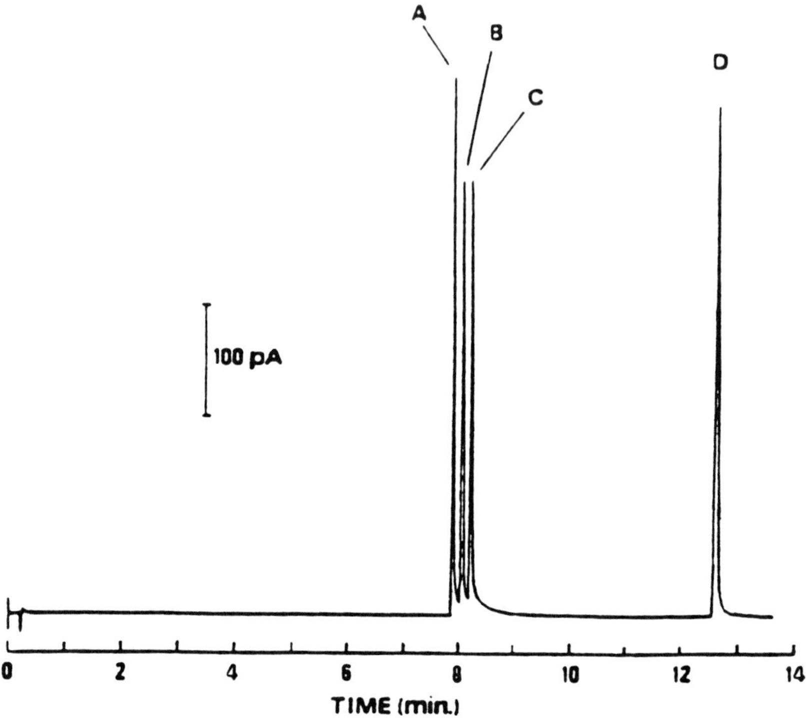

Figure 21 Electropherogram of catecholamines with electrochemical detection on a 26-μm–id column using a 10-μm carbon fiber electrode; 0.02 M MES buffer (pH 6.05); injection, 2 seconds at 5 kV; separation potential, 25 kV (2 μA); electrode potential, 0.7 V vs SSCE; A, dopamine; B, norepinephrine; C, epinephrine; D, catechol. (From Ref. 14)

amino acids; however, most of the peaks present were not identified. No attempts to quantitate the compounds in the snail neurons were made.

A considerable amount of quantitative data from single nerve cells has been obtained by the Jorgenson group using OTLC with LIF detection and electrochemical detection [22,52]. Individual *H. aspersa* neurons have been studied, using the sampling, precolumn derivatization, and injection techniques, described previously, with OTLC coupled to both LIF detection and electrochemical detection [22,52]. Amino acids and aminergic neurotransmitters have been quantitated using these techniques, and specific *H. aspersa* neurons have proved to be distinguishable, based on their chemical profiles. In addition, microcolumn LC with electrochemical detection has been used for the quantitative analysis of epinephrine and norepinephrine in individual cultured bovine adrenomedullary cells [66].

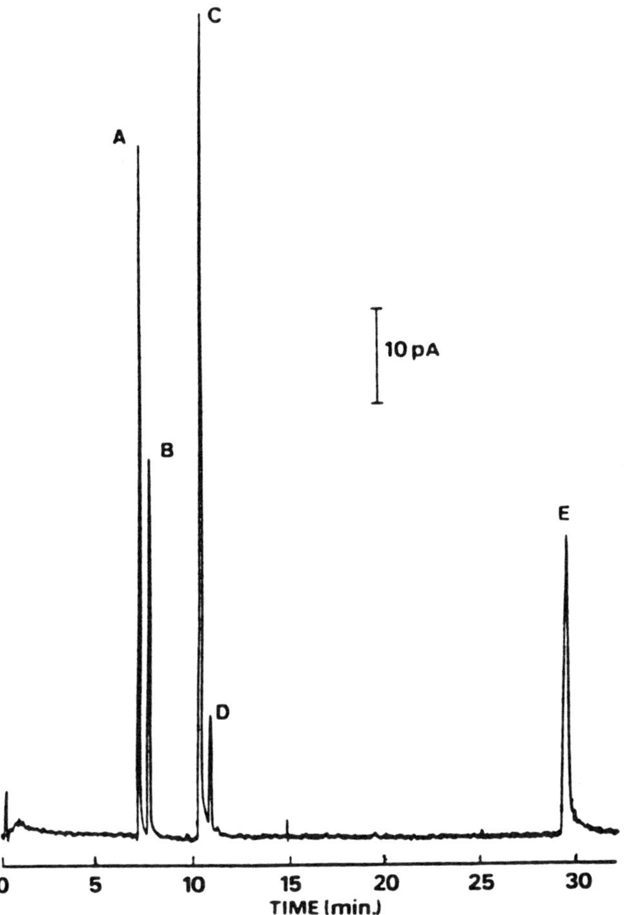

Figure 22 Capillary electropherogram of borate-complexed catechols and serotonin in a micellar solution: A, 1.0×10^{-5} M NE; B, 1.0×10^{-5} M E; C, 1.8×10^{-5} M DA; D, 1.0×10^{-5} M CAT; E, 2.0×10^{-5} M 5-HT. Electrochemical detection was used with a 10-μm carbon fiber electrode at 0.7 V vs SSCE. (From Ref. 18)

Ewing and co-workers have employed a different strategy for the analysis of whole neurons from the pond snail, *P. corneus*, using CE [42,65]. As described in the methods section, whole nerve cells can be pulled into the CE capillary by electroosmotic flow. When the dopamine cell of *P. corneus* is sampled and lysed (60 seconds of pH 5.65 MES buffer injected over the cell), some interesting results are

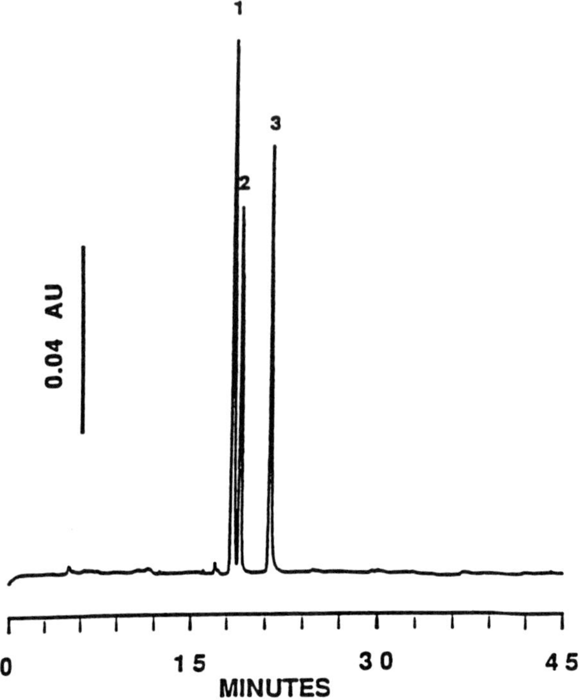

Figure 23 Electropherogram of a mixture of 1, c-AMP; 2, c-GMP; and 3, c-IMP. Samples were monitored at 210 nm. (From Ref. 31)

observed (Fig. 24). Peaks a, c, d, and e have been tentatively identified as dopamine, all electroactive neutrals, uric acid, and dihydroxyphenylacetic acid, respectively. Peak b is considerably broader than the others, and the size of this peak can be altered by bathing the cell with reserpine, which displaces dopamine from neurotransmitter vesicles. In fact, peak b nearly disappears following reserpine treatment, suggesting that this peak represents vesicular dopamine stores in the cell. The following model has been proposed for these data. Peak a apparently represents cellular dopamine from the cytoplasmic compartment and from vesicles that are easily released when the cell is stimulated by the lysing buffer. This is the easily released or "functional" compartment of dopamine. Peak b appears to represent dopamine present in vesicles located more internally in the cell and, hence, these are not lysed during the initial 60 seconds lyse time. Also when cells are injected onto the capillary and lysed for 5 minutes, peak b virtually disappears, and peak a increases by over 700%. These data are consistent with the hypothesis that neurotransmitters exist in two compartments, including a functional compartment and

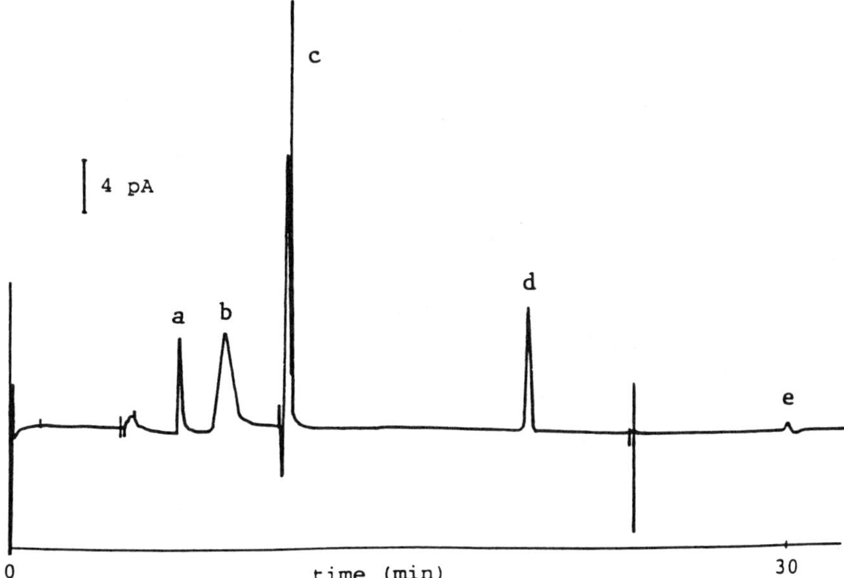

Figure 24 Electropherogram of a 75-μm–diameter dopamine cell from *P. corneus*. Conditions: 25-μm–id fused silica capillary with the injection end etched to approximately 60-μm–od; buffer, 25 mM MES, pH 5.65; 25-kV separation potential; 10-kV injection potential; detection, 11-μm diameter carbon fiber, 0.8 V vs SSCE.

nonfunctional (reserve) compartment [67,68]. However, these CE experiments represent the first data testing this hypothesis at the level of a single nerve cell.

Analysis of Cytoplasmic Samples From Individual Neurons

The analysis of single-cell compartments by CE has been taken a step farther by demonstrating the separation and detection of small portions of the cytoplasm sampled from single intact invertebrate neurons in vivo [28,30,41,43,53]. By using an etched capillary tip as an injector (as described in the methods section), 50- to 100-pL volumes of cytoplasm have been electrokinetically injected onto the capillary. This volume range represents approximately 1% of the total cell volume from these large cells. The overall injection system for acquisition of cytoplasmic samples is shown in Fig. 25. This research has primarily focused on the study of the giant dopamine cell found on the left pedal ganglion of the pond snail, *P. corneus*, [30,41,43]. Figure 26 shows an electropherogram from a 66-pL injection of cytoplasm from a

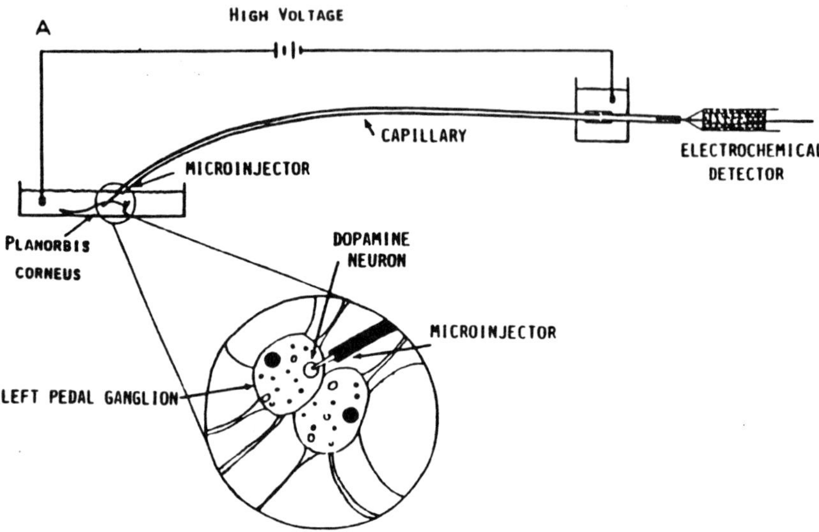

Figure 25 Schematic diagram of the capillary electrophoresis system used for the injection, separation, and detection of cytoplasmic samples. Also shown is an exploded view of the left and right pedal ganglia of *P. corneus* with the microinjector inserted directly into the dopamine neuron. (From Ref. 30)

giant dopamine cell. The first and third peaks were identified, respectively, as dopamine and dihydroxyphenylacetic acid (DOPAC), based on their calculated electrophoretic mobilities. The second peak was determined to be nonionic and, therefore, could not be identified; however, it can be used as an internal marker for electroosmotic flow. Quantitation of the dopamine peak provides a direct measurement of the cytoplasmic concentration of dopamine in a sampled neuron. Measurements with five snails (one measurement from the giant dopamine neuron per snail) provide a cytoplasmic dopamine concentration of 2.2×10^{-6} M with an error (SEM) of 0.52×10^{-6} M. The dopamine peak in Fig. 26 was determined to represent 105 amol.

Acquisition and separation of cytoplasmic samples with CE has been used together with intracellular voltammetry to estimate the amount of cytoplasmic dopamine versus vesicular dopamine present in the giant dopamine neuron in *P. corneus* [41]. Intracellular voltammetry with carbon ring microelectrodes ($r < 5$ μm) provides estimates of cytoplasmic dopamine below 2×10^{-6} M. Bathing of the cell with ethanol (300 μL, 50% solution) results in an increase of intracellular dopamine concentration to $9.7 \pm 0.9 \times 10^{-5}$ M as measured by voltammetry. These results suggest that the ethanol bathing promotes release of dopamine from vesicles in the neuron. Dopamine stored in vesicles is not accessible to voltammetric

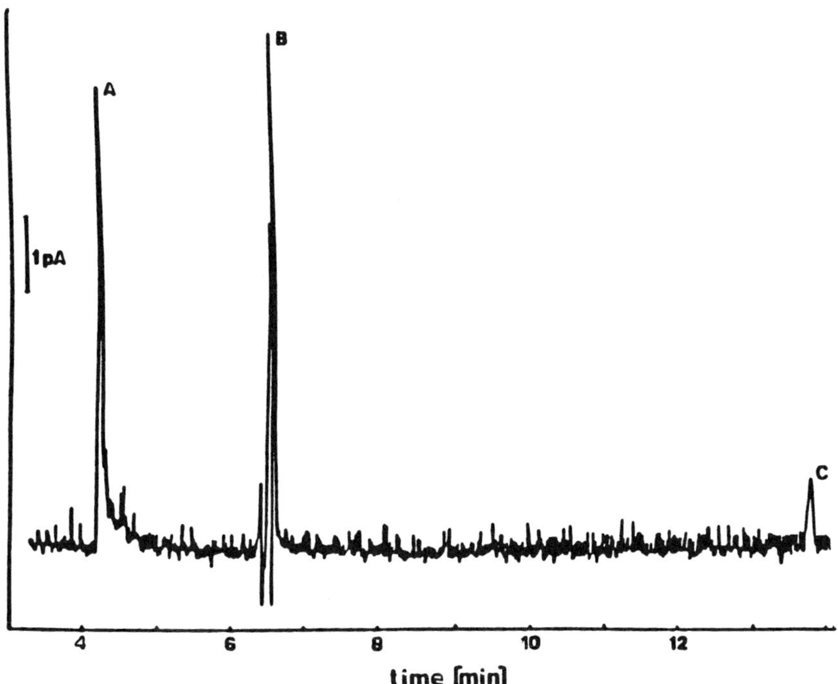

Figure 26 Capillary electrophoretic separation of solutes in a sample of cytoplasm removed from inside the giant dopamine neuron of *P. corneus*: electrophoresis capillary length, 66 cm; buffer, 0.025 M MES, pH, 5.65; injection, 5 seconds at 10 kV through a 700-μm–long, 10-μm–od microinjector; separation potential, 25 kV. Detection electrode, 0.7 V v SSCE. The injection volume was 66 pL, based on electroosmotic flow. Peaks A, B, and C have been identified as dopamine, a nonionic species, and dihydroxyphenylacetic acid, based on their calculated electrophoretic mobilities. The electrophoretic mobility of peak A is 2.40 $cm^2 V^{-1} s^{-1}$ and that of peak C is -2.28 $cm^2 V^{-1} s^{-1}$. (From Ref. 30)

measurement. Cytoplasmic sampling by CE confirms the voltammetric dopamine measurements after ethanol bathing. The voltammetric response has been identified as being due to dopamine, based on the shape of the voltammogram and half-wave potential, and CE confirms this identification; based on the electrophoretic mobility of dopamine. In addition, the dopamine concentration estimated after ethanol bathing is in the range from 4.7×10^{-5} to 1.4×10^{-5} M. The levels of dopamine for this study can only be estimated owing to the use of a microinjector that is less reliable than that developed later [28,40]. The results of this study suggest that at least 98% of the dopamine found in the giant dopamine neuron is bound and not directly accessible to the cytoplasm [41].

Measurements of cytoplasmic neurotransmitter levels have also been conducted for the large serotonin neuron on the right pedal ganglion of *P. corneus* [28]. The same sampling, separation, and detection methods used for studying the giant dopamine cell were used for studying the serotonin cell. Figure 27 compares an electropherogram from a serotonin cell and an electropherogram of a standard sample. The first peak has been identified as serotonin, based on calculated electrophoretic

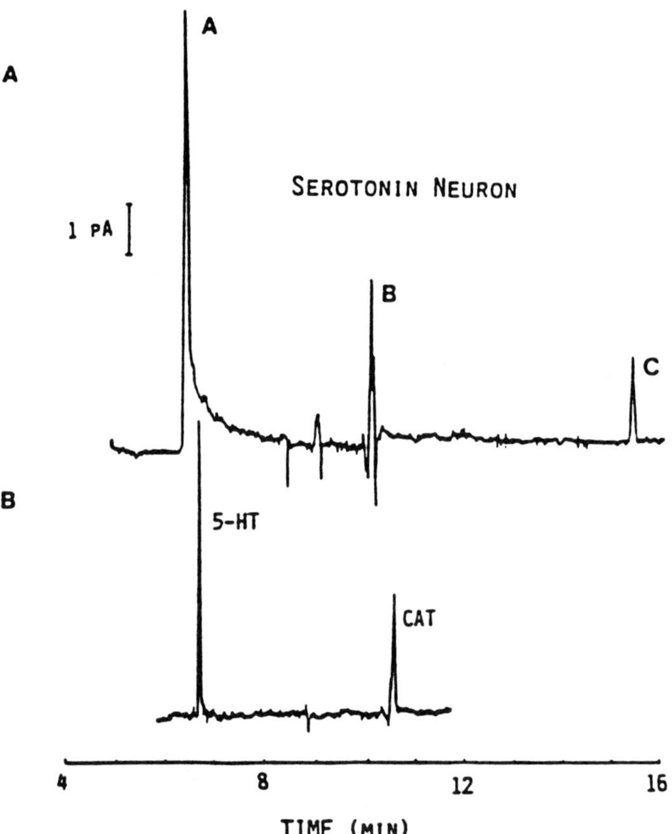

Figure 27 Comparison of electropherograms of a cytoplasmic sample obtained from a large serotonin neuron of *P. corneus* (part A) to an electropherogram of serotonin and catechol obtained from a standard solution (part B): separation capillary length, 77 cm; capillary id, 5 μm; buffer, 25 mM MES (pH 5.65); injection, 5 seconds at 10 kV (*i*, 9 nA); separation potential 25 kV (*i*, 21 nA). The electrophoretic mobilities of peaks A and C in part A correspond to the calculated electrophoretic mobilities of serotonin (basis for identification) and an unidentified anion, respectively. (From Ref. 28)

mobility, and the second peak is an unidentified neutral species marking electroosmotic flow. The third anionic peak has not been identified. The concentration of serotonin in this cell has been determined to be $3.1 \pm 0.57 \times 10^{-6}$ M (SEM) based on three experiments, each with a serotonin neuron from a different snail.

FUTURE PROSPECTS

The future prospects for the application of capillary electrophoresis to the study of neurochemistry promise to be both very rewarding and extremely challenging. All three of the important aspects for use of CE in neuroscience—sampling, separation, and detection—have been developed sufficiently to allow the analysis of neurochemicals at the single-cell level, but room is still available for substantial improvement in all three areas. The neurochemical studies accomplished to date with CE have been most interesting, yet technically difficult. However, the molecules and systems studied so far have been at the simplest and most accessible end of a spectrum of very complicated and comparably inaccessible neurochemical problems waiting to be examined.

Sample acquisition is the key to using CE for the analysis of extremely small volumes in heterogeneous neurochemical environments. Although this area has progressed to the extent that single whole cells and portions of single-cell cytoplasm can be sampled, room for improvement does exist. The snail neurons that have been sampled are very large and easily identified relative to mammalian neurons. In addition to sampling whole mammalian neurons and their cytoplasm, it would be extremely desirable to sample other smaller structures, such as vesicles, axons, and ultimately single synapses.

Sampling from these environments will require miniaturized instruments. The smallest sampling devices yet reported have been fabricated from 5-μm–id/150-μm–od capillaries and, in principle, smaller injectors could be made from the smallest commercially available capillaries with a 2-μm id [28]. The injection method used by Jorgenson and co-workers [22,66] cannot be used to sample intact cells or to sample from intact cells, but it does not suffer from the ionic bias of electrokinetic injection, and it allows the use of precolumn derivatization of samples. Finally, some type of new injection system could be developed for continuous small-volume sampling in neurochemical environments. It must also be remembered that small-volume injections from extremely small structures pushes microscopy and micromanipulation technology to their limits, and improvements in these areas can be as important as improvements to the actual sampling devices themselves.

Improvements in the CE separations will be important to future applications of CE in neurochemistry. Many important neurochemicals are very similar to each other in structure, as exemplified by catecholamines, neuropeptides, and proteins. Separating several closely related compounds found in the same environment can be difficult, and this requires optimizing separation conditions and finding new experimental conditions to improve resolution. One of the most critical aspects of improving CE separations of neuropeptides and proteins is minimizing adsorption of these species to the negatively charged capillary wall to reduce band broadening. As shown in the separation of bradykinin and lys-bradykinin, minimization of adsorption can have dramatic effects on a separation [21]. Also, when sampling from neurochemical environments, it is important that adsorption of molecules in the matrix (i.e., lipids and proteins) does not have too great an effect on the separation, column, and detector. Finally, the development of MECC separations of neurochemicals [14,43,53] is important to CE in neuroscience because it will allow sampling and separation advantages of CE to be applied to neutral species found in neurochemistry, such as the unidentified neutral species found in electropherograms from the cytoplasm of *P. corneus* neurons [30].

Improvements in detection may be the most important and most challenging factor in the future application of CE to neuroscience. To take advantage of the sampling capabilities of CE, a detector must be sensitive enough to detect trace compounds in picoliter samples. This translates to the need for attomole and zeptomole mass detection capabilities. Electrochemical detection, LIF detection, and radiometric detection, all can meet this requirement; however, all suffer from shortcomings [53]. Electrochemical and LIF detection will detect only a few compounds without derivatization, and radiometric detection will detect only radiolabeled molecules. Single-cell analysis has mainly focused on electroactive neurotransmitters and labeled amino acids, but neuropeptides, which are present at considerably lower concentrations, have not yet been examined at the single-cell level. Intracellular messengers, such as cyclic nucleotides and many general cell metabolites, have not been studied at the single-cell level because of a lack of suitable detection schemes.

An important aspect of using available electrochemical and LIF CE detectors for neuroscience is the development of effective derivatization schemes for compounds of neurochemical interest. Precolumn derivatization for both types of detection have already been developed and applied, and postcolumn derivatization schemes have been developed for general use. Development of effective postcolumn derivatization schemes for neurochemicals is most desirable because of minimization of sample-handling errors and simplified injection from ultrasmall environments.

One promising solution to CE detection for neuroscience is mass spectrometry. Mass spectrometry offers two detection advantages that make its continued development very important to CE in neuroscience. First it is a universal detector. Mass spectrometers with API interfaces are capable of analyzing most molecules, ranging in size from a molecular mass under 100 Da to proteins well over 100 kDa. Mass spectrometric detection also adds selectivity to CE analysis by providing mass spectra for eluting components that can be used to identify these components. However, attomole and subattomole detection limits for neurochemicals have yet to be attained with CE–MS.

In summary, the future of CE in neuroscience will depend on the continued development of sample injection, separation, and detection techniques. Sample injection techniques and detection sensitivity must both be improved to begin to study mammalian neurons and subcellular structures such as vesicles and synapses. Finally, analytical chemists and neuroscientists must begin to devote more resources toward applying CE techniques that are currently available toward problems in neuroscience that the methods are already capable of attacking.

ACKNOWLEDGMENTS

We would like to acknowledge the help of the many coworkers whose work is referenced in this chapter. The support of the National Science Foundation and the National Institutes of Health is gratefully acknowledged.

REFERENCES

1. L. L. Iversen, *Sci. Am.*, *241*:134 (1979).
2. R. A. Wall, *J. Chromatogr.*, *60*:195 (1971).
3. P. T. Kissinger, R. M. Riggin, R. L. Alcorn, and L.-D. Rau, *Biochem. Med.*, *13*:299 (1975).
4. G. Schwedt and H. H. Bussemas, *Chromatographia*, *9*:17 (1976).
5. J. H. Knox and J. Jurand, *J. Chromatogr.*, *125*:89 (1976).
6. R. Virtanen, *Acta Polytech. Scand.*, *123*:1 (1974).
7. F. E. P. Mikkers, F. M. Everaerts, and T. P. E. M. Verheggen, *J. Chromatogr.*, *169*:11 (1979).
8. J. W. Jorgenson and K. D. Lukacs, *J. Chromatogr.*, *218*:209 (1981).
9. J. W. Jorgenson and K. D. Lukacs, *Clin. Chem.*, *27*:1551 (1981).
10. J. W. Jorgenson and K. D. Lukacs, *Anal. Chem.*, *53*:1298 (1981).
11. J. W. Jorgenson and K. D. Lukacs, *Science*, *222*:266 (1983).
12. C. L. Rice and R. Whitehead, *J. Phys. Chem.*, *69*:4017 (1965).
13. E. D. Lee, W. Muck, J. D. Henion, and T. R. Covey, *J. Chromatogr.*, *458*:313 (1988).
14. R. A. Wallingford and A. G. Ewing, *Anal. Chem.*, *60*:258 (1988).
15. R. A. Wallingford, P. D. Curry, Jr., and A. G. Ewing, *J. Microcolumn Separations*, *1*:23 (1989).
16. N. A. Guzman, L. Hernandez, and B. G. Hoebel, *Biopharmaceutics, January* (1989).
17. P. D. Grossman, et al., *Anal. Chem.*, *61*:1186 (1989).
18. R. A. Wallingford and A. G. Ewing, *Anal. Chem.*, *61*:98 (1989).
19. K.-F. J. Chan and W. H. Chen, *Electrophoresis*, *11*:15 (1990).
20. W. M. Muck and J. D. Henion, *J. Chromatogr. Biomed. Appl.*, *495*:41 (1989).
21. M. A. Moseley, L. J. Deterding, K. B. Tomer, and J. W. Jorgenson, *Anal. Chem.*, *63*:109 (1991).
22. R. T. Kennedy, M. D. Oates, B. R. Cooper, B. Nickerson, and J. W. Jorgenson, *Science*, *246*:57 (1989).

23. N. A. Guzman, M. A. Trebilcock, and J. P. Advis, In *Techniques in Protein Chemistry II* (J. J. Villafranca, ed.), Academic Press, San Diego, pp. 37–54 (1991).
24. J. V. Sweedler, J. B. Shear, H. A. Fishman, R. N. Zare, and R. H. Scheller, In *International Conference on Scientific Optical Imaging (12/90 Georgetown, GCI)* (*Proc. SPIE*) pp. 37–46 (1992).
25. W. L. Caudill, A. G. Ewing, S. Jones, and R. M. Wightman, *Anal. Chem.*, *55*:1877 (1983).
26. J. G. White, R. L. St. Claire III, and J. W. Jorgenson, *Anal. Chem.*, *58*:293 (1986).
27. J. G. White and J. W. Jorgenson, *Anal. Chem.*, *58*:2992 (1986).
28. T. M. Olefirowicz and A. G. Ewing, *Anal. Chem.*, *62*:1872 (1990).
29. W. G. Kuhr, *Anal. Chem.*, *62*:403R (1990).
30. T. M. Olefirowicz and A. G. Ewing, *J. Neurosci. Methods*, *34*:11 (1990).
31. L. Hernandez, B. G. Hoebel, and N. A. Guzman, In *Analytical Biotechnology* (C. Horvath, and J. G. Nikelly, eds.), American Chemical Society, Washington, D.C., pp. 50–59 (1990).
32. U. Ungerstedt, In *Measurement of Neurotransmitter Release by Intracranial Dialysis* (C. A. Marsden, ed.), John Wiley & Sons, New York, pp. 81–105 (1984).
33. Q. J. Pittman, J. Disturnal, C. Riphagen, W. L. Veale, and L. Bauce, In *Neuromethods 1; General Neurochemical Techniques* (A. A. Boulton, G. B. Baker, eds.). Humana Press, Clifton, N.J., (1985).
34. R. D. Myers, *Physiol. Behav.*, *5*:243 (1970).
35. R. M. Caprioli, *Anal. Chem.*, *62*:477A (1990).
36. N. A. Guzman, L. Hernandez, and J. P. Advis, In *Current Research in Protein Chemistry* (J. J. Villafranca, ed.), Academic Press, New York (1990).
37. J. P. Advis, L. Hernandez, and N. A. Guzman, *Peptide Res.*, *2*:389 (1989).
38. L. Hernandez, N. A. Guzman, and B. G. Hoebel, *Psychopharmacology*, *105*:264 (1991).
39. T. W. Stone, *Microiontophoresis and Pressure Injection*, John Wiley & Sons, New York (1985).
40. R. A. Wallingford and A. G. Ewing, *Anal. Chem.*, *59*:678 (1987).
41. J. B. Chien, R. A. Wallingford, and A. G. Ewing, *J. Neurochem.*, *54*:633 (1990).
42. T. M. Olefirowicz and A. G. Ewing, *Chimia*, *45*:106 (1991).
43. R. A. Wallingford and A. G. Ewing, *Anal. Chem.*, *60*:1972 (1988).
44. R. M. McCormick, *Anal. Chem.*, *50*:2322 (1988).
45. H. H. Lauer and D. McManigill, *Anal. Chem.*, *58*:166 (1986).
46. J. S. Green and J. W. Jorgenson, *J. Chromatogr.*, *478*:63 (1989).
47. S. Hjerten, *J. Chromatogr.*, *347*:191 (1985).
48. J. K. Towns and F. E. Regnier, *Anal. Chem.*, *63*:1126 (1991).
49. R. T. Kennedy, R. L. St. Claire III, J. G. White, and J. W. Jorgenson, *Mikrochim. Acta [Wien]*, *2*:37 (1987).
50. J. W. Jorgenson, et al., *J. Res. NBS*, *93*:403 (1988).
51. R. T. Kennedy and J. W. Jorgenson, *Anal. Chem.*, *61*:436 (1989).
52. M. D. Oates, B. R. Cooper, and J. W. Jorgenson, *Anal. Chem.*, *62*:1573 (1990).
53. A. G. Ewing, R. A. Wallingford, and T. M. Olefirowicz, *Anal. Chem.*, *61*:292A (1989).
54. S. Wu and N. J. Dovichi, *Talanta*, *39*:173 (1992).

55. J. V. Sweedler, J. B. Shear, H. A. Fishman, R. N. Zare, and R. H. Scheller, *Anal. Chem.*, *63*:496 (1991).
56. J. P. Advis, K. Iqbal, A. W. Malick, and N. A. Guzman, In *Handbook of Hormonal Assay Techniques* (F. d. Pablo, C. G. Scanes, and B. D. Weintraub, eds.), Academic Press, San Diego, (in press; 1992).
57. R. D. Smith, J. A. Olivares, N. T. Nguyen, and H. R. Udseth, *Anal. Chem.*, *60*:436 (1988).
58. R. D. Smith, C. J. Barinaga, and H. R. Udseth, *Anal. Chem.*, *60*:1948 (1988).
59. S. A. Lammert and R. G. Cooks, *Rapid Commun. Mass Spectrom.*, *6*:75 (1992).
60. R. D. Smith, J. A. Loo, R. R. Ogorzalek Loo, M. Busman, and H. R. Udseth, *Mass Spectrom. Rev.*, *10*:359 (1991).
61. X. Huang, R. N. Zare, S. Sloss, and A. G. Ewing, *Anal. Chem.*, *63*:189 (1991).
62. S. Sloss and A. G. Ewing, *Anal. Chem.*, *65* (1993).
63. R. D. Smith, J. A. Loo, C. J. Barinaga, C. G. Edmonds, and H. R. Udseth, *J. Chromatogr.*, *480*:211 (1989).
64. F.-T. A. Chen, C.-M. Liu, Y.-Z. Hsieh, and J. C. Sternberg, *Clin. Chem.*, *37*:14 (1991).
65. H. Kristensen, Y. Y. Lau, and A. G. Ewing, *J. Neurosci.* (submitted; 1993).
66. B. R. Cooper, J. A. Jankowski, D. J. Leszczyszyn, R. M. Wightman, and J. W. Jorgenson, *Anal. Chem.*, *64*:691 (1992).
67. M. J. Besson, A. Cheramy, P. Feltz, and J. Glowinski, *Proc. Natl. Acad. Sci. USA*, *62*:741 (1969).
68. A. G. Ewing, J. C. Bigelow, and R. M. Wightman, *Science*, *221*:169 (1983).

29

Capillary Isoelectric Focusing of Peptides, Proteins, and Antibodies

J. R. Mazzeo* and Ira S. Krull

Northeastern University
Boston, Massachusetts

The performance of isoelectric focusing in capillary columns of 200-μm id or smaller began with the work of Hjerten in 1985 [1–2]. His approach of using polyacrylamide-coated capillaries to eliminate electroosmotic flow (EOF) and salt mobilization for detection of focused zones has been extensively developed and commercialized since [3–10]. Another group also reported using polyethylene glycol-coated capillaries and salt mobilization [11]. In 1991, Mazzeo and Krull reported the successful isoelectric focusing separation of model proteins in uncoated capillaries using methyl cellulose as an additive, with EOF as the force of mobilization [12,13]. Independently, in 1992, Thormann et al. also reported similar results with uncoated capillaries and electroosmotic zone displacement [14]. We can expect that other methods of performing capillary isoelectric focusing (CIEF) will soon be reported, possibly using other means of zone mobilization, such as application of a vacuum [15].

It now appears that no one method of performing CIEF is the winner, in terms of being able to separate species over wide- and narrow-range pH gradients with a high degree of resolution, reproducibility, and stability. In this chapter, we will discuss the key experimental variables of each CIEF method reported to date, as well as the advantages and limitations of each. We will also review the applications of CIEF that have been reported on peptides, proteins, and antibodies. The chapter will be divided into two sections: one on coated capillaries and salt mobilization,

**Current affiliation*: Millipore Corporation, Milford, Massachusetts the other on uncoated capillaries with EOF mobilization. Finally, we will conclude by discussing the problems that still need to be addressed in CIEF.

COATED CAPILLARIES AND SALT MOBILIZATION

Background

Early on, Hjerten believed that it was necessary to eliminate EOF for performing isoelectric focusing in silica capillaries [1]. Thus, his initial reports used capillaries permanently coated with methyl cellulose [1] or bonded with polyacrylamide [2–6], which essentially eliminated any EOF. The coated capillary was loaded with the sample–ampholyte mixture, and the focusing voltage applied, with acid at the anode (10 or 20 mM phosphoric acid) and base at the cathode (20 mM sodium hydroxide). The pH gradient was formed with Pharmalyte 3–10 as the ampholytes. Initially, when the focusing voltage was applied, the current was at a maximum value, then immediately began to decay in a logarithmic fashion. This decrease in the current is due to the depletion of charge carriers in the capillary as the focusing proceeds. Theoretically, when the zones have completely focused, they will have no net movement in the presence of the electric field, since they have no net charge and there is no EOF or other bulk flow to move them. Thus, a means of pushing the focused zones past a stationary detection point in the capillary was required.

In Hjerten's first paper, two means of mobilization were investigated [1]: One involved applying a mechanical pump to the capillary and generating hydrodynamic flow. For this, the focusing voltage was on during the pumping to avoid any zone broadening from the hydrodynamic flow. The disadvantage of this method of mobilization is the additional hardware required and the difficulty of applying a pump to a small-diameter capillary. The advantage is that protein zones can be recovered from the capillary quite easily in fractions. This method has not found any subsequent use in CIEF, probably owing to the extra hardware requirements, which would be difficult to implement with any of today's commercial CE instruments. Furthermore, the second-generation CE instruments now appearing on the market have fraction collection capabilities, so that the advantage for this method of mobilization can now be ascribed to others.

The second means of mobilization involved turning the focusing voltage off when focusing had been achieved, followed by replacement of the base at the cathode with acid or replacement of the acid at the anode with base. Next, the voltage was turned on, leading to mobilization toward the cathode if the base was replaced, or toward the anode if the acid was replaced. To explain this phenomenon, we will discuss only the case in which replacement of the base at the cathode with acid is performed, but a similar argument would apply for replacement of the acid at the anode with base.

When both the anode and cathode contain acid (phosphoric acid) solution and the voltage is reapplied to the focused pH gradient, protons enter the capillary at the anode end and phosphate anions enter at the cathode end. The incoming protons begin to titrate the ampholytes and focused proteins until their net charge is positive, causing them to move toward the cathode past the detection point. During the mobilization, the conductivity in the capillary increases owing to the incoming protons and phosphate ions, as evidenced by the increased current.

In Fig. 1, CIEF separations of hemoglobin and transferrin are shown using both anode mobilization and cathode mobilization. In Fig. 1a, base is present at both the anode and cathode, mobilization is toward the anode, and the most acidic protein, transferrin, migrates first, since it focuses closest to the anode. In Fig. 1b, acid is present at both anode and cathode, mobilization is toward the cathode, and the most basic protein, hemoglobin, migrates first.

As seen in Fig. 1, different separation patterns were achieved for each protein in the two different mobilizations. For both, resolution was highest for the protein that migrated first, indicative of loss of resolution for the later-migrating proteins, probably caused by a breakdown of the pH gradient. Thus, when using this type of mobilization, it appears that to obtain optimum resolution for both acidic and basic proteins, both anodic and cathodic mobilization must be performed.

In a later paper [3], Hjerten described the now popular salt mobilization technique. Here, rather than having acid or base at both the anode and cathode, salt (i.e., sodium chloride) was added to either the anolyte or catholyte. Because of electroneutrality, if salt was added to the anolyte, mobilization would be toward the anode, and if added to the catholyte, mobilization would be toward the cathode. Here, we will consider the case in which salt is added to the catholyte, called cathodic mobilization, but similar arguments can be made for addition of salt to the anode.

The concept of electroneutrality in IEF is that for every anion that enters the capillary, a corresponding cation must also enter. As a rough approximation, when acid is at the anode and base at the cathode, for every proton entering at the anode, there must be a hydroxyl ion entering at the cathode. If sodium chloride is added to the base at the cathode and voltage applied, for every proton entering at the anode, there will be an *anion* entering at the cathode. This anion can be either a hydroxyl or a chloride ion. For simplicity, let us consider that five protons have entered the capillary at the anode. Thus, owing to electroneutrality, five anions must enter at the cathode. Depending on the relative mobilities of the chloride ion and hydroxyl ion, it may be that two chloride ions and three hydroxyl ions will enter. In any event, the net effect is an excess of protons entering the capillary, leading to titration of the pH gradient, generating a net positive charge on the ampholytes and bulk movement toward the cathode.

The salt mobilization technique suffers because the mobilization efficiency is not the same throughout the pH gradient [10]. When cathodic mobilization is

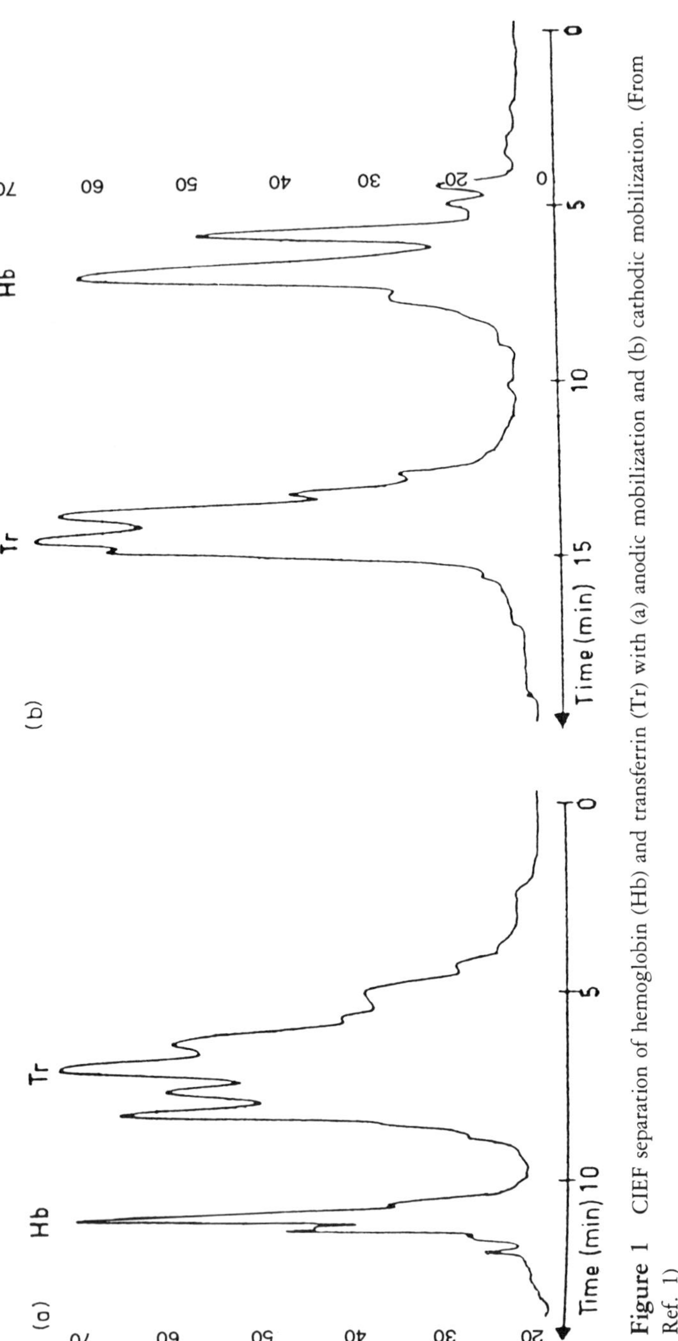

Figure 1 CIEF separation of hemoglobin (Hb) and transferrin (Tr) with (a) anodic mobilization and (b) cathodic mobilization. (From Ref. 1)

performed, the acidic region of the pH gradient is mobilized with less efficiency than the basic region. This has been overcome by using a zwitterion, instead of sodium chloride, for the mobilization. In Fig. 2**A**, CIEF of a standard protein mixture is shown with sodium chloride added to the cathode for salt mobilization. There was no peak seen for the acidic protein phycocyanin, pI 4.65. In Fig. 2**B**, a pI 3.22 zwitterion was added to the cathode instead of sodium chloride, and phycocyanin was detected. It was believed that the zwitterion entered the capillary and stopped in the pH gradient at its own pI, forming an ever-increasing zone width, thereby pushing the pH gradient above the zwitterion's pI out of the capillary. The concept

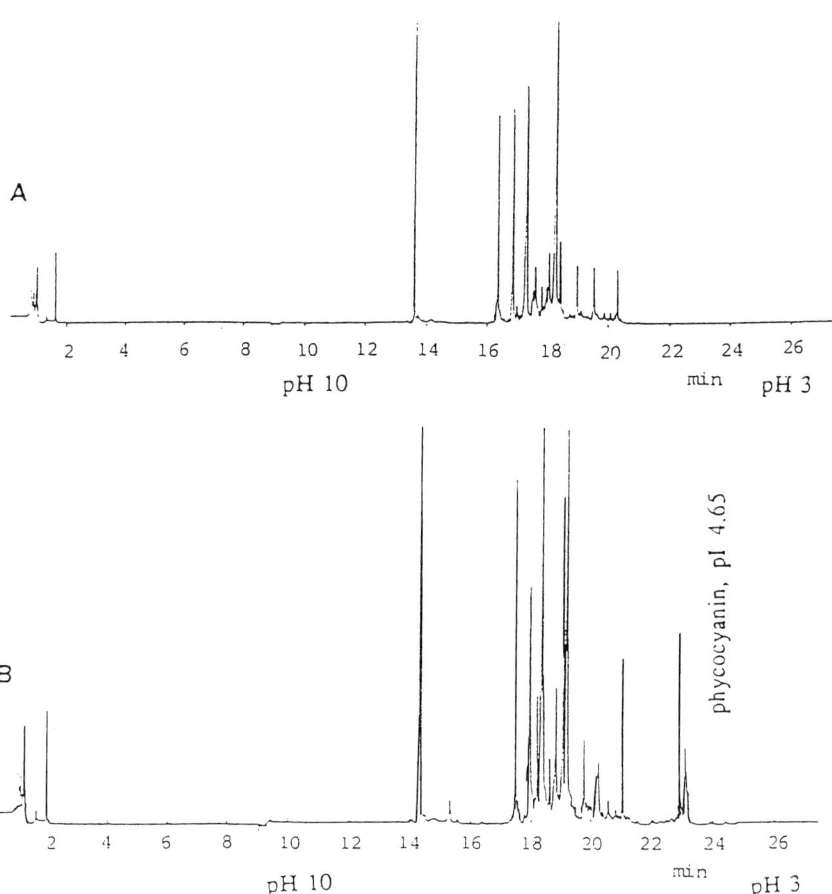

Figure 2 Mobilization of standard protein mix with sodium chloride in (A) the cathode (phycocyanin not detected) and (B) with a pI 3.22 zwitterion in the cathode. (From Ref. 10)

of using a zwitterion for mobilization can be used to selectively mobilize portions of the gradient. In Fig. 3a, the pI 3.22 zwitterion was used for the mobilization, whereas in Fig. 3b, a pI 6.90 zwitterion was used, leading to mobilization of all species with pIs above 3.22 in Fig. 3A, and only those species with pIs above 6.90 in Fig. 3B.

The clear disadvantage of the zwitterion mobilization is the nonlinear resolution generated throughout the gradient, as seen in Fig. 3a. Here, the resolution between the two hemoglobins, pI 7.5 and 7.1, was much larger than the resolution between the pI 7.1 hemoglobin and the phycocyanin, pI 4.65. It would seem that phycocyanin was mobilized with better efficiency than hemoglobin, leading to the poor resolution.

It appears at this point that no means of mobilization, whether salt, zwitterion, or having acid or base at both electrode reservoirs, can provide linear resolution throughout the entire pH gradient. This has to be considered disadvantageous, since one would ideally like to use CIEF to quickly determine pIs of proteins. Theoretically, this would be done by making a plot of migration time versus pI over the pH gradient, then estimating the pI of an unknown based on its migration time. This plot can be used over the entire pH gradient if there is linear resolution, but, this has yet to be achieved. However, as will be described later, one can make plots of migration time versus pI over short regions of the gradient at which they are linear, and these plots can be used to estimate pIs within the region.

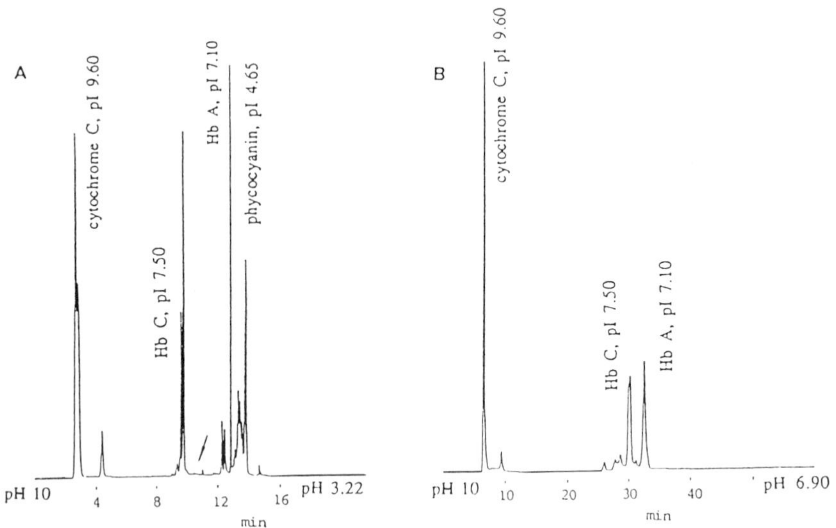

Figure 3 Mobilization of proteins with (A) pI 3.22 zwitterion and (B) pI 6.90 zwitterion added to cathode. (From Ref. 10)

Another means of estimating pIs by CIEF has been described by Kilar [9]. As discussed earlier, during the mobilization step the current increases. In Kilar's approach, a plot of observed current at the migration time of a peak versus its pI was constructed. However, the curve was not linear over a 1 pH-unit portion of the gradient, and it is not clear that this approach could be used over a wide range of pH.

Advantages

The general advantages of the coated capillary and salt mobilization approach of performing CIEF are fast run times, high resolution, and commercial availability. In terms of run times, the focusing step is usually about 5–10 minutes, and the mobilization step about 10–20 minutes, leading to total run times of 30 minutes or less. High resolution has been obtained, as evidenced by the separation of hemoglobins C, S, F, and A (Fig. 4). These proteins differ in pI by as little as 0.05 pH units (A and F). Finally, the entire approach has been commercialized by Bio-Rad (Richmond, California).

Limitations

During the development of CIEF, several problems and limitations were discovered and then minimized. One is that proteins tend to precipitate at their pIs, a well-known phenomenon in IEF, leading to sharp spikes in the separation pattern [4,8,10]. Addition of ethylene glycol or a suitable surfactant (i.e., Brij 35) to the separation media will minimize this problem. Furthermore, the method is very intolerant to salt in the protein sample, and all proteins must be desalted to a maximum of 10 mM before separation [10]. Again, intolerance to salt is well known in all forms of IEF, but seems to be especially problematic in CIEF. The presence of salt is easily noticed, as the initial focusing current will be much higher than usual, and the separation will be marked by very broad peaks and sharp, inconsistent spikes from precipitation.

Another problem is that the very basic portion of the gradient focuses past the detection point in the capillary, near the cathode, and cannot be detected unless mobilization is directed toward the anode. For instance, if a 14-cm capillary is used with a pH 3–10 gradient, there are 2 cm of capillary per pH unit. Thus, if the detection window is 10 cm away from the anode, the pH 8–10 region of the pH gradient will focus in the 4 cm of capillary past the detection point. This can be overcome by supplementation of the ampholytes with tetramethylethylenediamine (TEMED), a compound that extends the basic region of the pH gradient [4,10]. This causes the pH at the detection window to increase, forcing the pH 8–10 region of the gradient to focus before the window, and allows detection.

Detection is another point to consider. Theoretically, because CIEF is a concentrating technique, detectability should be greatly improved over free-solution

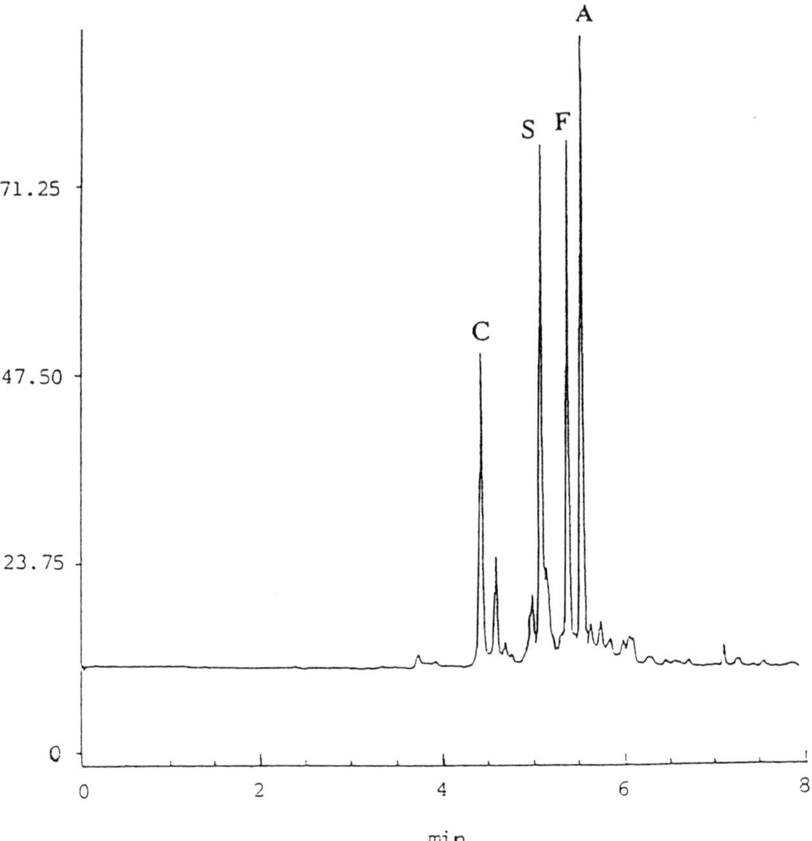

Figure 4 CIEF separation of hemoglobins C (pI 7.50), S (pI 7.25), F (pI 7.15), and A (pI 7.10). (From Ref. 10)

capillary electrophoresis (CZE). However, because the ampholytes used to establish the pH gradient absorb below 280 nm, detectability is compromised in comparison with CZE, for which wavelengths as low as 190 nm can be used. Recent reports of concentration gradient detection for CIEF may overcome this problem [16].

As discussed in the foregoing, a major limitation is the inability to obtain linear resolution throughout the entire pH gradient, owing to the lack of a mobilization scheme that can mobilize the entire pH gradient as a train. An elegant way around this problem may be to scan the entire capillary with the detector when the zones have been focused, either by moving the detector over the capillary, or by moving the capillary through the detector [17,18]. Problems with this approach are that the capillary requirements are strict in terms of shape and UV transparency, and the hardware problems in either moving the detector or the capillary. However, if the problems can be worked out, these approaches could find wide application, because, theoretically, linear resolution could be achieved as well as very fast run times. However, the existence of pH gradient drifts, whether anodic or cathodic, may still make linear resolution impossible.

The question also arises of when to stop the focusing voltage and begin mobilization. It has been said that it is best to stop at some minimum current value, or to stop when a plot of current versus time reaches a certain slope. It has also been reported that the migration time during the mobilization step is dependent on the focusing time [9]. One can imagine that it is critical that the focusing time be rigorously controlled to lead to reproducible migration times. Ideally, one would prefer not to have to perform mobilization in a separate step.

Finally, we come to the issue of the coating used to eliminate EOF; namely, the stability of the coating, especially at alkaline pH [19,20]. The polyacrylamide-coated capillary is now used exclusively, although there has been a report that used a polyethylene glycol (PEG) coated capillary [11]. In the PEG-coated capillary, EOF was not totally eliminated, so that methyl cellulose was added to the analyte–ampholyte mixture to further suppress it. However, there have been no subsequent reports on CIEF that used the PEG-coated capillary, so that the long-term stability of the coating is unknown.

In the polyacrylamide-coated capillary, anchoring to the silica wall is through a siloxane linkage, Si—O—Si, a bond that is known to be susceptible to base hydrolysis. The catholyte solution in CIEF is 20 mM sodium hydroxide, a condition basic enough to cause hydrolysis of the coating. A recent paper reported that when using the polyacrylamide-coated capillary for CIEF of antibodies in a pH 5–8 gradient, the column lifetime was five runs or fewer [19]. Clearly, the stability of the polyacrylamide-coated capillary, when used in CIEF, is an area of concern. Novotny recently described a polyacrylamide-coated capillary that is anchored to the silica wall by a Si—C—Si bond, which is much more pH-stable than the siloxane linkage [20]. This capillary was shown to have good stability at pHs 2 and 11 when run in the CZE mode for protein separations. Thus far, there have been no reports using this coating for CIEF.

It is our feeling that the larger issue is the stability of polyacrylamide itself. Specifically, it is well known that at high pH acrylamide is converted to acrylic acid. Thus, even though the coating may be anchored to the silica wall in a pH-stable manner, as described by Novotny, the acrylamide may be hydrolyzed to acrylic acid, leading to fixed negative charges on the wall and EOF in the cathode direction. It would seem that this problem would become magnified in CIEF, during which the pH in the capillary is not constant. This would lead to an inhomogeneous coating along the capillary. The answer may be to use vinyl acetate as the monomer for attachment to the coupling reagent of Novotny [20]. This would lead to a

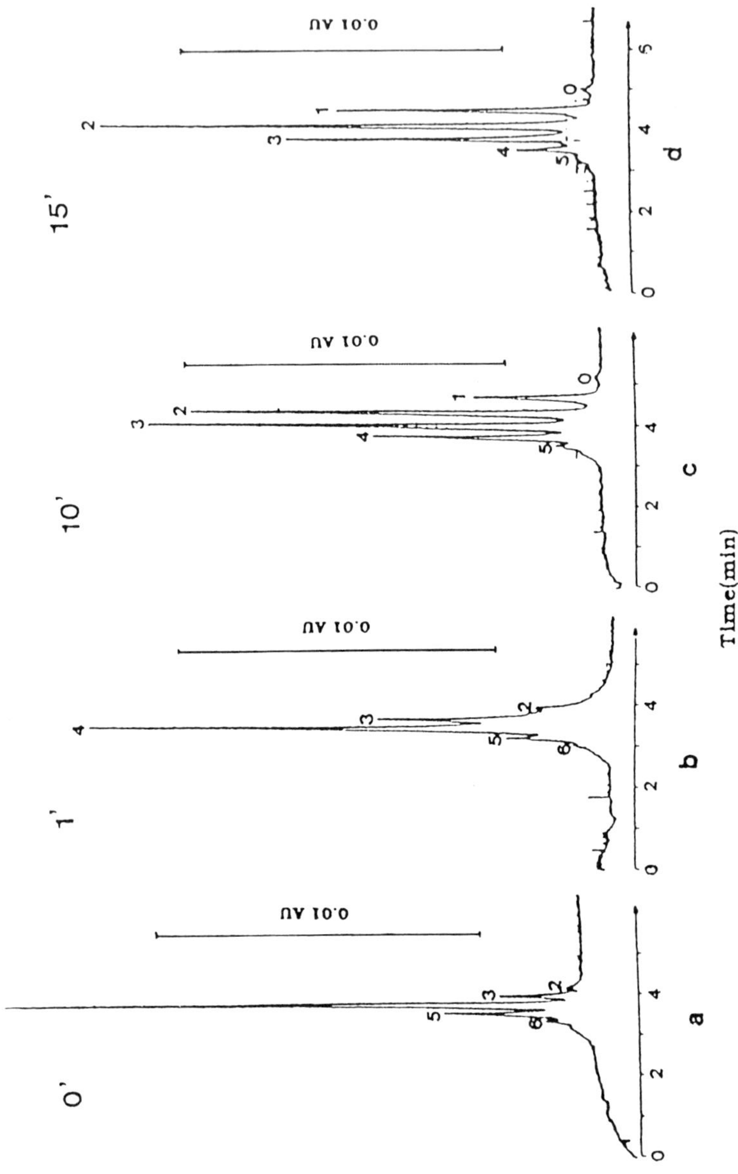
0'
1'
10'
15'
0.01 AU
0.01 AU
0.01 AU
0.01 AU
a
b
c
d
Time(min)

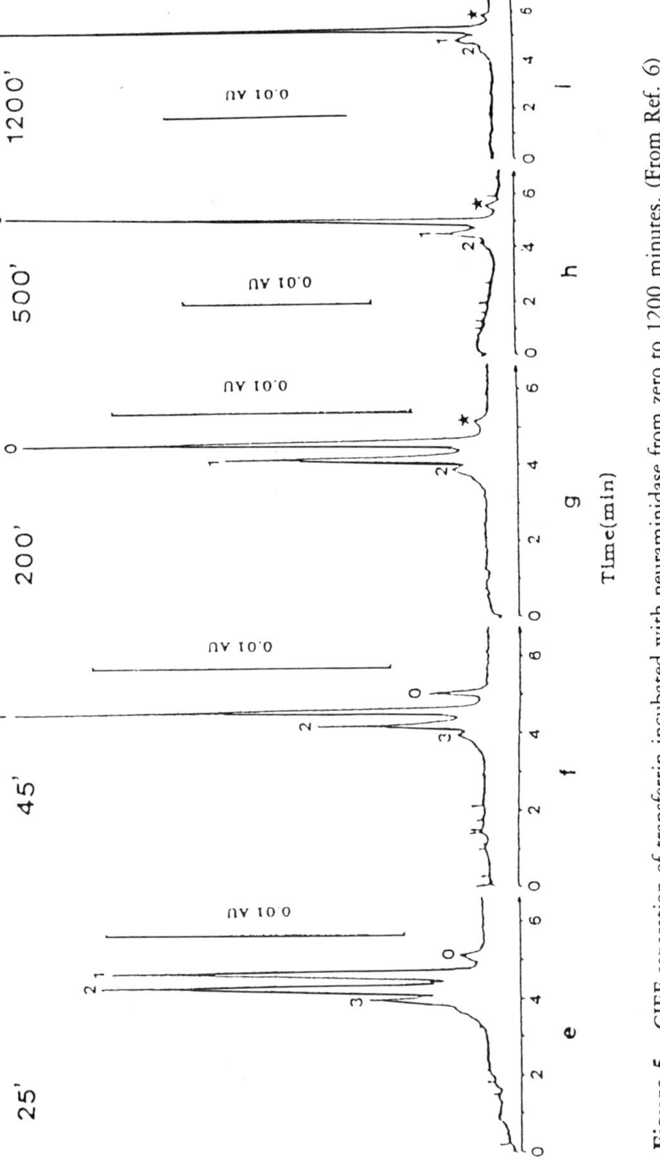

Figure 5 CIEF separation of transferrin incubated with neuraminidase from zero to 1200 minutes. (From Ref. 6)

polyvinyl acetate-coated capillary, which, when base-hydrolyzed, would lead to a polyvinyl alcohol coating. A polyvinyl alcohol coating should be pH stable over a wider range than the polyacrylamide-coated one.

Applications

The initial applications of CIEF with coated capillaries and salt mobilization were for hemoglobins and transferrin. As shown in Fig. 4, the separation of variant hemoglobins is easily realized. Two papers have been dedicated to the separation of transferrin [5,6]. In one [5], CIEF was used to separate transferrin bound to different numbers of iron atoms. When iron-free transferrin was subjected to CIEF separation, several peaks were obtained, which were believed to arise from different amounts of sialic acid bound to the transferrin [6]. When the iron-free transferrin was incubated with neuraminidase and the sample separated by CIEF over time, a clear pattern arose in terms of decreasing number of peaks (Fig. 5) presumably caused by release of sialic acids.

Another glycoprotein application has been reported by Yim [21]. The glycoforms of recombinant tissue plasminogen activator (rt-PA), were separated in a pH 6–8 gradient. Several peaks were seen, but incubation with neuraminidase led to only two major bands of higher pI, strongly suggesting that the different peaks corresponded to protein with different amounts of sialic acid. This example, and the foregoing transferrin, indicate that separating glycoforms with varying sialic acid content is a strong point of CIEF. Another potential application area of CIEF is for monitoring deamidation and proteolytic clipping of proteins. This has been reported for human growth hormone (hGH), for which CIEF was used to monitor the degradation of hGH to the deamidated and clipped forms [8].

Separation of antibodies is another potential strong point of CIEF. Silverman et al. have reported the separation of a recombinant antibody, MAb b72.3, by CIEF into five major and one minor bands in a pH 5–8 gradient (Fig. 6) [19]. In their study, a direct comparison was made between slab gel IEF, CIEF, and ion-exchange chromatography. In comparison with the other two techniques, CIEF gave the same resolution as slab gel IEF, better than ion exchange; CIEF was much faster than gel IEF; and the reproducibility of peak areas was much better by CIEF in comparison with gel IEF. However, the major limitations of CIEF were the poor reproducibility of migration time, as high as 33% relative standard deviation (RSD), and the instability of the capillary coating, with capillaries being good for only three to five runs.

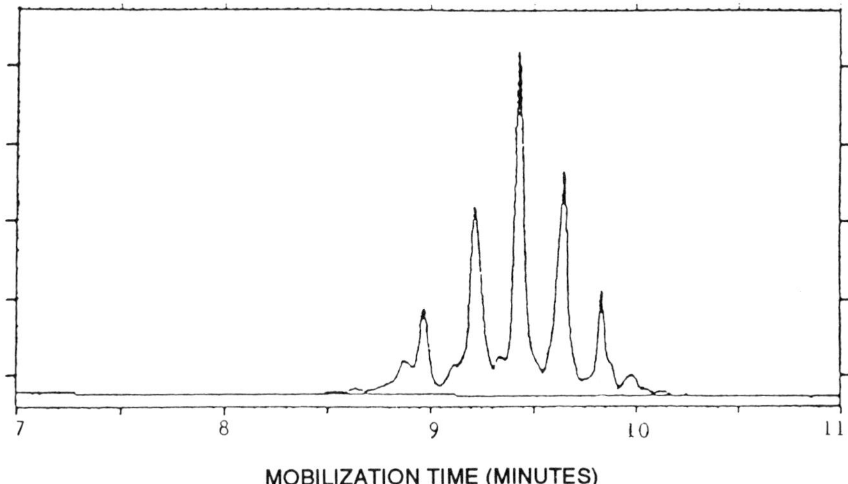

Figure 6 CIEF separation of a monoclonal antibody. (From Ref. 19)

UNCOATED CAPILLARIES WITH ELECTROOSMOTIC MOBILIZATION

Background

The problems with the coated capillary and salt mobilization approach to CIEF led us and others to explore alternative approaches for performing CIEF [12–14]. We felt that the solutions to some of the foregoing problems could be found through reexamination of the need to eliminate EOF. In theory, it is not necessary to eliminate EOF when performing CIEF, only that EOF be slow enough to allow bands to focus before migrating past the detection window [22]. Thus, by optimizing EOF, one could achieve successful focusing and resolution of protein bands, while also maintaining a bulk flow to mobilize zones past the flow cell for detection. This approach would then obviate the need for performing salt or other means of mobilization, and focusing, and elution could be done in one step.

It is possible to make coated capillaries that do not totally eliminate EOF. However, if one wished to be able to conveniently pick a certain EOF rate, the use of several different coated capillaries did not seem to be a viable approach. It had been demonstrated in CZE that EOF can be easily altered by using various concentrations of methyl cellulose [23]. We chose this approach to optimizing EOF, as did others [14].

Figure 7 shows the separation of four model proteins in an uncoated capillary, with 0.1% methyl cellulose to control EOF and 1% TEMED to allow detection of

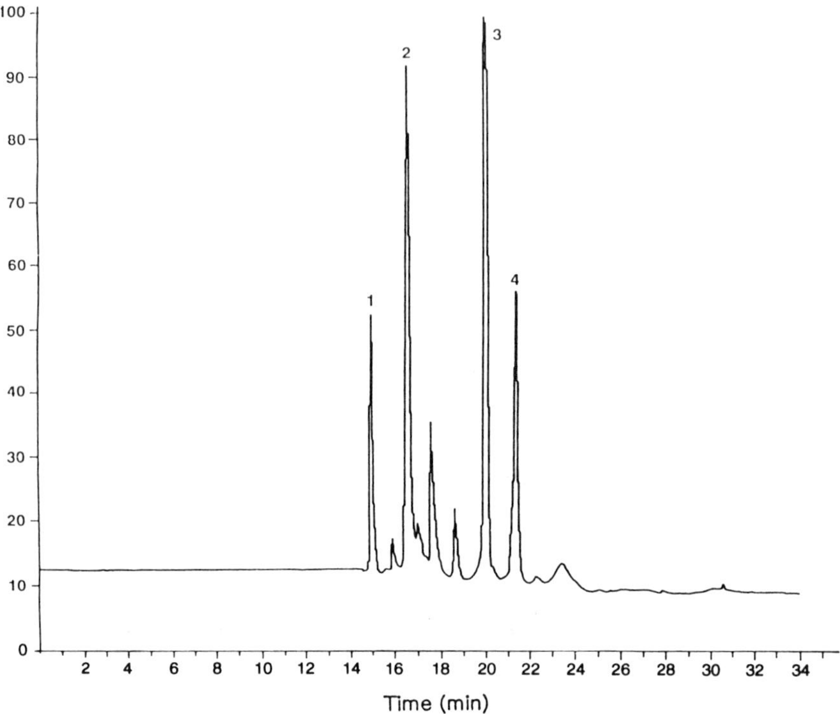

Figure 7 CIEF separation of four model proteins in an uncoated capillary with EOF mobilization. Peaks: 1, cytochrome *c*, pI 9.6; 2, chymotrypsinogen A, pI 9.1; 3, myoglobin, pI 7.2; and 4, myoglobin, pI 6.8. (From Ref. 13)

basic proteins. Note that the migration order of the proteins was high pI to low pI, since the direction of movement (i.e., the EOF) was toward the cathode. Good resolution of the proteins was achieved, and a plot of migration time versus pI for the peaks in this separation was linear, with a correlation coefficient of 0.999. Migration time reproducibility was 2.2% RSD.

The results in Fig. 7 clearly demonstrated that CIEF could be performed in uncoated capillaries, with EOF as the force of mobilization. When the concentration of methyl cellulose was increased, the migration times and peak widths of protein zones increased, owing to the decreased EOF. If the separation in Fig. 7 was performed with no TEMED, the cytochrome *c* and chymotrypsinogen A peaks were not detected, because they focused in the region past the detection point on the cathode side. If the separation was attempted without any methyl cellulose or TEMED, very broad peaks were obtained, with poor resolution, probably because of adsorption of protein to the capillary wall. Thus, the methyl cellulose served to both minimize EOF and adsorption of proteins to the wall, whereas TEMED seemed to do both equally well.

There are many variables that play a role in CIEF separations in uncoated capillaries with EOF mobilization, including the field, buffer concentrations, capillary length, concentration of methyl cellulose, and concentration of TEMED. A study was done on the effect of each one these variables, using the same model proteins as in Fig. 7 [24]. When all other variables were constant, longer column length led to better resolution and longer run time, as was expected. For methyl cellulose, up to 0.2% its effect was to decrease peak width and migration time, probably by decreased protein adsorption to the wall. At higher concentrations, the effect was to increase run time and peak width, as a result of decreased EOF.

Optimization of the TEMED concentration was required, since there are two competing factors. The concentration must be high enough to cause the basic proteins to focus before reaching the detection window. However, if the concentration is too high, EOF will decrease, leading to longer migration times and broader peaks. The TEMED will also affect the resolution of bands, since the more space in the capillary TEMED occupies, the less space for the pH gradient to extend. For conditions in Fig. 7, a TEMED concentration of 1–1.4% was appropriate.

It was also found that the TEMED concentration and applied electric field were dependent variables. Higher fields required lower TEMED concentrations for optimum separation of the model protein mix. This was believed to arise from a current limitation; that is, there appeared to be a maximum initial current that could be tolerated before the separation rapidly degraded, owing to precipitation of protein bands. An initial current of 40–50 μA was the limit. Since higher fields and higher TEMED concentrations lead to larger current, it is not surprising that higher fields required lower TEMED concentrations. This current limitation may be a joule-heating problem (i.e., the higher current led to higher powers and higher temperatures, which led to precipitation and variation of the EOF along the capillary diameter).

The total run time in Fig. 7 was 24 minutes, which we were interested in decreasing, if possible. That is, one would like to be able to perform very fast separations, 10 minutes or less, for the purpose of screening samples. This can be done by decreasing the separation distance. In Fig. 7, the total capillary length was 60 cm, 40 cm from the anode to the detection point. By reversing the polarity and switching the buffers, the distance from anode to detection point can be cut in half to 20 cm. When this was done, using all other conditions as in Fig. 7, the migration time of the pI 6.8 peak decreased from 21 minutes to 6 minutes [25]. However, this large decrease in run time came at the expense of lower resolution and broader peaks throughout the gradient, especially in the basic region. This was because the proteins were detected before they had completely focused. By increasing the concentration of TEMED to 1.2%, the migration time increased to 9 minutes, but the resolution was greatly improved (Fig. 8).

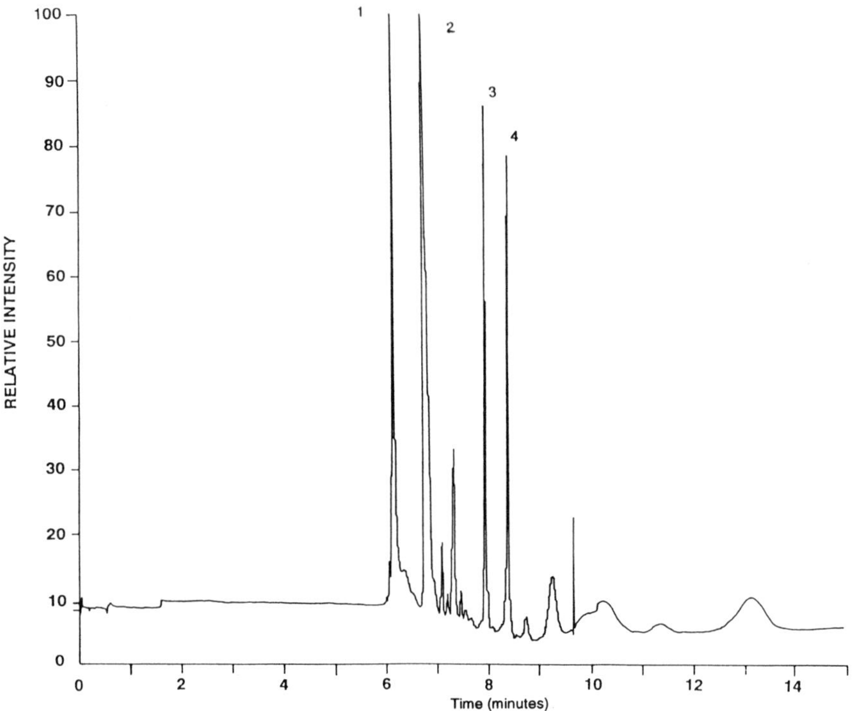

Figure 8 CIEF separation of four model proteins in 20-cm separation distance. Peaks as in Fig. 7. (From Ref. 25)

An easy way to shorten the separation distance would be to use a shorter capillary. However, for us, the minimum capillary length required by the instrument was about 60 cm. The TEMED concentration also influences the effective separation distance. At low TEMED concentrations, the separation distance is shorter than at high TEMED levels. However, one must also be aware that TEMED concentration will also influence the resolution if dramatic changes are made.

The effects of all the different variables may seem a bit confusing. In practice, we have found that a capillary length of 60 cm with a field of 400 V cm^{-1} is the best place to start. The sample should be dissolved in 5% Pharmalyte 3–10, with a methyl cellulose concentration of 0.1%. Then, various TEMED concentrations should be attempted in either a normal or reversed polarity mode until the desired separation is achieved.

Two areas not yet discussed are the effect of buffer concentrations and the separation of acidic species. In Fig. 8, β-lactoglobulin A, pI 5.1, was included in the protein–ampholyte mixture. Clearly, within the time frame monitored, no peak was seen for this acidic protein. It was believed that this was because the protein was moving at a slower rate than the other neutral and acidic species, because of the presence of anodic drift in the pH gradient. It is well known that both anodic and cathodic drift can occur in ampholyte gradients [26]. When EOF is in the direction of the cathode, as here, cathodic drift is not a real problem, since it will only lead to faster migration of basic species. However, anodic drift will cause problems for acidic species, since it is in the direction opposite that of EOF. Thus, acidic species will experience competing forces and may not move toward the cathode as the other species do. The solution to anodic drift was to increase the concentration of acid in the anode [26]. When the phosphoric acid concentration was increased from 10 to 20 mM using the conditions in Fig. 8, the separation in Fig. 9 was obtained, and β-lactoglobulin was detected.

Note that the peak for β-lactoglobulin was much broader than the peaks for the basic and neutral species. This was because the fastest an acidic species could be moving was the speed of EOF, whereas basic and neutral ones may move with

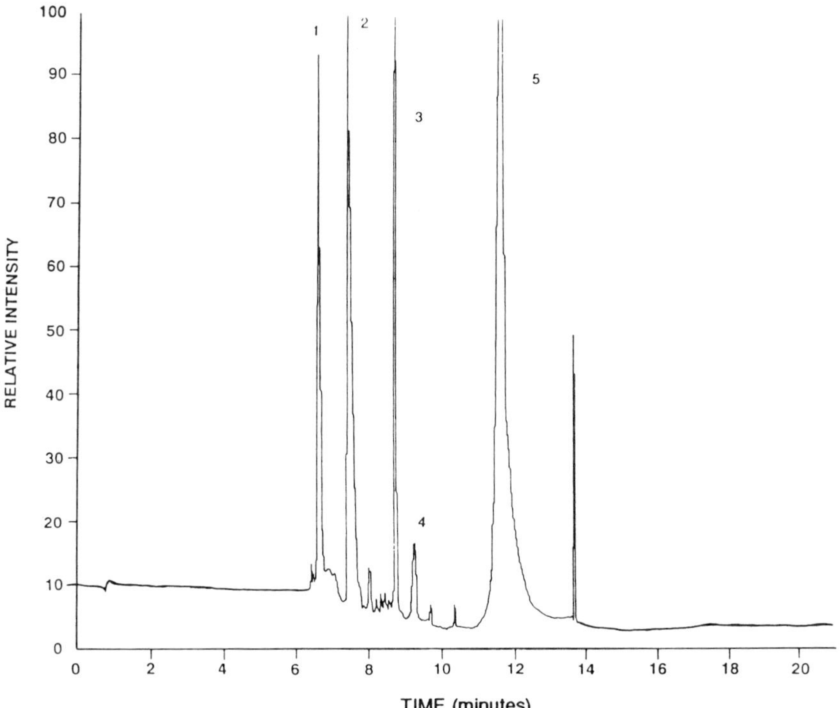

Figure 9 CIEF separation of model protein mixture with 20 mM phosphoric acid as anolyte. Peaks as in Fig. 7, plus 5, β-lactoglobulin A. (From Ref. 25)

both EOF and cathodic drift. This also assumed that EOF was constant during the entire run, which was probably not true. The EOF was a function of the average pH in the capillary, and as the CIEF separation proceeded, the average pH in the capillary decreased. This suggested that EOF also decreased during the run. Further evidence for this theory was that when a plot of migration time versus pI was made using all proteins in Fig. 9, the correlation coefficient was 0.949, owing to the larger-than-expected migration time for β-lactoglobulin. If β-lactoglobulin was excluded, the correlation coefficient was 0.999.

Thus, it would appear that, as with coated capillaries and salt mobilization, linear resolution throughout the entire pH gradient cannot be achieved. There are two things that could be tried to alleviate the problem with acidic species. The easiest, experimentally, is to use a capillary with EOF in the direction of the anode. Here, acidic species will migrate first, with the combined effect of EOF and anodic drift, leading to sharper peaks and better resolution; however, the basic species will be very broad and lose resolution, since cathodic drift and EOF would be in opposite directions. An experimentally more difficult solution would be to use an external electric field applied to the capillary to control the zeta potential and EOF, as has been reported for CZE [27]. Initially the external electric field would be low, to allow sufficient time for basic species to focus before elution. As the separation proceeded, the external electric field would be increased to increase the EOF and compensate for the reduced migration speed of acidic species. The focusing voltage would remain constant. Through proper optimization with known pI standards, linear resolution throughout the entire gradient should be achievable.

A similar method of performing IEF in uncoated capillaries with EOF mobilization has been reported by Thormann and associates [14]. They used hydroxypropyl methylcellulose to control EOF. The main difference from the method of Mazzeo and Krull was that only a small amount of the sample–ampholyte mixture was introduced into the capillary filled with the catholyte, rather than filling the entire capillary. The advantage to this approach was no TEMED was required for detection of basic proteins. Figure 10 shows a separation using this CIEF approach

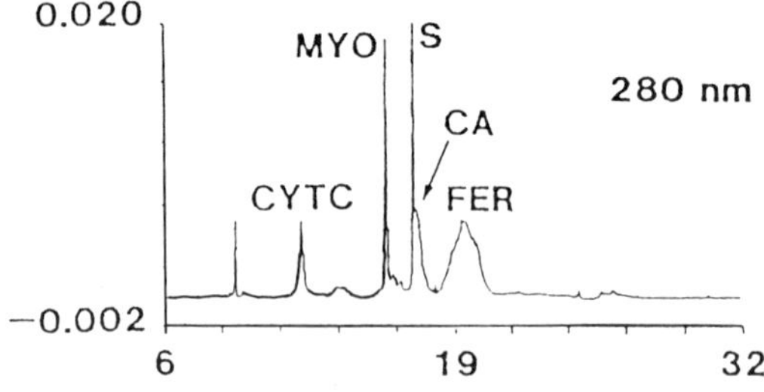

Figure 10 CIEF separation of model protein mix in uncoated capillary with electroosmotic zone mobilization. (From Ref. 14)

of cytochrome *c*, myoglobin, carbonic anhydrase, and ferritin. Run times were about 20 minutes. Note that the peak shape and resolution of the basic and neutral proteins—cytochrome *c*, myoglobin, and carbonic anhydrase—were quite good, but for the acidic protein—ferritin—the peak was very broad. Thus, it would appear that this method suffers from the same problems as that of Mazzeo and Krull, probably from anodic drift and changing EOF. In this example, multiwavelength detection was used, but its usefulness is less than in other forms of CE owing to the background absorption of the ampholytes at wavelengths below 280 nm.

Advantages

Many of the advantages of CIEF in coated capillaries with salt mobilization also apply to uncoated ones with EOF mobilization. These include fast run times and high resolution. There are some significant advantages of CIEF in uncoated capillaries with EOF mobilization over CIEF in coated capillaries with salt mobilization. One is the ease of use, as all that is required is uncoated capillaries, some methyl cellulose, Pharmalytes, and TEMED. For 300 dollars, one can purchase enough of these supplies to do CIEF separations for several months, making this form of CIEF much less expensive than the other, which requires the use of expensive coated capillaries. Focusing and mobilization are performed in one step, which should allow easier automation. Stability of the "coating" is not an issue, since it is continually renewed before every run. Capillaries can be rinsed with harsh solutions, such as 1 N NaOH, to remove bound material. This could not be done with a coated capillary. The lifetime of the capillaries that we have used is on the order of months, and they are rinsed after every run with 0.1 N NaOH. When a serious decrease in separation performance is seen, flushing with 1 N NaOH usually alleviates the problem. If this does not work, replacing the capillary is not an expensive ordeal.

Disadvantages

The main disadvantage of this form of CIEF, as with the other, is the inability to successfully separate and mobilize species over the entire range of the gradient. Here, the reasons are different, namely, nonconstant EOF and anodic drift. A further disadvantage is the possibility of adsorption to the wall. Since there is still some EOF present, there are negative charges on the wall, which could adsorb proteins during the focusing. This problem does not manifest itself in broad peak shapes of basic proteins, which would be expected to give the most adsorption. It may be that adsorbed material stays on the wall and does not come off. If adsorption does occur during the run, this could be another reason for the changing EOF.

Applications

Because this form of CIEF is less than a year old, fewer applications have been reported. It has been applied to the characterization of a cross-linked bovine hemoglobin sample [13]. When this sample was first run in the pH 3–10 gradient, a single peak was seen. However, as the sample was continually run over a 4-hour period, more and more peaks began to appear in the separation (Fig. 11). It was postulated that this was due to oxidation of one, two, three, or all four of the hemes present in the protein. When fresh sample was treated with carbon monoxide to complex with the heme, a single peak was seen by CIEF over a 1-day injection period. However, when the same sample was run in a narrow-range gradient, pH 6–8, two peaks were seen. This was attributed to the hemoglobin existing in two different forms differing by one amino acid.

It is rare that one would want to separate a real protein sample over a wide-range gradient, pH 3–10. Rather, for proteins, one is more interested in looking at deamidation, carbohydrate forms, and so on. In most instances, with the exception of proteins with a large amount of sialic acid, one would want to use narrow-range pH gradients to look for these things. One potential use of the wide-range pH gradients is for separating peptides from an enzymatic digestion of a protein, or for peptide mapping. We have investigated the potential of CIEF for peptide-mapping purposes [28]. Figure 12 shows the CIEF separation of the tryptic peptides from bovine cytochrome *c*. In Table 1a, the sequences and calculated pIs of the expected

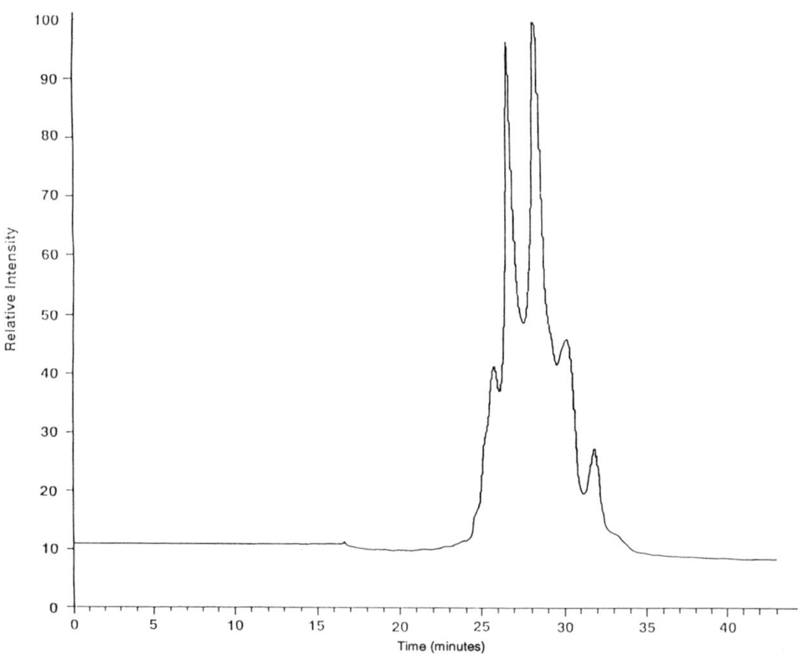

Figure 11 Resolution of different hemoglobin oxidation states by CIEF. (From Ref. 13)

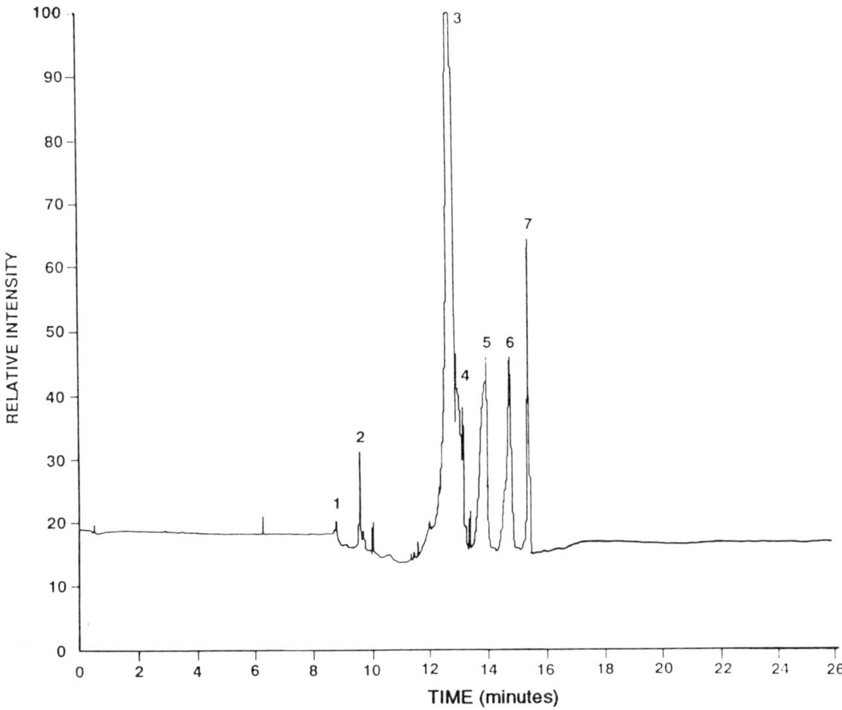

Figure 12 CIEF peptide map of the tryptic digest of bovine cytochrome *c*. See Table 1 for peak pIs. (From Ref. 28)

Table 1a Bovine Cytochrome *c* Expected Tryptic Peptides

Peptide no.	Sequence	pI Calculated
1	KYIPGTK	10.1
2	YIPGTK	9.5
3	KTGQAPGFSYTDANK	9.4
4	TGQAPGFSYTDANK	6.9
5	GITWGEETLMEYLENPKK	4.3
6	EDLIAYLK	4.1
7	GITWGEETLMEYLENPK	3.8
8	EDLIAY	3.0

Table 1b Bovine Cytochrome *c* Tryptic Map

Peak no.	T_m(s)	%RSD	Estimated pI
1	508	2.5	10.6
2	546	2.4	9.5
3	732	2.2	5.6
4	744	2.9	5.4
5	794	2.4	4.6
6	844	2.1	3.9
7	880	2.1	3.3

tryptic peptides are listed, as well as the estimated pIs (Table 1b) of each one of the peaks in Fig. 12. The correlation of calculated and measured pIs was reasonably good.

The pIs of the peaks from the CIEF tryptic map of bovine cytochrome *c* were estimated by running standard proteins of known pI and making plots of migration time versus pI. The migration times of the peaks were then plugged into the line equation to estimate their pIs. As discussed earlier, the problem is that when one includes the migration time and pI of both basic and acidic proteins, the plot is nonlinear. However, if separate plots are made for the basic species and the acidic species, the plots are linear. Thus, in Table 1, the pIs of the basic peptides were estimated using a plot derived from pI versus migration time for pI 9.6, 9.1, 7.2, and 6.8 standard proteins. The pIs of acidic peptides were estimated from a plot of pI versus migration time for pI 7.2, 6.8, and 4.7 standard proteins. This overcomes the problem of nonlinearity of migration time versus pI, but does not overcome the problem with acidic peptides being very broad (Fig. 12).

Currently, there are two problems that must be overcome before CIEF peptide mapping can be a stand-alone approach. The first is the problem with acidic peptides. The second is the issue of detectability. Since CIEF with ampholytes is constrained to working at 280 nm or above, only those peptides that contain either tyrosine or tryptophan will be detected. The solution is to derivatize the peptides with UV or FL active tags, which we are currently pursuing. We feel that it is worth pursuing CIEF peptide mapping, since it offers something that no other form of peptide mapping can; separation based on a calculable property of the peptides, their pIs.

Other applications of CIEF with uncoated capillaries and EOF mobilization will be forthcoming. We are investigating application to antibody separations. All those applications reported for the other CIEF method could be done here as well.

CONCLUSIONS

At the present time, there is not a perfect approach to performing CIEF. Each of the two methods have their advantages and disadvantages. Clearly the potential is there, but it has not been reached. The ability to separate both basic and acidic proteins with the same degree of resolution and peak shape would be highly desirable. Perhaps the approach of pulling the capillary through the detector may provide this in the case of coated capillaries (18), or using external fields to control EOF in uncoated ones. Separations in narrow range pH gradients need to be further explored and optimized, especially in terms of the catholytes and anolytes used. To date only narrow range gradients of 5-8 have been used. There are many areas of research that still need to be explored. We can expect that more and more researchers will be performing research in CIEF, which will help further advance the technology. We have come a long way since the pioneering work of Hjerten in 1985, but there is considerable distance to go.

ACKNOWLEDGMENT

Our own research was supported by a grant and CE instrument donations from Isco, Inc. We wish to thank K. Patt, J. Algaier, R. Allington, M. Gurkin, and J. Tehrani. We also acknowledge N. Guzman for the invitation to prepare this chapter on CIEF and for the opportunity to express our beliefs and ideas.

REFERENCES

1. S. Hjerten and M. Zhu, *J. Chromatogr.*, *346*:265 (1985).
2. S. Hjerten, *J. Chromatogr.*, *347*:191 (1985).
3. S. Hjerten, J. Liao, and K. Yao, *J. Chromatogr.*, *387*:127 (1987).
4. S. Hjerten, K. Elenbring, F. Kilar, J. Liao, J. Chen, C. Siebert, and M. Zhu, *J. Chromatogr.*, *403*:47 (1987).
5. F. Kilar and S. Hjerten, *Electrophoresis*, *10*:23 (1989).
6. F. Kilar and S. Hjerten, *J. Chromatogr.*, *480*:351 (1989).
7. M. Zhu, D. Hansen, S. Burd, and F. Gannon, *J. Chromatogr.*, *480*:311 (1989).
8. T. Wehr, M. Zhu, R. Rodriguez, D. Burke, and K. Duncan, *Am. Biotechnol. Lab.*, *8*:22 (1990).
9. F. Kilar, *J. Chromatogr.*, *545*:403 (1991).
10. M. Zhu, R. Rodriguez, and T. Wehr, *J. Chromatogr.*, *559*:479 (1991).
11. T. Kasper and M. Melera, Presented at 27th Eastern Analytical Symposium, New York, Oct. 2–7 (1988).
12. J. Mazzeo and I. S. Krull, *BioTechniques*, *10*:638 (1991).
13. J. Mazzeo and I. S. Krull, *Anal. Chem.*, *63*:2852 (1991).
14. W. Thormann, J. Caslavska, S. Moltenim, and J. Chmelik, *J. Chromatogr.*, *589*:321 (1992).
15. S. M. Chen, J. C. Colburn, and W. Wiktorowicz. Presented at High Performance Capillary Electrophoresis 1992, Amsterdam, Feb. 9–13, Abstr. M-39 (1992).
16. J. Wu and J. Pawliszyn, *Anal. Chem.*, *64*:219 (1992).
17. J. Wu and J. Pawliszyn, *Anal. Chem.*, *64*:224 (1992).
18. R. A. Hartwick, Presented at Pittsburgh Conference 1992. New Orleans, Mar. 9–12, Abstr. 367 (1992).

19. C. Silverman, M. Komar, K. Shields, G. Diegnan, and J. Adamovics, *J. Liq. Chromatogr.*, *15*:207 (1992).
20. K. Cobb, V. Dolnik, and M. Novotny, *Anal. Chem.*, *62*:2478 (1990).
21. K. W. Yim, *J. Chromatogr.*, *559*:401 (1991).
22. B. L. Karger, A. S. Cohen, and A. Guttman, *J. Chromatogr.*, *492*:85 (1989).
23. S. Hjerten, *Chromatogr. Rev.*, *9*:122 (1967).
24. J. Mazzeo and I. S. Krull, *J. Microcolumn Separations*, *4*:29 (1992).
25. J. Mazzeo and I. S. Krull, *J. Chromatogr.*, *606*(2), 291 (1992).
26. R. A. Mosher and W. Thormann, *Electrophoresis*, *11*:717 (1990).
27. C. T. Wu, T. Lopes, B. Patel, and C. S. Lee, *Anal. Chem.*, *64*:886 (1992).
28. J. Mazzeo and I. S. Krull, *Anal. Biochem.*, (in press; 1993).

30

Behavior of Ricin on Untreated and Treated Capillary Electrophoresis Columns

Harry B. Hines and Ernst E. Brueggemann

United States Army
Medical Research Institute
of Infectious Diseases
Fort Detrick, Maryland

INTRODUCTION

Capillary electrophoresis (CE) has been used extensively for analyzing both charged and uncharged species that possess widely different characteristics, such as molecular composition and weight. One of the most promising and potent aspects of CE is its ability to resolve complex mixtures of proteins. However, many applications require buffer or column modifications to overcome protein–silanol interactions that occur in bare or untreated fused silica columns. These interactions degrade column performance through band broadening that adversely affects peak symmetry and resolution. Exploiting solution chemistry to decrease surface interactions is an uncomplicated remedy to this problem, compared with the more complex approach of surface deactivation or modification. Nevertheless, both strategies have been used successfully for protein analyses.

Exploiting solution chemistry to decrease silanol–solute interactions depends on manipulating characteristics of the buffer(s). Buffer modifications generally involve changing composition, pH values, or the use of additives. Changing buffer composition can have a marked effect on peak shape in capillary electrophoresis. For example, if the analyte possesses appropriately oriented (*cis*) hydroxyl groups, borate will form a complex and inhibit binding to the silica surface. Phosphate buffer will not have this ability, so changing to the borate buffer would improve column performance.

Adjusting pH values is one of the simplest means of changing a buffer's characteristics and also improving column performance in capillary electrophoresis. Two commonly employed techniques involve using either acidic pH values or values that are approximately 2 units above the protein's pI. Such pH values act by changing the charge state of the protein(s) and silanols [1,2]. The repulsion of similar charges enhances peak symmetry, thereby enhancing resolution.

Another strategy to achieve the same goal employs buffer additives. Buffer additives have been used to dynamically coat columns [3], form ion pairs [4,5], or compete with analyte(s) for active silanol sites [6]. Additives may be incorporated in the sample buffer only, in the electrophoresis buffer only, or in both buffers. Regardless of how it is applied, a buffer additive is a reversible solution to the adsorption problem. This reversibility maintains flexibility, so many experiments can be performed with the same column under similar conditions.

Although it is more complex than altering solution chemistry, several techniques have also been employed for column treatment (deactivation). Column modifications range from reversibly to covalently bound coatings. Coatings may also be neutral or interactive [7 – 12]. Interactive coatings introduce another element into separations that may be used to exploit some characteristic of the analyte. For example, a hydrophobic coating will affect separations by binding hydrophobic sites on a peptide or protein. A hydrophilic protein would elute before the retained eluite. This approach may be used independently, or combinations of coatings and buffer modifications or additives may be used to achieve the desired result.

Each of these variations, whether it involves buffer or column modifications separately or portions of both, may require consideration when planning separation strategies for proteins. Depending on the chemical and physical properties of the analyte, it may be necessary to examine and change many variables to define the proper analytical system. For each analysis, specific analytical requirements direct the approach employed to meet the desired objective(s).

One objective of this laboratory is to develop instrumental analytical methods for detecting toxins in complex matrices. One toxin of particular interest is ricin. Ricin (RCA 60) is a heterodimeric, glycoprotein phytotoxin, with a relative molecular mass (M_r) of approximately 66,000 Da and a pI of 7.1 [13]. This toxin is produced by the castor bean plant, *Ricinus communis*, which grows globally in warm climates, including California and the southern United States. Most of the plant's ricin is found in the seed(s) [for review of early work on ricin, see Ref. 14]. Ricin is highly toxic to mammals and is considered to be one of the most toxic naturally occurring plant toxins known.

At the cellular level, ricin causes death by inhibiting protein synthesis [15,16]. In mammals, the ricin B chain mediates binding to galactosyl moieties on cell membranes [17]. The exact nature of the membrane receptor is currently unknown. Intact ricin is transported into the cell by pinocytosis. After dissociating from the B chain within the cell, the A chain, which possesses *N*-glycosidase activity, acts on the 28S ribosome by cleaving a specific adenosine glycosidic bond in rRNA [18–20]. This disrupts protein synthesis, and the cell dies. The precise mechanisms for A-chain dissociation, translocation, and transport to the ribosome are as yet unknown.

Capillary electrophoresis was selected for ricin analysis for several reasons. First, it has been used extensively for protein analysis, and many factors affecting its performance have been reported [21]. Second, instrumentation is portable and relatively easy to use. Third, little solvent is required, and little sample is needed. Fourth, automation has greatly improved the analytical precision of capillary electrophoresis by standardizing injection and analysis variables, such as temperature. Automation has also increased productivity by improving sample processing. Fifth, analysis times are generally shorter than those required when using high-performance liquid chromatography (HPLC). Previously, only HPLC and classic electrophoresis techniques were used to separate ricin. These techniques were typically restricted to ricin separation and purification for biological studies.

In this study, the electrophoretic behavior of ricin in fused silica capillaries was examined using several different operating parameters with untreated and treated fused silica capillary columns. The objective of the study was to define a system that conformed to the requirements just listed as closely as possible. Within these constraints, optimal analytical conditions were established to minimize adsorption and to maximize peak symmetry and column efficiency. The best systems were tested by analyzing unknown samples believed to contain ricin.

MATERIALS AND METHODS

Ricin was purchased from Vector Laboratories (Burlingame, California) and was used without further purification. Purity was reported to be a single protein, as determined by sodium dodecyl sulfate–polyacrylamide gel electrophoresis (SDS–PAGE) [22]. Some batches were desalted with 30,000 molecular weight-limit centrifuge filters (PGC Scientifics, Gaithersburg, Maryland). A modular CE unit from Spectrovision, Inc. (Chelmsford, Massachusetts) equipped with an ISCO CV^4 UV detector (Lincoln, Nebraska) and a Beckman P/ACE 2000 (Palo Alto, California) unit were employed in this study. Untreated fused silica capillary columns were purchased from Polymicro Technologies (Phoenix, Arizona) and Beckman Instrument Co. (Palo Alto, California). Treated fused silica capillary columns were purchased from Supelco, Inc. (Bellefonte, Pennsylvania). Tables 1 and 2 contain summaries of conditions and column dimensions for untreated and treated columns, respectively. Buffer components were purchased from Sigma Chemical Co. (St. Louis, Missouri) and were the highest purity available. Mesityl oxide, a compound used as a neutral marker for measuring electroosmotic flow, phosphoric acid, and putrescine were purchased from Aldrich Chemical Co. (Milwaukee, Wisconsin). Deionized water (18 MOhms)

Table 1 Partial Summary of CZE Conditions Used for Untreated Fused Silica Columns

Buffer	Column	kV	μA	UV	Injection
0.03 M sodium borate,[a] pH 8.5	75 μm × 35 cm[b]	20	80	214	Press,[c] 3 s
0.01 M/0.04 M Tricine, pH 9.0	75 μm × 50 cm	20	4	220	Press, 1 s
0.02 M K_2SO_4, 0.03 M putrescine, pH 7.0	50 μm × 57 cm	30	85	200	Press, 1 s
0.1 M trimethylammonium-propanesulfonate, 0.1 M KH_2PO_4, pH 9.0 (Accupure reagent)	50 μm × 57 cm	25	120	200	Press, 1 s
0.1 M KH_2PO_4, 4 M urea, pH 8.6	50 μm × 57 cm	20	95	200	Press, 1 s
0.04 M Tris/Tricine, 0.05 M SDS, 10% MeOH,[d] pH 8.0	50 μm × 57 cm	27	11	220	Press, 1 s
0.04 M Tris/Tricine, 0.005 M SDS, pH 8.0, sample treated	50 μm × 57 cm	20	3	220	Vac,[e] 1 s
0.03 M sodium borate, pH 8.0, ethylene glycol to sample	75 μm × 50 cm	20	80	214	Vac, 1 s
0.03 M sodium borate, pH 10.0, ethylene glycol to sample	75 μm × 50 cm	20	200	214	Vac, 1 s
0.03 M sodium borate, pH 8.5, 1% ethylene glycol to buffer	75 μm × 50 cm	20	82	214	Vac, 1 s
0.03 M sodium borate, pH 8.5, 3% ethylene glycol to buffer	75 μm × 50 cm	20	78	214	Vac, 1 s
0.03 M sodium borate, 6 M urea, pH 8.5	75 μm × 50 cm	20	150	214	Vac, 1 s
0.03 M KH_2PO_4, pH 2.5	75 μm × 50 cm	20	50	214	Vac, 1 s
0.1 M KH_2PO_4, pH 8.6	50 μm × 57 cm	20	114	200	Press, 1 s
0.1 M KH_2PO_4, 0.03 M putrescine, pH 3.5	50 μm × 57 cm	30	160	200	Press, 1 s
0.01 M Tris/Tricine, pH 9.0	50 μm × 57 cm	20	3	200	Press, 1 s
0.08 M Tris/Tricine, pH 9.0	50 μm × 57 cm	20	22	200	Press, 1 s
0.04 M Tris/0.01 M Tricine, pH 9.0	50 μm × 57 cm	20	8	200	Press, 1 s
0.04 M Tris/0.08 M Tricine, pH 9.0	50 μm × 57 cm	20	10	200	Press, 1 s
0.01 M Tris/0.04 M Tricine, pH 9.0	50 μm × 57 cm	20	10	200	Press, 1 s
0.08 M Tris/0.04 M Tricine, pH 9.0	50 μm × 57 cm	20	20	200	Press, 1 s
0.08 M Tris/0.04 M Tricine, pH 9.0	50 μm × 57 cm	10	11	200	Press, 1 s
0.16 M Tris/Tricine, pH 9.0	50 μm × 57 cm	20	47	200	Press, 1 s
0.005 M Tris/Tricine, pH 9.0	50 μm × 57 cm	20	2	200	Press, 1 s
0.04 M Tris/0.01 M Tricine, pH 9.0	50 μm × 57 cm	10	9	200	Press, 1 s
0.04 M Tris/Tricine, 0.01 M SDS,[f] pH 8.1	50 μm × 57 cm	20	6	200	Press, 1 s

Table 1 *(Continued)*

Buffer	Column	kV	μA	UV	Injection
0.04 M Tris/Tricine, 0.05 M SDS, pH 8.0	50 μm × 57 cm	20	11	200	Press, 1 s
0.04 M Tris/Tricine, 0.01 M SDS, pH 8.0	50 μm × 57 cm	20	18	200	Press, 1 s
0.01 M Tricine, 0.02 M KCl, 0.005 M putrescine, pH 8.2	50 μm × 57 cm	30	29	200	Press, 1 s
0.015 M sodium borate, pH 9.5	50 μm × 57 cm	20	23	200	Press, 1 s
0.015 M sodium borate, pH 8.0	50 μm × 57 cm	20	20	200	Press, 1 s
0.03 M sodium borate, pH 9.5	50 μm × 57 cm	20	45	200	Press, 1 s
0.03 M sodium borate, pH 8.0	50 μm × 57 cm	20	43	200	Press, 1 s
0.06 M sodium borate, pH 9.5	50 μm × 57 cm	20	100	200	Press, 1 s
0.06 M sodium borate, pH 8.0	50 μm × 57 cm	20	95	200	Press, 1 s
0.006 M sodium borate, pH 8.0	50 μm × 57 cm	20	7	200	Press, 1 s
0.03 M sodium borate, 0.025 HSA,[g] pH 9.35	50 μm × 57 cm	20	152	20	Press, 1 s
0.03 M sodium borate, 0.05 M HSA, pH 9.35	50 μm × 57 cm	20	170	200	Press, 1 s
0.03 M sodium borate, pH 10.0	75 μm × 50 cm	20	110	200	Press, 1 s
0.1 M CHES,[h] 0.125 M KH_2PO_4, pH 8.75	75 μm × 50 cm	15	170	200	Press, 1 s
0.02 M CHES, 0.01 M KCl, pH 8.0	75 μm × 50 cm	30	30	200	Press, 1 s
0.02 M CHES, 0.01 M KCl, 0.005 M putrescine, pH 9.0	50 μm × 50 cm	20	8	220	Vac, 1 s
0.02 M CHES, 0.005 M putrescine, pH 9.0	50 μm × 50 cm	20	3	220	Vac, 1 s
0.02 M CHES, 0.01 M KCl, pH 9.0	50 μm × 50 cm	20	5	220	Vac, 1 s
0.02 M CHES, 0.01 M KCl, 0.1% TEA,[i] pH 8.6	50 μm × 50 cm	20	5	220	Vac, 1 s
0.01 M Tricine, 0.0085 M putrescine, pH 9.3	50 μm × 50 cm	20	3	220	Vac, 1 s

[a]Unless otherwise indicated, concentrations are equal for both components.
[b]Column dimensions: inner diameter × length to detector.
[c]Pressure injection: 0.5 psi.
[d]MeOH: methanol.
[e]Vacuum injection: 20 in. mercury.
[f]SDS: sodium dodecyl sulfate.
[g]HSA: hexanesulfonic acid.
[h]CHES: 2-(*N*-cyclohexylamino)ethanesulfonic acid.
[i]TEA: triethylamine.

Table 2 Partial Summary of CZE Conditions Used with Treated Fused Silica Columns

Buffer	Column	kV	μA	UV	Injection
0.1 M sodium acetate, pH 3.5	50 μm × 57 cm Aminopropyl[b]	20	1	200	Press,[a] 1 s
0.01 M KH_2PO_4, pH 8.6	50 μm × 50 cm Supelco P150	20	15	200	Press, 1 s
0.005 M Na_2HPO_4, 0.005 M SDS,[c] pH 6.0	50 μm × 57 cm Supelco C_{18}	30	10	200	Press, 1 s
0.005 M KH_2PO_4, pH 6.0	50 μm × 50 cm Supelco C_1	30	10	200	Press, 1 s

[a]Press: pressure injection: 0.5 psi.
[b]Aminopropyl: aminopropyl phase coated in this laboratory.
[c]SDS: sodium dodecyl sulfate.

was used in all experiments. An aminopropyl-coated column was prepared according to the method of Mosely et al. [23].

RESULTS

Many CE variables were screened to determine which combination(s) would be optimal for conducting ricin analyses. Several combinations yielded promising results and were explored in more detail. Initial systems employed untreated fused silica capillary columns. Of the conditions tested, using 0.03 M borate buffer, pH 8.5, and a 35-cm column provided acceptable results on untreated columns when theoretical plate number and peak shape were considered. Figure 1 contains a representative electropherogram for ricin analyzed under the foregoing conditions. A relatively broad, tailing peak was evident ($N = 1550$). Tailing is indicative of either silanol binding or sample heterogeneity. Borate concentration and pH were varied both independently and concurrently to overcome adsorption. Neither plate number nor peak shape improved, regardless of changes made in pH and borate concentration. When a smaller inner-diameter untreated column (50 μm) and the same buffer were used, the peak shown in Fig. 1 began to split. The appearance of an additional peak indicated that more than one component was present. Additional analyses performed by classic denaturing electrophoresis confirmed that ricin samples contained only one stainable component. The presence of additional peaks in the capillary electropherogram components may result from glycoside heterogeneity of pure ricin, an impurity in the sample, ricin aggregation, or ricin chain rearrangement. Because component resolution was unsuccessful with borate buffer, a different buffer system was applied.

The zwitterionic nature of Tricine has the potential to enhance peak shape by interacting with both silanol groups on the column and amine moieties on the

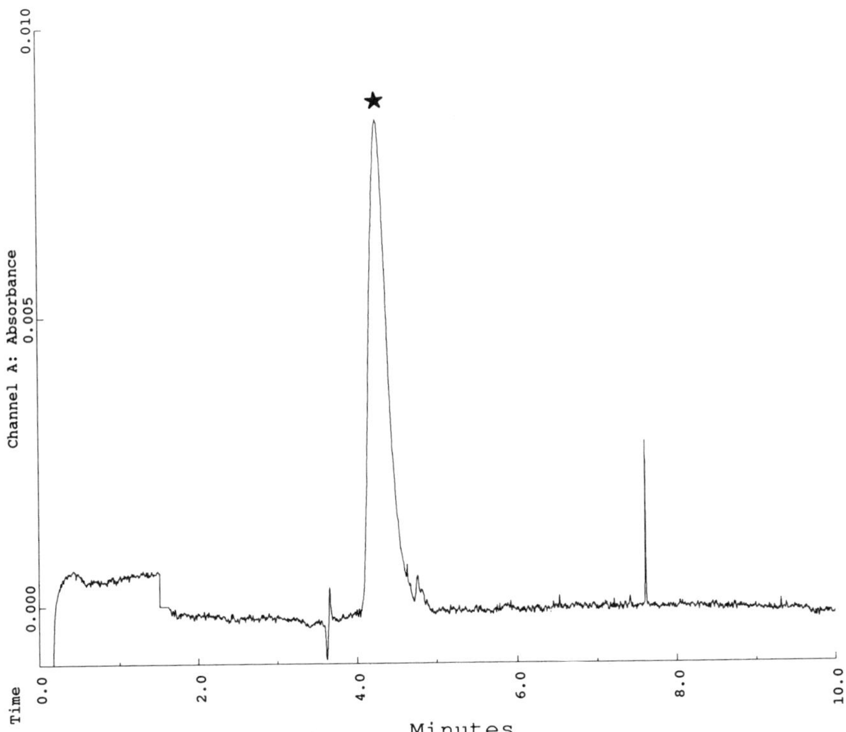

Figure 1 Representative electropherogram of ricin (★) using an untreated fused silica capillary column: 75 μm × 35 cm. Buffer: 0.03 M sodium borate, pH 8.5; 20 kV; 200 nm detection.

protein [11,12]. Figure 2 shows that concurrently increasing the concentration of Tris and Tricine from 0.01 M to 0.08 M improved component resolution slightly. Two peaks were clearly present with higher buffer concentrations, as shown in Fig. 2B. Varying Tris/Tricine concentration ratios resolved additional components in the ricin sample. Figure 3 shows that, by holding the Tris concentration constant and increasing Tricine levels, additional components were resolvable. The presence of four components constituted an improvement over classic electrophoresis, the borate buffer system, and equimolar concentrations of Tris and Tricine. Some peak tailing was still observed with these conditions, but it did not strongly affect resolution. From these results, Tricine emerged as the more important buffer component. This conclusion was strengthened when Tris concentration had little or no effect on resolution when it was increased and the Tricine concentration was held constant (Fig. 4). Decreasing the electrophoresis voltage from 20 to 10 kV with the same

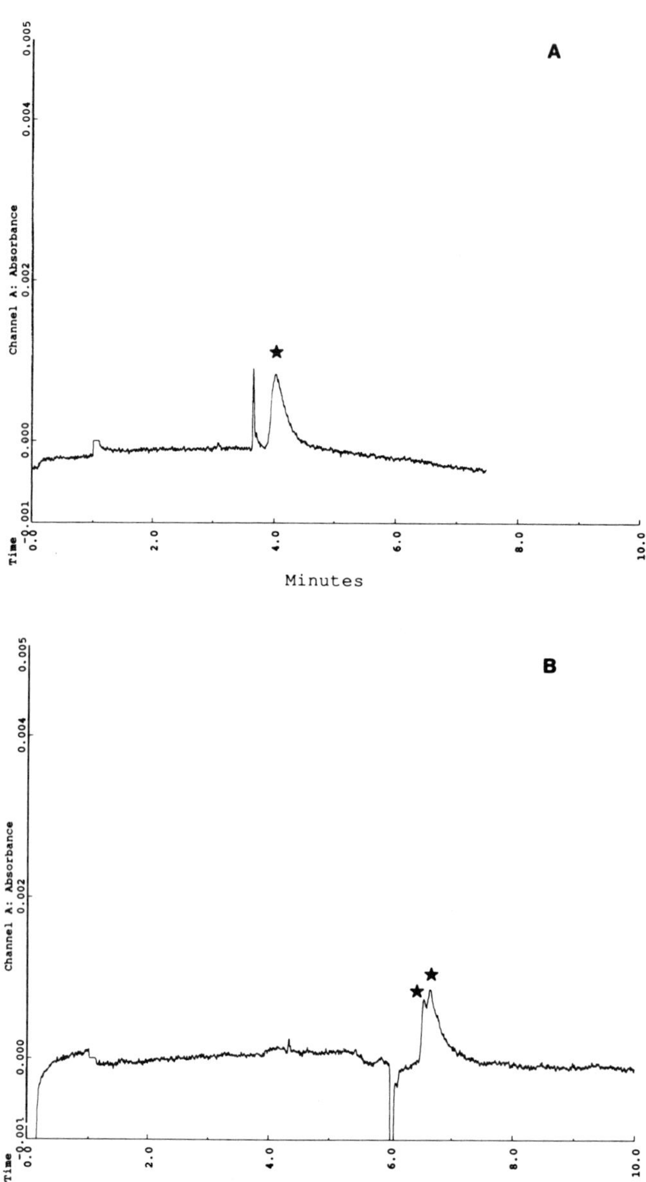

Figure 2 Electropherogram of ricin (★) depicting the effect of different concentrations of Tris/Tricine upon peak shape. (A) 0.01 M Tris and Tricine, pH 9.0. (B) 0.08 M Tris and Tricine, pH 9.0. Column: 75 μm × 50 cm; 20 kV; 200 nm detection.

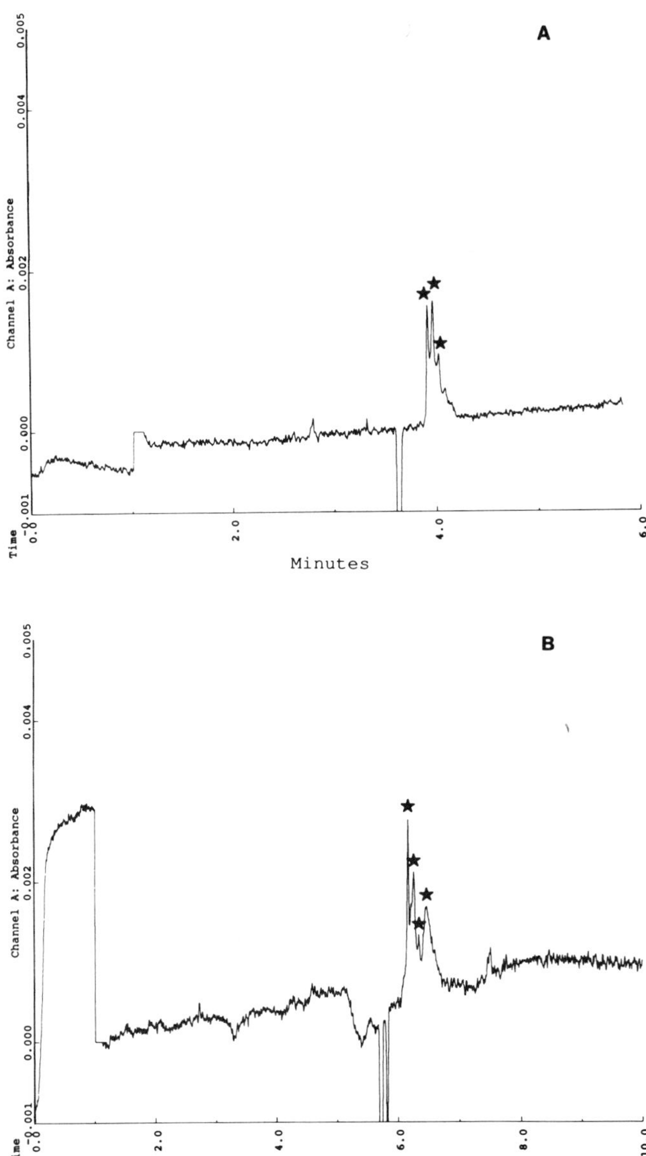

Figure 3 Electropherogram of standard ricin (★) showing enhanced resolution of sample components with changes in buffer composition. (A) 0.01 M Tris, 0.04 M Tricine, pH 9.0. (B) 0.01 M Tris, 0.01 M Tricine, pH 9.0. Column: Untreated, 75 μm × 50 cm; 20 kV; 200 nm detection.

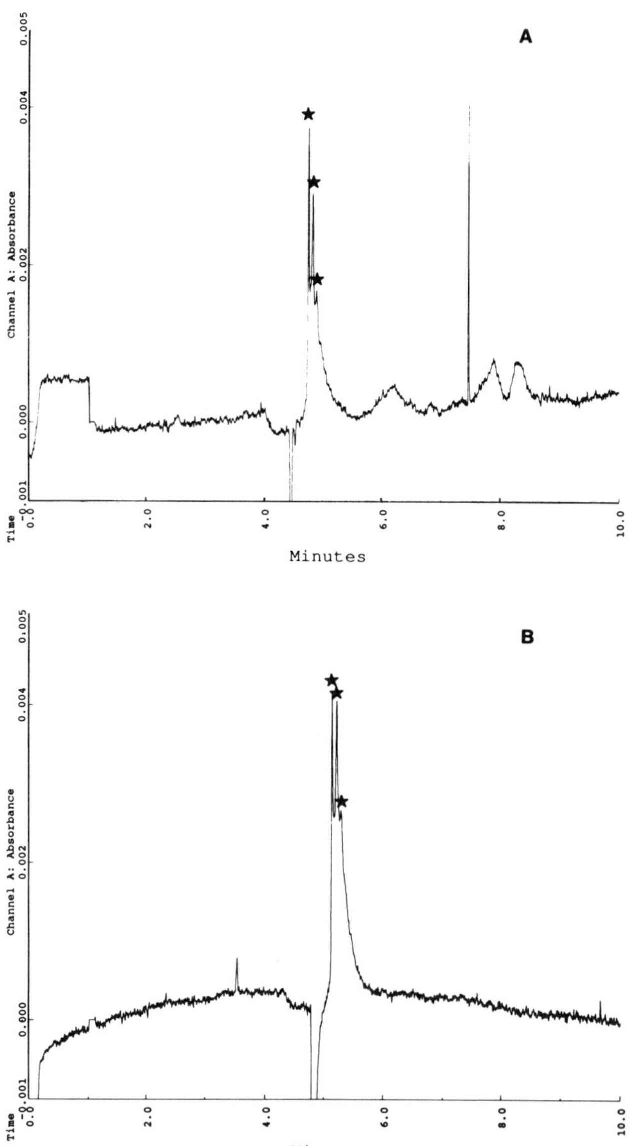

Figure 4 Electropherograms of standard ricin (★), demonstrating that changes in Tris concentration had little effect upon component separation. (A) 0.01 M Tris, 0.04 M Tricine, pH 9.0. (B) 0.08 M Tris, 0.04 M Tricine, pH 9.0. Column: untreated, 75 μm × 50 cm; 20 kV; 200 nm detection.

buffer conditions had little effect on resolution, but almost doubled retention time. Doubling Tris/Tricine concentrations, while maintaining a constant ratio, greatly compromised resolution, resulting in poor peak shapes and resolution. The same was also true when Tris/Tricine concentrations were decreased to 0.005 M. Tris was required for proper peak shape and resolution, however. When Tricine was used alone, it did not yield the resolution obtainable when Tris was included. It was concluded that a proper combination of these components was the most important characteristic necessary for optimal resolution. Changing untreated columns did not affect the reproducibility of this method. To test these conditions, a synthetic mixture of bovine serum albumin and ricin was prepared. Ricin and bovine serum albumin have similar M_r, but very different pI values (7.1 and 4.8, respectively). The optimal Tris/Tricine buffer and an untreated column separated these two proteins, as depicted

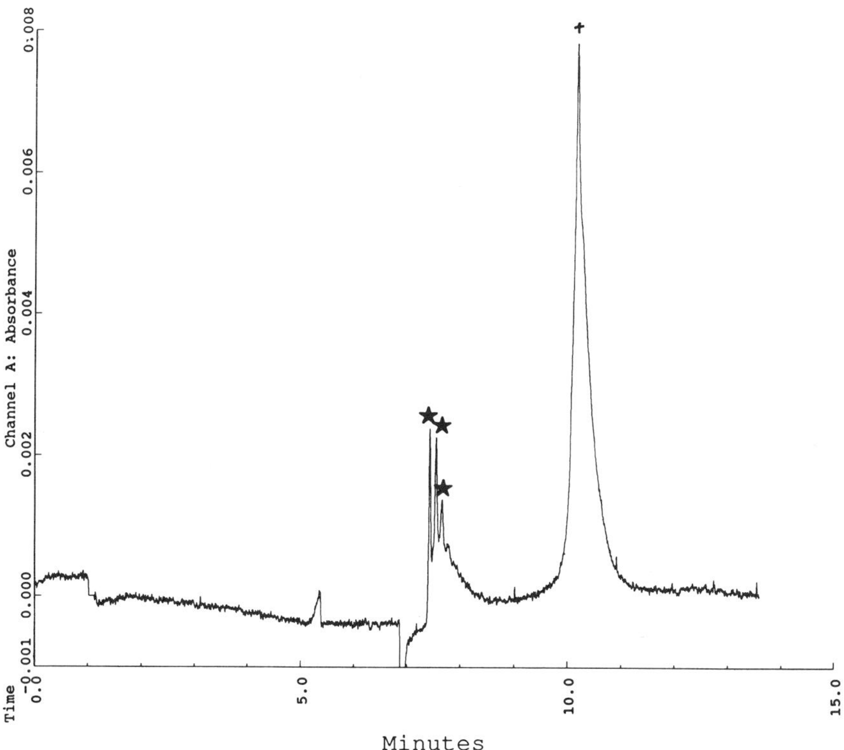

Figure 5 Electropherogram of standard ricin (★) and standard bovine serum albumin (+). Separation was achieved using the optimum Tris/Tricine concentration ratio. Buffer: 0.04 M Tris, 0.01 M Tricine, pH 9.0. Column: untreated, 75 μm × 50 cm; 20 kV; 200 nm detection.

in Fig. 5. This verified that the system had the potential to separate ricin from protein(s) of similar size, but differing pI values.

Because results obtained with Tricine were encouraging, other bifunctional compounds were also used as additives. Putrescine (1,4-diaminobutane) in potassium sulfate buffer [13] gave very good results for a mixture of three standard proteins (Fig. 6A). The mixture included lysozyme, typically a difficult protein to analyze by CE owing to its basic nature. Theoretical plate numbers were 138,300, 32,000, and 170,700 for lysozyme, cytochrome *c*, and chymotrypsinogen A, respectively. When ricin was analyzed under identical conditions, excellent resolution and peak shape were obtained (see Fig. 6B). These results paralleled those obtained with Tris/Tricine buffer. The theoretical plate number for the largest peak in the ricin electropherogram was 64,000.

Another zwitterionic compound, trimethylammoniumpropanesulfonate (Accupure Z-1 Methyl reagent, Waters Div., Millipore Corp., Milford, Massachusetts), was also used with potassium phosphate buffer. This buffer additive reduced peak width somewhat, but improved peak shape (Fig. 7). Three components were detectable with this system. Electropherograms were reproducible over several weeks. These results indicated that it would be possible to analyze an unknown sample with these conditions. Figure 8 contains the electropherogram for an unknown sample. It was possible to tentatively identify ricin in this mixture, which was confirmed to be a crude, acidic extract of castor beans. The presence of multiple ricin peaks complicated, but did not prevent, identification.

To investigate the origin of multiple peaks in the ricin sample, an attempt was made to dissociate any aggregates that might be present. A phosphate buffer system containing 6 M urea was used for this purpose. No evidence of aggregation was found (Fig. 9). Additional peaks should have disappeared if dissociable aggregates had been present. Impurities or microheterogeneity were the best possible explanations for the origin of multiple peaks.

We also examined the behavior of ricin on treated columns. This approach is the second general technique used to minimize protein–silanol attraction. Four treated columns were investigated in this study (see Table 2). As seen in Fig. 10, a commercial column coated with a proprietary hydrophilic phase (Supleco P150) gave peak symmetry results similar to those obtained with an untreated column when the same buffer system was used for both columns ($N = 4500$). Attempts to improve peak shape by varying buffer composition and pH values were unsuccessful. A C_{18} column also performed unsatisfactorily for ricin (Fig. 11). On the other hand, a C_1 column gave a relatively sharp peak ($N = 12{,}436$) when SDS was included in the buffer. A representative electropherogram is shown in Fig. 12. Although the results were promising for this system, the equilibration period was quite long (>12 hours), which diminished its usefulness. Furthermore, the column was unable to resolve components in the ricin sample. Better results for peak characteristics were obtained with an aminopropyl column ($N = 20{,}100$) prepared in this laboratory.

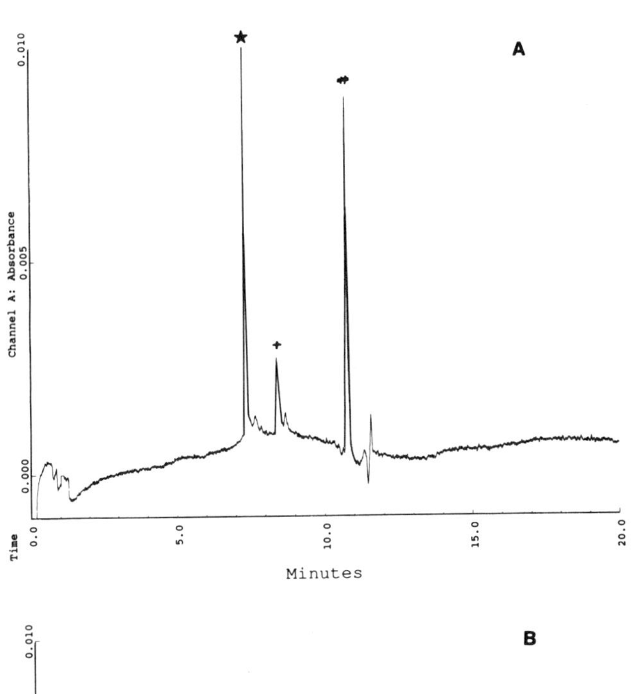

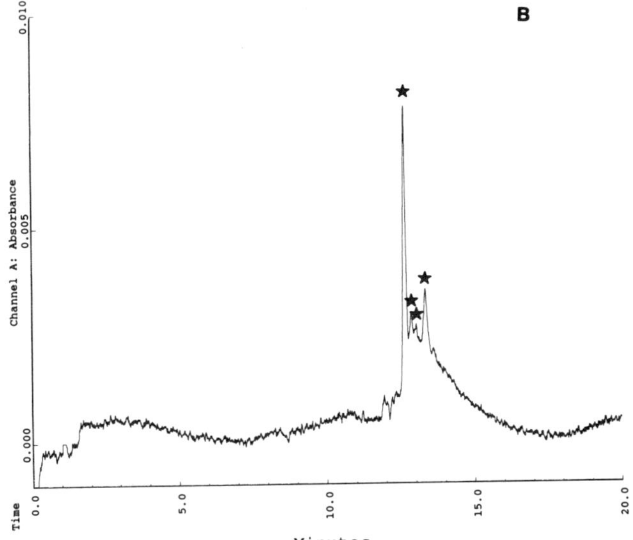

Figure 6 Electropherograms of (A) standard protein mixture containing chymotrypsinogen A (★), cytochrome *c* (+), and lysozyme (#); and (B) ricin (★), analyzed under the same conditions. Buffer: 0.02 M K_2SO_4, 0.03 M putrescine, pH 7.0. Column: untreated, 50 μm × 57 cm; 20 kV; 200 nm detection.

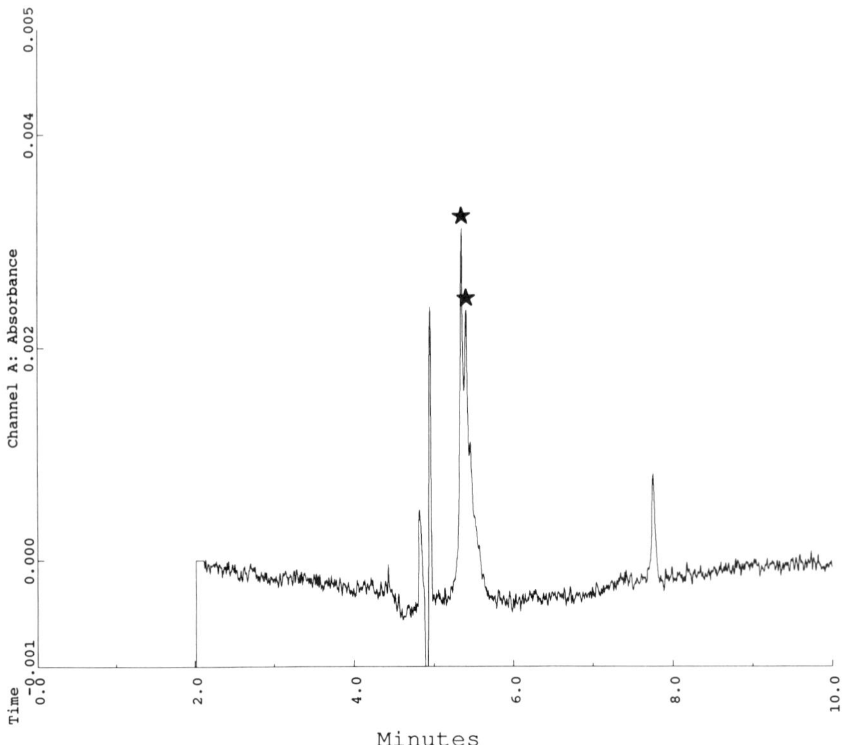

Figure 7 Electropherogram of standard ricin analyzed with 1 M Accupure Z1-Methyl reagent, 0.1 M KH_2PO_4, pH 8.0. Column: untreated fused silica, 50 μm × 57 cm; 25 kV; 200 nm detection.

Unfortunately, the coating was unstable (Fig. 13). The column became unusable when the efficiency dropped to 3500 theoretical plates after only 2 days of use. Approximately 30 injections were made during this period. The coating could be regenerated satisfactorily, but this made the column an unlikely choice for routine use.

DISCUSSION

To develop a useful CE method for ricin, many variables were systematically studied to minimize ricin adsorption and maximize peak symmetry, column efficiency, and reproducibility. Problems with attaining this goal originated in two areas. One area was protein adsorption, and the other area was sample impurity or heterogeneity.

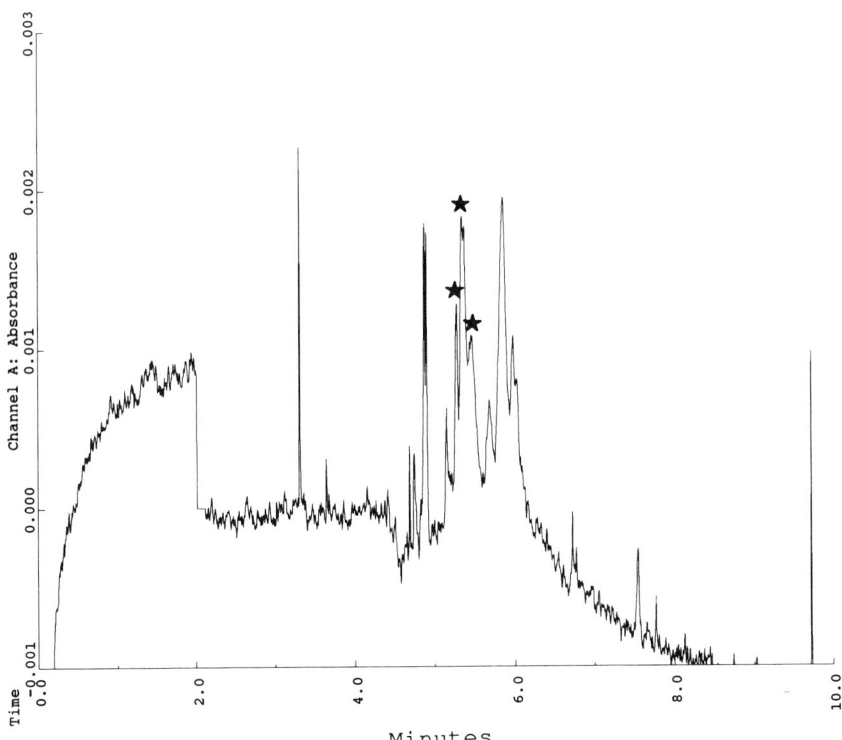

Figure 8 Electropherogram of a crude, acidic castor bean extract containing ricin (★). Buffer: 1 M Accupure Z-1 Methyl reagent, 0.1 M KH_2PO_4, pH 9.0. Column: untreated, 50 μm × 57 cm; 25 kV; 200 nm detection.

Efforts to solve the adsorption problem fell into three main categories: pH manipulation, buffer additives, and column inactivation (coating).

Coulombic repulsion has been used extensively to improve protein analyses. It was discovered early in protein CE analyses that inducing negative charges on protein(s) and the fused silica column would reduce protein adsorption and improve peak shape [1]. Because only pH adjustment was needed, it was also a simple modification. Of the many basic buffers tested in this study, only borate buffer, pH 8.5, gave acceptable results for ricin. Increasing the pH of this buffer above 8.5 had no additional beneficial effect on peak shape. It may be that ricin could not be sufficiently charged to effectively avoid silanol binding. Ricin, like other proteins, can undergo conformational changes at different pH values. Conformational changes may have altered ricin's physical characteristics. A change in ricin structure may have reduced charging or may have exposed other amino acids capable of

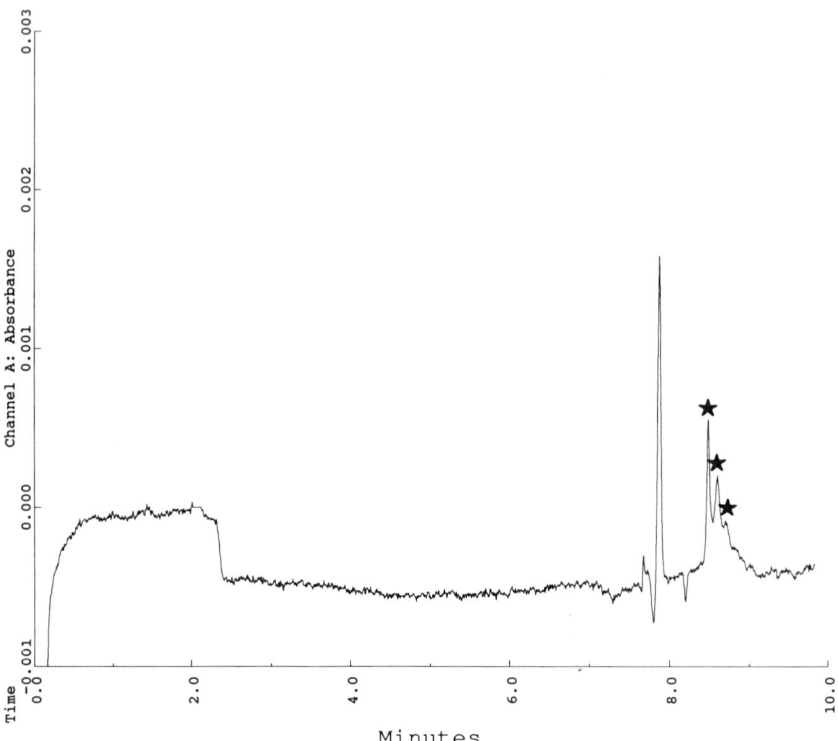

Figure 9 Electropherogram of ricin (★) demonstrating that urea failed to dissociate the components further. Buffer: 0.1 M KH_2PO_4, 4 M urea, pH 8.6. Column: untreated, 50 μm × 57 cm; 20 kV; 200 nm detection.

complexing silanols. Borate buffer may have also been more effective than other inorganic buffers used at the same pH values because borate may partially complex silanol groups. This possibility led to the use of zwitterionic buffers to enhance competition for silanol binding.

Organic zwitterions made a marked improvement in ricin's peak shape when compared with singly charged ions that had the same charge sign. Nitrogen buffers such Tris and additives such as triethylamine improved peak shape somewhat, but better results were obtained when Tricine was included with Tris. Sulfonate buffers and additives such as cyclohexylaminoethanesulfonic acid (CHES) and hexanesulfonic acid were also tested to determine if the acidic moiety would compete or pair with ricin, but no advantage was noted. Consequently, it was concluded that Tris and Tricine competed effectively for silanol groups and may have also paired with amino acids to further reduce silanol attractions. It was evident also that putrescine, a

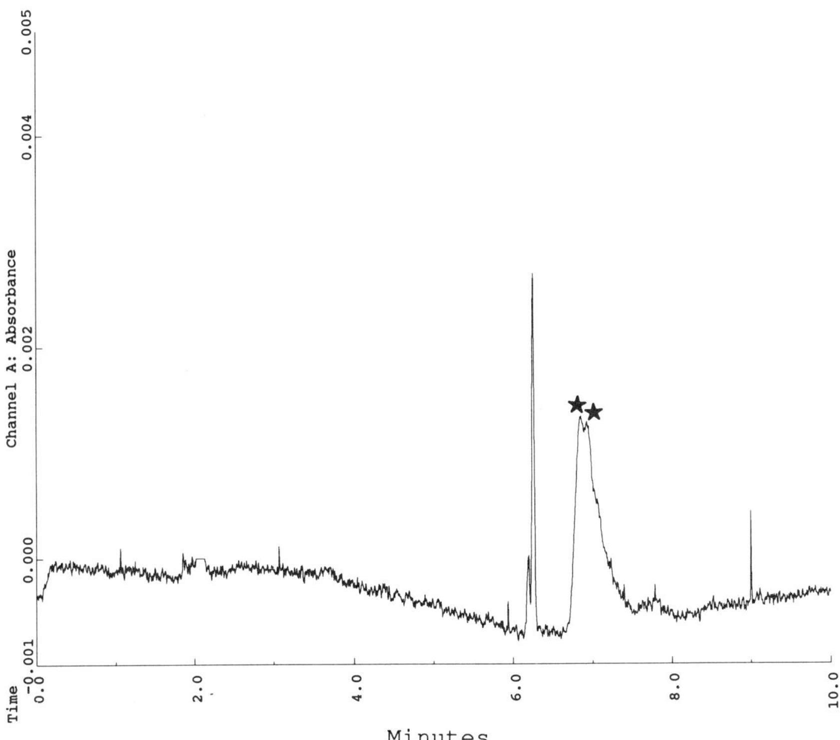

Figure 10 Electropherogram of ricin (★) on a Supelco P-150 column. Buffer: 0.01 M KH_2PO_4, pH 8.6. Column: 50 μm × 57 cm; 20 kV; 200 nm detection.

diamino compound, had a marked positive effect on ricin peak shape. The apparent need for a zwitterion in ricin analysis was strengthened by the successful use of Accupure Z-1 Methyl reagent prepared by Millipore Corp. The quaternary amine and sulfonate group seemed to effectively inhibit protein adsorption.

The third approach used in this study to enhance ricin analysis involved column modification to inactivate silanol groups. Several strategies were investigated. First, reversible column coatings were evaluated. It was reported that use of ethylene glycol added to the sample and 0.02 M borate elution buffer (pH 10) could reduce protein adsorption [3]. In this study, ethylene glycol was added to both sample and borate buffer in an attempt to further enhance performance, but no improvement was gained in ricin's peak shape. Another reversible coating agent, Micro-Coat (ABI, Foster City, California), was also tested. This proprietary coating did not improve ricin's performance either. Lack of success with reversible coatings prompted the examination of bound coatings.

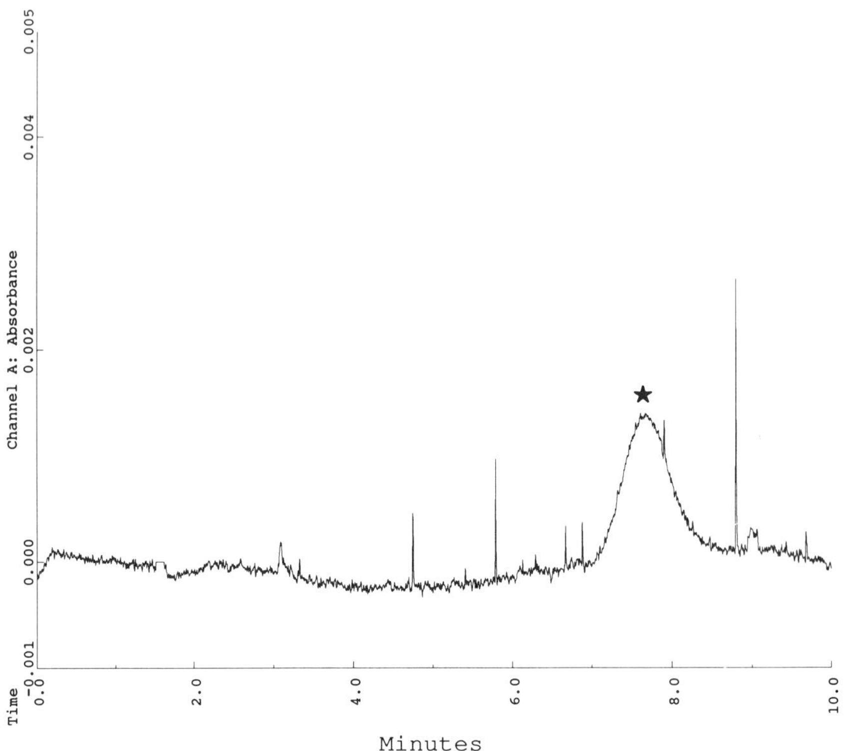

Figure 11 Electropherogram of ricin using a Supelco C_{18} column (50 μm × 57 cm); buffer: 0.005 M Na_2HPO_4, 0.005 M SDS, pH 6.0; 30 kV; 200 nm detection.

Three commercial columns, having different stationary phases, were tested. Supelco's P-150 coating is proprietary. Electroosmotic flow data supplied by the company indicated that its flow characteristic paralleled untreated fused silica columns when identical buffers were used. When the same buffer was used for both types of column, ricin results were similar. A broad, unresolved peak was obtained for both applications. Apparently, ricin adsorption was not decreased when this coating was used. It is possible that interaction occurred between the coating and ricin, as well as ricin and unreacted silanol groups. Coating coverage becomes an important consideration when these columns are used. A low percentage of covering will leave many active sites exposed, plus introduce the binding behavior of the coating. Mixed mechanisms result that complicate interpretation. Coverage was not investigated in this study.

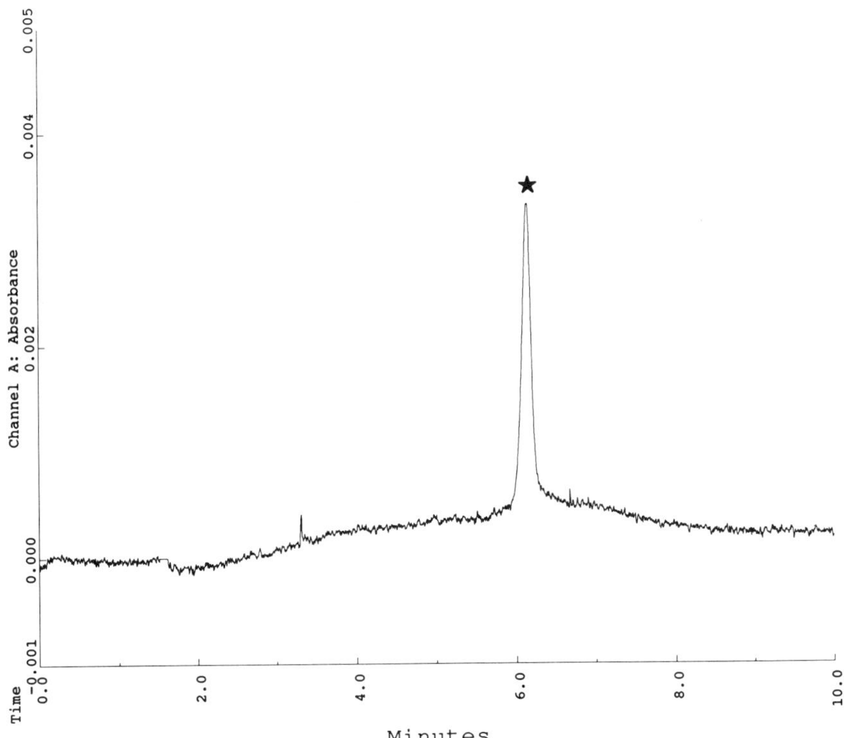

Figure 12 Electropherogram of ricin analyzed on a Supelco C_1 column (50 μm × 57 cm); buffer; 0.005 M Na_2HPO_4, 0.005 M SDS, pH 6.0; 30 kV; 200 nm detection.

Next, the performance of an octadecyl (C_{18}) coating was checked. Previous success with C_{18} reversed-phase HPLC in this laboratory indicated that ricin may elute as a well-defined peak with such a system. Results were disappointing, however. Again, the coating percentage may not have been sufficient to prevent binding, or ricin may not have eluted from the stationary phase. To overcome phase–protein binding, methanol, acetonitrile, or SDS was added to the buffer. None of these treatments significantly improved peak shape, however. Addition of detergent had a greater effect than either methanol or acetonitrile.

Finally, a C_1-coated column was investigated. This coating was chosen because the short-chain length would not contribute to phase–analyte interaction as much as the C_{18} coating. Ricin's peak shape improved, compared with the other two columns, but it was not equal to that obtained with an untreated column and Tris/Tricine buffer.

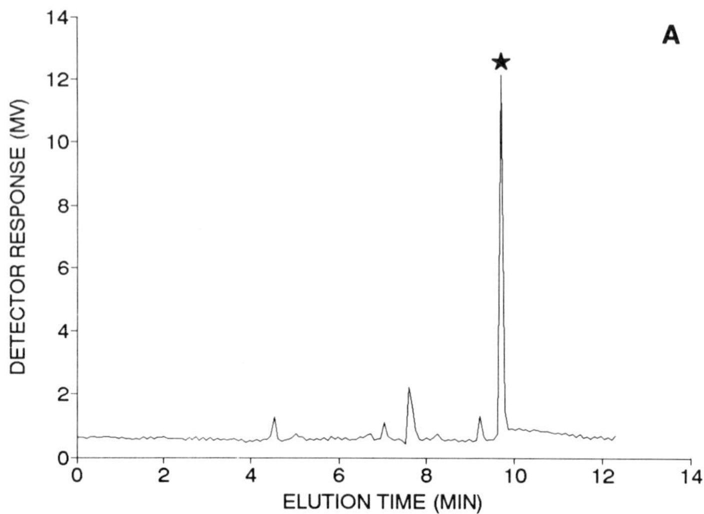

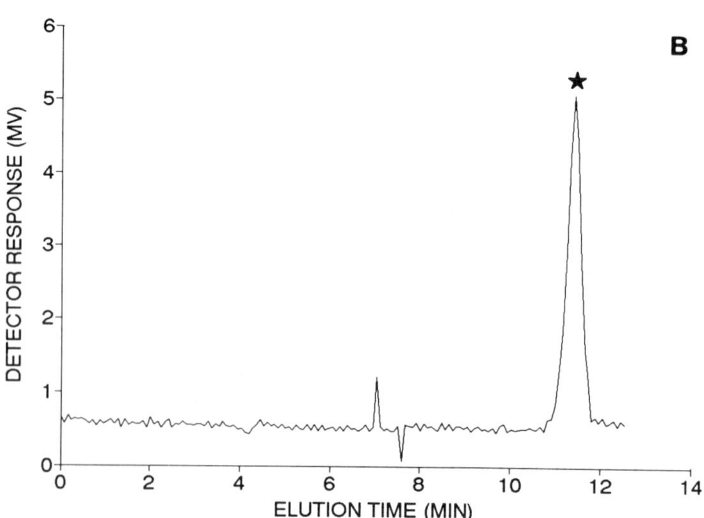

Figure 13 Electropherograms of ricin analyzed on an aminopropyl-coated column (50 μm × 57 cm). (A) Freshly prepared column. (B) Same column after 3 days use. Buffer: 0.01 M sodium acetate, pH 3.5; 20 kV; 200 nm detection.

Results obtained with the aminopropyl column that was prepared in this laboratory initially surpassed those acquired with the commercial columns. Peak shape was good for ricin and nearly matched that observed with Tris/Tricine and an uncoated column; however, sustained used caused rapid deterioration in performance.

Column stability is another important variable that must be examined when using coated columns. Using mesityl oxide as a marker for electroosmotic flow, it was found that P-150 and C_1 column performance deteriorated with use. The C_{18} column retained its performance longer, although no quantitative studies were conducted to characterize the lifetimes of these columns. Of the coated columns studied, the stability of the aminopropyl column required the most improvement. Other investigators have reportedly been more successful using this type of column, even interfaced with mass spectrometry [23]. The primary weakness with this coating is the silicon–carbon bond that is subject to hydrolysis.

Other surface modifications, including acrylamide-coated columns, must be explored for ricin analyses. It is especially important that these columns and buffers be compatible with mass spectrometric detection. Certain restraints are required to ensure compatibility. For example, acidic buffers, which are commonly used with coated columns, are currently required for interfacing CE and electrospray ionization mass spectrometry (ESIMS). Relative molecular mass and important structural information would then be available with ESIMS. Unfortunately, poor results were obtained in the present study when acidic buffers were used, regardless of the column type.

The second problem arose with the presence of extra peaks in the electropherogram. There are several possible explanations for the presence of these peaks. If the additional peaks represented ricin impurities, they possessed chemical characteristics, and possibly structures, that strongly resembled ricin itself. The commercial preparation used in this study was reportedly affinity-purified from castor beans, and only two components were present when analyzed by classic denaturing reducing electrophoresis techniques. Those components corresponded to ricin A and B chains. No ricin dimer was detected. Urea was used in a CE buffer to separate aggregates of such dimers, but no evidence was found for their presence. It seems more likely that the extra peaks have another origin.

Another possible explanation for the appearance of the extra peaks involves carbohydrate structure. Variability in the carbohydrate chains has been reported for ricin [24,25]. These chains are neutral, since they do not contain charged monosaccharides, such as sialic acid [25]. Their principal influences would involve changes in M_r and solution structure. Conformational changes may also affect the availability of surface charges. Of these two possibilities, the greatest influence will involve relative molecular mass. Deglycosylation and structural studies will be required to answer this question.

Finally, a third explanation for the appearance of these peaks involves disulfide bonds. The pH values of successful buffers were close to the pK value(s) of most disulfide bonds. Bond dissociation and reassociation would cause interchange that would alter conformation by rearranging protein structure [26]. Corresponding electrostatic and hydrodynamic changes would yield separate peaks with CE. Conformers would not be susceptible to dissociation with urea. Furthermore, because ricin chains are linked by a disulfide bond, scrambling may form three potential recombinants. These would be AA, BB, and the native AB chains. Although possible, this explanation for the extra peaks seemed unlikely, since no hydrogen donor equivalent to a thiol was included in the buffer. Nevertheless, additional studies concentrating upon preservation of disulfide bonds in ricin will be necessary to fully exclude this possibility.

The results of this study indicated that ricin samples are heterogeneous. Regardless of the origin of these peaks, we showed that, under appropriate conditions, capillary electrophoresis has the ability to partially resolve different protein forms of ricin or closely related impurities. It was not possible to differentiate between these two possibilities in this study.

CONCLUSIONS

Several reproducible CE methods were developed for ricin analysis. Untreated fused silica capillary columns used with zwitterionic buffers or doubly charged, basic additives yielded the best peak symmetry, column efficiency, and resolution for ricin. These buffers and additives may also have application to other glycoproteins.

Many buffers, acidic and basic, typically used for protein analyses with untreated fused silica capillary columns were not applicable to ricin analysis.

Columns with hydrophobic coatings were not useful for ricin analyses with the conditions employed in this study.

ACKNOWLEDGMENTS

The authors wish to thank Dr. Judith G. Pace and Robert W. Wannemacher for critically reviewing this manuscript. We also wish to thank Ms K. Kenyon for her assistance as USAMRIID editor. We acknowledge the assistance of Richard Dinterman for technical assistance in performing classic electrophoresis analysis of commercial ricin. The views of the authors do not purport to reflect the position of the Department of the Army or the Department of Defense.

REFERENCES

1. H. H. Lauer and D. McManigill, *Anal. Chem.*, *58*:166 (1986).
2. R. M. McCormick, *Anal. Chem.*, *60*:2322 (1988).
3. M. J. Gordon, K. Lee, A. A. Arias, and R. N. Zare, *Anal. Chem.*, *63*:69 (1991).
4. J. A. Bullock and L. Yuan, *J. Microcolumn Separation*, *3*:241 (1991).

5. M. M. Bushey and J. W. Jorgensen, *J. Chromatogr.*, *480*:301 (1989).
6. J. S. Green and J. W. Jorgensen, *J. Chromatogr.*, *478*:63 (1989).
7. S. Hjerten, *J. Chromatogr.*, *347*:191 (1985).
8. G. J. M. Bruin, J. P. Chang, R. H. Kuhlman, K. Zegers, J. C. Kraak, and H. Poppe, *J. Chromatogr.*, *471*:429 (1989).
9. K. A. Cobb, V. Dolnik, and M. Novotny, *Anal. Chem.*, *62*:2478 (1990).
10. S. A. Swedberg, *Anal. Biochem.*, *185*:51 (1990).
11. J. K. Towns and F. E. Regnier, *J. Chromatogr.*, *516*:69 (1990).
12. M. Zhu, R. Rodriguez, D. Hansen, and T. Wehr, *J. Chromatogr.*, *516*:123 (1990).
13. R. Hegde and S. K. Podder, *Eur. J. Biochem.*, *204*:155 (1992).
14. G. A. Balint, *Toxicology*, *2*:77 (1974).
15. L. Montanaro, S. Sperti, and F. Stirpe, *Biochem. J.*, *136*:677 (1973).
16. S. Olsnes, *Nature*, *328*:474 (1987).
17. S. Olsnes and A. Phil, *Biochemistry*, *12*:3121 (1973).
18. Y. Endo, K. Mitsui, M. Motizuki, and K. Tsurugi, *J. Biol. Chem.*, *262*:5908 (1987).
19. Y. Endo and K. Tsurugi, *J. Biol. Chem.*, *263*:8735 (1988).
20. Y. Endo and K. Tsurugi, *Nucleic Acids Res.*, *19*:139 (1988).
21. P. D. Grossman, J. C. Colburn, H. H. Lauer, R. G. Nielsen, R. M. Riggin, G. S. Sittampalam, and E. C. Rickard, *Anal. Chem.*, *11*:1186 (1989).
22. U. K. Laemmli, *Nature*, *227*:680 (1970).
23. M. A. Mosely, L. J. Deterding, and K. B. Tomer, *Anal. Chem.*, *63*:109 (1991).
24. Y. Kimura, H. Kusuoku, M. Tada, S. Takagi, and G. Funatsu, *Agric. Biol. Chem.*, *54*:157 (1990).
25. Y. Kimura, S. Hase, Y. Kobayashi, Y. Kyogoku, T. Ikenaka, and G. Funatsu, *J. Biochem.*, *103*:944 (1988).
26. T. E. Creighton, In *Protein Structure: A Practical Approach* (T. E. Creighton, ed.) IRL Press, New York, p. 155 (1989).

Index